Protein Intake in Health and Disease

Proteins are fundamental to human health, serving as building blocks for cellular structures, providing essential energy, and supporting tissue function. In *Protein Intake in Health and Disease*, the critical biomedical and nutritional roles of protein in the diet are explored in depth. Topics include protein quality, the impact of both low and high-protein diets, and the metabolic disorders and diseases linked to protein malnutrition. The book delves into protein's influence across the lifespan, from maternal health and early development to the nutritional needs of children, adolescents, and the aging population. Each chapter is enriched with practical insights for broader health applications and concise summary points to aid understanding.

Features:

- Contains a mini dictionary of terms and summary points in each chapter to facilitate clear understanding
- Highly illustrated with figures and multiple tables in each chapter
- Chapter contributors represent global coverage

This book is written for nutritionists, food scientists, and health care professionals, as well as research scientists and practitioners. It is also practically designed for policy makers and libraries.

Protein Intake in Health and Disease

Edited by
Victor R. Preedy

CRC Press
Taylor & Francis Group
Boca Raton London New York

CRC Press is an imprint of the
Taylor & Francis Group, an **informa** business

Designed cover image: Shutterstock

First edition published 2026
by CRC Press
2385 NW Executive Center Drive, Suite 320, Boca Raton FL 33431

and by CRC Press
4 Park Square, Milton Park, Abingdon, Oxon, OX14 4RN

CRC Press is an imprint of Taylor & Francis Group, LLC

ISBN: 9781032458274 (hbk)
ISBN: 9781032467580 (pbk)
ISBN: 9781003383116 (ebk)

DOI: 10.1201/b23328

Typeset in Times
by Deanta Global Publishing Services, Chennai, India

Contents

Preface

Dietary proteins are an important component of health and wellbeing. Their compositional amino acids provide biochemical precursors and contribute to structural components of tissue systems via macromolecule formation. Protein requirements vary across the lifespan and different needs must be met in scenarios of ill health. However, diets low in protein can have a detrimental impact, as can diets very high in protein. Furthermore, there are many different types of dietary proteins, and, across populations, their intake varies in terms of amount, source, and quality. Also, there are significant variations in the extent to which each protein type is consumed. There is an ongoing dialogue to discover new sources of dietary protein and determine what optimal intakes are. However, some material on protein intakes and different protein sources fails to address the academic diversity of the readers or include comparative information. Moreover, information on dietary protein is scattered throughout the literature. These limitations are met in *Protein Intake in Health and Disease*. Coverage includes introductory chapters, low-protein diets and intakes, metabolic disorders and diseases, protein quality, metabolism, effects on organs, high protein diets and intakes, the life span, childhood, aging and other areas. This serves as a comprehensive resource on the biomedical and nutritional aspects of dietary protein.

Protein Intake in Health and Disease has the following sections:

SECTION 1: SETTING THE SCENE AND INTRODUCTORY CHAPTERS

Chapter 1 Protein Quality in Consumer Products and Diets
Chapter 2 Protein Intake in the Mediterranean Diet
Chapter 3 Protein Intake and Exercise
Chapter 4 Dietary Proteins and Systemic Inflammation
Chapter 5 Protein Requirements in Different Metabolic Scenarios: Focusing on Endurance Athletes, Muscle, and Whole-Body Protein Metabolism
Chapter 6 Understanding the Nutritional Aspects of Dietary Proteins: Sources, Quality, Digestibility, and Beyond
Chapter 7 Comparing Animal and Plant Protein Diets in Relation to Chronic Cardiometabolic Risk Factors
Chapter 8 Alpha-Ketoglutarate, the Bacteriome, Protein Metabolism, and Longevity

SECTION 2: PROTEIN QUALITY AND DIGESTIBILITY

Chapter 9 Protein Quality: Features, Measurements, and Uses in Evaluating Diets
Chapter 10 Improving the Quality of Lentil Protein-Based Diets: A Focus on Metabolic Availability of Methionine
Chapter 11 Improving Micronutrient Density through Increasing Protein Quality: A Strategy to Improve Diet Quality
Chapter 12 Protein Digestibility of Bread: Impact of Some Breadmaking Processes and Flour Sources
Chapter 13 Protein Sources and Determining Ileal Amino Acid Digestibility

SECTION 3: METABOLIC DISORDERS AND DISEASES RESULTING FROM PROTEIN UNDERNUTRITION OR OTHER FORMS OF MALNUTRITION

Chapter 14 Kwashiorkor Malnutrition: Linking Sulfur Amino Acids to Signs of Kwashiorkor
Chapter 15 Malnutrition and Refeeding Syndrome

Chapter 16 Postoperative Protein Malnutrition: The Example of Bariatric Surgery
Chapter 17 The Paradox of Dietary Protein and Bone Physiology
Chapter 18 Addressing Protein-Energy Wasting in Hemodialysis Patients: The Case for High-Flux and Highly-Efficient Convective-Based Therapies
Chapter 19 The Importance of Protein Intake in Relation to Oral Health: A New Narrative
Chapter 20 Protein in the Diet
Chapter 21 Protein-Calorie Malnutrition in Cirrhosis: A Narrative
Chapter 22 Dietary Protein and Cancer
Chapter 23 Dietary Protein and Gut Microbiota
Chapter 24 How Protein Deficiency Impacts Immunity and Infection Response: The Example of Helminth Parasite Infection
Chapter 25 Linking Protein Intolerance, Immunity, and Disease: The Example of Celiac Disease
Chapter 26 Modeling Protein Malnutrition in Rodents: Implications for Clinical Investigations

SECTION 4: HIGH PROTEIN DIETS AND INTAKES

Chapter 27 High-Protein Diets and Their Impact on Hypercholesterolemia
Chapter 28 High-Protein Diets in the Prevention and Treatment of Type 2 Diabetes

SECTION 5: THE LIFE SPAN: MATERNAL, CHILDHOOD, ADOLESCENCE AND AGING ASPECTS

Chapter 29 Maternal Low-Protein Diets and Effects on Fetal Programming
Chapter 30 Effects of Maternal Diet and Body Composition on Human Milk Protein
Chapter 31 The Paradoxical Impact of Dietary Protein on Growth in Early Life: The Role of Different Protein Sources
Chapter 32 Protein Requirements and Quality in Infancy and Early Childhood
Chapter 33 Protein Quantity and Quality and Links with Mental Health in Adolescents
Chapter 34 Protein Intake and Weight Loss in Postmenopausal Women
Chapter 35 The Impact of Plant and Animal Proteins on the Diets of Older Individuals

SECTION 6: RESOURCES

Chapter 36 Recommended Resources for Protein Intake in Health and Disease

Each chapter has the following subsections:

1. Mini Dictionary of Terms
2. Summary Points

These subsections encompass unique features in the book. The "Mini Dictionary of Terms" aids the novice reader. The "Summary Points" encapsulate the entire chapter in succinct sentences.

Contributors include international and national experts from respected institutions. Leaders in the field and trendsetters are also contributors. Furthermore, emerging fields of protein nutrition are incorporated in *Protein Intake in Health and Disease.*

This book is essential reading for nutritionists, food scientists, healthcare professionals, research scientists, the medical profession, and other practitioners, policymakers, and institutional or governmental departments. It is also designed for teachers and lecturers, undergraduates, graduates, post-graduates, professors, and libraries.

The Editor

Acknowledgements

I wish to acknowledge the professionalism and dedication of Tom Connelly, who helped throughout the preparation and production of this book.

I am also very grateful to Randy Brehm for her guidance and support, from when the ideas for this book were first proposed to the completion of the project.

The Editor

About the Editor

Victor R. Preedy BSc, PhD, DSc, FRSB, FRSPH, FRCPath, FRSC is Honorary Professor in the Department of Clinical Biochemistry at King's College Hospital, Emeritus Professor in the Faculty of Life Sciences and Medicine at King's College London, and Visiting Professor at the University of Suffolk. Professor Preedy graduated with an Honours Degree in Biology and Physiology with Pharmacology. He gained his University of London PhD on the biochemical measurements in protein metabolism and also gained a second doctoral degree for his outstanding contribution to protein metabolism in health and disease. Professor Preedy was elected as a Fellow of the Royal Society of Medicine in 2022. He is also a Fellow of the Royal Society of Biology, The Royal College of Pathologists, The Royal Society for Public Health, and The Royal Society of Chemistry. He was the founding Director of the Genomics Centre at King's College London and held the post from 2006 to 2020. In his career, Professor Preedy has carried out research at the Cardiothoracic Institute, National Heart Hospital (part of Imperial College London), the School of Pharmacy (now part of University College London), and the MRC Centre at Northwick Park Hospital. He has collaborated with research groups in Finland, Japan, Australia, the US, and Germany. He is a leading expert on the science of health. To his credit, Professor Preedy has published over 750 articles, including peer-reviewed manuscripts based on original research, abstracts and symposium presentations, reviews, and numerous books and volumes.

Contributors

Ludovico Abenavoli
Department of Health Sciences
University "Magna Graecia"
Catanzaro, Italy

Sandra Abreu
School of Life Sciences and Environment
University of Trás-os-Montes and Alto Douro
Vila Real, Portugal

Bani Tamber Aeri
Department of Food and Nutrition, Institute of Home Economics
University of Delhi
New Delhi, India

Jussilene Alves Amorim
Postgraduate Programme in Food and Nutrition
Federal University of Piauí
Teresina, Brazil

Peter Uchenna Amadi
Department of Pediatrics
University of Alberta
Edmonton, Alberta

Maryam Amini
Department of Nutrition Research
National Nutrition and Food Technology Research Institute
and
Faculty of Nutrition Sciences and Food Technology
Shahid Beheshti University of Medical Sciences
Tehran, Iran

Liriane Andressa Alves da Silva
Federal University of Piauí
Teresina, Brazil

Rosemary Aparecida de Carvalho
Food Engineering Department
University of São Paulo
Pirassununga, Brazil

Nastaran Asri
Research Institute for Gastroenterology and Liver Diseases
Shahid Beheshti University of Medical Sciences
Tehran, Iran

Samuel Selorm Attu
Center for Human Nutrition, Department of Food Science
University of Arkansas System Division of Agriculture
Fayetteville, AR

Murali Badanthadka
NITTE (Deemed to be University)
NGSM Institute of Pharmaceutical Sciences
Department of NITTE University Centre for Animal Research and Experimentation
Karnataka, India

Pablo C. Baldi
Facultad de Farmacia y Bioquímica, Cátedra de Inmunología
Universidad de Buenos Aires
Buenos Aires, Argentina

Jamie I. Baum
Center for Human Nutrition, Department of Food Science
University of Arkansas System Division of Agriculture
Fayetteville, AR

Zulfiqar Ahmed Bhutta
Institute of Global Health and Development
Aga Khan University
Karachi, Pakistan

Chellakkan S. Blesson
Department of Obstetrics and Gynecology
Baylor College of Medicine and Texas Children's Hospital
Houston, Texas

Neusa Camila Cavalcante Andrade Oliveira
Postgraduate Programme in Food and Nutrition
Federal University of Piauí
Teresina, Brazil

Bernard Canaud
School of Medicine
University of Montpellier
Montpellier, France

Carmen Colaci
Digestive Diseases Resident
University of "Magna Graecia"
Catanzaro, Italy

Luiz Alberto Colnago
Brazilian Agricultural Research Corporation, Embrapa Instrumentation
São Carlos, Brazil

Glenda Courtney-Martin
Department of Nutritional Sciences
The Hospital for Sick Children
Toronto, Ontario

Jean-Paul Cristol
Department of Biochemistry and Hormonology
Lapeyronie University Hospital Center of Montpellier
Montpellier, France

Gangani Dharmarathne
Australian Laboratory Services – Water and Hydrographics
Hume, Australia

Maria do Carmo de Carvalho e Martins
Department of Biophysics and Physiology
Federal University of Piauí
Teresina, Brazil

Piotr Dobrowolski
Department of Functional Anatomy and Cytobiology
Maria Curie Skłodowska University
Lublin, Poland

Christina F. Dolopikou
Schools of Physical Education and Sports Science (Serres)
Faculty of Physical Education and Sport Science
Aristotle University of Thessaloniki
Thessaloniki, Greece

Janine Donaldson
School of Physiology
University of the Witwatersrand
Johannesburg, South Africa

Vinitha D'Souza
NITTE (Deemed to be University)
NGSM Institute of Pharmaceutical Sciences
Department of NITTE University Centre for Animal Research and Experimentation
Karnataka, India

Rajavel Elango
Department of Pediatrics, School of Population and Public Health
University of British Columbia
Vancouver, British Columbia

Sharon Erdrich
School of Pharmacy, Faculty of Medicine and Health
The University of Sydney
Sydney, Australia

Jessica Faccioli
Department of Translational and Precision Medicine
Sapienza University of Rome
Rome, Italy

Mary Fewtrell
Department of Population, Policy and Practice
UCL Great Ormond Street Institute of Child Health
University College London
London, United Kingdom

Merry C. Fitzpatrick
Feinstein International Center, Friedman School of Nutrition Science and Policy
Tufts University
Boston, MA

Stefania Gioia
Department of Translational and Precision Medicine
Sapienza University of Rome
Rome, Italy

Delaram Ghodsi
Department of Nutrition Research
National Nutrition and Food Technology Research Institute
and
Faculty of Nutrition Sciences
and
Food Technology
Shahid Beheshti University of Medical Sciences
Tehran, Iran

Maísa Guimarães Silva Primo
Federal University of Piauí
Teresina, Brazil

Aubree L. Hawley
Center for Human Nutrition
Department of Food Science, University of Arkansas System Division of Agriculture
Fayetteville, AR

Suzanne M. Hodgkinson
Riddet Institute
Massey University
Palmerston North, New Zealand

Hamidreza Houri
Research Institute for Gastroenterology and Liver Diseases
Shahid Beheshti University of Medical Sciences
Tehran, Iran

Thom Huppertz
School of Food and Nutritional Sciences
University College Cork
Cork, Ireland

Aamer Imdad
Stead Family Department of Pediatrics
University of Iowa
Iowa City, IA

İzzet İslamoğlu
Faculty of Yaşar Doğu Sport Sciences
University of Ondokuz Mayıs
Samsun, Turkey

Thilini Jayasinghe
The Charles Perkins Centre, School of Dentistry
The University of Sydney
Sydney, Australia

Yves José de Souza Santos
Food Engineering Department
University of São Paulo
Pirassununga, Brazil

Lucas Jurado-Fasoli
Department of Physiology, Faculty of Medicine
University of Granada
Granada, Spain

Harmeet Kaur
Department of Food and Nutrition
University of Delhi
New Delhi, India

Yıldırım Kayacan
Faculty of Yaşar Doğu Sport Sciences
University of Ondokuz Mayıs
Samsun, Turkey

Kaori Kinoshita
Department of Frailty Research
National Center for Geriatrics and Gerontology
Obu, Japan

Eirini Koidou
Schools of Physical Education and Sports Science (Serres)
Faculty of Physical Education and Sport Science
Aristotle University of Thessaloniki
Thessaloniki, Greece

Kulnipa Kittisakmontri
Department of Pediatrics
Faculty of Medicine, Chiang Mai University
Chiang Mai, Thailand

Moattar Latif
Department of Pediatrics
University of Alberta
Edmonton, Alberta

Hélène Leray-Moragues
AIDER Santé, Clinique Jacques Mirouze
Lapeyronie University Hospital Center of Montpellier
Montpellier, France

Laís Lima de Castro Abreu
Postgraduate Programme in Food and Nutrition
Federal University of Piauí
Teresina, Brazil

Wenjing Liu
School of Public Health (Shenzhen)
Sun Yat-sen University
Shenzhen, China

Madhura R. J.
NITTE (Deemed to be University)
NGSM Institute of Pharmaceutical Sciences
Department of NITTE University Centre for Animal Research and Experimentation
Karnataka, India

Faidon Magkos
Department of Nutrition, Exercise and Sports
University of Copenhagen
Copenhagen, Denmark

Sarah Malik
Norton College of Medicine
SUNY Upstate Medical University
Syracuse, NY

Margarida Liz Martins
Polytechnic University of Coimbra
Coimbra Health School
Coimbra, Portugal

Huicui Meng
School of Public Health (Shenzhen)
Sun Yat-sen University
Shenzhen, China

Daniel R. Moore
Faculty of Kinesiology and Physical Education
University of Toronto
Toronto, ON

Cecília Morais
School of Life Sciences and Environment
University of Trás-os-Montes and Alto Douro
Vila Real, Portugal

Marion Morena-Carrere
Department of Biochemistry and Hormonology
Lapeyronie University Hospital Center of Montpellier
Montpellier, France

Silvia Nardelli
Department of Translational and Precision Medicine
Sapienza University of Rome
Rome, Italy

Tirang R. Neyestani
Department of Nutrition Research
National Nutrition and Food Technology Research Institute
and
Faculty of Nutrition Sciences and Food Technology
Shahid Beheshti University of Medical Sciences
Tehran, Iran

Bahareh Nikooyeh
Department of Nutrition Research
National Nutrition and Food Technology Research Institute
and
Faculty of Nutrition Sciences and Food Technology
Shahid Beheshti University of Medical Sciences
Tehran, Iran

Rei Otsuka
Department of Epidemiology of Aging
National Center for Geriatrics and Gerontology
Obu, Japan

Isil Ozer
Department of Child Nutrition and Metabolism
University of Ondokuz Mayis
Samsun, Turkey

Anabel N. Pallaro
Facultad de Farmacia y Bioquímica, Cátedra de Nutrición
Universidad de Buenos Aires
Buenos Aires, Argentina

Alyssa Paoletti
The Hospital for Sick Children
Toronto, Ontario

Sousana Papadopoulou
Department of Nutritional Sciences & Dietetics
International Hellenic University, Thessaloniki
Thessaloniki, Greece

Vinood B. Patel
School of Life Sciences
University of Westminster
London, United Kingdom

Kateryna Pierzynowska
Department of Biology
Lund University
Lund, Sweden

Stefan G. Pierzynowski
Department of Biology
Lund University
Lund, Sweden

Samira Rabiei
Department of Nutrition Research
National Nutrition and Food Technology Research Institute
Faculty of Nutrition Sciences and Food Technology
Shahid Beheshti University of Medical Sciences
Tehran, Iran

Rajkumar Rajendram
Department of Medicine, King Abdulaziz Medical City
Ministry of National Guard Health Affairs
Riyadh, Saudi Arabia
and
College of Medicine
King Saud bin Abdulaziz University for Health Sciences
Riyadh, Saudi Arabia

Lorenzo Ridola
Department of Translational and Precision Medicine
Sapienza University of Rome
Rome, Italy

Carmen Rodríguez-García
Vitamins and Carcinogenesis Laboratory
Jean Mayer USDA Human Nutrition Research Center on Aging at Tufts University
Boston, MA

Chetanjot Kaur Romana
Center for Human Nutrition, Department of Food Science
University of Arkansas System Division of Agriculture
Fayetteville, AR

Mohammad Rostami-Nejad
Research Institute for Gastroenterology and Liver Diseases
Shahid Beheshti University of Medical Sciences
Tehran, Iran

Raiany Rosa Bergamo
Pediatrics Research Center (CIPED), Faculty of Medical Sciences (FCM)
University of Campinas (Unicamp)
Campinas, Brazil

María P. Saracino
Facultad de Farmacia y Bioquímica, Cátedra de Inmunología
Universidad de Buenos Aires
Buenos Aires, Argentina

Marialaura Scarcella
Department of Anesthesia, Intensive Care and Nutritional Science
Santa Maria Hospital
Terni, Italy

Emidio Scarpellini
Translational Research Center for Gastrointestinal Disorders (TARGID)
Gasthuisberg University Hospital
Leuven, Belgium

Amanda Suellenn da Silva Santos Oliveira
Postgraduate Program in Food and Nutrition
Federal University of Piauí
Teresina, Brazil

Blerina Shkembi
Food Quality and Design Group
Wageningen University & Research
Wageningen, The Netherlands

Axel Spahr
School of Dentistry
University of Sydney
Surry Hills, Australia

Yuxiang Sun
Department of Nutrition
Texas A&M University (TAMU)
College Station, Texas

Fani Thoma
Schools of Physical Education and Sports Science (Serres)
Faculty of Physical Education and Sport Science
Aristotle University of Thessaloniki
Thessaloniki, Greece

Fernanda M. Vanin
Food Engineering Department
University of São Paulo
Pirassununga, Brazil

Varsha A.
NITTE (Deemed to be University)
NGSM Institute of Pharmaceutical Sciences
Department of NITTE University Centre for Animal Research and Experimentation
Karnataka, India

Vipin A. Vidyadharan
Department of Obstetrics and Gynecology
Baylor College of Medicine
Houston, Texas

Cecilia C. Vila
Facultad de Farmacia y Bioquímica, Cátedra de Inmunología
Universidad de Buenos Aires
Buenos Aires, Argentina

Eric Williamson
Faculty of Kinesiology and Physical Education
University of Toronto
Toronto, ON

Man Wu
School of Public Health (Shenzhen)
Sun Yat-sen University
Shenzhen, China

Piotr Wychowański
Department of Head and Neck and Sensory Organs
Catholic University of the Sacred Heart
Rome, Italy
and
Department of Interventional Dentistry
Collegium Modicum, Nicolaus Copernicus University
Bydgoszcz, Poland

Chandra Yallampalli
Department of Obstetrics and Gynecology
Baylor College of Medicine
Houston, Texas

Zahra Yari
Department of Nutrition Research
National Nutrition and Food Technology Research Institute
Faculty of Nutrition Sciences and Food Technology
Shahid Beheshti University of Medical Sciences
Tehran, Iran

Hayrullah Yazar
Faculty of Medicine, Department of Biochemistry
University of Sakarya
Sakarya, Turkey

1 Protein Quality in Consumer Products and Diets

Blerina Shkembi and Thom Huppertz

ABBREVIATIONS

ANFs	Anti-nutritional factors
Arg	Arginine
DAAs	Dispensable amino acids
DIAAS	Digestible indispensable amino acid score
EAR	Estimated average requirement
EFSA	European Food Safety Authority
FAO	Food and Agriculture Organization
IAAs	Indispensable amino acids
Lys	Lysine
PDCAAS	Protein digestibility-corrected amino acid score
RDA	Recommended dietary allowance

INTRODUCTION

The Importance of Protein

Proteins are a critical component of the human diet as they play a key role in growth and other essential functions of the human body (Grigg and Sheffield 1995; Qin et al. 2022; Hoffman and Falvo 2004). The primary role of protein in human growth was exemplified by Grasgruber and colleagues (2016), who found that males consuming a low-protein diet were shorter in stature compared to those consuming a high-protein diet. Proteins are composed of amino acids linked together by peptide bonds (Wu 2016). There is a global variation in protein consumption, with regions and countries having higher and lower consumption (Reedy et al. 2019). Dietary proteins can be derived from animal or plant sources (Hoffman and Falvo 2004). As the world's population increases, there is also a continuous increase in the demand for protein supplies (Henchion et al. 2017; Sá et al. 2019).

Because the human body does not store supplies of proteins or amino acids and is subject to daily losses of nitrogen due to protein turnover, a daily supply of dietary protein is required. Dietary protein serves as a source of indispensable amino acids (IAAs) and also as a nitrogen source for the synthesis of dispensable amino acids (DAAs). While recommended intake values for proteins have been established, e.g., 0.83 g of protein per kg of body weight per day for adults, it is not only the quantity of protein but also the quality of protein that is a critical factor in determining whether protein intake is sufficient to meet human requirements.

Protein Quality

Within the context of protein quality, two key aspects play a role, i.e., amino acid composition and digestibility. In terms of amino acid composition, dietary proteins are a source of IAAs, i.e., those amino acids that the body cannot synthesize, for all of which specific requirements, based on age,

DOI: 10.1201/b23328-1

exist. Furthermore, dietary proteins contain not only IAAs but also DAAs, which are essential sources of nitrogen for amino acid synthesis in the body (Tessari 2018). In addition, digestibility of proteins is critical and may vary between different dietary protein sources. Such variability may arise from factors intrinsic to the protein, as well as extrinsic factors (Bos et al. 2000). Intrinsic factors in protein digestibility include both the amino acid sequence and the folding of the protein. Long sequences in proteins without peptide bonds susceptible to hydrolysis by gastrointestinal proteases and peptidases can lead to peptides too large for the brush border peptidases and hence prevent uptake. Furthermore, specific folding of proteins may lead to peptide bonds being inaccessible to gastrointestinal proteases and peptidases. Inaccessibility may also be linked to extrinsic factors, e.g., the inaccessibility of peptide bonds due to interactions with other molecules, or even inaccessibility in the food matrix. Furthermore, the food matrix may also contain anti-nutritional factors (ANFs) that complex with the digestive enzymes which should either hydrolyze the proteins or break down other aspects of the food matrix to make the proteins accessible to proteases and peptidases (Adhikari et al. 2022).

When considering protein quality in the context of the human diet, it is important to note that most foods consumed by humans have undergone one or more processing and preparation steps prior to consumption. While some of these steps are mild and can, e.g., include the washing or removal of inedible parts, in other cases they can include extensive heat treatments or mechanical treatments, which can have a notable effect on the protein structure and also result in amino acid modifications (Sá et al. 2019). Product processing and preparation varies widely, not only based on the type of product but also geographically as a result of availability, societal context, and economic status. As such, generic proxies for protein quality in human diets are difficult to establish, and a detailed examination is required, as, e.g., protein quality in boiled peas will differ from that of pea proteins in a pea protein concentrate or isolate which is subsequently applied in, e.g., an extruded meat replacer or a sterilized beverage.

In this chapter, we will consider the effects of transferring raw materials to consumer products on protein quality. First, we will briefly review the assessment of protein quality in dietary protein sources. Next, we will provide an overview of global differences in dietary protein sources. Finally, we will describe the effects of food processing and preparation on protein quality.

ASSESSMENT OF PROTEIN QUALITY

With the aforementioned importance of protein quality in human diets, accurate assessment of protein quality parameters is critical. A large body of work has focused on the methodology for protein quality assessment, with the area extensively reviewed by several groups (Adhikari et al. 2022; Wolfe et al. 2016). In general, assessment of both amino acid composition and digestibility is preferred. The digestible indispensable amino acid score (DIAAS) method, based on the determination of ileal digestible levels of individual amino acids, is the most favored methodology. The protein digestibility-corrected amino acid score (PDCAAS) method, based on fecal digestibility of total nitrogen, is less favored but also provides useful insights. In vitro methods have been widely researched as well, and recent methodology optimizations by Sousa et al. (2023) yield values comparable to DIAAS. Previous methodologies may not provide direct links to DIAAS or PDCAAS values but often still offer relevance for quantitatively assessing treatment impacts on protein digestibility. In this chapter, we will consider studies using DIAAS, PDCAAS, and in vitro techniques for evaluating protein quality in global diets and the impact of processing and preparation.

MAIN SOURCES OF DIETARY PROTEIN: A GLOBAL PERSPECTIVE

Total Protein Intake

According to the European Food Safety Authority (EFSA), the estimated average requirement (EAR) for healthy adults is 0.66 g protein/kg body weight per day, and the recommended dietary allowance (RDA) is 0.83 g protein/kg body weight per day (EFSA 2012). For an average global body weight of ~62 kg, the RDA would be just over 50 g of protein per person per day. It has been shown that many people in the world consume more protein than required, particularly in richer regions (Ranganathan et al. 2016). However, some consume less protein. Inadequate dietary protein intake has been associated with various health problems including anemia, physical weakness, edema, vascular dysfunction, and impaired growth and immunity (Wu 2016). Figure 1.1 shows protein intake, obtained from FAO food balance sheets, in 2020 for 220 countries and regions across the globe, indicating that protein intake varies widely from <30 to >100 g per person per day, a difference of more than three-fold. Only ten countries (nine African plus Haiti) did not meet 50 g protein intake per person per day, whereas in all other countries, based on these estimates, average protein consumption did meet requirements. However, it should be noted that these are of course estimates for average intake, and do not account for, e.g., unequal distribution of available protein between people within populations.

The main dietary protein suppliers are animal and plant food sources (Hoffman and Falvo 2004). In a study by Reddy et al. (2019), global animal protein intake was reported to be higher than plant protein intake, which is in line with trends shown in Figure 1.1. Furthermore, in 2015, total protein intake worldwide was 78.2 g/day, ranging from 61.0 g/day for the Middle East and North Africa to 92.5 g/day for Asia (Reedy et al. 2019), comparable to trends in Figure 1.1. Protein can also serve as a source of energy for the body, although not the primary choice (Carreiro et al. 2016; Adhikari et al. 2022; Hoffman and Falvo 2004).

Main Protein Sources

Dietary proteins derive from both animal and plant sources, varying in protein content and amino acid composition (Grigg and Sheffield 1995; Hoffman and Falvo 2004). Animal proteins come from meat, poultry, fish, eggs, and milk (Hoffman and Falvo 2004), while plant proteins come from foods of plant origin including whole grains, legumes, and nuts (Qin et al. 2022). The data in Figures 1.1 and 1.2 show proportional protein intake from animal- and plant-based protein sources, including, in Figure 1.2, some unclassified foods (e.g., mixed foods), in different countries. Figure 1.2 shows variability in animal and plant protein intake across countries. For instance, the United States, Latin

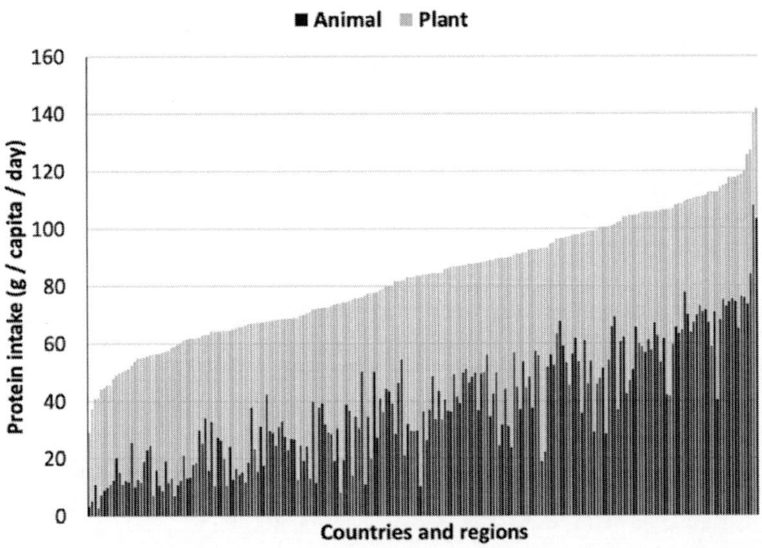

FIGURE 1.1 Protein intake across countries and regions. Estimated protein intake (in g per capita per day) from animal-based and plant-based food sources for 220 countries and regions in 2020 (data from FAO food balance sheets).

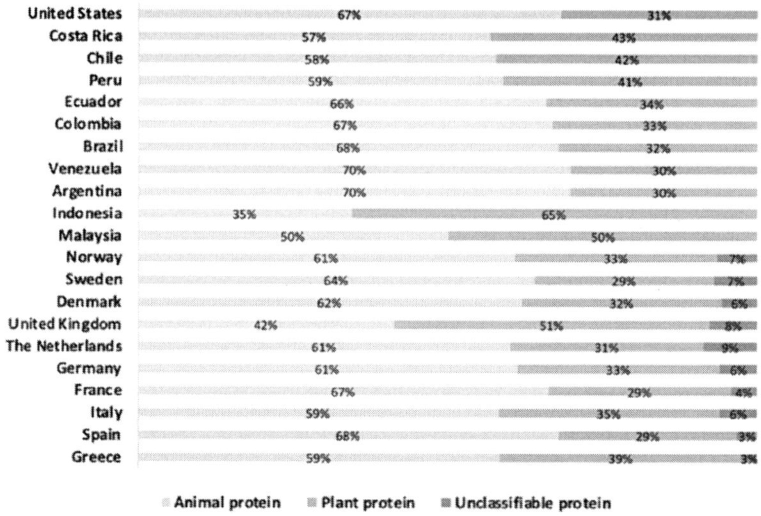

FIGURE 1.2 Global differences in protein intake from different sources. Calculated gender-averaged of total protein intake percentage (%) from animal, plant, and unclassifiable food sources among different countries. Data for European countries from Halkjaer et al. (2009), for South American countries from Herrera-Cuenca et al. (2023), for United States from Hoy et al. (2022), for Indonesia from Khusun et al. (2022), and for Malaysia from Drewnowski et al. (2020).

America, and Europe (except for the general population in the United Kingdom) showed the highest animal protein intake (57–70% of total protein) (Hoy et al. 2022; Halkjær et al. 2009; Herrera-Cuenca et al. 2023). In Malaysia, animal and plant protein intake were equal (50%) (Drewnowski et al. 2020). In Indonesia, 65% of dietary protein comes from plant sources, particularly cereals, and the rest from animal sources (35%) (Figure 1.2) (Khusun et al. 2022). A study by Reedy and colleagues (2019) reported that global animal protein intake was highest in Latin America and the Caribbean (42.8 g/day) and lowest in Sub-Saharan Africa (38.3 g/day). Bangladesh and Nepal had the lowest mean national animal protein intake (7.2 g/day and 10.1 g/day). Global plant protein intake was highest in the Middle East and North Africa (32.1 g/day) and lowest in the former Soviet Union (25.5 g/day) (Reedy et al. 2019). The choice of a specific protein may depend on factors such as education, ethnicity, and religion (Drewnowski et al. 2020).

Tables 1.1 and 1.2 indicate that meat and cereals are the main protein contributors in many diets. As shown in Table 1.1, dairy products are often considered the second source of animal protein after meat, while in some parts of Spain, fish is the second contributor. In some Eastern African countries, such as Burundi, Rwanda, Tanzania, and Uganda, fish contributes to notably more animal protein intake than the world average (Table 1.1) (Obiero et al. 2019). Table 1.2 shows that vegetables are the most important plant protein suppliers after cereals in most countries, followed by fruits, legumes, and potatoes. Cakes (2–7%) and non-alcoholic beverages (1–12%) also contributed to plant protein in several countries (Table 1.2) (Halkjær et al. 2009). In the Philippines, 19% of protein comes from plant sources such as rice (Table 1.2), followed by animal sources: 17% from pork, 17% from fish, and 5% from eggs and egg dishes (Table 1.1) (Angeles-Agdeppa and Custodio 2020).

TABLE 1.1
Main Dietary Animal Protein Sources in Different Regions

Country and Region	Dairy	Total Meat	Beef	Pork	Mutton	Red Meat	Poultry	Processed Meat	Fish	Eggs	References
Greece	34.7	39.7				24.6	9.9	3.4	21.5	2.7	Halkjær et al. 2009
Spain											Halkjær et al. 2009
Granada	27.2	40.9				15.9	12.3	12.1	24.6	4.3	
Murcia	22.6	47.8				17.4	15.6	12.9	22.1	4.8	
Navarra	19.0	50.4				21.6	12.1	14.7	22.4	5.8	
San Sebastian	17.7	51.1				26.2	12.3	10.3	23.9	6.1	
Asturias	22.4	47.5				21.6	10.6	13.4	22.7	5.4	
Italy											Halkjær et al. 2009
Ragusa	27.5	57.1				33.0	15.6	7.4	11.0	3.1	
Naples	35.6	47.1				28.0	9.6	8.9	10.0	3.0	
Florence	27.1	58.6				27.1	18.4	9.8	9.6	2.8	
Turin	29.0	57.1				26.6	17.1	11.6	8.8	3.2	
Varese	29.6	59.5				27.0	16.0	15.3	6.2	2.1	
France											Halkjær et al. 2009
South coast	35.4	44.7				18.0	13.8	9.3	14.4	3.0	
South	34.0	46.9				22.8	13.2	9.0	12.6	3.3	
Northeast	33.3	47.3				22.9	12.2	9.3	13.6	3.2	
Northwest	29.8	46.6	NA	NA	NA	20.0	14.5	10.2	17.8	3.3	
Germany											Halkjær et al. 2009
Heidelberg	27.8	58.3				25.2	10.7	21.4	6.4	3.2	
Potsdam	25.7	57.0				21.8	9.0	25.3	8.4	3.8	
The Netherlands											Halkjær et al. 2009

(Continued)

TABLE 1.1 (CONTINUED)
Main Dietary Animal Protein Sources in Different Regions

Country and Region	Dairy	Total Meat	Beef	Pork	Mutton	Red Meat	Poultry	Processed Meat	Fish	Eggs	References
Bilthoven	32.8	57.6				30.7	8.1	18.1	5.0	3.8	
Utrecht	41.6	49.0				26.0	8.8	13.9	4.6	3.9	
United Kingdom											Halkjær et al. 2009
General population	30.3	52.9				20.5	18.3	13.2	11.5	3.2	
Health-conscious	49.8	26.6				7.5	10.8	7.7	10.8	5.1	
Denmark											Halkjær et al. 2009
Copenhagen	31.3	52.1				31.6	9.4	10.1	11.7	4.1	
Aarhus	34.5	50.4				29.7	9.4	10.1	9.7	4.3	
Sweden											Halkjær et al. 2009
Malmö	33.7	48.6				26.5	6.7	14.4	11.6	4.0	
Umeå	39.3	43.8				20.3	4.3	14.3	10.1	3.8	
Norway											Halkjær et al. 2009
South and East	34.7	45.7				16.7	10.2	16.1	12.4	4.5	
North and West	32.7	41.2				18.4	4.6	16.8	18.8	4.1	
Malaysia	18.76 (including eggs)	19.43	1.47	1.41	0.30	NA	16.25	NA	12.12	NA	Drewnowski et al. 2020

(*Continued*)

TABLE 1.1 (CONTINUED)
Main Dietary Animal Protein Sources in Different Regions

Country and Region	Dairy	Total Meat	Beef	Pork	Mutton	Red Meat	Poultry	Processed Meat	Fish	Eggs	References
Indonesia	2.46	NA	3.44	0.13	0.04	NA	8.43	NA	9.81	10.32	Khusun et al. 2022
Philippines	NA	NA	NA	17	NA	NA	NA	NA	17	5 (eggs and egg dishes)	Angeles-Agdeppa and Custodio 2020
World									16.9		Obiero et al. 2019
Africa									18.4		Obiero et al. 2019
Burundi									28.1		Obiero et al. 2019
Ethiopia									1.6		Obiero et al. 2019
Kenya	NA	NA	NA	NA	NA	NA	NA	NA	7.3	NA	Obiero et al. 2019
Rwanda									25.1		Obiero et al. 2019
Tanzania									22.5		Obiero et al. 2019
Uganda									30.1		Obiero et al. 2019

Calculated gender-averaged percentage contribution of different product categories to dietary plant protein intake in different countries and regions.

TABLE 1.2
Main Dietary Plant Protein Sources in Different Regions

Country and Region	Potatoes	Vegetables	Legumes	Fruits	Rice	Cereals	Tubers	Cakes	Non-Alcoholic Drinks	References
Greece	3.1	13.8	7.1	8.1		59.6		3.2	2.1	Halkjær et al. 2009
Spain										Halkjær et al. 2009
Granada	5.8	11.2	7.9	12.3		47.6		3.0	2.3	
Murcia	5.8	18.8	7.2	14.8		39.8		6.2	1.3	
Navarra	4.8	11.9	9.7	11.9		47.0		4.1	1.7	
San Sebastian	6.5	12.7	13.7	11.3		42.3		3.1	1.4	
Asturias	8.0	7.0	14.3	11.4		43.9		4.8	2.0	
Italy										Halkjær et al. 2009
Ragusa	2.8	9.3	4.3	7.9		64.0		3.1	1.3	
Naples	2.9	13.0	3.7	7.4		60.7		3.5	2.7	
Florence	2.9	12.3	3.8	6.9		62.0		2.7	2.3	
Turin	3.1	15.2	1.6	8.0		60.5		3.2	2.2	
Varese	3.3	12.0	2.2	8.2		62.6		3.2	2.8	
France										Halkjær et al. 2009
South coast	2.5	15.5	2.9	10.4		52.8		3.8	3.1	
South	4.1	13.1	1.9	9.1	NA	56.2	NA	3.9	2.9	
Northeast	4.7	13.1	1.8	9.6		53.1		4.4	3.5	
Northwest	5.7	12.8	1.6	10.0		54.6		3.4	3.6	
Germany										Halkjær et al. 2009
Heidelberg	7.4	11.0	1.4	7.5		42.3		5.4	12.1	
Potsdam	8.8	11.0	1.2	8.7		41.2		5.9	10.1	
The Netherlands										Halkjær et al. 2009
Bilthoven	9.8	8.7	1.3	9.8		52.8		2.7	5.8	
Utrecht	8.7	9.6	1.3	11.9		51.2		3.2	6.7	

(Continued)

TABLE 1.2 (CONTINUED)
Main Dietary Plant Protein Sources in Different Regions

United Kingdom										Halkjær et al. 2009
General population	7.5	11.7	2.2	6.7		54.6		3.7	4.0	
Health-conscious	5.2	10.0	3.7	12.9		40.6		3.0	2.1	
Denmark										Halkjær et al. 2009
Copenhagen	7.8	10.1	0.3	5.6		58.2		2.4	8.5	
Aarhus	8.0	8.9	0.1	6.0		59.2		3.1	8.2	
Sweden										Halkjær et al. 2009
Malmö	9.1	7.9	1.4	5.6		53.7		4.6	4.7	
Umeå	9.1	6.2	0.8	4.8		58.4		7.2	3.4	
Norway										Halkjær et al. 2009
South and East	5.7	7.5	0.4	7.8		59.0		4.7	7.9	
North and West	7.2	6.7	0.3	7.8		58.6		5.4	6.8	
Indonesia	NA	NA	19.31	NA	NA	45.04	1.01	NA	NA	Khusun et al. 2022
Philippines	NA	NA	NA	NA	19	NA	NA	NA	NA	Angeles-Agdeppa and Custodio 2020

Calculated gender-averaged percentage contribution of different product categories to dietary plant protein intake in different countries and regions.

Protein Quality: Differences between Sources

When choosing a dietary protein, in addition to the quantity, it is very important to consider the quality as well (Adhikari et al. 2022). As discussed in Section 2, the quality of any protein can be determined by its amino acid composition and by its digestibility. Protein digestibility refers to the fraction of ingested protein that is hydrolyzed into amino acids, dipeptides, and tripeptides that can be readily absorbed by the organism (Adhikari et al. 2022). In general, animal proteins have a higher nutritional quality than plant proteins due to their amino acid profile and higher digestibility (Qin et al. 2022; Vieux et al. 2022; Hoffman and Falvo 2004; Adhikari et al. 2022). It has been reported that animal proteins have a DIAAS value higher than 90% or even higher than 100%, while most plant proteins have low quality with DIAAS values lower than 90%. The exception is soy-based products and mung beans, which have DIAAS values higher than 90% (Adhikari et al. 2022). Animal products like meat, poultry, eggs, milk, and fish provide the highest protein quality as they have complete protein with all the essential amino acids that the human body needs (Hoffman and Falvo 2004). Furthermore, among these animal products, fish is an important source of protein for different countries. In particular, in Africa, around 200 million people consume high-quality protein derived from fish (Obiero et al. 2019). Instead, plant products generally provide the lowest protein quality as they often contain incomplete proteins, thus being deficient in one or more essential amino acids (Hoffman and Falvo 2004). Hence, it has been suggested that to get all the essential amino acids from plant-based products, it is necessary to consume a variety of plant products (Hoffman and Falvo 2004).

Protein Quality: Differences between Diets

The impact of different protein sources in diets is also reflected in the overall protein quality of diets. It has, for instance, been reported that the true fecal digestibility of protein from diets in different parts of the world can vary from >90% to only just above 50% (Table 1.3), even without considering amino acid composition (Gilani et al. 2005). A similar variation was reported by Moughan (2021), who estimated DIAAS values for mixed diets in different countries in Asia and Africa. As shown in Figure 1.3, a wide variety was observed, with DIAAS values >90% observed for some developing countries (e.g., Thailand and Pakistan), whereas Nigeria and Ethiopia had estimated values <70%. This reflects large differences worldwide. In all cases, lysine was the first limiting amino acid in diets, typically reflecting high levels of plant-based protein sources in the diets. The differences in protein quality in global diets are important to consider. While data in Figure 1.1

TABLE 1.3
Protein Digestibility in Diets in Different Regions

Country or Area	Diet	Digestibility (%)
North America	Mixed	94
North America	Lactovegetarian	88
North America	Vegetarian	93
India	Rice, red gram dahl, milk powder, vegetables	75
India	Ragi, red gram, potatoes, milk powder, vegetables	54
Guatemala	Black beans, corn tortilla, rice, wheat rolls, cheese, eggs, vegetables	77
Nigeria	Cassava, rice, dried fish, vegetables	91
Sri Lanka	Rice, meat, fish, dairy products, breads, fruits, vegetables	87
Brazil	Rice, beans, meat, eggs, vegetables	78
Philippines	Mixed	88

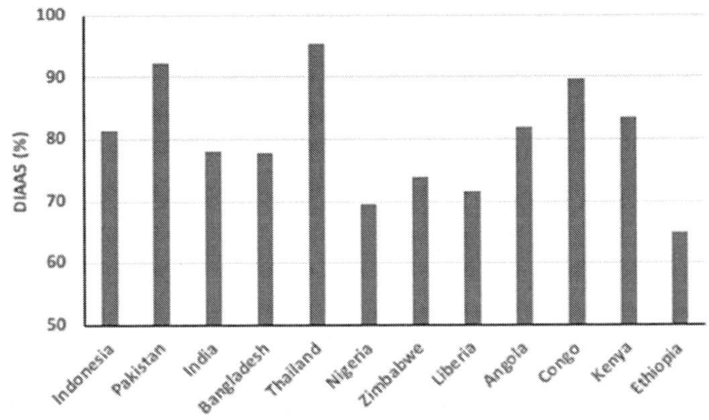

FIGURE 1.3 Protein quality in diets in different countries. Estimated DIAAS values of mixed diets from different countries (data from Moughan 2021).

indicate that protein intake in most countries could potentially be adequate, it is important to keep in mind that RDAs of 0.83 g protein per person per day are based on proteins with an ideal amino acid composition and complete digestibility, i.e., a DIAAS value of 100. Lower estimated DIAAS values for diets could mean that even though protein intake appears adequate, digestible amino acid requirements are not met.

The differences in protein quality between diets illustrated in Table 1.3 and Figure 1.3 highlight the key importance of considering both protein quantity and quality. While excess indispensable amino acids in some protein sources can compensate for shortfalls in others, shortfalls will remain in many diets. Even if recommended daily intake values are met, if protein quality is suboptimal, issues will still arise. Furthermore, Adhikari et al. (2022) argued that the complementarity of protein sources should be considered on a meal basis rather than a diet basis, which could lead to even lower values for protein quality on a diet basis if implemented. Similar arguments were put forward recently in consideration of nutrient-based frameworks for protein quality (Forester et al. 2023). Additionally, it is important to consider that many protein quality data are derived from studies on protein ingredients and not final consumer products. Further processing can impact protein quality, as described in Section 1.4.

IMPACT OF FOOD PROCESSING AND PREPARATION ON PROTEIN QUALITY: FROM RAW MATERIAL TO CONSUMER PRODUCT

When considering the consumption of dietary protein, it is important to note that virtually all major dietary protein sources consumed worldwide undergo various processing and preparation steps prior to consumption. For example, while raw milk is consumed by some people, the vast majority of milk produced worldwide is subjected to different treatments (e.g., heat treatment) or is even converted into different products, like cheese or yoghurt (Walstra et al. 2005). Likewise, wheat is not consumed as such but is typically first milled into flour and subsequently forms the basis of a wide range of food products, including bread, which undergoes substantial thermal processing during preparation (Shewry 2009). Therefore, while understanding protein quantities and quality parameters in raw materials is important, understanding the transitions of the proteins throughout the different stages from raw material to final product is equally critical, as are other factors affecting their digestibility. Within this field of process-induced protein modifications, the following three aspects are considered most important:

(1) Process-induced changes in amino acids
(2) Process-induced changes in protein structure
(3) Process-induced changes in anti-nutritional compounds

Process-Induced Changes in Amino Acids

During the processing of food, some process-induced changes in amino acids can occur. The most notable example is probably the glycation of Lys and Arg residues, which can occur as a first step in the Maillard reaction, between the amino group on the side chain of these amino acids and reducing sugars, such as glucose, fructose, maltose, and lactose (Martins et al. 2001). The reaction is promoted by the application of heat, occurs at a higher rate at elevated pH (because the reaction requires the non-protonated amino groups), and has a strong dependence on water activity, with maximum rates often found in the water activity range of 0.5–0.7 (Ames 1990). Examples of where these reactions occur include the baking or toasting of bread, processing of cereals, frying, and grilling of meat products. Protein glycation can lead to substantial losses in (bio)available lysine. For cereal products, more than 50% of lysine may be unavailable (Rutherfurd and Moughan 2007). However, common at-home preparation methods for meat (e.g., grilling, frying, etc.) have only a limited effect on bioavailable lysine in beef products (Hodgkinson et al. 2018). For dairy products, process-induced losses in available lysine are generally limited in most common dairy products, not exceeding 10% of total lysine; for some products, e.g., evaporated milk products, which typically undergo sterilization treatments, higher losses in available lysine have been reported (Mehta and Deeth 2016). Overall, process-induced changes in amino acids should be carefully considered, particularly given the importance of lysine as one of the most limiting IAAs globally.

Process-Induced Changes in Protein Structure

Proteins can be characterized by primary, secondary, tertiary, and sometimes even quaternary structures. These structures are subject to change based on changes in environmental conditions including, e.g., temperature, pH, and ionic strength. Such changes may be reversible or irreversible. Irreversible changes often occur as a result of, e.g., heat treatment, when denaturation of the protein occurs. As a result of denaturation, peptide bonds previously inaccessible to digestive proteases become accessible, which can lead to enhanced digestion. This has, for instance, been shown for the heat-induced denaturation of whey proteins (Peram et al. 2013) but also for pressure-induced denaturation of whey proteins (Zeece et al. 2008), both of which improved whey protein digestion in vitro. However, protein denaturation can also lead to (extensive) aggregation of proteins, which, in turn, can make parts of the protein poorly accessible and reduce digestibility. This, for instance, has been observed in the case of proso millet protein (Gulati et al. 2017) and sorghum protein (Duodu et al. 2003), where intermolecular formation of hydrophobic bonds and disulphide bonds occurred. Overall, processing-induced changes in protein conformation and their impact on protein digestibility are topics that require careful consideration. While, e.g., heat treatments are often necessary in food processing to inactivate undesirable micro-organisms and enzymes, they will also affect digestibility.

Process-Induced Changes in Anti-Nutritional Compounds

In addition to proteins and other nutrients, food products can also contain compounds that can negatively affect the digestion and absorption of nutrients, including those that affect protein digestion. Two different categories of effects can be considered: components that interact with the proteins and thereby affect digestibility, and components that affect protein digestibility via interactions with or effects on the digestive enzymes.

Trypsin Inhibitors

Enzyme inhibitors can impact digestibility of protein foods. The most common inhibitor class in this context are the so-called trypsin inhibitors, which affect not only trypsin but also enzymes such as chymotrypsin and elastase. Of the most common protein sources in the human diet, soybeans contain the highest levels of trypsin inhibitors. Defatting of soybeans to create soy flour does not change the protein: trypsin inhibitor ratio. However, due to their proteinaceous nature, trypsin inhibitors are susceptible to heat-induced inactivation, which can occur, e.g., during the toasting of soy flour, extrusion, or heat treatments, such as sterilization, applied to extracts of soy proteins or products in which they are used. In soy protein concentrate, soy protein isolate, tofu, or soy drink, trypsin inhibitor levels are typically approximately five- to ten-fold lower than in soybeans (Anderson and Wolf 1995). Autoclaving raw soybean flour has been shown to increase the ileal digestibility of amino acids approximately two-fold (Li et al. 1998). Trypsin inhibitors are also present in other legumes, including, e.g., many types of peas and beans, as well as cereals (Gilani et al. 2005). Reducing trypsin inhibitor activity is also warranted here to improve protein digestibility.

Tannins

Tannins are water-soluble polyphenolic compounds that occur naturally in a range of foods important for human consumption, including sorghum, millet, barley, and various beans and peas. Levels may reach tens of grams per kg of (dry) product. Their ability to interact with proteins and precipitate them from aqueous solutions, as well as to interact with digestive enzymes, can lead to reduced digestibility of food proteins. An example is shown in Figure 1.4 shows the effect of including a tannin-rich fraction of faba bean hull material on the digestion of casein in rats, where a linear decrease in digestibility was observed. Likewise, for faba bean protein, reduced digestibility has also been observed in high-tannin variants compared to low-tannin variants (Jansman et al. 1995). In addition to concentrations in products, tannin levels in meals and interactions between products should also be considered. Kashyap et al. (2019) illustrated this nicely by showing that co-ingestion with tea reduced the digestibility of egg protein compared to co-ingestion with water, presumably due to the tannins present in the tea. Common practices like dehulling, soaking and cooking can reduce tannin content in, e.g., legumes.

Phytate

Another common anti-nutritional factor in food products is phytate, which can affect protein digestibility in two ways: through chelation of metal ions required in the active site of digestive enzymes or via direct complexation with the proteins. Phytate is found in, e.g., nuts, seeds, legumes, and

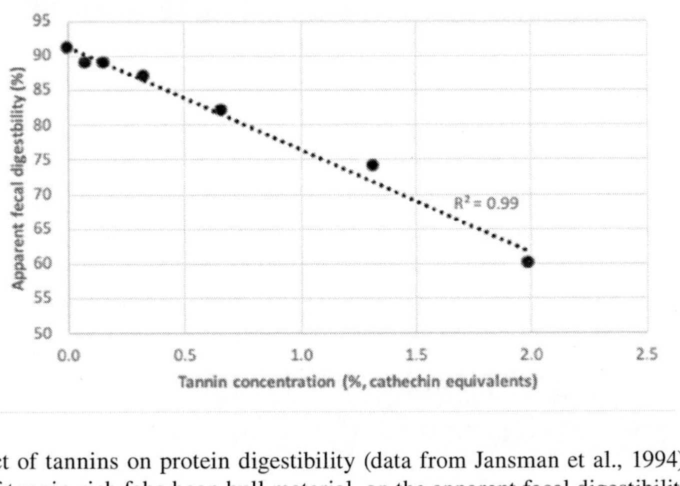

FIGURE 1.4 Effect of tannins on protein digestibility (data from Jansman et al., 1994). Effect of tannin, added in the form of tannin-rich faba bean hull material, on the apparent fecal digestibility of caseins in rats (data from Jansman et al. 1994).

grains. Phytate levels in cereals can be reduced through milling and refining because more of the phytate is present in the bran of wheat, the germ of corn, or the pericarp of rice, which are typically removed during the refining processes. In legumes, however, most phytate is found in the cotyledon, together with the protein, so this is largely retained during fractionation and refinement. Phytate is heat-stable but can be broken down by enzymes (phytase) and fermentation. Breakdown or removal of phytate has been shown to improve protein digestibility (Elliott et al. 2022), though the overall effects of phytate on protein digestibility are smaller than those of trypsin inhibitors and tannins.

MINI DICTIONARY OF TERMS

Amino acid digestibility: The fraction of amino acids that can be absorbed in the gastrointestinal tract after protein digestion.

Animal-based protein: Dietary protein derived from animal-based sources, such as meat, fish, dairy, and eggs.

Dispensable amino acids: Amino acids that can be synthesized by the human body.

Indispensable amino acids: Amino acids that cannot be synthesized by the human body.

Phytate: A phosphate ester present in plant-based foods and food ingredients that can impair protein digestion.

Plant-based protein: Dietary protein derived from plant-based sources, such as grains, cereals, legumes, nuts, and vegetables.

Protein digestion: The breakdown of proteins by gastrointestinal enzymes.

Protein glycation: Non-enzymatic covalent bond formation between amines in proteins (mainly lysine residues) and reducing carbohydrates.

Protein quality: The extent to which a protein is able to provide indispensable amino acids in a digestible form relative to a reference pattern.

Tannins: A class of water-soluble polyphenols present in plant-based foods and food ingredients that can impair protein digestion.

SUMMARY POINTS

- Dietary protein quality strongly differs between countries and regions.
- Higher animal protein intake is often related to higher dietary protein quality.
- The presence of anti-nutritional factors can strongly reduce protein quality.
- Protein glycation reduces protein digestibility and amino acid bioavailability.
- Tannins and phytates can reduce protein digestibility via complexation with proteins and digestive enzymes.
- Processing can reduce anti-nutritional factors but can also impair protein digestibility via glycation and protein aggregation.
- Consideration of protein quality at a consumer product level rather than at the raw material level is critical.

REFERENCES

Adhikari, Shiksha, Marijke Schop, Imke J. M. de Boer, and Thom Huppertz. 2022. 'Protein Quality in Perspective: A Review of Protein Quality Metrics and Their Applications'. *Nutrients*. doi:10.3390/nu14050947.

Ames, Jennifer M. 1990. 'Control of the Maillard Reaction in Food Systems'. *Trends in Food Science and Technology*. doi:10.1016/0924-2244(90)90113-D.

Anderson, R. L., and W. J. Wolf. 1995. 'Compositional Changes in Trypsin Inhibitors, Phytic Acid, Saponins and Isoflavones Related to Soybean Processing'. *Journal of Nutrition* 125. doi:10.1093/jn/125.3_Suppl.581S.

Angeles-Agdeppa, Imelda, and Ma Rosel S. Custodio. 2020. 'Food Sources and Nutrient Intakes of Filipino Working Adults'. *Nutrients* 12 (4). doi:10.3390/nu12041009.

Bos, Cécile, Claire Gaudichon, and Daniel Tomé. 2000. 'Nutritional and Physiological Criteria in the Assessment of Milk Protein Quality for Humans'. *Journal of the American College of Nutrition* 19. doi :10.1080/07315724.2000.10718068.

Carreiro, Alicia L., Jaapna Dhillon, Susannah Gordon, Kelly A. Higgins, Ashley G. Jacobs, Breanna M. McArthur, Benjamin W. Redan, Rebecca L. Rivera, Leigh R. Schmidt, and Richard D. Mattes. 2016. 'The Macronutrients, Appetite, and Energy Intake'. *Annual Review of Nutrition*. doi:10.1146/annurev-nutr-121415-112624.

Drewnowski, Adam, Elise Mognard, Shilpi Gupta, Mohd Noor Ismail, Norimah A. Karim, Laurence Tibère, Cyrille Laporte, et al. 2020. 'Socio-Cultural and Economic Drivers of Plant and Animal Protein Consumption in Malaysia: The SCRiPT Study'. *Nutrients* 12 (5). doi:10.3390/nu12051530.

Duodu, K. G., J. R. N. Taylor, P. S. Belton, and B. R. Hamaker. 2003. 'Factors Affecting Sorghum Protein Digestibility'. *Journal of Cereal Science*. Academic Press. doi:10.1016/S0733-5210(03)00016-X.

EFSA. 2012. 'EFSA Panel on Dietetic Products; Nutrition and Allergies (NDA). Scientific Opinion on Dietary Reference Values for Protein'. *EFSA Journal* 10 (2): 2557. doi:10.2903/j.efsa.2012.2557.

Elliott, Hannah, Patrick Woods, Brian D. Green, and Anne P. Nugent. 2022. 'Can Sprouting Reduce Phytate and Improve the Nutritional Composition and Nutrient Bioaccessibility in Cereals and Legumes?' *Nutrition Bulletin*. John Wiley and Sons Inc. doi:10.1111/nbu.12549.

Forester, Shavawn M., Emily M. Jennings-Dobbs, Shazia A. Sathar, and Donald K. Layman. 2023. 'Perspective: Developing a Nutrient-Based Framework for Protein Quality'. *The Journal of Nutrition*. doi:10.1016/j.tjnut.2023.06.004.

Gilani, G. Sarwar, Kevin A. Cockell, and Estatira Sepehr. 2005. 'Effects of Antinutritional Factors on Protein Digestibility and Amino Acid Availability in Foods'. *Journal of AOAC International* 88 (3). AOAC International: 967–87. doi:10.1093/jaoac/88.3.967.

Grasgruber, P., M. Sebera, E. Hrazdíra, J. Cacek, and T. Kalina. 2016. 'Major Correlates of Male Height: A Study of 105 Countries'. *Economics and Human Biology* 21 (May). Elsevier B.V.: 172–95. doi:10.1016/j.ehb.2016.01.005.

Grigg, David. 1995. 'The Pattern of World Protein Consumption'. *Geoforum* 26: 1–17.

Gulati, Paridhi, Aixia Li, David Holding, Dipak Santra, Yue Zhang, and Devin J. Rose. 2017. 'Heating Reduces Proso Millet Protein Digestibility via Formation of Hydrophobic Aggregates'. *Journal of Agricultural and Food Chemistry* 65 (9). American Chemical Society: 1952–59. doi:10.1021/acs.jafc.6b05574.

Halkjær, J., A. Olsen, L. J. Bjerregaard, G. Deharveng, A. Tjønneland, A. A. Welch, F. L. Crowe, et al. 2009. 'Intake of Total, Animal and Plant Proteins, and Their Food Sources in 10 Countries in the European Prospective Investigation into Cancer and Nutrition'. *European Journal of Clinical Nutrition* 63: S16–36. doi:10.1038/ejcn.2009.73.

Henchion, Maeve, Maria Hayes, Anne Maria Mullen, Mark Fenelon, and Brijesh Tiwari. 2017. 'Future Protein Supply and Demand: Strategies and Factors Influencing a Sustainable Equilibrium'. *Foods* 6 (7). MDPI AG: 1–21. doi:10.3390/foods6070053.

Herrera-Cuenca, Marianella, Martha Cecilia Yépez García, Lilia Yadira Cortés Sanabria, Pablo Hernández, Yaritza Sifontes, Guillermo Ramírez, Maura Vásquez, et al. 2023. 'Contribution of Proteins to the Latin American Diet: Results of the ELANS Study'. *Nutrients* 15 (3). MDPI. doi:10.3390/nu15030669.

Hodgkinson, Suzanne M., Carlos A. Montoya, Patty T. Scholten, Shane M. Rutherfurd, and Paul J. Moughan. 2018. 'Cooking Conditions Affect the True Ileal Digestible Amino Acid Content and Digestible Indispensable Amino Acid Score (DIAAS) of Bovine Meat as Determined in Pigs'. *Journal of Nutrition* 148 (10). Oxford University Press: 1564–69. doi:10.1093/jn/nxy153.

Hoffman, Jay R., and Michael J. Falvo. 2004. 'Protein - Which Is Best?' *Journal of Sports Science and Medicine* 3: 118–30.

Hoy, M. Katherine, Theophile Murayi, and Alanna J. Moshfegh. 2022. 'Diet Quality and Food Intakes Samong US Adults by Level of Animal Protein Intake, What We Eat in America, NHANES 2015-2018'. *Current Developments in Nutrition* 6 (5). doi:10.1093/cdn/nzac035.

Jansman, A. J. M., A. A. Frohlich, and R. R. Marquardt. 1994. 'Production of Proline-Rich Proteins by the Parotid Glands of Rats Is Enhanced by Feeding Diets Containing Tannins from Faba Beans (Vicia Faba L.)'. *Journal of Nutrition* 124 (2). doi:10.1093/jn/124.2.249.

Jansman, A. J. M., M. W. Verstegen, J. Huisman, and J. W. van den Berg. 1995. 'Effects of Hulls of Faba Beans (Vicia Faba L.) with a Low or High Content of Condensed Tannins on the Apparent Ileal and Fecal Digestibility of Nutrients and the Excretion of Endogenous Protein in Ileal Digesta and Feces of Pigs'. *Journal of Animal Science* 73 (1): 118–27. doi:10.2527/1995.731118x.

Kashyap, Sindhu, Nirupama Shivakumar, Aneesia Varkey, Thomas Preston, Sarita Devi, and Anura V. Kurpad. 2019. 'Co-Ingestion of Black Tea Reduces the Indispensable Amino Acid Digestibility of Hens' Egg in Indian Adults'. *Journal of Nutrition* 149 (8). Oxford University Press: 1363–68. doi:10.1093/jn/nxz091.

Khusun, Helda, Judhiastuty Februhartanty, Roselynne Anggraini, Elise Mognard, Yasmine Alem, Mohd Ismail Noor, Norimah Karim, et al. 2022. 'Animal and Plant Protein Food Sources in Indonesia Differ Across Socio-Demographic Groups: Socio-Cultural Research in Protein Transition in Indonesia and Malaysia'. *Frontiers in Nutrition* 9 (February). Frontiers Media S.A. doi:10.3389/fnut.2022.762459.

Li, Shaoyan, Willem C. Sauer, and William R. Caine. 1998. 'Response of Nutrient Digestibilities to Feeding Diets with Low and High Levels of Soybean Trypsin Inhibitors in Growing Pigs'. *Journal of the Science of Food and Agriculture* 76 (3). doi:10.1002/(SICI)1097-0010(199803)76:3<357::AID-JSFA955>3.0.CO;2-2.

Martins, S., W. Jongen, and M. A. J. S. van Boekel. 2001. 'A Review of Maillard Reaction in Food and Implications to Kinetic Modelling'. *Trends in Food Science and Technology* 11: 364–373.

Mehta, Bhavbhuti M., and Hilton C. Deeth. 2016. 'Blocked Lysine in Dairy Products: Formation, Occurrence, Analysis, and Nutritional Implications'. *Comprehensive Reviews in Food Science and Food Safety* 15 (1). Blackwell Publishing Inc.: 206–18. doi:10.1111/1541-4337.12178.

Moughan, Paul J. 2021. 'Population Protein Intakes and Food Sustainability Indices: The Metrics Matter'. *Global Food Security* 29 (June). Elsevier B.V. doi:10.1016/j.gfs.2021.100548.

Obiero, Kevin, Paul Meulenbroek, Silke Drexler, Adamneh Dagne, Peter Akoll, Robinson Odong, Boaz Kaunda-Arara, and Herwig Waidbacher. 2019. 'The Contribution of Fish to Food and Nutrition Security in Eastern Africa: Emerging Trends and Future Outlooks'. *Sustainability (Switzerland)* 11 (6). MDPI. doi:10.3390/su11061636.

Peram, Malleswara R., Simon M. Loveday, Aiqian Ye, and Harjinder Singh. 2013. 'In Vitro Gastric Digestion of Heat-Induced Aggregates of β-Lactoglobulin'. *Journal of Dairy Science* 96 (1): 63–74. doi:10.3168/jds.2012-5896.

Qin, Pingxu, Taoran Wang, and Yangchao Luo. 2022. 'A Review on Plant-Based Proteins from Soybean: Health Benefits and Soy Product Development'. *Journal of Agriculture and Food Research* 7. doi:10.1016/j.jafr.2021.100265.

Ranganathan, Janet, Richard Waite, Tim Searchinger, Daniel Vennard, Patrice Dumas, and Brian Lipinski. 2016. 'Shifting Diets for a Sustainable Food Future SUPREMA View Project Resource Watch'. https://Resourcewatch.Org/ View Project'. doi:10.13140/RG.2.1.3808.2961.

Reedy, Julia, Frederick Cudhea, Peilin Shi, Jianyi Zhang, Jennifer Onopa, Victoria Miller, Gitanjali Singh, Renata Micha, and Dariush Mozaffarian. 2019. 'Global Intakes of Total Protein and Sub-Types; Findings from the 2015 Global Dietary Database (P10-050-19)'. *Current Developments in Nutrition* 3. doi:10.1093/cdn/nzz034.p10-050-19.

Rutherfurd, Shane M., and Paul J. Moughan. 2007. 'Development of a Novel Bioassay for Determining the Available Lysine Contents of Foods and Feedstuffs'. *Nutrition Research Reviews* 20 (1): 3–16. doi:10.1017/S0954422407739124.

Sá, Amanda Gomes Almeida, Yara Maria Franco Moreno, and Bruno Augusto Mattar Carciofi. 2019. 'Food Processing for the Improvement of Plant Proteins Digestibility'. *Critical Reviews in Food Science and Nutrition*. doi:10.1080/10408398.2019.1688249.

Shewry, P. R. 2009. 'Wheat'. *Journal of Experimental Botany* 60 (6): 1537–53.

Sousa, Raquel, Isidra Recio, Dominique Heimo, Sébastien Dubois, Paul J. Moughan, Suzanne M. Hodgkinson, Reto Portmann, and Lotti Egger. 2023. 'In Vitro Digestibility of Dietary Proteins and in Vitro DIAAS Analytical Workflow Based on the INFOGEST Static Protocol and Its Validation with in Vivo Data'. *Food Chemistry* 404 (March). Elsevier Ltd. doi:10.1016/j.foodchem.2022.134720.

Tessari, P. 2018. 'Requirements of Non Essential Amino Acids for Protein Conservation in Humans: A Method of Estimation'. *Clinical Nutrition* 37. doi:10.1016/j.clnu.2018.06.1596.

Vieux, Florent, Didier Rémond, Jean Louis Peyraud, and Nicole Darmon. 2022. 'Approximately Half of Total Protein Intake by Adults Must Be Animal-Based to Meet Nonprotein, Nutrient-Based Recommendations, with Variations Due to Age and Sex'. *Journal of Nutrition* 152 (11). doi:10.1093/jn/nxac150.

Walstra, Pieter, Jan T. M. Wouters, and Tom J. Geurts. 2005. *Dairy Science and Technology*. CRC Press. doi:10.1201/9781420028010.

Wolfe, Robert R., Shane M. Rutherfurd, Il Young Kim, and Paul J. Moughan. 2016. 'Protein Quality as Determined by the Digestible Indispensable Amino Acid Score: Evaluation of Factors Underlying the Calculation'. *Nutrition Reviews* 74 (9). Oxford University Press: 584–99. doi:10.1093/nutrit/nuw022.

Wu, Guoyao. 2016. 'Dietary Protein Intake and Human Health'. *Food and Function*. Royal Society of Chemistry. doi:10.1039/c5fo01530h.

Zeece, Michael, Thom Huppertz, and Alan Kelly. 2008. 'Effect of High-Pressure Treatment on in-Vitro Digestibility of β-Lactoglobulin'. *Innovative Food Science and Emerging Technologies* 9 (1): 62–69. doi:10.1016/j.ifset.2007.05.004.

2 Protein Intake in the Mediterranean Diet

Ludovico Abenavoli, Carmen Colaci, Marialaura Scarcella, and Emidio Scarpellini

ABBREVIATIONS

ACE	Angiotensin-converting enzyme
AD	Alzheimer's disease
BCAAs	Branched-chain amino acids
BCAT1 gene	Branched-chain amino acid transaminase 1
CKD	Chronic kidney disease
CRC	Colorectal cancer
CVD	Cardiovascular disease
EVO (Oil)	Extra virgin olive (oil)
GFR	Glomerular filtration rate
eIF2α	Eukaryotic initiation factor 2
HCAs	Heterocyclic amines
IBD	Inflammatory bowel disease
IBS	Irritable bowel syndrome
IL-12	Interleukin-12
IL-17	Interleukin-17
LPS	Lipopolysaccharides
mTOR	Mechanistic target of rapamycin
NAFLD	Nonalcoholic fatty liver disease
ND	Neurological disorders
PAHs	Polycyclic aromatic hydrocarbons
PD	Parkinson's disease
SAM	S-adenosylmethionine
SAH	S-adenosylhomocysteine
SCFA	Short-chain fatty acids
T2DM	Type 2 diabetes mellitus
TLR4	Toll-like receptor 4
TMAO	Trimethylamine-N-oxide
UNESCO	United Nations Educational, Scientific and Cultural Organization
VEGF	Vascular endothelial growth factor
ZO-1	Zonula occludens-1

INTRODUCTION

The Mediterranean diet is named after its origin and diffusion among civilizations that populate the basin of the Mediterranean Sea, and it is one of the most studied and well-known nutritional models throughout the world. More extensively, the term Mediterranean diet is closely related not

DOI: 10.1201/b23328-2

FIGURE 2.1 The Mediterranean region. This figure shows a physical map of the Mediterranean area. The Mediterranean diet is a dietary scheme developed in Southern Italy, most of Greece and Crete, with variations in Morocco, other regions of Italy, Lebanon, Spain, Portugal, Tunisia, Turkey, and several other locations along the Mediterranean coast.

only to nutritional habits but also to the lifestyle and social profiles of the civilizations in this basin. Notably, the practice of sharing meals is one of the main aspects of the Mediterranean diet, where people of the same family or community eat together at the same table. Eating together is the basis of the cultural identity of the Mediterranean civilizations (Figure 2.1); it encompasses the values of hospitality, neighborhood, intercultural dialogue, and creativity. Agricultural practices and the transformation of products are historically deeply linked with the cultural and nutritional habits of this geographical region and go hand in hand with care and respect for the environment and biodiversity (Willett et al. 1995). The Mediterranean diet has been widely adopted for decades and, geographically, reflects food patterns typical of Southern Italy and most of Greece and Crete, with variations in Morocco, other regions of Italy, Lebanon, Spain, Portugal, Tunisia, Turkey, and several other locations along the coast of the Mediterranean basin (Willett et al. 1995).

The Mediterranean dietary model is mainly associated with areas where olives are traditionally cultivated and is strictly tied to extra virgin olive (EVO) oil production and consumption. Its alimentary constituents include the consumption of large quantities of fresh vegetables and fruit, whole grain cereals, legumes, nuts, and fish (Bach-Faig et al. 2011). The diet is characterized by low alcohol intake (mainly red wine) and low amounts of red meat and dairy products (Urquiaga et al. 2010; Morris and Bhatnagar 2016). From a nutritional intake perspective, the Mediterranean diet guarantees an adequate amount of proteins (10–15%), carbohydrates (55–60%), and fats (25%). The relative quantities of foods are well recognized in the Mediterranean diet. Cereals and whole grains represent the basis with 1–2 servings per day, as well as vegetables and fruit. Dairy products are present in 1–2 servings per day, preferably milk, low-fat yogurt, and low-fat cheese. Fish and white meat are allowed in 2–3 servings per week, while red meat, preferably lean, is consumed in less than two servings per week and no more than 2–4 eggs per week. Sweets, pastries, and sugar products are limited, and alcohol intake is no more than occasional (namely, one glass of red wine for female subjects and two glasses for males per day). EVO oil is the dressing of choice, and animal-derived fats are present in very low proportions (Serra-Majem et al. 2020). It has been widely demonstrated that adherence to the Mediterranean diet is associated with significant health benefits, including a lower risk of mortality from cardiovascular disease, metabolic disease, and cancer (Schwingshackl et al. 2019). This is related to the consumption of bioactive compounds such

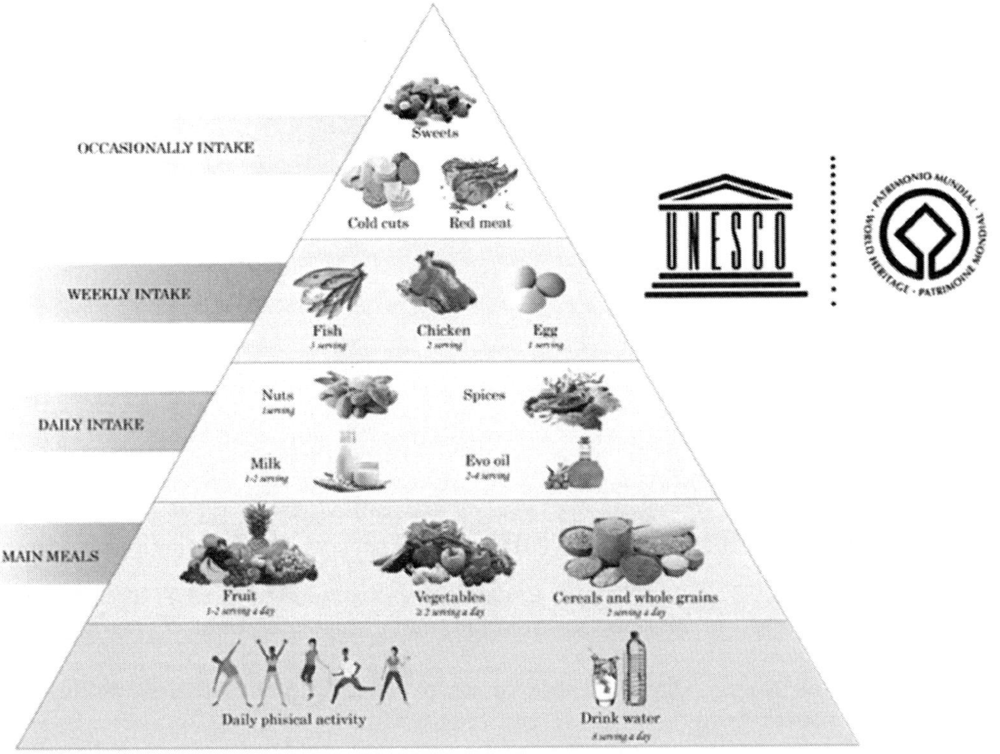

FIGURE 2.2 The Mediterranean diet pyramid. This figure shows the proportion of different foods in the Mediterranean diet, stratified by their intake frequency.

as polyphenols, monounsaturated and polyunsaturated fatty acids, and fibers present in large quantities in the Mediterranean diet (Abenavoli et al. 2021). These substances reduce blood lipids, inflammation, and oxidative stress, and improve insulin sensitivity and endothelial and antithrombotic function. The Mediterranean diet has both high scientific and humanistic value, and since 2010 has been identified as an intangible cultural heritage by UNESCO (Figure 2.2).

A KEY ROLE OF PROTEIN IN THE MEDITERRANEAN DIET: HEALTH AND DISEASE

Adequate protein intake is essential for nutritional health. The World Health Organization (WHO) suggests that the adequate daily intake of proteins is 0.83 g/kg/day (Ahnen et al. 2019). Proteins have long been considered able to help with weight management, increase high-density lipoprotein cholesterol levels, and improve thermogenesis and bone mineralization (Pasiakos et al. 2015). The total protein intake in the Mediterranean diet is 20% lower than in the typical Western diet, with the main sources being legumes, whole grains, and nuts (Figure 2.3).

Therefore, the main source of proteins in the Mediterranean diet is of vegetable origin, reaching an average intake of about 40 g per day. Animal-derived protein consumption is limited, at 50–60% lower than in the typical Western diet. Several studies have shown that plant-based protein intake can reduce cancer mortality compared to animal protein intake (Levine et al. 2014). A prospective cohort study found that replacing animal proteins (especially red meat) with plant proteins was associated with a reduced risk of colorectal cancer (CRC) (Liao et al. 2019). Multiple studies have linked increased CRC risk to excessive consumption of red meat and processed meat. While the pathogenetic mechanism remains unknown, several hypotheses have been proposed regarding the association between red meat and processed meat consumption and CRC. These include

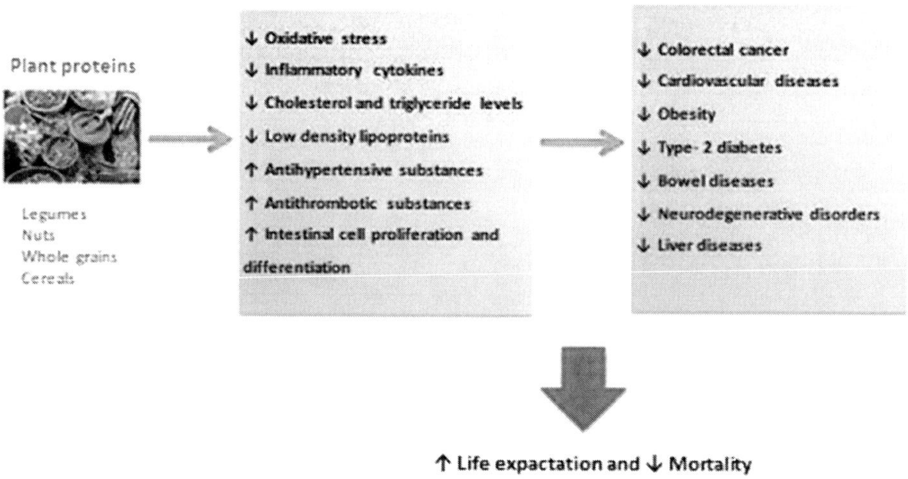

FIGURE 2.3 Effects of plant proteins on human health. This figure shows the main sources of plant-derived proteins in the Mediterranean diet and their pathophysiological pathways and effects on health and diseases.

heterocyclic amines (HCAs) and polycyclic aromatic hydrocarbons (PAHs), present in meat cooked at high temperatures, as well as heme-iron, nitrates, and nitrites used in meat processing (Carr et al. 2015).

Plant-based protein intake has also been shown to reduce the risk of cardiovascular disease (CVD), type 2 diabetes mellitus (T2DM), and obesity (Richter et al. 2015; Martín-Peláez et al. 2020). Additionally, many studies have shown a strong association between excessive consumption of animal proteins, especially from red meat, and chronic kidney disease (CKD). Replacing animal proteins with vegetable proteins has been shown to reduce the risk of CKD by 31–62.4% (Lew et al. 2016). This is because excessive intake of animal proteins can cause glomerular hyperfiltration, which is associated with albuminuria and a progressive reduction of the glomerular filtration rate (GFR) (Kamper and Strandgaard 2017).

Among plant sources of proteins, legumes provide high levels of protein without the high-fat content often associated with high-protein foods. The particularly high protein content of legumes makes them unique among vegetable protein sources, especially compared to cereals, which have half the protein content in proportion to the energy they provide (Boye et al. 2010). Legumes are also rich in bioactive co-nutrients, such as polyphenols, which play an important role in cancer prevention and total cholesterol reduction (Kalogeropoulos et al. 2010). Nuts are another protein source in the Mediterranean diet. Although nuts contain low amounts of essential amino acids, this is not a nutritional concern due to protein complementation. Furthermore, nuts have a low lysine/arginine ratio, which is inversely associated with the risk of developing hypercholesterolemia and atherosclerosis (Brufau et al. 2006). Cereals form the staple of the Mediterranean diet, providing carbohydrates, fiber, and bioactive peptides. Research has shown that the bioactive peptides in wheat, oats, barley, and rice, though present in smaller quantities than in legumes, have a protective role against chronic non-communicable diseases. These peptides inhibit the angiotensin-converting enzyme (ACE), producing an antihypertensive effect, and the enzyme dipeptidyl-peptidase, resulting in an antithrombotic action. A recent multi-center randomized and controlled study showed that the Mediterranean-style dietary pattern led to a significant reduction in systolic blood pressure and arterial stiffness (Jennings et al. 2019). Wheat and rice contain a higher proportion of proteins than other cereals, and anticancer amino acid sequences have been identified in these grains (Cavazos and De Mejia 2013).

Research shows that the Mediterranean diet is effective in reducing abdominal adiposity, particularly metabolically harmful visceral fat, independent of weight loss, making it a healthy dietary choice for individuals with obesity and overweight (Bendall et al. 2017). The reduced content of saturated fats associated with the increased consumption of vegetable proteins and cereals makes the Mediterranean diet an adequate nutritional approach to combat obesity. The elimination of visceral fat significantly reduces cardiovascular mortality and morbidity. A recent meta-analysis demonstrated that adherence to the Mediterranean diet was associated with a 19% lower risk of developing metabolic syndrome. Individual components, such as waist circumference and blood pressure, were also improved (Godos et al. 2016).

Recent studies link high protein intake with changes in the incidence of inflammatory bowel disease (IBD), suggesting that high protein intake from several sources, including red meat, fish, eggs, milk, cheese, and nuts, may be a factor in modulating IBD incidence. The mechanisms underlying the role of animal proteins in modulating the onset of IBD remain largely unknown. Researchers hypothesize that the degradation of animal proteins in the gut may produce substrates that promote pathobiont expansion or short-chain fatty acids (SCFAs) that modulate enterocyte function (Mentella et al. 2020).

GUT MICROBIOTA AND PROTEIN INTAKE

Gut microbiota consists of about 100 trillion microbes and includes mainly bacteria, viruses, fungi, and a small percentage of parasites. The most abundant bacterial species are *Prevotella*, *Ruminococcus*, *Bacteroidetes*, and *Firmicutes*. An adequate ratio between the *Bacteroidetes* and *Firmicutes* species plays an important role in human health (Adak and Khan 2018). Gut microbiota is deeply involved in pathologies such as IBD, nonalcoholic fatty liver disease (NAFLD), CVD, obesity, and neurological disorders (ND) (Solch et al. 2022). The interaction between diet and gut microbiota is bi-directional. While the microbiota acts on digested nutrients, food has a strong impact on intestinal microbial composition. A diet rich in fats, with high consumption of red meat and refined carbohydrates and low consumption of fish, vegetables, and fruit can directly affect the immune system, causing dysbiosis (Kaoutari et al. 2013). The typical Western diet, consisting of foods rich in fat, sugar, and animal proteins, has been linked to low-grade chronic inflammation, metabolic diseases, and obesity. Recent studies have shown that such a diet can alter the intestinal microbiota, increasing lipopolysaccharides (LPS) and trimethylamine-N-oxide (TMAO) while decreasing SCFAs, thus causing systemic inflammation.

SCFAs play an important role as fuels for intestinal epithelial cells (IEC). They modulate IEC proliferation and differentiation and the functions of subpopulations such as enteroendocrine cells, influencing intestinal motility and strengthening intestinal barrier function and host metabolism (Martin-Gallausiaux et al. 2020). The Mediterranean diet, which is based on the consumption of vegetable proteins, has been observed to modulate the microbiota (Singh et al. 2020). Adherence to this diet modifies microbial composition, increasing *Bacteroides*, *Prevotella*, and *Enterococci*, which are essential for maintaining health, and decreasing *Clostridium*, which is highly pathogenic (Lopez-Legarrea et al. 2014). Furthermore, adherence to the Mediterranean diet for just three months has been shown to increase SCFA levels and reduce inflammatory cytokines, such as interleukin-12 (IL-12), interleukin-17 (IL-17), and vascular endothelial growth factor (VEGF) (Pagliai et al. 2019).

Several studies have shown that excessive intake of animal proteins increases fermented products in the intestine, causing remodeling of the intestinal barrier and increased permeability ("leaky gut"). This increased intestinal permeability and reduced expression of zonula occludens-1 (ZO-1), claudin, and occludin, which form the tight junctions of the intestinal barrier, cause LPS translocation into the systemic level. LPS activates the Toll-like receptor 4 (TLR4) present on most cells and macrophages. This binding initiates a cascade of signals through the expression of cytokines, inducing an inflammatory response (Minihane et al. 2015). Chronic inflammatory response has been associated with insulin resistance and T2MD (Cani et al. 2007).

Altered intestinal permeability and chronic inflammation cause the onset and progression of IBD, irritable bowel syndrome (IBS), obesity, CRC, liver disease, inflammatory joint diseases, and neurological diseases (Mentella et al. 2020). Levels of some metabolites from the fermentation of animal proteins are increased in patients with ulcerative colitis compared to healthy subjects. Consequently, the mucus layer undergoes remodeling, with cell and mucus loss that increases paracellular permeability. Other metabolites from normal protein degradation may be harmful in conditions of altered microbiota and intestinal inflammation. Phenolic compounds, products of aromatic amino acid fermentation by *Bacteroidetes spp.* and some *Firmicutes*, have a detrimental effect on the mucosal barrier function in vitro, which depends on the presence of other nutrients in vivo. N-nitroso compounds have carcinogenic potential. Polyamines may influence the expression of a cotransporter for monocarboxylates such as lactate, pyruvate, and leucine, which contribute to the regulation of central metabolic pathways and insulin secretion (Gilbert et al. 2018).

Alzheimer's disease (AD) and Parkinson's disease (PD) are the two most prevalent neurodegenerative diseases, neither of which has prevention or cure. The Mediterranean diet may be neuroprotective by modulating gut microbiota. Conversely, the Western diet, which is low in fiber, may be associated with AD or PD. Gut dysbiosis has been proposed as a mechanism for neuroinflammation leading to these diseases. Adherence to the Mediterranean diet is associated with a lower risk of AD and PD development and, more generally, improved neurologic health. The health benefits of the Mediterranean diet have traditionally been attributed to its anti-inflammatory and antioxidant effects, mediated by foods rich in omega-3 fatty acids and polyphenols. Interestingly, polyphenols, abundant in vegetables, fruits, and wine, are poorly absorbed in the small intestine (~5–10% of intake), suggesting a secondary mechanism for their associated benefits. Accumulation of polyphenols in the colon allows for microbial fermentation and growth of beneficial microbes such as *Lactobacillus*, while decreasing pathogenic genera such as *Clostridium*, *Shigella*, and *Escherichia*. Additionally, fiber serves as the main energy substrate for gut microbiota (Figure 2.4).

The Mediterranean diet reduces the bacterial species that produce SCFAs. Lower levels of SCFA can increase gut permeability and alter gut pH, allowing for the growth of opportunistic pathogens such as *Shigella* and *Escherichia*, which are found at higher levels in both AD and PD. Specifically, there is evidence of a decrease in *Roseburia* in AD and *Roseburia*, *Blautia*, and *Faecalibacterium* in PD. Conversely, increases in SCFA-producing bacteria such as *Akkermansia* and *Bifidobacterium* in PD and *Blautia* and *Bifidobacterium* in AD have been reported. These genera are usually

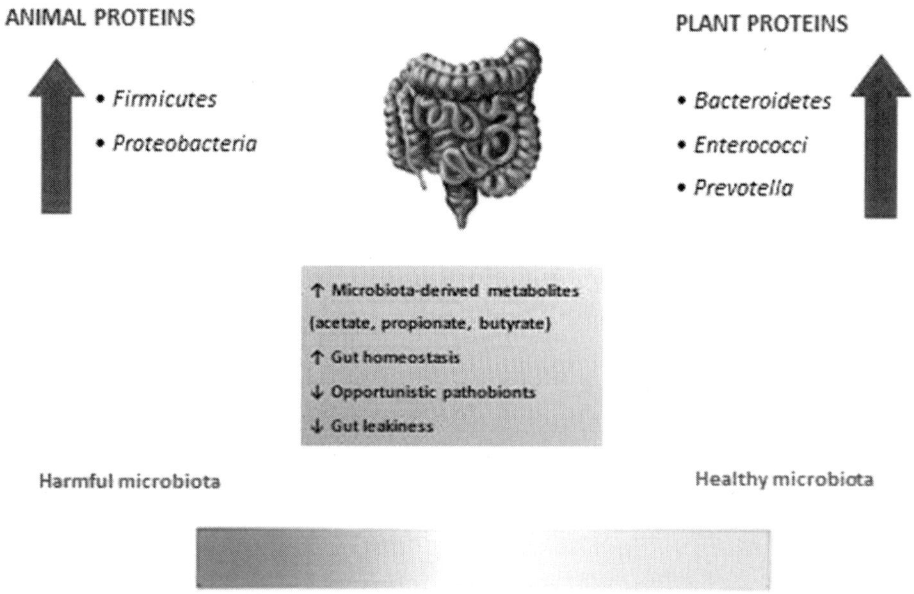

FIGURE 2.4 Protein source and gut microbiota. This figure shows the different effects of animal- vs plant-derived proteins on gut homeostasis and the principal physiopathological mechanisms involved. Arrows indicate an increase in bacterial phyla.

considered beneficial for human health and are being explored as probiotics (*Bifidobacterium*) or potential probiotic candidates (Solch et al. 2022).

BRANCHED-CHAIN AMINO ACIDS (BCAAS)

The different ratios between animal and plant proteins result in different contents of some essential amino acids in the Mediterranean and Western diets. For example, methionine intake is lower in the Mediterranean diet. Methionine is the first amino acid present in proteins encoded by the nucleus, as the methionine codon signals the beginning of protein translation. Methionine metabolism occurs predominantly in the liver and consists of two steps: the transmethylation cycle for the production of S-adenosylmethionine (SAMe) and its metabolism to homocysteine, and the transsulfuration pathway, which leads to the generation of several sulfur metabolites, including cysteine, glutathione, and the gaseous signaling molecule hydrogen sulfide (H_2S). Adequate levels of H_2S play a key cytoprotective role and reduce inflammation by activating different signaling pathways, such as the target of rapamycin (mTOR) pathway and eukaryotic initiation factor 2 (eIF2α) (Hine et al. 2015). However, elevated H_2S levels are present in cardiovascular disease, rheumatoid arthritis, and tumor growth (Hellmich et al. 2015).

Researchers have observed altered circulating BCAA levels in patients diagnosed with cancer. Retrospective studies of metabolomics have shown that elevated plasma levels of BCAAs are associated with more than double the risk of pancreatic cancer and precede clinical presentation by many years (Mayers et al. 2014). In different types of cancer, there is increased expression of the BCAT1 gene (branched-chain amino acid transaminase 1), which is important in the metabolism of BCAAs. Increased expression of this gene facilitates the synthesis of BCAAs, resulting in tumor growth (Zheng et al. 2016). Several studies, conducted on model organisms such as rats and mice, have shown that methionine dietary restriction results in increased lifespan and reduced chronic disease risk, including cancer, obesity, and liver steatosis (Brown-Borg and Buffenstein 2017).

Other essential amino acids, such as leucine, isoleucine, valine, and tryptophan, are also present in lower amounts in the Mediterranean diet compared to the traditional Western diet due to lower intake of meat, eggs, and dairy products (Figure 2.5). These BCAAs modulate insulin sensitivity, improving its function and playing a role in glucose metabolism. Elevated levels of BCAAs have been found in insulin-resistant humans and rodents (Cuomo et al. 2022). The increase in circulating BCAAs seems to affect heart function, although the mechanisms are still unclear. High levels of intracellular cardiac leucine can activate the mTOR pathway (mechanistic target of rapamycin), thus promoting cardiac insulin resistance and hypertrophy. Indeed, mTOR inhibition mitigates heart dysfunction in multiple models of heart failure (Völkers et al. 2013). Furthermore, increased BCAA concentrations in serum or plasma are linked to the onset and severity of NAFLD (Abenavoli et al. 2019; Hasegawa et al. 2020).

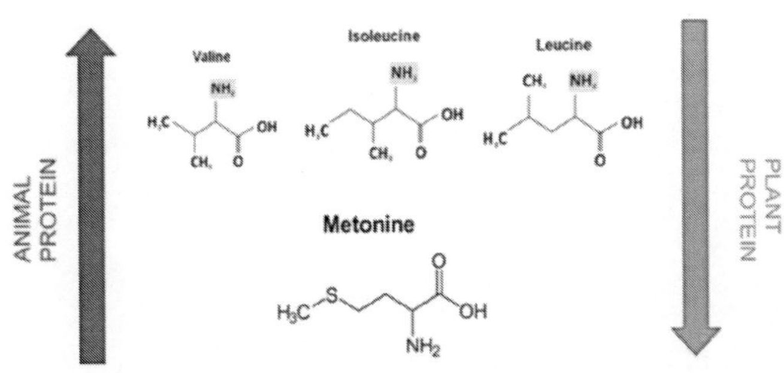

FIGURE 2.5 Proportion of different BCCAs in animal and plant proteins.

CONCLUSIONS

In summary, evidence suggests that the Mediterranean diet can improve life expectancy by modifying the incidence and prevalence of several chronic diseases, such as CVD, NAFLD, and cancer, which are reduced due to different intake of plant- and animal-derived proteins, as compared to the Western diet. Adherence to the Mediterranean diet is associated with reduced oxidative stress and increased antihypertensive and antithrombotic substances. Moreover, lower levels of BCAAs associated with reduced animal protein intake decrease the likelihood of occurrence and progression of chronic diseases, such as obesity, T2DM, and hypertension, liver disease, and neurological disorders. High levels of BCAAs appear to promote the growth of some cancers and worsen their prognosis. In addition, lower consumption of products derived from animal proteins promotes a favorable balance of gut microbiota, increasing the presence of healthy bacteria species while lowering harmful ones. This condition helps prevent dysbiosis and its related illnesses, such as NAFLD and IBD.

MINI DICTIONARY OF TERMS

Dysbiosis: Condition of imbalance in the composition and function of the microbiota.

Gut microbiota: Set of microorganisms that colonize the intestinal tract.

Leaky gut: Altered permeability of the intestinal mucosa.

Oxidative stress: Pathological condition caused by the loss of the physiological balance between the production and elimination of oxidizing chemical species.

Polyphenols: Family of natural, semi-natural, or synthetic organic molecules present in the plant kingdom, characterized by the presence of multiple phenolic groups.

Processed meat: Meat processed by salting, seasoning, fermentation, smoking, or other processes that increase its taste or improve its preservation.

SUMMARY POINTS

- The Mediterranean diet is a balanced diet based on the consumption of cereals, legumes, fruits and vegetables, nuts, olive oil, fish, small amounts of red meat, and alcohol, especially red wine.
- The primary source of protein in the Mediterranean diet is plant protein.
- Plant protein intake plays a protective role against various chronic pathologies such as CVD, neurological disorders, CKD, T2DM, IBD, NAFLD, and cancer.
- Plant proteins contain BCAAs in lower concentrations than animal proteins.
- Reduced BCAA levels have been shown to protect against oxidative stress and related diseases.
- Animal model studies have shown that reducing methionine levels can lower the risk of cancer, obesity, and liver steatosis.
- Lower levels of other BCAAs, such as isoleucine, leucine, valine, and tryptophan, improve insulin sensitivity and reduce the risk of T2DM.
- Plant proteins modulate gut microbiota, favoring some bacterial species, such as *Bacteroides*, *Prevotella*, and *Enterococci*, and reducing dysbiosis.
- A correct balance of gut microbiota modifies the onset and progression of non-communicable chronic diseases.
- Substituting animal proteins with plant proteins improves the quality of life and life expectancy.

REFERENCES

Abenavoli, L., L. Boccuto, A. Federico, M. Dallio, C. Loguercio, L. Di Renzo, and A. De Lorenzo. 2019. "Diet and Non-Alcoholic Fatty Liver Disease: The Mediterranean Way." *International Journal of Environmental Research and Public Health* 16 (17): 3011.

Abenavoli, L., T. Larussa, A. Corea, A. Procopio, L. Boccuto, M. Dallio, A. Federico, and F. Luzza. 2021. "Dietary Polyphenols and Non-Alcoholic Fatty Liver Disease." *Nutrients* 13 (2): 494.

Adak, A., and R. K. Mojibur. 2018. "An Insight into Gut Microbiota and Its Functionalities." *Cellular and Molecular Life Sciences* 76 (3): 473–93.

Ahnen, R. T., S. J. Satya, and J. L. Slavin. 2019. "Role of Plant Protein in Nutrition, Wellness, and Health." *Nutrition Reviews* 77 (11): 735–47.

Bach-Faig, A., E. M. Berry, D. Lairon, J. Reguant, A. Trichopoulou, S. Dernini, F. X. Medina, et al. 2011. "Mediterranean Diet Pyramid Today." *Science and Cultural*14 (12A): 2274–84.

Bendall, C., H. L. Mayr, R. Opie, M. Bes–Rastrollo, C. Itsiopoulos, and C. J. Thomas. 2017. "Central Obesity and the Mediterranean Diet: A Systematic Review of Intervention Trials." *Critical Reviews in Food Science and Nutrition* 58 (18): 3070–84.

Boye, J., F. Zare, and A. Pletch. 2010. "Pulse Proteins: Processing, Characterization, Functional Properties and Applications in Food and Feed". *Food Research International* 43: 414–31.

Brown-Borg, M. Holly, and R. Buffenstein. 2017. "Cutting Back on the Essentials: Can Manipulating Intake of Specific Amino Acids Modulate Health and Lifespan?" *Ageing Research Reviews* 39 (October): 87–95.

Brufau, G., J. Boatella, and M. Rafecas. 2006. "Nuts: Source of Energy and Macronutrients." *British Journal of Nutrition* 96 (S2): S24–28.

Cani, P. D., J. Amar, M. Iglesias, M. Poggi, C. Knauf, D. Bastelica, A. M. Neyrinck, et al. 2007. "Metabolic Endotoxemia Initiates Obesity and Insulin Resistance." *Diabetes* 56 (7): 1761–72.

Carr, P. R., V. Walter, H. Brenner, and M. Hoffmeister. 2015. "Meat Subtypes and Their Association with Colorectal Cancer: Systematic Review and Meta-Analysis." *International Journal of Cancer* 138 (2): 293–302.

Cavazos, A., and E. G. De Mejia. 2013. "Identification of Bioactive Peptides from Cereal Storage Proteins and Their Potential Role in Prevention of Chronic Diseases." *Comprehensive Reviews in Food Science and Food Safety* 12 (4): 364–80.

Cuomo, P., R. Capparelli, A. Iannelli, and D. Iannelli. 2022. "Role of Branched-Chain Amino Acid Metabolism in Type 2 Diabetes, Obesity, Cardiovascular Disease and Non-Alcoholic Fatty Liver Disease." *Journal of Molecular Sciences* 23 (8): 4325.

Gilbert, N. I., A. K. Kies, and S. W.C. Van Mil. 2018. "Protein Fermentation in the Gut; Implications for Intestinal Dysfunction in Humans, Pigs, and Poultry." *American Journal of Physiology-Gastrointestinal and Liver Physiology* 315 (2): G159–70.

Godos, J., G. Zappalà, S. Bernardini, I. Giambini, M. Bes–Rastrollo, and M. Á. Martínez-González. 2016. "Adherence to the Mediterranean Diet Is Inversely Associated with Metabolic Syndrome Occurrence: A Meta-Analysis of Observational Studies." *International Journal of Food Sciences and Nutrition* 68 (2): 138–48.

Hasegawa, T., Y. Takikawa, T. Endo, K. Mikami, M. Kimura, N. Sawada, S. Nakaji, and S. Fukuda. 2020. "Changed Amino Acids in NAFLD and Liver Fibrosis: A Large Cross-Sectional Study without Influence of Insulin Resistance." *Nutrients* 12 (5): 1450.

Hellmich Mark, R., C. Coletta, C. Chao, and C. Szabo. 2015. "The Therapeutic Potential of Cystathionine Β-Synthetase/Hydrogen Sulfide Inhibition in Cancer." *Antioxidants & Redox Signaling* 22 (5): 424–48.

Hine, C., E. Harputlugil, Y. Zhang, C. Ruckenstuhl, B. C. Lee, L. E. Brace, A. Longchamp, et al. 2015. "Endogenous Hydrogen Sulfide Production Is Essential for Dietary Restriction Benefits." *Cell* 160 (1–2): 132–44.

Jennings, A., A. M. Berendsen, L. C. P. G. M. De Groot, E. J. M. Feskens, A. Brzozowska, E. Sicińska, B. Pietruszka, et al. 2019. "Mediterranean-Style Diet Improves Systolic Blood Pressure and Arterial Stiffness in Older Adults." *Hypertension* 73 (3): 578–86.

Kalogeropoulos, N., A. Chiou, M. Ioannou, et al. 2010. "Nutritional Evaluation and Bioactive Microconstituents (Phytosterols, Tocopherols, Polyphenols, Triterpenic Acids) in Cooked Dry Legumes Usually Consumed in the Mediterranean Countries." *Food Chemistry* 121: 682–90.

Kamper, A. L., and S. Strandgaard. 2017. "Long-Term Effects of High-Protein Diets on Renal Function." *Annual Review of Nutrition* 37 (1): 347–69.

Kaoutari, A. E., F. Armougom, J. I. Gordon, D. Raoult, and B. Henrissat. 2013. "The Abundance and Variety of Carbohydrate-Active Enzymes in the Human Gut Microbiota." *Nature Reviews Microbiology* 11 (7): 497–504.

Levine, M. E., J. Suarez, S. Brandhorst, P. Balasubramanian, C. W. Cheng, F. Madia, L. Fontana, et al. 2014. "Low Protein Intake Is Associated with a Major Reduction in IGF-1, Cancer, and Overall Mortality in the 65 and Younger but Not Older Population." *Cell Metabolism* 19 (3): 407–17.

Lew Quan-Lan, J., T. H. Jafar, W. L. K Hiromi., A. Jin, K. Yew Chow, J. M. Yuan, and W. P. Koh. 2016. "Red Meat Intake and Risk of ESRD." *Journal of the American Society of Nephrology* 1: 304–12.

Liao, L M., E. Loftfield, A. Etemadi, B. L. Graubard, and R. Sinha. 2019. "Substitution of Dietary Protein Sources in Relation to Colorectal Cancer Risk in the NIH-AARP Cohort Study." *Cancer Causes & Control* 30 (10): 1127–35.

Lopez-Legarrea, P., R. F. Nicholas, M. A. Zulet, J. A. Martínez, and I. D. Caterson. 2014. "The Influence of Mediterranean, Carbohydrate and High Protein Diets on Gut Microbiota Composition in the Treatment of Obesity and Associated Inflammatory State." *Asia Pacific Journal of Clinical Nutrition* 23 (3): 360–68.

Mayers, J. R., C. Wu, C. B. Clish, P. Kraft, M. E. Torrence, et al. 2014. "Elevation of Circulating Branched-chain Amino Acids is An Early Event in Human Pancreatic Adenocarcinoma Development." *Nature Medicine* 20 (10): 1193–98.

Martin-Gallausiaux, C., L. Marinelli, H. M. Blottière, P. Larraufie, and N. Lapaque. 2020. "SCFA: Mechanisms and Functional Importance in the Gut." *Processing of the Nutrition Society* 80 (1): 37–49.

Martín-Peláez, S., M. Fitó, and O. Castañer. 2020. "Mediterranean Diet Effects on Type 2 Diabetes Prevention, Disease Progression, and Related Mechanisms. A Review." *Nutrients* 12 (8): 2236.

Mentella, M. C., F. Scaldaferri, M. Pizzoferrato, A. Gasbarrini, and G. A. Donato Miggiano. 2020. "Nutrition, IBD and Gut Microbiota: A Review." *Nutrients* 12 (4): 944.

Minihane, A. M., S. Vinoy, W. R. Russell, A. Baka, H. M. Roche, K. Tuohy, J. L. Teeling, et al. 2015. "Low-Grade Inflammation, Diet Composition and Health: Current Research Evidence and Its Translation." *British Journal of Nutrition* 114 (7): 999–1012.

Morris, L., and D. Bhatnagar. 2016. "The Mediterranean Diet." *Current Opinion in Lipidology* 27 (1): 89–91.

Pagliai, G., E. Russo, E. Niccolai, M. Dinu, V. Di Pilato, A. Magrini, G. Bartolucci, et al. 2019. "Influence of a 3-Month Low-Calorie Mediterranean Diet Compared to the Vegetarian Diet on Human Gut Microbiota and SCFA: The CARDIVEG Study." *European Journal of Nutrition* 59 (5): 2011–24.

Pasiakos, S. M., H. R. Lieberman, and V. L. Fulgoni. 2015. "Higher-Protein Diets Are Associated with Higher HDL Cholesterol and Lower BMI and Waist Circumference in US Adults." *Journal of Nutrition* 145 (3): 605–14.

Richter, C. K., A. C. Skulas-Ray, C. M. Champagne, and P. M. Kris-Etherton. 2015. "Plant Protein and Animal Proteins: Do They Differentially Affect Cardiovascular Disease Risk?" *Advances in Nutrition* 6 (6): 712–28.

Schwingshackl, L., J. Morze, and G. Hoffmann. 2019. "Mediterranean Diet and Health Status: Active Ingredients and Pharmacological Mechanisms." *Journal of Pharmacology* 177 (6): 1241–57.

Serra-Majem, L., L. Tomaino, S. Dernini, E. M. Berry, D. Lairon, J. Ngo, A. Bach-Faig, et al. 2020. "Updating the Mediterranean Diet Pyramid towards Sustainability: Focus on Environmental Concerns." *International Journal of Enviromental Research and Public Health* 17 (23): 8758.

Singh, R., D. A. Halaka, Z. Hayouka, and O. Tirosh. 2020. "High-Fat Diet Induced Alteration of Mice Microbiota and the Functional Ability to Utilize Fructooligosaccharide for Ethanol Production." *Frontiers in Cellular and Infection Microbiology* 10 (August 7): 376.

Solch, R. J., J. O. Aigbogun, A. G. Voyiadjis, G. Talkington, R. Darensbourg, S. O'Connell, K. Pickett, S. Perez, and D. M. Maraganore. 2022. "Mediterranean Diet Adherence, Gut Microbiota, and Alzheimer's or Parkinson's Disease Risk: A Systematic Review." *Journal of the Neurological Sciences* 434 (March): 120166.

Urquiaga, I., P. Strobel, D. Pérez, C. Martínez, A. Cuevas, O. Castillo, G. Marshall, J. Rozowski, and F. Leighton. 2010. "Mediterranean Diet and Red Wine Protect against Oxidative Damage in Young Volunteers." *Atherosclerosis* 211 (2): 694–99.

Völkers, M., H. Toko, S. Doroudgar, S. Din, P. Quijada, A. Joyo, L. Ornelas, et al. 2013. "Pathological Hypertrophy Amelioration by PRAS40-Mediated Inhibition of MTORC1." *Proceeding of the National Academy of Sciences of the United States of America* 110 (31): 12661–66.

Willett, W. C., F. M. Sacks, A. Trichopoulou, G. Drescher, A. Ferro-Luzzi, E. Helsing, and D. Trichopoulos. 1995. "Mediterranean Diet Pyramid: A Cultural Model for Healthy Eating." *The American Journal of Clinical Nutrition* 61 (6): 1402S–6S.

Zheng, Y., H. Weixin, C. Bicheng, T. H. Min Grahn, Z. Yanrong, H. L. Bao, Y. Xiao Zhu, and Q. Zhang. 2016. "BCAT1, a Key Prognostic Predictor of Hepatocellular Carcinoma, Promotes Cell Proliferation and Induces Chemoresistance to Cisplatin." *Liver International* 36 (12): 1836–47.

3 Protein Intake and Exercise

Yıldırım Kayacan, İzzet İslamoğlu, and Hayrullah Yazar

ABBREVIATIONS

1 RM	One-repetition maximum
5-HTP	Hydroxytryptophan
ADP	Adenosine diphosphate
AMPK	Adenosine monophosphate-activated protein kinase
ATP	Adenosine triphosphate
cAMP	Cyclic adenosine monophosphate
CBG	Corticosteroid-binding globulin
CK	Creatine kinase
GH	Growth hormone
IGF	Insulin-like growth factors
IP3	Inositol trisphosphate
ISSN	International Society of Sports Nutrition
MHC	Major histocompatibility complex
RDA	Recommended dietary allowance
T3	Triiodothyronine
T4	Thyroxine
WP	Whey protein isolate

INTRODUCTION

Proteins are large molecules formed by the arrangement of amino acids, which constitute fundamental components of living organisms and serve various crucial functions. Amino acids are the building blocks of proteins. Through the utilization of 20 distinct amino acids in various combinations, the body can synthesize thousands of diverse proteins. Amino acids have a structure containing an amino group (-NH2) and a carboxyl group (-COOH) (Wang et al., 2022).

Daily protein consumption varies based on individual needs, age, gender, activity level, body composition, and objectives (Koguchi, 2022). Generally, the protein intake necessary for sustaining health and supporting bodily functions is categorized by age group. For adults aged 19 and above, 10–35% of the daily total calorie intake should comprise protein; for children aged 4–18, this range is 10–30%; and for children aged 1–3, it is 5–22%. This range is commonly accepted as suitable for most of the population (Institute of Medicine, 2006).

Health authorities, such as the American Dietetic Association and the World Health Organization, typically recommend that an average adult's daily protein requirement is 0.8 grams per kilogram of body weight. Nevertheless, this quantity may differ for more active individuals, athletes aiming for muscle development, or those wanting to maintain or reduce body weight. The necessity for daily protein fluctuates depending on objectives and individual attributes. Individuals engaging in bodybuilding or resistance training consume more protein to facilitate muscle development, repair, and for energy. In older individuals, it is used to prevent muscle mass loss, while pregnant and lactating women need it to meet increased protein requirements (Stephens et al., 2015).

DOI: 10.1201/b23328-3

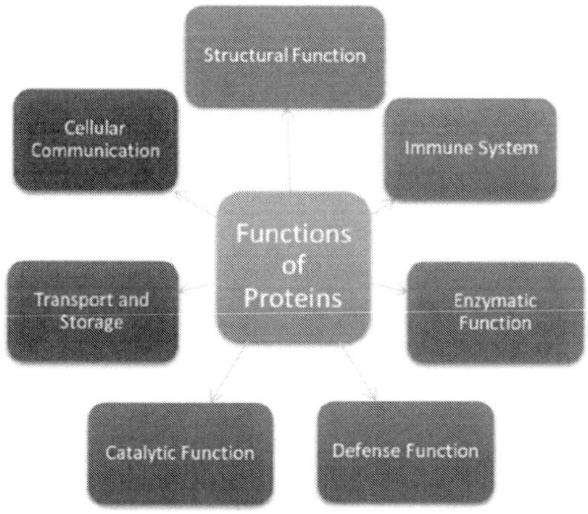

FIGURE 3.1 Functions of proteins. The diverse functions of proteins contribute to the regulation of numerous biochemical and physiological processes within the body. The structure and function of each protein have been specifically defined, and it is crucial for these proteins to be present in appropriate amounts for the body to function properly and maintain health.

However, excessive protein intake can also cause health complications. Particularly, individuals at risk of kidney disease or those with chronic health issues should meticulously manage their protein consumption. All of the protein functions outlined below are associated with exercise. The functions of proteins are shown in Figure 3.1.

Some of the basic functions of proteins are stated below.

Structural Function

Proteins form the basis of cellular structure. For instance, many physical characteristics, ranging from muscle structure to skin elasticity, are determined by proteins (Kustatscher et al., 2022). Proteins are used especially to support the structure of tissues like connective tissue and skeletal muscle. Proteins such as collagen, elastin, and keratin provide durability and elasticity to tissues like skin, bone, tendons, and others. Some proteins support the structure of the cell membrane and facilitate interactions with it. These proteins stabilize the cell membrane and facilitate molecular exchange between the interior and exterior of the cell. Other proteins catalyze cellular reactions. These enzymes regulate metabolic pathways and support cellular functions. Enzymes can also aid in the transport of molecules within the cell.

These structural proteins play a significant role in preserving and organizing the organism's physical structure. Furthermore, their functions are vital for establishing cellular organization and maintaining overall health. Exercise leads to microscopic tears in muscle fibers. The body utilizes proteins to repair these tears and strengthen the muscles. Proteins assist in muscle repair and growth. Additionally, regular exercise enhances muscle protein synthesis. Resistance exercises are potent stimuli for increasing protein assimilation in muscles, as they can improve the balance between protein synthesis and breakdown in muscles. However, for hypertrophy to occur, post-exercise recovery should be accompanied by dietary intake. Therefore, athletes should consider post-exercise protein supplementation to achieve a positive protein balance and maximize the skeletal muscle's adaptive response (Koopman et al., 2007).

Enzymatic Function

Enzymes are proteins that accelerate biochemical reactions. Numerous fundamental biological processes, such as metabolism, nutrient digestion, and DNA replication, occur through enzyme mediation (Schwartz, 2023). Enzymes encourage substrate molecules to engage in chemical reactions. These reactions alter the chemical bonds of substrates, resulting in product formation. Enzymes facilitate the transition of substrates into reaction transition states, thereby lowering activation energy. They lead to the formation of specific products by enabling substrates to undergo reactions. Each enzyme catalyzes a specific reaction, thus aiding in the regulation of cellular metabolism. Enzymes work with specific substrates and typically catalyze only a particular reaction. This specificity governs the intricate metabolic pathways of organisms, ensuring controlled execution of each reaction.

Enzymes regulate vital processes within organisms and are coded by genetic instructions. Each enzyme serves a distinct purpose and governs the complexity of cellular operations. Enzyme activity increases alongside physical training. Exercise boosts metabolic rates due to energy production and intense utilization. This, in turn, influences enzymatic reactions required for ATP (adenosine triphosphate) production during muscle contractions. Exercise results in lactic acid-lactate energy production, which is metabolized through supporting enzymes.

Immune System

Proteins play a critical role in many biological processes, including the immune system. They contribute to immune system functioning in various ways: Antibodies (immunoglobulins) assist in recognizing and neutralizing foreign microorganisms, toxins, and other foreign substances in the body. These specialized proteins constitute essential components of the immune response (Bungener et al., 2002). Cytokines are proteins that facilitate intercellular communication. Produced by the immune system, they play a role in processes such as inflammation and immune response regulation. Complement proteins within this system perform functions like breaking down cellular membranes and enhancing the effectiveness of antibodies.

These protein types are examples, and the effects of proteins on the immune system are quite extensive. During exercise, the population of white blood cells (leukocytes) can vary based on the intensity, severity, and duration of the activity. Moderate-intensity exercise prompts an inflammatory response, whereas high-intensity exercise tends to encourage an anti-inflammatory response to mitigate skeletal muscle damage (Terra et al., 2012).

Transport Function

Proteins serve various roles in cellular transport, including transporting proteins present in the cell membrane inside and outside the cell, carrying certain hormones (e.g., thyroid hormones), moving molecules between organelles within the cell, transporting minerals (e.g., iron) and trace elements (e.g., zinc), and incorporating ingested nutrients into the energy system. Proteins like hemoglobin play a role in transporting oxygen from the lungs to tissues and removing carbon dioxide from the body (Rensvold et al., 2022). Hemoglobin in erythrocytes (red blood cells) carries oxygen from the lungs to various body tissues of the body. During exercise, muscles require more oxygen, which can increase the amount of oxygen carried by hemoglobin. Proteins aid in nutrient transport through the digestive system and transport cellular signaling molecules like hormones. Hormones released during exercise regulate processes such as energy production, muscle growth, and metabolism. The process of muscle contraction, described by the sliding filament theory, involves mineral transport into or out of cells. Proteins play a role in this process. Increased muscle contraction during exercise leads to a greater need for calcium.

These examples illustrate the interaction between exercise and protein transport functions. The increased energy demand during exercise may necessitate more efficient transport of nutrients, oxygen, and other components. Consequently, protein transport functions become even more crucial during exercise to ensure the balanced functioning of the body.

Hormonal Function

Peptide and protein hormones regulate various bodily processes. For instance, insulin plays a role in blood sugar regulation (Vendruscolo and Fuxreiter, 2022). Peptide hormones, also known as protein hormones, are composed of amino acid chains. Examples include insulin, glucagon, and growth hormone (GH). Peptide hormones initiate cellular signaling by binding to receptors on the cell surface. Insulin regulates blood sugar levels, while GH controls growth and cellular renewal. Steroid hormones are synthesized from lipids like cholesterol. These hormones can pass through the cell membrane and affect cellular responses by binding to intracellular receptors. Examples of steroid hormones include estrogen, progesterone, testosterone, and cortisol. Estrogen and testosterone regulate gender-specific characteristics, while cortisol controls stress response and metabolism. Thyroid hormones are protein hormones synthesized in the thyroid gland, affecting metabolism, energy production, growth, and nervous system functions. Triiodothyronine (T3) and thyroxine (T4) are important examples of thyroid hormones.

During exercise, muscles utilize glucose for energy. This process requires insulin-glucose regulation. Acute exercise increases the release of certain hormones (e.g., cortisol, growth hormone, testosterone), which can lead to a healthy decrease in their secretion over the long term.

Cellular Communication

Certain proteins facilitate communication between cells. For instance, receptor proteins located on cell surfaces aid in transmitting signals into the cell (Wilk et al., 2023). These receptor proteins on the cell surface or within the cell detect chemical signals received by cells. These signals can include neurotransmitters, hormones, growth factors, and other signaling molecules. Receptors recognize signals and can transmit them to intracellular signaling pathways to initiate or modulate cellular responses. Some signaling molecules trigger other molecules called second messengers within the cell to affect signal transduction pathways. Examples include cyclic adenosine monophosphate (cAMP) and inositol trisphosphate (IP3), which play roles in intracellular signaling pathways. Signal transmission and signal transduction pathways lead to various responses within the cell. These responses can encompass cellular growth, division, differentiation, apoptosis (programmed cell death), and more. They are also essential for the organism to adapt to environmental changes and maintain homeostasis.

Cellular communication is vital for organisms to maintain internal balance and function harmoniously. Proteins constitute fundamental components of these communication mechanisms and are essential for organismal adaptation to the internal and external environment. Hormones released during exercise influence cellular communication. For example, exercise can impact cellular responses by increasing the release of stress hormones like cortisol. Cortisol exists in the blood both in free (biologically active) form and bound to proteins. Approximately 90% of cortisol binds to a protein called transcortin or corticosteroid-binding globulin (CBG). Literature indicates significant differences in cortisol awakening responses before or after exercise among athletes (Kayacan et al., 2021; Kayacan et al., 2023).

Catalytic Function

Proteins can catalyze chemical reactions and facilitate molecular transformations. For instance, lipid chaperones, a type of protein, reversibly bind lipids in the central nervous system by flipping them, playing a crucial role in lipid homeostasis. Dysfunctional lipid chaperones can disrupt lipid homeostasis, increasing the likelihood of neurological disorders (Scarfò et al., 2022). During exercise, the increased energy demand is met by the enzymatic breakdown of nutrients like carbohydrates, fats, and proteins to generate energy. For instance, the breakdown of glucose through glycolysis produces energy in cells. Muscle contraction and relaxation involve enzymes. Processes such as the entry and exit of calcium ions in muscle cells are regulated through enzymatic reactions. Exercise enhances protein synthesis in muscles. Protein synthesis is facilitated by ribosomes, enzymes that play a critical role in synthesizing protein structures.

During exercise, 5'-adenosine monophosphate-activated protein kinase (AMPK) is activated in skeletal muscles in response to changes in cellular energy status, such as an increase in the ratio of adenosine diphosphate (ADP), adenosine triphosphate (ATP), and creatine kinase (CK). AMPK activation is intensity-dependent and triggers a process of ATP inhibition. It activates pathways involved in carbohydrate and fatty acid metabolism to restore ATP levels to their baseline state. Recent evidence suggests that while AMPK plays a significant metabolic role during acute exercise periods, its ability to regulate muscle fuel depots and its sensitivity to exercise adaptations vary, making it a crucial component in the adaptation of skeletal muscles to endurance exercise (Aschenbach et al., 2004). Consequently, a multitude of biochemical reactions occur during exercise via enzyme-mediated mechanisms. These reactions critically govern processes such as energy production, tissue repair, muscle contraction, and other pivotal physiological functions.

Transport and Storage

Proteins like ferritin enable the storage of important minerals like iron within the body. Hemoglobin proteins carry oxygen from the lungs to other parts of the body within the bloodstream. Similarly, myoglobin proteins assist in storing oxygen within muscle cells, ensuring the availability of oxygen needed during muscle activity. During exercise, there is a need for amino acids to repair muscle damage and promote muscle growth. Hence, adequate amino acid storage in muscles is crucial for healthy muscle function. Additionally, energy-rich molecules like glycogen and fats cannot be stored without proteins. Glycogen is a carbohydrate form stored in the liver and muscles, serving as an energy source during prolonged exercise. Fats play a role in long-term energy storage within the body.

Proteins are essential components for maintaining the functioning of an organism. The body can either synthesize its own proteins by obtaining necessary amino acids through dietary sources or be supplemented with external protein intake when needed. Exercise science encompasses a wealth of research on protein utilization. However, research outcomes vary based on factors such as the studied population, age, gender, and the type of protein used, among others. Generally speaking, the relationship between exercise and protein is crucial for bodybuilding, muscle gain, performance enhancement, and overall health. Proteins play a significant role in reducing biological and physiological stress (oxidative stress, etc.), as well as in repairing and growing muscles in individuals who engage in exercise. In a previous study conducted with rats, we examined the effects of using a type of amino acid called 5-HTP (hydroxytryptophan) on the oxidation process in rats performing moderate-intensity running exercises. Our findings showed that protein intake alongside exercise significantly reduced oxidative damage (Kayacan et al., 2019). In another study, we found similar results in various protein sources, particularly in foods such as meat, fish, dairy products, eggs, legumes, and nuts, which contain L-tyrosine as an amino acid (Kayacan et al., 2018).

Adequate protein intake can expedite post-exercise recovery by supporting muscle repair and growth. Consuming an appropriate amount of protein is important to ensure an exercise program's effectiveness. When engaging in resistance training exercises, sufficient protein intake can contribute to increased muscle mass and strength development. Additionally, proteins can assist in sustaining the body during exercise and maintaining energy levels. Moreover, proteins can aid in replenishing glycogen stores after exercise. The digestion and metabolism of proteins require more energy, which leads to an increase in metabolic rate through a process known as thermogenesis (Choi et al., 2023). Therefore, a diet rich in protein can support an increase in metabolic rate. Figure 3.2 shows the use of protein for general health and exercise purposes.

Physiological and morphological adaptations resulting from resistance and endurance exercise show differences. Strength training often aims to increase muscle strength by performing fewer

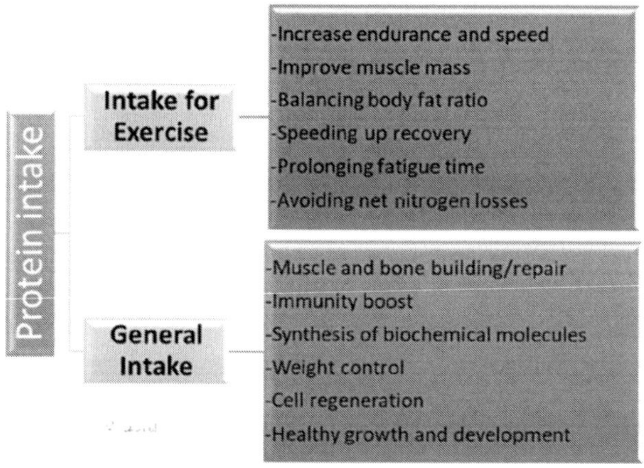

FIGURE 3.2 Protein intake for healthy people and athletes. For healthy individuals, proteins are preferred to maintain bodily functions, facilitate muscle and tissue repair, support hormone production, and sustain overall well-being. However, the protein needs of professional athletes are generally higher compared to regular individuals. This is due to the fact that athletes engage in more intense physical activities and require faster muscle repair and growth.

repetitions at a specific high weight. This training approach is known as "high-intensity training." In strength training, the concept of one-repetition maximum (1 RM) represents the maximum weight a person can lift for a single repetition in a particular exercise. This value reflects an individual's maximal muscle strength. High-intensity training is typically performed with loads ranging from 60% to 100%. This range can vary based on different goals and training stages.

Load between 60% and 75% is used to enhance strength endurance and support muscle development. This range often involves higher repetitions (8 to 12 repetitions). Load between 75% and 85% aims to increase strength. Commonly, this involves moderate repetitions (4 to 6 repetitions). Load between 85% and 100% is used to target maximal strength. Lower repetitions (1 to 3 repetitions) are often employed. These intensity ranges should be adjusted based on the training program's objective, individual fitness level, and training experience. However, high-intensity training is generally considered effective for increasing muscle strength and growth. Strong evidence in the literature suggests that protein consumption before and/or after exercise enhances muscle protein synthesis. This is a common nutritional strategy among athletes and fitness enthusiasts. Studies indicate that skeletal muscle protein synthesis is positively impacted after endurance-based workouts, and athletes have a greater amino acid capacity compared to sedentary individuals (Pikosky et al., 2006). Muscle protein synthesis is a critical process for muscle repair and growth. It becomes more active after exercise, highlighting the importance of consuming amino acids that muscles require. Hence, the recommendation to consume protein within an hour before and after exercise is frequently heard. However, recent research suggests that this time window can be flexible, and protein intake can remain effective even several hours after exercise. In this context, the relationship between exercise and protein usage appears to be a solid strategy for professional athletes. Academic studies examining exercise and protein use in recent years are listed in Table 3.1.

Most studies in the literature show that the use of protein is effective in increasing exercise performance, lean muscle mass, power and strength. This effect is higher in individuals with an athletic profile.

Using protein as part of a diet can increase the feeling of fullness and extend the time between meals, which can help reduce the risk of overeating and assist with weight management. Professional athletes also use protein supplementation to accelerate fat burning without losing muscle mass. A related study examined the effects of a low-protein diet (1.2 g/kg/day) versus a high-protein diet (2.4 g/kg/day) on body composition in recreationally active athletes. Participants underwent four weeks

TABLE 3.1
Some Research Findings in the Literature Regarding Exercise and Protein Intake

Campbell et al. (2018)	Investigated the impact of eight weeks of high and low protein intake (whey protein) through a resistance training program on body composition and maximal strength was investigated in 17 female athletes. Maximal strength significantly increased, with a notable enhancement in lean body mass.
Trommelen et al. (2023)	Studied the effects of casein and whey protein consumption on mitochondrial and myofibrillar protein synthesis in 33 sedentary males following endurance exercise. Both mitochondrial and myofibrillar protein synthesis rates increased during the recovery period.
Tinsley et al. (2019)	Assessed the impact of β-hydroxy β-methylbutyrate and whey intake during resistance training on intermittent fasting in a cohort of 24 sedentary women. Increased lean body mass, skeletal muscle hypertrophy, and enhanced muscle performance were achieved.
Ribeiro et al. (2022)	Examined the impact of protein consumption (meat, dairy products, and eggs) on body composition in 130 elderly women following a 24-week resistance training program. Low protein intake negatively affected body composition.
Grootswagers et al. (2020)	Investigated the effects of a four-week exercise regimen and protein supplementation (whey protein concentrate) on skeletal muscle in 17 women and men. The combined intervention of exercise and protein consumption led to muscle-related improvements in adults with low levels of physical activity.
Res et al. (2012)	Studied the impact of casein consumption on recovery, digestion and absorption kinetics, and protein metabolism in 16 sedentary men following resistance exercise. Casein was efficiently digested and absorbed, effectively stimulating muscle protein synthesis.
Ormsbee et al. (2018)	Examined the effects of a six-month strength and endurance training program with protein supplementation (whey) on muscle strength and body composition in 51 men and women. Both genders exhibited positive impacts on body composition, with men experiencing enhanced strength gains.
Jonvik et al. (2019)	Investigated the effects of a 12-week endurance exercise training program and protein supplementation (casein) on oxidative capacity and endurance performance in 60 men. Casein did not yield significant effects on these parameters.
Knuiman et al. (2019)	Assessed the changes in VO_2 max due to protein supplementation during a ten-week endurance training program in 44 men. Substantial improvements were observed in VO_2 max, and protein supplementation facilitated lean mass accumulation but did not enhance endurance performance.
Longland et al. (2016)	Studied the impact of a four-week intense exercise training program with dietary protein intake (whey isolate) on body composition in 40 men. The study revealed reductions in both lean body mass and fat mass.
Röhling et al. (2021)	Examined the effects of a 12-week protein supplementation (soy) regimen on endurance performance and physiological parameters in 30 individuals with an endurance training background participating in a marathon race. Protein supplementation led to reductions in muscle and inflammatory stress responses, contributing to improved endurance performance and physiological well-being during the marathon race.
Lamb et al. (2020)	Investigated the effects of a ten-week resistance training program with protein supplementation (peanut) on muscle, body composition, and strength in 30 men and women. The protein supplementation enhanced specific aspects of muscle hypertrophy and strength.
Yang et al. (2012)	Assessed the changes in myofibrillar protein synthesis following soy protein intake in 30 sedentary men after resistance exercise. The study reported that this protein type ineffectively stimulated myofibrillar protein synthesis.
Banaszek et al. (2019)	Studied the physiological adaptations following eight weeks of protein supplementation (pea and whey) in 15 trained men after high-intensity training. The study observed improvements in body composition, muscle thickness, and strength production, indicating that the protein supplementation supported these outcomes.

(Continued)

TABLE 3.1 (CONTINUED)
Some Research Findings in the Literature Regarding Exercise and Protein Intake

Xia et al. (2018)	Examined the protective role of oat protein against exercise-induced muscle damage and performance decline in 16 sedentary men over 19 days. Oat protein prevented damaging limb edema and mitigated the negative effects on muscle strength, knee joint range of motion, and vertical jump performance.
Hevia-Larraín et al. (2021)	Investigated the changes in muscle mass and strength due to a 12-week resistance training program and protein intake (soy protein isolate) in 19 sedentary men. A high-protein, plant-based diet supported muscle strength and mass development.
Huang et al. (2016)	Studied the relationship between plant protein intake (beans, vegetables, and plant-based staple foods) and low muscle mass in 327 sedentary women. Low plant protein intake was strongly associated with a heightened risk of low muscle mass.
Haub et al. (2002)	Assessed the impact of protein consumption (soy) over a 12-week period on muscle size, strength, and body composition in response to resistance training in 28 sedentary men. The consumed protein did not influence muscle strength, size, or body composition.

of plyometric exercises, resistance training, and high-intensity sprint training protocols six days a week. Both diets resulted in a 40% energy deficit for the athletes. The research findings showed no significant difference in total body weight loss between the groups. However, the high-protein diet group exhibited a greater reduction in fat mass compared to the low-protein group (−4.8 vs −3.5 kg) and experienced a significant increase in lean mass (+1.2 vs +0.1 kg) (Longland et al., 2016). These data suggest that high protein intake in conjunction with frequent intense training sessions positively impacts muscle mass and contributes to the success of fat loss strategies.

In summary, the relationship between exercise and protein is important for muscle repair, growth, energy provision, metabolism enhancement, weight control, and overall health. However, requirements can vary based on individual factors, exercise type, intensity, and personal goals. Consulting a nutritionist or healthcare professional about appropriate protein intake for specific exercise programs can be beneficial.

PROTEIN DIETS IN EXERCISE (HIGH OR LOW)

In healthy adults, the recommended safe protein intake ranges from approximately 0.65 to 0.83 g/kg/day for high-quality protein, often reported as 0.8 to 1.0 g/kg/day. The National Academy of Medicine (USA) recommends a dietary allowance (RDA) of 0.8 g per kilogram of ideal body weight. However, this intake can increase up to 1.2 g/kg/day with lower-quality dietary protein (Institute of Medicine, 2006). Daily energy intake also influences protein requirements. A healthy dietary pattern typically includes 10–15% of total energy intake from protein. For predominantly plant-based diets, this percentage is recommended at 10–13%, while for predominantly animal-based diets, it is suggested to be in the range of 13–15% (Lonnie and Johnstone, 2020). Figure 3.3 shows the amount of protein consumed according to exercise types.

The current literature lacks clear recommendations on the optimal dose and duration of protein supplementation for individuals engaging in exercise and sports. However, it is well established that individuals involved in exercise and sports require protein intake higher than 0.8 g/kg/day to prevent protein/amino acid oxidation. Exercise typically increases daily total energy requirements by around 20–30%, and a balanced diet can allocate 10–15% of this additional energy to meet protein needs (Westerterp-Plantenga et al., 2012). Some studies suggest that exceeding a total daily protein intake of 1.6 g/kg/day with protein supplementation may not provide additional benefits (Morton et al., 2018), and at least ten weeks of supplementation (3–5 times per week) is necessary to increase

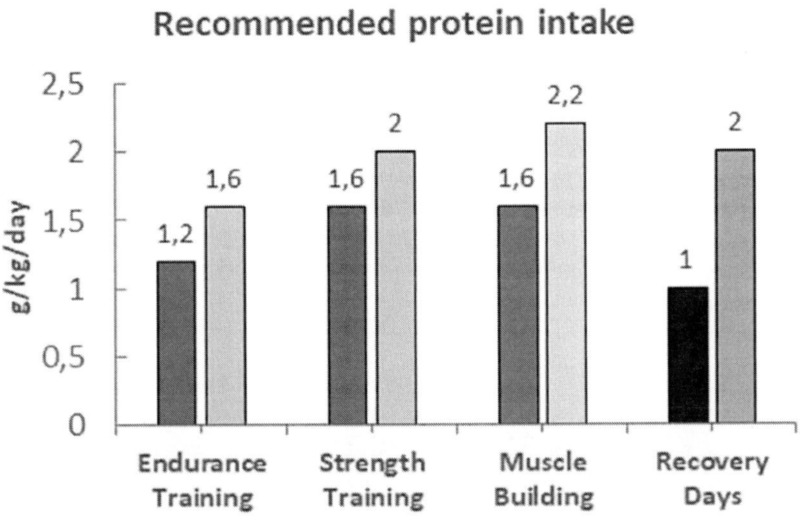

FIGURE 3.3 The amount of protein consumed according to exercise types.

muscle hypertrophy and strength (Stark et al., 2012). The International Society of Sports Nutrition (ISSN) recommends protein intake based on the type of exercise. For endurance exercises, recommendations range from 1.0 to 1.6 g/kg per day, depending on exercise intensity, duration, and individual experience. For strength and power exercises, recommendations range from 1.6 to 2.0 g/kg per day, considering the same factors. ISSN's general recommendation for exercise and sports participants is 1.4 to 2.0 g/kg. This range should be adjusted based on exercise type, with lower levels for endurance exercises, moderate levels for team sports (1.4–1.7 g/kg), and higher levels for strength exercises (Campbell et al., 2007).

MEAT- VS PLANT-BASED PROTEINS IN EXERCISE

Muscle contraction occurs when myosin and actin filaments interact via a sliding filament mechanism, shortening the sarcomeres. This contraction generates the force needed for various muscle functions, such as movement, maintaining posture, and generating heat. Each muscle fiber can contain numerous myofibrils, making them fundamental components of muscle tissue. Undoubtedly, myofibrils are composed of proteins, with research in the literature investigating the comparison between plant-based and animal-based proteins for enhancing these myofibrils through exercise. Animal-derived proteins are generally recognized as complete proteins, containing adequate amounts of all essential amino acids. This allows for a more balanced intake of the amino acids required by the body. Additionally, their typically higher biological value and digestion rate often support more efficient utilization of proteins. Plant-derived proteins may exhibit deficiencies in terms of amino acid profiles, but these deficiencies can be mitigated when various plant sources are consumed in combination. In a study, muscle development was compared between individuals following protein-rich vegan and omnivorous (plant and animal mixture) diets. The results of the research demonstrated similar muscle gains in both diet groups (Monteyne et al., 2023). This finding suggests comparable adaptive skeletal muscle responses during prolonged high-intensity resistance training, regardless of the protein source in the diet. While soy is a high-quality plant protein, it generally contains less leucine (50–60%) compared to animal proteins like milk, meat, and eggs. However, animal products can be metabolically complex and may lead to fatigue and weakness. Plants are rich in nutrients and low in fat. Meat contains cortisol, which impedes recovery, and arachidonic acid, an inflammatory unsaturated fat. Some athletes believe that plant-focused diets enhance endurance and performance (Rogers, 2020). Numerous studies have shown that consuming sufficient amounts of plant-based protein containing essential amino acids, primarily leucine, for 8–12 weeks can support adaptations similar to those observed with animal protein sources in professional sports performance. Further research is needed to determine whether longer supplementation periods maintain the balance between protein sources and to resolve variations in current findings. Several methods exist to enhance the anabolic capacity of plant protein sources, but more research is required to identify which, if any, provide noteworthy benefits. Given the positive effects of increased plant consumption on health and environmental sustainability, interest in increasing plant protein consumption will likely continue. The evidence base for plant protein in exercise has expanded significantly, with most results now showing that consuming an effective dose (often higher than animal protein) from a plant protein source can induce similar positive changes in acid-base status.

PROTEINS AS ENERGY SOURCES AND SPORTS PERFORMANCE

The history of dietary supplements dates back to ancient times. Today, protein supplements are commonly used by professional athletes and individuals engaging in regular exercise. These supplements are often formulated as powders, tablets, or liquids (Saxton and McDougal, 2021). Users consume these products to enhance endurance, speed, and skill, build muscle mass, maintain body fat balance, prevent protein and amino acid breakdown during exercise, and accelerate post-exercise recovery. Additionally, they extend the fatigue onset and prevent net loss of nitrogen. (Aljaloud and Ibrahim, 2013).

Protein consumption benefits biological and physiological capacity following intense exercise. Research reveals that different types of proteins enhance physical performance. For instance, post-exercise protein consumption can induce anti-inflammatory signaling by downregulating the nuclear factor kappa-β (NF Kβ) pathway. This can accelerate tissue repair and regeneration by reducing inflammatory cytokines levels (such as interleukin 6, interleukin 1, and tumor necrosis factor) (McKinlay, 2021).

Furthermore, protein consumption related to intense exercise in adults reduces muscle damage and provides energy. Nevertheless, previous studies have not re-evaluated performance to confirm the significance of improvements shown in studies displaying weak proinflammatory responses after high-protein training. Moreover, research on the acute effects of post-exercise protein intake on recovery has primarily focused on adult athletes, with limited attention to young athletes. For instance, a study of sedentary individuals engaging in intense aerobic exercise found that whey protein isolate (WP) at different doses (0.8 g/kg/day and 1.6 g/kg/day) significantly increased monocyte glutathione levels when consumed after strenuous exercise. However, this effect was absent in sedentary individuals (Middleton et al., 2004). In this context, WP appears advantageous compared to other protein supplements due to its effects on muscle protein synthesis, rapid absorption, rapid digestion, and water solubility. Increased blood amino acids resulting from rapid digestion can potentially prevent exercise-induced muscle protein breakdown in professional athletes (Miller et al., 2014).

A systematic review (Morton et al., 2018) examining 49 studies involving 1,863 participants indicated a significant difference in strength, maximum repetition, muscle size, muscle fiber cross-sectional area, and performance improvement with protein supplementation. This difference was more pronounced among trained athletes. Long-term endurance exercise regimens are common among healthy adults. However, it is crucial to determine when and how to consume protein supplements based on an individual's age, physical measurements, and exercise type.

MINI DICTIONARY OF TERMS

Actin: A round-shaped structural protein in the muscle.

Albumin: The most common protein found in the blood plasma of humans and other mammals.

Amino group (-NH2): A component included in the chemical formula of ammonia.

Antibodies: Substances with a protein structure formed by the body's immune system against foreign substances entering the body.

Anti-inflammatory: The common name for substances and treatments that reduce inflammation and edema.

Carboxyl group: A functional group of carboxylic acids.

Catalyst: The effect of a substance that does not seem to be undergoing a chemical reaction, causing a reaction to occur or a change in the rate of the reaction, without undergoing any change.

Cell junctions: Structures found in the tissues of some multicellular organisms.

Collagen: A protein that helps stimulate cartilage production.

Cytokines: Proteins that facilitate intercellular communication.

Elastin: A specialized protein found in connective tissues throughout the bodies of mammals, including humans.

Enzymes: Biomolecules that perform catalysis are called enzymes.

Immune system: A flawless defense system that protects the body against microbes through the mutual interaction and cooperation of various cells and proteins.

Inflammation: A protective response against injury, infection, or damage.

Integrins: Specific proteins that regulate cell-to-cell communication.

Keratin: A fibrous structural protein found in the epidermis (the outermost layer) of the skin, hair, nails, feathers, hooves, and horns of vertebrate animals, including humans.

Leukocytes: A group of white blood cells that protect the body against diseases by fighting infections and playing a preventive role against pathogens at the inflammatory and cellular levels.

Lymphocytes: Lymphocytes, like T cells and B cells, are key components of the immune system.

Microtubules: A crucial component of the cytoskeleton, a complex network of protein filaments found in the cells of eukaryotic organisms.

Peptides: Amino acid chains that support various elements of the skin, such as elastin and collagen.

Phagocytes: Cells that engulf and clear foreign substances entering the body.

Steroids: Lipids with a carbon skeleton consisting of four fused rings. Steroids are formed from the acetyl CoA biosynthesis pathway.

Substrate: The name given in biochemistry to substances processed in enzyme reactions.

Tyroxin: A non-essential α-amino acid that occurs mostly in the L-form in nature and is found in the structure of many proteins.

SUMMARY POINTS

- This chapter focuses on the relationship between the use of protein as a dietary source and exercise.
- Proteins are important compounds that perform many functions in our body.
- Current research on the use and effects of protein in exercise has been reviewed.
- The role of animal proteins in exercise performance seems to be more effective.
- In exercise, plant-based proteins can be taken together in different combinations to meet the needs of animal proteins.
- Attention should be paid to the amount of protein used to increase exercise performance and prevent negative consequences.

REFERENCES

Aljaloud, S. O., Ibrahim, S. A. 2013. Use of dietary supplements among professional athletes in Saudi Arabia. *Journal of Nutrition and Metabolism* 2013:245349.

Aschenbach, W. G., Sakamoto, K., Goodyear, L. J. 2004. 5'adenosine monophosphate-activated protein kinase, metabolism and exercise. *Sports Medicine* 34:91–103.

Banaszek, A., Townsend, J. R., Bender, D., Vantrease, W. C., Marshall, A. C., Johnson, K. D. 2019. The effects of whey vs. Pea protein on physical adaptations following 8-weeks of high-intensity functional training (HIFT): a pilot study. *Sports* 7:12.

Bungener, L., Huckriede, A., Wilschut, J., Daemen, T. 2002. Delivery of protein antigens to the immune system by fusion-active virosomes: a comparison with liposomes and ISCOMs. *Bioscience Reports* 22(2):323–338.

Campbell, B. I., Aguilar, D., Conlin, L., Vargas, A., Schoenfeld, B. J., Corson, A., Gai, C., Best, S., Galvan, E., Couvillion, K. 2018. Effects of high versus low protein intake on body composition and maximal strength in aspiring female physique athletes engaging in an 8-week resistance training program. *International Journal of Sport Nutrition and Exercise Metabolism* 28(6):580–585.

Campbell, B. I., Kreider, R. B., Ziegenfuss, T., La Bounty, P., Roberts, M., Burke, D., Antonio, J. 2007. International Society of Sports Nutrition position stand: protein and exercise. *Journal of the International Society of Sports Nutrition* 4(1):1–7.

Choi, K. M., Ko, C. Y., An, S. M., Cho, S. H., Rowland, D. J., Kim, J. H., Yoon, J. C. 2023. Regulation of beige adipocyte thermogenesis by the cold-repressed ER protein NNAT. *Molecular Metabolism* 69:101679.

Grootswagers, P., de Regt, M., Domić, J., Dronkers, J., Visser, M., Witteman, B., Hopman, M., Mensink, M. 2020. A 4-week exercise and protein program improves muscle mass and physical functioning in older adults - A pilot study. *Experimental Gerontology* 141:111094.

Haub, M. D., Wells, A. M., Tarnopolsky, M. A., Campbell, W. W. 2002. Effect of protein source on resistive-training-induced changes in body composition and muscle size in older men. *The American Journal of Clinical Nutrition* 76(3):511–517.

Hevia-Larraín, V., Gualano, B., Longobardi, I., Gil, S., Fernandes, A. L., Costa, L. A. R., Pereira, R. M. R., Artioli, G. G., Phillips, S. M., Roschel, H. 2021. High-protein plant-based diet versus a protein-matched omnivorous diet to support resistance training adaptations: A comparison between habitual vegans and omnivores. *Sports Medicine* 51(6):1317–1330.

Huang, R. Y., Yang, K. C., Chang, H. H., Lee, L. T., Lu, C. W., Huang, K. C. 2016. The association between total protein and vegetable protein intake and low muscle mass among the community-dwelling elderly population in Northern Taiwan. *Nutrients* 8(6):373.

Institute of Medicine. 2006. *Dietary Reference Intakes: The Essential Guide to Nutrient Requirements.* Washington, DC: The National Academies Press. https://doi.org/10.17226/11537.

Jin, T. Z., Yadav, M. P., Qi, P. X. 2023. Antimicrobial and physiochemical properties of films and coatings prepared from bio-fiber gum and whey protein isolate conjugates. *Food Control* 148:109666.

Jonvik, K. L., Paulussen, K. J. M., Danen, S. L., Ceelen, I. J. M., Horstman, A. M., Wardenaar, F. C., Van Loon, L. J. C., Van Dijk, J. W. 2019. Protein supplementation does not augment adaptations to endurance exercise training. *Medicine & Science in Sports & Exercise* 51(10):2041–2049.

Kayacan, Y., Derebasi, D. G., Ucar, C., Ozgocer, T., Yildiz, S. 2021. The hypothalamic-pituitary-adrenal axis activity in archers: cortisol release, stress, anxiety and success. *The Journal of Sports Medicine and Physical Fitness* 62(1):139–148.

Kayacan, Y., Makaracı, Y., Ucar, C., Amonette, W. E., Yıldız, S. 2023. Heart rate variability and cortisol levels before and after a brief anaerobic exercise in handball players. *Journal of Strength and Conditioning Research* 37(7):1479–1485.

Kayacan, Y., Yazar, H., Cerit, G., Ghojebeigloo, B. E. 2019. A new oxidative stress indicator: Effect of 5-hydroxytryptophan on thiol-disulfide homeostasis in exercise. *Nutrition* 63:114–119.

Kayacan, Y., Yazar, H., Kisa, E. C., Ghojebeigloo, B. E. 2018. A novel biomarker explaining the role of oxidative stress in exercise and l-tyrosine supplementation: thiol/disulphide homeostasis. *Archives of Physiology and Biochemistry* 124(3):232–236.

Knuiman, P., van Loon, L. J. C., Wouters, J., Hopman, M., Mensink, M. 2019. Protein supplementation elicits greater gains in maximal oxygen uptake capacity and stimulates lean mass accretion during prolonged endurance training: a double-blind randomized controlled trial. *The American Journal of Clinical Nutrition* 110(2):508–518.

Koguchi, T. 2022. Modification of macronutrient intake for prevention of gout in Japanese people in 2019: 2022 update. *American Journal of Health Research* 10:83–106.

Koopman, R., Saris, W. H., Wagenmakers, A. J., Van Loon, L. J. 2007. Nutritional interventions to promote post-exercise muscle protein synthesis. *Sports Medicine* 37:895–906.

Kustatscher, G., Collins, T., Gingras, A. C., Guo, T., Hermjakob, H., Ideker, T., Rappsilber, J. 2022. Understudied proteins: opportunities and challenges for functional proteomics. *Nature Methods* 19(7):774–779.

Lamb, D. A., Moore, J. H., Smith, M. A., Vann, C. G., Osburn, S. C., Ruple, B. A., Fox, C. D., Smith, K. S., Altonji, O. M., Power, Z. M., Cerovsky, A. E., Ross, C. O., Cao, A. T., Goodlett, M. D., Huggins, K. W., Fruge, A. D., Young, K. C., Roberts, M. D. 2020. The effects of resistance training with or without peanut protein supplementation on skeletal muscle and strength adaptations in older individuals. *Journal of the International Society of Sports Nutrition* 17(1):66.

Longland, T. M., Oikawa, S. Y., Mitchell, C. J., Devries, M. C., Phillips, S. M. 2016. Higher compared with lower dietary protein during an energy deficit combined with intense exercise promotes greater lean mass gain and fat mass loss: a randomized trial. *The American Journal of Clinical Nutrition* 103(3):738–746.

Lonnie, M., Johnstone, A. M. 2020. The public health rationale for promoting plant protein as an important part of a sustainable and healthy diet. *Nutrition Bulletin* 45(3):281–293.

McKinlay, B. 2021. *The Role of Protein Following Intense Exercise in Competitive Youth Athletes.* St. Catharines, On: Doctor of Philosophy, Health Biosciences, Brock University.

Middleton, N., Jelen, P., Bell, G. 2004. Whole blood and mononuclear cell glutathione response to dietary whey protein supplementation in sedentary and trained male human subjects. *International Journal of Food Sciences and Nutrition* 55(2):131–141.

Miller, P. E., Alexander, D. D., Perez, V. 2014. Effects of whey protein and resistance exercise on body composition: a meta-analysis of randomized controlled trials. *Journal of the American College of Nutrition* 33(2):163–175.

Monteyne, A. J., Coelho, M. O., Murton, A. J., Abdelrahman, D. R., Blackwell, J. R., Koscien, C. P., ... Wall, B. T. (2023). Vegan and omnivorous high protein diets support comparable daily myofibrillar protein synthesis rates and skeletal muscle hypertrophy in young adults. *The Journal of Nutrition*, 153(6): 1680–1695.

Morton, R. W., Murphy, K. T., McKellar, S. R., Schoenfeld, B. J., Henselmans, M., Helms, E., Phillips, S. M. 2018. A systematic review, meta-analysis and meta-regression of the effect of protein supplementation on resistance training-induced gains in muscle mass and strength in healthy adults. *British Journal of Sports Medicine* 52(6):376–384.

Oikawa, S. Y., Bahniwal, R., Holloway, T. M., Lim, C., McLeod, J. C., McGlory, C., Baker, S. K., Phillips, S. M. 2020. Potato protein isolate stimulates muscle protein synthesis at rest and with resistance exercise in young women. *Nutrients* 12(5):1235.

Ormsbee, M. J., Willingham, B. D., Marchant, T., Binkley, T. L., Specker, B. L., Vukovich, M. D. 2018. Protein supplementation during a 6-month concurrent training program: Effect on body composition and muscular strength in sedentary individuals. *International Journal of Sport Nutrition and Exercise Metabolism*: 28(6), 619–628.

Pikosky, M. A., Gaine, P. C., Martin, W. F., Grabarz, K. C., Ferrando, A. A., Wolfe, R. R., Rodriguez, N. R. 2006. Aerobic exercise training increases skeletal muscle protein turnover in healthy adults at rest. *The Journal of Nutrition* 136(2):379–383.

Rensvold, J. W., Shishkova, E., Sverchkov, Y., Miller, I. J., Cetinkaya, A., Pyle, A., Pagliarini, D. J. 2022. Defining mitochondrial protein functions through deep multiomic profiling. *Nature* 606(7913):382–388.

Res, P. T., Groen, B., Pennings, B., Beelen, M., Wallis, G. A., Gijsen, A. P., Senden, J. M., Van Loon, L. J. 2012. Protein ingestion before sleep improves postexercise overnight recovery. *Medicine & Science in Sports & Exercise* 44(8):1560–1569.

Ribeiro, A. S., Pereira, L. C., Schoenfeld, B. J., Nunes, J. P., Kassiano, W., Nabuco, H. C. G., Sugihara Junior, P., Fernandes, R. R., Antunes, M., Aguiar, A. F., Cyrino, E. S. 2022. Moderate and higher protein intakes promote superior body recomposition in older women performing resistance training. *Medicine and Science in Sports and Exercise* 54(5):807–813.

Rogers, D. 2020. *Plant Based High Protein Cookbook for Athletes: Maintain a Low-Carb Diet for Athletic Performance, Eating Vegan. A Nutrition Guide, 100+ Veggie Recipes and a 30-Day Meal Plan to Fuel Your Workout.* 978-1914164057. Rebirth Publishing Ltd. Brentford, United Kingdom.

Röhling, M., McCarthy, D., Berg, A. 2021. Continuous protein supplementation reduces acute exercise-induced stress markers in athletes performing marathon. *Nutrients* 13(9):2929.

Saxton, R., McDougal, O. M. 2021. Whey protein powder analysis by mid-infrared spectroscopy. *Foods* 10(5):1033.

Scarfò, G., Piccarducci, R., Daniele, S., Franzoni, F., Martini, C. (2022). Exploring the role of lipid-binding proteins and oxidative stress in neurodegenerative disorders: A focus on the Neuroprotective effects of Nutraceutical supplementation and physical exercise. *Antioxidants* 11(11):2116.

Schwartz, S. D. 2023. Protein dynamics and enzymatic catalysis. *The Journal of Physical Chemistry B* 127(12):2649–2660.

Stark, M., Lukaszuk, J., Prawitz, A., Salacinski, A. 2012. Protein timing and its effects on muscular hypertrophy and strength in individuals engaged in weight-training. *Journal of the International Society of Sports Nutrition* 9(1):1–8.

Stephens, T. V., Payne, M., Ball, R. O., Pencharz, P. B., Elango, R. 2015. Protein requirements of healthy pregnant women during early and late gestation are higher than current recommendations. *The Journal of Nutrition* 145(1):73–78.

Terra, R., Silva, S. A. G. D., Pinto, V. S., Dutra, P. M. L. 2012. Effect of exercise on immune system: response, adaptation and cell signaling. *Revista brasileira de medicina do esporte* 18:208–214.

Tinsley, G. M., Moore, M. L., Graybeal, A. J., Paoli, A., Kim, Y., Gonzales, J. U., Harry, J. R., VanDusseldorp, T. A., Kennedy, D. N., Cruz, M. R. 2019. Time-restricted feeding plus resistance training in active females: a randomized trial. *The American Journal of Clinical Nutrition* 110(3):628–640.

Trommelen, J., Kouw, I. W. K., Holwerda, A. M., Snijders, T., Halson, S. L., Rollo, I., Verdijk, L. B., van Loon, L. J. C. 2018. Presleep dietary protein-derived amino acids are incorporated in myofibrillar protein during postexercise overnight recovery. *American Journal of Physiology-Endocrinology and Metabolism* 314(5):E457–E467.

Trommelen, J., van Lieshout, G. A. A., Pabla, P., Nyakayiru, J., Hendriks, F. K., Senden, J. M., Goessens, J. P. B., van Kranenburg, J. M. X., Gijsen, A. P., Verdijk, L. B., de Groot, L. C. P. G. M., van Loon, L. J. C. 2023. Pre-sleep protein ingestion increases mitochondrial protein synthesis rates during overnight recovery from endurance exercise: a randomized controlled trial. *Sports Medicine* 53(7):1445–1455.

Vendruscolo, M., Fuxreiter, M. (2022). Protein condensation diseases: therapeutic opportunities. *Nature Communications* 13(1): 5550.

Wang, D., Ye, J., Shi, R., Zhao, B., Liu, Z., Lin, W., & Liu, X. 2022. Dietary protein and amino acid restriction: Roles in metabolic health and aging-related diseases. *Free Radical Biology and Medicine* 178:226–242.

Westerterp-Plantenga, M. S., Lemmens, S. G., Westerterp, K. R. 2012. Dietary protein–its role in satiety, energetics, weight loss and health. *British Journal of Nutrition* 108(2):105–112.

Wilk, A. J., Shalek, A. K., Holmes, S., Blish, C. A. 2023. Comparative analysis of cell–cell communication at single-cell resolution. *Nature Biotechnology*: 42(3), 470–483.

Xia, Z., Cholewa, J. M., Dardevet, D., Huang, T., Zhao, Y., Shang, H., Yang, Y., Ding, X., Zhang, C., Wang, H., Liu, S., Su, Q., Zanchi, N. E. 2018. et al. Effects of oat protein supplementation on skeletal muscle damage, inflammation and performance recovery following downhill running in untrained collegiate men. *Food Function* 9:4720–4729.

Yang, Y., Churchward-Venne, T. A., Burd, N. A., Breen, L., Tarnopolsky, M. A., Phillips, S. M. 2012. Myofibrillar protein synthesis following ingestion of soy protein isolate at rest and after resistance exercise in elderly men. *Nutrition & Metabolism* 9:57.

4 Dietary Proteins and Systemic Inflammation

Zahra Yari, Samira Rabiei, Bahareh Nikooyeh, and Tirang R. Neyestani

ABBREVIATIONS

AGEs	Advanced glycation end-products
ATI	Alpha-amylase/trypsin inhibitors
BMI	Body mass index
CCL	Chemokine ligand
CKD	Chronic kidney disease
COX-2	Cyclooxygenase 2
CRP	C-reactive protein
CVD	Cardiovascular disease
DAMPs	Damage-associated molecular patterns
eNOS	Endothelial nitric oxide synthase
GSH	Glutathione
hs-CRP	High-sensitivity C-reactive protein
IFN-γ	Interferon gamma
IL	Interleukin
iNOS	Inducible nitric oxide synthase
LGSI	Low-grade systemic inflammation
LPS	Lipopolysaccharide
MAPK	Mitogen-activated protein kinase
MCP-1	Monocyte chemoattractant protein-1
MetS	Metabolic syndrome
NCDs	Non-communicable diseases
NF-κB	Nuclear factor kappa B
NLRP3	Nod-like receptor protein 3
OS	Oxidative stress
PAMPs	Pathogen-associated molecular patterns
PGE2	Prostaglandin E2
PUFAs	Polyunsaturated fatty acids
ROS	Reactive oxygen species
SI	Systemic inflammation
T2D	Type 2 diabetes
Th	T helper
TLR	Toll-like receptor
TMA	Trimethylamine
TMAO	Trimethylamine N-oxide
TNF-α	Tumor necrosis factor-alpha

DOI: 10.1201/b23328-4

INTRODUCTION

Systemic inflammation (SI) is a multi-organ process triggered by the release of pro-inflammatory cytokines from immune-related cells and an increased number of circulating neutrophils and monocytes, among other parameters (Zotova, Chereshnev, and Gusev 2016). Extra fat mass contributes to SI by causing metabolic inflammation or "meta-inflammation" (Hotamisligil 2017). Low-grade systemic inflammation (LGSI) appears to be associated with higher risks of various complications, frailty, disability, and ultimately greater mortality (Furman et al. 2019).

Researchers have evaluated different nutritional strategies to manage SI (vel Szic et al. 2015). Growing evidence indicates that dietary components, including proteins, play a role in SI. A prospective study shows an association between different dietary protein sources and all-cause and cause-specific mortality rates (Sun et al. 2021). Considering SI's role in various pathologies (Furman et al. 2019), dietary proteins likely contribute to morbidities by affecting SI. Different macronutrients may trigger inflammatory responses through various mechanisms, partly related to gut microbiota. For instance, protein intake is associated with an increase in the intestinal population of *Bacteroidetes* but a decrease in *Firmicutes* (Wu et al. 2022). An imbalance in the intestinal microbial community or dysbiosis with a shift from Bacteroidetes to Firmicutes can lead to the activation of the toll-like receptors (TLR) signaling pathway and increased intestinal permeability to endotoxins, including lipopolysaccharides (LPS), the main component of Gram-negative bacterial cell walls (d'Hennezel et al. 2017). Absorption of LPS into systemic circulation may activate pro-inflammatory pathways, including nuclear factor kappa B (NF-κB) and mitogen-activated protein kinase (MAPK), leading to increased production of inflammatory mediators (Valacchi et al. 2018).

This chapter discusses current knowledge about plant and animal dietary proteins' association with SI and the underlying mechanisms. It also provides dietary recommendations and suggestions for future studies based on current data.

BIOMARKERS OF SYSTEMIC INFLAMMATION

Inflammation is a defense mechanism against agents the immune system recognizes as "non-self." As the major task of the immune system, it is crucial for survival. The most familiar form of inflammation is the acute inflammatory response, characterized by redness, swelling, pain, and occasionally loss of function. In the chronology of chronic disease progression, inflammation is initially subclinical, i.e., LGSI or "silent inflammation." LGSI is a slow process featuring up to a three-fold increase in circulating inflammatory mediators (Calçada et al. 2014). During an inflammatory response, some molecules (mostly proteins) may be upregulated, while others may be downregulated. These molecules can serve as the "biomarkers" of the inflammatory reaction. Some inflammatory biomarkers are disease-specific, while others are general. Early SI studies showed that postprandial increases in oxidative stress (OS) and inflammatory biomarkers are physiological responses to food intake. Decreased cellular glutathione (GSH) content and increased reactive oxygen species (ROS) production following energy substrate processing augment the redox state. Consequently, this activates certain transcription factors like NF-κB, which in turn induce pro-inflammatory cytokine genes, including interleukin (IL)-1, IL-6, tumor necrosis factor-alpha (TNF-α), interferon gamma (IFN-γ), and monocyte chemoattractant protein-1 (MCP-1). This process may be accompanied by upregulation of cyclooxygenase 2 (COX-2)—the key enzyme in the production of prostaglandins from arachidonic acid—and inducible nitric oxide synthase (iNOS) (Seyedsadjadi and Grant 2020). On the other hand, hyperplasia and hypertrophy of adipose tissue are associated with oxygen depletion and establishing an inflammatory condition, i.e., meta-inflammation, which leads to an increase in the aforementioned pro-inflammatory mediators (Hotamisligil 2017). This process triggers the production and release of C-reactive protein (CRP), an acute-phase protein. Measuring small amounts of circulating CRP, i.e., high-sensitivity (hs)-CRP, indicates subclinical chronic inflammation (Banait et al. 2022). Increased concentration of CRP reliably indicates cardiovascular mortality risk and is associated with higher risks of various diseases, including cancers and neurological disorders (Markozannes et al. 2021). CRP exerts pro-inflammatory effects on several cell lines, including endothelial cells, leukocytes, and platelets, by activating various signaling pathways (Luan and Yao 2018; Gasecka et al. 2020). Notwithstanding, no consensus exists on the most appropriate biomarkers for LGSI or how to differentiate between chronic systemic and acute inflammation. Consequently, nutrition researchers frequently use various cellular and soluble markers, including total and differential leukocyte count, cytokines (IL-1β, IL-6, TNF-α, IFN-γ), adipokines (leptin, adiponectin, resistin), chemokine ligands (CCL2, CCL3, CCL5), adhesion molecules (vascular adhesion molecule-1, intercellular adhesion molecule-1, E-selectin), and several biomarkers of OS (total antioxidant capacity, malondialdehyde, oxidized low-density lipoprotein, antioxidant enzymes, and circulating concentrations of antioxidants and pro-oxidants) (Minihane et al. 2015).

ROLE OF SYSTEMIC INFLAMMATION IN HEALTH AND DISEASE

Acute inflammatory response (AIR) is initiated by an infectious agent or damage and develops through interactions between pattern recognition receptors on immune cells and pathogen-associated molecular patterns (PAMPs) on pathogens or damage-associated molecular patterns (DAMPs) released from damaged cells (Furman et al. 2019). Following AIR, polyunsaturated fatty acids (PUFAs)-derived molecules—including resolvins, protectins, lipoxins, and maresins—contribute to alleviating inflammation (Ferreira et al. 2022). Chronic SI, on the other hand, can be initiated by both infectious and non-infectious factors that constantly provoke the immune system. Among the causes of chronic SI are infection, trauma, and surgery, all causing augmented imbalance between pro-oxidants and antioxidants in favor of pro-oxidants, known as "oxidative stress," which can damage tissues (Hertiš Petek et al. 2022). A huge body of evidence supports the etiologic role of chronic SI in a wide spectrum of human non-communicable diseases (NCDs), including cardiovascular disease (CVD), diabetes, and cancers (Nediani and Giovannelli 2020). Inflammation and oxidative stress are therefore considered as preventive and therapeutic targets in NCDs (Nediani and Dinu 2022). Chronic SI may accompany physiologic conditions, notably pregnancy and aging. Maternal LGSI may adversely affect intrauterine programming and fetal neurodevelopmental disorders (Parisi et al. 2021; Han et al. 2021). LGSI in aging, often accompanied by augmented oxidative stress and increased systemic pro-inflammatory cytokines (a condition called "inflammaging"), is associated with sarcopenia. Nutritional interventions, including dietary antioxidants and proteins, may slow down this process (Draganidis et al. 2016).

DIETARY PROTEINS AND SYSTEMIC INFLAMMATION

A growing body of evidence supports nutrition's role as a major determinant of health through modulation of LGSI (Hofseth and Hébert 2022). Dietary proteins are important macronutrients supporting both innate and adaptive arms of a healthy immune system, considering amino acids' essential roles in synthesizing immune proteins, cytokines, antibodies, and enzymes involved in the inflammatory response (Tourkochristou, Triantos, and Mouzaki 2021). One possible reason is amino acids' role in regulating certain metabolic pathways. For example, glutamine is an important modulator of lymphocyte proliferation, cytokine production, macrophage phagocytosis, and other immune mechanisms. Plasma glutamine concentration has an inverse relationship with inflammation and oxidative stress, which leads to impaired immune response, especially in critical illness (Newsholme 2001). Cysteine is another amino acid that plays a role in the synthesis of glutathione tripeptide, an essential antioxidant whose deficiency is associated with oxidative stress (Wu et al. 2004). Arginine is known as a precursor for nitric oxide synthesis, which, besides its antioxidant function, participates in many physiological processes, such as endothelial-mediated vasodilation and immune response, and its deficiency can exacerbate inflammatory conditions (Moncada and

Higgs 1993). Furthermore, the interplay between inflammation and muscle wasting underlies catabolic mechanisms in many diseases (Ferrando et al. 1999).

In general, total dietary protein may actually have an overall null effect on inflammation and oxidative stress, but depending on its source, it can potentially create a pro-inflammatory state and provide substrates for anti-inflammatory and antioxidative processes. Some studies report that high protein intake is associated with increased concentrations of specific circulating inflammatory biomarkers, such as CRP, which may relate to the dietary source of protein (Gögebakan et al. 2011). Additionally, results from observational studies confirm that plant-based diets are associated with a lower risk of elevated inflammatory biomarkers, while Western dietary patterns, characterized by excessive intake of animal proteins and processed meats, are associated with higher levels of inflammatory biomarkers (Silveira et al. 2018; Hart et al. 2021). For instance, among animal protein foods, dairy products may significantly lower serum concentrations of CRP (Moosavian et al. 2020), while other products, such as red meat (O'Connor et al. 2021) and eggs (Sajadi Hezaveh et al. 2019), have produced null results.

Certain animal protein sources are associated with pro-inflammatory responses due to their methylamine content. Methylamines include choline/lecithin (found in fish, eggs, and some dairy products), L-carnitine (found in red meat), and ergothioneine (found in some meat products, as well as in mushrooms and several types of beans). Methylamines are converted to trimethylamine (TMA) in the gut by microbiota, then absorbed, transported to the liver, and oxidized to TMA N-oxide (TMAO) (Thomas and Fernandez 2021). It has been seen that TMAO is associated with an increased risk of cardiovascular and neurodegenerative diseases due to the increased inflammatory state caused by NLRP3 inflammasome activation, caspase-1 activity, and inhibition of ATG16L1 gene expression (Janeiro et al. 2018; Yang et al. 2019). Regardless of dietary intake, TMAO production appears to strongly depend on gut microbiota profile (e.g., the *Firmicutes/Bacteroidetes* ratio) (Schoeler and Caesar 2019). Therefore, the effect of different dietary proteins on SI may vary depending on the protein source.

ANIMAL PROTEINS

Dairy Proteins, Including Whey and Casein

Studies on dairy products' effects on inflammation yield conflicting results. A study comparing a diet with moderate dairy products to a low-calorie diet containing high dairy products in obese subjects found no significant change in serum CRP levels (Thompson et al. 2005). Similarly, 12-week supplementation with a milk drink in adults with hypertension had no significant effect on the circulatory levels of inflammatory markers, including CRP and IL-6 (Lee et al. 2007). Another study showed that consumption of full-fat dairy products enriched with conjugated linoleic acid (CLA) did not affect systemic and vascular inflammatory markers in healthy middle-aged men (Tricon et al. 2006). In contrast, a cross-sectional study in Spanish adults reported that higher dairy consumption was associated with decreased serum concentrations of both systemic and vascular inflammatory markers, even after adjusting for BMI (Salas-Salvadó et al. 2008). Notwithstanding, the observed associations might partly result from the fat content of dairy products (Esmaillzadeh and Azadbakht 2010).

A systematic review showed that dairy consumption has no potentially adverse or beneficial effects on SI. This review appraised eight trials evaluating dairy proteins' effects on inflammatory markers. Seven trials reported no effect of dairy protein intervention on CRP (Nieman, Anderson, and Cifelli 2021).

In the past decade, there has been growing interest in the antioxidant and anti-inflammatory roles of dairy proteins, especially whey protein. Evidence from in vitro models indicates that whey protein may prevent hydrogen peroxide toxicity, reduce lipid peroxidation, and increase the antioxidant enzyme activity by decreasing intracellular calcium concentrations and suppressing mitochondrial apoptotic pathways (Jin et al. 2013). Similarly, an experimental study in a murine model showed that whey protein supplementation prevented the onset of oxidative stress and inflammatory reactions by increasing antioxidant enzyme activity, increasing glutathione availability, and maintaining mitochondrial integrity (Athira et al. 2013).

Meat, Egg, Poultry, and Fish

Although a recent meta-analysis of clinical trials lasting up to 16 weeks reported that total red meat consumption had no effect on CRP, IL-6, or TNF-α concentrations (O'Connor et al. 2021), some studies still indicate the pro-inflammatory effects of animal proteins, especially red meat. Several reasons have been proposed for this potential effect, including high contents of heme iron (de Oliveira Otto et al. 2011), saturated fat (Clarke et al. 2008), and advanced glycation end-products (AGEs) (Chen and Smith 2015). High animal protein intake, especially from red meat, may induce inflammatory reactions through dysbiosis, an imbalance in gut microbiota, and the consequent decrease in short-chain fatty acids (SCFAs) (Cai et al. 2022). Nevertheless, increasing body fat and resulting obesity (both general and central) may confound or mediate this relationship (Schlesinger et al. 2019).

It is worth noting that these findings cannot be generalized to all animal proteins. For example, egg consumption is not associated with significant changes in serum concentrations of inflammatory biomarkers such as hs-CRP, TNF-α, and IL-6 (Sajadi Hezaveh et al. 2019). Though eggs contain bioactive compounds with both anti- and pro-inflammatory properties—including certain proteins (ovotransferrin), lutein, zeaxanthin, phospholipids, and cholesterol—it is unlikely that egg consumption, at least at the recommended amount (i.e., 1–2 a day), has any provocative effect on inflammatory reactions in healthy people (Andersen 2015).

Evidence regarding the chicken protein's effect on inflammation is scarce and conflicting. An experimental study reported that high chicken or pork protein intake (40% of calories) in a high-fat diet induced an inflammatory response in the hippocampus of Wistar rats (Zhang et al. 2020). However, the composition of the studied diet was quite unusual for most, if not all, human populations. In contrast, another experimental study in male C57BL/6 mice on a high-fat diet found that supplementation with chicken protein hydrolysate significantly reduced the inflammatory reaction, as judged by circulating concentrations of IL-1β, IFN-γ, TNF-α, and MCP-1 (Aloysius et al. 2018).

Several studies have examined the effect of fish consumption on inflammatory biomarkers. However, the type of fish is a determining factor, as oily fish usually have anti-inflammatory effects mostly due to their ω-3 fatty acids (Costabile et al. 2021). To differentiate the effects of fish fats and proteins, an experimental study examined how supplementation with different fish protein isolates (bonito, herring, mackerel, and salmon) affected inflammatory and metabolic biomarkers in male Wistar rats fed a high-fat diet. After 28 days, TNF-α and IL-6 expression in visceral adipose tissue significantly decreased in all fish protein-supplemented groups, compared with casein (Pilon et al. 2011). Similarly, another study revealed the antioxidant and anti-inflammatory effects of protein hydrolysates derived from some less commonly consumed fish (Da Rocha et al. 2018). In contrast to these findings, a clinical trial investigated an energy-restricted diet's effect with either 15% or 30% of energy intake from protein (including fish) on inflammation in 96 adult subjects with obesity. After eight weeks, though the amount of weight loss was similar in both groups, the inflammation score based on circulating concentrations of hs-CRP, IL-6, TNF-α, and plasminogen activator inhibitor 1 was significantly lower in the low-protein group. Importantly, total protein intake was associated with inflammation, but vegetable- or fish-derived protein intake had no effect on inflammatory biomarkers (Lopez-Legarrea et al. 2014).

Fish protein's anti-inflammatory effects have been attributed to its bioactive peptides (Kemp and Kwon 2021). Mechanistically, bioactive peptides can inhibit inflammatory mediators, including NF-κB, and enhance the immune system. Some of these peptides can indirectly affect the inflammatory state through increased antioxidant activity (for example, inhibition of lipid peroxidation and metal chelating activities), antihypertensive activity (i.e., inhibitory activity of angiotensin-converting enzyme-1), antimicrobial activity, and intestinal microbiota imbalance regulatory activity (e.g., normalization of *Bacteroidetes* and *Firmicutes* populations) (Ryan et al. 2011).

PLANT PROTEINS

Plant foods commonly have higher energy-to-protein and energy-to-phosphorus ratios. In other words, they provide energy requirements but relatively low nitrogen and phosphorus. Furthermore, these foods contain low saturated fats. Plant foods may improve glycemic control compared with an omnivorous diet and reduce oxidative stress and inflammation. The following sections review certain plant proteins' effects on inflammatory reactions.

Soy Protein

Soybean contains 35–40% protein on a dry-weight basis. Its carbohydrate content is lower than in other legumes. Soy's specific characteristics (high protein and low carbohydrate) differentiate it from other legumes. Additionally, soybeans are a high-quality protein source, with amino acid scores similar to animal sources, such as eggs and milk (van den Berg et al. 2022). Soy's bioactive components are believed to have anti-inflammatory, anticoagulant, anticancer, and antioxidant properties. Daidzein, an isoflavone found in soybean and other legumes, can cause the downregulation of genes associated with inflammatory pathways, like toll-like receptors (Franca-Oliveira et al. 2023).

Findings on soy protein's effects on inflammatory biomarkers have been controversial. For example, a clinical trial reported that high soy isoflavone consumption was associated with increased IL-6 concentration in middle-aged women (Jenkins et al. 2002). On the other hand, some researchers suggest no beneficial effect on inflammatory biomarkers and vascular adhesion molecules after soy consumption (Dong et al. 2011; Rebholz et al. 2013; Mangano et al. 2013; Liu et al. 2013). Furthermore, a meta-analysis of 14 trials in postmenopausal women reported that soy food consumption does not influence serum CRP levels. Nevertheless, it has an inverse association with CRP levels only in women with pre-existing elevated CRP (Dong et al. 2011).

Some evidence indicates that soy protein may suppress inflammation via its antioxidant properties (Rizzo 2020). In vitro studies have shown that soy isoflavones, genistein, and daidzein can upregulate endothelial nitric oxide synthase (eNOS) and decrease the production of reactive oxygen species (ROS) in vascular endothelial cells (Zhang et al. 2016) through the microRNA-155/SOCS1-mediated repression of the NF-κB signaling pathway (Zhang et al. 2017). Notwithstanding, data obtained from clinical trials on the antioxidant effects of soy protein are conflicting (Celec et al. 2013; Sen et al. 2012).

Legume Protein

Legumes contain high amounts of components with direct or indirect (like reducing body weight and blood glucose) anti-inflammatory properties, including bioactive peptides, phenolic compounds, and fiber.

Legume peptides are natural antioxidants and anti-inflammatory agents. Their antioxidant and anti-inflammatory bioactivity stems mainly from their hydrophobic amino acids (Tawalbeh, Al-U'datt M, and Wan Ahmad 2023). Bioactive peptides in legumes can modulate inflammatory mediators such as PGE2, NO, iNOS, cyclooxygenase 2 (COX2), cytokines, and chemokines (Kiersnowska and Jakubczyk 2022). Lupin, a protein hydrolysate from legumes, can reduce NO and ROS levels and downregulate pro-inflammatory cytokines in a Caco-2 monolayer culture medium (Montserrat-de la Paz and Villanueva 2021). Lupin may also modulate the number of T helper (Th) cell subpopulations, including Th1, Th9, and Th17, that produce pro-inflammatory cytokines (Tawalbeh, Al-U'datt M, and Wan Ahmad 2023).

Notably, legumes are good sources of prebiotics that are metabolized into immune-enhancing and anti-inflammatory molecules by the gut microbiota. Therefore, some immunomodulatory and anti-inflammatory actions of legumes may be attributable to these microbial metabolites (Aranda-Olmedo and Rubio 2020).

Wheat Protein

The effect of wheat and other cereal grains on inflammatory reactions is conflicting. Gliadin, an alcohol-soluble fraction of gluten, contains immunomodulatory peptides that can be recognized by T-cells, thus triggering an inflammatory reaction that may lead to gluten sensitivity and celiac disease in genetically susceptible individuals (Gupta et al. 2018). Gliadin peptides increase intestinal permeability through the release of zonulin, the regulator protein of paracellular intestinal permeability. This process interacts with the immune system, leading to inflammation through the paracellular translocation of gliadin and microbial antigens (Shimada and Tanigawa 2019).

An experimental study on human keratinocyte culture reported that an acid-hydrolyzed wheat protein provokes an inflammatory reaction through NF-κB pathway activation (Kuroda et al. 2020). Wheat and related cereals contain a non-gluten alpha-amylase/trypsin inhibitors (ATI) family that are strong inducers of the innate immune response. ATI family members can activate the TLR4–MD2–CD14 complex and create strong innate immune effects after oral or systemic challenge (Junker et al. 2012).

Activated myeloid cells may mediate ATI-induced intestinal inflammation (Zevallos et al. 2017).

Conversely, wheat contains phytochemicals with anti-inflammatory properties that can offset any potentially pro-inflammatory activities of gluten in wheat (Fardet 2010). An experimental study revealed that at least one bioactive component of wheat gluten hydrolysate could downregulate the expression of pro-inflammatory cytokines IL-1β, IL-6, IL-17, and IFN-γ in human monocyte culture medium, suggesting it could be an anti-inflammatory and cardioprotective food component (Montserrat-de la Paz et al. 2020). Other dietary components, genetic makeup, and gut microbiota of the host likely contribute to the eventual pro- or anti-inflammatory effects of wheat and other dietary proteins in the body (Caminero and Verdu 2019).

THE IMPORTANCE OF PLANT PROTEINS IN CERTAIN CLINICAL SETTINGS

Diabetes and Metabolic Syndrome (MetS)

Inflammatory mechanisms are involved in the pathogenesis of diabetes mellitus and its long-term complications (Tsalamandris et al. 2019). In a prospective study of 37 adult subjects with T2D, a high protein diet (30% of energy intake) from either animal (dairy and meat; n=18) or plant (mainly legume; n=19) sources for six weeks caused a significant reduction in hepatic fat, necro-inflammation, and insulin resistance independent of body weight (Markova et al. 2017). Both diets also equally decreased pro-inflammatory adipokines chemerin and progranulin (Markova et al. 2020). In contrast, findings from combined data from two large prospective studies, the Women's Health Initiative and the UK Biobank, indicated that replacing animal protein with plant protein reduced T2D risk mainly via decreasing obesity-induced inflammation (Li et al. 2022).

Renal Disease

Traditionally, patients with chronic kidney disease (CKD) are advised to limit the intake of several nutrients, including protein, potassium, and phosphorous. This may lead to a low intake of fruits, vegetables, and plant proteins, which are often lower quality than animal proteins. However, these recommendations have been challenged (Carrero et al. 2020). A meta-analytical study found an inverse association between serum CRP concentrations and plant protein, as compared with animal protein, consumption in non-dialysis patients. Among animal proteins, the effect of eggs and red meat on raising CRP was more prominent than whey protein. Nevertheless, some of the studies included in the meta-analysis were non-randomized controlled trials, raising the possibility of bias (Aycart et al. 2021). Furthermore, plant dietary sources usually contain many phytochemicals with antioxidant and anti-inflammatory properties, including flavonoids, polyphenols, and cordycepin, with potential benefits for CKD (Rapa et al. 2019). Consequently, the reported anti-inflammatory effect of plant proteins in CKD patients is difficult to attribute solely to the plant protein itself.

Cardiovascular Diseases

Evidence increasingly suggests that a plant-based diet is consistent with lower CVD risk and all-cause mortality in the general population (Kim et al. 2019). Nonetheless, the contribution of plant protein to these associations and its possible effect on inflammatory CVD biomarkers remains unclear. Analysis of data from 2061 participants of the Framingham Heart Study Offspring cohort (both sexes, aged 60±8.8 years) revealed that total dietary protein intake was inversely associated with oxidative stress and inflammatory indicators, with a stronger association for plant proteins compared to animal proteins (Hruby and Jacques 2019). Here too, the observed effect might be mostly due to the phytochemicals in plant dietary sources. A recent meta-analysis of thirteen randomized clinical trials and seven cohort studies failed to obtain concrete evidence supporting a protective effect of plant protein against CVD mortality and T2D risk. The authors considered the observed association to be limited-suggestive (Lamberg-Allardt et al. 2023).

CONCLUDING REMARKS

Inflammation is a natural immune system reaction to non-self invaders, essential for survival. However, LGSI is associated with a higher risk of various diseases and related mortalities (Furman et al. 2019). Proteins are essential dietary elements that may have antioxidant properties either in their intact form or as post-hydrolysis peptide derivatives (Corrochano et al. 2018; Nwachukwu and Aluko 2019). Adequate protein intake, especially from plant sources, is inversely associated with circulating inflammatory biomarkers (Hruby and Jacques 2019). Yet, there is no concrete evidence favoring plant proteins over animal proteins in suppressing SI. The final effect of dietary protein on SI likely depends on interactions among whole dietary components, including macro- and micronutrients, phytochemicals, and fiber, with the intestinal microbiota. The host's sex, age, body composition (percent of body fat mass), and genetic makeup also contribute to these complex interactions. Nevertheless, consumption of high amounts of animal proteins, especially meats and high-fat dairy, is not recommended in healthy dietary guidelines (Fischer et al. 2020). Meanwhile, plant protein sources are favored for sustainable nutrition (Langyan et al. 2022).

Consequently, a well-balanced diet combined with weight control and an active lifestyle is crucial for suppressing SI and its adverse consequences (Figure 4.1). This diet consists of abundant antioxidant-rich vegetables and sufficient amounts of fruits, whole grains, and plant and healthy animal proteins (fish, poultry, lean meat, eggs, and low-fat dairy) while minimizing the intake of saturated and trans fats and red and processed meats. In this context, it is unlikely that the proportion of plant or animal protein is a determinant in the provocation of SI and related pathologies. However, this notion needs to be investigated by well-designed studies.

The findings of future studies on the effects of dietary proteins on inflammation and their possible clinical outcomes will be exciting. Nevertheless, it is noteworthy that a statistically significant decrease in a given inflammatory biomarker like hs-CRP (especially within its reference interval) does not necessarily mean significant improvement in health or clinical outcomes. Future investigations on the effect of dietary proteins on inflammation should take into account as many of the aforementioned determinants as possible, both in designing the study and in data interpretation.

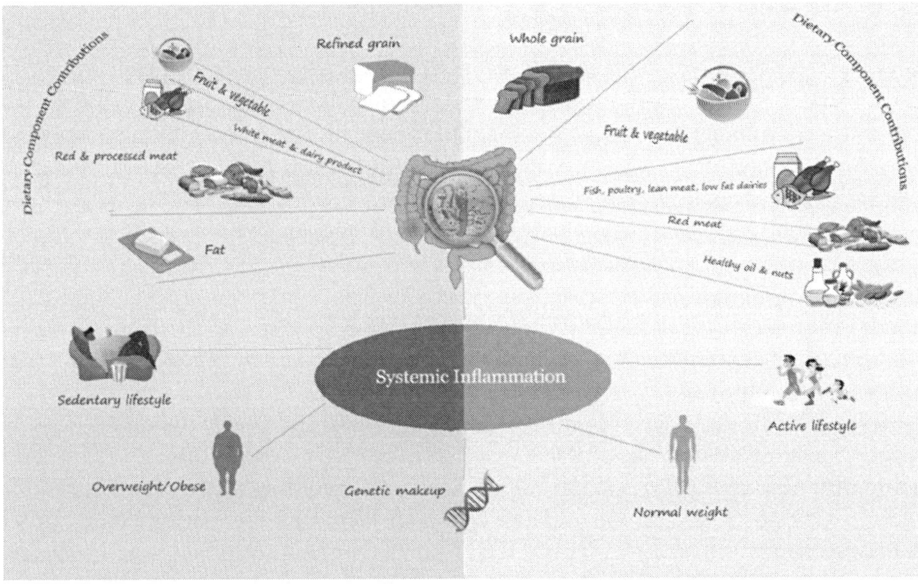

FIGURE 4.1 The intertwined interactions of all dietary components, lifestyle, genetic makeup, and gut microbiota determine the provocation or suppression of systemic inflammation. A healthy lifestyle, including weight control, regular workouts, and a balanced diet consisting of whole grains, plenty of vegetables and fruits, legumes, nuts, low-fat dairy, poultry, and lean meats, and the avoidance of saturated fats and processed meats, is consistent with SI suppression (right side). Whereas a sedentary lifestyle and an unhealthy diet comprising red and processed meats, refined grains, high amounts of unhealthy (saturated and trans) fats, and low intake of vegetables and fruits will bring about increased whole body and visceral fat mass, augmented oxidative stress, provoked SI, and consequent adverse health outcomes (left side).

MINI DICTIONARY OF TERMS

Acute inflammatory response (AIR): A rapid and intense inflammatory response typically triggered by an infection or injury.

Advanced glycation end-products (AGEs): Harmful compounds formed when sugars react with proteins or fats in food, particularly in high-temperature cooking.

Alpha-amylase/trypsin inhibitors (ATI): Proteins in wheat and related cereals that stimulate the innate immune system.

Celiac disease: An autoimmune disorder triggered by gluten consumption, leading to small intestine damage.

Chronic systemic inflammation: Long-term, low-level inflammation caused by various factors, including infections, trauma, and metabolic conditions.

C-reactive protein (CRP): An acute-phase protein that increases in the blood in response to inflammation and serves as a marker of systemic inflammation.

Gut microbiota: The collection of microorganisms in the digestive tract that can influence health.

Isoflavones: Naturally occurring compounds in soybeans and other legumes that have antioxidant and potentially anti-inflammatory properties.

Lipopolysaccharides (LPS): Molecules in the cell walls of certain bacteria, particularly Gram-negative bacteria, which can trigger an immune response.

Low-grade Systemic Inflammation (LGSI): A chronic, low-level inflammation associated with an increased risk of numerous chronic diseases, including cardiovascular disease, type 2 diabetes, frailty, disability, and mortality.

Meta-inflammation: A chronic, low-grade systemic inflammation induced by excess body fat mass.

Mitogen-activated protein kinase (MAPK): A family of proteins involved in signaling pathways that regulate various cellular processes, including inflammation.

Neutrophils and monocytes: Types of white blood cells involved in the immune response.

Nuclear factor kappa B (NF-κB): A protein complex that plays a role in the regulation of immune response and inflammation.

Oxidative stress (OS): An imbalance favoring harmful pro-oxidants over the body's ability to neutralize them with antioxidants.

Phytochemicals: Naturally occurring compounds in plants with various health-promoting properties.

Pro-inflammatory cytokines: Signaling molecules produced by immune cells that promote inflammation in the body.

Sustainable nutrition: A diet that is environmentally friendly and maintainable long-term without depleting natural resources.

Systemic inflammation (SI): A multi-organ process triggered by the release of pro-inflammatory cytokines from immune-related cells and increased circulating neutrophils and monocytes.

Toll-like receptors (TLR): Proteins crucial in the innate immune response for recognizing molecules associated with pathogens.

SUMMARY POINTS

- Systemic inflammation (SI) is a multi-organ process triggered by pro-inflammatory cytokines and other factors. Extra fat mass can lead to metabolic inflammation or "meta-inflammation." Low-grade systemic inflammation (LGSI) is linked to various complications, frailty, disability, and increased mortality.
- Various nutritional strategies have been explored to manage SI. Dietary proteins play a role in SI modulation. Different macronutrients can trigger an inflammatory response, partly related to gut microbiota composition.
- Dietary proteins have complex effects on inflammation. Plant-based diets are associated with lower inflammatory markers, while "westernized" diets with excessive animal proteins are linked to increased inflammation. Effects of different animal protein sources vary, with some being pro-inflammatory due to components like methylamines.
- Suppressing systemic inflammation (SI) and its associated adverse effects primarily hinges on maintaining a well-balanced diet, managing body weight, and leading an active lifestyle. This entails a dietary pattern rich in antioxidants from vegetables, adequate servings of fruits, whole grains, and a mix of plant-based and lean animal proteins, while reducing saturated and trans fats and limiting the consumption of red and processed meats.
- The impact of plant-to-animal protein ratio of SI and related health issues remains uncertain, necessitating further investigation through rigorous research studies.

REFERENCES

Aloysius, T. A., A. K. Carvajal, R. Slizyte, J. Skorve, R. K. Berge, and B. Bjørndal. 2018. "Chicken protein hydrolysates have anti-inflammatory effects on high-fat diet induced obesity in mice." *Medicines (Basel)* 6 (1). doi: 10.3390/medicines6010005.

Andersen, Catherine J. 2015. "Bioactive egg components and inflammation." *Nutrients* 7 (9):7889–7913.

Aranda-Olmedo, Isabel, and Luis A Rubio. 2020. "Dietary legumes, intestinal microbiota, inflammation and colorectal cancer." *Journal of Functional Foods* 64:103707.

Athira, S, Bimlesh Mann, Rajan Sharma, and Rajesh Kumar. 2013. "Ameliorative potential of whey protein hydrolysate against paracetamol-induced oxidative stress." *Journal of dairy Science* 96 (3):1431–1437.

Aycart, Danielle Francesca, Sofía Acevedo, Lucía Eguiguren-Jimenez, and Jeanette Mary Andrade. 2021. "Influence of plant and animal proteins on inflammation markers among adults with chronic kidney disease: a systematic review and meta-analysis." *Nutrients* 13 (5):1660.

Banait, Tanvi, Anil Wanjari, Vedika Danade, Shashank Banait, and Jyoti Jain. 2022. "Role of high-sensitivity C-reactive protein (hs-CRP) in non-communicable diseases: a review." *Cureus* 14 (10).

Cai, J., Z. Chen, W. Wu, Q. Lin, and Y. Liang. 2022. "High animal protein diet and gut microbiota in human health." *Critical Reviews in Food Science and Nutrition* 62 (22):6225–6237. doi: 10.1080/10408398.2021.1898336 [doi].

Calçada, Dulce, Dario Vianello, Enrico Giampieri, Claudia Sala, Gastone Castellani, Albert de Graaf, Bas Kremer, Ben van Ommen, Edith Feskens, and Aurelia Santoro. 2014. "The role of low-grade inflammation and metabolic flexibility in aging and nutritional modulation thereof: a systems biology approach." *Mechanisms of Ageing and Development* 136:138–147.

Caminero, Alberto, and Elena F. Verdu. 2019. "Metabolism of wheat proteins by intestinal microbes: Implications for wheat related disorders." *Gastroenterología y Hepatología (English edition)* 42 (7):449–457.

Carrero, Juan J., Ailema González-Ortiz, Carla M. Avesani, Stephan J. L. Bakker, Vincenzo Bellizzi, Philippe Chauveau, Catherine M. Clase, Adamasco Cupisti, Angeles Espinosa-Cuevas, Pablo Molina, Karine Moreau, Giorgina B. Piccoli, Adrian Post, Siren Sezer, and Denis Fouque. 2020. "Plant-based diets to manage the risks and complications of chronic kidney disease." *Nature Reviews Nephrology* 16 (9):525–542. doi: 10.1038/s41581-020-0297-2.

Celec, P., J. Hodosy, R. Pálffy, R. Gardlík, L. Halčák, and D. Ostatníková. 2013. "The short-term effects of soybean intake on oxidative and carbonyl stress in men and women." *Molecules* 18 (5):5190–200. doi: 10.3390/molecules18055190.

Chen, Gengjun, and J. Scott Smith. 2015. "Determination of advanced glycation endproducts in cooked meat products." *Food Chemistry* 168:190–195.

Clarke, Robert, Martin Shipley, Jane Armitage, Rory Collins, and William Harris. 2008. "Plasma phospholipid fatty acids and CHD in older men: whitehall study of London civil servants." *British Journal of Nutrition* 102 (2):279–284.

Corrochano, Alberto R, Vitaly Buckin, Phil M Kelly, and Linda Giblin. 2018. "Invited review: whey proteins as antioxidants and promoters of cellular antioxidant pathways." *Journal of Dairy Science* 101 (6):4747–4761.

Costabile, G., G. Della Pepa, C. Vetrani, P. Vitaglione, E. Griffo, R. Giacco, M. Vitale, D. Salamone, A. A. Rivellese, G. Annuzzi, and L. Bozzetto. 2021. "An oily fish diet improves subclinical inflammation in people at high cardiovascular risk: a randomized controlled study." *Molecules* 26 (11). doi: molecules26113369 [pii] molecules-26-03369 [pii] 10.3390/molecules26113369 [doi].

Da Rocha, Meritaine, Ailén Alemán, Gyselle Chrystina Baccan, M. Elvira López-Caballero, Carmen Gómez-Guillén, Pilar Montero, and Carlos Prentice. 2018. "Anti-inflammatory, antioxidant, and antimicrobial effects of underutilized fish protein hydrolysate." *Journal of Aquatic Food Product Technology* 27 (5):592–608.

de Oliveira Otto, Marcia C. C. Alvaro Alonso, Duk-Hee Lee, George L. Delclos, Nancy S. Jenny, Rui Jiang, Joao A. Lima, Elaine Symanski, David R. Jacobs Jr, and Jennifer A. Nettleton. 2011. "Dietary micronutrient intakes are associated with markers of inflammation but not with markers of subclinical atherosclerosis." *The Journal of Nutrition* 141 (8):1508–1515.

d'Hennezel, Eva, Sahar Abubucker, Leon O. Murphy, and Thomas W. Cullen. 2017. "Total lipopolysaccharide from the human gut microbiome silences toll-like receptor signaling." *Msystems* 2 (6):e00046–17.

Dong, J. Y., P. Wang, K. He, and L. Q. Qin. 2011. "Effect of soy isoflavones on circulating C-reactive protein in postmenopausal women: meta-analysis of randomized controlled trials." *Menopause* 18 (11):1256–1262. doi: 10.1097/gme.0b013e31821bfa24.

Draganidis, Dimitrios, Leonidas G. Karagounis, Ioannis Athanailidis, Athanasios Chatzinikolaou, Athanasios Z. Jamurtas, and Ioannis G. Fatouros. 2016. "Inflammaging and skeletal muscle: can protein intake make a difference?" *The Journal of Nutrition* 146 (10):1940–1952.

Esmaillzadeh, Ahmad, and Leila Azadbakht. 2010. "Dairy consumption and circulating levels of inflammatory markers among Iranian women." *Public Health Nutrition* 13 (9):1395–1402.

Fardet, A. 2010. "New hypotheses for the health-protective mechanisms of whole-grain cereals: what is beyond fibre?" *Nutrition Research Reviews* 23 (1):65–134. doi: 10.1017/s0954422410000041.

Ferrando, Arny A., Charles A. Stuart, Melinda Sheffield-Moore, and Robert R. Wolfe. 1999. "Inactivity amplifies the catabolic response of skeletal muscle to cortisol." *The Journal of Clinical Endocrinology & Metabolism* 84 (10):3515–3521.

Ferreira, I., F. Falcato, N. Bandarra, and A. P. Rauter. 2022. "Resolvins, protectins, and maresins: DHA-derived specialized pro-resolving mediators, biosynthetic pathways, synthetic approaches, and their role in inflammation." *Molecules* 27 (5). doi: 10.3390/molecules27051677.

Fischer, Nicole Mercado, Vincent A. Pallazola, Helen Xun, Miguel Cainzos-Achirica, and Erin D. Michos. 2020. "The evolution of the heart-healthy diet for vascular health: a walk through time." *Vascular Medicine* 25 (2):184–193.

Franca-Oliveira, Giselle, Adolfo J. Martinez-Rodriguez, Esperanza Morato, and Blanca Hernández-Ledesma. 2023. "Contribution of proteins and peptides to the impact of a soy protein isolate on oxidative stress and inflammation-associated biomarkers in an innate immune cell model." *Plants* 12 (10):2011.

Furman, David, Judith Campisi, Eric Verdin, Pedro Carrera-Bastos, Sasha Targ, Claudio Franceschi, Luigi Ferrucci, Derek W. Gilroy, Alessio Fasano, Gary W. Miller, Andrew H. Miller, Alberto Mantovani, Cornelia M. Weyand, Nir Barzilai, Jorg J. Goronzy, Thomas A. Rando, Rita B. Effros, Alejandro Lucia, Nicole Kleinstreuer, and George M. Slavich. 2019. "Chronic inflammation in the etiology of disease across the life span." *Nature Medicine* 25 (12):1822–1832. doi: 10.1038/s41591-019-0675-0.

Gasecka, Aleksandra, Dominika Siwik, Magdalena Gajewska, Miłosz J Jaguszewski, Tomasz Mazurek, Krzysztof J. Filipiak, Marek Postuła, and Ceren Eyileten. 2020. "Early biomarkers of neurodegenerative and neurovascular disorders in diabetes." *Journal of Clinical Medicine* 9 (9):2807.

Gögebakan, Özlem, Angela Kohl, Martin A. Osterhoff, Marleen A. van Baak, Susan A. Jebb, Angeliki Papadaki, J. Alfredo Martinez, Teodora Handjieva-Darlenska, Petr Hlavaty, and Martin O. Weickert. 2011. "Effects of weight loss and long-term weight maintenance with diets varying in protein and glycemic index on cardiovascular risk factors: the diet, obesity, and genes (DiOGenes) study: a randomized, controlled trial." *Circulation* 124 (25):2829–2838.

Gupta, Kunj Bihari, Shishir Upadhyay, Ram Gopal Saini, Anil K. Mantha, and Monisha Dhiman. 2018. "Inflammatory response of gliadin protein isolated from various wheat varieties on human intestinal cell line." *Journal of Cereal Science* 81:91–98. https://doi.org/10.1016/j.jcs.2018.03.012.

Han, Velda X., Shrujna Patel, Hannah F. Jones, Timothy C. Nielsen, Shekeeb S. Mohammad, Markus J. Hofer, Wendy Gold, Fabienne Brilot, Samantha J. Lain, and Natasha Nassar. 2021. "Maternal acute and chronic inflammation in pregnancy is associated with common neurodevelopmental disorders: a systematic review." *Translational Psychiatry* 11 (1):71.

Hart, Michael J., Susan J. Torres, Sarah A. McNaughton, and Catherine M. Milte. 2021. "Dietary patterns and associations with biomarkers of inflammation in adults: a systematic review of observational studies." *Nutrition Journal* 20 (1):1–14.

Hertiš Petek, Tjaša, Tadej Petek, Mirjam Močnik, and Nataša Marčun Varda. 2022. "Systemic inflammation, oxidative stress and cardiovascular health in children and adolescents: a systematic review." *Antioxidants* 11 (5):894.

Hofseth, Lorne J., and James R. Hébert. 2022. "Chapter 3 - Diet and acute and chronic, systemic, low-grade inflammation." In *Diet, Inflammation, and Health*, edited by James R. Hébert and Lorne J. Hofseth, 85–111. Academic Press.

Hotamisligil, Gökhan S. 2017. "Inflammation, metaflammation and immunometabolic disorders." *Nature* 542 (7640):177–185. doi: 10.1038/nature21363.

Hruby, Adela, and Paul F. Jacques. 2019. "Dietary protein and changes in biomarkers of inflammation and oxidative stress in the framingham heart study offspring cohort." *Current Developments in Nutrition* 3 (5):nzz019. https://doi.org/10.1093/cdn/nzz019.

Janeiro, Manuel H., María J. Ramírez, Fermin I. Milagro, J. Alfredo Martínez, and Maite Solas. 2018. "Implication of trimethylamine N-oxide (TMAO) in disease: potential biomarker or new therapeutic target." *Nutrients* 10 (10):1398.

Jenkins, D. J., C. W. Kendall, P. W. Connelly, C. J. Jackson, T. Parker, D. Faulkner, and E. Vidgen. 2002. "Effects of high- and low-isoflavone (phytoestrogen) soy foods on inflammatory biomarkers and pro-inflammatory cytokines in middle-aged men and women." *Metabolism* 51 (7):919–924. doi: 10.1053/meta.2002.33352.

Jin, Man-Man, Li Zhang, Hui-Xin Yu, Jun Meng, Zhen Sun, and Rong-Rong Lu. 2013. "Protective effect of whey protein hydrolysates on H2O2-induced PC12 cells oxidative stress via a mitochondria-mediated pathway." *Food Chemistry* 141 (2):847–852.

Junker, Y., S. Zeissig, S. J. Kim, D. Barisani, H. Wieser, D. A. Leffler, V. Zevallos, T. A. Libermann, S. Dillon, T. L. Freitag, C. P. Kelly, and D. Schuppan. 2012. "Wheat amylase trypsin inhibitors drive intestinal inflammation via activation of toll-like receptor 4." *Journal of Experimental Medicine* 209 (13):2395–2408. doi: 10.1084/jem.20102660.

Kemp, David C., and Jung Yeon Kwon. 2021. "Fish and shellfish-derived anti-inflammatory protein products: properties and mechanisms." *Molecules* 26 (11):3225.

Kiersnowska, K., and A. Jakubczyk. 2022. "Bioactive peptides obtained from legume seeds as new compounds in metabolic syndrome prevention and diet therapy." 11 (20). doi: 10.3390/foods11203300.

Kim, Hyunju, Laura E. Caulfield, Vanessa Garcia-Larsen, Lyn M. Steffen, Josef Coresh, and Casey M. Rebholz. 2019. "Plant-based diets are associated with a lower risk of incident cardiovascular disease, cardiovascular disease mortality, and all-cause mortality in a general population of middle-aged adults." *Journal of the American Heart Association* 8 (16):e012865.

Kuroda, Y., T. Yuki, Y. Takahashi, H. Sakaguchi, K. Matsunaga, and H. Itagaki. 2020. "An acid-hydrolyzed wheat protein activates the inflammatory and NF-κB pathways leading to long TSLP transcription in human keratinocytes." *Journal of Toxicological Sciences* 45 (6):327–337. doi: 10.2131/jts.45.327 [doi].

Lamberg-Allardt, C., L. Bärebring, E. K. Arnesen, B. I. Nwaru, B. Thorisdottir, A. Ramel, F. Söderlund, J. Dierkes, and A. Åkesson. 2023. "Animal versus plant-based protein and risk of cardiovascular disease and type 2 diabetes: a systematic review of randomized controlled trials and prospective cohort studies." *Food & Nutrition Research* 67. doi: 9003 [pii] 10.29219/fnr.v67.9003 [doi].

Langyan, Sapna, Pranjal Yadava, Fatima Nazish Khan, Zahoor A. Dar, Renu Singh, and Ashok Kumar. 2022. "Sustaining protein nutrition through plant-based foods." *Frontiers in Nutrition* 8. doi: 10.3389/fnut.2021.772573.

Lee, Yu-Mi, Thomas Skurk, Michael Hennig, and Hans Hauner. 2007. "Effect of a milk drink supplemented with whey peptides on blood pressure in patients with mild hypertension." *European Journal of Nutrition* 46:21–27.

Li, Jie, Andrea J. Glenn, Qingling Yang, Ding Ding, Lingling Zheng, Wei Bao, Jeannette Beasley, Erin LeBlanc, Kenneth Lo, JoAnn E. Manson, Lawrence Philips, Lesley Tinker, and Simin Liu. 2022. "Dietary protein sources, mediating biomarkers, and incidence of type 2 diabetes: findings from the women's health initiative and the UK biobank." *Diabetes Care* 45 (8):1742–1753. doi: 10.2337/dc22-0368.

Liu, Z. M., S. C. Ho, Y. M. Chen, and J. Woo. 2013. "Effect of soy protein and isoflavones on blood pressure and endothelial cytokines: a 6-month randomized controlled trial among postmenopausal women." *Journal of Hypertension* 31 (2):384–392. doi: 10.1097/HJH.0b013e32835c0905.

Lopez-Legarrea, Patricia, Rocio de la Iglesia, Itziar Abete, Santiago Navas-Carretero, J. Alfredo Martinez, and M. Angeles Zulet. 2014. "The protein type within a hypocaloric diet affects obesity-related inflammation: the RESMENA project." *Nutrition* 30 (4):424–429.

Luan, Ying-yi, and Yong-ming Yao. 2018. "The clinical significance and potential role of C-reactive protein in chronic inflammatory and neurodegenerative diseases." *Frontiers in Immunology* 9:1302.

Mangano, K. M., H. L. Hutchins-Wiese, A. M. Kenny, S. J. Walsh, R. H. Abourizk, R. S. Bruno, R. Lipcius, P. Fall, A. Kleppinger, L. Kenyon-Pesce, K. M. Prestwood, and J. E. Kerstetter. 2013. "Soy proteins and isoflavones reduce interleukin-6 but not serum lipids in older women: a randomized controlled trial." *Nutrition Research* 33 (12):1026–1033. doi: 10.1016/j.nutres.2013.08.009.

Markova, M., Liselot Koelman, Silke Hornemann, Olga Pivovarova, Stephanie Sucher, Juergen Machann, Natalia Rudovich, Ralph Thomann, Rosemarie Schneeweiss, and Sascha Rohn. 2020. "Effects of plant and animal high protein diets on immune-inflammatory biomarkers: A 6-week intervention trial." *Clinical Nutrition* 39 (3):862–869.

Markova, M., O. Pivovarova, S. Hornemann, S. Sucher, T. Frahnow, K. Wegner, J. Machann, K. J. Petzke, J. Hierholzer, R. Lichtinghagen, C. Herder, M. Carstensen-Kirberg, M. Roden, N. Rudovich, S. Klaus, R. Thomann, R. Schneeweiss, S. Rohn, and A. F. Pfeiffer. 2017. "Isocaloric diets high in animal or plant protein reduce liver fat and inflammation in individuals with type 2 diabetes." *Gastroenterology* 152 (3):571–585.e8. doi: 10.1053/j.gastro.2016.10.007.

Markozannes, Georgios, Charalampia Koutsioumpa, Sofia Cividini, Grace Monori, Konstantinos K. Tsilidis, Nikolaos Kretsavos, Evropi Theodoratou, Dipender Gill, John PA Ioannidis, and Ioanna Tzoulaki. 2021. "Global assessment of C-reactive protein and health-related outcomes: an umbrella review of evidence from observational studies and Mendelian randomization studies." *European Journal of Epidemiology* 36:11–36.

Minihane, A. M., S. Vinoy, W. R. Russell, A. Baka, H. M. Roche, K. M. Tuohy, J. L. Teeling, E. E. Blaak, M. Fenech, D. Vauzour, H. J. McArdle, B. H. Kremer, L. Sterkman, K. Vafeiadou, M. M. Benedetti, C. M. Williams, and P. C. Calder. 2015. "Low-grade inflammation, diet composition and health: current research evidence and its translation." *British Journal of Nutrition* 114 (7):999–1012. doi: 10.1017/s0007114515002093.

Moncada, Salvador, and Annie Higgs. 1993. "The L-arginine-nitric oxide pathway." *New England Journal of Medicine* 329 (27):2002–2012.

Montserrat-de la Paz, Sergio, Noelia M. Rodriguez-Martin, Alvaro Villanueva, Justo Pedroche, Ivan Cruz-Chamorro, Francisco Millan, and Maria C. Millan-Linares. 2020. "Evaluation of anti-inflammatory and atheroprotective properties of wheat gluten protein hydrolysates in primary human monocytes." *Foods* 9 (7):854.

Montserrat-de la Paz, S., and A. Villanueva. 2021. "Antioxidant and anti-inflammatory properties of bioavailable protein hydrolysates from lupin-derived agri-waste." 11 (10). doi: 10.3390/biom11101458.

Moosavian, Seyedeh Parisa, Mehran Rahimlou, Parvane Saneei, and Ahmad Esmaillzadeh. 2020. "Effects of dairy products consumption on inflammatory biomarkers among adults: A systematic review and meta-analysis of randomized controlled trials." *Nutrition, Metabolism and Cardiovascular Diseases* 30 (6):872–888.

Nediani, Chiara, and Monica Dinu. 2022. *Oxidative Stress and Inflammation as Targets for Novel Preventive and Therapeutic Approaches in Non-Communicable Diseases II*. MDPI.

Nediani, Chiara, and Lisa Giovannelli. 2020. *Oxidative Stress and Inflammation as Targets for Novel Preventive and Therapeutic Approaches in Non-Communicable Diseases*. MDPI.

Newsholme, Philip. 2001. "Why is L-glutamine metabolism important to cells of the immune system in health, postinjury, surgery or infection?" *The Journal of Nutrition* 131 (9):2515S–2522S.

Nieman, Kristin M., Barbara D. Anderson, and Christopher J. Cifelli. 2021. "The effects of dairy product and dairy protein intake on inflammation: a systematic review of the literature." *Journal of the American College of Nutrition* 40 (6):571–582.

Nwachukwu, Ifeanyi D., and Rotimi E. Aluko. 2019. "Structural and functional properties of food protein-derived antioxidant peptides." *Journal of Food Biochemistry* 43 (1):e12761.

O'Connor, Lauren E., Jung Eun Kim, Caroline M. Clark, Wenbin Zhu, and Wayne W. Campbell. 2021. "Effects of total red meat intake on glycemic control and inflammatory biomarkers: a meta-analysis of randomized controlled trials." *Advances in Nutrition* 12 (1):115–127.

Parisi, Francesca, Roberta Milazzo, Valeria M Savasi, and Irene Cetin. 2021. "Maternal low-grade chronic inflammation and intrauterine programming of health and disease." *International Journal of Molecular Sciences* 22 (4):1732.

Pilon, Geneviève, Jérôme Ruzzin, Laurie-Eve Rioux, Charles Lavigne, Phillip J. White, Livar Frøyland, Hélène Jacques, Piotr Bryl, Lucie Beaulieu, and André Marette. 2011. "Differential effects of various fish proteins in altering body weight, adiposity, inflammatory status, and insulin sensitivity in high-fat–fed rats." *Metabolism* 60 (8):1122–1130.

Rapa, Shara Francesca, Biagio Raffaele Di Iorio, Pietro Campiglia, August Heidland, and Stefania Marzocco. 2019. "Inflammation and oxidative stress in chronic kidney disease—potential therapeutic role of minerals, vitamins and plant-derived metabolites." *International Journal of Molecular Sciences* 21 (1):263.

Rebholz, C. M., K. Reynolds, M. R. Wofford, J. Chen, T. N. Kelly, H. Mei, P. K. Whelton, and J. He. 2013. "Effect of soybean protein on novel cardiovascular disease risk factors: a randomized controlled trial." *European Journal of Clinical Nutrition* 67 (1):58–63. doi: 10.1038/ejcn.2012.186.

Rizzo, G. 2020. "The antioxidant role of soy and soy foods in human health." *Antioxidants (Basel)* 9 (7). doi: 10.3390/antiox9070635.

Ryan, Joseph Thomas, Reynolds Paul Ross, Declan Bolton, Gerald F Fitzgerald, and Catherine Stanton. 2011. "Bioactive peptides from muscle sources: meat and fish." *Nutrients* 3 (9):765–791.

Sajadi Hezaveh, Zohreh, Masoumeh Khalighi Sikaroudi, Mohammadreza Vafa, Zachary Stephen Clayton, and Sepideh Soltani. 2019. "Effect of egg consumption on inflammatory markers: a systematic review and meta-analysis of randomized controlled clinical trials." *Journal of the Science of Food and Agriculture* 99 (15):6663–6670.

Salas-Salvadó, J., A. Garcia-Arellano, R. Estruch, F. Marquez-Sandoval, D. Corella, M. Fiol, E. Gómez-Gracia, E. Viñoles, F. Arós, and C. Herrera. 2008. "Components of the Mediterranean-type food pattern and serum inflammatory markers among patients at high risk for cardiovascular disease." *European Journal of Clinical Nutrition* 62 (5):651–659.

Schlesinger, Sabrina, Manuela Neuenschwander, Carolina Schwedhelm, Georg Hoffmann, Angela Bechthold, Heiner Boeing, and Lukas Schwingshackl. 2019. "Food groups and risk of overweight, obesity, and weight gain: a systematic review and dose-response meta-analysis of prospective studies." *Advances in Nutrition* 10 (2):205–218.

Schoeler, Marc, and Robert Caesar. 2019. "Dietary lipids, gut microbiota and lipid metabolism." *Reviews in Endocrine and Metabolic Disorders* 20:461–472.

Sen, C., Y. Morimoto, S. Heak, R. V. Cooney, A. A. Franke, and G. Maskarinec. 2012. "Soy foods and urinary isoprostanes: results from a randomized study in premenopausal women." *Food Function* 3 (5):517–521. doi: 10.1039/c2fo10251j.

Seyedsadjadi, Neda, and Ross Grant. 2020. "The potential benefit of monitoring oxidative stress and inflammation in the prevention of non-communicable diseases (NCDs)." *Antioxidants* 10 (1):15.

Shimada, S., and T. Tanigawa. 2019. "Involvement of gliadin, a component of wheat gluten, in increased intestinal permeability leading to non-steroidal anti-inflammatory drug-induced small-intestinal damage." 14 (2):e0211436. doi: 10.1371/journal.pone.0211436.

Silveira, Brenda Kelly Souza, Thatianne Moreira Silva Oliveira, Patricia Amaro Andrade, Helen Hermana Miranda Hermsdorff, Carla de Oliveira Barbosa Rosa, and Sylvia do Carmo Castro Franceschini. 2018. "Dietary pattern and macronutrients profile on the variation of inflammatory biomarkers: scientific update." *Cardiology Research and Practice* 2018:4762575.

Sun, Yangbo, Buyun Liu, Linda G Snetselaar, Robert B. Wallace, Aladdin H. Shadyab, Candyce H. Kroenke, Bernhard Haring, Barbara V. Howard, James M. Shikany, and Carolina Valdiviezo. 2021. "Association of major dietary protein sources with all-cause and cause-specific mortality: Prospective cohort study." *Journal of the American Heart Association* 10 (5):e015553.

Tawalbeh, D., H. M. Al-U'datt, and W. A. N. Wan Ahmad. 2023. "Recent Advances in In Vitro and In Vivo Studies of Antioxidant, ACE-Inhibitory and Anti-Inflammatory Peptides from Legume Protein Hydrolysates." 28 (6). doi: 10.3390/molecules28062423.

Thomas, Minu S., and Maria Luz Fernandez. 2021. "Trimethylamine N-oxide (TMAO), diet and cardiovascular disease." *Current Atherosclerosis Reports* 23:1–7.

Thompson, Warren G., Nicole Rostad Holdman, Denise J. Janzow, Jeffrey M. Slezak, Kristin L. Morris, and Michael B. Zemel. 2005. "Effect of energy-reduced diets high in dairy products and fiber on weight loss in obese adults." *Obesity Research* 13 (8):1344–1353.

Tourkochristou, Evanthia, Christos Triantos, and Athanasia Mouzaki. 2021. "The influence of nutritional factors on immunological outcomes." *Frontiers in Immunology* 12. doi: 10.3389/fimmu.2021.665968.

Tricon, Sabine, Graham C. Burdge, Emma L. Jones, Jennifer J. Russell, Soraya El-Khazen, Emmanuelle Moretti, Wendy L. Hall, Andrew B. Gerry, David S. Leake, and Robert F. Grimble. 2006. "Effects of dairy products naturally enriched with cis-9, trans-11 conjugated linoleic acid on the blood lipid profile in healthy middle-aged men." *The American Journal of Clinical Nutrition* 83 (4):744–753.

Tsalamandris, Sotirios, Alexios S. Antonopoulos, Evangelos Oikonomou, George-Aggelos Papamikroulis, Georgia Vogiatzi, Spyridon Papaioannou, Spyros Deftereos, and Dimitris Tousoulis. 2019. "The role of inflammation in diabetes: current concepts and future perspectives." *European Cardiology Review* 14 (1):50.

Valacchi, Giuseppe, Fabio Virgili, Carlo Cervellati, and Alessandra Pecorelli. 2018. "OxInflammation: From subclinical condition to pathological biomarker." *Frontiers in Physiology* 9:858.

van den Berg, Lisa A., Jurriaan J. Mes, Marco Mensink, and Anne J. Wanders. 2022. "Protein quality of soy and the effect of processing: A quantitative review." *Frontiers in Nutrition* 9:2148.

vel Szic, Katarzyna Szarc, Ken Declerck, Melita Vidaković, and Wim Vanden Berghe. 2015. "From inflammaging to healthy aging by dietary lifestyle choices: is epigenetics the key to personalized nutrition?" *Clinical Epigenetics* 7:1–18.

Wu, Guoyao, Yun-Zhong Fang, Sheng Yang, Joanne R. Lupton, and Nancy D. Turner. 2004. "Glutathione metabolism and its implications for health." *The Journal of Nutrition* 134 (3):489–492.

Wu, Shujian, Zuhaib F. Bhat, Rochelle S. Gounder, Isam A. Mohamed Ahmed, Fahad Y. Al-Juhaimi, Yu Ding, and Alaa E-DA Bekhit. 2022. "Effect of dietary protein and processing on gut microbiota—a systematic review." *Nutrients* 14 (3):453.

Yang, Shengjie, Xinye Li, Fan Yang, Ran Zhao, Xiandu Pan, Jiaqi Liang, Li Tian, Xiaoya Li, Longtao Liu, and Yanwei Xing. 2019. "Gut microbiota-dependent marker TMAO in promoting cardiovascular disease: inflammation mechanism, clinical prognostic, and potential as a therapeutic target." *Frontiers in Pharmacology* 10:1360.

Zevallos, Victor F., Verena Raker, Stefan Tenzer, Carolina Jimenez-Calvente, Muhammad Ashfaq-Khan, Nina Rüssel, Geethanjali Pickert, Hansjörg Schild, Kerstin Steinbrink, and Detlef Schuppan. 2017. "Nutritional wheat amylase-trypsin inhibitors promote intestinal inflammation via activation of myeloid cells." *Gastroenterology* 152 (5):1100–1113. e12.

Zhang, H. P., J. H. Zhao, H. X. Yu, and D. X. Guo. 2016. "Genistein ameliorated endothelial nitric oxidase synthase uncoupling by stimulating sirtuin-1 pathway in ox-LDL-injured HUVECs." *Environ Toxicol Pharmacol* 42:118–124. doi: 10.1016/j.etap.2016.01.011.

Zhang, H. P., Z. Zhao, X. Pang, J. Yang, H. Yu, Y. Zhang, H. Zhou, and J. Zhao. 2017. "Genistein protects against Ox-LDL-induced inflammation through MicroRNA-155/SOCS1-mediated repression of NF-κB signaling pathway in HUVECs." *Inflammation* 40 (4):1450–1459. doi: 10.1007/s10753-017-0588-3.

Zhang, M., S. Song, D. Zhao, J. Shi, X. Xu, G. Zhou, and C. Li. 2020. "High intake of chicken and pork proteins aggravates high-fat-diet-induced inflammation and disorder of hippocampal glutamatergic system." *Journal of Nutritional Biochemistry* 85:108487. doi: 10.1016/j.jnutbio.2020.108487.

Zotova, N. V., V. A. Chereshnev, and E. Yu Gusev. 2016. "Systemic inflammation: methodological approaches to identification of the common pathological process." *PLoS One* 11 (5):e0155138.

5 Protein Requirements in Different Metabolic Scenarios

Focusing on Endurance Athletes, Muscle, and Whole-Body Protein Metabolism

Eric Williamson and Daniel R. Moore

ABBREVIATIONS

ACSM	American College of Sports Medicine
AMDR	Acceptable macronutrient distribution range
EAR	Estimated average requirement
IAAO	Indicator amino acid oxidation
IOM	Institute of Medicine
RDA	Recommended dietary allowance
RMR	Resting metabolic rate
WHO	World Health Organization

INTRODUCTION

Nutrition guidelines for endurance athletes have historically centered on carbohydrates to fuel their activity and fluid to prevent dehydration, with relatively little emphasis on protein (American College of Sports Medicine et al. 2000; Jeukendrup 2011). However, with increased awareness of the importance of dietary protein in providing the requisite building blocks to support training recovery and adaptation (Moore et al. 2014) more recent reviews (Vitale and Getzin 2019) and position stands (Thomas et al. 2016; Jäger et al. 2017) acknowledge the importance of protein for endurance training. This chapter will summarize the current understanding of endurance exercise's impact on body protein metabolism and how it affects the requirement for dietary protein in these athletes. We will outline contemporary research into the impact of dietary protein amount and timing on endurance athletes' ability to recover, adapt, and perform.

PROTEIN METABOLISM AND ENDURANCE EXERCISE

Body protein (including skeletal muscle) constantly undergoes "turnover" through the processes of breakdown (into its constituent amino acids) and synthesis. This turnover maintains tissue quality and quantity and is the fundamental process that underpins training adaptations. As protein synthesis is energetically expensive, body protein synthesis decreases during endurance exercise, contributing to a negative protein balance (Mazzulla et al. 2017). This negative protein balance is further exacerbated because amino acids liberated by protein breakdown (especially within skeletal muscle) can be deaminated, and their carbon skeletons oxidized as a source of energy or diverted

DOI: 10.1201/b23328-5

to gluconeogenesis. Research has demonstrated that the oxidation of the branched-chain amino acid leucine correlates positively with intensity measured as a percentage of VO_2 max (Lemon et al. 1982; Millward et al. 1982). The contribution of amino acid oxidation to energy is approximately 2–10%, depending on duration, intensity, and nutritional status (e.g., muscle glycogen levels) (Howarth et al. 2010). When essential amino acids are oxidized, these losses require replacement through post-exercise protein ingestion (Tarnopolsky 2004; Mazzulla et al. 2017). Therefore, endurance exercise results in substantial perturbation to whole-body proteostasis that contributes to a negative protein balance (Figure 5.1).

Endurance exercise involving prolonged repetitive contractions (especially weight-bearing, such as running) can also contribute to tissue protein damage, as evidenced by microstructure damage of muscle tissue (Koller et al. 1998; Chantler et al. 2021), even in elite athletes (Simmons et al. 2021). The gastrointestinal tract may also experience cell stress and damage due to splanchnic ischemia during activity (Costa et al. 2017), which may compromise its integrity and function. This is further exacerbated by performing endurance exercise in the heat (Snipe et al. 2018; Osborne et al. 2019) and/or with low carbohydrate availability during the activity (Snipe et al. 2017; Jonvik et al. 2019). Furthermore, post-endurance exercise protein ingestion may support increased remodeling of oxidative and fatigue-resistant muscle fibers (i.e., primarily type I, type IIa), which is observed with training (Andersen and Henriksson 1977; Gehlert et al. 2012) as well as mitochondrial protein synthesis (Breen et al. 2011; Moore and Stellingwerff 2012). Thus, protein intake may contribute to adaptations related to aerobic capacity (VO_2 max), anaerobic threshold, and running economy (Wilkinson et al. 2008; Breen et al. 2011): factors that have been identified as differentiating elite from non-elite athletes (Lorenz et al. 2013).

Following endurance exercise, protein breakdown increases to maintain the plasma amino acid pool (Williamson et al. 2022). Although protein synthesis is also elevated in this period (Sheffield-Moore et al. 2004), whole-body protein balance does not shift towards a net positive (i.e., synthesis

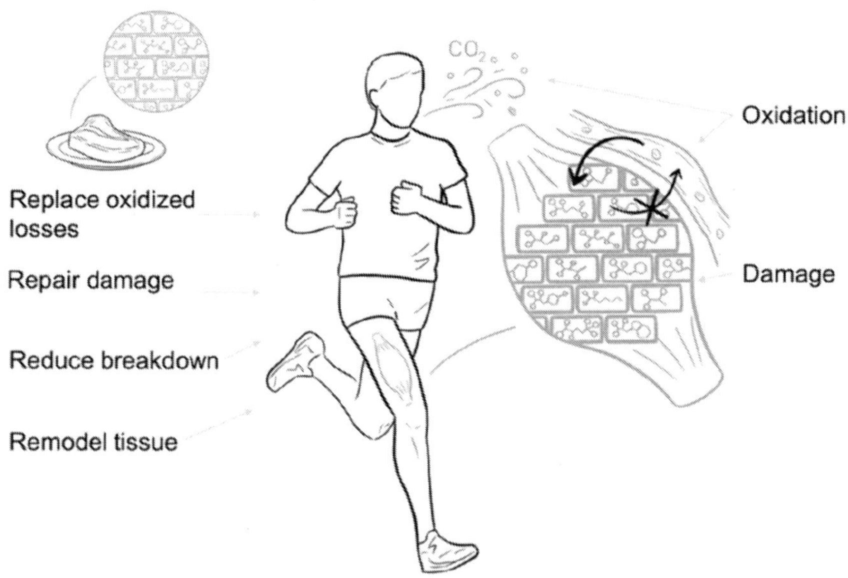

FIGURE 5.1 The impact of endurance exercise on protein metabolism. The right of the illustration shows the impact of endurance exercise on tissue proteins, while the left shows the subsequent effects of post-exercise protein ingestion. During endurance exercise, damage is incurred to tissues such as muscle, and amino acids are oxidized as a fuel source. Amino acid ingestion from dietary protein following exercise supports the replacement of the oxidized amino acids, repair of damaged tissue, reductions in protein breakdown, and remodeling of tissue proteins. From Williamson (2022), with permission.

> breakdown) anabolic environment until dietary protein is consumed (Churchward-Venne et al. 2020). Therefore, recovery of oxidized amino acids does not begin, and repair of damaged tissue is likely not maximized until the consumption of dietary protein. This is due to dietary protein's role in both enhancing protein synthesis and mitigating protein breakdown (Churchward-Venne et al. 2020).

PROTEIN REQUIREMENTS

The World Health Organization (WHO) defines protein requirements as "the minimum intake that will allow nitrogen equilibrium" (i.e., intake equal to excretion), with the recommended dietary allowance (RDA) being 0.8 $g \cdot kg^{-1} \cdot d^{-1}$ for all ages of adulthood, regardless of physical activity intensity and duration (WHO 2007). This recommendation is based on studies utilizing nitrogen balance methodology, which has limitations including the likelihood of overestimating nitrogen intake and underestimating nitrogen excretion (e.g., due to delays in metabolic adjustments to changes in nitrogen intake); collectively, this can result in an underestimation of nitrogen balance (Forbes 1973; Millward 1998). More modern techniques using amino acid tracers have led to an increasingly prominent definition of the term "requirement" as the minimum intake to maximize protein synthesis and protein/nitrogen turnover rates (Elango et al. 2008; Phillips et al. 2016), particularly for athletes (Tipton and Witard 2007; Phillips 2012). This will be our defined use of the term "requirement" for the remainder of this chapter. This more optimal protein intake is thought to allow for improved tissue quality through the construction of new proteins and the replacement of damaged and dysfunctional ones. This could include, for example, the synthesis of plasma albumin (i.e., increased plasma volume expansion) (Goto et al. 2010), sarcoplasmic proteins (i.e., metabolic enzymes) (Oikawa et al. 2020), collagen (e.g., tendon and bone) (Miller et al. 2005), vasculature (e.g., endothelial and smooth muscle cells within capillaries) (Prior et al. 2004) and neutrophils (Costa R et al. 2011). Thus, while protein remodeling may occur in a variety of body tissues of varying sizes and rates of protein turnover, methods that determine maximal whole-body protein synthesis (an average of all body protein pools) could provide a suitable estimate of the protein intake that supports the repair and synthesis of most/all body proteins. For example, acute studies have shown that muscle protein synthesis after endurance exercise is maximized at a lower protein intake than whole-body protein synthesis (Levenhagen et al. 2001; Churchward-Venne et al. 2020).

The most recent (2016) American College of Sports Medicine (ACSM) position recommends a protein intake of 1.2–2.0 $g \cdot kg^{-1} \cdot d^{-1}$ across different forms of exercise, including endurance training, to "maximize metabolic adaptations to training" (Thomas et al. 2016). Research using non-invasive whole-body tracer methodology has shown that male endurance athletes had an estimated average protein requirement (EAR) and safe intake (as determined by the upper 95% confidence interval) of 1.65 and 1.83 $g \cdot kg^{-1} \cdot d^{-1}$, respectively, to maximize whole-body protein synthesis on a day involving 20 km running training (Kato et al. 2016). Using a similar methodology but in a free-living setting, Williamson et al. (2023) recently demonstrated that the safe intake for male and female athletes during the luteal phase was similar at ~1.89 $g \cdot kg^{-1} \cdot d^{-1}$, further emphasizing that this protein intake would support body protein remodeling in similarly trained endurance athletes regardless of sex (Williamson et al. 2023). To explore the ecological validity of this protein recommendation, Williamson et al. (2019) expanded on these findings and demonstrated that 1.83 $g \cdot kg^{-1} \cdot d^{-1}$ outperformed the low end (i.e., 1.2 $g \cdot kg^{-1} \cdot d^{-1}$) of the ACSM's recommendation in achieving a net positive protein balance and maintaining performance over a four-day intensified training period. Specifically, post-training performance testing revealed moderate and low effect sizes for maximal voluntary contraction and a 5 km running time trial for 1.83 $g \cdot kg^{-1} \cdot d^{-1}$ compared to 0.9 and 1.2 $g \cdot kg^{-1} \cdot d^{-1}$, respectively. Although performance differences were relatively small and not statistically significant due to variability associated with performance testing, from a practical perspective, improvements of <1% are relevant to athletes (Hopkins et al. 1999). Thus, consuming protein below this threshold to maximize whole-body protein synthesis resulted in metabolic and performance decrements in only four days of relatively high volume (55 km) training. The primary reason for these protein requirements may be to replenish amino acid oxidative losses, given that the branched-chain amino acids (leucine, isoleucine, valine) were determined to be the primary limiting dietary amino acids for endurance athletes (Kato et al. 2018).

It is important to note that the training stimulus in the above studies (Kato et al. 2016; Williamson et al. 2023) was comparable to the average training session of marathoners who have a completion time of <4 h (i.e., ~16.0 km·session^{-1}, ~5.52 min·km^{-1}·session^{-1}, ~96 min·session^{-1}, ~61% VO_{2max}) (Gordon et al. 2017). Although middle-distance endurance athletes tend to incorporate more high-intensity as well as strength and power training, they are known to have high-volume periods and regularly incorporate >15 km runs into their training (Tjelta 2016; Haugen et al. 2021), suggesting these findings may be applicable to this population as well. However, as discussed above, amino acid oxidation makes up a large portion (>40%) (Kato et al. 2016; Williamson et al. 2023) of the enhanced requirements for post-endurance exercise protein ingestion. Given that there is a wide range of endurance sports that differ in their training, it is likely that amino acid oxidation rates will differ as well, which will impact their post-exercise protein needs. For example, while the work of Kato et al. (2016) and Williamson et al. (2023) represents the average training volume of marathon runners, other athletes are known to exercise at similar average intensities for longer durations, such as cyclists (~160 min) (van Erp et al. 2019; Gallo et al. 2022). Considering both energy requirements and amino acid oxidation rates increase with intensity and duration of exercise, it is currently recommended that those exercising for longer durations and/or greater intensities follow the protein requirement expressed as a proportion of caloric needs, which was also determined by Williamson et al. (2023) at 15% of total calories from protein.

While many studies report endurance athletes consume mean habitual intakes in the range of 1.5–2.0 g·kg^{-1}·d^{-1} (Burke et al. 2003; Tarnopolsky 2004; Gillen et al. 2017) or ~14% of calories, others report intakes of ~1.2 g·kg^{-1}·d^{-1} or below (Tarnopolsky 2004) (Figure 5.2). This highlights that

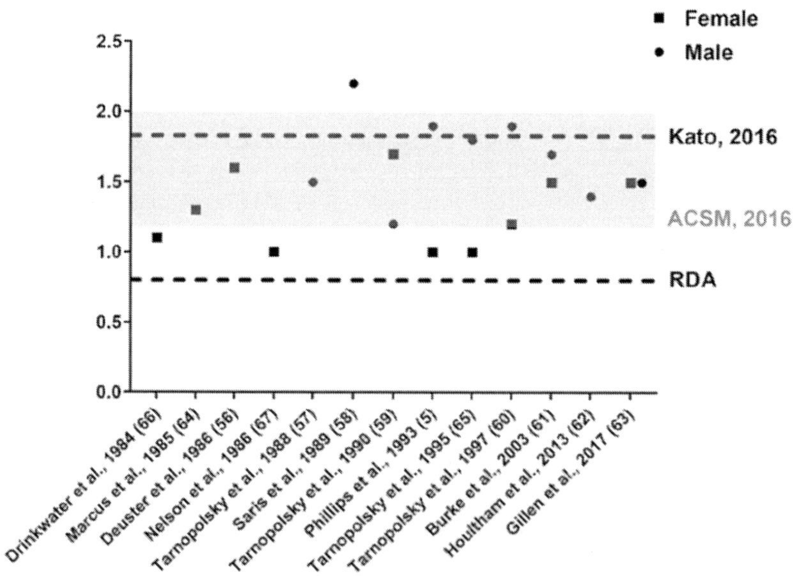

FIGURE 5.2 Protein intakes reported by endurance athletes. This figure shows the protein intake of female and male endurance athletes across various studies compared to recommendations by the WHO (dashed line; i.e., RDA of 0.8 g·kg^{-1}·d^{-1}), ACSM (shaded area; i.e., 1.2–2.0 g·kg^{-1}·d^{-1}) and using the Indicator Amino Acid Oxidation (IAAO) method (dotted line; i.e., 1.8 g·kg^{-1}·d^{-1}). Intakes generally exceed the RDA and fall within the range of the ACSM guidelines, but some individuals report falling short of that which would maximize whole-body protein synthesis rates. From Williamson (2022), with permission.

many athletes are meeting recommendations, while others do not. The recommendation expressed as a percentage of energy is similar to that reported by the general population (i.e., ~15% of total energy intake) (Berryman et al. 2018) and is within the acceptable macronutrient distribution range (AMDR) of 10–35% of energy put forth by the Institute of Medicine (IOM) (Institute of Medicine, 2005) based on a balanced diet. Accordingly, a prudent recommendation for athletes, coaches, and sports dietitians may be to prioritize adequate energy intake, as this may inherently meet our protein recommendation, which is supported by previous research in high-performance athletes (Burke et al. 2003; Gillen et al. 2017).

DAILY PROTEIN DISTRIBUTION

It is clear that dietary protein ingestion stimulates whole-body and muscle protein synthesis up to a plateau, after which dietary amino acids are diverted to energy metabolic pathways (Moore et al. 2009; Mazzulla et al. 2018). Thus, it has been argued that "efficient" protein ingestion could be viewed through the lens of meal intakes that maximize protein synthesis but minimize amino acid oxidation (Barnes et al. 2023). Although meal protein intake may be related to a variety of food-matrix factors that may influence protein metabolism (Burd et al. 2019), meal protein targets of ~0.3 $g{\cdot}kg^{-1}$ generally support maximal rates of muscle protein synthesis (Moore et al. 2015; Moore 2019). The repeated ingestion of moderate protein-containing meals (~0.25 $g{\cdot}kg^{-1}$ every 3 h) supports greater rates of muscle protein synthesis and whole-body anabolism (i.e., net balance) after resistance exercise (Moore et al. 2012; Areta et al. 2013). Although the effect of the distribution of protein intake throughout the day has yet to be assessed following endurance exercise, it is notable that endurance athletes have been reported to consume at least four meals per day (Burke et al. 2003), which would ostensibly help them meet their relatively high daily energy requirements. Following a distribution pattern of 4–6 meals per day separated by about 3–5 hours would maximize MPS while minimizing MPB and amino acid oxidation (Williamson and Moore 2021) to support optimal recovery and adaptation following endurance exercise. A recent investigation determined that a protein dose of ~0.49 $g{\cdot}kg^{-1}$ maximally stimulates myofibrillar protein synthesis immediately following 90 minutes of intense endurance exercise (Churchward-Venne et al. 2020), although this is likely related primarily to the need to replenish exercise-induced amino acid oxidative losses and is unlikely to persist throughout the remainder of the day (Moore 2020). Therefore, the most prudent advice for endurance athletes may be to meet their energy requirements (and thus, daily protein requirement) and to spread this over their typical 4–6 meals/snacks per day (Burke et al. 2003; Gillen et al. 2017), which would ultimately result in an efficient daily consumption of 1.85–$g{\cdot}kg^{-1}{\cdot}d^{-1}$.

ENERGY AVAILABILITY

Negative energy balance and low energy availability can lead to lower net protein balance within as little as four days (Carbone et al. 2019; Oxfeldt et al. 2023). Although this effect has not been studied directly in endurance athletes, researchers have observed it in military personnel performing concurrent training (Pasiakos et al. 2013; Margolis et al. 2016) and low-moderate intensity ski march training (Margolis et al. 2014). These conditions may result in under-recovery, poorer performance adaptations, and loss of muscle mass (Carbone et al. 2019). Studies in resistance-trained and inactive populations have demonstrated that this negative net balance can be mitigated through elevated protein intakes (Josse et al. 2011; Pasiakos et al. 2013). Protection of fat-free mass has been demonstrated with elevated protein intakes (1.6–2.4 $g{\cdot}kg^{-1}{\cdot}d^{-1}$) in individuals engaging in concurrent training (Josse et al. 2011; Pasiakos et al. 2013) and in a case study of a female road cyclist consuming ~2.4 $g{\cdot}kg^{-1}{\cdot}d^{-1}$ of protein (Haakonssen et al. 2013).

For the general athlete population, protein recommendations during weight loss to spare lean mass depend on the extent of energy deficiency and range from 1.6 $g{\cdot}kg^{-1}{\cdot}d^{-1}$ for a deficit of 10% of energy requirements to 2.4 $g \cdot kg^{-1} \cdot d^{-1}$ for deficits ≥30% (Hector and Phillips 2018). Current recommendations suggest a rate of weight loss of 0.5–1.0% of body weight per week to limit the catabolic impact of caloric deficiency (Aragon et al. 2017; Hector and Phillips 2018), corresponding to an approximate energy deficit of 20% (Garthe et al. 2011; Hector and Phillips 2018). This scenario implies a protein intake of at least 2.0 $g \cdot kg^{-1} \cdot d^{-1}$ or 20% of total energy balance requirements (Garthe et al. 2011; Hector and Phillips 2018) from protein. Further research is warranted in the endurance athlete population to refine this requirement.

Athletes should reserve low energy availability for planned periods when body weight loss is necessary to improve performance. However, in certain cases (e.g., relative energy deficiency in sport syndrome), low energy availability may occur in the absence of body weight loss due to metabolic accommodations (e.g., reproductive suppression) that contribute to lower resting metabolic rate (RMR) and the overall detriment of the athlete (Melin et al. 2023). In these conditions, experts recommend improving energy availability and meeting general protein recommendations of 1.85 $g \cdot kg^{-1} \cdot d^{-1}$ or 15% of energy needs. Regardless, carbohydrates should remain a priority in the athlete's diet when training quality and performance need to remain high. Thus, athletes are discouraged from consuming a higher protein intake at the expense of carbohydrates in both energy balance and low energy availability, which is likely to have a greater overall detriment to recovery and performance (Jarvis et al. 2002; Macdermid and Stannard 2006).

CARBOHYDRATE AVAILABILITY

The contribution of amino acids to total energy expenditure during endurance exercise depends on glycogen levels (Howarth et al. 2010) and therefore has implications for athletes who train with low carbohydrate availability (i.e., "training low"). In recent years, researchers have increasingly incorporated this type of training into nutrition periodization recommendations for endurance athletes to facilitate aerobic adaptations and promote more efficient utilization of fat as an energy substrate (Jeukendrup 2017; Podlogar and Wallis 2022). Gillen et al. (2019) explored the effects of training low on protein metabolism in a crossover design study where endurance-trained participants completed an afternoon glycogen-depleting training session followed by either a "train-low" protocol with minimal carbohydrate intake (0.2 $g \cdot kg^{-1}$) or a "train-high" protocol with adequate carbohydrate intake (5.0 $g \cdot kg^{-1}$) before a 10 km run at ~80% of max heart rate the next morning. The researchers estimated that this low carbohydrate availability training increases relative protein requirements by ~0.12 $g \cdot kg^{-1} \cdot d^{-1}$ (i.e., ~7%). Considering amino acid oxidation scales with energy expenditure (Rennie et al. 2006), this requirement may be greater in athletes exercising for longer durations and/or higher intensities. This increased requirement may be important in facilitating the protein remodeling that is the basis of this training (Margolis and Pasiakos 2013) and should be considered for athletes who utilize this type of training within their periodization.

SEX

Female athletes comprise a large portion of high-performance athletics but are unfortunately underrepresented in exercise physiology research (Costello et al. 2014; Devries and Jakobi 2021) despite known metabolic differences, including in protein metabolism (Devries and Jakobi 2021). For example, the female hormones estrogen and progesterone may have opposing impacts on amino acid metabolism, with estrogen having a protein-sparing effect (i.e., reduced amino acid oxidation) (Kriengsinyos et al. 2004; Hamadeh et al. 2005) and progesterone demonstrating a catabolic effect (i.e., enhanced amino acid oxidation) (Landau and Lugibihl 1961, 1967; Kriengsinyos et al. 2004). Unlike males, the fluctuations in hormones (i.e., the estrogen:progesterone ratio) during the menstrual cycle (averaging 28 days) (Bull et al. 2019) may result in variations in metabolism and nutrient requirements (Draper et al. 2018). The mid-luteal phase, characterized by higher levels of sex hormones and a low estrogen:progesterone ratio, is suggested to be the phase in which amino

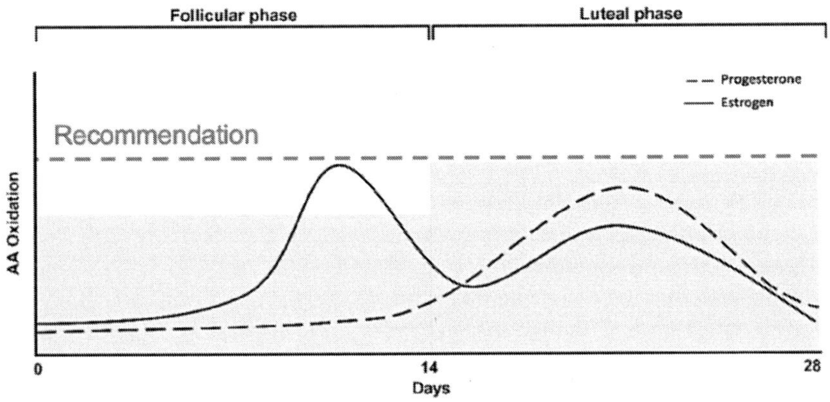

FIGURE 5.3 The impact of menstrual cycle phase on amino acid oxidation.This graph illustrates hormonal changes over a 28-day menstrual cycle. Day 1 begins on the first day of menstruation. The shaded bars represent amino acid oxidation rates. During the follicular phase, progesterone levels remain low and estrogen levels surge prior to ovulation (day 14), resulting in lower amino acid oxidation rates. Following ovulation, progesterone and estrogen levels rise, with progesterone exceeding estrogen, resulting in greater amino acid oxidation. The recommended protein intake of 1.8 $g \cdot kg^{-1} \cdot d^{-1}$ was determined during the luteal phase (Williamson et al. 2023) and thus would presumably cover requirements across the entire cycle. From Williamson (2022), with permission.

acid oxidation is elevated both at rest and during exercise (Lamont et al. 1987; Lariviere et al. 1994). This is relative to the follicular phase, when the estrogen:progesterone ratio is high (i.e., estrogen is higher than progesterone) and lower amino acid oxidation rates are observed both at rest and during exercise (Lamont et al. 1987; Lariviere et al. 1994) (Figure 5.3).

Previous studies have demonstrated lower amino acid oxidation rates in females during the follicular phase of the menstrual cycle compared to males (Phillips et al. 1993; Horton et al. 1998). This finding led to the hypothesis that protein requirements might be lower in female endurance athletes than in current recommendations based primarily on studies of male athletes (Tarnopolsky 2004; Mercer et al. 2020). However, Williamson et al. (2023) found no difference in protein requirements between sexes when they studied females studied in the luteal phase. As the hormonal environment of the luteal phase likely represents the menstrual phase with the highest rates of amino acid oxidation due to the low estrogen-to-progesterone ratio, female athletes who consume the safe protein intake (i.e., ~1.85 $g \cdot kg^{-1} \cdot d^{-1}$) would minimize the chance of consuming suboptimal protein throughout the entire menstrual cycle (Figure 5.3). However, practitioners should be aware that female athletes are more likely to consume less than recommended protein intake compared to males (Moss et al. 2023).

OLDER AGE

Sarcopenia is prevalent among older adults (i.e., >60 years) (Yuan and Larsson 2023) and refers to an age-related decline in muscle mass and strength (Cruz-Jentoft et al. 2010). The etiology of sarcopenia is partly related to a blunted protein synthetic response to protein ingestion, although low activity may primarily mediate and/or exacerbate this effect (Oikawa et al. 2019). Older individuals who maintain regular endurance training demonstrate greater physical function and sensitivity to protein compared to their inactive counterparts (Moore 2021). Thus, increases in protein intake may not be required in the endurance training population discussed here (Doering et al. 2016b; Moore 2021). From a practical lens, many older endurance athletes report inadequate daily protein intakes as well as a suboptimal dose of protein in their first post-exercise meal (Doering et al. 2016a). Daily distribution of protein intake within the general older adult population is also suboptimal

(Cardon-Thomas et al. 2017; Gaytán-González et al. 2020). Therefore, while per-meal and daily requirements are similar for the older adult population, practitioners should aim to encourage optimal distribution and per-meal doses of protein, which are likely to subsequently meet daily protein requirements, as demonstrated in the general athlete population (Gillen et al. 2017)

CONCLUSION

Protein intake is an important component of the endurance athlete's diet for recovery and adaptation to training. Current recommendations suggest an intake of 1.85 $g{\cdot}kg^{-1}{\cdot}d^{-1}$ or 15% of energy needs distributed equally into 4–6 meals to maximize whole-body protein synthesis. This recommendation applies to older adults and female athletes with similar training regimens to young males; however, conditions of reduced energy and/or carbohydrate availability increase protein requirements. Athletes should avoid increasing protein intake at the expense of carbohydrate requirements. Although many athletes meet current daily recommendations, practitioners should recognize that athletes consume suboptimal levels, and many athletes, particularly those of older age, could benefit from improved distribution of their current daily protein intake.

MINI DICTIONARY OF TERMS

Amino acid oxidation: The catabolism of amino acids to produce energy for metabolic processes (e.g., skeletal muscle contraction), resulting in carbon being excreted primarily through the breath as carbon dioxide, hydrogen excreted primarily as H_2O in urine, and nitrogen excreted primarily as urea and ammonia in urine.

Gluconeogenesis: The de novo synthesis of glucose from other substrates such as gluconeogenic amino acids.

Nitrogen balance: The mathematical relationship between nitrogen consumed (e.g., through dietary protein) and nitrogen excreted (e.g., through sweat, urine, and feces).

Protein breakdown: The catabolism of protein into individual amino acids.

Protein synthesis: The incorporation of amino acids into complete proteins.

Sarcopenia: An age-related decline in muscle mass and strength.

SUMMARY POINTS

- This chapter examines protein requirements for endurance exercise.
- Protein recommendations following endurance exercise are 1.85 $g{\cdot}kg^{-1}{\cdot}d^{-1}$, which is near the upper range of consensus guidelines of 1.2–2.0 $g{\cdot}kg^{-1}{\cdot}d^{-1}$ for athletes, or 15% of total energy requirements.
- Distributing total daily protein intake into 4–6 feedings separated by 3–5 hours may further enhance daily protein synthesis rates and, thus, recovery and adaptations to training.
- Low energy availability may increase protein requirements to preserve lean mass, with a 20% energy deficit requiring 2.0 $g{\cdot}kg^{-1}{\cdot}d^{-1}$ or 20% of total energy balance requirements.
- Training with low carbohydrate availability may raise protein requirements by 0.12 $g{\cdot}kg^{-1}{\cdot}d^{-1}$ or an added 7% of current protein recommendations.
- Sex influences protein metabolism, as the follicular phase of the menstrual cycle is characterized by lower amino acid oxidation during exercise. However, luteal phase requirements are similar, and therefore, protein recommendations are similar to those of male athletes to avoid deficiency across the menstrual cycle.
- Older age results in a blunted protein synthetic response to protein ingestion, which is mitigated by regular endurance exercise. Older adults engaging in regular endurance exercise have protein requirements similar to those of their younger counterparts.

- In summary, current recommendations suggest an intake of 1.85 $g \cdot kg^{-1} \cdot d^{-1}$ or 15% of energy needs distributed equally into 4-6 feedings to support recovery and adaptations to endurance exercise when training with adequate energy and carbohydrate availability.

REFERENCES

American College of Sports Medicine, American Dietetic Association, Dietitians of Canada (2000) Joint Position Statement: nutrition and athletic performance. American College of Sports Medicine, American Dietetic Association, and Dietitians of Canada. *Med Sci Sports Exerc* 32:2130–2145. https://doi.org/10.1097/00005768-200012000-00025

Andersen P, Henriksson J (1977) Training induced changes in the subgroups of human type II skeletal muscle fibres. *Acta Physiol Scand* 99:123–125. https://doi.org/10.1111/j.1748-1716.1977.tb10361.x

Aragon AA, Schoenfeld BJ, Wildman R, et al. (2017) International society of sports nutrition position stand: Diets and body composition. *J Int Soc Sports Nutr* 14.

Areta JL, Burke LM, Ross ML, et al. (2013) Timing and distribution of protein ingestion during prolonged recovery from resistance exercise alters myofibrillar protein synthesis. 591:2319–2331. https://doi.org/10.1113/jphysiol.2012.244897

Barnes TM, Deutz MT, Zupančič Ž, et al. (2023) Protein quality and the food matrix: defining optimal versus maximal meal-based protein intakes for stimulating muscle protein synthesis. *Appl Physiol Nutr Metab Physiol Appl Nutr Metab* 48:340–344. https://doi.org/10.1139/apnm-2022-0373

Berryman CE, Lieberman HR, Fulgoni VL, Pasiakos SM (2018) Protein intake trends and conformity with the dietary reference intakes in the United States: Analysis of the national health and nutrition examination survey, 2001–2014. *Am J Clin Nutr* 108:405–413. https://doi.org/10.1093/ajcn/nqy088

Breen L, Philp A, Witard OC, et al. (2011) The influence of carbohydrate-protein co-ingestion following endurance exercise on myofibrillar and mitochondrial protein synthesis. *J Physiol* 589:4011–4025

Bull JR, Rowland SP, Scherwitzl EB, et al. (2019) Real-world menstrual cycle characteristics of more than 600,000 menstrual cycles. *NPJ Digit Med* 2. https://doi.org/10.1038/s41746-019-0152-7

Burd NA, Beals JW, Martinez IG, et al. (2019) Food-first approach to enhance the regulation of post-exercise skeletal muscle protein synthesis and remodeling. *Sports Med Auckl NZ* 49:59–68. https://doi.org/10.1007/s40279-018-1009-y

Burke LM, Slater G, Broad EM, et al. (2003) Eating patterns and meal frequency of elite Australian athletes. *Int J Sport Nutr Exerc Metab* 13:521–538

Carbone JW, McClung JP, Pasiakos SM (2019) Recent advances in the characterization of skeletal muscle and whole-body protein responses to dietary protein and exercise during negative energy balance. *Adv Nutr Bethesda Md* 10:70–79. https://doi.org/10.1093/advances/nmy087

Cardon-Thomas DK, Riviere T, Tieges Z, Greig CA (2017) Dietary protein in older adults: Adequate daily intake but potential for improved distribution. *Nutrients* 9:184. https://doi.org/10.3390/nu9030184

Chantler S, Griffiths A, Matu J, et al. (2021) The effects of exercise on indirect markers of gut damage and permeability: A systematic review and meta-analysis. *Sports Med Auckl NZ* 51:113–124. https://doi.org/10.1007/s40279-020-01348-y

Churchward-Venne TA, Pinckaers PJM, Smeets JSJ, et al. (2020) Dose-response effects of dietary protein on muscle protein synthesis during recovery from endurance exercise in young men: A double-blind randomized trial. *Am J Clin Nutr* 112:303–317. https://doi.org/10.1093/ajcn/nqaa073

Costa RJS, Snipe RMJ, Kitic CM, Gibson PR (2017) Systematic review: Exercise-induced gastrointestinal syndrome-implications for health and intestinal disease. *Aliment Pharmacol Ther* 46:246–265. https://doi.org/10.1111/apt.14157

Costa RJS, Walters R, Bilzon J LJ, Walsh NP (2011) Effects of immediate postexercise carbohydrate ingestion with and without protein on neutrophil degranulation. *Int J Sport Nutr Exerc Metab* 21:205–213. https://doi.org/10.1123/ijsnem.21.3.205

Costello JT, Bieuzen F, Bleakley CM (2014) Where are all the female participants in sports and exercise medicine research? *Eur J Sport Sci* 14:847–851. https://doi.org/10.1080/17461391.2014.911354

Cruz-Jentoft AJ, Baeyens JP, Bauer JM, et al. (2010) Sarcopenia: European consensus on definition and diagnosis: Report of the European working group on sarcopenia in older people. *Age Ageing* 39:412–423. https://doi.org/10.1093/ageing/afq034

Devries MC, Jakobi JM (2021) Importance of considering sex and gender in exercise and nutrition research. *Appl Physiol Nutr Metab* 46:. https://doi.org/10.1139/apnm-2021-0298

Doering TM, Reaburn PR, Cox G, Jenkins DG (2016a) Comparison of postexercise nutrition knowledge and postexercise carbohydrate and protein intake between Australian masters and younger triathletes. *Int J Sport Nutr Exerc Metab* 26:338–346. https://doi.org/10.1123/ijsnem.2015-0289

Doering TM, Reaburn PR, Phillips SM, Jenkins DG (2016b) Postexercise dietary protein strategies to maximize skeletal muscle repair and remodeling in masters endurance athletes: A review. *Int J Sport Nutr Exerc Metab* 26:168–178. https://doi.org/10.1123/ijsnem.2015-0102

Draper CF, Duisters K, Weger B, et al. (2018) Menstrual cycle rhythmicity: Metabolic patterns in healthy women. *Sci Rep* 8. https://doi.org/10.1038/s41598-018-32647-0

Elango R, Ball RO, Pencharz PB (2008) Indicator amino acid oxidation: Concept and application. *J Nutr* 138:243–246. https://doi.org/10.1093/jn/138.2.243

Forbes GB (1973) Another source of error in the metabolic balance method. *Nutr Rev* 31:297–300

Gallo G, Leo P, March MM, et al. (2022) Differences in training characteristics between junior, under 23 and professional cyclists. *Int J Sports Med* 43:1183–1189. https://doi.org/10.1055/a-1847-5414

Garthe I, Raastad R, Refsnes PE, et al. (2011) Effect of two different weight-loss rates on body composition and strength and power-related performance in elite athletes. *Int J Sport Nutr Exerc Metab* 21:97–104

Gaytán-González A, Ocampo-Alfaro M de J, Torres-Naranjo F, et al. (2020) Dietary protein intake patterns and inadequate protein intake in older adults from four countries. *Nutrients* 12:3156. https://doi.org/10.3390/nu12103156

Gehlert S, Weber S, Weidmann B, et al. (2012) Cycling exercise-induced myofiber transitions in skeletal muscle depend on basal fiber type distribution. *Eur J Appl Physiol* 112:2393–2402. https://doi.org/10.1007/s00421-011-2209-4

Gillen JB, Trommelen J, Wardenaar FC, et al. (2017) Dietary protein intake and distribution patterns of well-trained dutch athletes. *Int J Sport Nutr Exerc Metab* 27:105–114.

Gillen JB, West DWD, Williamson EP, et al. (2019) Low-carbohydrate training increases protein requirements of endurance athletes. *Med Sci Sports Exerc* 51. https://doi.org/10.1249/MSS.0000000000002036

Gordon D, Wightman S, Basevitch I, et al. (2017) Physiological and training characteristics of recreational marathon runners. *Open Access J Sports Med* 8:231–241. https://doi.org/10.2147/OAJSM.S141657

Goto M, Okazaki K, Kamijo Y, et al. (2010) Protein and carbohydrate supplementation during 5-day aerobic training enhanced plasma volume expansion and thermoregulatory adaptation in young men. *J Appl Physiol Bethesda Md* 1985 109:1247–1255. https://doi.org/10.1152/japplphysiol.00577.2010

Haakonssen EC, Martin DT, Burke LM, Jenkins DG (2013) Increased lean mass with reduced fat mass in an elite female cyclist returning to competition: Case study. *Int J Sports Physiol Perform* 8:699–701. https://doi.org/10.1123/ijspp.8.6.699

Hamadeh MJ, Devries MC, Tarnopolsky MA (2005) Estrogen supplementation reduces whole body leucine and carbohydrate oxidation and increases lipid oxidation in men during endurance exercise. *J Clin Endocrinol Metab* 90:3592–3599. https://doi.org/10.1210/jc.2004-1743

Haugen T, Sandbakk Ø, Enoksen E, et al. (2021) Crossing the golden training divide: The science and practice of training world-class 800- and 1500-m runners. *Sports Med Auckl NZ* 51:1835–1854. https://doi.org/10.1007/s40279-021-01481-2

Hector AJ, Phillips SM (2018) Protein recommendations for weight loss in elite athletes: A focus on body composition and performance. *Int J Sport Nutr Exerc Metab* 28:170–177.

Hopkins WG, Hawley JA, Burke LM (1999) Design and analysis of research on sport performance enhancement. *Med Sci Sports Exerc* 3:472–485. https://doi.org/10.1097/00005768-199903000-00018

Horton TJ, Pagliassotti MJ, Hobbs K, Hill JO (1998) Fuel metabolism in men and women during and after long-duration exercise. *J Appl Physiol Bethesda Md 1985* 85:1823–1832. https://doi.org/10.1152/jappl.1998.85.5.1823

Howarth KR, Phillips SM, MacDonald MJ, et al. (2010) Effect of glycogen availability on human skeletal muscle protein turnover during exercise and recovery. *J Appl Physiol* 109:431–438. https://doi.org/10.1152/japplphysiol.00108.2009

Institute of Medicine. Panel on M, Institute of Medicine. Standing Committee on the Scientific Evaluation of Dietary Reference I (2005) *Dietary reference intakes for energy, carbohydrate, fiber, fat, fatty acids, cholesterol, protein, and amino acids*. National Academies Press.

Jäger R, Kerksick CM, Campbell BI, et al. (2017) International society of sports nutrition position stand: Protein and exercise. *J Int Soc Sports Nutr* 14:1–25. https://doi.org/10.1186/s12970-017-0177-8

Jarvis M, McNaughton L, Seddon A, Thompson D (2002) The acute 1-week effects of the Zone diet on body composition, blood lipid levels, and performance in recreational endurance athletes. *J Strength Cond Res* 16:50–57.

Jeukendrup AE (2011) Nutrition for endurance sports: Marathon, triathlon, and road cycling. *J Sports Sci* 29 Suppl 1:S91–S99. https://doi.org/10.1080/02640414.2011.610348

Jeukendrup AE (2017) Periodized nutrition for athletes. *Sports Med Auckl NZ* 47:51–63. https://doi.org/10.1007/s40279-017-0694-2

Jonvik KL, Lenaerts K, Smeets JSJ, et al. (2019) Sucrose but not nitrate ingestion reduces strenuous cycling-induced intestinal injury. *Med Sci Sports Exerc* 51:436–444. https://doi.org/10.1249/MSS.0000000000001800

Josse AR, Atkinson SA, Tarnopolsky MA, Phillips SM (2011) Increased consumption of dairy foods and protein during diet- and exercise-induced weight loss promotes fat mass loss and lean mass gain in overweight and obese premenopausal women. *J Nutr* 141:1626–1634. https://doi.org/10.3945/jn.111.141028

Kato H, Suzuki K, Bannai M, Moore DR (2016) Protein requirements are elevated in endurance athletes after exercise as determined by the indicator amino acid oxidation method. *PLoS One* 11:e0157406. https://doi.org/10.1371/journal.pone.0157406

Kato H, Suzuki K, Bannai M, Moore DR (2018) Branched-chain amino acids are the primary limiting amino acids in the diets of endurance-trained men after a bout of prolonged exercise. *J Nutr* 1–7. https://doi.org/10.1093/jn/nxy048

Koller A, Mair J, Schobersberger W, et al. (1998) Effects of prolonged strenuous endurance exercise on plasma myosin heavy chain fragments and other muscular proteins. Cycling vs running. *J Sports Med Phys Fitness* 38:10–17.

Kriengsinyos W, Wykes LJ, Goonewardene LA, et al. (2004) Phase of menstrual cycle affects lysine requirement in healthy women. *Am J Physiol - Endocrinol Metab* 287:489–496. https://doi.org/10.1152/ajpendo.00262.2003

Lamont LS, Lemon PW, Bruot BC (1987) Menstrual cycle and exercise effects on protein catabolism. *Med Sci Sports Exerc* 19:106–110

Landau RL, Lugibihl K (1961) The effect of progesterone on amino acid metabolism. *J Clin Endocrinol Metab* 21:1355–1363. https://doi.org/10.1210/jcem-21-11-1355

Landau RL, Lugibihl K (1967) The effect of progesterone on the concentration of plasma amino acids in man. *Metabolism* 16:1114–1122. https://doi.org/10.1016/0026-0495(67)90057-1

Lariviere F, Moussalli R, Garrel DR (1994) Increased leucine flux and leucine oxidation during the luteal phase of the menstrual cycle in women. *Am J Physiol* 267:E422–E428. https://doi.org/10.1152/ajpendo.1994.267.3.E422

Lemon PW, Nagle FJ, Mullin JP, Benevenga NJ (1982) In vivo leucine oxidation at rest and during two intensities of exercise. *J Appl Physiol* 53:947–954. https://doi.org/10.1152/jappl.1982.53.4.947

Levenhagen DK, Gresham JD, Carlson MG, et al. (2001) Postexercise nutrient intake timing in humans is critical to recovery of leg glucose and protein homeostasis. *Am J Physiol Endocrinol Metab* 280:E982–E993. https://doi.org/10.1152/ajpendo.2001.280.6.E982

Lorenz DS, Reiman MP, Lehecka BJ, Naylor A (2013) What performance characteristics determine elite versus nonelite athletes in the same sport? *Sports Health* 5:542–547. https://doi.org/10.1177/1941738113479763

Macdermid PW, Stannard SR (2006) A whey-supplemented, high-protein diet versus a high-carbohydrate diet: Effects on endurance cycling performance. *Int J Sport Nutr Exerc Metab* 16:65–77. https://doi.org/10.1123/ijsnem.16.1.65

Margolis LM, Murphy NE, Martini S, et al. (2014) Effects of winter military training on energy balance, whole-body protein balance, muscle damage, soreness, and physical performance. *Appl Physiol Nutr Metab Physiol Appl Nutr Metab* 39:1395–1401. https://doi.org/10.1139/apnm-2014-0212

Margolis LM, Murphy NE, Martini S, et al. (2016) Effects of supplemental energy on protein balance during 4-d arctic military training. *Med Sci Sports Exerc* 48:1604–1612. https://doi.org/10.1249/MSS.0000000000000944

Margolis LM, Pasiakos SM (2013) Optimizing intramuscular adaptations to aerobic exercise: Effects of carbohydrate restriction and protein supplementation on mitochondrial biogenesis. *Adv Nutr Bethesda Md* 4:657–664. https://doi.org/10.3945/an.113.004572

Mazzulla M, Parel JT, Beals JW, et al. (2017) Endurance exercise attenuates postprandial whole-body leucine balance in trained men. *Med Sci Sports Exerc* 49:2585–2592. https://doi.org/10.1249/mss.0000000000001394

Mazzulla M, Volterman KA, Packer JE, et al. (2018) Whole-body net protein balance plateaus in response to increasing protein intakes during post-exercise recovery in adults and adolescents. *Nutr Metab* 15. https://doi.org/10.1186/s12986-018-0301-z

Melin AK, Areta JL, Heikura IA, et al. (2023) Direct and indirect impact of low energy availability on sports performance. *Scand J Med Sci Sports*. https://doi.org/10.1111/sms.14327

Mercer D, Convit L, Condo D, et al. (2020) Protein requirements of pre-menopausal female athletes: Systematic literature review. *Nutrients* 12:1–20. https://doi.org/10.3390/nu12113527

Miller BF, Olesen JL, Hansen M, et al. (2005) Coordinated collagen and muscle protein synthesis in human patella tendon and quadriceps muscle after exercise. *J Physiol* 567:1021–1033. https://doi.org/10.1113/jphysiol.2005.093690

Millward DJ (1998) Metabolic demands for amino acids and the human dietary requirement: Millward and Rivers (1988) revisited. *J Nutr* 128. https://doi.org/10.1093/jn/128.12.2563S

Millward DJ, Davies CT, Halliday D, et al. (1982) Effect of exercise on protein metabolism in humans as explored with stable isotopes. *Fed Proc* 41:2686–2691.

Moore DR (2019) Maximizing post-exercise anabolism: The case for relative protein intakes. *Front Nutr* 6. https://doi.org/10.3389/fnut.2019.00147

Moore DR (2020) One size doesn't fit all: Postexercise protein requirements for the endurance athlete. *Am J Clin Nutr* 112:249–250. https://doi.org/10.1093/ajcn/nqaa144

Moore DR (2021) Protein requirements for master athletes: Just older versions of their younger selves. *Sports Med Auckl NZ* 51:13–30. https://doi.org/10.1007/s40279-021-01510-0

Moore DR, Areta J, Coffey VG, et al. (2012) Daytime pattern of post-exercise protein intake affects whole-body protein turnover in resistance-trained males. *Nutr Metab* 9:91–95. https://doi.org/10.1186/1743-7075-9-91

Moore DR, Camera DM, Areta JL, Hawley JA (2014) Beyond muscle hypertrophy: Why dietary protein is important for endurance athletes. *Appl Physiol Nutr and Metab* 39(9):987–997.

Moore DR, Churchward-Venne TA, Witard O, et al. (2015) Protein ingestion to stimulate myofibrillar protein synthesis requires greater relative protein intakes in healthy older versus younger men. *J Gerontol - Ser Biol Sci Med Sci* 70:57–62. https://doi.org/10.1093/gerona/glu103

Moore DR, Robinson MJ, Fry JL, et al. (2009) Ingested protein dose response of muscle and albumin protein synthesis after resistance exercise in young men. *Am J Clin Nutr* 89:161–168. https://doi.org/10.3945/ajcn.2008.26401

Moore DR, Stellingwerff T (2012) Protein ingestion after endurance exercise: The "evolving" needs of the mitochondria? *J Physiol* 590:1785–1786. https://doi.org/10.1113/jphysiol.2011.224188

Moss K, Kreutzer A, Graybeal AJ, et al. (2023) Nutrient adequacy in endurance athletes. *Int J Environ Res Public Health* 20:5469. https://doi.org/10.3390/ijerph20085469

Oikawa SY, Holloway TM, Phillips SM (2019) The impact of step reduction on muscle health in aging: Protein and exercise as countermeasures. *Front Nutr* 6. https://doi.org/10.3389/fnut.2019.00075

Oikawa SY, Macinnis MJ, Tripp TR, et al. (2020) Lactalbumin, not collagen, augments muscle protein synthesis with aerobic exercise. *Med Sci Sports Exerc* 52:1394–1403. https://doi.org/10.1249/MSS.0000000000002253

Osborne JO, Stewart IB, Beagley KW, Minett GM (2019) The effect of cycling in the heat on gastrointestinal-induced damage and neuromuscular fatigue. *Eur J Appl Physiol* 119:1829–1840. https://doi.org/10.1007/s00421-019-04172-z

Oxfeldt M, Phillips SM, Andersen OE, et al. (2023) Low energy availability reduces myofibrillar and sarcoplasmic muscle protein synthesis in trained females. *J Physiol* 601:3481–3497. https://doi.org/10.1113/JP284967

Pasiakos SM, Cao JJ, Margolis LM, et al. (2013) Effects of high-protein diets on fat-free mass and muscle protein synthesis following weight loss: A randomized controlled trial. *FASEB J* 27:3837–3847. https://doi.org/10.1096/fj.13-230227

Phillips SM (2012) Dietary protein requirements and adaptive advantages in athletes. *Br J Nutr* 108 Suppl 2:S158–S167. https://doi.org/10.1017/S0007114512002516

Phillips SM, Atkinson SA, Tarnopolsky MA, MacDougall JD (1993) Gender differences in leucine kinetics and nitrogen balance in endurance athletes. *J Appl Physiol* 75:2134–2141. https://doi.org/10.1152/jappl.1993.75.5.2134

Phillips SM, Chevalier S, Leidy HJ (2016) Protein "requirements" beyond the RDA: Implications for optimizing health. *Appl Physiol Nutr Metab* 41:565–572. https://doi.org/10.1139/apnm-2015-0550

Podlogar T, Wallis GA (2022) New horizons in carbohydrate research and application for endurance athletes. *Sports Med Auckl NZ* 52:5–23. https://doi.org/10.1007/s40279-022-01757-1

Prior BM, Yang HT, Terjung RL (2004) What makes vessels grow with exercise training? *J Appl Physiol Bethesda Md* 1985 97:1119–1128. https://doi.org/10.1152/japplphysiol.00035.2004

Rennie MJ, Bohé J, Smith K, et al. (2006) Branched-chain amino acids as fuels and anabolic signals in human muscle. *J Nutr* 136:264S–268S. https://doi.org/10.1093/jn/136.1.264S

Sheffield-Moore M, Yeckel CW, Volpi E, et al. (2004) Postexercise protein metabolism in older and younger men following moderate-intensity aerobic exercise. *Am J Physiol - Endocrinol Metab* 287:513–522. https://doi.org/10.1152/ajpendo.00334.2003

Simmons R, Doma K, Sinclair W, et al. (2021) Acute effects of training loads on muscle damage markers and performance in semi-elite and elite athletes: A systematic review and meta-analysis. *Sports Med Auckl NZ* 51:2181–2207. https://doi.org/10.1007/s40279-021-01486-x

Snipe RMJ, Khoo A, Kitic CM, et al. (2017) Carbohydrate and protein intake during exertional heat stress ameliorates intestinal epithelial injury and small intestine permeability. *Appl Physiol Nutr Metab Physiol Appl Nutr Metab* 42:1283–1292. https://doi.org/10.1139/apnm-2017-0361

Snipe RMJ, Khoo A, Kitic CM, et al. (2018) The impact of mild heat stress during prolonged running on gastrointestinal integrity, gastrointestinal symptoms, systemic endotoxin and cytokine profiles. *Int J Sports Med*. https://doi.org/10.1055/s-0043-122742

Tarnopolsky M (2004) Protein requirements for endurance athletes. *Nutrition* 20:662–668. https://doi.org/10.1016/j.nut.2004.04.008

Thomas DT, Erdman KA, Burke LM (2016) American college of sports medicine joint position statement. Nutrition and athletic performance. *Med Sci Sports Exerc* 48:543–568. https://doi.org/10.1249/MSS.0000000000000852

Tipton KD, Witard OC (2007) Protein requirements and recommendations for athletes: Relevance of ivory tower arguments for practical recommendations. *Clin Sports Med* 26:17–36. https://doi.org/10.1016/j.csm.2006.11.003

Tjelta LI (2016) The training of international level distance runners. *Int J Sports Sci Coach* 11:122–134. https://doi.org/10.1177/1747954115624813

van Erp T, Sanders D, de Koning JJ (2019) Training characteristics of male and female professional road cyclists: A 4-year retrospective analysis. *Int J Sports Physiol Perform*:1–7. https://doi.org/10.1123/ijspp.2019-0320

Vitale K, Getzin A (2019) Nutrition and supplement update for the endurance athlete: Review and recommendations. *Nutrients* 11:1289. https://doi.org/10.3390/nu11061289

WHO (2007) *Protein and amino acid requirements in human nutrition: Report of a joint WHO/FAO/UNU expert consultation*. World Health Organization

Wilkinson SB, Phillips SM, Atherton PJ, et al. (2008) Differential effects of resistance and endurance exercise in the fed state on signalling molecule phosphorylation and protein synthesis in human muscle. *J Physiol* 586:3701–3717. https://doi.org/10.1113/jphysiol.2008.153916

Williamson E (2022) *Protein requirements and amino acid metabolism following endurance exercise*. University of Toronto.

Williamson E, Fung HJW, Adams C, et al. (2023) Protein requirements are increased in endurance-trained athletes but similar between females and males during postexercise recovery. *Med Sci Sports Exerc* 55:1866–1875. https://doi.org/10.1249/MSS.0000000000003219

Williamson E, Kato H, Volterman KA, et al. (2019) The effect of dietary protein on protein metabolism and performance in endurance-trained males. *Med Sci Sports Exerc* 51:352–360. https://doi.org/10.1249/MSS.0000000000001791

Williamson E, Kato H, Volterman KA, et al. (2022) Greater plasma essential amino acids and lower 3-methylhistidine with higher protein intake during endurance training: A randomised control trial. *Amino Acids*. https://doi.org/10.1007/s00726-022-03210-z

Williamson E, Moore DR (2021) A muscle-centric perspective on intermittent fasting: A suboptimal dietary strategy for supporting muscle protein remodeling and muscle mass? *Front Nutr* 8:640621. https://doi.org/10.3389/fnut.2021.640621

Yuan S, Larsson SC (2023) Epidemiology of sarcopenia: Prevalence, risk factors, and consequences. *Metabolism* 144:155533. https://doi.org/10.1016/j.metabol.2023.155533

6 Understanding the Nutritional Aspects of Dietary Proteins

Sources, Quality, Digestibility, and Beyond

Moattar Latif and Peter Uchenna Amadi

ABBREVIATIONS

BV	Biological value
EAR	Estimated average requirement
FAO	Food and Agricultural Organization
FDA	US Food and Drug Administration
PDCAAS	Protein digestibility-corrected amino acid score
PER	Protein efficiency ratio
RDA	Recommended dietary allowance
WHO	World Health Organization

INTRODUCTION

Protein is a dietary macronutrient that meets a wide range of nutritional and physiological needs within the human body (Das Suchintita et al. 2022). It stands as one of the three fundamental macronutrients, along with carbohydrates and fats. Protein plays a pivotal role in a well-balanced diet, being indispensable for various bodily functions. According to the World Health Organization (WHO), adults are recommended to consume 0.8 grams of protein per kilogram of body weight daily, with a protein digestibility-corrected amino acid score value of 1.0 for optimal nutrition. Amino acids, often termed the "building blocks of life," are the components of proteins. There are 20 different types of amino acids, as shown in Figures 6.1 and 6.2. Nine of these, illustrated in Figure 6.1, are designated as essential because the body cannot synthesize them and must derive them from dietary sources. A complete protein source contains all nine essential amino acids in ample proportions, whereas incomplete sources lack one or more of these vital amino acids. High-quality dietary protein generally has an efficiency of 75–85% in terms of its ability to utilize truly digestible protein for maintaining the body and depositing protein in it (Wu 2021; Bandyopadhyay et al. 2022). Proteins are found in a wide variety of foods, coming from both animal and plant sources. Animal-based protein sources include meat, poultry, fish, eggs, and dairy products. Plant-derived protein sources encompass legumes such as beans, lentils, and chickpeas, along with tofu, tempeh, nuts, seeds, and whole grains.

A balanced diet typically incorporates a blend of these protein sources. The quality of protein depends on its amino acid composition and how easily it can be digested. Animal-based proteins are generally considered high-quality as they typically offer all essential amino acids in the right proportions and exhibit exceptional digestibility. Plant-based proteins can also be high-quality when thoughtfully combined, such as pairing beans with rice. Nevertheless, certain plant sources may

DOI: 10.1201/b23328-6

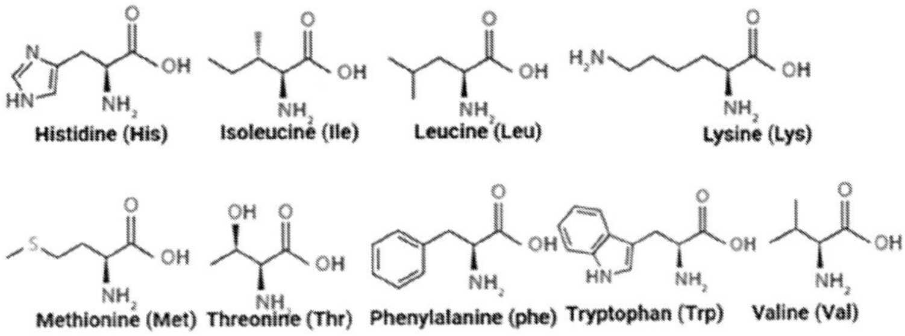

FIGURE 6.1 Essential amino acids.

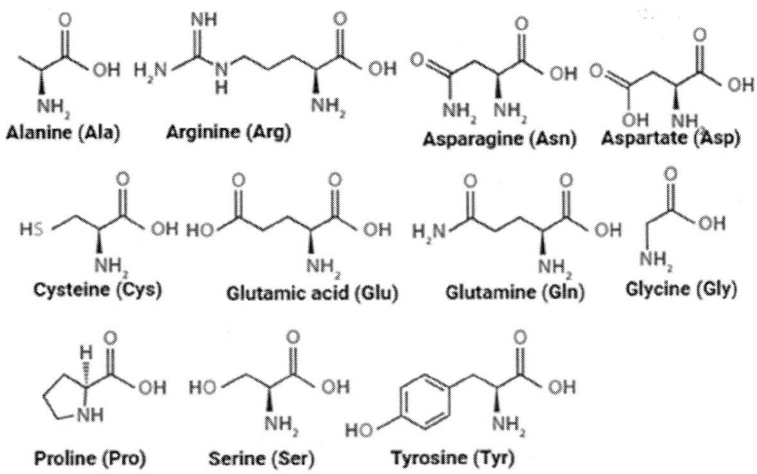

FIGURE 6.2 Non-essential amino acids.

lack specific essential amino acids (Hertzler et al. 2020). Certain proteins can create gels, which find application in a diverse range of products, including desserts like gelatin molds and panna cotta. Whey protein (WP), derived from the liquid residue of cheese production, is a mixture of globular proteins with distinctive functional qualities. Egg white proteins are particularly valuable in the food industry due to their ability to create gels, dissolve easily, emulsify, and retain water. Gelatin, derived from the chemical and structural conversion of collagen, serves as a colorless, flavorless protein with extensive utility across diverse segments of the food industry.

The food industry has conventionally employed synthetic surfactants or animal-based proteins, primarily sourced from milk or eggs, to stabilize the interfaces in emulsions and foams. Proteins present in egg whites and casein exhibit the remarkable ability to generate robust foams, which is an essential element in crafting delectable treats such as meringues, soufflés, and various baked goods that rely on incorporating air. This property is also leveraged in manufacturing products like mayonnaise and salad dressings (Bertsch et al. 2021). In processed foods, proteins serve as binding agents, uniting various ingredients. This role is evident in the preparation of items like meatballs and sausages, where proteins play a key part in preserving the desired texture. Additionally, certain proteins also play a role in enhancing the flavor of foods. For example, the Maillard reaction, an intricate process involving the interaction of amino acids (found in proteins) and sugars during cooking, contributes to the creation of intricate and enticing flavors in roasted meats and baked goods.

Within food processing, proteins find application in crafting appealing coatings, glazes, and browning effects, as exemplified in the golden crust on bread or pastries. Proteins also wield a profound influence on the mouthfeel of food, imparting attributes such as creaminess (as in the case of ice cream) or a satisfying chewy texture (found in specific plant-based meat substitutes). The precise impact of proteins on the structure and attributes of a food item is contingent on a multitude of variables, encompassing the type of protein, its concentration, the methods employed in processing, and the additional ingredients featured in the recipe. A thorough understanding of these dynamics is of utmost significance for food scientists and culinary experts striving to attain the intended food textures, flavors, and visual presentations. The properties of food proteins depend on their origin, the methods used for extraction, the degree of purification, and any modifications introduced during food processing. This knowledge holds utmost importance in the food industry to ensure the production of high-quality products.

The demand for protein, whether from plant or animal sources, is on the rise, reflecting consumers' growing recognition of protein's significance in their diets. Research, particularly concerning older adults, has been accumulating to substantiate the health advantages of sufficient protein intake. Some of these advantages include the potential to preserve or enhance muscle mass, a critical consideration for older individuals to counteract age-related muscle decline, known as sarcopenia. Protein consumption can also contribute to enhanced physical capabilities, such as increased leg strength and walking speed, which are pivotal for maintaining mobility and independence in the elderly. Furthermore, protein, being an integral component of bone tissue, has been linked to improved bone health and density.

EXISTING PROTEIN SOURCES

Plant-Based Protein

In recent times, there has been a growing interest in alternative sources of food protein, with the potential to replace traditional animal-based proteins. Multiple compelling factors are driving this dietary transformation, underpinned by concerns related to sustainability, health, and ethical values. As we confront the emerging challenge of climate change and strive to safeguard food security for our increasingly populous planet, the shift toward a predominantly plant-based diet may be an inescapable necessity (Poore and Nemecek 2018; Willett et al. 2019). Cereal proteins constitute a significant share of the world's dietary protein intake and hold vital importance for both animals and humans. Their prominence is particularly notable in the diets of developing nations, and wheat stands as the most prevalent plant protein source in Western diets (Van et al. 2013). While recognizing the significant importance of cereals and pulses, plant leaves are often deprived of worthy recognition, despite being rich in all essential amino acids.

Differing in function from plant seeds, mature plant leaves are mainly responsible for photosynthesis, a process leading to harnessing sunlight to supply carbohydrates and energy to the plant. Plant leaves, especially from leafy vegetables, contain sufficient amino acids comparable to animal proteins. Examples of leafy vegetables rich in proteins include spinach and broccoli. These leafy vegetables are predominantly made up of ribulose 1,5-bisphosphate carboxylase/oxygenase (RUBISCO), a polypeptide complex responsible for the atmospheric fixation of carbon. This protein is the predominant protein in the leaves, is located within the chloroplast, and has been shown to supply all the essential amino acids sufficiently in accordance with WHO recommendations.

Pulses

Pulses stand out with their diverse and challenging-to-match nutritional profile. They serve as an excellent source of protein, fiber, B vitamins, iron, zinc, copper, and a range of other essential nutrients. In comparison to alternative food sources, pulses exhibit lower energy content, and reduced fat and sodium levels, rendering them a notably nutrient-dense food option. Notably, pulses offer a substantial protein source, accompanied by well-balanced amino acid profiles (Derbyshire and

Delange 2021). Pulse proteins are currently being explored for their potential as a valuable source of nutritional and functional proteins, with comparisons to well-known protein sources like soy and whey. When combined with cereal proteins, they hold promise for the future of nutrition. Pulse proteins are emerging as a strong alternative to soy proteins. Notably, they are free from Genetically Modified Organisms (GMOs) concerns and are generally considered a less significant allergen source, though this remains a topic of debate. Pulses are the dry, edible seeds of various legumes, which encompass green and yellow peas, faba beans, lentils, chickpeas, adzuki beans, and mung beans. These pulses also have the advantage of possessing very low glycemic index (GI) values, making them valuable in regulating blood sugar and lipid levels in both diabetic and non-diabetic individuals. Furthermore, pulses contain an array of bioactive components that can be extracted and harnessed for their advantageous health properties.

Within pulses, several bioactive components are present, including lectins, which are carbohydrate-binding proteins, protease inhibitors, and an angiotensin-converting enzyme inhibitor. This inhibitor, found in various human tissues, has the potential to raise blood pressure by converting angiotensin I into angiotensin II, leading to vasoconstriction. These anti-nutritional compounds (ANCs) can induce discomfort, especially when consuming raw pulses, resulting in symptoms like bloating, vomiting, and pancreas enlargement. (Roy et al. 2010). Hence, when evaluating the nutritional content of dishes featuring pulses, it's crucial to factor in these anti-nutritional compounds (ANCs). While pulses do contain compounds with potential health benefits, it's equally important to acknowledge that they can also harbor anti-nutritional compounds, which can counteract their health advantages. ANCs are substances that interfere with the digestive processes, especially when pulses are consumed in their raw form.

Dry Peas

Dry peas (*Pisum sativum* L.) are considered the second most significant pulse crop, occupying a substantial portion of the total land area dedicated to dry pulses, specifically 34.2%. The protein content in pea seeds can range from 23% to 31%, with carbohydrates making up 60% to 65% of their composition, and a fat content of about 1% to 2%. (Bogahawaththa et al. 2019; Rempel, Geng, and Zhang 2019). Pea protein has gained significant interest as a promising replacement for conventional protein sources such as animal proteins and soy protein. This stems from its low allergenicity, non-genetically modified status, high nutritional value, ready availability, and origin from a sustainable crop. Furthermore, pea protein can be classified as a high-quality protein due to its well-balanced amino acid profile. It contains all essential amino acids except for methionine, making it capable of meeting FAO/WHO recommendations. (Gorissen et al. 2018).

Quinoa

Quinoa (*Chenopodium quinoa*, Willd.) is not classified as a true grain; rather, it is a pseudocereal indigenous to the Andes region (Abugoch 2009). The protein content of quinoa seeds, expressed as g/100 g of edible matter, falls within the range of 13.1% to 16.7%. These protein levels in quinoa exceed those found in rice, barley, corn, and rye and approach the protein content of wheat. The primary storage proteins in quinoa, accounting for 35% and 37% of the total protein content, are albumins and globulins. Prolamins, on the other hand, are present in relatively low concentrations. Apart from their substantial quantity, quinoa proteins are distinguished by their high quality, characterized by a well-balanced profile of essential amino acids (Vilcacundo and Hernandez-Ledesma 2017). Quinoa's cultivation has expanded to include regions in Africa, Asia, Australia, Europe, and North America, owing to its remarkable adaptability to diverse agroecological conditions.

Quinoa is often regarded as a complete food due to its exceptional protein quantity and quality, along with its rich content of vitamins, minerals, phytosterols, flavonoids, omega-6 fatty acids, and vitamin E. In addition to its rich protein, lipid, fiber, vitamin, and mineral content, and a well-balanced essential amino acid profile, quinoa has been discovered to possess a diverse array of phytochemicals. These include saponins, phytosterols, phytoecdysteroids, phenolics, and bioactive peptides. These compounds hold the potential to exert positive effects on metabolic, cardiovascular, and gastrointestinal health. This review focuses on quinoa's nutritional and functional significance, emphasizing the substantiated evidence from both animal and clinical studies. One of the key attributes of quinoa seeds is their demonstrated antioxidant activity, which is closely associated with their abundant phenolic compound content. In addition to its high nutritional value and gluten-free nature, quinoa has been reported to confer beneficial effects on specific high-risk consumer groups, including children, the elderly, those with lactose intolerance, and individuals dealing with conditions such as anemia, diabetes, obesity, dyslipidemia, and celiac disease.

Animal-Based Protein

In recent decades, there has been a notable global rise in the consumption and production of animal-based products, and this trend is expected to continue. While there is a widely accepted consensus regarding the benefits of including animal-based protein in one's diet for maintaining muscle mass during the aging process, the consumption of animal-based protein has recently sparked significant debate within both scientific literature and popular media. There is growing concern that animal-based proteins may be linked to chronic diseases, including heart disease, colorectal cancer, breast cancer, prostate cancer, and bone demineralization. It is generally accepted that animal protein offers superior nutritional quality compared to plant-based protein. Moreover, animal-based proteins are typically easier to digest than plant-based proteins, thereby enhancing their potential to stimulate muscle protein synthesis (Ajomiwe et al. 2024).

Animal protein's superior nutritional quality refers to its amino acid profile, ease of digestion, and capacity to facilitate the absorption of essential nutrients like calcium and iron. Furthermore, its technological versatility in terms of abilities such as gelling, emulsification, and foaming, which contribute to the desirable texture and sensory characteristics of food, is generally regarded as superior to plant-based proteins (Kim, Wang, and Selomulya 2020).

Animal-based proteins have been linked to chronic diseases like heart disease, colorectal cancer, breast cancer, prostate cancer, and bone demineralization. Additionally, the increased demand for animal protein production can result in soil nutrient depletion and environmental pollution. (Menzi et al. 2010). Egg protein has frequently served as a reference for assessing the biological value of other protein sources. Lean beef, in particular, holds significant appeal for older adults. There are 29 beef cuts that meet the criteria for being labeled as lean or extra lean (containing five grams or less of fat and two grams or less of saturated fat). The majority (20%) of animal-based protein consumption comes from red meat, with 14% attributed to beef and the remainder stemming from pork, lamb, game, or a combination of beef and pork with other meats. An additional 18% is sourced from dairy products (including milk and cheese, with yogurt comprising less than 2%), 14% from poultry (primarily chicken), and the rest is distributed among seafood, fish, and eggs. As a result, beef, pork, and chicken constitute the predominant sources of animal-based protein intake in the United States. The latest National Health and Nutrition Examination Survey (NHANES) data, from 2015 to 2016, indicate an upsurge in poultry consumption (47 grams per week), a decline in unprocessed red meat consumption (from 340 grams per week in 1999 to 2000 to 284 grams per week in 2015 to 2016), and no significant shift in consumption patterns for processed red meats (182 grams per week).

The nutritional quality of animal-source protein foods varies, despite their excellent digestibility in the human digestive system. For instance, when it comes to individuals looking to build muscle and using protein powder after exercise, whey protein is often regarded as superior to casein. Several potential reasons account for this distinction. Firstly, whey protein contains higher levels of essential amino acids, such as arginine, leucine, lysine, and sulfur-containing amino acids like methionine and cysteine, compared to casein. Secondly, whey protein releases individual amino acids more rapidly, earning it the classification of a "fast-release" protein, which is digested within an hour in the small intestine. This rapid release of amino acids from whey protein immediately supports muscle protein synthesis following exercise. As a result, the consumption of whey protein

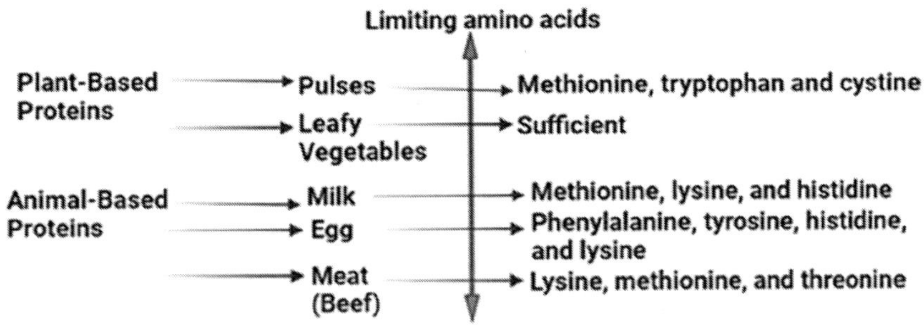

FIGURE 6.3 Limiting amino acids of common protein sources for human nutrition.

is more effective than casein or casein hydrolysate in stimulating post-meal muscle protein growth in elderly men (Pennings et al. 2011). In a study of postmenopausal women who engaged in weight training and supplemented their diet with whey protein (0.15 grams per kilogram of body weight per day) for 24 weeks, researchers observed a consistent 9% improvement in skeletal muscle strength (Holm et al. 2008). On the other hand, elderly men who took a casein supplement (20 grams per session, three sessions per week) immediately before and after resistance training showed varying outcomes, including instances with no change in muscle strength (Verdijk et al. 2009). Nonetheless, it remains uncertain whether this regimen of casein supplementation effectively maintains elevated amino acid levels in the bloodstream necessary to trigger muscle protein synthesis.

Animal-based proteins, particularly dairy proteins, play a crucial role in delivering essential nutrition for human well-being, especially for infants, who rely on them for cognitive and physical development. An abundance of reports and recommendations support the inclusion of animal protein sources in food aid products (Li Day et al. 2022). Figure 6.3 presents a summary of limiting amino acids in popular protein sources.

New and Emerging Sources of Protein

In a broad sense, alternative proteins serve as substitutes for traditional animal-based proteins. These alternatives include sources such as plants (e.g., grains, legumes, pulses, and nuts), fungi (e.g., mushrooms), algae, insects, and lab-grown (cultured) meat (Bekhit, Riley, and Hussain 2022). Many alternative protein sources have long been staples in certain regions but may not be readily embraced in other parts of the world where they are unfamiliar. For instance, insects are a common food source in numerous Asian and African countries but are regarded as novel by most Western consumers.

Most alternative proteins, except for canola protein and cultured meat, have been part of the human diet for centuries. Humans have developed methods to incorporate these sources into their diets, often involving minimal processing, such as sun-drying, roasting, boiling, steaming, or frying. These foods are typically consumed in their whole form without extensive separation into components. Whole-food alternatives include algae, edible insects, and mushrooms. Sometimes, particular plant materials are finely ground and further processed to enhance their nutritional content and flavor, as seen in cereals and traditional soy products like soy milk and tofu. Moreover, these alternative proteins are frequently incorporated into a variety of food products, typically following appropriate processing steps, such as the production of edible insect flour and pulse flour (Onwezen et al. 2021).

Current Dietary Protein Recommendations

Dietary guidelines offer daily protein intake recommendations that may differ according to factors such as age, gender, and physical activity. These guidelines aim to help individuals maintain their overall well-being. In essence, protein is a vital nutrient that contributes to human health in various ways. Consumer dietary choices are evolving, and research increasingly substantiates the health advantages associated with sufficient protein consumption, especially among older individuals.

ASSESSMENT OF PROTEIN QUALITY

Protein quality refers to the amount of indispensable amino acids and their digestibility sufficient to meet the general dietary protein requirements for humans (Hoffman and Falvo 2004). Indispensable or essential amino acids cannot be synthesized in the body or are not produced in sufficient quantities to meet metabolic demands; therefore, they must be acquired from dietary sources. The nine essential amino acids are histidine, valine, phenylalanine, methionine, lysine, threonine, tryptophan, leucine, and isoleucine (Figure 6.1). Thus, assessing protein quality from a dietary source primarily involves considering the amount of indispensable amino acids that meets the growth needs of a target group (Institute of Medicine 1998).

Two perspectives are taken into account in assessing general dietary protein requirements: the estimated average requirement (EAR) and the recommended dietary allowance (RDA). The EAR is the daily intake value estimated to meet the requirement of 50% of healthy individuals in a target group, gender, or life stage, regardless of whether the dietary needs of the other half of individuals in the age group are met (Institute of Medicine 1998). The EAR for adults is 0.66 g/kg body weight/day, but increases during pregnancy and lactation (NDA 2012). The RDA is the average daily intake level for a nutrient adequate to meet the nutrient requirement for 97% to 98% of the target population (Adhikari et al. 2022). It is usually calculated as the EAR plus two times the standard deviation. Given these perspectives, there is no need to define upper limits for protein intake or restrict certain amino acids. Infants between six months and one year old have the highest requirements for RDA values, which decrease with age, reflecting the central role of essential amino acids in the growth and maintenance of the human body.Furthermore, several principles guide the assessment of protein quality, including amino acid composition, digestibility, and the amounts supplied versus required by the target group. The target group requirements are assessed using a so-called reference protein specific to distinct age groups, which ideally contains all essential amino acids in the desired amounts with 100% digestibility (Millward and Alan 2004). Digestibility refers to a protein's potential to be hydrolyzed into absorbable amino acids or small peptides. Several methods have been appraised for determining protein quality, each adopting different approaches to assess or categorize the source of the protein quality. Figure 6.4 summarizes the amino acid scoring patterns for toddlers, children, adolescents, and adults.

PROTEIN DIGESTIBILITY-CORRECTED AMINO ACID SCORE (PDCAAS)

The PDCAAS assesses protein quality based on the requirement and digestibility of a food source's amino acid components (Schaadsma 2012). This method ranks protein quality by comparing its amino acid content to that fof a reference protein, with the highest attainable score being 1.0 (Rutherfurd 2015). This score implies that the protein, after digestion, provides 100% or more of the required amino acids per unit of protein. A PDCAAS <1 implies that one or more essential amino acids are lacking, while a score >1 indicates that no amino acid is limiting in the protein (Marinangeli and House 2017). Usually, PDCAAS scores above 100% are rounded to 100% as excess amino acids are typically catabolized (Castellanos 2006).

In 1993, the US FDA and the Food and Agriculture Organization of the United Nations/World Health Organization (FAO/WHO 1990) endorsed the PDCAAS as "the preferred 'best' method to determine protein quality." They considered it more appropriate for meeting human dietary requirements because, unlike other protein assessment methods based on the amino acid requirements of animals, it was based on human amino acid requirements. Despite its efficiency, the PCDAAS has several recognized drawbacks. For instance, it does not account for the point of protein digestion

			His	Ile	Leu	Lys	SAA	AAA	Thr	Trp	Val
Tissue amino acid pattern (mg/g protein)[1]			27	35	75	73	35	73	42	12	49
Maintenance amino acid pattern (mg/g protein)[2]			15	30	59	45	22	38	23	6	39
	Protein requirements (g/kg/d)										
Age (yr)	**Maintenance**	**Growth[3]**	*amino acid requirements (mg/kg/d)[4]*								
0.5	0.66	0.46	22	36	73	63	31	59	35	9.5	48
1-2	0.66	0.20	15	27	54	44	22	40	24	6	36
3-10	0.66	0.07	12	22	44	35	17	30	18	4.8	29
11-14	0.66	0.07	12	22	44	35	17	30	18	4.8	29
15-18	0.66	0.04	11	21	42	33	16	28	17	4.4	28
>18	0.66	0.00	10	20	39	30	15	25	15	4.0	26
			scoring pattern mg/g protein requirement[5]								
0.5			20	32	66	57	27	52	31	8.5	43
1-2			18	31	63	52	25	46	27	7	41
3-10			16	30	61	48	23	41	25	6.6	40
11-14			16	30	61	48	23	41	25	6.6	40
15-18			16	30	60	47	23	40	24	6.3	40
>18			15	30	59	45	22	38	23	6.0	39

His, histidine; Ile, isoleucine; Leu, leucine; SAA, sulphur amino acids; AAA, aromatic amino acids, Thr, threonine, Trp, tryptophan; Val, valine

[1] Amino acid composition of whole-body protein.

[2] Adult maintenance pattern.

[3] Calculated as average values for the age range: growth adjusted for protein utilization of 58%.

[4] Sum of amino acids contained in the dietary requirement for maintenance (maintenance protein x the adult scoring pattern) and growth (tissue deposition adjusted for a 58% dietary efficiency of utilization x the tissue pattern).

[5] Amino acid requirements/protein requirements for the selected age groups. Note that these values, some of which are slightly amended from the 2007 report, are the correctly calculated values. In the published report, the value for the SAA requirement for children aged 3-10 is incorrect (18mg/kg/d) as are the SAA patterns for infants preschool and school children up to 10, (28, 26 and 24 mg/g protein).

FIGURE 6.4 Amino acid scoring patterns for toddlers, children, adolescents, and adults (amended values from the 2007 WHO/FAO/UNU report), adapted from the 2011 report of an FAO Expert Consultation.

and generally assumes incorrectly that lost or non-bioavailable amino acids have been digested. In the same way, amino acids that traverse beyond the terminal ileum are usually not absorbable or used for protein synthesis, thus exiting the system or being metabolized by bacteria. At a WHO/FAO meeting, it was agreed that proteins scoring above 1.0 should be leveled to 1.0 for fairness (FAO/WHO 1990). Table 6.1 summarizes the PDCAAS of common protein sources in the human diet.

Protein Efficiency Ratio

The protein efficiency ratio (PER) was the first developed method for assessing the protein quality of human food. It is referred to as the growth-based method of protein quality assessment because it calculates how efficiently a dietary protein is utilized for body growth. The PER is based on the amino acid requirements of growing rats, which differ extensively from those of humans. It is a standardized method whereby a test protein, also containing a known amount of casein, is fed to an experimental subject, after which the weight gain is divided by the amount of protein consumed

TABLE 6.1
PDCAAS of Protein Sources of Human Nutrition

Protein Source	PDCAAS
Animal Proteins	
Beef	0.92
Casein	1
Egg white	1
Branded cricket	0.65
Yellow mealworm	0.79–0.82
Plant Proteins	
Soybean protein (concentrate)	0.99
Rapeseed protein (concentrate)	0.93
Soybean protein (isolate)	0.92
Potato protein	1
Lupin	0.4–0.8
Brown rice	0.59
Pea protein (concentrate)	0.72
Pea flour	0.69
Peanut meal	0.52
Whole wheat	0.40
Sunflower protein (isolate)	0.37
Wheat gluten	0.24

Adapted from János-István et al. 2016; Zeng et al. 2022

to estimate the protein efficiency ratio. Table 6.2 shows the derived PERs of protein sources in the human diet.

Researchers make necessary adjustments to the PER values when comparing PER values across different foods and laboratories. Usually, they adjust the PER values to an assumed value of 2.5 for a casein control (McLaughlan 1974). Currently, experts dispute the acceptability of the PER as an effective tool for estimating protein quality due to issues with proportionality, reproducibility, and precision (Boling-Frankenbach et al. 2001). However, its main criticism is that the PER fails to account for certain proteins that help maintain the human body. The demerits of this assessment method worsen with proteins of lower quality. Therefore, there is a complete lack of proportionality between protein quality and the PER.

Biological Value

Biological value (BV) simply measures the amount of protein from food the body retains for growth and maintenance, estimated through the proportion of absorbed nitrogen. The biological value measures how much of the protein from the food is incorporated into the organism's protein. By doing so, it measures how well the body can use digested proteins from foods for protein synthesis. Researchers measure the biological value of a protein by comparing the amount of consumed nitrogen to the amount excreted. Table 6.3 presents common protein sources and their biological values.

Researchers feed nitrogen-free diets to subjects to determine the obligatory excretion of nitrogen through feces or urine, then administer the test protein, after which they determine the amount of urine and fecal nitrogen. In other words, BV is a measure of protein usability calculated from the ratio of incorporated nitrogen over the absorbed nitrogen. The biological value does not account for the unabsorbed dietary protein that appears in the feces, which means that unlike some other

TABLE 6.2
PERs of Common Protein Sources in Human Nutrition

Protein Source	PER
Almonds	0.4
Cheese, cheddar	2.5
Corn, whole	1.4
Dried whey	2.6
Egg white	3
Egg, whole	3.1
Fish	2.7
Gelatin or hydrolyzed collagen	0
Kidney beans	1.55
Lentils, whole green	1.3
Liver, beef	2.7
Milk	2.5
Pea flour	1.2
Pinto beans	1.64
Pork, ham	2.7
Poultry	2.7
Rice	1.5
Rice-wheat gluten	0.2
Rye	1.3
Shellfish	2.7
Soy protein	2
Sunflower seed	1.2
Wheat, whole	0.8

Adapted from the Canadian Food Inspection Agency's "Elements within the Nutrition Facts Table"

TABLE 6.3
Biological Value of Common Protein Sources for Human Nutrition

Protein Type	Biological Value
Beef	80
Casein	77
Egg	100
Milk	91
Soy protein	74
Wheat gluten	64
Whey protein	104

Adapted from US Dairy Export Council, *Reference Manual for U.S. Whey Products*, 2nd edition, 1999

metrics used to measure the usability of proteins, the biological value does not identify the digestibility or absorbability of the protein. The higher the biological value, the more it indicates the compatibility of the protein with the body's requirements. However, when one indispensable amino acid is lacking, the biological value usually approaches zero. If the biological value occurs as a fraction, researchers may express this value as a percentage by multiplying it by 100 (Hambraeus 2014). Two scales are most frequently adopted when using the BV method, which include the percentage utilization and the relative percentage utilization to a readily utilizable source of protein. The variability of the biological value among different foods is enormous, which is mostly due to differences in food preparation or even the recent food consumed by the organism. This factor limits the reliable use of the biological value for protein quality estimation, in which case experts recommend fasting before assessing biological value.

To address the limitations of existing protein quality assessment methods, the Monsonego Ornan and Reifen (2022) group has proposed a novel and sensitive way of measuring the biological contribution of proteins to the growth and development of the human body. To address the pitfalls of other protein assessment methods, they developed a sensitive preclinical model based on postnatal growth, which accurately and reliably addresses the disparities in accounting for proteins from different sources. By mitigating the variabilities of nutrition and lifestyle, this method measures postnatal growth through the changes in body weight and size at a growing age. This specific growth period, using rats as an experimental model, typifies the growth requirements during skeletal development. The main reasons for this choice of growth period and model are its sensitivity to slight changes in diet components, as skeletal development is particularly sensitive to dietary changes, and the short timeframe of approximately ten weeks, which allows timely and accelerated preclinical testing of the food to implement a successful clinical study. Another strength of this model is the close similarity between the nutrition and development of rats and humans, which makes them an acceptable preclinical model.

MINI DICTIONARY OF TERMS

Biological value: The amount of protein from a food retained in the body for growth.
Protein digestibility: Potential of a protein to hydrolyze to amino acids.
Protein efficiency ratio: Measures how well a protein is utilized for growth.
Protein quality: Amount and bioavailability of amino acids in a protein source.
Pulses: Edible leguminous seeds.

SUMMARY POINTS

- This review provides an overview of plant- and animal-based sources of dietary proteins and protein quality assessment methods.
- Animal sources of protein provide sufficient amounts of indispensable amino acids with high digestibility.
- Pulses are the closest plant source to providing adequate dietary protein to meet the recommended dietary allowance.
- Dietary integration of alternative sources of proteins, including canola and mushrooms, provides sufficient amino acids.
- The limitations of the PDCAAS, PER, and BV necessitate a combination of these three methods to provide a comprehensive method of protein quality assessment.

REFERENCES

Abugoch, James L. E. "Quinoa (Chenopodium quinoa Willd.): Composition, chemistry, nutritional, and functional properties." *Adv Food Nutr Res.* 58 (2009): 1–31.

Adhikari, Shiksha, Marijke Schop, Imke J. M. de Boer, and Thom Huppertz. "Protein quality in perspective: A review of protein quality metrics and their applications." *Nutrients* 14, no. 5 (2022): 947.

Ajomiwe N., M. Boland, S. Phongthai, M. Bagiyal, J. Singh, L. Kaur. "Protein nutrition: Understanding structure, digestibility, and bioavailability for optimal health." *Foods* 13, no. 11 (2024):1771.

Bandyopadhyay S., S. Kashyap, J. Calvez, S. Devi, D. Azzout-Marniche, D. Tomé, A. V. Kurpad, and C. Gaudichon. "Evaluation of protein quality in humans and insights on stable isotope approaches to measure digestibility – a review." *Adv Nutr.* 13, no. 4 (2022):1131–1143.

Bekhit, Alaa El-Din A., William W. Riley, and Malik A. Hussain, eds. *Alternative Proteins: Safety and Food Security Considerations.* CRC Press, 2022.

Bertsch, Pascal, Lukas Böcker, Alexander Mathys, and Peter Fischer. "Proteins from microalgae for the stabilization of fluid interfaces, emulsions, and foams." *Trends in Food Science & Technology* 108 (2021): 326–342.

Bogahawaththa, Dimuthu, Nguyen Hoang Bao Chau, Jigar Trivedi, Muditha Dissanayake, and Todor Vasiljevic. "Impact of selected process parameters on solubility and heat stability of pea protein isolate." *LWT* 102 (2019): 246–253.

Boling-Frankenbach, S. D., C. M. Peter, M. W. Douglas, J. L. Snow, C. M. Parsons, and D. H. Baker. "Efficacy of phytase for increasing protein efficiency ratio values of feed ingredients." *Poultry Science* 80, no. 11 (2001): 1578–1584.

Castellanos, Victoria Hammer, Mary D. Litchford, and Wayne W. Campbell. "Modular protein supplements and their application to long-term care." *Nutrition in Clinical Practice* 21, no. 5 (2006): 485–504.

Das Suchintita, Rahel, Brijesh K. Tiwari, Farid Chemat, and Marco Garcia-Vaquero. "Impact of ultrasound processing in alternative protein systems: Protein extraction, nutritional effects and associated challenges." *Ultrasonics Sonochemistry* 91 (2022): 106234.

Day, Li, Julie A. Cakebread, and Simon M. Loveday. "Food proteins from animals and plants: Differences in the nutritional and functional properties." *Trends in Food Science & Technology* 119 (2022): 428–442.

Derbyshire, Emma J., and Joanne Delange. "Fungal protein–what is it and what is the health evidence? A systematic review focusing on mycoprotein." *Frontiers in Sustainable Food Systems* 5 (2021): 581682.

EFSA Panel on Dietetic Products, Nutrition and Allergies (NDA). "Scientific opinion on dietary reference values for protein." *EFSA Journal* 10, no. 2 (2012): 2557.

FAO/WHO. *Expert Consultation on Protein Quality Evaluation.* Food and Agriculture Organization of the United Nations, 1990.

Gorissen, Stefan H. M., Julie J. R. Crombag, Joan M. G. Senden, W. A. Huub Waterval, Jörgen Bierau, Lex B. Verdijk, and Luc J. C. van Loon. "Protein content and amino acid composition of commercially available plant-based protein isolates." *Amino Acids* 50 (2018): 1685–1695.

Hambræus, Leif. "Protein and amino acids in human nutrition." *Reference Module in Biomedical Sciences* 3 (2014): 1–13.

Hertzler, Steven R., Jacqueline C. Lieblein-Boff, Mary Weiler, and Courtney Allgeier. "Plant proteins: assessing their nutritional quality and effects on health and physical function." *Nutrients* 12, no. 12 (2020): 3704.

Hoffman, Jay R., and Michael J. Falvo. "Protein–which is best?." *Journal of Sports Science & Medicine* 3, no. 3 (2004): 118.

Holm, Lars, Jens L. Olesen, Keitaro Matsumoto, Tatsuya Doi, Masao Mizuno, Thomas J. Alsted, Abigail L. Mackey, Peter Schwarz, and Michael Kjaer. "Protein-containing nutrient supplementation following strength training enhances the effect on muscle mass, strength, and bone formation in postmenopausal women." *Journal of Applied Physiology* 105, no. 1 (2008): 274–281.

Institute of Medicine (US). Food, and Nutrition Board. *Dietary Reference Intakes: A Risk Assessment Model for Establishing Upper Intake Levels for Nutrients.* National Academies Press, 1998.

János-István Petrusán, Harshadrai Rawel, and Gerd Huschek. "Protein-rich vegetal sources and trends in human nutrition: A review." *Current Topics in Peptide & Protein Research* 17 (2016): 1–19.

Kim, Woojeong, Yong Wang, and Cordelia Selomulya. "Dairy and plant proteins as natural food emulsifiers." *Trends in Food Science & Technology* 105 (2020): 261–272.

Marinangeli, Christopher P. F., and James D. House. "Potential impact of the digestible indispensable amino acid score as a measure of protein quality on dietary regulations and health." *Nutrition Reviews* 75, no. 8 (2017): 658–667.

McLaughlan, J. M. "Evaluation of standard rat assays. In *Nutrients in Processed Foods Proteins*, edited by P. L White and D. C. Fletcher, 69. Publishing Sciences Group, Inc., 1974.

Menzi, Harald, Oene Oenema, Colin Burton, Oleg Shipin, Pierre Gerber, Tim Robinson, and Gianluca Franceschini. "Impacts of intensive livestock production and manure management on the environment." *Livestock in a changing landscape* 1 (2010): 139–163.

Millward, D. Joe, and Alan A. Jackson. "Protein/energy ratios of current diets in developed and developing countries compared with a safe protein/energy ratio: implications for recommended protein and amino acid intakes." *Public Health Nutrition* 7, no. 3 (2004): 387–405.

Monsonego Ornan, Efrat, and Ram Reifen. "Revisiting protein quality assessment to include alternative proteins." *Foods* 11, no. 22 (2022): 3740.

Onwezen, Marleen C., Emily P. Bouwman, Machiel J. Reinders, and Hans Dagevos. "A systematic review on consumer acceptance of alternative proteins: Pulses, algae, insects, plant-based meat alternatives, and cultured meat." *Appetite* 159 (2021): 105058.

Pennings, Bart, Yves Boirie, Joan M. G. Senden, Annemie P. Gijsen, Harm Kuipers, and Luc J. C. van Loon. "Whey protein stimulates postprandial muscle protein accretion more effectively than do casein and casein hydrolysate in older men." *The American Journal of Clinical Nutrition* 93, no. 5 (2011): 997–1005.

Poore, Joseph, and Thomas Nemecek. "Reducing food's environmental impacts through producers and consumers." *Science* 360, no. 6392 (2018): 987–992.

Rempel, Curtis, Xin Geng, and Yachuan Zhang. "Industrial scale preparation of pea flour fractions with enhanced nutritive composition by dry fractionation." *Food Chemistry* 276 (2019): 119–128.

Roy, Ambrish, Alper Kucukural, and Yang Zhang. "I-TASSER: a unified platform for automated protein structure and function prediction." *Nature Protocols* 5, no. 4 (2010): 725–738.

Rutherfurd, Shane M., Aaron C. Fanning, Bruce J. Miller, and Paul J. Moughan. "Protein digestibility-corrected amino acid scores and digestible indispensable amino acid scores differentially describe protein quality in growing male rats." *The Journal of Nutrition* 145, no. 2 (2015): 372–379.

Schaafsma, Gertjan. "Advantages and limitations of the protein digestibility-corrected amino acid score (PDCAAS) as a method for evaluating protein quality in human diets." *British Journal of Nutrition* 108, no. S2 (2012): S333–S336.

Van der Spiegel, M., M. Y. Noordam, and H. J. Van der Fels-Klerx. "Safety of novel protein sources (insects, microalgae, seaweed, duckweed, and rapeseed) and legislative aspects for their application in food and feed production." *Comprehensive Reviews in Food Science and Food Safety* 12, no. 6 (2013): 662–678.

Verdijk, Lex B., Richard A. M. Jonkers, Benjamin G. Gleeson, Milou Beelen, Kenneth Meijer, Hans H. C. M. Savelberg, Will K. W. H. Wodzig, Paul Dendale, and Luc J. C. van Loon. "Protein supplementation before and after exercise does not further augment skeletal muscle hypertrophy after resistance training in elderly men." *The American Journal of Clinical Nutrition* 89, no. 2 (2009): 608–616.

Vilcacundo, Ruben, and Blanca Hernández-Ledesma. "Nutritional and biological value of quinoa (Chenopodium quinoa Willd.)." *Current Opinion in Food Science* 14 (2017): 1–6.

Willett, Walter, Johan Rockström, Brent Loken, Marco Springmann, Tim Lang, Sonja Vermeulen, Tara Garnett et al. "Food in the anthropocene: the EAT–lancet commission on healthy diets from sustainable food systems." *The Lancet* 393, no. 10170 (2019): 447–492.

Wu, Guoyao. *Amino Acids: Biochemistry and Nutrition*. CRC Press, 2021.

Zeng, Y., E. Chen, X. Zhang, D. Li, Q. Wang, and Y. Sun. "Nutritional value and physicochemical characteristics of alternative protein for meat and dairy-a review." *Foods* 11, no. 21 (October 23, 2022): 3326.

7 Comparing Animal and Plant Protein Diets in Relation to Chronic Cardiometabolic Risk Factors

Wenjing Liu, Man Wu, and Huicui Meng

ABBREVIATIONS

BMI	Body mass index
CHD	Coronary heart disease
CMD	Cardiometabolic disease
CMRF	Cardiometabolic risk factor
CRP	C-reactive protein
CVD	Cardiovascular disease
HbA1c	Hemoglobin A1C
HDL-C	High-density lipoprotein cholesterol
hs-CRP	High-sensitivity C-reactive protein
IL-6	Interleukin-6
IR	Insulin resistance
LBP	Lipopolysaccharide-binding protein
LDL-C	Low-density lipoprotein cholesterol
MetS	Metabolic syndrome
RCT	Randomized controlled trial
TC	Total cholesterol
TG	Triglyceride
TMAO	Trimethylamine-N-oxide
TNF-α	Tumor necrosis factor-α
T2DM	Type 2 diabetes mellitus
WC	Waist circumference

INTRODUCTION

Cardiometabolic diseases (CMDs), including cardiovascular disease (CVD), type 2 diabetes mellitus (T2DM), stroke, and metabolic syndrome (MetS), remain the leading causes of mortality and morbidity globally (WHO 2020). Despite decades of efforts to reduce cardiometabolic risk, the prevalence of CMDs has been increasing dramatically and continues to be a leading public health challenge worldwide (GBD 2020). Cardiometabolic risk factors (CMRFs) contribute to the increased risk of CMDs and CMD-specific mortality. Traditional CMRFs include overweight and obesity, dyslipidemia, glucose intolerance, and low-grade chronic inflammation, and emerging CMRFs include elevated serum uric acid concentration or hyperuricemia (Rahimi-Sakak et al.

DOI: 10.1201/b23328-7

2019). Adhering to healthy lifestyle behaviors, especially following healthy dietary patterns and eating behaviors, are key strategies for improving CMRFs and preventing CMDs (Wang et al. 2023).

Dietary protein, essential for overall health maintenance, can be categorized as animal or plant protein. Animal proteins are considered complete proteins as they contain all the essential amino acids necessary for optimal body function. The amino acid compositions and patterns are close to the needs of the human body, which makes animal proteins highly bioavailable and of high quality. In contrast, most plant proteins are incomplete proteins, except for soy protein. These plant proteins are low in essential amino acids and rich in non-essential amino acids, resulting in lower nutritional value compared to animal proteins.

Global animal protein consumption is increasing (Bajracharya et al. 2023), coinciding with rising CMD incidence and prevalence. This chapter focuses on the long-term associations between animal and plant protein intake and their impact on CMD risk and CMRFs, aiming to provide evidence-based dietary guidance for optimal protein selection.

PROTEIN INTAKE AND RISK OF CVD

Table 7.1 summarizes the associations between dietary animal and plant protein intake and CVD risk. Several large-scale prospective cohort studies have reported positive associations between higher intake of animal protein or isocaloric substitutions of animal protein for plant protein or carbohydrate with higher CVD risk (Hosseini-Esfahani et al. 2020). However, other studies and a recent meta-analysis have reported no association between animal protein intake and risk of coronary heart disease (CHD) (Mousavi et al. 2022) or total CVD (Kelemen et al. 2005). Differences in animal protein sources may account for these discrepancies. Most cohort studies and meta-analyses show higher CVD risk associated with red meat intake, particularly processed red meat (Bechthold et al. 2019; Zhong et al. 2020). In contrast, white meat intake, particularly poultry or fish, is inversely associated with CVD risk (Zhong et al. 2020; Zhang et al. 2020). Data on the associations between dairy or egg intake and CVD risk have been inconsistent (Zhong et al. 2019; Pan et al. 2022).

Although an earlier prospective cohort study and a recent meta-analysis have reported no association between plant protein intake and CHD risk (Mousavi et al. 2022; Haring et al. 2014), other studies have found inverse associations between high plant protein intake and risk of total CVD in Dutch (Voortman et al. 2021) and US (Halton et al. 2006) populations. Substituting high-quality plant protein, such as legumes or soy (Yan et al. 2017; Zuo et al. 2023), for animal protein is associated with a lower risk of CVD.

In conclusion, dietary approaches that limit the consumption of animal-derived foods, especially processed meat, and shift the sources of dietary protein from animal-derived to plant-derived foods may be better strategies than dietary protein restriction to reduce CVD in individuals either healthy or with high cardiometabolic risk.

PROTEIN INTAKE AND RISK OF STROKE

Table 7.2 presents associations between dietary animal and plant protein intake and stroke risk. Several prospective cohort studies (Ozawa et al. 2017; Larsson et al. 2012) and a meta-analysis (Zhang et al. 2014) demonstrate that intake of certain animal foods, including eggs, dairy products, and fish, is associated with a reduced risk of intracerebral hemorrhagic stroke. In contrast, several meta-analyses report positive associations between higher intake of red meat, but not white meat, and increased risk of total stroke and ischemic stroke (Yang et al. 2016; Kim et al. 2017), but not hemorrhagic stroke (Yang et al. 2016). However, other studies report no significant association between animal protein intake and the risk of stroke (Preis et al. 2010; Haring et al. 2015).

Only one study has reported an inverse association between intake of plant protein and the risk of total or ischemic stroke (Ozawa et al. 2017), while other studies (Larsson et al. 2012; Preis et al. 2010; Haring et al. 2015) and a meta-analysis (Zhang et al. 2014) have reported no association.

TABLE 7.1
Summary of Prospective Cohort Studies and Meta-Analyses that Investigated the Associations between Dietary Animal and Plant Protein Intake and Risk of Cardiovascular Diseases

	Prospective Cohort Studies			Meta-Analyses		
Source of Protein	**Positive Association**	**No Significant Association**	**Inverse Association**	**Positive Association**	**No Significant Association**	**Inverse Association**
Total animal protein	Hosseini-Esfahani et al. (2020)	Kelemen et al. (2005), Haring et al. (2014), Voortman et al. (2021), Halton et al. (2006)	–	–	Mousavi et al. (2022)	–
Dairy products	–	Haring et al. (2014)	–	–	Bechthold et al. (2019)	–
Eggs	Zhong et al. (2019)	Haring et al. (2014)	–	–	Bechthold et al. (2019)	–
Fish	–	Zhong et al. (2020), Haring et al. (2014)	–	–	–	Bechthold et al. (2019), Zhang et al. (2020)
Poultry	Zhong et al. (2020)	–	–	–	–	–
Red and processed meat	Zhong et al. (2020)	Haring et al. (2014)	–	Bechthold et al.(2019), Shi et al. (2023)	–	–
Unprocessed meat	Zhong et al. (2020)	–	–	Shi et al. (2023)	–	–
Total plant protein	–	Hosseini-Esfahani et al. (2020), Haring et al. (2014)	Kelemen et al. (2005), Voortman et al. (2021), Halton et al. (2006)	–	Mousavi et al. (2022)	–
Legumes	–	Haring et al. (2014)	–	–	–	Bechthold et al. (2019)

(Continued)

TABLE 7.1 (CONTINUED)
Summary of Prospective Cohort Studies and Meta-Analyses that Investigated the Associations between Dietary Animal and Plant Protein Intake and Risk of Cardiovascular Diseases

	Prospective Cohort Studies			Meta-Analyses		
Source of Protein	**Positive Association**	**No Significant Association**	**Inverse Association**	**Positive Association**	**No Significant Association**	**Inverse Association**
Soy	–	–	–	–	–	Yan et al. (2017), Zuo et al. (2023)
Soy protein	–	–	–	–	Yan et al. (2017)	–
Soy milk	–	–	–	–	Zuo et al. (2023)	–
Tofu	–	–	–	–	Yan et al. (2017)	Zuo et al. (2023)
Miso	–	–	–	–	Yan et al. (2017), Zuo et al. (2023)	–
Natto	–	–	–	–	–	Zuo et al. (2023)
Nuts	–	Haring et al. (2014)	–	–	Bechthold et al. (2019)	–

TABLE 7.2
Summary of Prospective Cohort Studies and Meta-Analyses that Investigated the Associations between Dietary Animal and Plant Protein Intake and Risk of Stroke

	Prospective Cohort Studies			Meta-Analyses		
Sources of Protein	**Positive Association**	**No Significant Asociation**	**Inverse Association**	**Positive Association**	**No Significant Association**	**Inverse Association**
Total animal protein	–	Ozawa et al. (2017), Preis et al. (2010), Haring et al. (2015)	Larsson et al. (2012)	Yang et al. (2016)	–	Zhang et al. (2014)
Dairy products	–	Haring et al. (2015)	–	–	Bechthold et al. (2019)	–
Eggs	–	Haring et al. (2015)	–	–	Bechthold et al. (2019)	–
Fish	–	Haring et al. (2015)	–	–	–	Bechthold et al. (2019)
Poultry	–	Haring et al. (2015)	–	–	–	Kim et al. (2017)
Red and processed meat	Haring et al. (2015)	–	–	Bechthold et al. (2019), Kim et al. (2017), Shi et al. (2023)	–	–
Unprocessed meat	–	–	–	–	Shi et al. (2023)	–
Total meat	–	–	–	Kim et al. (2017)	–	–
Total plant protein	–	Larsson et al. (2012), Preis et al. (2010), Haring et al. (2015)	Ozawa et al. (2017)	–	Zhang et al. (2014)	–
Legumes	–	–	–	–	Bechthold et al. (2019)	–
Soy	–	–	–	–	–	Yan et al. (2017), Zuo et al. (2023)
Soy milk	–	–	–	–	Zuo et al. (2023)	–
Tofu	–	–	–	–	Zuo et al. (2023)	–
Miso	–	–	–	–	Zuo et al. (2023)	–
Natto	–	–	–	–	–	Zuo et al. (2023)
Nuts	–	–	–	–	Bechthold et al. (2019)	–

Future studies should investigate the associations between the isocaloric substitution of plant protein for animal protein and the risk of stroke in large, diverse populations.

PROTEIN INTAKE AND RISK OF T2DM

The acute effect of dietary protein blunting postprandial glycemic responses has been reported in previous studies (Hutchison et al. 2015; Lan-Pidhainy and Wolever 2010). The mechanism underpinning this effect has been attributed to delayed gastric emptying rates mediated by the sustained release of gut-derived hormones, and to the direct stimulation of insulin secretion from pancreatic β cells by certain amino acids such as leucine, valine, and lysine (Hutchison et al. 2015; Lan-Pidhainy and Wolever 2010). Despite these beneficial short-term effects, the hypothesis that high protein intake may improve glycemic control in T2DM patients remains unproven. Long-term associations between total, animal, and plant protein intake and T2DM risk have been investigated in a considerable body of work with equivocal results (Table 7.3). Some observational studies and meta-analyses have identified positive associations between animal protein intake and T2DM risk (Ye et al. 2019; Fan et al. 2019; Gu et al. 2023; Shi et al. 2023), and inverse associations between plant protein intake and T2DM risk (Fan et al. 2019; Gu et al. 2023; Virtanen et al. 2017). However, other studies have reported no significant association (Halton et al. 2008; Alhazmi et al. 2012). This inconsistency may be due to variability in genetic background, dietary patterns, and food sources of animal or plant proteins in various cohorts. Most available studies have focused on Western populations with animal-based dietary patterns (Fan et al. 2019). More studies are needed on the associations between dietary animal and plant protein intake and T2DM risk in Asian populations, who usually follow plant-based dietary patterns while bearing an increasing prevalence of T2DM.

PROTEIN INTAKE AND RISK OF METS

The associations between dietary animal and plant protein intake and the risk of MetS are presented in Table 7.4. Cohort studies have reported heterogenous findings on the associations between intake of animal protein and incidence of MetS. Although some studies have found inverse associations between intake of animal protein and incidence of MetS (Jamshidi et al. 2022; Hajihashemi et al. 2021), others have reported positive (Vasbinder et al. 2021) or no (Chung et al. 2020) association. These inconsistent findings may be attributed to differences in food sources of animal protein. Higher red meat and processed red meat consumption has been associated with a significantly higher prevalence or incidence of MetS (Shang et al. 2017; Kim and Je 2018). Most cohort studies and meta-analyses have shown that higher intake of white meat (Kim and Je 2018), especially fish (Zaribaf et al. 2014; Kim et al. 2015), are associated with a lower risk of MetS. Data on the associations between dairy (Mena-Sánchez et al. 2019) or egg (Hajihashemi et al. 2021) intake and MetS risk have been inconsistent.

Despite two studies reporting no significant association between plant protein intake and MetS risk (Vasbinder et al. 2021; Chung et al. 2020), other studies have found inverse associations (Jamshidi et al. 2022; Hajihashemi et al. 2021), particularly for high-quality plant proteins such as soy and nuts (Li et al. 2018; Chalvon-Demersay et al. 2017).

In light of the literature's heterogenous findings, dietary approaches that minimize the consumption of animal-derived foods, especially processed meat, and prioritize plant-derived protein sources may represent the most promising strategy for reducing MetS risk in both healthy individuals and those with elevated cardiometabolic risk.

TABLE 7.3
Summary of Prospective Cohort Studies and Meta-Analyses that Investigated the Associations between Dietary Animal and Plant Protein Intake and Risk of Type 2 Diabetes Mellitus

	Prospective Cohort Studies			Meta-Analyses		
Sources of Protein	**Positive Association**	**No Significant Association**	**Inverse Association**	**Positive Association**	**No Significant Association**	**Inverse Association**
Total Animal Protein	–	Virtanen et al. (2017), Halton et al. (2008)	–	Ye et al. (2019), Fan et al. (2019)	Alhazmi et al. (2012)	–
Dairy products	–	Virtanen et al. (2017)	Gu et al. (2023)	Fan et al. (2019)	Fan et al. (2019)	Fan et al. (2019)
Eggs	–	–	Virtanen et al. (2017)	Fan et al. (2019)	–	–
Fish	–	Virtanen et al. (2017)	–	–	Fan et al. (2019)	–
Poultry	–	–	–	–	Fan et al. (2019)	–
Red and processed meat	Gu et al. (2023)	Virtanen et al. (2017)	–	Fan et al. (2019), Shi et al. (2023)	–	–
Unprocessed meat	Gu et al. (2023)	Virtanen et al. (2017)	–	Shi et al. (2023)	–	–
Total meat	–	Virtanen et al. (2017)	–	–	–	–
Total Plant Protein	–	–	Virtanen et al. (2017), Halton et al. (2008)	–	Ye et al. (2019), Alhazmi et al. (2012)	Fan et al. (2019)
Soy	–	–	–	–	Fan et al. (2019)	Zuo et al. (2023)
Soy milk	–	–	–	–	Zuo et al. (2023)	–
Tofu	–	–	–	–	Zuo et al. (2023)	–
Miso	–	–	–	–	Zuo et al. (2023)	–
Nuts and legumes	–	–	Gu et al. (2023)	–	–	–

TABLE 7.4
Summary of Observational Studies and Meta-analyses that Investigated the Associations between Dietary Animal and Plant Protein Intake and Risk of Metabolic Syndrome

	Observational Studies			Meta-Analyses		
Sources of Protein	**Positive Association**	**No Significant Association**	**Inverse Association**	**Positive Association**	**No Significant Association**	**Inverse Association**
Total animal protein	Vasbinder et al. (2021), Shang et al. (2017), Chung et al. (2020) (♂)	Jamshidi et al. (2022) (♀), Chung et al. (2020) (♀)	Hajihashemi et al. (2021), Jamshidi et al. (2022) (♂)	–	–	–
Dairy products	–	Hajihashemi et al. (2021)	–	–	Mena-Sánchez et al. (2019)	Mena-Sánchez et al. (2019)
Eggs	–	–	Hajihashemi et al. (2021)	–	–	–
Fish	–	Hajihashemi et al. (2021)	Zaribaf et al. (2014)	–	–	Kim et al. (2015)
White meat	Shang et al. (2017)	–	Hajihashemi et al. (2021), Kim and Je (2018)	–	–	Kim and Je (2018)
Red and processed meat	Kim and Je (2018), Shang et al. (2017)	Hajihashemi et al. (2021), Kim and Je (2018)	Hajihashemi et al. (2021)	Kim and Je (2018)	–	–
Total meat	–	–	–	Kim and Je (2018)	–	–
Total plant protein	–	Vasbinder et al. (2021), Chung et al. (2020)	Hajihashemi et al. (2021), Shang et al. (2017), Jamshidi et al. (2022)	–	–	–
Legumes and nuts	–	Shang et al. (2017)	–	–	–	–
Legumes and soy	–	Hajihashemi et al. (2021)	–	–	–	–
Soy	–	–	–	–	–	Chalvon-Demersay et al. (2017)
Nuts	–	–	–	–	–	Li et al. (2018)
Nuts and seeds	–	–	Hajihashemi et al. (2021)	–	–	–

♂ male; ♀ *female.*

DIETARY PROTEIN INTAKE AND CMRFS

Overweight and Obesity

Several studies have investigated the associations between dietary animal and plant protein intake and overweight and obesity (Figures 7.1 and 7.2). In populations with habitual high carbohydrate and low protein intake, substitution of animal protein for carbohydrate is inversely correlated with overweight or obesity (Hemler et al. 2022). In generally healthy individuals, the intake of animal protein is positively associated with body mass index (BMI) (Moon et al. 2021; Lin et al. 2011), waist circumference (WC) (Lin et al. 2011), and weight gain (Halkjær et al. 2011), as well as increased prevalence of overweight and obesity (Bujnowski et al. 2011), especially abdominal obesity (Chung et al. 2020). When examining specific food sources, higher consumption of red meat and processed meat is positively associated with overweight or obesity (Muga et al. 2017; Rouhani et al. 2014).

In contrast to animal protein, intake of plant protein, especially nuts and soy, has been reported to be inversely associated with body weight (Tran et al. 2020), WC (Lin et al. 2011), BMI (Lin et al. 2011), and prevalence of overweight or obesity (Lin et al. 2011; Bujnowski et al. 2011; Navas-Carretero et al. 2019). Randomized controlled trials (RCTs) have reinforced the beneficial effect of plant protein intake on promoting favorable body composition and weight loss, which may possibly be due to lower intake of leucine and histidine in these diets (Kahleova et al. 2018). Overall, dietary approaches that limit protein intake from animal-derived foods and shift the sources of dietary protein from animal-derived to plant-derived foods may be beneficial for weight management.

	Cross-sectional studies		Prospective cohort studies	
	Animal protein	Plant protein	Animal protein	Plant protein
Overweight or obesity	- (Hemler. 2022)	NS (Hemler. 2022)	+ (Bujnowski. 2011)	- (Bujnowski. 2011)
BMI	+ (Moon. 2021, Lin. 2011)	NS (Moon. 2021)		
		- (Lin. 2011)		
Weight			+ (Halkjær. 2011)	NS (Halkjær. 2011)
Waist circumference	+ (Lin. 2011)	- (Lin. 2011)	NS (Halkjær. 2011)	NS (Halkjær. 2011)
Lipid or lipoprotein profiles				
TC	+ (Meng. 2021)	NS (Meng. 2021)	NS (Hruby and Jacques 2018)	NS (Hruby and Jacques. 2018)
TG	NS (Meng. 2021, Azemati. 2021)	NS (Meng. 2021, Azemati. 2021)		
LDL-C	+ (Meng. 2021)	NS (Meng. 2021, Azemati. 2021)	NS (Hruby and Jacques 2018)	NS (Hruby and Jacques. 2018)
	NS (Azemati. 2021)			
HDL-C	NS (Meng. 2021, Azemati. 2021)	- (Meng. 2021)	NS (Hruby and Jacques. 2018)	NS (Hruby and Jacques. 2018)
		NS (Azemati. 2021)		
Blood glucose homeostasis biomarkers				
Fasting glucose	NS (Meng. 2021)	NS (Meng. 2021)	+ (Chung. 2020)	NS (Chung. 2020)
		+ (Alsulami. 2021)		
HbA1c	+ (Berryman. 2016)	+ (Alsulami. 2021)		
		NS (Berryman. 2016)		
insulin resistance	+ (Berryman. 2016)	NS (Berryman. 2016)	+ (Chen. 2020)	NS (Chen. 2020)
Inflammatory biomarkers				
hs-CRP	NS (Meng. 2021)	NS (Meng. 2021)		
	- (Matsunaga. 2021)	- (Matsunaga. 2021)		
IL-6	NS (Azzini. 2022)	NS (Azzini. 2022)		
TNF-α	NS (Azzini. 2022)	- (Azzini. 2022)		
Uric acid	+ (Meng. 2021)	- (Meng. 2021, Limirio. 2021)		
	NS (Limirio. 2021)			

FIGURE 7.1 Summary of observational studies that investigated the associations between dietary animal and plant protein intake and cardiometabolic risk factors. NS: no significant association; + positive association; – inverse association; BMI: body mass index; HbA1c: hemoglobin A1C; HDL-C: high-density lipoprotein cholesterol; hs-CRP: high-sensitivity C-reactive protein; IL-6: interleukin-6; LDL-C: low-density lipoprotein cholesterol; TC: total cholesterol; TNF-α: tumor necrosis factor-α; TG: triglyceride.

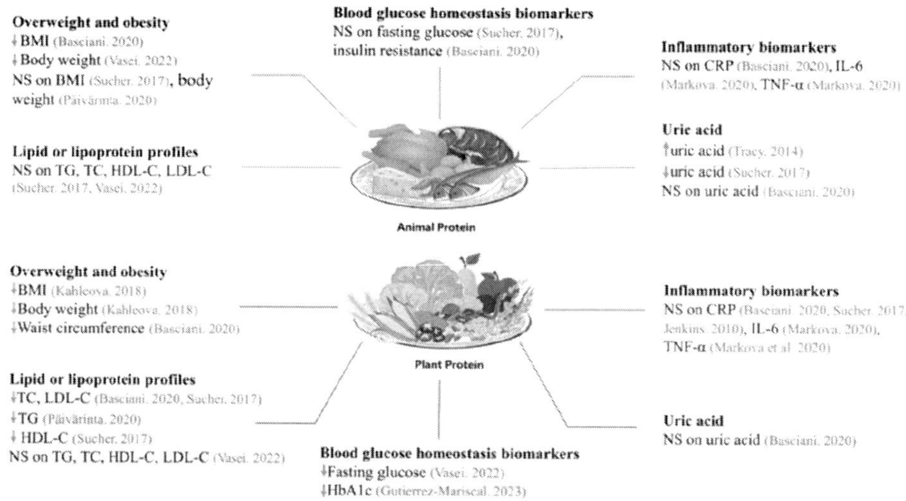

FIGURE 7.2 Effects of dietary animal and plant protein intake on cardiometabolic risk factors based on randomized controlled trials. ↓: negative effect; ↑: positive effect; BMI: body mass index; CRP: C-reactive protein; HbA1c: hemoglobin A1C; HDL-C: high-density lipoprotein cholesterol; IL-6: interleukin-6; LDL-C: low-density lipoprotein cholesterol; NS: no significant effect; TC: total cholesterol; TNF-α: tumor necrosis factor-α; TG: triglyceride.

Lipid or Lipoprotein Profiles

Despite many studies measuring the relationships between dietary animal protein and lipid and lipoprotein profiles, the results are controversial (Figures 7.1 and 7.2). While several cross-sectional studies have found positive associations between dietary animal protein intake and higher circulating concentrations of total cholesterol (TC) (Meng et al. 2021), and low-density lipoprotein cholesterol (LDL-C), other studies have reported contradictory (Akhavan et al. 2020) or null (Azemati et al. 2021; Hruby and Jacques 2018) associations. In RCTs, two studies have demonstrated that increased animal protein consumption for 6–7 weeks resulted in lower concentrations of circulating triglyceride (TG), TC, LDL-C, and high-density lipoprotein cholesterol (HDL-C) in individuals with obesity (Basciani et al. 2020) or T2DM (Sucher et al. 2017) when compared to the baseline. However, these reductions were not significantly different when compared to plant protein consumption, which aligns with a recent RCT (Vasei et al. 2022). Collectively, current RCTs have consistently reported no significant effect of increased animal protein consumption on circulating lipid and lipoprotein profiles.

Divergent findings have been reported regarding the relationships between intake of plant protein and lipid and lipoprotein profiles (Figures 7.1 and 7.2). Cross-sectional (Azemati et al. 2021) or prospective cohort studies (Hruby and Jacques 2018) have found no significant correlations between plant protein intake and concentrations of circulating TG, TC, LDL-C, and HDL-C. Additionally, several RCTs (Basciani et al. 2020; Sucher et al. 2017; Vasei et al. 2022; Jenkins et al. 2010) have corroborated these null effects. Conversely, several RCTs and meta-analyses have indicated that replacing animal protein with plant protein or increasing plant protein results in lower concentrations of circulating TC, LDL-C, or TG in both healthy individuals (Päivärinta et al. 2020; Lamberg-Allardt et al. 2023) and those with elevated cardiometabolic risk (Basciani et al. 2020; Sucher et al. 2017; Zhao et al. 2020) when compared to higher animal protein consumption or lower plant protein consumption. Further mechanistic studies are required to explore the potential mechanisms underlying the differential effects of animal and plant protein intake on circulating lipid and lipoprotein profiles.

Blood Glucose Homeostasis Biomarkers

Cross-sectional or prospective cohort studies have consistently demonstrated positive associations between animal protein intake and concentrations of fasting glucose (Chung et al. 2020), hemoglobin A1c (HbA1c) (Berryman et al. 2016), as well as insulin resistance (IR) (Berryman et al. 2016; Chen et al. 2020) (Figure 7.1). Based on these studies, the beneficial effects of animal protein on glucose homeostasis biomarkers may be attributed to its ability to induce the release of intestinal hormones, such as glucagon-like peptide 1, which further delays gastric emptying rate (Hutchison et al. 2015) and stimulates insulin secretion by pancreatic β cells (Wang et al. 2020). Nonetheless, RCTs have consistently demonstrated that higher consumption of animal protein for 6–8 weeks results in no significant differences in concentrations of fasting glucose and insulin, as well as insulin resistance index in individuals with obesity (Basciani et al. 2020) or T2DM (Sucher et al. 2017) compared to higher plant protein consumption (Figure 7.2).

Despite the heterogenous findings regarding the associations between intake of plant protein with glucose homeostasis biomarkers in observational studies (Meng et al. 2021; Chen et al. 2020; Alsulami et al. 2021), two RCTs have demonstrated lowering effects of increased plant protein consumption on concentrations of fasting glucose (Vasei et al. 2022) and HbA1c (Gutierrez-Mariscal et al. 2023) compared to higher animal protein or lower plant protein consumption groups (Figures 7.1 and 7.2). Additional mechanistic studies are required to explore the potential mechanisms underlying the differential effects of animal and plant protein intake on blood glucose homeostasis biomarkers.

Inflammatory Biomarkers

An initial cross-sectional study reported an inverse association between animal protein intake and serum high-sensitivity C-reactive protein (hs-CRP) concentration (Matsunaga et al. 2021) (Figure 7.1). From a mechanistic perspective, animal protein consumption results in lower levels of serum lipopolysaccharide-binding protein (LBP) compared to plant protein consumption in a murine model, subsequently inhibiting NF-κB-regulated inflammatory signaling pathways (Zhu et al. 2015). Despite these findings, other studies have reported either no significant association (Meng et al. 2021; Azzini et al. 2022) or no notable effect (Basciani et al. 2020; Markova et al. 2020) between animal protein intake and circulating concentrations of hs-CRP as well as other pro-inflammatory biomarkers, including interleukin-6 (IL-6) and tumor necrosis factor-α (TNF-α) (Figures 7.1 and 7.2).

Regarding plant protein, observational studies have presented mixed findings. Some reported an inverse association between plant protein intake and serum hs-CRP (Matsunaga et al. 2021) or TNF-α (Azzini et al. 2022) concentrations, while others found no significant correlations with hs-CRP (Meng et al. 2021) or IL-6 (Azzini et al. 2022) (Figure 7.1). RCTs have also reported no significant effect of plant protein consumption on circulating CRP (Basciani et al. 2020; Sucher et al. 2017, Jenkins et al. 2010), IL-6 (Markova et al. 2020), or TNF-α (Markova et al. 2020) concentrations (Figure 7.2). Consequently, additional RCTs and mechanistic studies are required to further compare the differential effects of animal and plant protein on hs-CRP, CRP, IL-6, and TNF-α, as well as other pro- or anti-inflammatory biomarkers, and explore the potential underlying mechanisms.

Uric Acid

Studies that measured how intake of animal protein could affect serum uric acid concentrations have yielded inconsistent results (Figures 7.1 and 7.2). In healthy individuals, a cross-sectional study (Meng et al. 2021) reported a positive association between higher animal protein intake and serum uric acid concentrations. One RCT demonstrated higher serum uric acid concentrations in healthy individuals (Tracy et al. 2014) following fish protein compared to chicken or beef protein. The potential mechanistic explanation may be attributed to higher levels of exogenous purines in foods enriched in red meat (Zhang et al. 2012). However, another RCT showed a greater reduction in serum uric acid concentrations with animal protein consumption compared to plant protein in T2DM patients (Sucher et al. 2017), while a separate RCT reported no significant effects in obese individuals (Basciani et al. 2020). These discrepant results may be attributed to the different health status of study participants.

Inverse associations between plant protein and serum uric acid concentration have been documented in cross-sectional studies (Meng et al. 2021; Limirio et al. 2021). However, the only available RCT has demonstrated a null effect of higher plant protein consumption on uric acid compared to higher animal protein consumption (Basciani et al. 2020). Additional RCTs and mechanistic studies are required to further compare the differential effects of animal and plant protein on uric acid concentrations, as well as prevalence of hyperuricemia and gout, and explore the potential underlying mechanisms.

CONCLUSIONS AND FUTURE DIRECTIONS

The existing literature presents mixed findings regarding the long-term associations between animal and plant protein intake with CMD risk and CMRFs, which have been investigated in a considerable body of work with equivocal results. While some studies have identified positive associations between animal protein intake and elevated CMD risk and unfavorable CMRFs, as well as inverse associations between plant protein intake with CMD risk and CMRFs, other research has reported no significant association. This heterogeneity may stem from the variability in genetic background, differential dietary patterns of participants, and different food sources of animal or plant proteins in various cohorts. Despite the inconsistent results, current evidence suggests that dietary approaches that limit the consumption of animal-derived foods, especially processed meat, and shift the sources of dietary protein from animal-derived to plant-derived foods may be better strategies than dietary protein restriction to reduce CMD risk and unfavorable CMRFs in both healthy individuals and those with high cardiometabolic risk.

One of the underlying mechanisms for the different associations between animal and plant proteins with CMD risk and CMRFs may be partially attributed to their distinct amino acid compositions (Mirmiran et al. 2017). Compared to plant proteins, animal proteins contain more essential amino acids, such as methionine and lysine, which have been reported to induce elevated concentrations of TC and LDL-C, and IR in experimental animals (Giroux et al. 1999). An additional underlying mechanism is the difference in gut microbial composition following the consumption of animal and plant protein. Higher consumption of animal protein, compared to plant protein, leads to changes in gut microbial composition with concurrent changes in lipoprotein profiles and other CMRFs (Basciani et al. 2020; Ijaz et al. 2018). Carnitine, which is derived from animal protein, interacts with the gut microbiota to produce trimethylamine-N-oxide (TMAO), a compound that has been reported to be associated with dyslipidemia and increased CMD risk (Richter et al. 2015).

In summarizing the literature to date on associations between animal and plant protein intake with CMD risk and the comparable effects of animal and plant protein intake in modulating CMRFs, it is clear that additional studies are required to clarify the following issues:

1. Future studies should focus on investigating the associations between isocaloric substitutions of plant protein for animal protein and the risk of CMDs in large, diverse populations to directly compare their differential roles in CMD prevention.
2. Additional RCTs and mechanistic studies are required to directly compare the differential effects of animal and plant protein consumption on established and emerging CMRFs and to explore the potential mediators that may link animal and plant protein with changes in CMRFs.
3. Research should aim to identify the optimal ratio of animal and plant protein in a diet needed to reduce unfavorable CMRF concentrations and further reduce CMD risk.

4. The majority of the available studies have focused on Western populations who follow habitual animal-based dietary patterns. More studies are needed with a focus on the associations between dietary animal and plant protein intake with CMD risk, as well as the role of animal and plant protein consumption in modulating CMRFs in Asian populations, who usually follow plant-based dietary patterns while bearing an increasing prevalence of CMDs.

MINI DICTIONARY OF TERMS

Cardiometabolic disease (CMD): CMDs refer to a group of diseases that affect the function of the heart and blood vessels, which are often associated with metabolic disorders such as cardiovascular disease, stroke, type 2 diabetes mellitus, and metabolic syndrome.

Cardiometabolic risk factors (CMRFs): CMRFs refer to a group of risk factors that contribute to the onset and progression of CMDs. Traditional CMRFs include overweight, obesity, dyslipidemia, glucose intolerance, insulin resistance, and low-grade systemic chronic inflammation. Emerging CMRFs include high serum uric acid concentration or hyperuricemia.

Metabolic syndrome (MetS): MetS is a medical diagnosis that is characterized by a cluster of metabolic abnormalities and physiological changes, such as hypertension, hyperglycemia, hyperlipidemia, and central obesity. MetS is often associated with an increased risk of CMDs and other health complications.

Trimethylamine-N-oxide (TMAO): Choline of animal origin is synthesized by gut microbiota using choline-trimethylamine lyase to generate trimethylamine, which then enters the hepatic portal circulation and undergoes oxidation by flavin monooxygenase to form TMAO. Consequently, TMAO is a gut microbiota-derived metabolite that has emerged as a potential risk factor for CMDs.

Tumor necrosis factor-α (TNF-α): TNF-α is a cytokine involved in the inflammatory process and is primarily produced by immune cells such as macrophages, lymphocytes, and dendritic cells, as well as other cell types such as adipocytes and endothelial cells. It has significant impacts on biological processes such as immune regulation, inflammatory response, cell proliferation, and apoptosis.

SUMMARY POINTS

- The current review focuses on the long-term associations between animal and plant protein intake and CMD risk and CMRFs.
- There is heterogeneity in the literature regarding the associations between animal and plant protein intake and CMD risk and CMRFs, which may be attributed to the variability in genetic background, differential dietary patterns of participants, and different food sources of animal or plant proteins in various cohorts.
- Overall, the current literature has demonstrated that dietary approaches that limit the consumption of animal-derived foods, especially processed meat, and shift the sources of dietary protein from animal-derived to plant-derived foods may be more effective strategies than dietary protein restriction to reduce CMD risk and unfavorable CMRFs in healthy individuals and those with elevated cardiometabolic risk.
- The underlying mechanisms for the differential associations between animal and plant proteins with CMD risk and CMRFs may be attributed to differences in their amino acid compositions and gut microbial compositions following consumption.
- Future studies should focus on investigating the associations between isocaloric substitutions of plant protein for animal protein and CMD risk, as well as directly comparing the differential effects of animal and plant protein consumption on established and emerging CMRFs in large, diverse populations.

REFERENCES

Akhavan, N. S., S. Pourafshar, S. A. Johnson, et al. 2020. The relationship between protein intake and source on factors associated with glycemic control in individuals with prediabetes and type 2 diabetes. *Nutrients* 12 (7):2031.

Alhazmi, A., E. Stojanovski, M. McEvoy, et al. 2012. Macronutrient intakes and development of type 2 diabetes: a systematic review and meta-analysis of cohort studies. *J Am Coll Nutr* 31 (4):243–258.

Alsulami, S., D. Bodhini, V. Sudha, et al. 2021. Lower dietary intake of plant protein is associated with genetic risk of diabetes-related traits in urban Asian Indian adults. *Nutrients* 13 (9):3064.

Azemati, B., S. Rajaram, K. Jaceldo-Siegl, et al. 2021. Dietary animal to plant protein ratio is associated with risk factors of metabolic syndrome in participants of the AHS-2 calibration study. *Nutrients* 13 (12):4296.

Azzini, E., I. Peluso, F. Intorre, et al. 2022. Total and plant protein consumption: the role of inflammation and risk of non-communicable disease. *Int J Mol Sci* 23 (14):8008.

Bajracharya, R., V. Katzke, T. Mukama, et al. 2023. Effect of iso-caloric substitution of animal protein for other macro nutrients on risk of overall, cardiovascular and cancer mortality: prospective evaluation in EPIC-heidelberg cohort and systematic review. *Nutrients* 15 (3):794.

Basciani, S., E. Camajani, S. Contini, et al. 2020. Very-low-calorie ketogenic diets with whey, vegetable, or animal protein in patients with obesity: a randomized pilot study. *J Clin Endocrinol Metab* 105 (9):dgaa336.

Bechthold, Angela, Heiner Boeing, Carolina Schwedhelm, et al. 2019. Food groups and risk of coronary heart disease, stroke and heart failure: a systematic review and dose-response meta-analysis of prospective studies. *Crit Rev Food Sci Nutr* 59 (7):1071–1090.

Berryman, C. E., S. Agarwal, H. R. Lieberman, et al. 2016. Diets higher in animal and plant protein are associated with lower adiposity and do not impair kidney function in US adults. *Am J Clin Nutr* 104 (3):743–749.

Bujnowski, D., P. Xun, M. L. Daviglus, et al. 2011. Longitudinal association between animal and vegetable protein intake and obesity among men in the United States: the Chicago Western electric study. *J Am Diet Assoc* 111 (8):1150–1155.

Chalvon-Demersay, T., D. Azzout-Marniche, J. Arfsten, et al. 2017. A systematic review of the effects of plant compared with animal protein sources on features of metabolic syndrome. *J Nutr* 147 (3):281–292.

Chen, Z., O. H. Franco, S. Lamballais, et al. 2020. Associations of specific dietary protein with longitudinal insulin resistance, prediabetes and type 2 diabetes: the rotterdam study. *Clin Nutr* 39 (1):242–249.

Chung, S., M. Y. Chung, H. K. Choi, et al. 2020. Animal protein intake is positively associated with metabolic syndrome risk factors in middle-aged korean men. *Nutrients* 12 (11):3415.

Fan, M., Y. Li, C. Wang, et al. 2019. Dietary protein consumption and the risk of type 2 diabetes: a dose-response meta-analysis of prospective studies. *Nutrients* 11 (11):2783.

GBD. 2020. Global burden of 369 diseases and injuries in 204 countries and territories, 1990–2019: a systematic analysis for the global burden of disease study 2019. *Lancet* 396 (10258):1204–1222.

Giroux, I., E. M. Kurowska, and K. K. Carroll. 1999. Role of dietary lysine, methionine, and arginine in the regulation of hypercholesterolemia in rabbits. *J Nutr Biochem* 10 (3):166–171.

Gu, X., J. P. Drouin-Chartier, F. M. Sacks, et al. 2023. Red meat intake and risk of type 2 diabetes in a prospective cohort study of United States females and males. *Am J Clin Nutr* 118 (6):1153–1163.

Gutierrez-Mariscal, F. M., J. F. Alcalá-Diaz, G. M. Quintana-Navarro, et al. 2023. Changes in quantity plant-based protein intake on type 2 diabetes remission in coronary heart disease patients: from the CORDIOPREV study. *Eur J Nutr* 62 (4):1903–1913.

Hajihashemi, P., R. Hassannejad, F. Haghighatdoost, et al. 2021. The long-term association of different dietary protein sources with metabolic syndrome. *Sci Rep* 11 (1):19394.

Halkjær, J., A. Olsen, K. Overvad, et al. 2011. Intake of total, animal and plant protein and subsequent changes in weight or waist circumference in European men and women: the diogenes project. *International Journal of Obesity* 35 (8):1104–1113.

Halton, T. L., S. Liu, J. E. Manson, et al. 2008. Low-carbohydrate-diet score and risk of type 2 diabetes in women. *Am J Clin Nutr* 87 (2):339–346.

Halton, T. L., W. C. Willett, S. Liu, et al. 2006. Low-carbohydrate-diet score and the risk of coronary heart disease in women. *N Engl J Med* 355 (19):1991–2002.

Haring, B., N. Gronroos, J. A. Nettleton, et al. 2014. Dietary protein intake and coronary heart disease in a large community based cohort: results from the Atherosclerosis Risk in Communities (ARIC) study [corrected]. *PLoS One* 9 (10):e109552.

Haring, B., J. R. Misialek, C. M. Rebholz, et al. 2015. Association of dietary protein consumption with incident silent cerebral infarcts and stroke: the atherosclerosis risk in communities (ARIC) study. *Stroke* 46 (12):3443–3450.

Hemler, E. C., S. Bromage, A. W. Tadesse, et al. 2022. Associations of percentage energy intake from total, animal and plant protein with overweight/obesity and underweight among adults in Addis Ababa, Ethiopia. *Public Health Nutr* 25 (11):3107–3120.

Hosseini-Esfahani, F., G. Koochakpoor, Z. Tahmasebinejad, et al. 2020. The association of dietary macronutrients composition with the incidence of cardiovascular disease, using iso-energetic substitution models: Tehran lipid and glucose study. *Nutr Metab Cardiovasc Dis* 30 (12):2186–2193.

Hruby, A., and P. F. Jacques. 2018. Dietary protein and changes in markers of cardiometabolic health across 20 years of follow-up in middle-aged Americans. *Public Health Nutr* 21 (16):2998–3010.

Hutchison, A. T., D. Piscitelli, M. Horowitz, et al. 2015. Acute load-dependent effects of oral whey protein on gastric emptying, gut hormone release, glycemia, appetite, and energy intake in healthy men. *Am J Clin Nutr* 102 (6):1574–1584.

Ijaz, M. U., M. I. Ahmed, X. Zou, et al. 2018. Beef, casein, and soy proteins differentially affect lipid metabolism, triglycerides accumulation and gut microbiota of high-fat diet-fed C57BL/6J mice. *Front Microbiol* 9:2200.

Jamshidi, A., M. Farjam, M. Ekramzadeh, et al. 2022. Evaluating type and amount of dietary protein in relation to metabolic syndrome among Iranian adults: cross-sectional analysis of Fasa Persian cohort study. *Diabetol Metab Syndr* 14 (1):42.

Jenkins, D. J., K. Srichaikul, J. M. Wong, et al. 2010. Supplemental barley protein and casein similarly affect serum lipids in hypercholesterolemic women and men. *J Nutr* 140 (9):1633–1637.

Kahleova, H., R. Fleeman, A. Hlozkova, et al. 2018. A plant-based diet in overweight individuals in a 16-week randomized clinical trial: metabolic benefits of plant protein. *Nutr Diabetes* 8 (1):58.

Kelemen, L. E., L. H. Kushi, D. R. Jacobs, Jr., et al. 2005. Associations of dietary protein with disease and mortality in a prospective study of postmenopausal women. *Am J Epidemiol* 161 (3):239–249.

Kim, K., J. Hyeon, S. A. Lee, et al. 2017. Role of total, red, processed, and white meat consumption in stroke incidence and mortality: a systematic review and meta-analysis of prospective cohort studies. *J Am Heart Assoc* 6 (9):e005983.

Kim, Y. S., and Y. Je. 2018. Meat consumption and risk of metabolic syndrome: results from the korean population and a meta-analysis of observational studies. *Nutrients* 10 (4):390.

Kim, Y. S., P. Xun, and K. He. 2015. Fish consumption, long-chain omega-3 polyunsaturated fatty acid intake and risk of metabolic syndrome: a meta-analysis. *Nutrients* 7 (4):2085–2100.

Lamberg-Allardt, C., L. Bärebring, E. K. Arnesen, et al. 2023. Animal versus plant-based protein and risk of cardiovascular disease and type 2 diabetes: a systematic review of randomized controlled trials and prospective cohort studies. *Food Nutr Res* 67 (10):29219.

Lan-Pidhainy, X., and T. M. Wolever. 2010. The hypoglycemic effect of fat and protein is not attenuated by insulin resistance. *Am J Clin Nutr* 91 (1):98–105.

Larsson, S. C., J. Virtamo, and A. Wolk. 2012. Dietary protein intake and risk of stroke in women. *Atherosclerosis* 224 (1):247–251.

Li, H., X. Li, S. Yuan, et al. 2018. Nut consumption and risk of metabolic syndrome and overweight/obesity: a meta-analysis of prospective cohort studies and randomized trials. *Nutr Metab (Lond)* 15:46.

Limirio, L. S., H. O. Santos, A. S. Dos Reis, et al. 2021. Association between dietary intake and serum uric acid levels in kidney transplant patients. *J Ren Nutr* 31 (6):637–647.

Lin, Y., S. Bolca, S. Vandevijvere, et al. 2011. Plant and animal protein intake and its association with overweight and obesity among the Belgian population. *British Journal of Nutrition* 105 (7): 1106–1116.

Markova, M., L. Koelman, S. Hornemann, et al. 2020. Effects of plant and animal high protein diets on immune-inflammatory biomarkers: a 6-week intervention trial. *Clin Nutr* 39 (3):862–869.

Matsunaga, M., E. L. Hurwitz, and D. Li. 2021. Adequate intake of plant protein foods and moderate intake of animal protein foods are inversely associated with C-reactive protein in US adults with diabetes: a cross-sectional study with national health and nutrition examination survey. *Nutrition* 89:111276.

Mena-Sánchez, G., N. Becerra-Tomás, N. Babio, et al. 2019. Dairy product consumption in the prevention of metabolic syndrome: a systematic review and meta-analysis of prospective cohort studies. *Adv Nutr* 10 (suppl_2):s144–s153.

Meng, S., Z. Cui, M. Li, et al. 2021. Associations between dietary animal and plant protein intake and cardiometabolic risk factors-a cross-sectional study in china health and nutrition survey. *Nutrients* 13 (2):336.

Mirmiran, P., Z. Bahadoran, A. Ghasemi, et al. 2017. Contribution of dietary amino acids composition to incidence of cardiovascular outcomes: a prospective population-based study. *Nutr Metab Cardiovasc Dis* 27 (7):633–641.

Moon, K., C. Krems, T. Heuer, et al. 2021. Association between body mass index and macronutrients differs along the body mass index range of German adults: results from the German national nutrition survey II. *J Nutr Sci* 10:e8.

Mousavi, S. M., A. Jayedi, Y. Jalilpiran, et al. 2022. Dietary intake of total, animal and plant proteins and the risk of coronary heart disease and hypertension: a systematic review and dose-response meta-analysis of prospective cohort studies. *Crit Rev Food Sci Nutr* 62 (5):1336–1349.

Muga, M. A., P. O. Owili, C. Y. Hsu, et al. 2017. Dietary patterns, gender, and weight status among middle-aged and older adults in Taiwan: a cross-sectional study. *BMC Geriatr* 17 (1):268.

Navas-Carretero, S., R. San-Cristobal, K. M. Livingstone, et al. 2019. Higher vegetable protein consumption, assessed by an isoenergetic macronutrient exchange model, is associated with a lower presence of overweight and obesity in the web-based Food4me European study. *Int J Food Sci Nutr* 70 (2):240–253.

Ozawa, M., D. Yoshida, J. Hata, et al. 2017. Dietary protein intake and stroke risk in a general Japanese population: the hisayama study. *Stroke* 48 (6):1478–1486.

Päivärinta, E., S. T. Itkonen, T. Pellinen, et al. 2020. Replacing animal-based proteins with plant-based proteins changes the composition of a whole nordic diet-a randomised clinical trial in healthy finnish adults. *Nutrients* 12 (4):943.

Pan, L., L. Chen, J. Lv, et al. 2022. Association of egg consumption, metabolic markers, and risk of cardiovascular diseases: a nested case-control study. *Elife* 11:e72909.

Preis, S. R., M. J. Stampfer, D. Spiegelman, et al. 2010. Lack of association between dietary protein intake and risk of stroke among middle-aged men. *Am J Clin Nutr* 91 (1):39–45.

Rahimi-Sakak, F., M. Maroofi, J. Rahmani, et al. 2019. Serum uric acid and risk of cardiovascular mortality: a systematic review and dose-response meta-analysis of cohort studies of over a million participants. *BMC Cardiovasc Disord* 19 (1):218.

Richter, C. K., A. C. Skulas-Ray, C. M. Champagne, et al. 2015. Plant protein and animal proteins: do they differentially affect cardiovascular disease risk? *Adv Nutr* 6 (6):712–728.

Rouhani, M. H., A. Salehi-Abargouei, P. J. Surkan, et al. 2014. Is there a relationship between red or processed meat intake and obesity? A systematic review and meta-analysis of observational studies. *Obes Rev* 15 (9):740–748.

Shang, X., D. Scott, A. Hodge, et al. 2017. Dietary protein from different food sources, incident metabolic syndrome and changes in its components: An 11-year longitudinal study in healthy community-dwelling adults. *Clin Nutr* 36 (6):1540–1548.

Shi, W., X. Huang, C. M. Schooling, et al. 2023. Red meat consumption, cardiovascular diseases, and diabetes: a systematic review and meta-analysis. *Eur Heart J* 44 (28):2626–2635.

Sucher, S., M. Markova, S. Hornemann, et al. 2017. Comparison of the effects of diets high in animal or plant protein on metabolic and cardiovascular markers in type 2 diabetes: A randomized clinical trial. *Diabetes Obes Metab* 19 (7):944–952.

Tracy, C. R., S. Best, A. Bagrodia, et al. 2014. Animal protein and the risk of kidney stones: a comparative metabolic study of animal protein sources. *J Urol* 192 (1):137–141.

Tran, E., H. F. Dale, C. Jensen, et al. 2020. Effects of plant-based diets on weight status: a systematic review. *Diabetes Metab Syndr Obes* 13:3433–3448.

Vasbinder, A., L. F. Tinker, M. L. Neuhouser, et al. 2021. Risk of metabolic syndrome and metabolic phenotypes in relation to biomarker-calibrated estimates of energy and protein intakes: an investigation from the Women's Health Initiative. *Am J Clin Nutr* 113 (3):706–715.

Vasei, M. H., S. Hosseinpour-Niazi, E. Ainy, et al. 2022. Effect of dietary approaches to stop hypertension (DASH) diet, high in animal or plant protein on cardiometabolic risk factors in obese metabolic syndrome patients: A randomized clinical trial. *Prim Care Diabetes* 16 (5):634–639.

Virtanen, H. E. K., T. T. Koskinen, S. Voutilainen, et al. 2017. Intake of different dietary proteins and risk of type 2 diabetes in men: the Kuopio ischaemic heart disease risk factor study. *Br J Nutr* 117 (6):882–893.

Voortman, T., Z. Chen, C. Girschik, et al. 2021. Associations between macronutrient intake and coronary heart disease (CHD): the rotterdam study. *Clin Nutr* 40 (11):5494–5499.

Wang, W., Y. Liu, Y. Li, et al. 2023. Dietary patterns and cardiometabolic health: clinical evidence and mechanism. *MedComm* 4 (1):e212.

Wang, Y., S. Nair, and J. Gagnon. 2020. Herring milt and herring milt protein hydrolysate are equally effective in improving insulin sensitivity and pancreatic beta-cell function in diet-induced obese- and insulin-resistant mice. *Mar Drugs* 18 (12):635.

WHO. 2020. *The Top 10 Causes of Death*. World Health Organization https://www.who.int/news-room/fact-sheets/detail/the-top-10-causes-of-death. (accessed August 26, 2023).

Yan, Z., X. Zhang, C. Li, et al. 2017. Association between consumption of soy and risk of cardiovascular disease: A meta-analysis of observational studies. *Eur J Prev Cardiol* 24 (7):735–747.

Yang, Cuili, Lei Pan, Chengcao Sun, et al. 2016. Red meat consumption and the risk of stroke: a dose–response meta-analysis of prospective cohort studies. *J Stroke Cerebrovasc Dis* 25 (5):1177–1186.

Ye, J., Q. Yu, W. Mai, et al. 2019. Dietary protein intake and subsequent risk of type 2 diabetes: a dose-response meta-analysis of prospective cohort studies. *Acta Diabetol* 56 (8):851–870.

Zaribaf, F., E. Falahi, F. Barak, et al. 2014. Fish consumption is inversely associated with the metabolic syndrome. *Eur J Clin Nutr* 68 (4):474–480.

Zhang, B., K. Xiong, J. Cai, et al. 2020. Fish consumption and coronary heart disease: a meta-analysis. *Nutrients* 12 (8):2278.

Zhang, Y., C. Chen, H. Choi, et al. 2012. Purine-rich foods intake and recurrent gout attacks. *Ann Rheum Dis* 71 (9):1448–1453.

Zhang, Z., G. Xu, F. Yang, et al. 2014. Quantitative analysis of dietary protein intake and stroke risk. *Neurology* 83 (1):19–25.

Zhao, H., A. Song, C. Zheng, et al. 2020. Effects of plant protein and animal protein on lipid profile, body weight and body mass index on patients with hypercholesterolemia: a systematic review and meta-analysis. *Acta Diabetol* 57 (10):1169–1180.

Zhong, V. W., L. Van Horn, M. C. Cornelis, et al. 2019. Associations of dietary cholesterol or egg consumption with incident cardiovascular disease and mortality. *JAMA* 321 (11):1081–1095.

Zhong, V. W., L. Van Horn, P. Greenland, et al. 2020. Associations of processed meat, unprocessed red meat, poultry, or fish intake with incident cardiovascular disease and all-cause mortality. *JAMA Intern Med* 180 (4):503–512.

Zhu, Y., X. Lin, F. Zhao, et al. 2015. Meat, dairy and plant proteins alter bacterial composition of rat gut bacteria. *Sci Rep* 5:15220.

Zuo, X., R. Zhao, M. Wu, et al. 2023. Soy consumption and the risk of type 2 diabetes and cardiovascular diseases: a systematic review and meta-analysis. *Nutrients* 15 (6):1358.

8 Alpha-Ketoglutarate, the Bacteriome, Protein Metabolism, and Longevity

Kateryna Pierzynowska, Piotr Wychowański, Piotr Dobrowolski, Janine Donaldson, and Stefan G. Pierzynowski

ABBREVIATIONS

2KA	2-ketoglutaric acid = α-ketoglutaric acid
2-OA	2-oxoglutaric acid = α-ketoglutaric acid
2-OP	2-oxopentanedioic acid = α-ketoglutaric acid
2-OG	2-oxoglutarate = α-ketoglutaric acid (not specified salt)
AKG	Alpha-ketoglutarate (not specified salt)
α-KG	Alpha-ketoglutaric acid
AGC	Aspartate/glutamate carrier
BMMSCs	Bone marrow-derived mesenchymal stem cells
CIC	Citrate carrier
GALT	Gut-associated lymphatic tissue
GC	Glutamate carrier
GDH:	Glutamate dehydrogenase
GOT	Glutamate oxaloacetate transaminase
Gx	Gastrectomy
HA	Healthy aging
HL	Happy longevity
IDH	Isocitrate dehydrogenase
MDH	Malate dehydrogenase
MG	Mongolian gerbils
$-NH_2$	Amine group
NH_4	Ammonium
OG	2-oxaloglutamate
OGC	2-oxoglutarate carrier
PC	Pyruvate carboxylase
PDH	Pyruvate dehydrogenase
PLEM	Pancreatic-like enzymes of microbial origin
SCFA	Short-chain fatty acids

INTRODUCTION

Alpha-ketoglutarate (AKG), an intermediary metabolite in the Krebs cycle, is a molecule involved in multiple pleiotropic metabolic activities, mainly related to protein metabolism (Zdzisińska, Żurek,

DOI: 10.1201/b23328-8

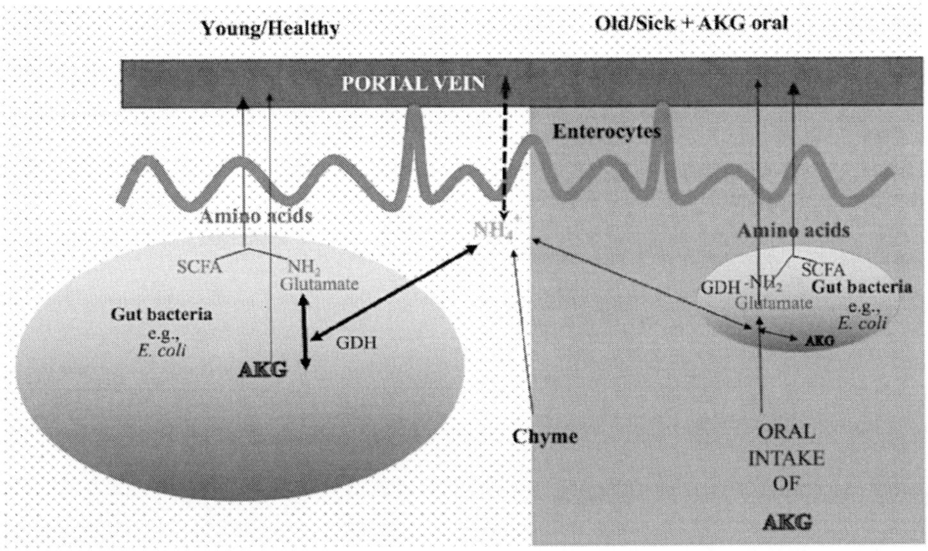

FIGURE 8.1 Summary of gut bacteria pathways for nitrogen fixation to AKG in a young/healthy and in an old/sick host. Note: Broken line indicates passive diffusion of NH4, AKG – alpha-ketoglutarate; SCFA – short-chain fatty acid; NH4 – ammonium; enterocytes – gut mucosa cells; -NH2 – amine group; chyme – gut content (from Pierzynowski and Pierzynowska 2022, Creative Commons CC-BY license).

and Kandefer-Szerszeń 2017; Bayliak and Lushchak 2021). This chapter focuses on parenteral AKG and AKG originating from the gut bacteriome (Figure 8.1) – a key factor in improving, for example, blood circulation in elderly mice (Niemiec et al. 2011) and promoting longevity. As highlighted in the review by Naeini et al., AKG plays a role in host protein metabolism, with particular relevance to malnutrition, neurodegenerative diseases, and osteoporosis – the silent, long-lasting, slow illnesses (Naeini et al. 2023). By eliminating these illnesses, longevity could transform from a period of often painful, unwanted 'final hours' to the 'happy hours' of life.

Animal proteins are indispensable structural and functional components essential for the maintenance of life. The remaining macronutrients, e.g., lipids (storage energy) and carbohydrates (disposable energy), can be created from proteins via host metabolic processes. The availability of proteins, as well as other macronutrients in nature, is extremely variable during periods of winter/drought vs. green/rainy periods. Thus, host metabolic loss of protein, or, to be more specific, the loss of amine groups, should essentially be minimized to keep the organism alive during periods of low or no dietary protein. Nature has partially 'solved' the issue of low dietary protein through gut microbes, but research concerning the role of the gut microbiota in protein metabolism is often overlooked.

The topic is especially disregarded in ruminant animals, which can survive in extreme conditions and procreate without any dietary proteins, thanks to the production of bacterial and animal proteins (since protozoa belong to the animal kingdom (Jouany and Ushida 1999)) by their rumen microflora. However, rumen bacteria are also food for rumen protozoa and are able to produce amino acids and then protein from non-organic nitrogen and carbohydrates, as proven by the 1945 Nobel Prize winner, Artturi Virtanen (1945, n.d.). One should thus remember that AKG is the only molecule that can assimilate non-organic nitrogen and convert it to glutamate, which is the first step of bacterial amino acid synthesis.

Thus, the enzyme glutamate dehydrogenase (GDH), which catalyzes this step, should be considered a 'creator of life.' In human nutrition (but not in that of wildlife), a lack of dietary protein (hunger) only occurs sporadically and is, for the time being, not really a problem for human existence – excluding some regions where political reasons have created hunger.

On the contrary, overfeeding with protein has become a health issue in Western civilization. The primary function and role of gut bacteria is to 'spare' host protein as a 'buffer,' allowing organisms to survive times of hunger. The host microbiome is recognized by scientists as a vehicle for bacteria, which then, in turn, 'govern' these bacteria. Current research is investigating the important effects of the gut microbiome on, e.g., cognition, longevity, obesity, osteoporosis, etc. Some of these effects may be obvious; however, they must be interpreted in the context of a surplus of AKG production and in terms of the availability of AKG for the host to 'upgrade' the aging metabolism.

IMPACT OF AKG ORIGINATING FROM THE GUT MICROBIOTA ON PROTEIN METABOLISM

Gut microbiota are involved in host protein production and preservation through alpha-ketoglutarate (AKG), a key metabolite known by multiple names, including α-ketoglutaric acid (α-KG), 2-keto-glutaric acid (2KA), 2-oxoglutaric acid (2-OA), 2-oxoglutarate (2-OG), 2-oxaloglutamate (OG), and 2-oxopentanedioic acid (2-OP) (Pierzynowski and Pierzynowska 2022; Yao et al. 2012; Helling 1994; Allison, Robinson, and Baetz 1979; Yuan et al. 2009). In bacterial prokaryotes, AKG is produced in the Krebs cycle in the cytoplasm, unlike in eukaryotes, where it is generated in the mitochondria. This cytoplasmic placement allows AKG to be more accessible for metabolic processes beyond energy production. In eukaryotes, AKG needs to be accessed from the mitochondria (Figure 8.2).

Unfortunately, very often, researchers overlook the critical role of AKG – the molecule of life – in gut bacteria metabolism. Even respected authors often focus on short-chain fatty acids (SCFAs) when discussing the origin of bacterial amino acids or proteins. While SCFAs are obviously an important bacterial product, they should be considered secondary to AKG. We should also not forget the origin of the amine group needed for the complete amino acid structure. Amine groups

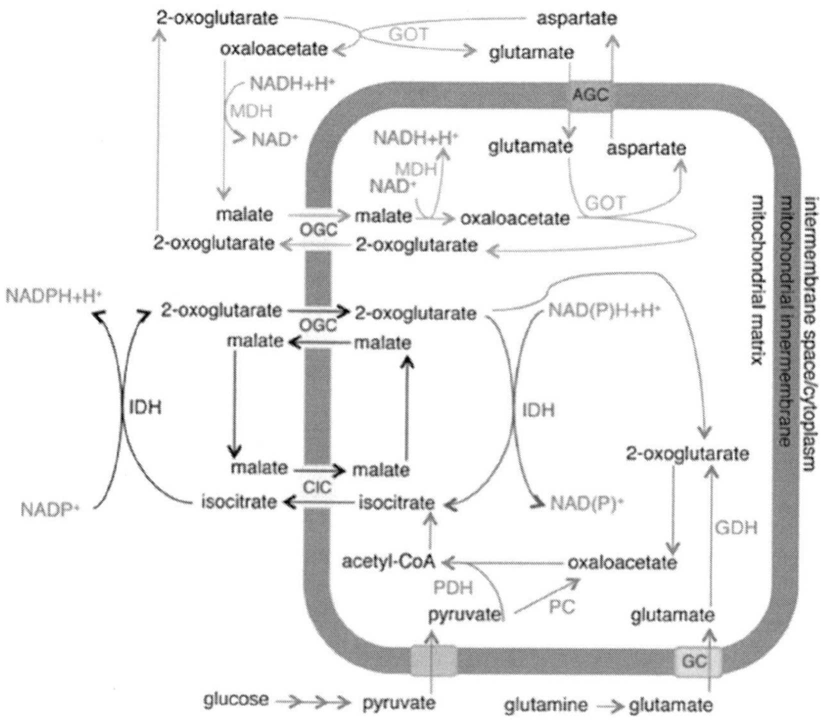

FIGURE 8.2 Physiological role of 2-oxoglutarate carrier (OGC). Note: The malate-aspartate shuttle (grey arrows), the oxoglutarate-isocitrate shuttle (blue and black arrows) and the pyruvate-isocitrate cycle (green and black arrows) are shown with the enzymes malate dehydrogenase (MDH), glutamate oxaloacetate trans-aminase (GOT), isocitrate dehydrogenase (IDH), pyruvate carboxylase (PC), pyruvate dehydrogenase (PDH) and glutamate dehydrogenase (GDH); and the carriers for oxoglutarate (OGC), citrate (CIC), aspartate/ gluta-mate (AGC), and glutamate (GC) (Monné et al. 2013, Copyright License # 5718190734214).

can obviously only be provided by glutamate, via its creation from ammonium – the end product of metabolized host or dietary amino acids – and AKG, under the control of glutamate dehydrogenase (GDH) (den Besten et al. 2013; Neis, Dejong, and Rensen 2015). Thus, AKG serves as a scavenger for ammonia by incorporating ammonia into glutamate by glutamate dehydrogenase, decreasing glutamate and Gln oxidation, and/or reducing urea production (Rutten et al. 2005; Radzki, Bieńko, and Pierzynowski 2009). This reaction is the most important reason for the maintenance and preservation of gut microbiota in the gastrointestinal tract throughout evolutionary history. While SCFA and bacterial vitamin production are important, they remain secondary to the production of amino acids and the protection of host nitrogen balance.

Gut microbes evolved as important 'helpers and protectors' of host proteins, strategically providing the correct amino acids and peptides at precise metabolic moments.

One must recognize that the total number of gut bacteria in healthy individuals typically exceeds the total number of human cells. The fundamental yet frequently overlooked role of gut bacteria is protecting and preserving host protein and nitrogen. The only mechanism to prevent nitrogen loss is to convert the ammonium group back to an amine group. How does this occur? By coupling ammonium to AKG to form glutamate, which is considered the first amino acid. Subsequently, the amine group from glutamate/glutamine can be transferred to different carbon skeletons of bacterial SCFA to create new amino acids (Walker and van der Donk 2016) (Figure 8.3).

All gut bacteria can produce AKG, but *Escherichia Coli* – a frequently occurring gut bacteria often recognized as a pathogen – is the master of AKG production. *E. Coli* likely produces AKG in amounts exceeding its own metabolic requirements and those of neighboring bacteria. *E. Coli* possesses two independent metabolic pathways for glutamate synthesis, thus requiring double the amount of AKG (Helling 1994). Due to the increased amount of AKG synthesized, the AKG can pass through the *E. Coli* membrane, using the NaDC-1 transporter (Lo and Bewick 1978).

AKG produced by gut bacteria or obtained from food, when not utilized by microbes, can be absorbed by intestinal enterocytes (Buddington et al. 2004) and metabolized in the enterocytes into other amino acids, such as proline (Kristensen et al. 2002), which ultimately crosses the blood-brain barrier. Research by Kristensen et al. (2002), Lambert et al. (2002, 2006), and Junghans et al. (2006) revealed that only half of the enteral AKG, in contrast to glutamine or glutamate, is converted to energy during its first pass through the enterocytes; the rest is converted to amino acids – including exogenous amino acids (Lambert et al. 2002, 2006) – remarkably lowering energy expenditure (Junghans et al. 2006). Moreover, previous studies have clearly demonstrated that AKG increases the levels of serum and intramuscular free amino acids and improves protein metabolism in pig muscles under protein-restricted conditions (Chen et al. 2018).

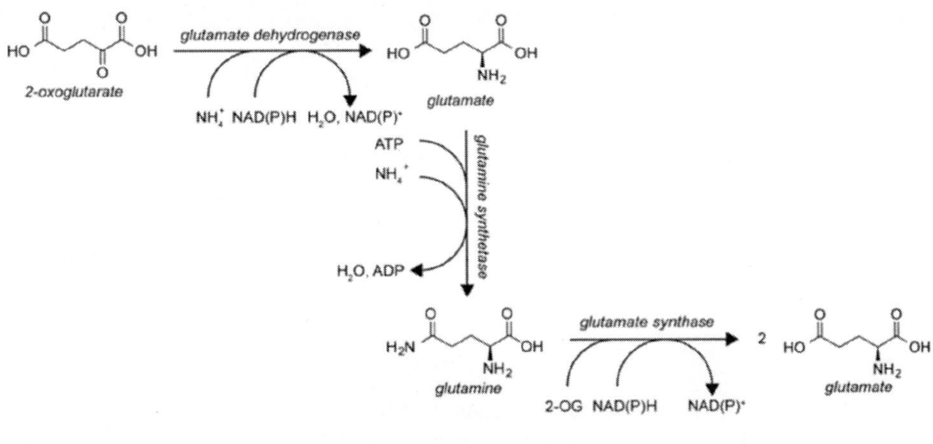

FIGURE 8.3 Glutamate biosynthetic pathways. Note: Glutamate is made from the citric acid cycle intermediate 2-oxoglutarate (2-OG) by reductive amination with either ammonium or glutamine as the nitrogen source (Walker and van der Donk 2016, Copyright License No. 5718191284069).

AKG participates in anabolic pathways in peripheral tissues, with up to 20% of dietary AKG recoverable in the blood (Kristensen et al. 2002). However, the origin and role of plasma AKG in whole-body metabolism, especially in the gut, remain largely unclear. The potential mechanisms of host mitochondrial AKG migration to the plasma also remain unknown. Theoretically, the average person produces more AKG in the mitochondria than their total body weight. Remarkably, the availability and transportation of this mitochondrial intermediate out of the mitochondria depend on the inflow of subsequent Krebs cycle intermediates, which simultaneously provide the AKG transporter on the mitochondrial membrane. This process is illustrated in Figure 8.2, where malate inflow into the mitochondria, regulated by the 2-oxoglutarate carrier (OGC), is a critical and limiting step for AKG outflow (Monné et al. 2013).

To date, AKG has been shown to participate in more than 3000 bodily reactions, positioning it as a potential 'molecule of life' that may ensure longevity. Numerous dicarboxylic acid sodium-dependent transporters (NaDC-1) are found within the gut (Buddington et al. 2004). However, most of the food we ingest daily contains little to no AKG. Why then do AKG transporters exist within the gut mucosa? Are they waiting for a miracle or for enterally produced AKG? Notably, the NaDC-1 transporter and AKG receptor GPR99, typically found in AKG-responsive cells (He et al. 2004), are present in the brain, kidneys, and other tissues. Is there perhaps another source of AKG from the gut?

During aging, gut bacteria change drastically (Ghosh, Shanahan, and O'Toole 2022; O'Toole, Paul, and Claesson 2010). Some of the signs of aging like malnutrition, osteoporosis, or intellectual disability – all related to reduced protein synthesis – are often overlooked in animal studies examining enteral AKG. It is compelling to postulate that age-related changes to the bacterial community may be responsible for decreased bacterial production of AKG and its availability for the host. In our own studies, we have documented that the concentration of AKG in the blood decreases ten-fold between the ages of 40 and 80 (Harrison and Pierzynowski 2008; Pierzynowski and Pierzynowska 2022), which leads to a deterioration of all the processes in which AKG is involved.

AKG VS. PROTEIN MALNUTRITION AND PANCREATIC ENZYMES

Malnutrition caused by maldigestion and/or malabsorption is recognized as one of the signs of aging (Rémond et al. 2015). Among other factors, the decreased stomach and pancreatic enzyme secretion emerges as the primary factor causing malnutrition (Rémond et al. 2015). Feldman postulated that between 70 and 90 years of age, basal pepsin secretion can diminish up to four-fold (Feldman et al. 1996). Studies conducted in humans showed that compared to young controls, subjects above 65–70 years old had significantly reduced bicarbonate and enzyme secretions, due to both a decrease in secreted volume and enzyme concentrations (Vellas et al. 1988; Ishibashi et al. 1991; Laugier et al. 1991; Sato et al. 2012). In a series of experiments by our lab, we showed that AKG applied enterally stimulates trypsin production in a pig model (Pierzynowski and Pierzynowska – data not published).

So, AKG acts indirectly as an anti-malnutrition factor, improving pancreatic enzyme production and supporting proteolysis of dietary protein to amino acids. At the same time, AKG improves pancreatic enzyme secretion, which ensures the plastein reaction-dependent synthesis of digestive peptides (Northrop 1947; Gong et al. 2015). Recent studies have proven that dietary proteins in young creatures are absorbed up to 80% in the form of peptides (Shen and Matsui 2019).

POSSIBLE IMPLICATIONS OF AKG AND PANCREATIC ENZYMES FOR DEGENERATIVE BRAIN DISEASES

If one combines the above-described age-dependent reduction in blood AKG levels – probably because of lowered AKG production by the gut bacteria – and the discussed deterioration in stomach and pancreas enzyme secretion during aging as a causative factor of malnutrition, it would be possible to accept the following:

1. Aging bacterial communities produce limited AKG for host usage.
2. Reduced blood levels of AKG can negatively affect thousands of metabolic reactions in which AKG is usually involved, including neurogenesis, protein metabolism, bone collagen synthesis (osteoporosis), etc.
3. Enterally applied, exogenous AKG can fulfill AKG host requirements and improve gut enzymology, which also slows down during aging.

These three points are summarized in the experiment performed by Goncharova (now Pierzynowska) et al. (2015), in which an excess of enterally supplied AKG was explored in aging Mongolian gerbils, in order to ensure, e.g., neuronal regeneration and behavior (Goncharova et al. 2015) (Figure 8.4). Neuronal regeneration is the goal of many treatments for neurodegenerative diseases.

The mix of alpha-ketoglutaric acid calcium and sodium salts (1:1) and a mixture of pancreatic-like enzymes of microbial origin (PLEM) consisting of a protease, a lipase, and an amylase were tested as feed additives on aging Mongolian gerbils. Following a six-month study, the animals were sacrificed, and the hippocampi were removed and analyzed to quantify hippocampal neurogenesis and ultrastructure.

Researchers used nestin as a marker for neural stem/progenitor cells, while NeuN was used as a marker for neurons. The investigation revealed two main populations of pyramidal neurons in the CA1 hippocampal area. One population was only NeuN-positive, whereas the second demonstrated both NeuN and nestin-positive staining. The number of nestin/NeuN-positive neurons was significantly less than the number of NeuN-positive cells and varied between experimental groups. The double-positive staining for nestin and NeuN and structural similarities with pyramidal neurons allowed the classification of this type of cell as newly generated neurons. In some cells, long processes that seemed to be dendrites extended into deeper layers of the hippocampus, and these processes were nestin-positive as well.

The data on synaptic plasticity obtained after ultrastructural and biochemical analysis of the CA1 hippocampal area demonstrated that diet supplementation with AKG and/or PLEM significantly reversed the age-related changes in the pool of excitatory synapses and brought the synaptic parameter values closer to those observed in the young control group of gerbils.

During the treatment, the animals were subjected to a T-maze spontaneous alternation test assessing learning and memory abilities. Intriguingly, the results perfectly correlated with the data obtained during the morphological brain investigation, suggesting that changes in neurogenesis and synaptic plasticity yielded an improved learning capacity. Irrespective of the mechanism, the experiment shows that AKG, PLEM, and AKG + PLEM improved cognitive function in old animals.

The results indicated that both AKG and PLEM supplementation synergistically increased hippocampal neurogenesis and improved synaptic plasticity compared to age-matched controls (old controls) (see Figure 8.4). Notably, the combination of AKG + PLEM was even more effective, resulting in a synergistic increase in neurogenesis (the AKG + PLEM group was significantly different from both the AKG and PLEM groups). The level of neurogenesis achieved by combination therapy brought the level of neurogenesis in the old animals to the level of the young animals.

In conclusion, since both regeneration and synaptic terminal formation are predominantly dependent on new protein synthesis, we suggest that the observed synergistic effect of AKG and enzymes is mostly dependent on the proteinase and AKG used in the experiment. In some ways, the described morphological effects observed are manifested in the results of the T-maze cognitive test. Thus, protease and AKG have an additive action on neuron regeneration and cognitive performance, approaching values obtained in untreated young gerbils. AKG produced by gut bacteria and pancreatic proteases are both important factors regulating the participation of dietary protein in cognition and protecting the brain from degenerative diseases.

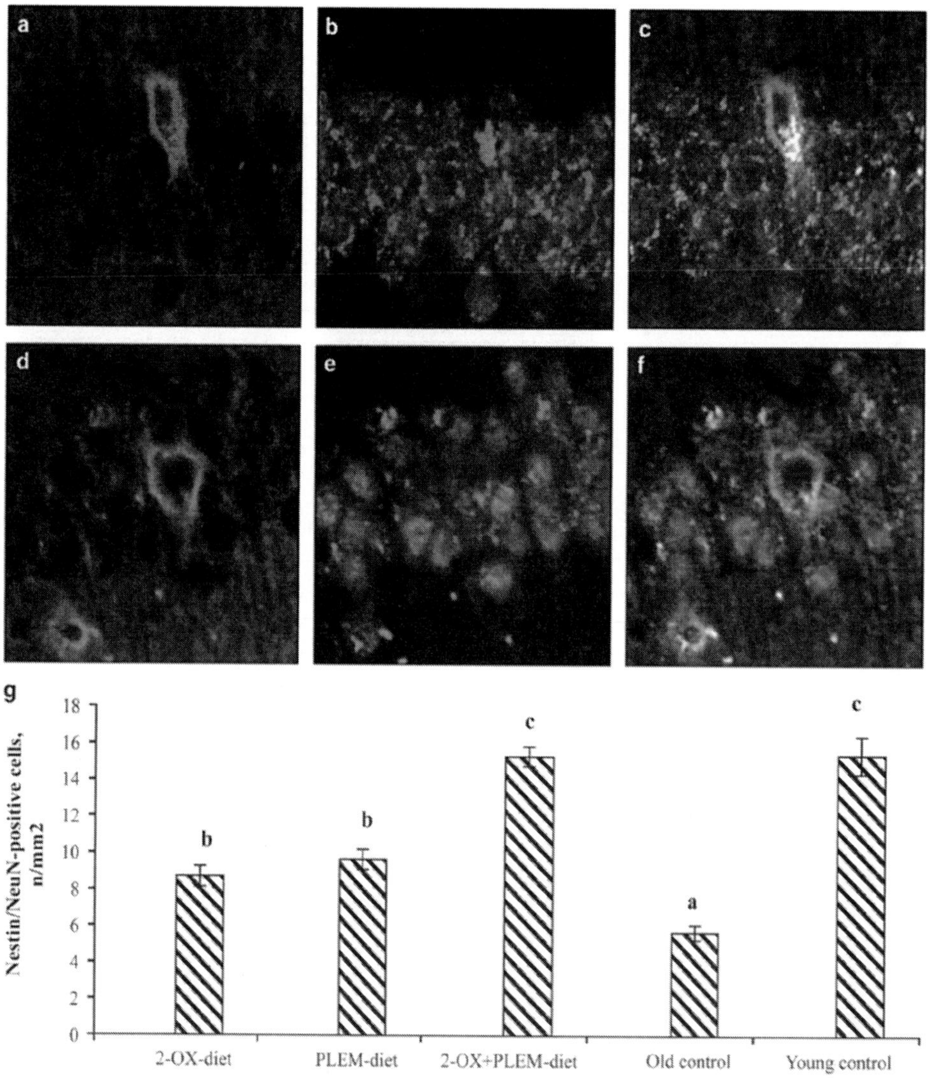

FIGURE 8.4 Immunohistochemical analysis of hippocampal CA1 area. Note: (a–f) Consistency of nestin (violet) and NeuN (green) staining in the CA1 hippocampal area of MGs. (a–c) Old control group. (d–f) 2-OX+PLEM group. (a and d) Nestin-positive cells (purple). (b and e) NeuN-positive cells (green), (c and f) composite merged images. (g) Number of Nestin/NeuN-positive cells per 1 mm2 of stratum pyramidale in hippocampal CA1 area. 2-OX-diet group – old MGs treated with 2-OX-enriched diet (n = 14); PLEM-diet group – old MGs treated with PLEM-enriched diet (n = 14); 2-OX+PLEM group – old MGs treated with both 2-OX and PLEM (n = 17); old control group – control old MGs (n = 18); young control group – young MGs (6 months; n = 6). Data presented as mean ± s.d. a,b,c Mean values with unlike letters were significantly different between the groups ($P \leqslant 0.05$) – from (Goncharova et al. 2015, Creative Commons CC-BY license).

CURATIVE EFFECTS OF AKG ON OSTEOPOROSIS AND BONE DEGENERATION DURING THE PLACEMENT OF DENTAL IMPLANTS

AKG is a potentially beneficial agent for addressing osteoporosis and bone degeneration (Tatara et al. 2003; Filip et al. 2007; Dobrowolski et al. 2008). This is particularly significant in cases where the body's capacity to support bone tissue is impaired, such as in diseases or disruptions in homeostasis. Consequently, AKG supplementation is believed to be more effective in augmenting the body's ability to maintain and repair bone tissues.

AKG plays a critical role in collagen stability and formation, especially under osteoporotic conditions where bone density and integrity are compromised. It acts as a precursor for amino acids such as proline and lysine, which are essential for collagen's structural integrity. This aspect of AKG function is particularly important in the context of dental implantation, in which bone quality has a significant impact on the success of the procedure. In addition, energy metabolism facilitated by AKG supports overall protein synthesis, which is critical for bone regeneration and healing after implantation.

Furthermore, AKG's interaction with the body's protein metabolism extends to its influence on cellular energy pathways, proving particularly valuable when normal homeostasis is disrupted. During illness or recovery from surgery, AKG supplementation can provide a much-needed boost to the natural processes of bone maintenance and repair. In conditions of compromised health, AKG is a more potent agent that counteracts the effects of bone degeneration and promotes recovery, making it an essential component in the treatment of osteoporosis and enhancing the success of dental implant procedures.

Overall, AKG supplementation is emerging as an important strategy in the management of bone health, particularly in osteoporosis and dental implant procedures. Its multifaceted role in improving bone quality and aiding recovery during periods of physical stress or disease underscores its potential as a therapeutic agent in these critical areas of healthcare.

Filip et al. (2007) demonstrated the reversal of osteopenia when Ca-AKG was given *per os* in studies on menopausal women (Filip et al. 2007).

Tatara et al. (2003), in studies on turkeys, showed that enterally administered AKG, offered as feed additives, speeds up the process of healing in broken bones (Tatara et al. 2003).

Dobrowolski et al. (2008) showed that calvariae of rats subjected to gastrectomy and treated with either AKG or vehicle exhibited far better bone regeneration compared to untreated animals (Figure 8.5) (Dobrowolski et al. 2008).

The most spectacular effect of AKG as a feed additive was observed in an experiment performed on osteoporotic mice, examining the osseointegration of titanium implants with and without AKG supplementation in drinking water during a 21-day study (Liu et al. 2023). The authors showed

FIGURE 8.5 Transillumination photos of calvariae of rats subjected to gastrectomy (Gx) or sham operation (Sham) and treated with either alpha-ketoglutarate (AKG) or vehicle. Note: Experimental groups: Sham Vehicle, Gx Vehicle, Sham AKG, Gx AKG. Notice the presence of lacunae (rich in hematopoietic tissue) in the calvaria of the Gx rat (from Dobrowolski et al. 2008, Copyright License No. 5718200658254).

that AKG supplementation prevented implant failure due to enhanced osseointegration and inhibited osteoclastogenesis. AKG was identified as one of the main regulators of bone marrow-derived mesenchymal stem cell (BMMSC) proliferation and differentiation (Yang et al. 2018), which are considered crucial in the process of osseointegration. In their study, Liu et al. demonstrated that AKG promotes BMMSC proliferation and osteogenic differentiation, enhances the expression of osteogenic genes, and inhibits cell apoptosis (Liu et al. 2023).

These results are confirmed by our own studies on rats where different models of titanium screws were tested for their osseointegration (Wychowański et al. – in press). Osseointegration of

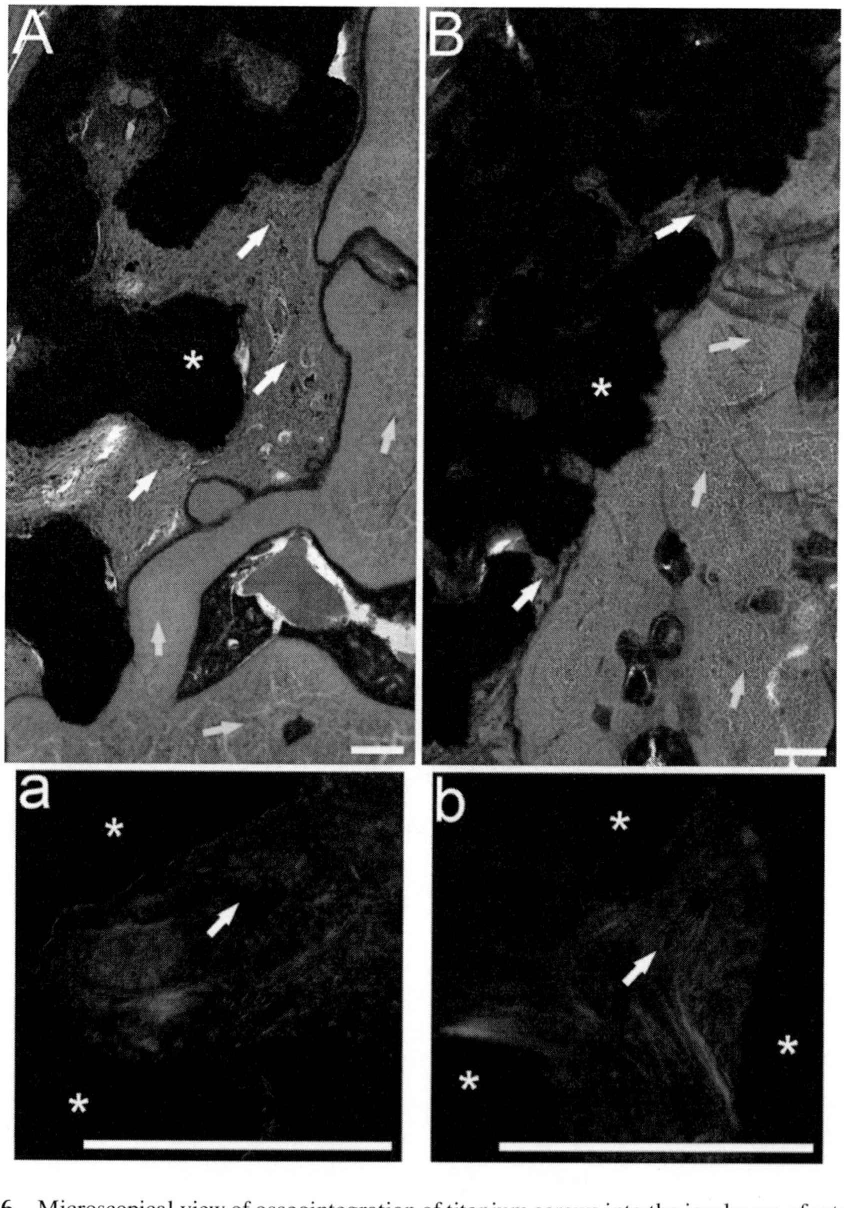

FIGURE 8.6 Microscopical view of osseointegration of titanium screws into the jaw bones of rats. Note: (A) Example of osseointegration of titanium screw with bone in rat not treated with AKG. (B) Example of osseointegration of titanium screw with bone in rat treated with AKG. (a) Example of fibrous stroma of forming bone in contact with titanium screw in rat not treated with AKG. (b) Example of fibrous stroma of forming bone in contact with titanium screw in rat treated with AKG.

the titanium screws was far better in the rats receiving 1% Na_2AKG additives in their drinking water for two months, as compared to controls (Figure 8.6). We revealed an active healing process and bone ingrowth, as well as a more mature collagen fiber structure, and therefore suggest that AKG supplementation not only accelerates the healing process but also aids in the formation of mature load-bearing connective tissue structures (Figure 8.6).

Osseointegration with titanium screws in rats receiving AKG is characterized by a significantly more mature structure of collagen fibers, and the preparations show a more directed arrangement of collagen fiber bundles. This suggests the directions of the acting force vectors in the tissue surrounding and connecting to the implant. It seems that the percentage of fine-fibrillar collagen in rats not receiving AKG is higher and, ergo, less mature connective tissue. Most likely, AKG accelerates the healing and shaping processes of connective tissue – shaping towards mature structures that can bear the load.

MINI DICTIONARY OF TERMS

Alpha-ketoglutarate: An intermediate in the Krebs cycle.

Gut microbiota/gut microbiome/gut bacteriome/gut flora: The community of microorganisms, including bacteria, archaea, fungi, and viruses, that live in the digestive tracts of animals.

Hippocampal neurogenesis: A remarkable – sometimes controversial – form of brain structural plasticity in adult mammals, in which new functional neurons are generated from adult neural stem cells.

Neurodegenerative disease: A condition characterized by the progressive loss of structure or function of neurons, potentially leading to cell death.

Osteoporosis: A systemic skeletal disorder characterized by low bone mass and microarchitectural deterioration of trabecular bone tissue, leading to an increased risk of bone fracture.

Synaptic plasticity: The ability of neurons to modify their connections, especially following different types of brain damage.

T-maze spontaneous alternation test: A physiological test for measuring exploratory behavior and memory, especially in rodent models for Central Nervous System (CNS) disorders. Rodents are naturally willing to explore a new environment.

FUTURISTIC EPILOGUE

The evolution from a caecum to an appendix in hominids most probably took place because brain-dependent conscious cognition impacted the evolutionary processes. This reduced the evolutionary impact of classical unconscious automatic metabolic cognition, resulting in reduced gut size and, as a consequence, a reduction in gut bacteria number and type.

One can postulate that from that specific moment in hominids, the brain/awareness began regulating the body's response to dietary protein availability. This may explain why seasonal changes in protein availability – such as during dry vs. wet seasons or in spring vs. winter – govern gut bacteria communities in wildlife both qualitatively and quantitatively via classic metabolic cognition. Thus, the main driving force for gut bacteria evolution in humans, domesticated husbandry animals, and synanthropic animals is not natural selection but ideas/awareness which, together with unconscious metabolic cognition, govern the development of a nitrogen metabolism overfed with proteins.

Healthy aging (HA) is, in fact, a contradictory concept – aging is a degenerative sickness. Any form of life must eventually die. Sickness, illness, and the 'wear and tear' of living body parts speed up that process. Thus, HA is simply not possible and is somewhat misleading. However, quite a good alternative to the contradictory HA slogan is that of a 'happy longevity' (HL) and its dependency on the 'molecule of life,' AKG.

SUMMARY

1. Alpha-ketoglutarate (AKG) is the indispensable precursor of amino acid *de novo* synthesis, both in the plant and animal kingdom, as it is a scavenger of ammonium ions, which in turn come from protein metabolism or indirectly from air.
2. The biological action of AKG originating from the gut can prevent nervous system and bone degeneration as well as other degenerative sicknesses, e.g., cancer.
3. Integration of titanium teeth/dental implants into bone, as well as collagen production for bone regeneration, is enhanced by enteral AKG.
4. AKG produced in the cytoplasm of the bacteriome is probably available for host metabolism.
5. Longevity can be ensured by the maintenance of a 'healthy' bacteriome community, as well as through the supplementation of exogenous AKG.

REFERENCES

Allison, M. J., I. M. Robinson, and A. L. Baetz. 1979. "Synthesis of Alpha-Ketoglutarate by Reductive Carboxylation of Succinate in Veillonella, Selenomonas, and Bacteriodes Species." *Journal of Bacteriology* 140 (3): 980–86. doi:10.1128/jb.140.3.980-986.1979.

Artturi Virtanen. n.d. "The Biological Fixation of Nitrogen and the Preservation of Fodder in Agriculture, and Their Importance to Human Nutrition." Nobel Lecture 1945. https://www.nobelprize.org/uploads/2018/06/virtanen-lecture.pdf NobelPrize.Org. Nobel Prize Outreach AB 2023. Tuesday, 26 December 2023.

Bayliak, Maria M., and Volodymyr I. Lushchak. 2021. "Pleiotropic Effects of Alpha-Ketoglutarate as a Potential Anti-Ageing Agent." *Ageing Research Reviews* 66 (March): 101237. doi:10.1016/j.arr.2020.101237.

Besten, Gijs den, Karen van Eunen, Albert K. Groen, Koen Venema, Dirk-Jan Reijngoud, and Barbara M. Bakker. 2013. "The Role of Short-Chain Fatty Acids in the Interplay between Diet, Gut Microbiota, and Host Energy Metabolism." *Journal of Lipid Research* 54 (9): 2325–40. doi:10.1194/jlr.R036012.

Buddington, Randal K., Ana Pajor, Karyl K. Buddington, and Stefan Pierzynowski. 2004. "Absorption of Alpha-Ketoglutarate by the Gastrointestinal Tract of Pigs." *Comparative Biochemistry and Physiology. Part A, Molecular & Integrative Physiology* 138 (2): 215–20. doi:10.1016/j.cbpb.2004.03.007.

Chen, Jiashun, Wenxuan Su, Baoju Kang, Qian Jiang, Yurong Zhao, Chenxing Fu, and Kang Yao. 2018. "Supplementation with α-Ketoglutarate to a Low-Protein Diet Enhances Amino Acid Synthesis in Tissues and Improves Protein Metabolism in the Skeletal Muscle of Growing Pigs." *Amino Acids* 50 (11): 1525–37. doi:10.1007/s00726-018-2618-3.

Dobrowolski, Piotr J., Tomasz Piersiak, Vikas V. Surve, Danuta Kruszewska, Antoni Gawron, Paulina Pacuska, Rolf Håkanson, and Stefan G. Pierzynowski. 2008. "Dietary α-Ketoglutarate Reduces Gastrectomy-Evoked Loss of Calvaria and Trabecular Bone in Female Rats." *Scandinavian Journal of Gastroenterology* 43 (5): 551–58. doi:10.1080/00365520701824951.

Feldman, M., B. Cryer, K. E. McArthur, B. A. Huet, and E. Lee. 1996. "Effects of Aging and Gastritis on Gastric Acid and Pepsin Secretion in Humans: A Prospective Study." *Gastroenterology* 110 (4): 1043–52. doi:10.1053/gast.1996.v110.pm8612992.

Filip, Pierzynowski, Lindegard, Wernerman, Haratym-Maj, and Podgurniak. 2007. "Alpha-Ketoglutarate Decreases Serum Levels of C-Terminal Cross-Linking Telopeptide of Type I Collagen (CTX) in Postmenopausal Women with Osteopenia: Six-Month Study." *International Journal for Vitamin and Nutrition Research* 77 (2): 89–97. doi:10.1024/0300-9831.77.2.89.

Ghosh, Tarini Shankar, Fergus Shanahan, and Paul W. O'Toole. 2022. "The Gut Microbiome as a Modulator of Healthy Ageing." *Nature Reviews Gastroenterology & Hepatology* 19 (9): 565–84. doi:10.1038/s41575-022-00605-x.

Goncharova, K., G. Skibo, T. Kovalenko, I. Osadchenko, G. Ushakova, M. Vovchanskii, and S. G. Pierzynowski. 2015. "Diet-Induced Changes in Brain Structure and Behavior in Old Gerbils." *Nutrition & Diabetes* 5 (6): e163–63. doi:10.1038/nutd.2015.13.

Gong, Min, Aishwarya Mohan, Angus Gibson, and Chibuike C. Udenigwe. 2015. "Mechanisms of Plastein Formation, and Prospective Food and Nutraceutical Applications of the Peptide Aggregates." *Biotechnology Reports* 5 (March): 63–69. doi:10.1016/j.btre.2014.12.003.

Harrison, A. P., and S. G. Pierzynowski. 2008. "Biological Effects of 2-Oxoglutarate with Particular Emphasis on the Regulation of Protein, Mineral and Lipid Absorption/Metabolism, Muscle Performance, Kidney Function, Bone Formation and Cancerogenesis, All Viewed from a Healthy Ageing Perspective State of the Art--Review Article." *Journal of Physiology and Pharmacology : An Official Journal of the Polish Physiological Society* 59 Suppl 1 (August): 91–106.

He, Weihai, Frederick J.-P. Miao, Daniel C.-H. Lin, Ralf T. Schwandner, Zhulun Wang, Jinhai Gao, Jin-Long Chen, Hui Tian, and Lei Ling. 2004. "Citric Acid Cycle Intermediates as Ligands for Orphan G-Protein-Coupled Receptors." *Nature* 429 (6988): 188–93. doi:10.1038/nature02488.

Helling, R. B. 1994. "Why Does Escherichia Coli Have Two Primary Pathways for Synthesis of Glutamate?" *Journal of Bacteriology* 176 (15): 4664–68. doi:10.1128/jb.176.15.4664-4668.1994.

Ishibashi, Tadaaki, Shuji Matsumoto, Hideo Harada, Koji Ochi, Juntaro Tanaka, Toshinobu Seno, Hiroo Oka, Hirofumi Miyake, and Ikuro Kimura. 1991. "Aging and Exocrine Pancreatic Function Evaluated by the Recently Standardized Secretin Test." *Nippon Ronen Igakkai Zasshi. Japanese Journal of Geriatrics* 28 (5): 599–605. doi:10.3143/geriatrics.28.599.

Jouany, J. P., and K. Ushida. 1999. "The Role of Protozoa in Feed Digestion - Review." *Asian-Australasian Journal of Animal Sciences* 12 (1): 113–28. doi:10.5713/ajas.1999.113.

Junghans, Peter, Michael Derno, Stefan Pierzynowski, Ulf Hennig, Paul Eberhard Rudolph, and Wolfgang B. Souffrant. 2006. "Intraduodenal Infusion of α-Ketoglutarate Decreases Whole Body Energy Expenditure in Growing Pigs." *Clinical Nutrition* 25 (3): 489–96. doi:10.1016/j.clnu.2005.11.003.

Kristensen, N. B., H. Jungvid, J. A. Fernández, and S. G. Pierzynowski. 2002. "Absorption and Metabolism of **α** -ketoglutarate in Growing Pigs." *Journal of Animal Physiology and Animal Nutrition* 86 (7–8): 239–45. doi:10.1046/j.1439-0396.2002.00380.x.

Lambert, Barry D., Barbara Stoll, Harri Niinikoski, Douglas G. Burrin, and Stefan Pierzynowski. 2002. "Net Portal Absorption of Enterally Fed α-Ketoglutarate Is Limited in Young Pigs." *The Journal of Nutrition* 132 (11): 3383–86. doi:10.1093/jn/132.11.3383.

Lambert, Barry D., Rafal Filip, Barbara Stoll, Peter Junghans, Michael Derno, Ulf Hennig, Wolfgang B. Souffrant, Stefan Pierzynowski, and Douglas G. Burrin. 2006. "First-Pass Metabolism Limits the Intestinal Absorption of Enteral α-Ketoglutarate in Young Pigs." *The Journal of Nutrition* 136 (11): 2779–84. doi:10.1093/jn/136.11.2779.

Laugier, R., J. P. Bernard, P. Berthezene, and P. Dupuy. 1991. "Changes in Pancreatic Exocrine Secretion with Age: Pancreatic Exocrine Secretion Does Decrease in the Elderly." *Digestion* 50 (3–4): 202–11. doi:10.1159/000200762.

Liu, Ruojing, Yi Gao, Li Huang, Bing Shi, Xing Yin, and Shujuan Zou. 2023. "Alpha-ketoglutarate Up-regulates Autophagic Activity in Peri-implant Environment and Enhances Dental Implant Osseointegration in Osteoporotic Mice." *Journal of Clinical Periodontology* 50 (5): 671–83. doi:10.1111/jcpe.13784.

Lo, T. C., and M. A. Bewick. 1978. "The Molecular Mechanisms of Dicarboxylic Acid Transport in Escherichia Coli K12. The Role and Orientation of the Two Membrane-Bound Dicarboxylate Binding Proteins." *The Journal of Biological Chemistry* 253 (21): 7826–31.

Monné, Magnus, Daniela Valeria Miniero, Faustino Bisaccia, and Giuseppe Fiermonte. 2013. "The Mitochondrial Oxoglutarate Carrier: From Identification to Mechanism." *Journal of Bioenergetics and Biomembranes* 45 (1–2): 1–13. doi:10.1007/s10863-012-9475-7.

Naeini, Saghi Hakimi, Laleh Mavaddatiyan, Zahra Rashid Kalkhoran, Soroush Taherkhani, and Mahmood Talkhabi. 2023. "Alpha-Ketoglutarate as a Potent Regulator for Lifespan and Healthspan: Evidences and Perspectives." *Experimental Gerontology* 175 (May): 112154. doi:10.1016/j.exger.2023.112154.

Neis, Evelien, Cornelis Dejong, and Sander Rensen. 2015. "The Role of Microbial Amino Acid Metabolism in Host Metabolism." *Nutrients* 7 (4): 2930–46. doi:10.3390/nu7042930.

Niemiec, T., J. Sikorska, A. Harrison, M. Szmidt, E. Sawosz, E. Wirth-Dzieciolowska, J. Wilczak, and S. Pierzynowski. 2011. "Alpha-Ketoglutarate Stabilizes Redox Homeostasis and Improves Arterial Elasticity in Aged Mice." *Journal of Physiology and Pharmacology: An Official Journal of the Polish Physiological Society* 62 (1): 37–43.

Northrop, John H. 1947. "Plastein From Pepsin and Trypsin." *Journal of General Physiology* 30 (5): 377–78. doi:10.1085/jgp.30.5.377.

O'Toole, Paul W., and Marcus J. Claesson. 2010. "Gut Microbiota: Changes throughout the Lifespan from Infancy to Elderly." *International Dairy Journal* 20 (4): 281–91. doi:10.1016/j.idairyj.2009.11.010.

Pierzynowski, Stefan, and Kateryna Pierzynowska. 2022. "Alpha-Ketoglutarate, a Key Molecule Involved in Nitrogen Circulation in Both Animals and Plants, in the Context of Human Gut Microbiota and Protein Metabolism." *Advances in Medical Sciences* 67 (1): 142–47. doi:10.1016/j.advms.2022.02.004.

Radzki, Radoslaw P., Marek Bieńko, and Stefan G. Pierzynowski. 2009. "Effect of Dietary Alpha-ketoglutarate on Blood Lipid Profile during Hypercholesterolaemia in Rats." *Scandinavian Journal of Clinical and Laboratory Investigation* 69 (2): 175–80. doi:10.1080/00365510802464633.

Rémond, Didier, Danit R. Shahar, Doreen Gille, Paula Pinto, Josefa Kachal, Marie-Agnès Peyron, Claudia Nunes Dos Santos, et al. 2015. "Understanding the Gastrointestinal Tract of the Elderly to Develop Dietary Solutions That Prevent Malnutrition." *Oncotarget* 6 (16): 13858–98. doi:10.18632/oncotarget.4030.

Rutten, Erica P. A., Marille P. K. J. Engelen, Annemie M. W. J. Schols, and Nicolaas E. P. Deutz. 2005. "Skeletal Muscle Glutamate Metabolism in Health and Disease: State of the Art." *Current Opinion in Clinical Nutrition and Metabolic Care* 8 (1): 41–51. doi:10.1097/00075197-200501000-00007.

Sato, Tomohiro, Katsuyoshi Ito, Tsutomu Tamada, Teruki Sone, Yasufumi Noda, Atsushi Higaki, Akihiko Kanki, Daigo Tanimoto, and Hiroki Higashi. 2012. "Age-Related Changes in Normal Adult Pancreas: MR Imaging Evaluation." *European Journal of Radiology* 81 (9): 2093–98. doi:10.1016/j.ejrad.2011.07.014.

Shen, Weilin, and Toshiro Matsui. 2019. "Intestinal Absorption of Small Peptides: A Review." *International Journal of Food Science & Technology* 54 (6): 1942–48. doi:10.1111/ijfs.14048.

Tatara, Marcin R., Grzegorz S. Pierzynowski, Piotr Majcher, Witold Krupski, Agata Wawrzyniak-Gacek, Rafał Filip, Piotr Silmanowicz, and Tadeusz Studziński. 2003. "The Influence of Alpha-Ketoglutarate (AKG) on Mineralisation, Mechanical and Structural Properties of Ulna in Turkey under Conditions of Osteotomy and Denervation." *Ortopedia, Traumatologia, Rehabilitacja* 5 (5): 666–72.

Vellas, B., D. Balas, J. Moreau, M. Bouisson, F. Senegas-Balas, M. Guidet, and A. Ribet. 1988. "Exocrine Pancreatic Secretion in the Elderly." *International Journal of Pancreatology: Official Journal of the International Association of Pancreatology* 3 (6): 497–502. doi:10.1007/BF02788208.

Walker, Mark C., and Wilfred A. van der Donk. 2016. "The Many Roles of Glutamate in Metabolism." *Journal of Industrial Microbiology and Biotechnology* 43 (2–3): 419–30. doi:10.1007/s10295-015-1665-y.

Yang, Ruili, Tingting Yu, Xiaoxing Kou, Xiang Gao, Chider Chen, Dawei Liu, Yanheng Zhou, and Songtao Shi. 2018. "Tet1 and Tet2 Maintain Mesenchymal Stem Cell Homeostasis via Demethylation of the P2rX7 Promoter." *Nature Communications* 9 (1): 2143. doi:10.1038/s41467-018-04464-6.

Yao, Kang, Yulong Yin, Xilong Li, Pengbin Xi, Junjun Wang, Jian Lei, Yongqing Hou, and Guoyao Wu. 2012. "Alpha-Ketoglutarate Inhibits Glutamine Degradation and Enhances Protein Synthesis in Intestinal Porcine Epithelial Cells." *Amino Acids* 42 (6): 2491–500. doi:10.1007/s00726-011-1060-6.

Yuan, Jie, Christopher D. Doucette, William U. Fowler, Xiao-Jiang Feng, Matthew Piazza, Herschel A. Rabitz, Ned S. Wingreen, and Joshua D. Rabinowitz. 2009. "Metabolomics-driven Quantitative Analysis of Ammonia Assimilation in *E. Coli*." *Molecular Systems Biology* 5 (1). doi:10.1038/msb.2009.60.

Zdzisińska, Barbara, Aleksandra Żurek, and Martyna Kandefer-Szerszeń. 2017. "Alpha-Ketoglutarate as a Molecule with Pleiotropic Activity: Well-Known and Novel Possibilities of Therapeutic Use." *Archivum Immunologiae et Therapiae Experimentalis* 65 (1): 21–36. doi:10.1007/s00005-016-0406-x.

9 Protein Quality
Features, Measurements, and Uses in Evaluating Diets

Kaori Kinoshita and Rei Otsuka

ABBREVIATIONS

CH_4	Methane
CO_2	Carbon dioxide
DIAAS	Digestible indispensable amino acid score
FAO	Food and Agriculture Organization of the United Nations
GHGE	Greenhouse gases
N_2O	Nitrous oxide
PDCAAS	Protein digestibility-corrected amino acid score
UNU	United Nations University
WHO	World Health Organization

INTRODUCTION

Based on a systematic review of data from 195 countries, it is estimated that, in 2017, 11 million deaths and 255 million disability-adjusted life years were attributable to dietary factors (GBD 2017 Diet Collaborators 2019). A nutritionally balanced diet that provides essential nutrients in human-appropriate amounts, without inducing nutrient overload, is important for improving health.

In recent years, food production has been recognized as a cause of global environmental degradation. The EAT-*Lancet* Commission has called for sustainable food production to protect the global environment (Willett et al. 2019). Sustainable food production mandates no additional land use, protection of existing biodiversity, reduction of water consumption, and reduction of nitrogen and phosphorus pollution, without any increase in greenhouse gases (GHGE) (Willett et al. 2019). GHGE include carbon dioxide (CO_2), methane (CH_4), and nitrous oxide (N_2O), wherein CH_4 and N_2O values are converted to CO_2 equivalents and the total value is expressed as CO_2 emissions. The environmental impact of food production is defined according to the amount of CO_2 emissions associated with the generation of food calories from farming, processing, transport, and, ultimately, consumer use (i.e., GHGE/kcal) (de Vries and de Boer 2010). These gases are produced by ruminant animals such as cattle. Several studies have shown that, in farms, GHGE emissions from plants are lower than those from animals (Tilman and Clark 2014; Drewnowski et al. 2015). Sustainable diets recommend shifting the human diet from animal- to plant-based foods (Willett et al. 2019). However, plant-based foods generally have low quantities of essential nutrients, such as vitamin B_{12} and iodine (Selinger et al. 2019; Weikert et al. 2020, Bakaloudi et al. 2021) and contain phytic and oxalic acids that inhibit calcium and zinc absorption (Bakaloudi et al. 2021; Hansen et al. 2018; Weaver and Plawecki 1994; Saunders et al. 2013; Hunt 2003). Thus, plant-based foods may be deficient in certain nutrients and there is concern about the protein quality of plant-derived foods.

DOI: 10.1201/b23328-9

Protein quality (i.e., bioavailability) is evaluated based on amino acid composition and digestibility (Joint WHO/FAO/UNU Expert Consultation 2007). High-quality proteins possess both high digestibility and an adequate amino acid composition to meet human nutritional requirements. It is well known that the bioavailability of protein varies widely among food sources, as both amino acid composition and digestibility differ (Herreman et al. 2020; Moughan and Wolfe 2019). Proteins in plant-based foods are limited by one or more indispensable amino acids such as lysine, methionine, tryptophan, threonine, and leucine (i.e., limiting amino acids) (Wu et al. 2014). Protein-rich plant-based foods, such as legumes, contain fiber, which inhibits protein digestion and absorption. In contrast, animal-based foods contain high levels of indispensable amino acids, such as lysine, methionine, tryptophan, threonine, and leucine, which constitute limiting amino acids in plant-based food. Combining foods can compensate for the indispensable amino acids; however, their combination with other nutrients and cooking can affect the digestibility of proteins and alter their bioavailability. However, insufficient data are available to assess the bioavailability of mixed diets. Thus, assessing the dietary protein quality for mixed foods and its implementation is not easy. In this chapter, we review the methods for evaluating protein quality, current issues, and prospects.

INDISPENSABLE AMINO ACID MALNUTRITION AND EVALUATION OF DIETARY PROTEIN QUALITY

According to the Food and Agriculture Organization of the United Nations (FAO), chronic undernutrition among children is widespread in sub-Saharan Africa and Southern Asia, with diets in these regions heavily relying on cereals, such as sorghum, wheat, rice, and maize. (Joint FAO/IFAD/UNICEF/WFP/WHO Consultation 2018; Cervantes-Pahm et al. 2014; Shaheen et al. 2016; Abelilla et al. 2018). These cereals are deficient in essential amino acids, such as lysine (i.e., limiting amino acids). Moreover, the digestibility of proteins and amino acids in these cereals further decreases during the refining process (Duodu et al. 2002; FAO 2013). In 1933, Cicely Williams identified kwashiorkor as a serious protein deficiency-associated disease in children caused by a maize-based baby diet (Williams 1933). Against this background, the FAO and World Health Organization (WHO) have proposed a method for assessing the quality of food proteins. In 2012, the risk of protein deficiency in different countries was calculated by Ghosh et al. based on estimates of total protein and utilizable protein (Ghosh et al. 2012). Furthermore, in 2016, a relationship between chronically undernourished children or stunted children and low serum levels of nine indispensable amino acids was reported by Semba and colleagues (Semba et al. 2016). Dietary protein quality can be improved by mixed meals, which combine foods with limited amino acids and those that compensate for those limitations. Therefore, improving the diet by combining foods containing sufficient amounts of indispensable amino acids is crucial. In fact, the FAO has recognized the need for protein-quality evaluation to help the global population to meet their protein requirements and avoid malnutrition in order to ensure health and well-being (FAO 2013).

PROTEIN DIGESTION AND ABSORPTION

The digestibility of proteins in food influences protein quality. Digestibility is defined in terms of the amino acid balance through the small intestine (i.e., ileal digestibility) or through the entire intestine (i.e., fecal digestibility) and is calculated as the difference between intake and loss, which measures the extent to which food proteins are digested and absorbed as amino acids in the gastrointestinal tract and utilized by the body (Joint WHO/FAO/UNU Expert Consultation 2007).

Actual digestibility is complex because of nitrogen exchange in the form of proteins, amino acids, and urea in the systemic pool and in the intestinal lumen (Figure 9.1). As shown in Figure 9.2, proteins are hydrolyzed to amino acids and peptides by pepsin in gastric juice, trypsin, chymotrypsin, and elastase in pancreatic juice, and peptidase in intestinal juice. Peptides and amino acids resulting from enzymatic hydrolysis are absorbed through the small intestinal mucosa and

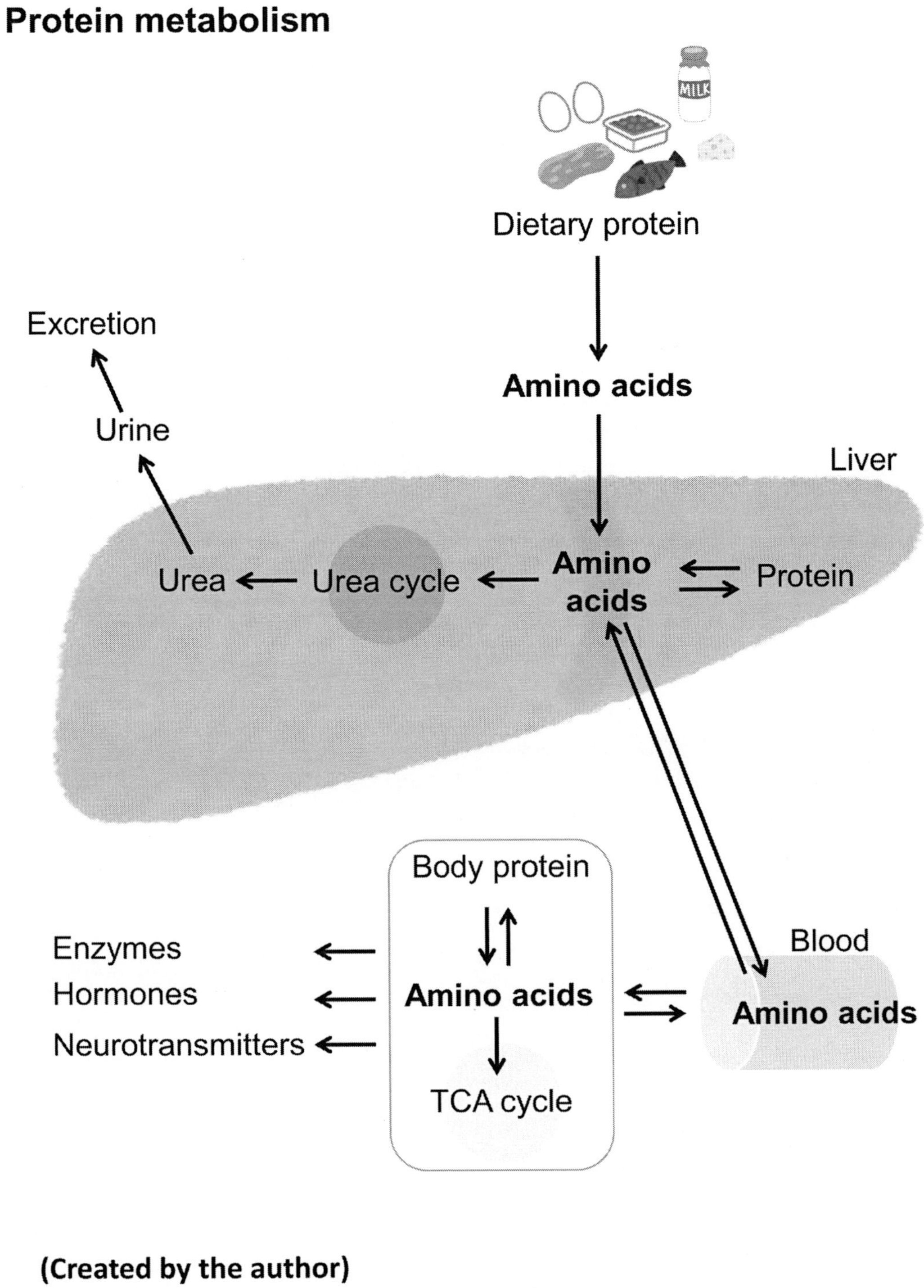

FIGURE 9.1 Protein metabolism. This figure illustrates protein metabolism. Dietary proteins are absorbed as amino acids, some of which are free amino acids. A certain amount of free amino acids is stored in the body. Amino acids are mainly used for protein synthesis, but when more than the required amount of amino acids is ingested, or when carbohydrates cannot be used as an energy source, amino acids are used as an energy source. As toxic ammonia is produced in this process, the liver converts ammonia to urea and excretes it.

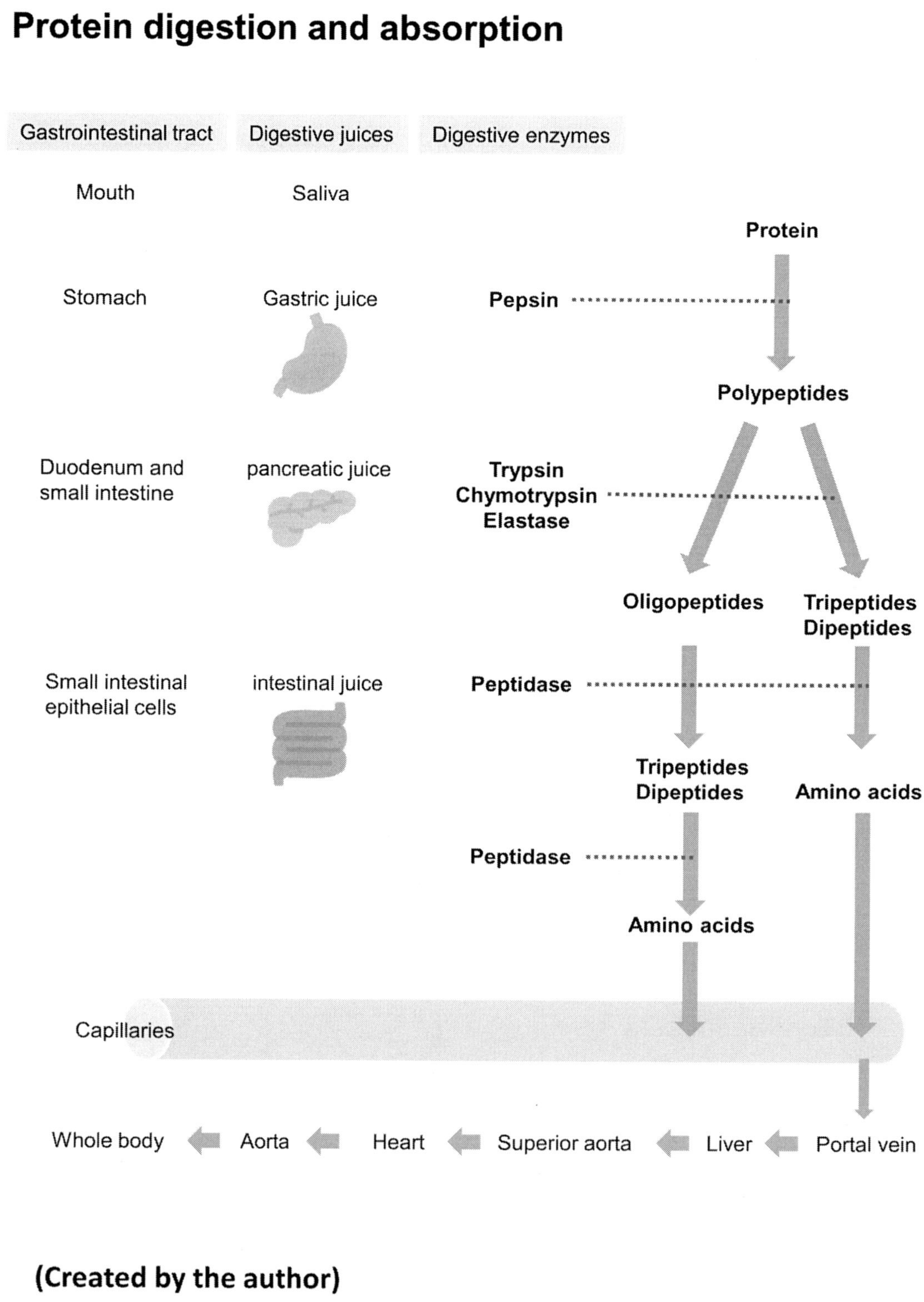

FIGURE 9.2 Protein digestion and absorption. This figure illustrates digestion and absorption of dietary proteins. Proteins are broken down by enzymes, such as pepsin, into amino acids and di- and tripeptides, which are rapidly absorbed through the small intestinal epithelium.

are generally considered to be almost completely absorbed at the end of the small intestine (i.e., terminal ileum) (Jahan-Mihan et al. 2011). Peptides and amino acids not absorbed in the terminal ileum are transported to the large intestine, where they are consumed by the gut microbiota. The gut microbiota produces some amino acids in the large intestine are produced during the digestion and fermentation of undigested proteins and peptides (van der Wielen et al. 2017). Thus, the gut microbiota may remove or add amino acids that pass through the terminal ileum in the large intestine.

Nonetheless, whether the absorption of peptides and amino acids is almost completed in the small intestine or continues in the large intestine depends on the food, resulting in variations in protein digestibility measurements between samples taken from the terminal ileum and those taken from the large intestine. These are important factors to consider when evaluating protein quality, as discussed below.

PROTEIN-QUALITY EVALUATION

The purpose of protein-quality evaluation is to determine whether individual foods and food-mixed diets meet the metabolic requirements of amino acids and nitrogen in the human body. Researchers have developed methods for determining protein quality, and commonly used principles of protein quality describe the ability to supply the body with sufficient indispensable amino acids. Therefore, protein quality indicates protein bioavailability (Joint WHO/FAO/UNU Expert Consultation 2007). Protein utilization is explained by its biological value (the extent to which the amino acids absorbed from the diet meet the required levels) and digestibility (the amount that can be effectively used after digestion and absorption) (Joint WHO/FAO/UNU Expert Consultation 2007).

Mitchell and Block reported that all amino acids must be supplied simultaneously for protein synthesis and indicated the concept of a chemical score as a measure of the nutritional value of a protein (Mitchell and Block 1946). The calculation of the chemical score is based on the indispensable amino acid content of the protein sources and their requirements, which are referred to as reference patterns. The absolute percentage of amino acids with the least amount of amino acids relative to the reference pattern is the chemical score, which is called the amino acid score. The least abundant amino acid limits body protein synthesis, and is therefore referred to as the first limiting amino acid. The amino acid score is calculated using the following formula (Joint WHO/FAO/UNU Expert Consultation 2007):

Amino acid score = amino acid (mg) in test protein (1g) ÷ amino acid (mg) in the reference pattern

The requirements for indispensable amino acids are summarized in a reference pattern containing all the required levels of indispensable amino acids per gram of protein based on 100% digestibility (Joint WHO/FAO/UNU Expert Consultation 2007). Reference patterns for the nine indispensable amino acids are provided for each age group to calculate amino acid scores (Joint WHO/FAO/UNU Expert Consultation 2007). Table 9.1 shows the reference patterns for the nine indispensable amino acids for calculating amino acid scores (WHO/FAO/UNU 2007). Both amino acid score and digestibility can be used to determine the quality of food proteins. The data used for digestibility include fecal protein digestibility and ileal amino acid digestibility.

THE PROTEIN DIGESTIBILITY-CORRECTED AMINO ACID SCORE (PDCAAS)

The PDCAAS is a protein-quality evaluation method that combines amino acid scores and fecal protein digestibility, first published by the FAO/WHO Expert Consultation in 1989. A subsequent WHO/FAO/United Nations University Expert Consultation on protein requirements updated and endorsed the PDCAAS methods (Joint WHO/FAO/UNU Expert Consultation 2007). The PDCAAS can be calculated using the following formula (Joint WHO/FAO/UNU Expert Consultation 2007):

TABLE 9.1
Reference Patterns for the Nine Indispensable Amino Acids for Calculating Amino Acid Scores

	Scoring Pattern mg/g Protein Requirement								
Age (yr)	His	Ile	Leu	Lys	SAA	AAA	Thr	Trp	Val
0.5	20	32	66	57	27	52	31	8.5	43
1–2	18	31	63	52	25	46	27	7.0	41
3–10	16	30	61	48	23	41	25	6.6	40
11–14	16	30	61	48	23	41	25	6.6	40
15–18	16	30	60	47	23	40	24	6.3	40
>18	15	30	59	45	22	38	23	6.0	39

His, Histidine; Ile, Isoleucine; Leu, Leucine; SAA, Sulfur amino acids; AAA, Aromatic amino acids; Thr, Threonine; Trp, Tryptophan; Val, Valine (WHO/FAO/UNU 2007).

$$PDCAAS = amino\ acid\ score \times fecal\ protein\ digestibility$$

The PDCAAS was proposed as a method for evaluating protein quality, both for proteins in meals with mixed foods and for individual protein sources (Joint WHO/FAO/UNU Expert Consultation 2007). It uses protein digestibility rather than individual indispensable amino acid digestibility. (The digestible indispensable amino acid score (DIAAS), described below, calculates the amino acid score by multiplying the digestibility of each indispensable amino acid.)

To evaluate the protein quality of mixed meals using PDCAAS, the amino acid score of the food mixture is calculated from the weighted average digestible amino acid content. The specific procedures are as follows (Joint FAO/IFAD/UNICEF/WFP/WHO Consultation 2018):

1) Calculate the total protein and amino acid content of the food mixture.
2) Convert the amount of each indispensable amino acid (mg) to the amount per gram of protein (mg/g) for each food source to calculate the pattern value of each indispensable amino acid (mg/g) in the protein source.
3) Multiply the protein content of each food source by the fecal protein digestibility of the food source to calculate the amount of digestible protein in the food source.
4) Calculate the total digestible amino acid content (total content in mg per food source) of each amino acid by multiplying the digestible protein value by the calculated amino acid pattern (mg/g) value.
5) Calculate the weighted digestibility, weighted by the protein contribution of each food source. This assumes the digestibility of food does not change when consumed in a mixture. (Note: as discussed further, digestibility can be affected by food processing and antinutrient factors).
6) Calculate the amount of each amino acid per gram of digestible protein (mg/g protein). Divide this value by the reference pattern to calculate the amino acid score for amino acids below the recommended amount.
7) Identify the amino acid with the lowest score as the limiting amino acid. Calculate the PDCAAS by multiplying the limiting amino acid score and weighted digestibility.
8) If the PDCAAS exceeds 100, round it down to 100.

Fecal digestibility values are available for various food sources because measuring fecal digestibility is easier than measuring ileal digestibility. Thus, PDCAAS is widely used to determine the protein quality of food products.

THE DIGESTIBLE INDISPENSABLE AMINO ACID SCORE

As mentioned in the "Protein Digestion and Absorption" section, most peptides and amino acids hydrolyzed by digestive enzymes from dietary proteins are absorbed in the terminal ileum; however, if ileal absorption is incomplete, the peptides and amino acids pass into the large intestine and are either consumed or synthesized by the gut microbiota. The extent of absorption in the terminal ileum versus the large intestine depends on the food source (e.g., milk is mostly absorbed in the ileal terminal). Therefore, using protein digestibility measured from fecal samples may lead to errors in the evaluation of protein quality. Some studies comparing ileal digestibility (i.e., amino acid balance in the small intestine) and fecal digestibility (i.e., amino acid balance in the entire intestine) in pigs and rats have shown that fecal digestibility is higher than ileal digestibility for most foods (Bailey and Stein 2020; Mathai et al. 2017; Rutherfurd et al. 2015). On the other hand, other studies have shown that fecal digestibility is lower than ileal digestibility (Mathai et al. 2017).

These differences between ileal and fecal digestibility have been reported, particularly for dietary proteins with low digestibility, which show a weak correlation between ileal and fecal digestibility (Hendriks et al. 2012). Furthermore, digestibility varies among amino acids in different protein sources (FAO 2013). Therefore, the DIAAS has been proposed to address these digestibility issues based on amino acid scores and ileal digestibility.

The DIAAS is calculated based on the concentration of each digestible amino acid (per gram of protein) and ileal digestibility. These digestible indispensable amino acid values are compared with the reference pattern. The digestible amino acid with the lowest score relative to the reference pattern is considered the first limiting amino acid. This first limiting amino acid score becomes the DIAAS score of the protein. Thus, the DIAAS can be calculated as follows (FAO 2013):

DIAAS % = 100 × lowest value [(mg of digestible dietary indispensable amino acid in 1 g of the dietary protein) ÷ (mg of the same dietary indispensable amino acid in 1 g of the reference protein)]

Table 9.2 shows the amino acid pattern of the reference protein for DIAAS calculations by age (FAO 2013).

The DIAAS is considered superior to the PDCAAS because it evaluates ileal, rather than fecal, digestibility and the digestibility of individual amino acids rather than proteins (FAO 2013). For the measurement of ileal digestibility to evaluate protein quality, the FAO indicates that human digests are the best, followed by pig and rat digests (FAO 2018). However, the only way to assess ileal digestibility in humans is to collect samples by inserting a tube through the nose into the ileal cavity or after ileostomy surgery. Therefore, the majority of reported DIAAS values to date have been assessed using animal models, and there is insufficient data on true ileal amino acid digestibility in humans (Deglaire et al. 2009). Thus, the implementation of the DIAAS is currently limited.

TABLE 9.2
Amino Acid Pattern of the Reference Protein for the DIAAS Calculations by Age

	Scoring Pattern mg/g Protein Requirement								
Age Group	His	Ile	Leu	Lys	SAA	AAA	Thr	Trp	Val
Infant (birth to 6 months)	21	55	96	69	33	94	44	17	55
Child (6 months to 3 years)	20	32	66	66	27	52	31	8.5	43
Older child, adolescent, and adult	16	30	61	61	23	41	25	6.6	40

His, Histidine; Ile, Isoleucine; Leu, Leucine; SAA, Sulfur amino acids; AAA, Aromatic amino acids; Thr, Threonine; Trp, Tryptophan; Val, Valine (FAO 2013).

EVALUATION OF THE PROTEIN QUALITY OF THE DIET (MIXED MEALS): EFFECTS OF NUTRIENT INTERACTIONS

Evaluating the protein quality of mixed meals is important because people generally eat a diet containing a mixture of foods. The protein quality of mixed meals can be determined by the additive DIAAS values calculated for the individual ingredients (FAO 2013). For example, peas are limited in sulfur amino acids but rich in lysine. White rice has a high concentration of sulfur-containing amino acids, but lysine is its limiting amino acid. Combining these two foods may therefore result in a balanced amino acid composition relative to the amino acid reference pattern and thereby improve the dietary protein quality (Rutherfurd et al. 2015).

Nonetheless, as the available DIAAS values are currently limited, mixing foods and changing their nutrient compositions can affect food digestibility. For example, the simultaneous consumption of protein and fat improves protein digestibility and amino acid absorption, which in turn improves protein quality. This has been demonstrated in growing pig trials (Cervantes-Pahm and Stein 2008), where fat slows down the rate of gastric emptying and thereby increases the duration that the ingested protein is exposed to proteolytic enzymes, improving protein digestibility. In addition, the rate of passage into the small intestine is reduced, thus prolonging the time for amino acid absorption. As protein quality is determined by amino acid content and digestibility, changes in digestibility because of food mixtures are likely to affect the protein quality of the mixed meal. Besides, in humans, increased branched-chain amino acids, which are found in animal products, have been shown to reduce the uptake of phenylalanine and tryptophan (Fernstrom 2005). With the established "protein-sparing effect of non-protein energy substrates," carbohydrate and fat intake may affect individual essential amino acid requirements (Joint WHO/FAO/UNU Expert Consultation 2007). However, neither PDCAAS nor DIAAS considers such nutrient interactions.

EVALUATION OF DIETARY PROTEIN QUALITY (FOR MIXED MEALS): EFFECTS OF COOKING AND FOOD PROCESSING

Cooking and processing can affect dietary protein quality (Williams et al. 2006). As shown in Table 9.3, different DIAAS values for proteins have been reported depending on the cooking method used. Most dietary proteins are heat-treated before consumption. Cooking with an internal temperature higher than 70°C improves protein digestibility via protein denaturation (Evenepoel et al. 1998, Williams et al. 2006). However, one study reported that the DIAAS (i.e., protein bioavailability, which is determined by ileal amino acid digestibility and indispensable amino acid contents) was superior in raw beef compared to grilled beef (Bailey and Stein 2020). Thus, heat treatment may affect the amino acid content and overall nutritional quality of food products.

Cooking and processing reduce the bioavailability of lysine by producing antinutrient factors such as Maillard-type reaction products (FAO 2014). When evaluating the protein quality of a diet, the impact on the bioavailability of amino acids due to reactions between various components, such as sugars, starch, and minerals, as well as amino acids, should be considered (FAO 2014). Furthermore, as processed foods are subjected to various treatments, such as lowering/adding certain components or heating, their dietary protein evaluation may be difficult. Nevertheless, meals are usually prepared by boiling, grilling, frying, steaming, or including processed foods. Thus, the evaluation of the protein quality of meals in which foods are mixed and cooked is complex.

Craddock et al. stated that the PDCAAS is the most suitable tool for evaluating protein quality, especially in diets containing plant-based foods (Craddock et al. 2021). The DIAAS does not consider differences in nitrogen-to-protein conversion factors between plant- and animal-based foods, and data on plant-based foods are limited. Thus, PDCAAS may continue to be used until an adequate database of ileal digestibility and DIAAS values for commonly consumed human foods is completed.

TABLE 9.3
DIAAS Values for Cooked Food Protein

Cooked Food	DIAAS Value (%) Reference Protein Pattern (Limiting Amino Acids)		
	Infants	Child	Older Child, Adolescent, and Adult
Beef			
Topside steak (raw)[a]	–	97 (Val)	
Topside steak (boiled)[a]	–	99 (Val)	
Topside steak (grilled)[a]	–	80 (Val)	
Topside steak (pan-fried)[a]	–	98 (Val)	
Topside steak (roasted)[a]	–	91 (Val)	
Ground beef (raw)[b]	–	111 (Trp)	121 (Leu)
Ground beef (cooked, fully browned)[b]	–	92 (Leu)	99 (Leu)
Ribeye roast (56°C)[b]	–	104 (Val)	111 (Val)
Ribeye roast (64°C)[b]	–	121 (Val)	130 (Val)
Ribeye roast (72°C)[b]	–	99 (Leu)	107 (Val)
Pork			
Raw[c]	–	111 (Val)	119 (Val)
Bacon (smoked)[c]	–	109 (Val)	117 (Val)
Bacon (smoked-cooked with a CM)[c]	–	126 (Trp)	142 (Val)
Ham (non-cured)[c]	–	115 (Val)	124 (Val)
Ham (alternatively cured)[c]	–	123 (Val)	133 (Val)
Ham (conventionally cured)[c]	–	117 (Val)	126 (Val)
Loin (63°C)[c]	–	129 (Val)	139 (Val)
Loin (68°C)[c]	–	109 (Val)	118 (Val)
Loin (72°C)[c]	–	109 (Val)	117 (Val)

Created by the author
[a] Hodgkinson SM et al., *J Nutr.* 2018; [b] Bailey HM et al., *Br J Nutr.* 2020; [c] Bailey HM et al., *J Nutr.* 2020. CM, commercial microwave; Leu, Leucine; Trp, Tryptophan; Val, Valine

DAILY DISTRIBUTION OF PROTEIN CONSUMPTION AND PROTEIN QUALITY

The evaluation of protein quality should be considered on a meal-by-meal basis. Blood amino acid levels increase postprandially and return to basal levels after a few hours. After meal consumption, amino acids are released into the blood and used for the biosynthesis of body proteins or bioactive substances, such as catecholamines and serotonin. However, excess amino acids not utilized in the biosynthesis of body proteins and bioactive substances are oxidized, releasing ammonia. Ammonia is toxic and is converted into urea in the hepatic urea cycle. Therefore, uneven distribution of protein intake throughout the day may reduce the efficiency of protein utilization (i.e., leading to reduced muscle protein synthesis after consuming a low-protein diet or excessive amino acid oxidation after consuming a high-protein diet). Despite the same amount of total daily protein intake, evenly distributing protein intake between breakfast, lunch, and dinner results in a higher rate of skeletal muscle synthesis than concentrating protein intake biased at dinner (Mamerow et al. 2014). However, protein consumption is generally biased toward dinner and less toward breakfast.

Therefore, improving the quality of breakfast protein is important. This aspect can be considered from the perspective of chrono-nutrition, which has received increased attention in recent years.

The light–dark cycle is a 24-hour cycle, whereas the human circadian rhythm is slightly longer at approximately 25 hours. Therefore, synchronizing the biological clock with the circadian cycle is necessary, with diet and nutritional components considered important synchronization factors (Tahara and Shibata 2014). The processes of digestion, absorption, and metabolism follow a circadian rhythm. Murine studies have shown circadian rhythm disruption can impair metabolic functions. In protein metabolism, the ability to absorb certain peptides in the intestinal tract changes throughout the day (Tahara and Shibata 2016; Saito et al. 2008). Amino acid metabolism varies with the time of day, and clock genes expressed in skeletal muscle may regulate muscular biological function (Takahashi et al. 2018; Aoyama and Shibata 2017).

BREAKFAST PROTEIN QUALITY AND HEALTH OUTCOMES

Kinoshita et al. reported findings from studies on breakfast protein quality and health in older adults. As mentioned earlier, it is currently difficult to completely evaluate the protein quality of meals containing mixed foods subjected to various cooking and food-processing treatments. Although the study had limitations in that it may not have accurately evaluated daily dietary protein quality, a longitudinal study suggested the importance of a high-quality protein diet at breakfast for good health in older adults (Kinoshita et al. 2022). The study's participants were community-dwelling older adults aged 60 years and older who were followed for a maximum of 9.2 years. To investigate the association between breakfast protein quality and the development of muscle weakness, participants were classified into three groups based on the tertiles of breakfast protein quality at baseline (low-, middle-, and high-quality groups). The results showed that the odds ratio for muscle weakness was 50% lower in the high-quality group than in the low-quality group (Kinoshita et al. 2022). The results are illustrated in Figure 9.3. Moreover, when analyzing only participants whose breakfast protein intake was below the recommended amount, the odds ratio was even lower (Kinoshita et al. 2022). This finding emphasizes the importance of increasing protein quality in individuals with lower protein intake. In another longitudinal study by Kinoshita and colleagues using cognitive decline as an outcome measure, the authors found a positive association between breakfast protein quality and cognitive function (Kinoshita et al. 2022).

In these studies, researchers measured the protein quality of mixed meals using the PDCAAS. They set digestibility at 90% based on the human fecal digestibility when consuming a diet corresponding to the average Japanese food consumption (i.e., a diet with a variety of mixed and prepared foods), as reported by Kaneko and colleagues (Kaneko et al. 1988). In 2007, the Joint WHO/FAO/United Nations University (UNU) expert consultation published true digestibility values for various protein sources in humans (Joint WHO/FAO/UNU Expert Consultation 2007). This report indicated the true digestibility of mixed meals in several countries. While the data for true digestibility of Japanese mixed diets are not provided, those of the American mixed diet are given as 95%, the Chinese mixed diet as 96%, and the Filipino mixed diet as 88% (Joint WHO/FAO/UNU Expert Consultation 2007). Thus, the 90% fecal digestibility rate used in the study is reasonable.

PROSPECTS

The FAO has recognized that there is insufficient data on ileal digestibility to apply true ileal amino acid digestibility in calculating the DIAAS. In 2014, the FAO proposed a research protocol and methodology for studying the ileal amino acid digestibility of human foods, using both human and animal models (FAO 2014; Lee et al. 2016), summarized as follows:

1. Accumulation of sufficient digestibility data on humans to implement the DIAAS

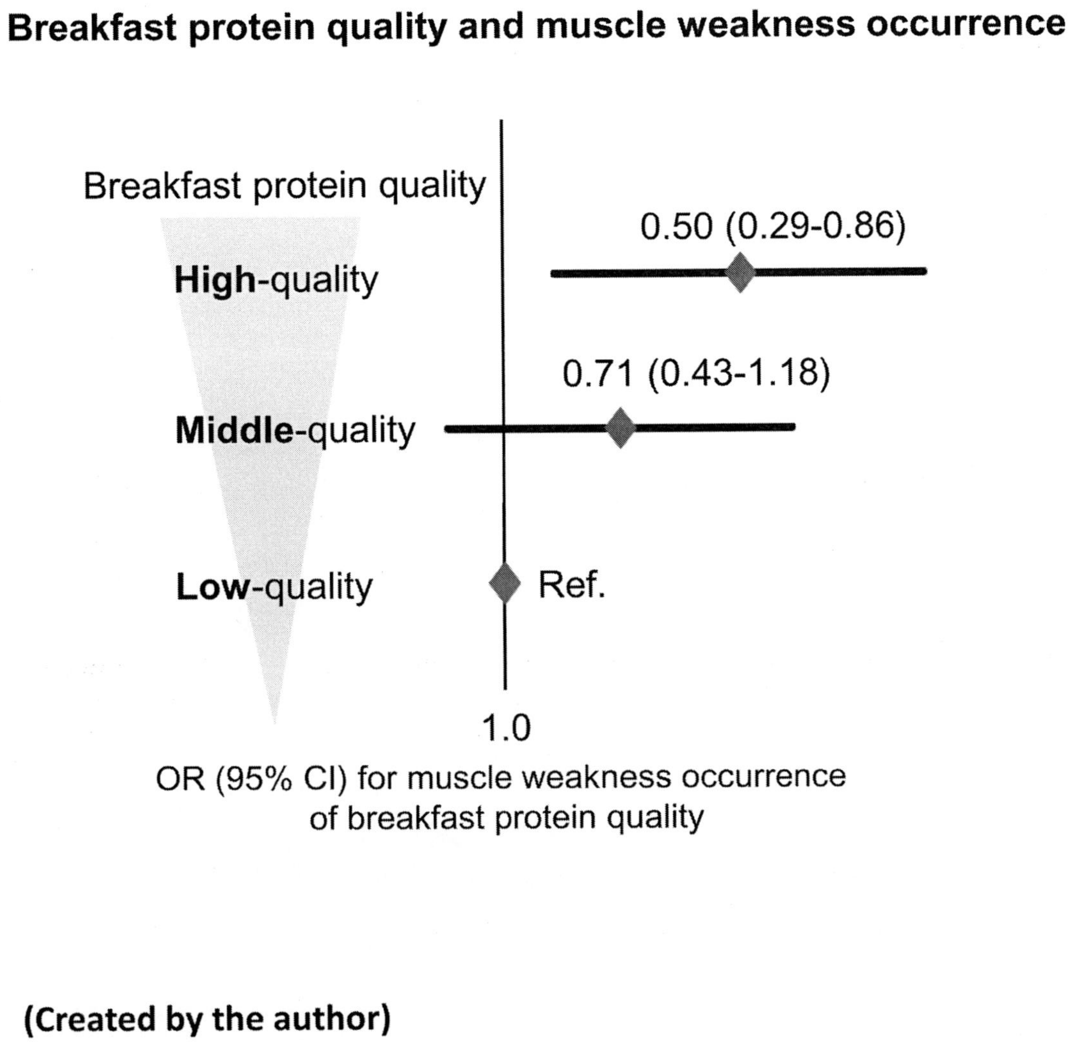

FIGURE 9.3 Breakfast protein quality and muscle weakness occurrence. This figure shows the multivariate adjusted association between breakfast PDCAAS scores and low grip strength during the follow-up period. Breakfast protein quality evaluated using PDCAAS. Muscle weakness defined as low grip strength based on sarcopenia definition (<28 kg in male and <18 kg in female). The odds ratios and 95% confidence intervals are shown and were estimated using the generalized estimating equation after adjusting for sex, age, follow-up time, body mass index, total physical activity, cognitive function, education, smoking status, annual household income, medical history (hypertension, dyslipidemia, diabetes mellitus, and ischemic heart disease), PDCAAS values for lunch and dinner (low to high at each meal), energy and protein intake at the three regular meals, and grip strength at baseline.

There is a need to develop a completely accessible and robust database of amino acid digestibility of foods and diets in various regions worldwide. This requires focusing on the diets consumed by vulnerable people of different ages in low-income countries. Multiple studies using the same methodology need to be conducted in various countries (FAO 2014).

2. Verification of the practical effects on public health policies and programs

The practical effects of using the DIAAS, instead of the PDCAAS, on public health must be examined before making a final decision to adopt the DIAAS (FAO 2014).

3. Assessment of protein requirements for vulnerable populations

Protocols are needed for assessing protein requirements for vulnerable populations whose requirements increase due to physiological changes during pregnancy or chronic inflammation, as well as from the effects of sanitation and hygiene. Furthermore, amino acid reference patterns for these individuals must be developed and applied to the DIAAS (FAO 2014).

4. Collaboration with agricultural research institutions worldwide

Collaboration with agricultural research institutions is necessary to introduce stable isotopes into plants, and this requires funding. It would be beneficial to give young researchers from developing countries opportunities to participate in research on this topic (FAO 2014).

5. Assessment of the DIAAS additivity in mixed meals and prepared multi-ingredient food products

In the long term, the goal is to develop methods that government agencies can use to regulate the protein quality of specific foods (FAO 2014).

It is not easy to properly evaluate the quality of dietary protein in humans. However, a diet with high protein quality significantly contributes to avoiding malnutrition and preserving the global environment. Further studies should be conducted in this regard.

MINI DICTIONARY OF TERMS

Digestible indispensable amino acid score: A method used to evaluate protein quality, calculated based on the concentration of each digestible indispensable amino acid (per gram of protein) and its ileal digestibility.

Fecal protein digestibility: Digestibility defined in terms of the amino acid balance throughout the entire intestine.

Greenhouse gases: Emissions including CO_2, CH_4, and N_2O. CH_4 and N_2O values are converted to CO_2 equivalents, and the total value is expressed as CO_2 emissions.

Ileal amino acid digestibility: Digestibility defined in terms of the amino acid balance throughout the small intestine.

Limiting amino acids: An indispensable amino acid present at the lowest value in a food protein compared to a reference food protein.

Protein digestibility-corrected amino acid score: A method for evaluating protein quality that combines the amino acid score and fecal protein digestibility, as published by the FAO/WHO Expert Consultation.

SUMMARY POINTS

- Protein quality (i.e., bioavailability) is determined by its biological value (i.e., the extent to which dietary amino acids meet required levels) and digestibility (i.e., the amount effectively used after digestion and absorption).
- The nutritional value of proteins is measured using chemical score concept, as all amino acids must be supplied simultaneously for protein synthesis.
- Digestibility is assessed in the small intestine (ileal amino acid digestibility) or entire intestine (fecal protein digestibility) and calculated as the difference between intake and loss.
- Peptides and amino acids that are not absorbed in the terminal ileum are transported to the large intestine and are removed or added by the gut microbiota, so using fecal protein digestibility can lead to errors in protein-quality evaluation.

- Protein-quality evaluation methods include the protein digestibility-corrected amino acid score (PDCAAS) and the digestible indispensable amino acid score (DIAAS).
- The PDCAAS can be calculated as follows: PDCAAS = amino acid (mg) in 1 g of dietary protein ÷ amino acid (mg) in the reference pattern × fecal protein digestibility.
- The DIAAS can be calculated as follows: DIAAS = digestible dietary indispensable amino acids (mg) in 1 g of dietary protein ÷ dietary indispensable amino acids (mg) in 1 g of reference protein.
- The DIAAS is considered superior to the PDCAAS because it uses ileal digestibility rather than fecal digestibility, and considers individual amino acid digestibility rather than overall protein digestibility.
- Challenges in protein-quality evaluation include: (i) insufficient data on ileal amino acid digestibility in humans, and (ii) digestibility variations due to protein interactions with other nutrients and the effects of cooking and processing.

REFERENCES

Abelilla, J. J., Liu, Y., & Stein, H. H. 2018. The digestible indispensable amino acid score (DIAAS) and protein digestibility-corrected amino acid score (PDCAAS) of the oat protein concentrate were measured in 20- to 30-kilogram pigs. *Journal of the Science of Food and Agriculture* 98:410–414.

Aoyama, S. and Shibata, S. 2017. Role of circadian rhythms in muscular and osseous physiology and their regulation by nutrition and exercise. *Frontiers of Neuroscience* 11:63.

Bailey, H. M. and Stein, H. H. 2020. Raw and roasted pistachio nuts (Pistacia vera L.) are 'good' sources of protein based on their digestible indispensable amino acid scores in pigs. *Journal of the Science of Food and Agriculture* 100:3878–3885.

Bakaloudi, D. R., Halloran, A., Rippin, H. L., Oikonomidou, A. C., Dardavesis, T. I., Williams, J., Chourdakis, M. 2021. Intake and adequacy of the vegan diet. Systematic review of evidence. *Clinical Nutrition* 40:3503–3521.

Cervantes-Pahm, S. K., Liu, Y., & Stein, H. H. 2014. Digestible indispensable amino acid scores and digestible amino acids in eight cereal grains. *British Journal of Nutrition* 111:1663–1672.

Cervantes-Pahm, S. K., Stein, H. H. 2008. Effects of dietary soybean oil and protein concentrations on the concentration of digestible amino acids in soybean products fed to growing pigs. *Journal of Animal Science* 86:1841–1849.

Craddock, J. C., Genoni, A., Strutt, E. F., Goldman, D. M. 2021. Limitations of the Digestible Indispensable Amino Acid Score (DIAAS) with special attention to plant-based diets: a review. *Current Nutrition Report* 10:93–98.

de Vries, M., & de Boer, I. J. M. 2010. Comparison of the environmental impacts of livestock products: A review of life cycle assessments. *Livestock Science* 128:1–11.

Deglaire, A., Bos, C., Tomé, D., & Moughan, P. J. 2009. Ileal digestibility of dietary proteins in growing and adult pigs. *British Journal of Nutritions* 102:1752–1759.

Drewnowski, A., Rehm, C. D., Martin, A., Verger, E. O., Voinnesson, M., & Imbert, P. 2015. Energy and nutrient density of foods in relation to carbon footprints. *American Journal of Clinical Nutrition* 101:184–191.

Duodu, K. G., Nunes, A., Delgadillo, I., Parker, M. L., Mills, E. N. C., Belton, P. S., Taylor, J. R. N. 2002. Effects of grain structure and cooking on the in vitro Protein Digestibility of sorghum and maize. *Journal of Cereal Science* 35:161–174.

Evenepoel, P., Geypens, B., Luypaerts, A., Hiele, M., Ghoos, Y., & Rutgeerts, P. (1998). Digestibility of cooked and raw egg proteins in humans assessed using stable isotope techniques. *Journal of Nutrition* 128:1716–1722.

Fernstrom, J. D. 2005. Branched-chain amino acids and brain function. *Journal of Nutrition* 135:1539s–1546s.

Food and Agriculture Organization of the United Nations. 2013. Dietary protein quality evaluation in human nutrition: Report of an FAO expert consultation. *FAO Food and Nutrition Paper* 2013;92:1–66.

Food and Agriculture Organization of the United Nations. 2014. Research approaches and methods for evaluating the protein quality of human foods. Report of a FAO Expert Working Group, 2–5 March 2014, Bangalore, India. https://www.fao.org/documents/card/es/c/5fa39b04-68bf-4dd7-bc44-3d1f249a6c5e/ (accessed June 23, 2023).

Food and Agriculture Organization of the United Nations. 2018. Protein quality assessment in follow-up formula for young children and ready to use therapeutic foods. Report of the FAO Expert Working Group, Rome, 6–9 November 2017.

GBD 2017 Diet Collaborators. 2019. Health effects of dietary risks in 195 countries, 1990–2017: A systematic analysis for the global burden of disease study 2017. *The Lancet* 393:1958–1972.

Ghosh, S., Suri, D. & Uauy, R. 2012. Assessment of protein adequacy in developing countries: Quality issues. *British Journal of Nutrition* 108:S77–S87.

Hansen, T. H., Madsen, M. T. B., rgensen, N. R., Cohen, A. S., Hansen, T., Vestergaard, H., & Allin, K. H. 2018. Bone turnover, calcium homeostasis, and vitamin D status of Danish vegans. *European Journal of Clinical Nutrition* 72:1046–1054.

Hendriks, W. H., van Baal, J., & Bosch, G. 2012. Ileal and fecal protein digestibility measurements in humans and other non-ruminants: A comparative species view. *British Journal of Nutrition* 108:S247–S257.

Herreman, L., Nommensen, P., Pennings, B., Laus, M. C. 2020. Comprehensive overview of the quality of plant- and animal-sourced proteins based on digestible indispensable amino acid scores. *Food Science and Nutrition* 8:5379–5391.

Hunt, J. R. 2003. Bioavailability of iron, zinc, and other trace minerals in vegetarian diets. *American Journal of Clinical Nutrition* 78:633S–639S.

Jahan-Mihan, A., Luhovyy, B. L., El Khoury, D., & Anderson, G. H. 2011. Dietary proteins determine the metabolic and physiological functions of the gastrointestinal tract. *Nutrients* 3:574–603.

Joint FAO/IFAD/UNICEF/WFP/WHO Consultation. 2018. The State of Food Security and Nutrition in the World 2018: Building climate resilience for food security and nutrition: Building climate resilience for food security and nutrition. Rome, FAO. https://www.unicef.org/reports/state-food-security-and-nutrition-world-2018 (accessed June 23, 2023)

Joint WHO/FAO/UNU Expert Consultation. 2007. Protein and amino acid requirements in human nutrition. *World Health Organization Technical Report Series* 2007:1–265.

Kaneko, K., Ishikawa, K., Setoguchi, K. & Koike, G. 1988. Utilization and requirement of dietary proteins considering dermal and miscellaneous nitrogen loss in Japanese women. *Journal of Nutritional Science and Vitaminology (Tokyo)* 34:459–467. doi:10.3177/jnsv.34.459

Kinoshita, K., Otsuka, R., Nishita, Y., Tange, C., Tomida, M., Zhang, S. & Arai, H. 2022. Breakfast protein quality and muscle strength in older Japanese adults: A community-based longitudinal study. *Journal of the American Medical Directors Association* 25(5):729–735.e2.

Kinoshita, K., Otsuka, R., Takada, M., Tsukamoto-Yasui, M., Nishita, Y., Tange, C. & Arai, H. 2022. Low amino acid score of breakfast is associated with the incidence of cognitive impairment in older japanese adults: A community-based longitudinal study. *Journal of Prevention of Alzheimers Disease* 9:151–157.

Lee, W. T., Weisell, R., Albert, J., Tome, D., Kurpad, A. V., & Uauy, R. 2016. In 2014, the food and agriculture organization expert working group proposed research approaches and methods for evaluating the protein quality of human foods. *Journal of Nutrition* 146:929–932.

Mamerow, M. M., Mettler, J. A., English, K. L., Casperson, S. L., Arentson-Lantz, E., Sheffield-Moore, M., & Paddon-Jones, D. 2014. Dietary protein distribution positively influences 24-h muscle protein synthesis in healthy adults. *Journal of Nutrition* 144:876–880.

Mathai, J. K., Liu, Y., & Stein, H. H. 2017. Values of digestible indispensable amino acid scores (DIAAS) for some dairy and plant proteins may describe protein quality better than those calculated using the protein digestibility-corrected amino acid score (PDCAAS). *British Journal Nutrition* 117:490–499.

Mitchell, H. H., & Block, R. J. 1946. Relationships between the amino acid content of proteins and their nutritional value in rats. *Journal of Biological Chemistry* 163:599–620.

Moughan, P. J., & Wolfe, R. R. 2019. Determination of the digestibility of dietary amino acids in humans. *Journal of Nutrition* 149:2101–2109.

Neis, E. P., Dejong, C. H., & Rensen, S. S. 2015. The role of microbial amino acid metabolism in host metabolism. *Nutrients* 7:2930–2946.

Rutherfurd, S. M., Fanning, A. C., Miller, B. J., & Moughan, P. J. 2015. Protein digestibility-corrected amino acid scores and digestible indispensable amino acid scores differentially describe protein quality in growing male rats. *Journal of Nutrition* 145:372–379.

Saito, H., Terada, T., Shimakura, J., Katsura, T., & Inui, K. 2008. Regulatory mechanism governing the diurnal rhythm of intestinal H+/peptide cotransporter 1 (PEPT1). *American Journal of Physiology-Gastrointestinal and Liver Physiology* 295:G395–402.

Saunders, A. V., Craig, W. J., & Baines, S. K. 2013. Zinc and vegetarian diets. *Medical Journal of Australia* 199:S17–21.

Selinger, E., Kühn, T., Procházková, M., Anděl, M., & Gojda, J. 2019. Vitamin B12 deficiency is prevalent among Czech vegans who do not use vitamin B12 supplements. *Nutrients* 11(12): 3019.

Semba, R. D., Shardell, M., Sakr Ashour, F. A., Moaddel, R., Trehan, I., Maleta, K. M., Isabel Ordiz, M., Kraemer, K., Khadeer M. A., Ferrucci, L., Manary, M. J. 2016. Child stunting is associated with low circulating essential amino acids. *EBioMedicine* 6:246–252.

Shaheen, N., Islam, S., Munmun, S., Mohiduzzaman, M., & Longvah, T. 2016. Amino acid profiles and digestible indispensable amino acid scores of proteins from the prioritized key foods in Bangladesh. *Food Chemistry* 213:83–89.

Tahara, Y., & Shibata, S. 2014. Chrono-biology, chrono-pharmacology, and chrono-nutrition. *Journal of Pharmacological Science* 124:320–335.

Tahara, Y., & Shibata, S. 2016. Circadian rhythms of liver physiology and disease: experimental and clinical evidence. *Nature Reviews of Gastroenterology and Hepatology* 13:217–226.

Takahashi, M., Ozaki, M., Kang, M. I., Sasaki, H., Fukazawa, M., Iwakami, T., Jean Lim, P., Kim, H. K., Aoyama, S., & Shibata, S. 2018. Effects of meal timing on postprandial glucose metabolism and blood metabolites in healthy adults. *Nutrients* 10(11):1763.

Tilman, D., & Clark, M. 2014. Global diets link environmental sustainability and human health. *Nature* 515:518–522.

van der Wielen, N., Moughan, P. J., & Mensink, M. 2017. Amino acid absorption in the large intestine of humans and porcine models. *Journal of Nutrition* 147:1493–1498.

Weaver, C. M., & Plawecki, K. L. 1994. Dietary calcium: adequacy of a vegetarian diet. *American Journal of Clinical Nutrition* 59:1238s–1241s.

Weikert, C., Trefflich, I., Menzel, J., Obeid, R., Longree, A., Dierkes, J., Meyer, K., Aeberli, I. H., Mai, K., Stangl, G. I., Muller, S. M., Schwerdtle, T., Lampen, A ., Abraham, K. 2020. Vitamin and mineral status in a vegan diet. *Deutsches A¨rzteblatt International* 117:575–582.

Willett, W., Rockström, J., Loken, B., Springmann, M., Lang, T., Vermeulen, S., Garnett, T., Tilman, D., DeClerck, F., Wood, A., Jonell, M., Clark, M., Gordon, L. J., Fanzo, J., Hawkes, C., Zurayk, R., Rivera, J. A., Vries, W. D., Sibanda, L. M., Afshin, A., Chaudhary, A., Herrero, M., Agustina, R., Branca, F., Lartey, A., Fan, A., Crona, B., Fox, E., Bignet, V., Troell, M., Lindahl, T., Singh, S., Cornell, S. E., Reddy, K. S., Narain, S., Nishtar, S., Murray, C. J. L. 2019. Food in the Anthropocene: the EAT-Lancet Commission on healthy diets from sustainable food systems. *The Lancet* 393:447–492.

Williams, C. D. 1933. A nutritional disease of childhood associated with a maize diet. *Archives of Disease in Childhood* 8:423–433.

Williams, P. A., Hodgkinson, S. M., Rutherfurd, S. M., & Hendriks, W. H. 2006. Lysine content in canine diets can be severely heat damaged. *Journal of Nutrition* 136:1998s–2000s.

Wu, G., Fanzo, J., Miller, D. D., Pingali, P., Post, M., Steiner, J. L., & Thalacker-Mercer, A. E. 2014. Production and supply of high-quality food protein for human consumption: sustainability, challenges, and innovations. *Annals of the New York Academy of Sciences* 1321:1–19.

10 Improving the Quality of Lentil Protein-Based Diets

A Focus on Metabolic Availability of Methionine

Alyssa Paoletti, Rajavel Elango, and Glenda Courtney-Martin

ABBREVIATIONS

AA	Amino acid(s)
DIAAS	Digestible indispensable amino acid score
FAO	Food and Agricultural Organization
IAAO	Indicator amino acid oxidation
MPS	Muscle protein synthesis
WHO	World Health Organization
UNICEF	United Nations International Children's Emergency Fund

INTRODUCTION

Lentil (*Lens culinaris*) is a major pulse crop, and global production of lentils has increased by 93% since 2000 (Dhull, Kinabo, and Uebersax 2023). Lentils are native to North America and Western Asia, with Canada being the leading producer, followed by India (FAO 2021). The steady growth in lentil production parallels the global shift toward plant-based diets (Willett et al. 2019; Health Canada 2019), which offer various health and environmental benefits (Gorissen et al. 2018; Willett et al. 2019; Grasso et al. 2021). Pulses, including lentils, provide complex carbohydrates, protein, dietary fiber, vitamins, and minerals (De Almeida Costa et al. 2006). Relative to cereal grains, pulses contain more protein and fewer carbohydrates, making them an important substitute for animal protein foods (Dhull, Kinabo, and Uebersax 2023).

There are many lentil varieties, with compositions uniquely determined by genetic and environmental factors (Grusak 2009). As reviewed by Grusak (2009), environmental factors include agroecology (i.e., soil type, temperature, rainfall), production practice (i.e., fertilizer and pesticide use), and growing conditions (i.e., biotic and abiotic stress). Reported nutrient ranges for lentils include: protein (15.9–31.4%), carbohydrates (3.4– 74.9%), fat (0.3–3.5%), total fiber (5.1–26.6%), and ash (2.2–6.4%). The nutrient composition of lentil (per 100 g), raw, is shown in Table 10.1.

Lentils also contain anti-nutrients, including trypsin inhibitors, phytic acid, and tannins, which reduce nutrient bioavailability (Bhatty, Christison, and Centre 1984; Nosworthy et al. 2017). These components, particularly lentils' high tannin content, can bind to the amino acids (AA) lysine and methionine, rendering them unavailable for body protein synthesis (Davis 1981). Additionally, lentils are first limiting in the indispensable AA methionine, which makes them a low-quality protein source compared to animal proteins (Temba et al. 2016; Khazaei et al. 2019). Therefore, understanding methionine bioavailability is essential for improving the protein quality of a lentil-based diet.

DOI: 10.1201/b23328-10

TABLE 10.1
Nutrient Composition of Raw Lentils

Composition[1]	Unit	Raw	
		Green	Red
Energy	Kcal	352	358
Protein	g	24.6	23.9
Fat	g	1.06	2.17
Carbohydrate	g	63.4	63.1
Fiber	g	10.7	10.8
Calcium	mg	35	48
Iron	mg	6.51	7.39
Magnesium	mg	47	59
Potassium	mg	677	668
Zinc	mg	3.27	3.6
Vitamin C	mg	4.5	1.7
Folate	μg	479	204
Choline	mg	96.4	–
Vitamin A, RAE	μg	2	3
Alanine	mg/g protein	42.4	43.6
Arginine		79.1	80.6
Asparagine		57	57.7
Aspartic Acid		57	57.7
Cysteine		22.2	27.4
Glutamine		77.9	80.9
Glutamic Acid		77.9	80.9
Glycine		40.8	42.4
Histidine		26.9	29.4
Isoleucine		42.6	45.1
Leucine		73.1	75.7
Lysine		69.9	72.8
Methionine		8.1	8.9
Phenylalanine		49.8	51.4
Proline		38.8	43.6
Serine		48.9	48.1
Threonine		36.9	37.4
Tryptophan		8.1	9.3
Tyrosine		26.7	27.9
Valine		48.3	51.8

[1] Lentil composition per 100 g of lentil. Adapted from Dhull, Kinabo, and Uebersax (2023). Thiamin, riboflavin, vitamin B6/B12, and K are all <1.0 mg. Amino acid composition of green lentils obtained from Rafii et al. (2022), and red from the Canadian Nutrient File (Food code: 4869).

THE LIMITING AMINO ACID

The protein content of foods varies in both the amount and composition of AA. As such, dietary proteins can be categorized as either high-quality or low-quality, based on the quantity of AA they provide relative to their corresponding AA requirements. Most animal proteins (e.g., eggs, meat, fish) provide all indispensable AA in concentrations adequate to meet human requirements, whereas indispensable AA in plant proteins (methionine in lentils, lysine in cereal grains) may not meet human requirements under all conditions (IOM 2006). Thus, in the quantity normally consumed, plant proteins like lentils may fail to fulfill the requirement for methionine, which is denoted as the limiting AA (Block and Mitchell 1946). For those consuming a strict plant/lentil-based diet, the methionine concentration might be inadequate to meet the body's metabolic needs, particularly because AA from plant sources are relatively less bioavailable (IOM 2006). This is important because when an indispensable AA is deficient in the diet, all other AA appear in relative excess and are oxidized, since there is no storage of AA in the body (Elango, Ball, and Pencharz 2008). Over the long term, this may negatively impact whole-body protein turnover. For example, adult males who received three levels of leucine intake (from deficient to excess) over a three-week period (Young et al. 1987) experienced a decline in leucine oxidation and flux at the deficient intakes from weeks 1 to 3 (Young et al. 1987). This decrease was observed despite improvements in leucine balance, demonstrating a decreased rate of whole-body protein turnover. As addressed by Waterlow (Waterlow 1984), this decline in whole-body protein turnover may be clinically significant because it parallels alterations in skeletal muscle protein turnover. Since skeletal muscle plays an important role in the body's response to stress (i.e., trauma or infection), a lower rate of skeletal protein turnover may compromise the host's ability to mount an adequate stress response, increasing susceptibility to infection/disease (Young and Marchini 1990). Therefore, protein quality is important to ensure individuals' dietary AA needs are met when consuming adequate protein.

PROTEIN QUALITY

In addition to AA content, protein quality is determined by the digestibility and bioavailability of AA (Elango et al. 2012). The Food and Agricultural Organization (FAO) recommends treating each AA as an individual nutrient when studying protein quality and including bioavailability data in food tables, since the limiting AA determines the utilization of all other AA for protein synthesis. Digestibility is defined as the net absorption of AA following digestion (FAO 2013), while bioavailability refers to the amount of AA available metabolically at the cellular level for whole-body protein synthesis, which is fundamental to protein quality (Batterham 1992).

Protein intake can be assessed in terms of quantity alone or quantity adjusted for quality. Quality adjustment is particularly important for plant-based diets. A meta-analysis of 200 countries found estimates of total energy, total protein, and total utilizable protein (protein intake adjusted for protein quality) were found to be independently, significantly, and adversely associated with the prevalence of stunting (Ghosh, Suri, and Uauy 2012). Specifically, total energy, total protein, and total utilizable protein accounted for 41%, 34%, and 43% of variations in stunting, respectively (Ghosh, Suri, and Uauy 2012). Furthermore, the meta-analysis showed that the risk of dietary protein inadequacy measured using total protein (uncorrected for protein quality) compared to total protein utilization (corrected for protein quality) significantly underestimated the risk of protein inadequacy. Inadequacy was most prevalent in East and Southern Africa and South and Southeast Asia, where the diet is predominantly plant-based. These findings demonstrate the importance of controlling for protein quality when planning and assessing the capacity of diets to meet AA needs, particularly when the diet is plant-based.

Data on methionine digestibility in lentils as consumed by humans is lacking. Using the indicator amino acid oxidation (IAAO) method, Rafii et al. found the metabolic availability of methionine (the first limiting AA) in cooked lentils to be 69% in healthy young adult males (Rafii et al. 2022). This indicates that in addition to being limiting in lentils, methionine's metabolic availability in lentils is also low. The term "metabolic availability" was coined during the validation of the IAAO method to study protein quality (Moehn et al. 2005). It refers to the digestibility, absorption, and bioavailability of AA for protein synthesis, accounting for losses of dietary AA during digestion, absorption, and cell metabolism, thus providing a more precise measure of protein quality compared with digestibility (FAO 2014). A number of strategies are available to improve the protein quality of lentil-based diets, including increasing total protein intake (quantity), biofortification, and protein complementation (i.e., protein blends or mixed-meal complementation), which are discussed below.

PROTEIN QUANTITY

The protein quality of plant-based diets can be improved by increasing the volume of plant foods consumed. This has been demonstrated in healthy younger and older adult populations. In a randomized control trial with healthy young adult females, researchers tested the effect of supplementing potato protein isolate on muscle protein synthesis (MPS), with and without resistance exercise (Oikawa et al. 2020). The intervention group received 25 g of potato protein twice daily, increasing the total protein intake to 1.6 g/kg/d compared to 0.8 g/kg/d in the control group. The results showed increased MPS rates in the intervention group, both with and without exercise, compared to the controls (Oikawa et al. 2020). This suggests that the limiting AA intake can be increased by consuming a larger magnitude of total protein from plant food.

Similar results were observed in older adults. In healthy males (mean age 71 years), the ingestion of whey, casein, and wheat protein hydrolysates on MPS rates were compared (Gorissen et al. 2016). While 35 g of wheat protein hydrolysate resulted in a lower MPS rate than the same amount of casein and whey protein, increasing the wheat protein hydrolysate intake to 60 g significantly increased MPS rates (Gorissen et al. 2016). In both studies, participants were fed isolated proteins, which lack the anti-nutritional factors naturally found in intact plant proteins. Therefore, in the context of intact plant proteins, an even greater volume of food may be required to attain the observed benefits.

While increasing total protein intake is a simple solution, it reduces diet diversity, resulting in the risk of inadequate intake of other essential nutrients. Including a variety of protein sources, including plant and animal proteins, could help ensure individuals meet the requirements for AA and all other essential dietary nutrients. Moreover, increasing the total volume of plant foods may result in excessive energy consumption, resulting in excessive weight gain. This is especially relevant to older adults, who have lower energy requirements (FAO/WHO/UNU 2005), making them more susceptible to obesity. For example, the digestible indispensable AA (DIAAS) score of a lentil-based diet that meets the protein needs (RDA = 0.8 g/kg/d) of an 80 kg older adult male >60 years is 42%, with methionine as the limiting AA (Table 10.2). To improve the DIAAS score, the total protein intake and quantity of lentils were doubled, yet the DIAAS increased only by 2%, while energy intake rose by 50%. Thus, alternative strategies, such as plant protein biofortification, may be necessary.

BIOFORTIFICATION

In experimental settings, AA fortification has proven to be successful (Bressani et al. 1958, 1960, 1963; Hussain et al. 2004; Ghosh et al. 2010; Sikka and Johari 1979). In the 1970s, three seminal AA fortification trials were conducted: (1) in Thailand, preschool children were fed fortified rice with lysine, threonine, thiamine, riboflavin, vitamin A, and iron; (2) in Tunisia, malnourished populations received wheat fortified with lysine, iron, and vitamins; and (3) in Guatemala, subjects consumed corn fortified with lysine, minerals, and vitamins. Researchers measured anthropometry, plasma concentrations of hemoglobin and hematocrit, and collected information on morbidity and mortality. Initially, the trials found no associated benefits of AA fortification on the outcomes. However, rigorous assessment of these trials revealed major methodological and analytical issues (NRC and NAS 1988), leaving the effectiveness of AA fortification unresolved. Much later, lysine fortification of wheat flour in China and Pakistan was shown to be effective, with children receiving fortified wheat gaining significantly more weight and height compared to controls (no fortification) (Hussain et al. 2004; Zhao et al. 2004).

TABLE 10.2
The Digestible Indispensable Amino Acid Score of a Hypothetical Lentil-Based Diet for an Older Adult Male

Food	Weight[2] (g)	Digestible Indispensable Amino Acid Reference Ratio[3]			DIAAS[4]
Diet #1 Composition[1]					
		Lysine	Sulphur amino acid	Leucine	
Lentil	255	1.74	0.42	0.64	42
Orange	100				
	100				
Apple	100				
Green pepper					
Oil	35				
Diet #2 Composition[1]					
Food	**Weight[2] (g)**	**Digestible Indispensable Amino Acid Reference Ratio[3]**			**DIAAS[4]**
		Lysine	Sulphur amino acid	Leucine	
Lentil	575	1.80	0.44	0.66	44
Orange	100				
Apple	100				
Carrots	100				
Green pepper	100				
Oil	35				

(1) Food composition data were obtained from the United States Department of Agriculture Food Data Central.

(2) Diet #1 was devised to meet the current recommended dietary allowance for protein (0.8 g/kg/d), and diet #2 was devised to meet twice this recommendation.

(3) The reference ratio was calculated using the FAO's three- to ten-year-old reference pattern for lysine, the indicator amino acid oxidation-derived requirement for leucine (Szwiega et al. 2020), and total sulfur amino acid requirements (Paoletti et al. 2023).

(4) The DIAAS for the mixed diet is expressed as a percentage, representing the lowest value among the digestible indispensable amino acid reference ratios.

Successful augmentation of pulse methionine content has been achieved by overexpressing sulphur-rich proteins in Narbon bean, lupins, and forage alfalfa (Bagga et al. 2005; Saalbach et al. 1995). In poultry, lower amounts of supplemental methionine were required when fed transgenic lupin seeds (i.e., high methionine content) compared to control lupin seeds (Ravindran et al. 2002). Unfortunately, these transgenic lines were allergenic (Pastorello et al. 2001) which reduces their applicability to human nutrition.

Conventional breeding improves highly heritable traits that are regulated by a small number of genes (Salaria et al. 2022). Protein content – a trait with low heritability and high environmental effects – does not respond well to this method (Salaria et al. 2022). As a result, for lentils, this technique leads to higher protein concentration at the expense of yield (Qureshi et al. 2013) rendering it an inefficient strategy. An alternative method is hybridization-based breeding. One study comparing various lentil species identified a high-protein strain called ILWL 47 (Bhatty 1986). Another lentil cultivar, IC317520, was also found to be high in protein, sugar, and starch (Tripathi et al. 2019). Using these identified lentil cultivars, hybridization breeding may be able to increase the protein content of lentils (Salaria et al. 2022). Yet, there is a dearth of lentil hybrids because pollination takes place in unopened flowers, and there are no male sterile lines that can be used for large-scale production (Tan et al. 2022). Mutation breeding is another option; for example, the lentil mutant (NIA-MASOOR-5), created by gamma irradiation, has higher protein content and is both high yield and resistant to diseases (Fazal Ali, Arain, and Shaikh 2010).

These strategies are theoretical, and transgenic lines will require empirical tests directly in humans to assess whether increased methionine content translates into increased availability for protein synthesis. Since the bioavailability of methionine is unknown in these cultivars, the appropriate levels of fortification in the diet remain unresolved. Furthermore, despite the existence of various transgenic biofortified crops, commercialization has been limited due to low consumer acceptability and regulatory issues (Shahzad et al. 2021).

PROTEIN COMPLEMENTATION: PROTEIN BLENDS AND MIXED-MEAL COMPLEMENTATION

Combining two protein sources with complementary AA profiles has proven efficient in increasing the content of the limiting AA in the diet (IOM 2006). This concept, known as protein complementation, is a traditional and practical strategy used to improve the protein quality of plant-based diets (IOM 2006). Thus, an ideal method for protein quality assessment should be able to evaluate the effect of complementation to inform recommendations on appropriate ratios of intact plant proteins (FAO 2014). Current guidelines for protein complementation are nonexistent due to the scarcity of human data on AA bioavailability of foods and the impact of protein complementation on overall mixed-meal protein quality (FAO 2014).

Protein Blends

Protein blends combine different proteins in a single product, such as a cereal-legume blend. Cereals are limited in the AA lysine, which is abundant in pulses, whereas pulses are limited in methionine, which is abundant in cereals. Thus, these two sources of protein may be combined to form a composite, weaning, or supplementary food mix, or ready-to-use flour. Protein blends were recognized by the World Health Organization (WHO) and UNICEF in 1998 as a potential strategy to alleviate protein-energy malnutrition in developing regions (Brown, Dewey, and Allen 1998). Various composite, weaning, and supplementary food mixes have been researched *in vitro*, exploring functional, nutritional, and sensory qualities. Composite products of wheat and lentils have been created that contain a superior AA profile (Turfani et al. 2017; Teterycz et al. 2020; Portman et al. 2018; Fathelrahman, Kheri, and Mohamed Ahmed 2015). However, these studies do not assess the *in vivo* protein/AA digestibility or bioavailability. The effects of protein blends on clinical outcomes are addressed in the following section.

Effect of Protein Blends on Clinical Outcomes

Researchers have investigated the effectiveness of protein blends on linear growth in children. One study supplemented complementary diets of rural Malawian children (6–23 months old) for 6–12 months with common bean or cowpea flour and compared them to corn-soy flour (control) (Stephenson et al. 2017). The results revealed that cowpea flour added to corn flour improved linear growth compared to the common bean or corn-soy flour blend (Stephenson et al. 2017).

Another study in Ethiopia compared anthropometrics between children (6–18 months at baseline) whose mothers received an intervention (six months of nutrition education promoting the use of pulses in complementary feeds) and controls (nutrition education that did not address the use of pulses) (Mulualem et al. 2016). In the intervention group, mothers' knowledge and attitudes concerning pulse-based complementary foods increased, resulting in increased feeding of pulse-containing porridge rather than cereal-based porridge. As a result, weight, wasting, and underweight indices improved significantly in children in the intervention group compared to the control group. This demonstrates the improved protein quality of cereal-pulse blends compared to cereal-based foods alone. In addition to blends, researchers have studied mixed-meal feeding using the IAAO method to improve protein quality.

APPLICATION OF THE IAAO METHOD TO ASSESS METABOLIC AVAILABILITY OF METHIONINE IN LENTILS AND THE EFFECT OF PROTEIN COMPLEMENTATION ON PROTEIN SYNTHESIS

The IAAO method was successfully adapted to study protein quality directly in humans by measuring the bioavailability of the most limiting AA in plant foods (Elango et al. 2012). Briefly, the IAAO method is based on the physiological principle that when one indispensable AA is limiting for protein synthesis, all other AA, including the indicator AA, will be preferentially oxidized (Elango, Ball, and Pencharz 2008). The food under study is first analyzed for its protein and AA composition to identify the first limiting AA. When fed below the limiting AA requirement, a decrease in IAAO will be linear with increasing intakes of the limiting AA (Moehn et al. 2005). This patterned IAAO response can be used to measure the change in net protein synthesis with increasing intakes of a proteinaceous food containing a limiting AA. The IAAO response is combined with slope-ratio principles, which are considered the gold standard against which other bioavailability methods are judged (Batterham 1992). The IAAO response to intakes of the test limiting AA from a food is measured and compared to a reference protein (crystalline AA). The true digestibility of crystalline AA in pigs is 100% (Chung and Baker 1992). Therefore, the relative ratio of the IAAO response from food compared to the reference protein is proportional to the whole-body bioavailability of the limiting AA, capturing both absorption and availability of the AA for metabolic processes (i.e., protein synthesis). The IAAO method also permits direct evaluation of the effect of protein complementation (pulse-cereal combinations) in humans (Paoletti et al. 2022; Fakiha et al. 2020; Rafii et al. 2020) (Figure 10.1).

Knowledge of the limiting AA content and its metabolic availability from plant proteins determined using the IAAO method permits the quantification of a more balanced mixture of plant foods that will fulfill dietary AA needs.

Mixed-Meal Complementation: Rice-Chickpea and Millet/Sorghum-Lentils

The IAAO method was first applied to the study of protein complementation by combining chickpeas and rice in a mixed meal. In that study, the separately determined metabolic availability of methionine from steamed rice (metabolic availability = 100%) and cooked chickpeas (metabolic

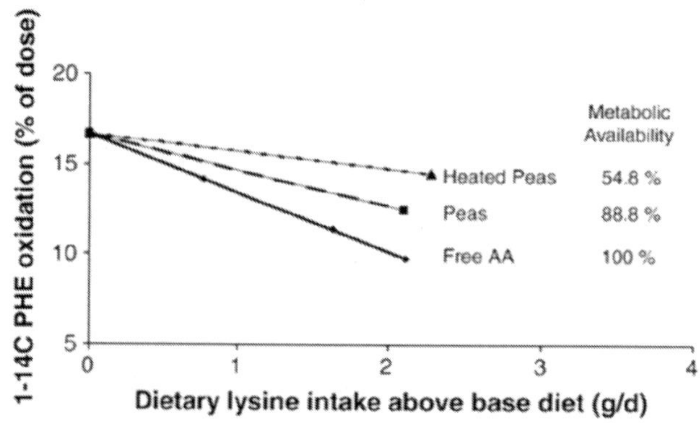

FIGURE 10.1 The validation of the application of the slope-ratio assay to the indicator amino acid oxidation method in an animal study. Note: Adapted with permission from Levesque et al. (2010). Abbreviations: AA, amino acid; Phe, phenylalanine.

TABLE 10.3
Effects of Cereal-Legume Protein Complementation Assessed Using the Combined Slope-Ratio and IAAO Method

Protein Complement	Limiting Amino Acid	Percent Change in Indicator Amino Acid Oxidation Response[1]	Reference
Chickpea + rice	Methionine	6–14	(Rafii et al. 2020)
Rice + lentil	Methionine	6–16	(Rafii et al. 2022)
Millet + lentils	Lysine	27	(Fakiha et al. 2020)
Sorghum + lentils	Lysine	19	(Paoletti et al. 2022)

[1] Percent change in indicator amino acid oxidation response refers to the decrease in the IAAO response of the mixed meal compared to the IAAO response of chickpea, millet, and sorghum proteins fed alone.

availability = 63%) was used to design a mixed meal of cooked rice plus chickpeas (Table 10.3) (Rafii et al. 2020). The aim was to augment the metabolically available methionine intake from chickpea protein by adding methionine from rice protein. The impact of complementation was assessed by comparing IAAO responses at each methionine intake from the cooked chickpeas alone to those of the IAAO responses at the same methionine intakes from cooked chickpeas plus rice after adjusting for the metabolic availability of methionine in chickpeas. This produced a decline in IAAO (6–14%) at each methionine intake compared to IAAO responses of methionine intakes from cooked chickpeas alone (Rafii et al. 2020) (Table 10.3). Another study observed a reduction in IAAO when cooked rice and lentils were fed in a mixed-meal format (Rafii et al. 2022). These studies demonstrate an increase in net protein synthesis due to increased intake of methionine from complementing two protein sources, reflecting the superior protein quality of the mixed meal.

Researchers have also assessed the effect of protein complementation by combining the cereal grains (millet or sorghum), which are limiting in lysine but sufficient in methionine, with lentils (Table 10.3) (Fakiha et al. 2020; Paoletti et al. 2022). Complementing cooked millet with stewed lentils resulted in a 27% reduction in IAAO response. Similarly, researchers observed a 19% reduction in the IAAO response when they combined cooked sorghum with stewed lentils in a complementary format. These studies demonstrate an improvement in the protein quality of the mixed meal, achieved by increasing the intake of the limiting AA from the complementary protein sources.

The results from the series of IAAO studies suggest that a more balanced ratio of 1:1 to 1:2 of lentils to cereal grain would be ideal to meet the demands of the limiting AA in plant-based diets.

CONCLUSION

Various strategies exist to enhance the protein quality of lentil-based diets. Among these strategies (increased protein intake, biofortification, and protein complementation), knowledge of the bioavailability or digestibility of the first limiting AA – methionine – is a necessity. The IAAO method has been used to assess the metabolic availability of methionine in cooked lentils individually and to evaluate the effect of protein complementation (cooked lentil plus rice) in improving lentil protein quality. The data demonstrate that protein complementation is an effective strategy to improve the protein quality of lentil-based diets using ratios of 1:1 to 1:2 of lentil to cereal grains.

MINI DICTIONARY OF TERMS

Bioavailability: The body's ability to use an amino acid for the synthesis of body proteins.

Biofortification: The addition of an amino acid or a protein rich in certain amino acids that are otherwise found in low amounts in foods.

Limiting amino acid: The quantity of an indispensable amino acid present in a food that is limiting, compared to other amino acids, for protein synthesis.

Protein quality: The ability of a food protein source to supply the body with an adequate pattern of amino acids needed for body protein synthesis.

SUMMARY POINTS

- This chapter focuses on strategies to improve the protein quality of lentil-based diets.
- Increasing protein quantity, biotechnology, protein blends, and protein complementation are discussed.
- Increasing plant protein quantity without focusing on protein quality results in reduced diet diversity and the risk of excessive calorie provision, especially in the elderly.
- Biotechnology offers a way to increase the protein and amino acid content of lentils; however, further research is required to test translation *in vivo.*
- Protein blends may be useful; however, they do not directly assess the effect on AA digestibility or bioavailability in humans.
- Protein complementation has been used to assess the improvement of a lentil-based diet directly in humans.
- Protein complementation is the most practical and effective strategy to date to improve the protein quality of lentil-based diets.

REFERENCES

Almeida Costa, Giovana Ermetice De, Keila Da Silva Queiroz-Monici, Soely Maria Pissini Machado Reis, and Admar Costa De Oliveira. 2006. "Chemical Composition, Dietary Fibre and Resistant Starch Contents of Raw and Cooked Pea, Common Bean, Chickpea and Lentil Legumes." *Food Chemistry* 94 (3): 327–30. https://doi.org/10.1016/j.foodchem.2004.11.020.

Bagga, S., C. Potenza, J. Ross, M. N. Martin, T. Leustek, and C. Sengupta-Gopalan. 2005. "A Transgene for High Methionine Protein Is Posttranscriptionally Regulated by Methionine." *In Vitro Cellular & Developmental Biology* 41: 731–41.

Batterham, E. S. 1992. "Availability and Utilization of Amino Acids for Growing Pigs." *Nutrition Research Reviews* 5: 1–18. https://doi.org/10.1079/NRR19920004.

Bhatty, R. S. 1986. "Protein Subunits and Amino Acid Composition of Wild Lentil." *Phytochemistry* 25 (3): 641–44. https://doi.org/10.1016/0031-9422(86)88015-3.

Bhatty, R. S., G. I. Christison, and C. D. Centre. 1984. "Laboratory Analysis Proximate Composition of the Three Meals, Protein Concentrates and Isolates." *Plant Foods for Human Nutrition* 34: 41–51.

Block, R. J., and H. H. Mitchell. 1946. "The Correlation of the Amino-Acid Composition of Proteins with Their Nutritive Value." *Nutrition Abstract Review* 16: 249–78.

Bressani, R., N. S. Scrimshaw, M. Béhar, and F. Viteri. 1958. "Supplementation of Cereal Proteins with Amino Acids. II. Effect of Amino Acid Supplementation of Corn-Masa at Intermediate Levels of Protein Intake on the Nitrogen Retention of Young Children." *The Journal of Nutrition* 66 (4): 501–13. https://doi.org/10.1093/jn/66.4.501.

Bressani, R., D. L. Wilson, M. Béhar, and N. S. Scrimshaw. 1960. "Supplementation of Cereal Proteins with Amino Acids. III. Effect of Amino Acid Supplementation of Wheat Flour as Measured by Nitrogen Retention of Young Children." *The Journal of Nutrition* 70 (2): 176–86. https://doi.org/10.1093/jn/70.2.176.

Bressani, R., D. L. Wilson, M. Chung, M. Behar, and N. S. Scrimshaw. 1963. "Supplementation of Cereal Proteins with Amino Acids VI. Effect of Amino Acid Supplementation of Rolled Oats as Measured by Nitrogen Retention in Young Children." *The Journal of Nutrition* 81 (4): 399–404. https://doi.org/10.1093/jn/81.4.399.

WHO (1998). "Complementary feeding of young children in developing countries: A review of current scientific knowledge." World Health Organization. https://iris.who.int/handle/10665/65932.

Chung, T. K., and D. H. Baker. 1992. "Apparent and True Amino Acid Digestibility of a Crystalline Amino Acid Mixture and of Casein: Comparison of Values Obtained with Ileal-Cannulated Pigs and Cecectomized Cockerels." *Journal of Animal Science* 70 (12): 3781–90. https://doi.org//1992.70123781x.

Davis, K. R. 1981. "Effect of Processing on Composition and Tetrahymena Relative Nutritive Value on Green and Yellow Peas, Lentils and White Pea Beans." *Cereal Chemistry* 58 (5): 454–60.

Dhull, Sanju Bala, Joyce Kinabo, and Mark A. Uebersax. 2023. "Nutrient Profile and Effect of Processing Methods on the Composition and Functional Properties of Lentils (Lens Culinaris Medik): A Review." *Legume Science* 5 (1): 1–14. https://doi.org/10.1002/leg3.156.

Elango, Rajavel, Ronald O. Ball, and Paul B. Pencharz. 2008. "Indicator Amino Acid Oxidation: Concept and Application." *The Journal of Nutrition* 138 (2): 243–46. https://doi.org/10.1093/jn/138.2.243.

Elango, Rajavel, C. Levesque, R. O. Ball, and P. B. Pencharz. 2012. "Available versus Digestible Amino Acids – New Stable Isotope Methods." *British Journal of Nutrition* 108 (S2): S306–14. https://doi.org/10.1017/S0007114512002498.

Fakiha, A., Z. Tul-Noor, A. Paoletti, P. B. Pencharz, R. O. Ball, C. L. Levesque, R. Elango, and G. Courtney-Martin. 2020. "Bioavailable Lysine, Assessed in Healthy Young Men Using Indicator Amino Acid Oxidation, Is Greater When Cooked Millet and Stewed Canadian Lentils Are Combined." *The Journal of Nutrition* 150 (10): 2729–37. https://doi.org/10.1093/jn/nxaa086.

FAO. 2013. *Dietary Protein Quality Evaluation in Human Nutrition*. Rome: FAO. http://www.nutrinfo.com/biblioteca/libros_digitales/fao_protein_quality.pdf.

FAO. 2014. *Research Approaches and Methods for Evaluating the Protein Quality of Human Foods*. Rome: FAO. https://doi.org/10.1017/CBO9781107415324.004.

FAO. 2021. "Crop Production and Trade Data." http://www.fao.org/faostat/en/#home.

FAO/WHO/UNU, Joint. 2005. "Human Energy Requirements. Scientific Background Papers from the Joint FAO/WHO/UNU Expert Consultation. October 17–24, 2001. Rome, Italy." *Public Health Nutrition* 8 (7A): 929–1228. http://www.ncbi.nlm.nih.gov/pubmed/19856424.

Fathelrahman, N. A., N. A. Kheri, and I. A. Mohamed Ahmed. 2015. "Development of Weaning Food from Wheat Flour Supplemented with Defatted Sesame Flour." *Journal of Biological & Agricultural Health* 16: 9–20.

Fazal Ali, J. A., M. A. Arain, and N. A. Shaikh. 2010. "Genetic Manipulation of Lentil through Induced Mutations." *Pakistan Journal of Botany* 42 (5): 3449–55.

Ghosh, S., M. Smriga, F. Vuvor, D. Suri, H. Mohammed, S. M. Armah, and N. S. Scrimshaw. 2010. "Effect of Lysine Supplementation on Health and Morbidity in Subjects Belonging to Poor Peri-Urban Households in Accra, Ghana." *American Journal of Clinical Nutrition* 92 (4): 928–39. https://doi.org/10.3945/ajcn.2009.28834.

Ghosh, S., D. Suri, and R. Uauy. 2012. "Assessment of Protein Adequacy in Developing Countries: Quality Matters." *British Journal of Nutrition* 108 (S2): S77–87. https://doi.org/10.1177/156482651303400217.

Gorissen, Stefan H. M., Astrid M. H. Horstman, Rinske Franssen, Julie J. R. Crombag, Henning Langer, Jörgen Bierau, Frederique Respondek, and Luc J. C. van Loon. 2016. "Ingestion of Wheat Protein Increases in Vivo Muscle Protein Synthesis Rates in Healthy Older Men in a Randomized Trial." *Journal of Nutrition* 146 (9): 1651–59. https://doi.org/10.3945/jn.116.231340.

Gorissen, Stefan H. M., Julie J. R. Crombag, Joan M. G. Senden, W. A. Huub Waterval, Jörgen Bierau, Lex B. Verdijk, and Luc J. C. van Loon. 2018. "Protein Content and Amino Acid Composition of Commercially Available Plant-Based Protein Isolates." *Amino Acids* 50 (12): 1685–95. https://doi.org/10.1007/s00726-018-2640-5.

Grasso, Alessandra C., Margreet R. Olthof, Corné van Dooren, Roline Broekema, Marjolein Visser, and Ingeborg A. Brouwer. 2021. "Protein for a Healthy Future: How to Increase Protein Intake in an Environmentally Sustainable Way in Older Adults in the Netherlands." *Journal of Nutrition* 151 (1): 109–19. https://doi.org/10.1093/jn/nxaa322.

Grusak, M. A. 2009. "Nutritional and Health-Beneficial Quality." In *The Lentil: Botany, Production and Uses*, edited by W. Erskine, F. J. Muehlbauer, A. Sarker, and B. Sharma, 368–90. CABI Publishing. https://doi.org/https://doi.org/10. 1079/9781845934873.0368.

Health Canada. 2019. *Canada's Dietary Guidelines for Health Professionals and Policy Makers*. Ottawa: Health Canada. https://food-guide.canada.ca/static/assets/pdf/CDG-EN-2018.pdf.

Hussain, T., S. Abbas, M. A. Khan, and N. S. Scrimshaw. 2004. "Lysine Fortification of Wheat Flour Improves Selected Indices of the Nutritional Status of Predominantly Cereal-Eating Families in Pakistan." *Food and Nutrition Bulletin* 25 (2): 114–22. https://doi.org/10.1177/156482650402500202.

IOM. 2006. *Dietary Reference Intakes: The Essential Guide to Nutrient Requirements*. Edited by J. J. Otten, J. Pitzi Hellwig, and L. D. Meyers. Washington, DC: The National Academies Press. https://doi.org/10.1016/S0002-8223(02)90177-X.

Khazaei, Hamid, Maya Subedi, Mike Nickerson, Cristina Martínez-Villaluenga, Juana Frias, and Albert Vandenberg. 2019. "Seed Protein of Lentils: Current Status, Progress, and Food Applications." *Foods* 8 (9): 1–23. https://doi.org/10.3390/foods8090391.

Levesque, C. L., S. Moehn, P. B. Pencharz, and R. O. Ball. 2010. "Review of Advances in Metabolic Bioavailability of Amino Acids." *Livestock Science* 133: 4–9. https://doi.org/10.1016/j.livsci.2010.06.013.

Moehn, Soenke, Robert F. P. Bertolo, Paul B. Pencharz, and Ronald O. Ball. 2005. "Development of the Indicator Amino Acid Oxidation Technique to Determine the Availability of Amino Acids from Dietary Protein in Pigs." *Journal of Nutrition* 135 (12): 2866–70. https://doi.org/10.1093/jn/135.12.2866.

Mulualem, Demmelash, Carol J. Henry, Getenesh Berhanu, and Susan J. Whiting. 2016. "The Effectiveness of Nutrition Education: Applying the Health Belief Model in Child-Feeding Practices to Use Pulses for Complementary Feeding in Southern Ethiopia." *Ecology of Food and Nutrition* 55 (3): 308–23. https://doi.org/10.1080/03670244.2016.1161617.

Nosworthy, Matthew G., Jason Neufeld, Peter Frohlich, Gina Young, Linda Malcolmson, and James D. House. 2017. "Determination of the Protein Quality of Cooked Canadian Pulses." (January): 896–903. https://doi.org/10.1002/fsn3.473.

NRC, and NAS. 1988. *Amino Acid Fortification of Cereals: Results and Interpretation of Trials in Three Countries: A Report of the Task Force on Amino Acid Fortification of Cereals, Committee on International Nutrition Programs, Food and Nutrition Board, Commission on Life Sc.* Ithaca, NY: Cornell University Program in International Nutrition.

Oikawa, Sara Y., Ravninder Bahniwal, Tanya M. Holloway, Changhyun Lim, Jonathan C. McLeod, Chris McGlory, Steven K. Baker, and Stuart M. Phillips. 2020. "Potato Protein Isolate Stimulates Muscle Protein Synthesis at Rest and with Resistance Exercise in Young Women." *Nutrients* 12 (5): 1–13. https://doi.org/10.3390/nu12051235.

Paoletti, A., A. Fakiha, Z. Tul-Noor, P. B. Pencharz, C. L. Levesque, Ronald O. Ball, D. Kong, R. Elango, and G. Courtney-Martin. 2022. "Bioavailable Lysine Assessed Using the Indicator Amino Acid Oxidation Method in Healthy Young Males Is High When Sorghum Is Cooked by a Moist Cooking Method." *The Journal of Nutrition* 152 (3): 770–78. https://doi.org/10.1093/jn/nxab410.

Paoletti, A., P. B. Pencharz, R. O. Ball, D. Kong, L. Xu, R. Elango, and G. Courtney-Martin. 2023. "The Dietary Requirement for Total Sulphur Amino Acids in Adults ≥60 Years Appears Higher in Males than in Females." *American Journal of Clinical Nutrition* 118 (3): 538–48. https://doi.org/10.1016/j.ajcnut.2023.06.015.

Pastorello, E. A., C. Pompei, V. Pravettoni, O. Brenna, L. Farioli, C. Trambaioli, and A. Conti. 2001. "Lipid Transfer Proteins and 2S Albumins as Allergens." *Allergy* 56: 45–47.

Portman, D., C. Blanchard, P. Maharjan, L. S. McDonald, J. Mawson, M. Naiker, and J. F. Panozzo. 2018. "Blending Studies Using Wheat and Lentil Cotyledon Flour—Effects on Rheology and Bread Quality." *Cereal Chemistry* 95 (6): 849–60.

Qureshi, A., S. Wani, A. Lone, Z. Dar, S. Wani, and F. Nehvi. 2013. "Breeding for Quality Traits in Grain Legumes." In *Conventional and Non-Conventional Interventions in Crop Improvement*, edited by C. Malik, G. Sanghera, and S. Wani, 1st ed. New Delhi: M D Publishers.

Rafii, Mahroukh, P. B. Pencharz, R. O. Ball, C. Tomlinson, R. Elango, and G. Courtney-Martin. 2020. "Bioavailable Methionine Assessed Using the Indicator Amino Acid Oxidation Method Is Greater When Cooked Chickpeas and Steamed Rice Are Combined in Healthy Young Men." *The Journal of Nutrition* 150 (7): 1834–44. https://doi.org/10.1093/jn/nxaa086.

Rafii, Mahroukh, Paul B. Pencharz, Kaylem Boileau, Ronald O. Ball, Christopher Tomlinson, Rajavel Elango, and Glenda Courtney-Martin. 2022. "Metabolic Availability of Methionine Assessed Using Indicator Amino Acid Oxidation Method, Is Greater When Cooked Lentils and Steamed Rice Are Combined in the Diet of Healthy Young Men." *Journal of Nutrition* 152 (6): 1467–75. https://doi.org/10.1093/jn/nxac049.

Ravindran, V., L. M. Tabe, L. Molvig, T. J. V. Higgins, and W. L. Bryden. 2002. "Nutritional Evaluation of Transgenic High-Methionine Lupins (Lupinus Angustifolius L) with Broiler Chickens." *Journal of the Science of Food and Agriculture* 82: 280–85.

Saalbach, I., D. Waddell, T. Pickardt, O. Schieder, and K. Müntz. 1995. "Stable Expression of the Sulphur-rich 2S Albumin Gene in Transgenic Vicia Narbonensis Increases the Methionine Content of Seeds." *Journal of Plant Physiology* 145: 674–81.

Salaria, Sonia, Jon Lucas Boatwright, Pushparajah Thavarajah, Shiv Kumar, and Dil Thavarajah. 2022. "Protein Biofortification in Lentils (Lens Culinaris Medik.) Toward Human Health." *Frontiers in Plant Science* 13 (April). https://doi.org/10.3389/fpls.2022.869713.

Shahzad, Rahil, Shakra Jamil, Shakeel Ahmad, Amina Nisar, Sipper Khan, Zarmaha Amina, Shamsa Kanwal, Hafiz Muhammad Usman Aslam, Rafaqat Ali Gill, and Weijun Zhou. 2021. "Biofortification of Cereals and Pulses Using New Breeding Techniques: Current and Future Perspectives." *Frontiers in Nutrition* 8 (October): 1–23. https://doi.org/10.3389/fnut.2021.721728.

Sikka, K. C., and R. P. Johari. 1979. "Comparative Nutritive Value and Amino Acid Content of Different Varieties of Sorghum and Effect of Lysine Fortification." *Journal of Agricultural and Food Chemistry* 27 (5): 962–65. https://doi.org/10.1021/jf60225a039.

Stephenson, Kevin B., Sophia E. Agapova, Oscar Divala, Yankho Kaimila, Kenneth M. Maleta, Chrissie Thakwalakwa, M. Isabel Ordiz, Indi Trehan, and Mark J. Manary. 2017. "Complementary Feeding with Cowpea Reduces Growth Faltering in Rural Malawian Infants: A Blind, Randomized Controlled Clinical Trial." *American Journal of Clinical Nutrition* 106 (6): 1500–7. https://doi.org/10.3945/ajcn.117.160986.

Szwiega, Sylwia, Paul B. Pencharz, Mahroukh Rafii, Mackenzie Lebarron, Jessica Chang, Ronald O. Ball, Dehan Kong, Libai Xu, Rajavel Elango, and Glenda Courtney-Martin. 2020. "Dietary Leucine Requirement of Older Men and Women Is Higher than Current Recommendations." *The American Journal of Clinical Nutrition*, 410–19. https://doi.org/10.1093/ajcn/nqaa323.

Tan, J., L. Wu, J. Peacock, and E. S. Dennis. 2022. "Capturing Hybrid Vigor for Lentil Breeding." *Crop Science* 62: 1787–96. https://doi.org/https://doi.org/10.1002/csc2.20777.

Temba, Makumba C., Patrick B. Njobeh, Oluwafemi A. Adebo, Adetola O. Olugbile, and Eugenie Kayitesi. 2016. "The Role of Compositing Cereals with Legumes to Alleviate Protein Energy Malnutrition in Africa." *International Journal of Food Science and Technology* 51 (3): 543–54. https://doi.org/10.1111/ijfs.13035.

Teterycz, Dorota, Aldona Sobota, Piotr Zarzycki, and Agnieszka Latoch. 2020. "Legume Flour as a Natural Colouring Component in Pasta Production." *Journal of Food Science and Technology* 57 (1): 301–9. https://doi.org/10.1007/s13197-019-04061-5.

Tripathi, Kuldeep, Padmavati G. Gore, Anjula Pandey, Rakesh Bhardwaj, Neeta Singh, Gautam Chawla, and Ashok Kumar. 2019. "Seed Morphology, Quality Traits and Imbibition Behaviour Study of Atypical Lentil (Lens Culinaris Medik.) from Rajasthan, India." *Genetic Resources and Crop Evolution* 66 (3): 697–706. https://doi.org/10.1007/s10722-019-00745-1.

Turfani, Valeria, Valentina Narducci, Alessandra Durazzo, Vincenzo Galli, and Marina Carcea. 2017. "Technological, Nutritional and Functional Properties of Wheat Bread Enriched with Lentil or Carob Flours." *Lwt* 78: 361–66. https://doi.org/10.1016/j.lwt.2016.12.030.

Waterlow, J. C. 1984. "Protein Turnover With Special Reference To Man." *Quarterly Journal of Experimental Physiology* 69 (3): 409–38. https://doi.org/10.1113/expphysiol.1984.sp002829.

Willett, Walter, Johan Rockström, Brent Loken, Marco Springmann, Tim Lang, Sonja Vermeulen, Tara Garnett, et al. 2019. "Food in the Anthropocene: The EAT–Lancet Commission on Healthy Diets from Sustainable Food Systems." *The Lancet* 393 (10170): 447–92. https://doi.org/10.1016/S0140-6736(18)31788-4.

Young, V. R., C. Gucalp, W. M. Rand, D. E. Matthews, and D. M. Bier. 1987. "Leucine Kinetics during Three Weeks at Submaintenance-to-Maintenance Intakes of Leucine in Men: Adaptation and Accommodation." *Human Nutrition: Clinical Nutrition* 41: 1–18.

Young, V. R., and J. S. Marchini. 1990. "Mechanisms and Nutritional Significance of Metabolic Responses to Altered Intakes of Protein and Amino Acids, with Reference to Nutritional Adaptation in Humans." *American Journal of Clinical Nutrition* 51 (2): 270–89. https://doi.org/10.1093/ajcn/51.2.270.

Zhao, W., F. Zhai, D. Zhang, Y. An, Y. Liu, Y. He, K. Ge, and N. S. Scrimshaw. 2004. "Lysine-Fortified Wheat Flour Improves the Nutritional and Immunological Status of Wheat-Eating Families in Northern China." *Food and Nutrition Bulletin* 25 (2): 123–29. https://doi.org/10.1177/156482650402500203.

11 Improving Micronutrient Density through Increasing Protein Quality

A Strategy to Improve Diet Quality

Delaram Ghodsi, Maryam Amini, Bahareh Nikooyeh, and Tirang R. Neyestani

ABBREVIATIONS

BV	Biological value
CHD	Coronary heart disease
CVD	Cardiovascular disease
DAA	Dispensable amino acid
DEAA	Digestible essential amino acid
DHA	Docosahexaenoic acid
DIAAS	Digestible indispensable amino acid score
EPIC	European Prospective Investigation into Cancer and Nutrition
EAA	Essential amino acid
EPA	Eicosapentaenoic acid
FAO	Food and Agriculture Organization
FCT	Food composition table
HR	Hazard ratio
LAA	Limiting amino acid
NPPU	Net postprandial protein utilization
OR	Odds ratio
PDCAAS	Protein digestibility-corrected amino acid score
PER	Protein efficiency ratio
PUFA	Polyunsaturated fatty acid
Q	Quartile
RR	Relative risk
SD	Standard deviation
SES	Socioeconomic status
SFA	Saturated fatty acid
SUCRA	Surface under the cumulative ranking
T	Tertile
TEI	Total energy intake
USDA	United States Department of Agriculture
WHO	World Health Organization

DOI: 10.1201/b23328-11

INTRODUCTION

Proteins are the most significant components of the body, with different structures and functions that make life possible. To develop a healthy dietary pattern, considering the quantity and quality of proteins is important. The dietary quality of proteins, which are polymers of amino acids, indicates to what extent they meet the body's requirements and depends on the content and bioavailability of their amino acids (Shevkani and Chourasia 2021).

Recent dietary guidelines recommend a shift toward more consumption of plant-based foods to support environmental sustainability, improve health, and reduce the risk of diseases,. Sustainable diets that aim to protect our planet and simultaneously meet the nutritional needs of the world's population are a new topic that is also relevant to the quality of dietary proteins (Langyan et al. 2021). Although plant-based dietary patterns may be preferable, dietary animal proteins, as compared with plant proteins, are more digestible. In addition, plant dietary proteins have less potency to stimulate muscle protein synthesis (Berrazaga et al. 2019).

Combining animal and plant protein is a strategy to complement amino acid profiles and enhance protein quality. A mixture of different types of plant protein can also complement amino acid compositions and produce high-quality protein. Therefore, diets that include staple foods (i.e., wheat, maize, rice, etc.) and other plants often have high protein quality as their amino acid profiles complement one another. Mixing different dietary protein sources in suitable proportions can improve amino acid profiles and yield high-quality protein. However, this complementary approach can affect both the micronutrient and energy density of the resulting food (Comerford et al. 2021). Notwithstanding, the shift toward the consumption of diverse types of proteins may not be easy and needs to be facilitated in some populations (Pikosky, Ragalie-Carr, and Miller 2022). In this chapter, the association between the quality of different protein sources and mixtures with micronutrient density and its potential health implications are comprehensively discussed.

THE IMPORTANCE OF DIETARY PROTEINS

The initiation of life is owed to the function of enzymes, and all enzymes are proteins. Proteins, as hormones, cytokines, and other mediators, make cellular cross-talk possible. Blood stabilization is achieved by secreting proteins from the liver, which makes plasma a solution with suitable viscosity and osmolarity. These plasma proteins also serve as carriers of different types of compounds in the blood. The most abundant protein deposit in the body is in muscle. Mobility without muscle contraction is not possible. Furthermore, muscle contraction is vital for pumping oxygen and nutrients throughout the body, for inhalation and exhalation by our lungs, and for movement. The body needs a constant protein supply to sustain all these vital functions in a homeostatic manner. The human body is equipped with synthetic machinery to produce the protein compounds needed from amino acids (AAs), the building blocks. Though some amino acids can be generated in the body (non-essential or dispensable amino acids; NEAAs or DAAs), there are nine amino acids that the body cannot generate (essential or indispensable amino acids; EAAs or IAAs), including phenylalanine, tryptophan, threonine, lysine, leucine, isoleucine, valine, methionine, and histidine. A high-quality or complete protein contains all EAAs in sufficient amounts to meet body needs. Animal proteins like meat, chicken, egg, and milk provide high-quality proteins. Those EAAs that are present in a protein in quantities below minimal body requirements are referred to as "limiting amino acids" (LAAs). Plant proteins commonly have one or more LAAs, and hence are considered incomplete proteins (Lopez and Mohiuddin 2023).

To establish a healthy eating pattern, ingesting a variety of high-quality proteins that fulfill our requirements and ensure adequate nutrient density is highly recommended. The intake of plant proteins is associated with lower all-cause and cardiovascular mortality and increased longevity (Naghshi et al. 2020). In addition, plant-based foods are considered much more sustainable than foods from animal sources (Langyan et al. 2021). The importance of sustainability becomes more salient when considering that the global population will soon surpass ten billion. Consequently, new dietary guidelines recommend the intake of plenty of vegetables and fruits to maximize the advantages of plants, including plant proteins (Binns et al. 2021).

DIETARY PROTEIN QUALITY

The ability of a food item to fulfill the body's protein requirements depends on the composition of its amino acids and their bioavailability. The quantity of amino acids in a food product may not necessarily reflect the amount that can be used by the body for its metabolic needs after consumption. Protein quality refers to the bioavailability and digestibility of protein and its ability to supply sufficient amounts of EAAs and nitrogen in absorbable forms to fulfill the body's protein requirement (Katz et al. 2019).

This issue is particularly relevant in developing countries where many people may have lower protein intake than their body requires and the majority of their protein intake comes from foods with low protein quality. In communities where protein intake is low and plant-based proteins such as cereals and legumes are the major dietary protein sources, the protein quality of the diet may be low due to poor digestion and the presence of high amounts of insoluble fiber and anti-nutritional factors (Cominelli et al. 2022). This issue should be considered for dietary recommendations, especially for vulnerable people who are at risk of malnutrition.

Generally, animal protein sources are considered high-quality proteins since they provide all essential amino acids (EAAs) in the right proportions, whereas plant proteins are known as low-quality proteins as they lack one or more EAAs (limiting amino acids; LAAs) in sufficient amounts compared with a reference protein like egg white. However, this traditional concept has recently been challenged (Mariotti 2017). In addition, plant foods are more environmentally friendly due to their lower greenhouse gas emissions per gram of protein and their overall effects on health. It is well known that lysine in cereals and methionine in legumes are LAAs. Consequently, these two food groups can be combined as a mixed food to increase protein quality through enhancement of the resulting amino acid profile, a process known as protein complementation. It should be noted that not all combinations of two plant protein sources give the desired quality (Adhikari et al. 2022a).

FACTORS AFFECTING PROTEIN QUALITY

All foods undergo some sort of pre-consumption processing, from gentle practices like washing and peeling to more invasive heating methods like boiling, baking, frying, grilling, and steaming, all of which can potentially influence the protein quality in the food. Processing can affect the digestibility of the protein and, hence, alter the bioavailability of the EAAs (Adhikari et al. 2022a). At high temperatures, amino acids may undergo chemical reactions that render them nutritionally unavailable. The Maillard reaction is a common example, where lysine reacts with reducing sugars to form blocked (or glycated) lysine that has poor bioavailability (Seiquer et al. 2006; Adhikari et al. 2022a).

Processing methods, duration, and storage conditions can have different effects on the amount of blocked lysine and overall bioavailability of amino acids. When heat and alkaline treatments are involved, the digestibility of not only lysine but also cysteine, threonine, and total protein can be extremely decreased. These processes can cause the conversion of L-amino acids into racemic forms and the creation of interconnected peptides, which not only decrease amino acid availability but also impede enzyme activity, resulting in reduced digestibility of amino acids. On the other hand, some processing methods can enhance protein digestibility by modifying the non-protein elements in the food. An example of this is the breakdown of cell walls, which enhances the protein's accessibility to digestive enzymes and consequently improves its digestibility. In general, it is evident that the processing of foods can significantly influence protein quality by affecting both the amino acids and the digestibility of the proteins (Adhikari et al. 2022a).

Certain components found in foods can also affect the quality of the food protein. For instance, legumes, cereals, potatoes, and tomatoes may contain inhibitors of proteolytic enzymes like trypsin, chymotrypsin, carboxypeptidases, and elastase, thereby decreasing protein digestibility and quality (Losso 2008). Phytic acid, found in cereals and legumes, is considered an anti-nutritional compound due to its negative charge that allows it to bind some minerals and proteins, thus reducing their bioavailability for absorption and utilization (Wang and Guo 2021). Phenolic compounds, such as tannins and saponins found in foods of plant origin, can form complexes with amino acids, acting as undesirable biological components (Kaspchak et al. 2020). However, adequate washing and heating processes, including steaming, may significantly decrease the amount of saponins in food (Rachwa-Rosiak, Nebesny, and Budryn 2015).

ASSESSMENT TOOLS

Different methods have been employed to evaluate the quality of proteins in different foods. Here, we briefly review some selected methods.

Protein Efficiency Ratio (PER)

This rat-based bioassay examines growth in relation to nitrogen retained from food products, comparing results to standard casein protein to provide insights into the biological value (BV) of the food product. BV indicates the quantity of nitrogen retained in the body, which is believed to correspond to nitrogen utilized for protein synthesis (Mansilla et al. 2020). However, PER is limited by its focus on a single level of protein intake and its specificity to rats. Therefore, new methods were necessary to determine protein quality for humans, considering the bioavailability of individual amino acids as well.

Protein Digestibility-Corrected Amino Acid Score (PDCAAS)

PDCAAS considers both the nutritional value of individual amino acids and the digestion of dietary protein. It is expressed as a percentage, comparing the amount of the most limiting amino acid in the test food to a reference amino acid pattern. The recommended reference pattern for all age groups, except infants, is based on the essential amino acid requirements of preschool-age children. Two factors are multiplied together to determine a food's PDCAAS: the amino acid score of the dietary protein, which is obtained by dividing indispensable amino acid requirements by the minimum requirement of high-quality protein, and its true fecal nitrogen digestibility (Mansilla et al. 2020).

While PDCAAS addressed some limitations of PER by incorporating human-specific amino acid requirements, it still has drawbacks. These include reliance on fecal nitrogen digestibility, which can be affected by gut microbiota salvaging, and the assumption that it represents the digestibility of individual amino acids (Monsonego-Ornan and Reifen 2022). Therefore, further enhancements were necessary to determine protein quality.

Digestible Indispensable Amino Acid Score

At the 2011 FAO Expert Committee meeting in New Zealand, it was agreed that dietary amino acids should be considered individual nutrients, and data on their digestibility and bioavailability should be provided individually (FAO 2013). This decision was based on the recognition that fecal digestibility measurement used in PDCAAS pertains to the whole digestive tract, which may overestimate the quality value of the tested protein. Instead, preference was given to determining ileal digestibility (Monsonego Ornan and Reifen 2022). These considerations led to the development of a new protein quality measure called the digestible indispensable amino acid score (DIAAS)

(Marinangeli and House 2017). Nevertheless, DIAAS also has limitations, especially when evaluating plant proteins (Monsonego Ornan and Reifen 2022). DIAAS is calculated using the following equation:

DIAAS% = 100 × [(mg of digestible dietary EAA in 1 g of the dietary protein) / (mg of the same dietary EAA in 1 g of the reference protein)].

Protein quality assessment tools that consider the concentrations and digestibility of individual amino acids are generally preferable. These methods enable us to rank different proteins and evaluate the complementarity of various protein sources. However, it is important to note that this assessment should be done at the meal level rather than the diet level. Additionally, when integrating protein quality into overall diet evaluation, measurements on an individual amino acid basis are recommended. Protein quality data play a crucial role in developing healthy and sustainable diets (Adhikari et al. 2022b).

Net Postprandial Protein Utilization (NPPU)

In this method, dietary proteins labeled with ^{15}N are given to subjects to track how nitrogen is processed in various metabolic pools, including serum protein, blood urea, urinary urea, and ileal digesta, over an eight-hour period. The retention of nitrogen in each pool is then used to determine the net postprandial protein utilization (NPPU). The total nitrogen flow rate can be calculated by multiplying the nitrogen content measured in the freeze-dried ileal sample by the dry matter of the sample and the flow rate of ileal digestion. The exogenous nitrogen flow rate is determined by adjusting the total nitrogen flow rate based on the ratio of ^{15}N enrichment excess in the waste products to the ^{15}N content in the initial meal. Using NPPU, a recent study has shown that pea isolate can be considered a high-quality protein (Guillin et al. 2022).

COMPARISON OF NUTRIENT DENSITY IN DIFFERENT DIETARY PROTEINS: ANIMAL VS. PLANT

Current evidence shows that nutrient intake might be different among those consuming animal vs. plant-based proteins. Plant-based proteins such as dried beans are rich sources of total and soluble fiber, as well as resistant starch. These components contribute to the low glycemic index of these foods. Additionally, they contain abundant polyphenols, which are powerful antioxidants (Singh et al. 2021). In the European Prospective Investigation into Cancer and Nutrition (EPIC)-Oxford study, individuals classified as "meat eaters" had higher energy intake and a different nutrient profile compared to vegetarians. Meat eaters had higher intakes of saturated fatty acids (SFAs) but lower intakes of fiber and polyunsaturated fatty acids (PUFAs). They also consumed more vitamin B_{12}, vitamin D, zinc, and iodine than vegetarians (Sobiecki et al. 2016).

In the United States, researchers have observed that animal-based protein foods such as meat contribute more to protein intake and several "positive" nutrients (notably zinc, vitamin B12, and highly bioavailable iron), but also to several "negative" nutrients (cholesterol and SFAs) compared to plant-based protein foods. Conversely, plant-based protein foods contributed more to the intake of other beneficial nutrients (such as dietary fiber, vitamin E, magnesium, and PUFAs) (Phillips et al. 2015).

It is important to note that both animal and plant proteins are consumed along with other nutrients and substances, forming a "protein package." Therefore, plant and animal proteins come with specific nutrient clusters (Mariotti 2019). Understanding this concept is crucial for long-term health considerations regarding protein. Combining protein from both animal and plant sources can help address potential deficiencies in micronutrients and ensure the attainment of the necessary micronutrient profile. Along the same lines, analysis of 141 nutrition surveys from North America, Europe, and South and East Asia revealed that plant-based diets are accompanied by lower intakes of protein, eicosapentaenoic acid (EPA), docosahexaenoic acid (DHA), vitamin B_{12}, iron, zinc, iodine, calcium, and lower bone turnover biomarkers. On the other hand, meat consumers may be at higher risk of inadequate intakes of fiber, folate, calcium, magnesium, and PUFA, including α-linoleic acid (ALA) (Neufingerl and Eilander 2021).

Though many believe that a plant-based diet is more consistent with general health, lower production of greenhouse gases, and thus lesser environmental damage and greater sustainability, the amount and sources of proteins are axial determinants of a healthy plant-based diet (Perraud et al. 2022). A recent study of 1,125 French adults documented that the percentage of plant protein in the diet is associated with overall nutrient adequacy, long-term health effects, and environmental impacts, with high plant-protein diets being more sustainable (Fouillet et al. 2023). Additionally, replacing animal proteins with legumes can enhance the intake of folic acid, high-quality carbohydrates (dietary fiber), and various other nutrients. Consequently, growing evidence highlights the need to redefine the notion of protein quality, incorporating considerations of superior nutritional value. In fact, some researchers suggest that a well-balanced combination of plant protein sources can result in greater reductions in saturated fat intake and increased intake of fiber, vitamin A, vitamin C, folate, magnesium, and potassium compared to animal sources (de la O et al. 2023; Katz et al. 2019).

PROTEIN COMPLEMENTATION AND MICRONUTRIENT DENSITY

Evidence suggests a direct association between protein quality and the micronutrient content of foods. A cross-sectional study of healthy young adults reported that a high protein diet is accompanied by higher dietary quality and micronutrient intake, including iron, zinc, calcium, magnesium, and vitamins A, D, E, C, B_1, B_2, B3, and B_{12} (Gwin et al. 2019). Similarly, a recent study that analyzed dietary intake from 17,535 participants revealed a direct association between dietary protein quality and adequate intake of micronutrients, including iron, zinc, magnesium, calcium, selenium, potassium, and various B vitamins, as well as vitamins A, C, and E (de la O et al. 2023).

Although most studies have revealed a positive association between dietary protein quality and micronutrient density, the key question is whether protein complementation can enhance the micronutrient density of mixed food. Low income and socioeconomic status (SES) usually shift food choices toward cereals, including rice, bread, and pasta (Sobhani, Eini-Zinab, and Rezazadeh 2022). Legumes, on the other hand, are good sources of protein and micronutrients (Didinger and Thompson 2021). Consequently, combining cereals and legumes not only improves the resultant protein quality but also substantially enhances the micronutrient content of the food. By consuming complementary proteins, individuals can obtain a wider range of vitamins and minerals, thus improving their overall nutrient intake.

Though legumes and pulses can be good substitutes for animal protein and are commonly consumed by households with lower income, some evidence indicates legume consumption is inversely associated with SES in certain communities (Sobhani, Eini-Zinab, and Rezazadeh 2022; Skórska, Grajeta, and Zabłocka-Słowińska 2021).

In Iran, the price of legumes, though much cheaper than meats, is one of the major determinants of their consumption. Nevertheless, adding even a small portion of legumes to a cereal-based diet can improve its micronutrient content dramatically. For instance, one common Iranian cuisine involves mixing lentils with rice or beans with rice at a ratio of 1 to 3. Another common practice is consuming bread together with beans. Similar combinations are used in cookery in other parts of the world, including South America. The nutritional values provided by these combinations surpass those of each individual food item when consumed separately (Table 11.1). These foods, when eaten in a usual portion size (~250–300 g), can provide a sizable amount of micronutrients (Figure 11.1).

TABLE 11.1
The Amount of Energy and some Selected Nutrients in 100 g of some Cereals (Cooked Rice, Pita Bread and Cooked Macaroni) and 30 g of some Cooked Legumes Alone and also in 100 grams of Various Combinations of Cereals and Legumes in the Ratio of 100:30 g

	Rice (100 g)	Pita Bread (100 g)	Macaroni (100 g)	Lentil (30 g)	Soy (30 g)	Kidney Red Bean (30 g)	Rice & Lentil (100 g)	Rice & Kidney Red Bean (100 g)	Pita Bread & Kidney Red Bean (100 g)	Macaroni & Soy (100 g)
Energy (kcal)	123	275	158	35	52	38.1	121.5	123.9	240.84	161.53
Protein (g)	2.91	9.1	5.80	2.71	4.99	2.60	4.32	4.23	9	8.3
Lysine (mg)	82	219	133	189	332	178	208	200	305	357
Methionine (mg)	70	160	65	23	67	39	71	83	153	101
Total lipid (g)	0.37	1.20	0.93	2.69	2.69	0.15	2.35	0.40	1.04	2.78
Carbohydrate (g)	26.05	6.03	30.86	2.51	2.98	6.75	21.96	25.2	48.03	26.03
Fiber (g)	0.9	2.37	1.80	1.8	1.8	2.22	2.07	2.4	3.4	2.76
Fe (mg)	0.24	1.4	0.50	1	1.54	0.882	1.37	0.86	1.75	1.57
Zn (mg)	0.37	0.84	0.51	0.38	0.34	0.32	0.57	0.53	0.89	0.653
Ca (mg)	19	86	7	6	31	8.4	19.23	21.07	72.61	29.23
Mg (mg)	9	26	18	11	26	13.5	15.38	17.3	30.38	33.84
Phosphorus (mg)	55	97	58	54	74	43	75.38	75.3	75.38	101.53
Vitamin A (IU)	–	0	0	2	3	0	1.5	0	0	2.30
Vitamin E (mg)	0.01	–	0.06	0.03	0.11	0.01	0.03	0.015	0.007	0.130
Vitamin K (µg)	0	–	0	0.5	5.8	2.5	0.38	1.92	1.92	4.46
B_1 (mg)	0.07	0.26	0.02	0.05	0.05	0.05	0.09	0.09	0.23	0.05
B_2 (mg)	0.02	0.10	0.02	0.02	0.08	0.017	0.03	0.15	1.81	0.08
B_6 (mg)	0.16	0.03	0.05	0.05	0.07	0.04	0.16	0.15	0.05	0.09
Niacin (mg)	2.31	2.14	0.40	0.32	0.12	0.17	2.02	1.91	1.78	0.40
Folate (µg)	3	24	7	54	16	39	43.80	32.30	48.46	17.69

Food composition data were obtained from United States Department of Agriculture (USDA) food composition tables (FCTs). Nutrients with no data in FCTs are indicated using "–" and considered as zero in calculations for mixed foods.

Cereals, including rice, bread, and pasta, are staple foods in many countries, especially in households with low socioeconomic status. They provide major parts of energy, macro-, and micronutrient requirements for consumers. Nevertheless, cereal protein is not considered "high quality" due to limited amounts of some amino acids, primarily lysine. Adding a small portion of legumes, whose limiting amino acid is methionine, to the cereals not only yields a final high-quality protein in terms of amino acid profile (complementation) but also improves micronutrient density in the mixed food. The colored right side of Table 11.1 shows selected combinations of cereals and legumes and their compositions per 100 grams of the mixed food. These combinations are the basis of some cuisines in Iran.

(a) (b) (c)

FIGURE 11.1 Combining cereals and legumes not only enhances protein quality through the complementary effect but also improves the micronutrient density in the resulting mixed food. These combinations form the basis of delicious cuisines in many countries, including Iran. (a) *Addas-Polo* (addas = lentil, polo = cooked rice); (b) *Loobia-Polo* (loobia = bean, polo = cooked rice). Both dishes may be eaten alone or with meat, chicken, fried egg, and/or yogurt, all of which add to the nutritional value. *Addas-Polo* is usually served with either roasted raisins or dates; (c) macaroni with soy. The crispy bottom layer of rice or macaroni in the pot, called *Tah-Deeg* (tah = bottom, deeg = pot), is popular among Iranians. Many Iranian cooks put flat bread or round sliced potatoes (seen in all three pictures) at the bottom of the pot to make a more delicious *Tah-Deeg*.

PROTEIN QUALITY AND HUMAN HEALTH

Research findings on the association of dietary proteins and health outcomes have been controversial to date (Table 11.2). A meta-analytical study failed to show any significant association between a high-protein diet and cardiovascular outcomes (Mantzouranis et al. 2023). However, this study did not discriminate between different sources of dietary proteins. In contrast, another meta-analysis of cohort studies reported a positive association between total protein intake and all-cause mortality, mainly due to the relationship between animal protein intake and CVD death. An inverse association between plant protein intake and CVD and all-cause mortality was observed (Chen et al. 2020). Similar findings were reported from a cohort study in Iran (Haghighatdoost et al. 2023).

Interestingly, this association may be reversed in elderly adults (≥85 years), where protein intake was found to be associated with a reduced risk of all-cause mortality (Kurata et al. 2023). One considerable limitation of these studies is that all animal-sourced proteins, including eggs, dairy products, poultry, fish, meats, and processed meats, were categorized as "animal protein." While plant proteins share very similar characteristics in terms of fat, fiber, sodium, and phosphorus, animal-sourced proteins differ significantly. A recent umbrella study of systematic reviews and meta-analyses reported that the strength of evidence for the association of dietary protein and most health outcomes was limited (Lv et al. 2022). Future studies must address the possible association of different animal proteins with health outcomes and mortality.

CONCLUDING REMARKS

Feeding the increasing global population with a healthy diet, including high-quality protein, in a sustainable way is a serious concern. Animal proteins are commonly considered high-quality proteins. Nevertheless, they are usually more expensive than plant protein sources and may be less affordable for people with lower SES. The recent coronavirus pandemic has had a great impact on dietary patterns (including dramatically decreased consumption of high-quality protein food sources and micronutrient-rich fruits) that may have persisted in some affected communities due to economic reasons (Nikooyeh et al. 2022a, 2022b, Amini et al. 2023). Even supplement use may not address the nutritional needs of low-income households (Ghodsi et al. 2023). Moreover, the sustainability of animal-sourced proteins is of great concern (Ferrari et al. 2022).

Current evidence strongly suggests a direct association between dietary protein quality and health outcomes. This association may be influenced partly by the amino acid profile and partly by the other components in the dietary protein source, including fatty acids, fiber, phytochemicals, and micronutrients (Ferrari et al. 2022). To reiterate, the complementation of different plant-sourced proteins, even at the cereal-to-legume ratio of 3:1, can enhance the micronutrient content of the mixed food. This issue must be considered in nutrition education for people with lower SES to make the most of the accessible food sources. Nevertheless, this is just a stepping stone for future research on the bioavailability of the micronutrients, especially the key trace elements, including iron, zinc, and calcium from different complementary combinations. Furthermore, vegans are at greater risk of micronutrient deficiencies and their consequent health outcomes, including fractures (Ogilvie et al. 2022). Though high-quality protein sources may meet the amino acid needs of vegans (Rojas Conzuelo et al. 2022), the micronutrient bioavailability and adequacy of these food sources warrant future studies.

TABLE 11.2
Characteristics of Prospective Studies on the Association between Protein Quality and Mortality

Author, Year	Study/ Population	Country	Follow-Up	Sample Size	Comparison	Outcome	OR, RR, HR
Bates, 2010 (Bates et al. 2010)	The community-living population of mainland Britain	United Kingdom	14	1,100	All-cause mortality	Protein (g)	0.86 (0.77, 0.97)
					Vascular disease		0.79 (0.67, 0.94)
Budhathoki, 2019 (Budhathoki et al. 2019)	Japan Public Health Center-based Prospective Cohort (JPHC) Study	Japan	18	70,696	All-cause mortality	Total protein: Quintile 5 vs. 1	0.99 (0.9, 1.09)
						Animal protein: Quintile 5 vs. 1	0.98 (0.88, 1.08)
						Plant protein: Quintile 5 vs. 1	0.87 (0.78, 0.96)
Chan, 2012 (Chan et al. 2012)	–	China	2.5	127	All-cause mortality	Isoflavone intake: >6.9 vs. <6.9 mg/d	OR 0.30 (0.05, 1.58)
Chen, 2020 (Chen et al. 2020)	The Rotterdam Study	Netherlands	13	7,786	All-cause mortality	Total Protein: Q4 vs. Q1	1.12 (1.01, 1.25)
						Animal protein: Q4 vs. Q1	1.18 (1.05, 1.31)
						Plant protein: Q4 vs. Q1	1.06 (0.92, 1.21)
Conroy, 2013 (Conroy et al. 2013)	Multiethnic Cohort	United States	14	3,842	All-cause mortality	Soy: T3 vs T1	1.03 (0.81, 1.33)
Haghighatdoost, 2023 (Haghighatdoost et al. 2023)	Isfahan cohort study (ICS)	Iran	11	5,431	All-cause mortality	Total protein: Quintile 5 vs. 1	1.22 (0.89, 1.67)
						Animal protein: Quintile 5 vs. 1	1.55 (1.15, 2.09)
						Plant protein: Quintile 5 vs. 1	0.67 (0.48, 0.95)
Kang, 2012 (Kang et al. 2012)	–	China	6	288	Breast cancer death	Soy protein: >15.7 vs. <4.5 mg/d	0.38 (0.17, 0.86)

(*Continued*)

TABLE 11.2 (CONTINUED)
Characteristics of Prospective Studies on the Association between Protein Quality and Mortality

Author, Year	Study/ Population	Country	Follow-Up	Sample Size	Comparison	Outcome	OR, RR, HR
Kurihara, 2019 (Kurihara et al. 2019)	The National Integrated Project for Prospective Observation of Non-communicable Disease and Its Trends in the Age 1990 (NIPPONDATA90)	Japan	13.9	7,744	Cardiovascular disease (CVD) death	Vegetable protein: Q4 vs Q1	0.8 (0.55, 1.16)
					Coronary heart disease (CHD) death		0.76 (0.31, 1.86)
					Stroke		0.61 (0.35, 1.05)
					Cerebral infection		0.77 (0.38,1.57)
					Cerebral hemorrhage		0.29 (0.08, 1.06)
Langsetmo, 2021 (Langsetmo et al. 2020)	Multicenter Osteoporotic Fractures in Men (MrOS) study	United States	10	5,790	All-cause mortality	1 HR unit = 2.9% TEI decrease for all protein intake variables	1.09 (1.04, 1.15)
					Non-dairy animal protein		1.08 (1.02, 1.15)
					Plant protein		1.07 (1.01, 1.13)
Leo QJN, 2016 (Leo et al. 2016)	Multiethnic Cohort	Japan	4.5	2,339	All-cause mortality	Soy: T3 vs. T1	0.93 (0.76, 1.14)
						Dairy: T3 vs. T1	1.14, 95 (1.00, 1.31)
						Legume: T2 vs. T1	0.83, (0.71, 0.98)
Levine, 2014 (Levine et al. 2014)	NHANES III	United States	13.1	3,039	Diabetes	Protein: high intake vs. low intake	2.77 (0.24, 31.73)
					All-cause mortality		1.18 (0.60, 2.31)
					CVD mortality		0.55 (0.19, 1.62)
					Cancer mortality		3.19 (1.21, 8.35)
McKenzie, 2022 (McKenzie et al. 2022)	UK Biobank	United Kingdom	11.1	120,963	All-cause mortality	Total Protein: upper vs. lowest third	0.82 (0.76, 0.89)
Meroño, 2022 (Meroño et al. 2022)	InCHIANTI Study	Italy	20	1,453	All-cause death	Animal protein intake: per 1% of total energy	0.96 (0.93, 0.99)
Nagata 2017 (Nagata et al. 2017)	Takayama Study	Japan	16	29,079	Stroke	Soy: Q4 vs. Q1	0.75 (0.57, 0.99)
					Ischemic heart disease		0.81 (0.53, 1.24)

(Continued)

TABLE 11.2 (CONTINUED)
Characteristics of Prospective Studies on the Association between Protein Quality and Mortality

Author, Year	Study/ Population	Country	Follow-Up	Sample Size	Comparison	Outcome	OR, RR, HR
Nechuta, 2012 (Nechuta et al. 2012)	Shanghai Breast Cancer Survival Study	China	7.4	4,856	All-cause mortality	Isoflavone: >10 vs. <4 mg/d	0.87 (0.7, 1.1)
Sun, 2021 (Sun et al. 2021)	Women's Health Initiative (WHI)	United States	24	102,521	All-cause death	Vegetable protein: Quintile 5 vs. 1	0.91 (0.86, 0.96)
						Animal protein: Quintile 5 vs. 1	0.99 (0.95, 1.04)
Talaei, 2014 (Talaei et al. 2014)	Singapore Chinese Health Study	China	14.7	63,257	CVD mortality	Soy: Q4 vs. Q1	1.06 (0.97, 1.17)
Tharrey, 2018 (Tharrey et al. 2018)	The Adventist Health Study 2 (AHS-2)	UnitedS and Canada	9.4	81,337	CVD death	Meat: Quintile 5 vs. Quintile 1	1.46 (0.98, 2.18)
Virtanen, 2019 (Virtanen et al. 2019)	Kuopio Ischemic Heart Disease Risk Factor Study	Finland	22.3	2,641	All-cause death	Total protein: Q4 vs. Q1	1.17 (0.99, 1.39)
						Animal protein: Q4 vs. Q1	1.13 (0.95, 1.35)
						Plant protein: Q4 vs. Q1	0.98 (0.76, 1.26)
Yang, 2013 (Yang et al. 2013)	Shanghai Women's Health Study	China	3	444	Lung cancer mortality	Soy: 90th vs 50th percentile	0.89 (0.68, 1.16)
Yamasaki, 2015 (Yamasaki, Kayaba, and Ishikawa 2015)	Jichi Medical School Cohort Study	Japan	11.4	4,309	All-cause mortality	Soy: almost daily vs. 1–2 times/ week	1.55 (1.19, 2.03)
Zhang, 2017 (Zhang et al. 2017)	Breast Cancer Family Registry	United States, Canada, and Australia	9.4	6,235	All-cause mortality	Dietary isoflavone: Q4 vs. Q1	0.79 (0.64, 0.97)

Abbreviations: HR: hazard ratio; OR: odds ratio; RR: relative risk; SD: standard deviation; T: tertile; TEI: total energy intake; Q: quartile.

MINI DICTIONARY OF TERMS

Amino acids (AAs): The building blocks of proteins.
Animal proteins: Proteins derived from animal sources like meat, chicken, eggs, and milk.
Bioavailability: The extent and rate at which a nutrient is absorbed and utilized by the body.
Dietary protein quality refers to how well proteins, as polymers of amino acids, meet the body's requirements depending upon amino acid content and bioavailability.
Digestible indispensable amino acid score (DIAAS): A measure of protein quality that considers the digestibility and bioavailability of amino acids.
Essential amino acids (EAAs): Amino acids that the body cannot produce and must be obtained from the diet.
Greenhouse gas emissions: The release of gases contributing to the greenhouse effect, often associated with environmental impact.
Limiting amino acid (LAA): An amino acid present in a protein in a quantity below the minimal body requirement.
Macronutrient density: The concentration of macronutrients (proteins, fats, and carbohydrates) in a food item.
Micronutrient density: The concentration of micronutrients (vitamins and minerals) in a food item.
Non-essential amino acids (NEAAs): Amino acids that the body can produce.
Plant proteins: Proteins derived from plant sources.
Protein complementation: Combining different protein sources to complement amino acid profiles and enhance protein quality.
Protein digestibility-corrected amino acid score (PDCAAS): A method to assess protein quality by considering amino acid requirements and true fecal nitrogen digestibility.
Sustainable diet: Diets that aim to protect the planet and meet the nutritional needs of the population.
Synthetic machinery: The body's ability to produce proteins from amino acids.

SUMMARY POINTS

- Proteins, comprising enzymes, hormones, and structural components, play a crucial role in vital bodily functions such as cellular communication, blood stabilization, and muscle contraction, emphasizing the necessity of a constant protein supply.
- Recent dietary guidelines recommend a shift toward plant-based diets for environmental sustainability and health benefits. While plant proteins are associated with a lower environmental impact, animal proteins are considered more digestible.
- Various methods, including protein efficiency ratio (PER), protein digestibility-corrected amino acid score (PDCAAS), digestible indispensable amino acid score (DIAAS), and net postprandial protein utilization, are employed to assess protein quality, each with its limitations and advantages.
- Animal and plant proteins come with specific nutrient clusters, and their combination can help address potential deficiencies in micronutrients, ensuring a well-balanced nutrient intake.
- Complementing different plant proteins, particularly by combining cereals and legumes, enhances both protein quality and micronutrient content in the diet. This strategy is crucial for populations with lower income and socioeconomic status.

REFERENCES

Adhikari, S., M. Schop, I. J. M. de Boer, and T. Huppertz. 2022a. "Protein quality in perspective: a review of protein quality metrics and their applications. " *Nutrients* 14 (5). doi: nu14050947 [pii] nutrients-14-00947 [pii] 0.3390/nu14050947 [doi].

Adhikari, Shiksha, Marijke Schop, Imke J. M. de Boer, and Thom Huppertz. 2022b. "Protein quality in perspective: a review of protein quality metrics and their applications. " *Nutrients* 14 (5):947.

Amini, Maryam, Azam Doustmohammadian, Bahareh Nikooyeh, Cain CT Clark, Samira Rabiei, Delaram Ghodsi, Hamid Rasekhi, Zahra Abdollahi, Mina Minaie, and Farzaneh Sadeghi Ghotbabadi. 2023. "Considerable decreased fruits and snack consumption in Iran population during COVID-19 lockdown: a cross-sectional web-based survey National Food and Nutrition Surveillance. " *Agriculture & Food Security* 12 (1):21.

Bates, Christopher J., Mohammed A. Mansoor, Kristina D. Pentieva, Mark Hamer, and Gita D. Mishra. 2010. "Biochemical risk indices, including plasma homocysteine, that prospectively predict mortality in older British people: the national diet and nutrition survey of people aged 65 years and over." *British journal of nutrition* 104 (6):893–899.

Berrazaga, I., V. Micard, M. Gueugneau, and S. Walrand. 2019. "The role of the anabolic properties of plant- versus animal-based protein sources in supporting muscle mass maintenance: a critical review." *Nutrients* 11 (8). doi: 10.3390/nu11081825.

Binns, Colin W., Mi Kyung Lee, Bruce Maycock, Liv Elin Torheim, Keiko Nanishi, and Doan Thi Thuy Duong. 2021. "Climate change, food supply, and dietary guidelines." *Annual Review of Public Health* 42:233–255.

Budhathoki, Sanjeev, Norie Sawada, Motoki Iwasaki, Taiki Yamaji, Atsushi Goto, Ayaka Kotemori, Junko Ishihara, Ribeka Takachi, Hadrien Charvat, and Tetsuya Mizoue. 2019. "Association of animal and plant protein intake with all-cause and cause-specific mortality in a Japanese cohort." *JAMA Internal Medicine* 179 (11):1509–1518.

Chan, Y. -H., K. -K. Lau, K. -H. Yiu, C. -W. Siu, H. -T. Chan, S. -W. Li, S. Tam, T. -H. Lam, C. -P. Lau, and Hung-Fat Tse. 2012. "Prospective observational study of isoflavone and the risk of stroke recurrence: potential clinical implications beyond vascular function." *The Journal of Nutrition, Health & Aging* 16:383–388.

Chen, Zhangling, Marija Glisic, Mingyang Song, Hamid A. Aliahmad, Xiaofang Zhang, Alice C. Moumdjian, Valentina Gonzalez-Jaramillo, Niels van der Schaft, Wichor M. Bramer, and Mohammad Arfan Ikram. 2020. *Dietary Protein Intake and All-cause and Cause-specific Mortality: Results from the Rotterdam Study and a Meta-analysis of Prospective Cohort Studies*. Springer.

Chen, Z., M. Glisic, M. Song, H. A. Aliahmad, X. Zhang, A. C. Moumdjian, V. Gonzalez-Jaramillo, N. van der Schaft, W. M. Bramer, M. A. Ikram, and T. Voortman. 2020. "Dietary protein intake and all-cause and cause-specific mortality: results from the rotterdam study and a meta-analysis of prospective cohort studies." 35 (5):411–429. doi: 10.1007/s10654-020-00607-6.

Comerford, K. B., G. D. Miller, W. Reinhardt Kapsak, and K. A. Brown. 2021. "The complementary roles for plant-source and animal-source foods in sustainable healthy diets." *Nutrients* 13 (10). doi: 10.3390/nu13103469.

Cominelli, Eleonora, Francesca Sparvoli, Silvia Lisciani, Chiara Forti, Emanuela Camilli, Marika Ferrari, Cinzia Le Donne, Stefania Marconi, Barend Juan Vorster, and Anna-Maria Botha. 2022. "Antinutritional factors, nutritional improvement, and future food use of common beans: a perspective." *Frontiers in Plant Science* 13:992169.

Conroy, Shannon M., Gertraud Maskarinec, Song-Yi Park, Lynne R. Wilkens, Brian E. Henderson, and Laurence N. Kolonel. 2013. "The effects of soy consumption before diagnosis on breast cancer survival: the multiethnic cohort study." *Nutrition and Cancer* 65 (4):527–537.

de la O, Víctor, Itziar Zazpe, Carmen de la Fuente-Arrillaga, Susana Santiago, Leticia Goni, Miguel Ángel Martínez-González, and Miguel Ruiz-Canela. 2023. "Association between a new dietary protein quality index and micronutrient intake adequacy: a cross-sectional study in a young adult Spanish Mediterranean cohort." *European Journal of Nutrition* 62 (1):419–432.

Didinger, C., and H. J. Thompson. 2021. "Defining nutritional and functional niches of legumes: a call for clarity to distinguish a future role for pulses in the dietary guidelines for Americans." *Nutrients* 13 (4). doi: nu13041100 [pii] nutrients-13-01100 [pii] 10.3390/nu13041100 [doi].

FAO. 2013. "Dietary protein quality evaluation in human nutrition. Report of an FAO expert consultation." 31 March–2 April, 2011, Auckland, New Zealand. Food and Nutrition Paper 92. Rome, Italy: FAO.

Ferrari, L., S. A. Panaite, A. Bertazzo, and F. Visioli. 2022. "Animal- and plant-based protein sources: a scoping review of human health outcomes and environmental impact." *Nutrients* 14 (23). doi: nu14235115 [pii] nutrients-14-05115 [pii] 10.3390/nu14235115 [doi].

Fouillet, Hélène, Alison Dussiot, Elie Perraud, Juhui Wang, Jean-François Huneau, Emmanuelle Kesse-Guyot, and François Mariotti. 2023. "Plant to animal protein ratio in the diet: nutrient adequacy, long-term health and environmental pressure." *Frontiers in Nutrition* 10. doi: 10.3389/fnut.2023.1178121.

Ghodsi, Delaram, Bahareh Nikooyeh, Maryam Amini, Samira Rabiei, Azam Doustmohammadian, Zahra Abdollahi, Mina Minaie, Farzaneh Sadeghi, Cain C. T. Clark, and Tirang R. Neyestani. 2023. "Dietary supplement use among iranian households during COVID-19 epidemic lockdown: less access in those who may need more national food and nutrition surveillance." *International Journal of Preventive Medicine* 14, no. 7 (July 27):14–99. doi: 10.4103/ijpvm. ijpvm_207_22.

Guillin, Florence M., Claire Gaudichon, Laetitia Guérin-Deremaux, Catherine Lefranc-Millot, Gheorghe Airinei, Nadezda Khodorova, Robert Benamouzig, Pierre-Henri Pomport, Juliette Martin, and Juliane Calvez. 2022. "Real ileal amino acid digestibility of pea protein compared to casein in healthy humans: a randomized trial." *The American Journal of Clinical Nutrition* 115 (2):353–363.

Gwin, J. A., J. P. Karl, L. J. Lutz, E. Gaffney-Stomberg, J. P. McClung, and S. M. Pasiakos. 2019. "Higher protein density diets are associated with greater diet quality and micronutrient intake in healthy young adults." *Frontiers in Nutrition* 6:59. doi: 10.3389/fnut.2019.00059.

Haghighatdoost, Fahimeh, Noushin Mohammadifard, Parisa Zakeri, Jamshid Najafian, Masoumeh Sadeghi, Hamidreza Roohafza, and Nizal Sarrafzadegan. 2023. "Differences in all-cause mortality risk associated with animal and plant dietary protein sources consumption." *Scientific Reports* 13 (1):3396. doi: 10.1038/s41598-023-30455-9.

Kang, Hong-Bin, Ya-Feng Zhang, Jin-Dun Yang, and Kuan-Liang Lu. 2012. "Study on soy isoflavone consumption and risk of breast cancer and survival." *Asian Pacific Journal of Cancer Prevention* 13 (3):995–998.

Kaspchak, Elaine, Andrea Briones Gonçalves Bonassoli, Patrícia Kanczewski Iwankiw, Cíntia Tiemi Misugi Kayukawa, Luciana Igarashi-Mafra, and Marcos R. Mafra. 2020. "Interactions of antinutrients mixtures with bovine serum albumin and its influence on in vitro protein digestibility." *Journal of Molecular Liquids* 315:113699. doi: https://doi.org/10.1016/j.molliq.2020.113699.

Katz, David L., Kimberly N. Doughty, Kate Geagan, David A. Jenkins, and Christopher D. Gardner. 2019. "Perspective: the public health case for modernizing the definition of protein quality." *Advances in Nutrition* 10 (5):755–764. doi: https://doi.org/10.1093/advances/nmz023.

Kurata, H., S. Meguro, Y. Abe, T. Sasaki, K. Asakura, Y. Arai, and H. Itoh. 2023. "Dietary protein intake and all-cause mortality: results from The Kawasaki Aging and Wellbeing Project." *BMC Geriatrics* 23 (1):479. doi: 10.1186/s12877-023-04173-w.

Kurihara, Ayako, Tomonori Okamura, Daisuke Sugiyama, Aya Higashiyama, Makoto Watanabe, Nagako Okuda, Aya Kadota, Naoko Miyagawa, Akira Fujiyoshi, and Katsushi Yoshita. 2019. "Vegetable protein intake was inversely associated with cardiovascular mortality in a 15-year follow-up study of the general Japanese population." *Journal of Atherosclerosis and Thrombosis* 26 (2):198–206.

Langsetmo, Lisa, Stephanie Harrison, Satya Jonnalagadda, Suzette L. Pereira, James M. Shikany, Samaneh Farsijani, Nancy E. Lane, Jane A. Cauley, Katie Stone, and Peggy M. Cawthon. 2020. "Low protein intake irrespective of source is associated with higher mortality among older community-dwelling men." *The Journal of Nutrition, Health & Aging* 24 (8):900–905.

Langyan, S., P. Yadava, F. N. Khan, Z. A. Dar, R. Singh, and A. Kumar. 2021. "Sustaining Protein Nutrition Through Plant-Based Foods." *Frontiers in Nutrition* 8:772573. doi: 10.3389/fnut.2021.772573.

Leo, Qi Jie Nicholas, Nicholas J. Ollberding, Lynne R. Wilkens, Laurence N. Kolonel, Brian E. Henderson, Loic Le Marchand, and Gertraud Maskarinec. 2016. "Nutritional factors and non-Hodgkin lymphoma survival in an ethnically diverse population: the multiethnic cohort." *European Journal of Clinical Nutrition* 70 (1):41–46.

Levine, Morgan E., Jorge A. Suarez, Sebastian Brandhorst, Priya Balasubramanian, Chia-Wei Cheng, Federica Madia, Luigi Fontana, Mario G. Mirisola, Jaime Guevara-Aguirre, and Junxiang Wan. 2014. "Low protein intake is associated with a major reduction in IGF-1, cancer, and overall mortality in the 65 and younger but not older population." *Cell Metabolism* 19 (3):407–417.

Lopez, M. J., and S. S. Mohiuddin. 2023. "Biochemistry, essential amino acids." In *StatPearls [Internet]*. Treasure Island, FL: StatPearls Publishing. Accessed March 13, 2023. https://www.ncbi.nlm.nih.gov/books/NBK557845/

Losso, Jack N. 2008. "The biochemical and functional food properties of the Bowman-Birk inhibitor." *Critical Reviews in Food Science and Nutrition* 48 (1):94–118.

Lv, Jia-Le, Qi-Jun Wu, Xin-Yu Li, Chang Gao, Ming-Zhu Xu, Juan Yang, Si-Tian Zang, Jie Luan, Dong-Zhen Cai, Qing Chang, and Yu-Hong Zhao. 2022. "Dietary protein and multiple health outcomes: an umbrella review of systematic reviews and meta-analyses of observational studies." *Clinical Nutrition* 41 (8):1759–1769. doi: https://doi.org/10.1016/j.clnu.2022.06.005.

Mansilla, Wilfredo D., Christopher P. F. Marinangeli, Cara Cargo-Froom, Adam Franczyk, James D. House, Rajavel Elango, Daniel A. Columbus, Elijah Kiarie, Mike Rogers, and Anna K. Shoveller. 2020. "Comparison of methodologies used to define the protein quality of human foods and support regulatory claims." *Applied Physiology, Nutrition, and Metabolism* 45 (9):917–926.

Mantzouranis, Emmanouil, Eleftheria Kakargia, Fotis Kakargias, George Lazaros, and Konstantinos Tsioufis. 2023. "The impact of high protein diets on cardiovascular outcomes: a systematic review and meta-analysis of prospective cohort studies." *Nutrients* 15 (6):1372.

Marinangeli, Christopher P. F., and James D. House. 2017. "Potential impact of the digestible indispensable amino acid score as a measure of protein quality on dietary regulations and health." *Nutrition Reviews* 75 (8):658–667.

Mariotti, François. 2017. "35 - plant protein, animal protein, and protein quality." In *Vegetarian and Plant-Based Diets in Health and Disease Prevention*, edited by François Mariotti, 621–642. Cambridge, MA: Academic Press.

Mariotti, François. 2019. "Animal and plant protein sources and cardiometabolic health." *Advances in Nutrition* 10 (Supplement_4):S351–S366.

McKenzie, Briar L., Katie Harris, Sanne A. E. Peters, Jacqui Webster, and Mark Woodward. 2022. "The association of energy and macronutrient intake with all-cause mortality, cardiovascular disease and dementia: findings from 120 963 women and men in the UK Biobank." *British Journal of Nutrition* 127 (12):1858–1867.

Meroño, Tomás, Raúl Zamora-Ros, Nicole Hidalgo-Liberona, Montserrat Rabassa, Stefania Bandinelli, Luigi Ferrucci, Massimiliano Fedecostante, Antonio Cherubini, and Cristina Andres-Lacueva. 2022. "Animal protein intake is inversely associated with mortality in older adults: the InCHIANTI study." *The Journals of Gerontology: Series A* 77 (9):1866–1872.

Monsonego Ornan, Efrat, and Ram Reifen. 2022. "Revisiting protein quality assessment to include alternative proteins." *Foods* 11 (22):3740.

Nagata, Chisato, Keiko Wada, Takashi Tamura, Kie Konishi, Yuko Goto, Sachi Koda, Toshiyuki Kawachi, Michiko Tsuji, and Kozue Nakamura. 2017. "Dietary soy and natto intake and cardiovascular disease mortality in Japanese adults: the Takayama study." *The American Journal of Clinical Nutrition* 105 (2):426–431.

Naghshi, Sina, Omid Sadeghi, Walter C. Willett, and Ahmad Esmaillzadeh. 2020. "Dietary intake of total, animal, and plant proteins and risk of all cause, cardiovascular, and cancer mortality: systematic review and dose-response meta-analysis of prospective cohort studies." *BMJ* 370:m2412. doi: 10.1136/bmj.m2412.

Nechuta, Sarah J., Bette J. Caan, Wendy Y. Chen, Wei Lu, Zhi Chen, Marilyn L. Kwan, Shirley W. Flatt, Ying Zheng, Wei Zheng, and John P. Pierce. 2012. "Soy food intake after diagnosis of breast cancer and survival: an in-depth analysis of combined evidence from cohort studies of US and Chinese women." *The American Journal of Clinical Nutrition* 96 (1):123–132.

Neufingerl, N., and A. Eilander. 2021. "Nutrient intake and status in adults consuming plant-based diets compared to meat-eaters: a systematic review." *Nutrients* 14 (1). doi: nu14010029 [pii] nutrients-14-00029 [pii] 10.3390/nu14010029 [doi].

Nikooyeh, Bahareh, Samira Rabiei, Maryam Amini, Delaram Ghodsi, Hamid Rasekhi, Azam Doustmohammadian, Zahra Abdollahi, Mina Minaie, Farzaneh Sadeghi, and Tirang R. Neyestani. 2022a. "COVID-19 epidemic lockdown-induced changes of cereals and animal protein foods consumption of Iran population: the first nationwide survey." *Journal of Health, Population and Nutrition* 41 (1):1–9.

Nikooyeh, Bahareh, Samira Rabiei, Maryam Amini, Delaram Ghodsi, Hamid Rasekhi, Azam Doustmohammadian, Zahra Abdollahi, Mina Minaie, Farzaneh Sadeghi, and Tirang R. Neyestani. 2022b. "COVID-19 epidemic lockdown-induced remarkable decrease in dairy products consumption of Iran population: does it really matter? National food and nutrition surveillance." *BMC Nutrition* 8 (1):1–8.

Ogilvie, Anna R., Brandon D. McGuire, Lingqiong Meng, and Sue A. Shapses. 2022. "Fracture risk in vegetarians and vegans: the role of diet and metabolic factors." *Current Osteoporosis Reports* 20 (6):442–452. doi: 10.1007/s11914-022-00754-7.

Perraud, Elie, Juhui Wang, Marion Salomé, Jean-François Huneau, Nathanaël Lapidus, and François Mariotti. 2022. "Plant and animal protein intakes largely explain the nutritional quality and health value of diets higher in plants: a path analysis in french adults." *Frontiers in Nutrition* 9. doi: 10.3389/fnut.2022.924526.

Phillips, Stuart M., Victor L. Fulgoni III, Robert P. Heaney, Theresa A. Nicklas, Joanne L. Slavin, and Connie M. Weaver. 2015. "Commonly consumed protein foods contribute to nutrient intake, diet quality, and nutrient adequacy." *The American Journal of Clinical Nutrition* 101 (6):1346S–1352S.

Pikosky, Matthew A., Jean Ragalie-Carr, and Gregory D. Miller. 2022. "Recognizing the importance of protein quality in an era of food systems transformation." *Frontiers in Sustainable Food Systems* 6:1012813.

Rachwa-Rosiak, Danuta, Ewa Nebesny, and Grażyna Budryn. 2015. "Chickpeas—composition, nutritional value, health benefits, application to bread and snacks: a review." *Critical Reviews in Food Science and Nutrition* 55 (8):1137–1145.

Rojas Conzuelo, Zaray, Natalie S. Bez, Steffen Theobald, and Katrin A. Kopf-Bolanz. 2022. "Protein quality changes of vegan day menus with different plant protein source compositions." *Nutrients* 14 (5):1088.

Seiquer, Isabel, José Díaz-Alguacil, Cristina Delgado-Andrade, Magdalena López-Frías, Antonio Muñoz Hoyos, Gabriel Galdó, and María Pilar Navarro. 2006. "Diets rich in Maillard reaction products affect protein digestibility in adolescent males aged 11–14 y." *The American Journal of Clinical Nutrition* 83 (5):1082–1088.

Shevkani, Khetan, and Shivani Chourasia. 2021. "Dietary proteins: functions, health benefits and healthy aging." In *Nutrition, Food and Diet in Ageing and Longevity*, edited by Suresh I. S. Rattan and Gurcharan Kaur, 3–37. Cham: Springer International Publishing.

Singh, Maleeka, Annamalai Manickavasagan, Shanmugam Shobana, and Viswanathan Mohan. 2021. "Glycemic index of pulses and pulse-based products: A review." *Critical Reviews in Food Science and Nutrition* 61 (9):1567–1588.

Skórska, K. B., H. Grajeta, and K. A. Zabłocka-Słowińska. 2021. "Frequency of legume consumption related to sociodemographic factors, health status and health-related variables among surveyed adults from Poland." *Public Health Nutrition* 24 (7):1895–1905. doi: S1368980020002116 [pii] 10.1017/S1368980020002116 [doi].

Sobhani, Seyyed Reza, Hassan Eini-Zinab, and Arezoo Rezazadeh. 2022. "Socioeconomic status and changes in iranian household food basket using national household budget and expenditure survey data, 1991–2017." *Iranian Journal of Public Health* 51 (4):919.

Sobiecki, Jakub G., Paul N. Appleby, Kathryn E. Bradbury, and Timothy J. Key. 2016. "High compliance with dietary recommendations in a cohort of meat eaters, fish eaters, vegetarians, and vegans: results from the European prospective investigation into cancer and nutrition–Oxford study." *Nutrition Research* 36 (5):464–477.

Sun, Yangbo, Buyun Liu, Linda G. Snetselaar, Robert B. Wallace, Aladdin H. Shadyab, Candyce H. Kroenke, Bernhard Haring, Barbara V. Howard, James M. Shikany, and Carolina Valdiviezo. 2021. "Association of major dietary protein sources with all-cause and cause-specific mortality: prospective cohort study." *Journal of the American Heart Association* 10 (5):e015553.

Talaei, Mohammad, Woon-Puay Koh, Rob M. van Dam, Jian-Min Yuan, and An Pan. 2014. "Dietary soy intake is not associated with risk of cardiovascular disease mortality in Singapore Chinese adults." *The Journal of Nutrition* 144 (6):921–928.

Tharrey, Marion, François Mariotti, Andrew Mashchak, Pierre Barbillon, Maud Delattre, and Gary E. Fraser. 2018. "Patterns of plant and animal protein intake are strongly associated with cardiovascular mortality: the adventist health study-2 cohort." *International Journal of Epidemiology* 47 (5):1603–1612.

Virtanen, Heli E. K., Sari Voutilainen, Timo T. Koskinen, Jaakko Mursu, Petra Kokko, Maija P. T. Ylilauri, Tomi-Pekka Tuomainen, Jukka T. Salonen, and Jyrki K. Virtanen. 2019. "Dietary proteins and protein sources and risk of death: the Kuopio Ischaemic heart disease risk factor study." *The American Journal of Clinical Nutrition* 109 (5):1462–1471.

Wang, Ruican, and Shuntang Guo. 2021. "Phytic acid and its interactions: Contributions to protein functionality, food processing, and safety." *Comprehensive Reviews in Food Science and Food Safety* 20 (2):2081–2105.

Yamasaki, Kyoko, Kazunori Kayaba, and Shizukiyo Ishikawa. 2015. "Soy and soy products intake, all-cause mortality, and cause-specific mortality in Japan: the Jichi medical school cohort study." *Asia Pacific Journal of Public Health* 27 (5):531–541.

Yang, Gong, Xiao-Ou Shu, Hong-Lan Li, Wong-Ho Chow, Wanqing Wen, Yong-Bing Xiang, Xianglan Zhang, Hui Cai, Bu-Tian Ji, and Yu-Tang Gao. 2013. "Prediagnosis soy food consumption and lung cancer survival in women." *Journal of Clinical Oncology* 31 (12):1548.

Zhang, Fang Fang, Danielle E. Haslam, Mary Beth Terry, Julia A. Knight, Irene L. Andrulis, Mary B. Daly, Saundra S. Buys, and Esther M. John. 2017. "Dietary isoflavone intake and all-cause mortality in breast cancer survivors: The breast cancer family registry." *Cancer* 123 (11):2070–2079.

12 Protein Digestibility of Bread

Impact of Some Breadmaking Processes and Flour Sources

Yves José de Souza Santos, Rosemary Aparecida de Carvalho, Luiz Alberto Colnago, and Fernanda M. Vanin

ABBREVIATIONS

DW	Dry weight
GFBs	Gluten-free breads
PD	Protein digestibility

INTRODUCTION

Wheat serves as a key staple food in many diets and contributes significantly to daily caloric intake (Cappelli, Lupori, and Cini 2021). The popularity of wheat products is largely due to wheat protein. The efficiency of protein absorption in nutrition is determined by its intestinal absorption capacity according to protein digestibility (PD) tests (FAO 2011).

Bakery products are consumed after wheat processing, which involves milling, fermentation, and baking. Some studies have proposed that these different processes can alter the characteristics of wheat proteins and, therefore, PD. This has been observed for bread through fermentation (Osman 2004) and baking (Abdel-Aal 2008; Smith et al. 2015) conditions, as well as indirectly by influencing interactions between proteins and other food components (Santos et al. 2023).

Beyond their popularity, bakery products are widely recognized as allergenic foods due to celiac disease syndrome, which occurs because of gluten proteins. Food allergens are associated with different characteristics, such as resistance to digestion (Bøgh and Madsen 2016). In general, when proteins undergo digestion, the amount of protein and/or amino acids and/or free amino released after a predetermined period is analyzed. As a result, when proteins reach the gastrointestinal tract mucosal immune system in their original form or as large fragments, they exhibit resistance to digestion. In the case of gluten protein, a decrease in PD has been linked to complex autoimmune responses that underlie celiac disease (Bøgh and Madsen 2016).

Other aspects related to PD in bakery products include wheat cultivar age (Shewry 2018) and organic cultivars. It is commonly believed that products made from spelt wheat may be more soothing and easily digestible than those made from common wheat. Similarly, foods derived from ancient wheat are thought to be healthier and more resistant to disease than those made from modern wheat.

Additionally, several studies have proposed substituting wheat flour with other non-conventional flour to improve the nutritional aspects of the final products or to address specific markets such as those related to "healthier" foods.

However, although the use of unconventional fruits, legumes, and vegetables allows the formulation of bakery products with improved nutritional composition, such as higher levels of resistant starch, dietary fiber, bioactive compounds, and minerals, the effect of these materials on PD has not been thoroughly evaluated.

DOI: 10.1201/b23328-12

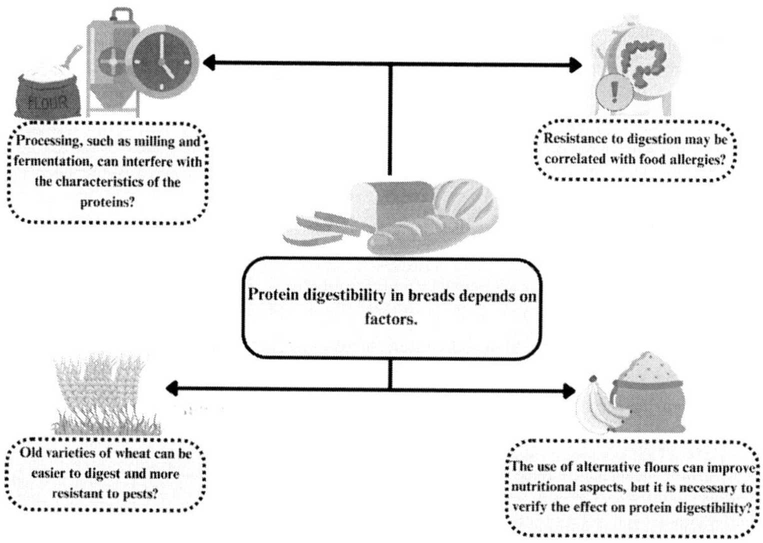

FIGURE 12.1 Nutritional, processing, variability, and enrichment aspects that can affect the digestibility of raw materials and bakery products (elaborated by the authors).

Therefore, bread PD is a crucial aspect that must be evaluated to enhance its nutritional value. In this context, PD may be related to the ease of digestion and absorption of bread proteins. This chapter presents the main findings related to bread PD observed during fermentation and baking, as well as the flour source, as illustrated in Figure 12.1.

BREADMAKING CONDITIONS

Bread processing conditions can significantly affect the PD of bread, including longer fermentation time, type of culture, particle size, and other factors. These conditions will be addressed in this study and are summarized in Table 12.1.

Fermentation

Bread fermentation is a crucial step in determining bread quality. Sourdough, one of the earliest fermentation methods, was widely employed as a leavening and acidifying agent for bread dough until the late 19th century (Arora et al. 2021).

Increased in vitro PD is a key benefit of sourdough (Arora et al. 2021; Graça et al. 2022; Rizzello et al. 2019). Graça et al. (2022) evaluated the effect of yogurt as a starter in sourdough fermentation to improve the PD of sourdough-wheat bread. Compared with yeast dough, yogurt sourdough fermentation significantly improved bread PD. For yeast bread, the PD was in the range of 78.9%, for sourdough bread, the values were 84.0% and 88.2% for bread prepared with only wheat flour and with a mixture of wheat-whole grain flour, respectively.

Similarly, Rizzello et al. (2019) compared the in vitro PD of yeast bread and two sourdough breads, one fermented for 4 hours and the other for 24 hours. The results showed that yeast bread presented an in vitro PD of 63.7%, while sourdough bread values were 71.6% and 79.8% for bread fermented for 4 and 24 hours, respectively. The authors noted that the degree of proteolysis was proportional to the duration of fermentation.

Kulathunga, Whitney, and Simsek (2023) analyzed the effect of flour particle size on the in vitro PD of sourdough bread produced with rye or wheat starter. Regardless of the starter culture, the lower the particle size, the higher the PD of the sourdough bread. Moreover, sourdoughs made with

TABLE 12.1
Summary of Aspects Evaluated in Different Types of Bread and the Most Relevant Findings of Each Study

Baking Product	Aspect Evaluated	Main Findings	Reference
Traditional wheat bread	Effect of baking conditions	1. Protein digestibility content ranged from 63.49% to 133.5%, according to modified baking parameters. 2. Baking process can increase protein digestibility of breads. 3. Use of steam during bread baking contributed to protein digestibility.	(Bredariol et al. 2020)
Wheat bread and wheat + whole grain bread (50:50%; flour weight)	Effect of yogurt incorporation as a starter in sourdough fermentation	1. Higher degree of protein proteolysis in blended bread fermented by yogurt. 2. Yogurt-sourdough improves technological characteristics, like breadcrumb softness. 3. Enhanced protein digestibility and free amino acids bioavailability.	(Graça et al. 2022)
Traditional wheat bread	Effect of baking temperature and duration	1. No influence of baking temperature and duration on peptide release profile and maximum peptide concentrations. 2. Baking altered the profile of immunogenic gluten peptides derived from α-gliadin.	(Ogilvie et al. 2021)
Sourdough bread	Effect of particle size and different starter culture	1. Higher protein digestibility found mostly in samples fermented by rye starter culture. 2. Protein digestibility showed positive correlation with smaller particle size. 3. Higher proteolytic activity from rye starter culture.	(Kulathunga, Whitney, and Simsek 2023)

the rye starter culture had greater PD values than those made with the wheat starter culture. The authors attributed these effects to higher proteolysis in flour with smaller particle size due to easier access of proteolytic enzymes to proteins provided by the large surface and to the fact that lactic acid bacteria in the rye starter culture have a more efficient proteolytic arrangement compared to those in the wheat starter culture.

Gao et al. (2022) evaluated the effects of yeast and sourdough fermentation on Chinese bread baked using steam or conventional baking. Regardless of the baking method used, sourdough Chinese bread showed a higher in vitro PD. Yeast and sourdough bread baked conventionally presented PD of 74.84% and 84.33%, respectively, while for bread baked with steam, the values were 83.20% and 91.12%, respectively. The authors proposed that when sourdough fermentation is used, a larger amount of acetic acid is produced, which in turn favors higher enzyme activity and higher gluten degradation into smaller proteins due to easier access to enzymes.

Osman (2004) analyzed three different varieties of sorghum cereal used in the production of Khamir bread with regard to parameters considered antinutritional, phytic acid, and tannin content and their probable correlation with trypsin inhibitory activity (a peptidase) and PD in vitro. In general, the author demonstrated that the higher the fermentation time (0 to 24h), the lower the

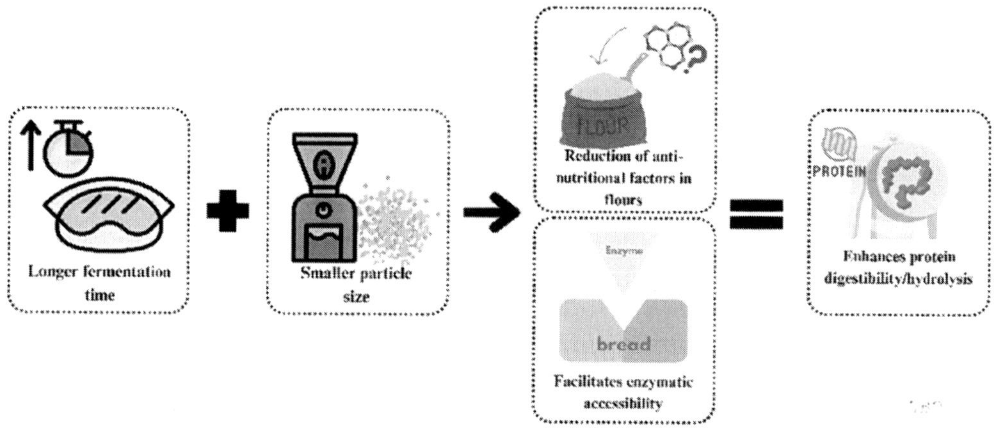

FIGURE 12.2 Effect of fermentation and particle size on the nutritional properties of bread and, consequently, the increase in protein digestibility (elaborated by the authors).

peptidase inhibitory activity, as well as the content of antinutritional compounds, which in turn enhanced PD. Therefore, it was concluded that the reduction in phytic acid and tannin content may improve the nutritional quality of sorghum grains, leading to the conclusion that longer bread fermentation times (24h) had better effects on protein bioavailability.

The increased proteolytic activity during sourdough fermentation, which helps to increase protein solubilization and makes the polypeptides more prone to partial degradation into small peptides and easily digestible, is likely the cause of the improved PD found in sourdough bread. These findings suggest that sourdough fermentation is a pre-digestion process that enhances PD and amino acid bioavailability, as illustrated in Figure 12.2.

Baking

During bread baking, large temperature gradients occur between the surface and the thermal center of the bread (Bredariol and Vanin 2022). Such thermal gradients result in different heating rates, which in turn can impact the chemical reactions during baking. Despite this, only a small number of studies have evaluated the effects of baking conditions on protein digestion.

Abdel-Aal (2008) compared the PD of different bakery products (bread, biscuit, cookie, and muffin) produced with organic and common wheat flour based on two methods of digestion in vitro. The author found significant differences in PD after baking for all products evaluated (increase in PD) as well as the type of wheat product. Therefore, this study demonstrated that the baking conditions associated with the production of different products are more important than the source of wheat used for baking products, whether organic or conventional cultivars.

Bredariol, Carvalho, and Vanin (2020) also demonstrated the importance of the baking conditions for bread PD. The authors evaluated the use of different temperatures, times, and vapor baking conditions during bread baking and their relationship with bread PD. Similar to Abdel-Aal and Hucl (2002), the authors verified that baking enhanced bread PD. However, for the evaluated conditions (160, 190, and 220°C, for 9, 12, 15, and 20 minutes, with or without vapor), a linear correlation was not observed.

Alvarez-Ramirez et al. (2018) analyzed the PD at different locations in white wheat bread slices. The results showed that the greater the distance from the bread surface, the higher the protein digestion.

Pasini et al. (2001) assessed the PD of baked bread by SDS-PAGE. Dough, crumb, and crust samples were freeze-dried before the protein digestion tests. The study concluded that uncooked bread dough had a higher PD than the crumb and crust samples. According to the authors, the

SDS-PAGE pattern of the undigested crumb was somewhat comparable to that of the undigested dough, but with poorer band definition, which was especially noticeable for the digested samples. Thorough solubilization of nitrogen from the samples related to the bread crust was not possible.

Another study reported that the proteins in the crust aggregated so intensely that it was not possible to insert them into the pores of the electrophoretic gel. The authors considered that one of the effects of baking, at the crust level, is the inhibition of the complete proteolytic degradation of allergenic proteins, causing a probable passage of large protein structures through the gastrointestinal tract and their subsequent allergic response in the body (Simonato et al. 2001).

Smith et al. (2015) assessed the digestibility of gluten proteins in flour samples and breads prepared using the same flour. These findings showed that baking decreases the digestibility of wheat gluten proteins. According to the authors, flour is more easily digestible than bread because it has more particles and a larger surface area, which makes it easier for digestive enzymes to reach proteins or carbohydrates. The authors also speculated that protein digestion might be hampered by gluten-starch complexes formed during baking. They verified that the presence of amylases accelerates protein breakdown in bread. The study proposed that amylase enzymes likely made gluten proteins more accessible by breaking down starch and amylose granules. Disintegration of the gluten network, into which starch granules are embedded, may increase amylase availability to starch granule shells, which may explain why proteolysis of gluten proteins also increases starch digestion. These findings demonstrate that the digestive process can be synergistic, utilizing several digestive enzymes to break down the composite environment. Considering this aspect related to starch-gluten interactions and their effects on PD, it should be noted that the use of freeze-drying before digestion analyses, as observed in the study by Pasini et al. (2001), can result in different gluten-starch interactions, which could have synergistic effects on PD.

Therefore, the literature is inconsistent regarding whether baking enhances bread protein digestion.

FLOUR SOURCES

In general, there are several differences between the experimental procedures used to evaluate PD. Figure 12.3 provides a quantitative example of the total number of studies used in this section, as well as the total number of studies for each type of material, the number of digestion phases used, and the quantification used. This information helps to elucidate the influences that the methodology can have on the results. Notably, in some studies, different analyses were performed to characterize the digested proteins.

WHEAT

Wheat Cultivars

Previous studies have suggested that over a century of wheat breeding, proteins and other quality indicators of wheat may have changed. For example, as the prevalence of disorders associated with wheat proteins, such as celiac disease, has increased in recent years, various theories have been developed to explain this trend. One theory is related to the technological wheat factors that have driven wheat selection for decades, resulting in wheat with enhanced technological properties (Dinu et al. 2018; Shewry 2018).

To verify this hypothesis, researchers have evaluated historical and modern wheat cultivars to determine whether modern wheat cultivars could be associated with lower wheat PD. However, although some authors have emphasized that the results are still inconsistent, recent studies have demonstrated that this trend cannot be supported (Dinu et al. 2018).

Abdel-Aal and Hucl (2002) used multi-enzyme one-step digestion and found that non-organic spelt and common wheat products had comparable in vitro PDs. Later, Abdel-Aal (2008) analyzed

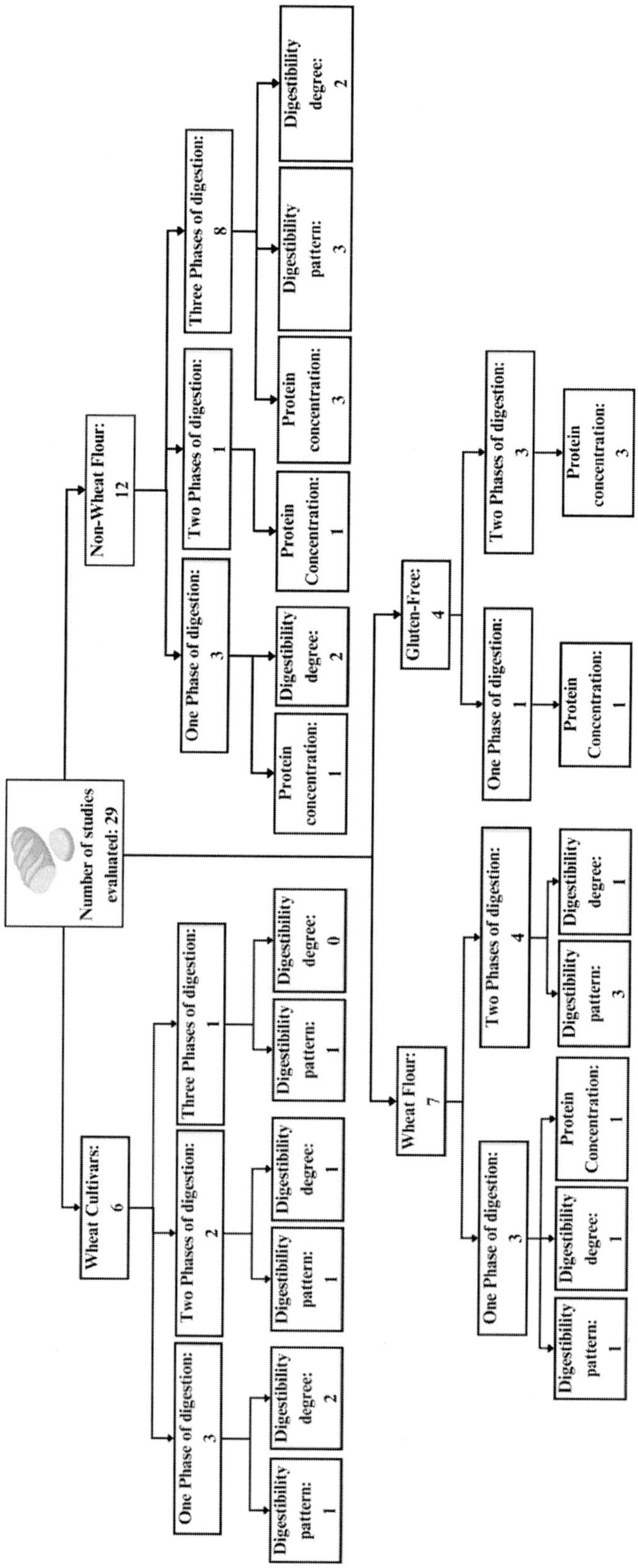

FIGURE 12.3 Number of studies related to bread protein digestibility considering four classifications: different wheat cultivars, commercial wheat, non-wheat, and gluten-free bread. The number of studies evaluated considers studies that carried out more than one analysis to characterize the digested protein; therefore, there may be a difference between the number shown in the figure and the number of studies presented below (elaborated by authors).

spelt wheat, which was mostly replaced by modern wheat cultivars and was the main wheat until the twentieth century. In this study, the in vitro PD of several baked goods (bread, biscuits, cookies, and muffins) made from organic spelt and common wheat flour was evaluated using two different methods: two-enzyme, two-step digestion and three-enzyme, one-step digestion. According to the author, the two-step digestion method was more accurate in identifying differences between the tested wheat products, whereas the one-step digestion method yielded results equivalent to in vivo (rat) methods. In both methods, the PD of the organic or common wheat products presented slight differences, while significant differences in PD were observed after baking for all products evaluated (increase in PD), as well as among the types of wheat products (bread, biscuit, cookie, or muffin).

Gulati et al. (2020) evaluated the PD of breads prepared from 21 different wheat varieties cultivated in the USA between 1870 and 2013 using in vitro analysis. The protein content in the flour was also measured. Based on these results, the protein concentration in wheat flour varied significantly depending on the year of wheat cultivar release. However, it was not possible to verify the effect of the year of planting on bread PD, even though the three older wheat cultivars presented significantly lower PD.

Ma and Baik (2021) also examined the impact of modern (soft red winter, hard winter, hard red spring, club, and durum) and historic (spelt, emmer, and einkorn) wheat classes and varieties on the in vitro PD of cooked flours. Based on the results, the authors concluded that PD was significantly influenced only by kernel hardness and albumin concentration in flour, rather than wheat classes or variety.

Lavoignat et al. (2022) evaluated the PD of bread produced from wheat grown in two environments from old and modern wheat cultivars (before or after 1960). The authors measured the proteolytic extension after 2 hours of digestion, and the results obtained for the PD of breads produced with old or modern wheat cultivars were not significantly different.

As a result, it is possible to conclude that more variables than just the "age" of the wheat affect PD. Studies have shown that even with years of difference between grains, the effect on digestibility varied more depending on other factors, such as grain hardness and protein and albumin concentrations.

Wheat Flour

Many studies on wheat flour have focused on the baking and fermentation aspects, as discussed previously. This section explores other aspects.

The effect of using simulated gastric fluids with varying pepsin concentrations (63 units or 182 units per mg of purified protein/flour/bread) on wheat flour products was examined (Smith et al. 2015). The authors found that purified gluten proteins can elucidate the mechanisms of in vitro digestion; however, these proteins are not reliable indicators for predicting digestibility in complex food matrices such as bread. The study concluded that, even at high pepsin concentrations, bread proteins were poorly accessible to enzymes, possibly due to aggregation or chemical modifications.

Researchers have also investigated the influence of bread formulated with wheat flour from different processes using in vitro systems that simulate adult and elderly conditions. When using soft, durum, and whole wheat flours for bread production, large variations in protein hydrolysis were observed in both systems (Moretton et al. 2023). Although whole wheat bread had a higher crude protein content, the presence of fiber may have prevented proteolytic enzymes from accessing their substrates in both adult and elderly systems. In addition, protein hydrolysis was more effective in the intestinal phase, and since breads are complex food matrices rich in carbohydrates, mainly starch, the greater degradation of this component in the intestinal phase may have aided protein digestion, as hydrolytic events break down this type of carbohydrate in the duodenal phase, making proteins more accessible.

Another area of study is the influence of nutritional composition on PD. Using the standard INFOGEST protocol (Minekus et al. 2014), researchers evaluated the use of refined (white) and whole wheat flour in bread production (Wu et al. 2017). The authors found that protein:protein or protein:carbohydrate associations are strong enough to remain bound during fermentation and/or swelling of polysaccharides are most likely to affect protein digestion. After baking for 20 minutes, the concentration of free amino was higher in bread containing white wheat flour than in bread containing whole wheat flour, suggesting that the fiber present in whole wheat flour interfered with protease activity.

These results indicate that factors such as interaction with other macronutrients significantly influence the PD/protein hydrolysis of bread made from different types of wheat flour. It is worth emphasizing that the use of enzymes complementary to proteolytic enzymes can help break down the structure of complex foods such as bread, increasing the concentration of hydrolyzed proteins (Freitas, Gómez-Mascaraque, and Brodkorb 2022).

Non-Wheat Flours

As mentioned previously, incorporation new raw material sources into bakery products is highly promising. However, caution should be exercised when using non-conventional raw materials without prior analysis. In terms of their cellular and structural complexity, plant sources may contain compounds commonly referred to as anti-nutrients, be deficient in essential amino acids, and have low digestibility (Sá, Moreno, and Carciofi 2019). Antinutritional factors include polyphenols (tannins and lignins), protease and amylase inhibitors, phytic and oxalic acids, and glycosides (Gilani, Cockell, and Sepehr 2005).

The incorporation of vegetables, fruits, cereals, and plants, as well as parts of these materials that are normally discarded, have been applied into bread formulations. For example, onion peels were used for bread production at concentrations of 1–4% as a substitute for wheat flour (Świeca et al. 2013). Although onion peel is rich in flavonoids, a compound with functional potential, the study highlighted that the presence of this bioactive compound reduced the digestibility of bread; even at low concentrations (1%), a reduction of 7% compared to control bread was observed. In contrast, the protein content in the reaction system increased as the onion peel incorporation increased, ranging from 3.94 to 8.53 mg/g of DW. The authors carried out other analyses, such as liquid chromatography and SDS-PAGE, and found a significant reduction in free amino group levels in the enriched bread and an increase in the chromatogram area compared with the control bread. These results indicate the formation of a protein-flavonoid complex, which significantly influences the bioaccessibility of proteins in vitro as well as a decrease in antioxidant potential, supporting the hypothesis of complex formation and providing another point of view regarding the issue of thermal deactivation of bioactive compounds. Researchers observed the same behavior with the incorporation of dried quinoa leaves in bread, with the formation of a protein-phenolic complex (Świeca et al. 2014).

Similarly, researchers have evaluated the inclusion of cereals and pseudo-cereals (gluten-containing and gluten-free flours) in bread formulations, using different breadmaking conditions (variation of the time-temperature binomial during baking) with constant protein content (Wu et al. 2017). According to these results, independent of the process, no significant variation was observed in the in vitro protein hydrolysis of breads formulated with whole wheat flour, buckwheat, rye, and chickpea (the latter, except for fermentation). However, bread formulated with refined wheat flour, lupin, and amaranth showed a decreased PD after baking. The authors attributed this effect mainly to two parameters, phenolic content and carbohydrates (fiber), which are used in non-enzymatic browning reactions.

Researchers have also used the peel of Amazonian fruits to formulate whole bread enriched with dietary fiber and phytochemicals (Salgado et al. 2011). In this study, cupuaçu peel was transformed into flour and incorporated in proportions of 0% (control), 3%, 6%, and 9% into whole bread. The study found an indirect relationship between the levels of tannins, phytic acid, and total phenolics and PD. In other words, the higher the concentration of cupuaçu peel flour in the bread, the higher the concentration of bioactive compounds and the lower the PD. Researchers observed similar behavior when incorporating complexes of different dietary fibers and prebiotics into bread, where, regardless of the type, PD was reduced, and the effect was more pronounced when incorporating gluco-oligosaccharide as a prebiotic agent (Angioloni and Collar 2011).

Researchers also studied the optimization of the protein quality of wheat bread, finding that the essential amino acid index and PD were not significantly affected by the incorporation of 30% native fava bean flour, although the content of peptides and free amino acids was significantly higher (Coda et al. 2017). One possible explanation for the lack of change in PD is the low content of tannins, flavonoids, and phenolics in fava beans (max. of 0.3 mg catechin/g DW, 0.540 mg catechin/g DW, and 1.22 mg gallic acid/g DW, respectively) (Rajhi et al. 2022). Similarly, an older study reported in vitro digestibility values of 75.6% and 75.3% for bread made with white and reddish-white sorghum, respectively. Furthermore, chemical composition analysis showed that the tannin content present in the flour was eliminated during processing, and no concentrations were determined in either type of bread (Khalil et al. 1984).

Recently, a research group demonstrated that fermentation with the fungus *Pleurotus ostreatus* reduced the concentration of tannins in black bean, red bean, and oat flours (Espinosa-Páez et al. 2017) and, when applied to bread, fermented red bean, and oat flour, improved PD compared to the same product without fermentation(Espinosa-Páez et al. 2023)

In general, the incorporation of new raw materials into bread, regardless of their origin, mainly involves the interaction of proteins with other compounds, particularly bioactive compounds such as tannins, flavonoids, and phenolics. An efficient way to counteract this problem is to optimize the formulation using flour that is rich in protein but has a lower bioactive content or by using treatment/processing techniques that improve enzymatic activity and, consequently, increase protein bioaccessibility.

Gluten-Free Flours

As described for traditional bread made with wheat flour, researchers have explored the incorporation of different raw materials to improve the nutritional aspects of gluten-free breads (GFBs).

Sciarini et al. (2017) evaluated the effects of oat bran fiber (insoluble fiber), inulin (soluble fiber), and resistant starch (insoluble fiber) on bread formulated with rice flour, cassava starch, and soy flour were evaluated to verify their effects on in vitro PD. In general, two interesting aspects were observed: (1) the incorporation of fibers in GFBs increased PD compared to the control GFBs, and (2) PD was higher in breads with 5% added fibers than in breads with 10%. The authors suggested that the incorporation of the fibers disturbs the formation of the bread structure, making soy proteins more suitable for enzymatic digestion. However, when the fiber concentration was higher (10%), a physical or chemical barrier could be responsible for the lower levels of PD observed.

When pea and egg white proteins were used for protein enrichment (30%) of GFB (made with starch), with constant hydration level and/or specific volume, a direct effect of the type of protein was observed. The control bread had a PD of 69.25%, while those enriched with pea protein had a PD of 79.03% (Sahagún et al. 2020). The GFB made with egg protein had the lowest PD among all samples, with a mean of 76.53%. This finding demonstrates that regardless of the mode of preparation, protein accessibility remained similar in other protein-enriched GFBs.

Montemurro, Pontonio, and Rizzello (2021) optimized the formulation to obtain clean-labeled GFB with increased protein content based on a blend of rice and corn flour. They performed protein enrichment using quinoa flour, psyllium, chia, linseed, and fermented chestnut paste. The results showed that the enriched bread had a PD of 76.9%, whereas for the standard GFB, this value was only 21.2%. Modifications to semolina to transform it into gluten-free raw material using a fermentation process and then applying it to GFBs also showed a significant increase in the digested protein concentration, with an increase of approximately 52% compared to a standard loaf of bread containing gluten (Rizzello, Montemurro, and Gobbetti 2016).

Therefore, in the case of GFBs, the difficulty in obtaining technological characteristics requires the use of alternative and/or modified treatments. Although nutritional improvement is not the main objective, it does improve nutritional aspects (Skendi, Papageorgiou, and Varzakas 2021), such as protein digestion, which can also limit the use of additives, as in the case of clean-label products.

METHODS USED TO EVALUATE PD IN BREAD

Digestion and digestibility are distinct analyses. The first refers to in vitro assays and bioassays that simulate the human gastrointestinal tract, focusing on the bioaccessibility of compounds of interest. On the other hand, PD is an analysis performed to measure the amount of protein and/or amino acids and/or free amino released after the period established for enzymatic hydrolysis.

Currently, complex digestive processes can be simulated using two distinct models: static and dynamic (Mackie, Mulet-Cabero, and Torcello-Gomez 2020). Such models have been formulated to overcome problems that occur with in vivo models, which are labor-intensive, expensive, and pose ethical concerns (Brodkorb et al. 2019). However, several digestion methodologies have been developed to study nutrient digestibility, such as those developed by Adler-Nissen (1979), Gauthier, Vachon, and Savoie (1686), Pedersen and Eggum (1983), Saunders et al. (1973), and, more recently, Minekus et al. (2014).

The methodologies used for the sample preparation and digestion stages are shown in Figure 12.4. It is important to note that the methods, whether they refer to the sample preparation or analysis stage, can be adapted according to the research objective, as in the case of Cattaneo et al. (2019), who aimed to evaluate the digestion of infant biscuits, and Gao et al. (2019), who found that the mode and level of mechanical disruption of the bread bolus, as well as homogenization, have an impact on digestion.

It is clear that the variety of in vitro protocols makes it difficult to choose the most appropriate, and coupled with the differences in the sample preparation stage, the comparison of results between studies is a complex process (Figure 12.4).

To minimize these differences, Brodkorb et al. (2019) developed a standardized protocol called INFOGEST 2.0. The main characteristic of this protocol is that it can be simultaneously performed using different samples. For a better understanding of the method and its specificities, readers are strongly encouraged to read the methodology in INFOGEST 2.0.

After digestion, the hydrolyzed material proceeds to the next step, which is the verification of PD. As with the digestion process, there are several methods to evaluate digestibility patterns (qualitative methods) and estimate the degree of hydrolysis and protein content (quantitative methods). For more details on these aspects, refer to the studies by Orlien et al. (2023), Silvestre (1997), and Zhou, Tan, and McClements (2023). However, among the quantitative methods, food PD analysis is essentially a measure of vulnerability to proteolysis and is closely linked to the structure of the protein, the intensity of processing, whether related to the material or the product itself, and the presence of antinutritional factors (Sá, Moreno, and Carciofi 2019).

CONCLUSION

Most studies related to bread PD consider only one or two compositional or processing parameters; therefore, it is difficult to understand the links between the chemical changes in proteins that occur during the breadmaking process. A review of the literature reveals that, in most cases, bioactive compounds, interactions between macronutrients, Maillard reaction products, baking, and fermentation are the main factors affecting PD in bread (Figure 12.5).

In addition, there is a growing consumer demand and, consequently, an increasing number of studies related to GFB, which have already shown interesting results, including those with non-wheat flour sources. However, in the latter case, great care must be taken regarding antinutritional factors that can negatively impact the nutritional quality of the products produced.

Additionally, methodologies need to be standardized, and the development of green technologies could aid in determining PD parameters related to both quantitative or qualitative aspects.

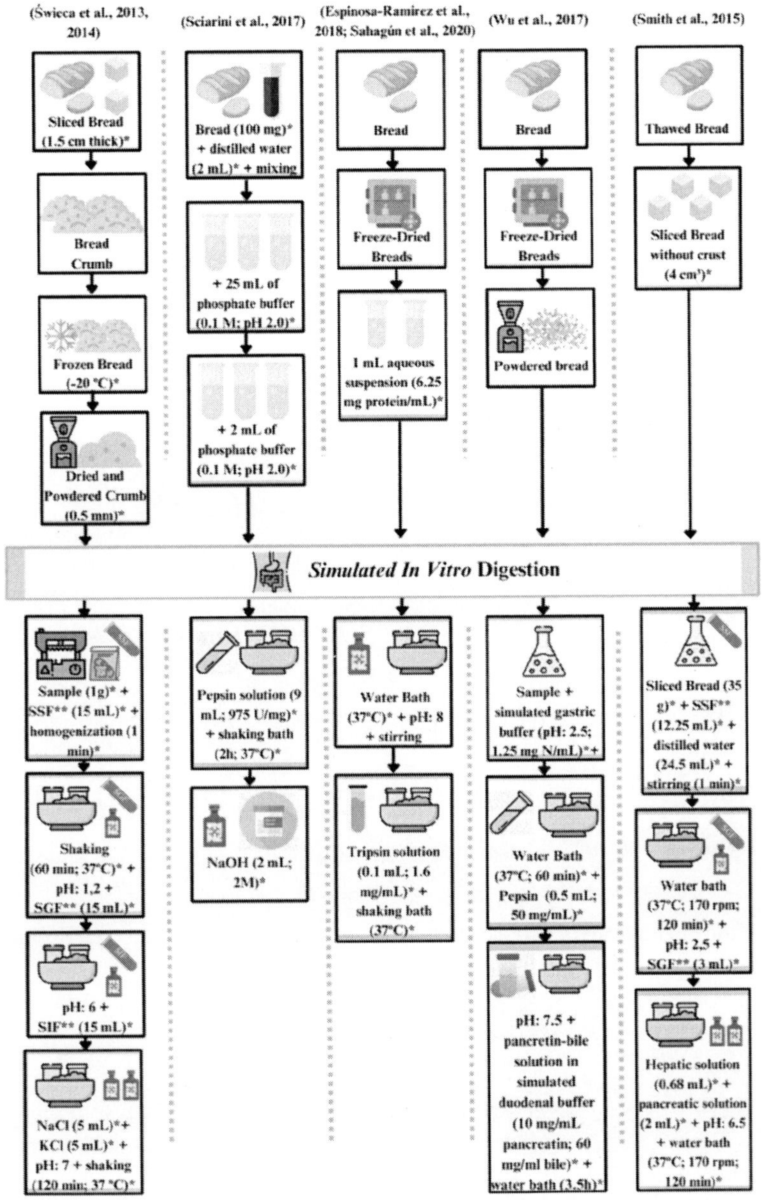

FIGURE 12.4 Illustrative overview of the differences between methodologies, both in the sample preparation stage and simulated digestion (elaborated by the authors).

MINI DICTIONARY OF TERMS

Bread processing conditions: Any change in the breadmaking process conditions related to fermentation time, type of yeast, flour particle size, or baking conditions.

Green technologies: Technologies that have little or no negative effects on the environment, society, and handlers.

INFOGEST 2.0: An internationally recognized in vitro digestion methodology.

Phenolic compounds: Secondary metabolites of plants and vegetables.

Tannins: Secondary metabolites of plants and vegetables.

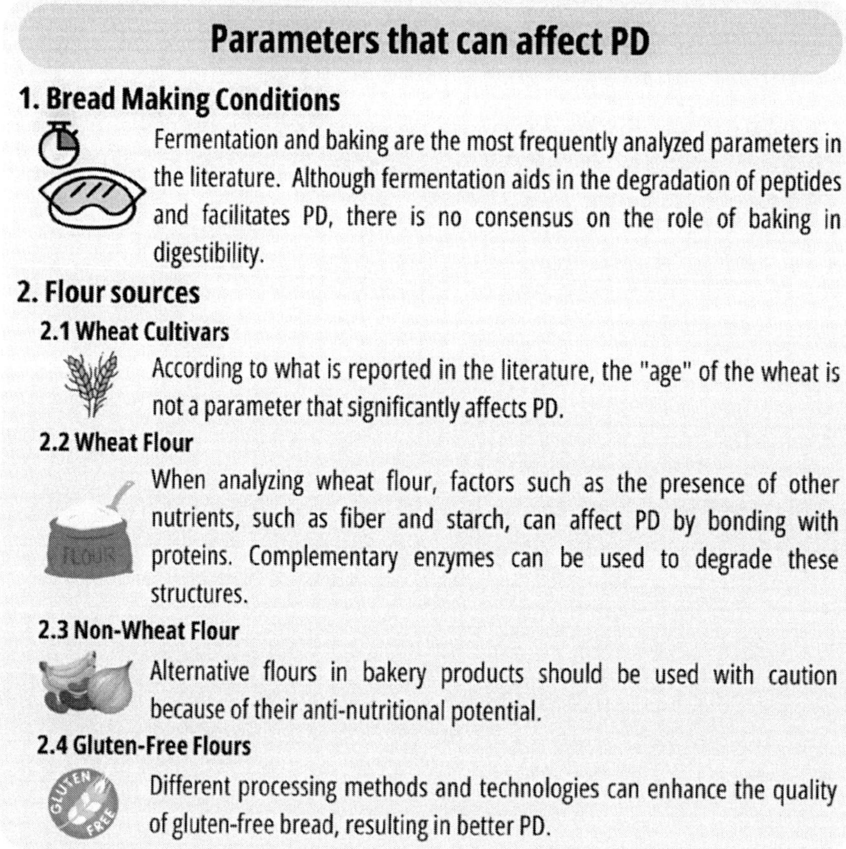

FIGURE 12.5 Illustrates the main parameters that can influence PD and considerations that should be taken into account when working with each factor (elaborated by the authors).

SUMMARY POINTS

- This chapter focuses on general aspects that can influence bread protein digestibility.
- Studies on fermentation and baking processes generally show that fermentation can enhance protein digestibility (PD); however, there is no consensus on the influence of the baking process.
- In non-wheat flour bread formulations, the interaction of components such as bioactive compounds can negatively affect PD.
- Chemical interactions between the different ingredients used in bread formulations strongly influence PD.
- A standardized PD protocol for bread using faster methods and sustainable techniques is essential.

ACKNOWLEDGMENTS

The authors thank the free icons available at Canva, Flaticom, and Freepik platforms used to create all illustrations in this chapter.

REFERENCES

Abdel-Aal, E. S. M. 2008. "Effects of Baking on Protein Digestibility of Organic Spelt Products Determined by Two in Vitro Digestion Methods." *Lwt* 41 (7): 1282–88. doi:10.1016/j.lwt.2007.07.018.

Abdel-Aal, E. S. M., and P. Hucl. 2002. "Amino Acid Composition and in Vitro Protein Digestibility of Selected Ancient Wheats and Their End Products." *Journal of Food Composition and Analysis* 15 (6): 737–47. doi:10.1006/jfca.2002.1094.

Adler-Nissen, J. 1979. "Determination of the Degree of Hydrolysis of Food Protein Hydrolysates by Trinitrobenzenesulfonic Acid." *Journal of Agricultural and Food Chemistry* 27 (6): 1258–62.

Alvarez-Ramirez, J., E. Rodriguez-Huezo, M. Meraz, S. Garcia-Diaz, P. C. Flores-Silva, and L. Mondragon-Reinoso. 2018. "Spatial Variation of In Vitro Starch and Protein Digestibility in White Wheat Bread." *Starch/Staerke* 70 (5–6): 1–9. doi:10.1002/star.201800025.

Angioloni, A., and C. Collar. 2011. "Physicochemical and Nutritional Properties of Reduced-Caloric Density High-Fibre Breads." *Lwt* 44 (3). Elsevier Ltd: 747–58. doi:10.1016/j.lwt.2010.09.008.

Arora, K., H. Ameur, A. Polo, R. Di Cagno, C. G. Rizzello, and M. Gobbetti. 2021. "Thirty Years of Knowledge on Sourdough Fermentation: A Systematic Review." *Trends in Food Science and Technology* 108 (November 2020). Elsevier Ltd: 71–83. doi:10.1016/j.tifs.2020.12.008.

Bøgh, K. L., and C. B. Madsen. 2016. "Food Allergens: Is There a Correlation between Stability to Digestion and Allergenicity?" *Critical Reviews in Food Science and Nutrition* 56 (9): 1545–67. doi:10.1080/10408398.2013.779569.

Bredariol, P., R. A. Carvalho, and F. M. Vanin. 2020. "The Effect of Baking Conditions on Protein Digestibility, Mineral and Oxalate Content of Wheat Breads." *Food Chemistry* 332 (December). Elsevier Ltd. doi:10.1016/j.foodchem.2020.127399.

Bredariol, P., and F. M. Vanin. 2022. "Bread Baking Review: Insight into Technological Aspects to Preserve Nutrition." *Food Reviews International* 38 (S1). Taylor & Francis: 651–68. doi:10.1080/87559129.2021.1878211.

Brodkorb, A., L. Egger, M. Alminger, P. Alvito, R. Assunção, S. Ballance, T. Bohn, et al. 2019. "INFOGEST Static in Vitro Simulation of Gastrointestinal Food Digestion." *Nature Protocols* 14 (4): 991–1014. doi:10.1038/s41596-018-0119-1.

Cappelli, A., L. Lupori, and E. Cini. 2021. "Baking Technology: A Systematic Review of Machines and Plants and Their Effect on Final Products, Including Improvement Strategies." *Trends in Food Science and Technology*. Elsevier Ltd. doi:10.1016/j.tifs.2021.06.048.

Cattaneo, S., F. Masotti, T. Silvetti, A. Hidalgo, and I. De-Noni. 2019. "Effect of Dairy Ingredients on the Heat Damage and the in Vitro Digestibility of Infant Biscuits." *European Food Research and Technology* 245 (11). Springer Verlag: 2489–97. doi:10.1007/s00217-019-03368-z.

Coda, R., J. Varis, M. Verni, C. G. Rizzello, and K. Katina. 2017. "Improvement of the Protein Quality of Wheat Bread through Faba Bean Sourdough Addition." *LWT* 82 (September). Academic Press: 296–302. doi:10.1016/j.lwt.2017.04.062.

Dinu, M., A. Whittaker, G. Pagliai, S. Benedettelli, and F. Sofi. 2018. "Ancient Wheat Species and Human Health: Biochemical and Clinical Implications." *Journal of Nutritional Biochemistry* 52. Elsevier Inc.: 1–9. doi:10.1016/j.jnutbio.2017.09.001.

Espinosa-Páez, E., M. G. Alanis-Guzmán, C. E. Hernández-Luna, J. G. Báez-González, C. A. Amaya-Guerra, and A. M. Andrés-Grau. 2017. "Increasing Antioxidant Activity and Protein Digestibility in Phaseolus Vulgaris and Avena Sativa by Fermentation with the Pleurotus Ostreatus Fungus." *Molecules* 22 (12). doi:10.3390/molecules22122275.

Espinosa-Páez, E., C. E. Hernández-Luna, S. Longoria-García, C. Torres-Alvarez, C. Velez-Argumedo, and B. E. González-Martínez. 2023. "Improving Nutritional and Functional Quality Characteristics in Bread by Using Flours Obtained from Fermentation of Kidney Beans and Oats with Pleurotus Ostreatus." *CyTA - Journal of Food* 21 (1): 151–58. doi:10.1080/19476337.2023.2173305.

FAO, Food and Agriculture Organization. 2011. *Dietary Protein Quality Evaluation in Human Nutrition. Report of an FAQ Expert Consultation*. FAO Food and Nutrition Paper. Vol. 92.

Freitas, D., L. G. Gómez-Mascaraque, and A. Brodkorb. 2022. "Digestion of Protein and Toxic Gluten Peptides in Wheat Bread, Pasta and Cereal and the Effect of a Supplemental Enzyme Mix." *Frontiers in Nutrition* 9. doi:10.3389/fnut.2022.986272.

Gao, J., S. Lin, X. Jin, Y. Wang, J. Ying, Z. Dong, and W. Zhou. 2019. "In Vitro Digestion of Bread: How Is It Influenced by the Bolus Characteristics?" *Journal of Texture Studies* 50 (3). Blackwell Publishing Ltd: 257–68. doi:10.1111/jtxs.12391.

Gao, S., J. Hong, C. Liu, X. Zheng, L. Li, and X. Tian. 2022. "Comparative Study of Different Fermentation and Cooking Methods on Dough Rheology and the Quality of Chinese Steamed/Baked Bread." *Journal of Food Processing and Preservation* 46 (2): 1–8. doi:10.1111/jfpp.16221.

Gauthier, S. F., C. Vachon, and L. Savoie. 1686. "Enzymatic Conditions of an In Vitro Method to Study Protein Digestion." *Journal of Food Science* 51 (4): 960–64.

Gilani, G. S., K. A. Cockell, and E. Sepehr. 2005. "Effects of Antinutritional Factors on Protein Digestibility and Amino Acid Availability in Foods." *Journal of AOAC International* 88 (3): 967–87. doi:10.1093/jaoac/88.3.967.

Graça, C., M. Edelmann, A. Raymundo, I. Sousa, R. Coda, T. Sontag-Strohm, and X. Huang. 2022. "Yoghurt as a Starter in Sourdough Fermentation to Improve the Technological and Functional Properties of Sourdough-Wheat Bread." *Journal of Functional Foods* 88. Elsevier Ltd: 104877. doi:10.1016/j.jff.2021.104877.

Gulati, P., S. Brahma, R. A. Graybosch, Y. Chen, and D. J. Rose. 2020. "In Vitro Digestibility of Proteins from Historical and Modern Wheat Cultivars." *Journal of the Science of Food and Agriculture* 100 (6): 2579–84. doi:10.1002/jsfa.10283.

Khalil, J. K., W. N Sawaya, W. J. Safi, and H. M Al-Mohammad. 1984. "Chemical Composition and Nutritional Quality of Sorghum Flour and Bread." *Qualitas Plantarum Plant Foods for Human Nutrition* 34 (2): 141–50. doi:10.1007/BF01094842.

Kulathunga, J., K. Whitney, and S. Simsek. 2023. "Impact of Starter Culture on Biochemical Properties of Sourdough Bread Related to Composition and Macronutrient Digestibility." *Food Bioscience* 53 (December 2022). Elsevier Ltd: 102640. doi:10.1016/j.fbio.2023.102640.

Lavoignat, M., S. Denis, A. Faye, L. Halupka, S. Perrochon, L. Rhazi, P. Giraudeau, et al. 2022. "Differences in Bread Protein Digestibility Traced to Wheat Cultivar Traits." *Journal of Cereal Science* 107 (July). doi:10.1016/j.jcs.2022.103533.

Ma, F., and B. K. Baik. 2021. "Influences of Grain and Protein Characteristics on in Vitro Protein Digestibility of Modern and Ancient Wheat Species." *Journal of the Science of Food and Agriculture* 101 (11): 4578–84. doi:10.1002/jsfa.11100.

Mackie, A., A. I. Mulet-Cabero, and A. Torcello-Gomez. 2020. "Simulating Human Digestion: Developing Our Knowledge to Create Healthier and More Sustainable Foods." *Food and Function* 11 (11). Royal Society of Chemistry: 9397–9431. doi:10.1039/d0fo01981j.

Minekus, M., M. Alminger, P. Alvito, S. Ballance, T. Bohn, C. Bourlieu, F. Carrière, et al. 2014. "A Standardised Static in Vitro Digestion Method Suitable for Food-an International Consensus." *Food and Function* 5 (6): 1113–24. doi:10.1039/c3fo60702j.

Montemurro, M., E. Pontonio, and C. G. Rizzello. 2021. "Design of a 'Clean-Label' Gluten-Free Bread to Meet Consumers Demand." *Foods* 10 (2): 462. doi:10.3390/foods10020462.

Moretton, M., M. Alongi, S. Melchior, and M. Anese. 2023. "Adult and Elderly in Vitro Digestibility Patterns of Proteins and Carbohydrates as Affected by Different Commercial Bread Types." *Food Research International* 167 (March). Elsevier Ltd: 112732. doi:10.1016/j.foodres.2023.112732.

Ogilvie, O., Roberts, S., Suttons, K., Gerrard, J., Larsen, J., and Domigan, L. 2021. "The effect of baking time and temperature on gluten protein structure and celiac peptide digestibility." *Food Research International* 140: 109988. doi: 10.1016/j.foodres.2020.109988.

Orlien, V., K. Aalaei, M. M. Poojary, D. S. Nielsen, L. Ahrné, and J. R. Carrascal. 2023. "Effect of Processing on in Vitro Digestibility (IVPD) of Food Proteins." *Critical Reviews in Food Science and Nutrition*. Taylor and Francis Ltd. doi:10.1080/10408398.2021.1980763.

Osman, M. A. 2004. "Changes in Sorghum Enzyme Inhibitors, Phytic Acid, Tannins and in Vitro Protein Digestibility Occurring during Khamir (Local Bread) Fermentation." *Food Chemistry* 88 (1): 129–34. doi:10.1016/j.foodchem.2003.12.038.

Pasini, G., B. Simonato, M. Giannattasio, A. D. B. Peruffo, and A. Curioni. 2001. "Modifications of Wheat Flour Proteins during in Vitro Digestion of Bread Dough, Crumb, and Crust: An Electrophoretic and Immunological Study." *Journal of Agricultural and Food Chemistry* 49 (5): 2254–61. doi:10.1021/jf0014260.

Pedersen, B., and B. O. Eggum. 1983. "Prediction of Protein Digestibility by an in Vitro Enzymatic PH-stat Procedure." *Zeitschrift Für Tierphysiologie Tierernährung Und Futtermittelkunde* 49 (1–5): 265–77. doi:10.1111/j.1439-0396.1983.tb00808.x.

Rajhi, I., M. Boulaaba, B. Baccouri, F. Rajhi, J. Hammami, F. Barhoumi, G. Flamini, and H. Mhadhbi. 2022. "Assessment of Dehulling Effect on Volatiles, Phenolic Compounds and Antioxidant Activities of Faba Bean Seeds and Flours." *South African Journal of Botany* 147 (July). Elsevier B.V.: 741–53. doi:10.1016/j.sajb.2022.03.010.

Rizzello, C. G., M. Montemurro, and M. Gobbetti. 2016. “Characterization of the Bread Made with Durum Wheat Semolina Rendered Gluten Free by Sourdough Biotechnology in Comparison with Commercial Gluten-Free Products.” *Journal of Food Science* 81 (9): H2263–72. doi:10.1111/1750-3841.13410.

Rizzello, C. G., P. Portincasa, M. Montemurro, D. M. di Palo, M. P. Lorusso, M. de Angelis, L. Bonfrate, B. Genot, and M. Gobbetti. 2019. “Sourdough Fermented Breads Are More Digestible than Those Started with Baker’s Yeast Alone: An in Vivo Challenge Dissecting Distinct Gastrointestinal Responses.” *Nutrients* 11 (12). doi:10.3390/nu11122954.

Sá, A. G. A., Y. M. F. Moreno, and B. A. M. Carciofi. 2019. “Food Processing for the Improvement of Plant Proteins Digestibility.” *Critical Reviews in Food Science and Nutrition*. Taylor and Francis Inc. doi:10.1080/10408398.2019.1688249.

Sahagún, M., Y. Benavent-Gil, C. M. Rosell, and M. Gómez. 2020. “Modulation of in Vitro Digestibility and Physical Characteristics of Protein Enriched Gluten Free Breads by Defining Hydration.” *Lwt* 117 (August 2019). Elsevier: 108642. doi:10.1016/j.lwt.2019.108642.

Salgado, J. M., B. S. Rodrigues, C. M. Donado-Pestana, C. T. Santos-Dias, and M. C. Morzelle. 2011. “Cupuassu (Theobroma Grandiflorum) Peel as Potential Source of Dietary Fiber and Phytochemicals in Whole-Bread Preparations.” *Plant Foods for Human Nutrition* 66 (4): 384–90. doi:10.1007/s11130-011-0254-0.

Santos, Y. J. S., W. M. Facchinatto, A. L. Rochetti, R. A. Carvalho, S. Le Feunteun, H. Fukumasu, M. Morzel, L. A. Colnago, and F. M. Vanin. 2023. “Systemic Characterization of Pupunha (Bactris Gasipaes) Flour with Views of Polyphenol Content on Cytotoxicity and Protein in Vitro Digestion.” *Food Chemistry* 405 (PA). Elsevier Ltd: 134888. doi:10.1016/j.foodchem.2022.134888.

Saunders, R. M., M. A. Connor, A. N. Booth, E. M. Bickoff, and G. O. Kohler. 1973. “Measurement of Digestibility of Alfalfa Protein Concentrates by in Vivo and in Vitro Methods.” *The Journal of Nutrition* 103 (4): 530–35. doi:10.1093/jn/103.4.530.

Sciarini, L. S., M. C. Bustos, M. B. Vignola, C. Paesani, C. N. Salinas, and G. T. Pérez. 2017. “A Study on Fibre Addition to Gluten Free Bread: Its Effects on Bread Quality and in Vitro Digestibility.” *Journal of Food Science and Technology* 54 (1). Springer India: 244–52. doi:10.1007/s13197-016-2456-9.

Shewry, P. R. 2018. “Do Ancient Types of Wheat Have Health Benefits Compared with Modern Bread Wheat?” *Journal of Cereal Science* 79. Elsevier Ltd: 469–76. doi:10.1016/j.jcs.2017.11.010.

Silvestre, M. P. C. 1997. “Review of Methods for the Analysis of Protein Hydrolysates.” *Food Chemistry* 60 (2): 263–71. doi:10.1016/S0308-8146(96)00347-0.

Simonato, B., G. Pasini, M. Giannattasio, A. D. B. Peruffo, F. De Lazzari, and A. Curioni. 2001. “Food Allergy to Wheat Products: The Effect of Bread Baking and in Vitro Digestion on Wheat Allergenic Proteins. A Study with Bread Dough, Crumb, and Crust.” *Journal of Agricultural and Food Chemistry* 49 (11): 5668–73. doi:10.1021/jf0104984.

Skendi, A., M. Papageorgiou, and T. Varzakas. 2021. “High Protein Substitutes for Gluten in Gluten-Free Bread.” *Foods* 10 (9): 1997. doi:10.3390/foods10091997.

Smith, F., X. Pan, V. Bellido, G. A. Toole, F. K. Gates, M. S. J. Wickham, P. R. Shewry, S. Bakalis, P. Padfield, and E. N. C. Mills. 2015. “Digestibility of Gluten Proteins Is Reduced by Baking and Enhanced by Starch Digestion.” *Molecular Nutrition and Food Research* 59 (10). Wiley-VCH Verlag: 2034–43. doi:10.1002/mnfr.201500262.

Świeca, M., U. Gawlik-Dziki, D. Dziki, B. Baraniak, and J. Czyz. 2013. “The Influence of Protein-Flavonoid Interactions on Protein Digestibility in Vitro and the Antioxidant Quality of Breads Enriched with Onion Skin.” *Food Chemistry* 141 (1). Elsevier Ltd: 451–58. doi:10.1016/j.foodchem.2013.03.048.

Świeca, M., Ł. Sęczyk, U. Gawlik-Dziki, and D. Dziki. 2014. “Bread Enriched with Quinoa Leaves - The Influence of Protein-Phenolics Interactions on the Nutritional and Antioxidant Quality.” *Food Chemistry* 162 (November). Elsevier Ltd: 54–62. doi:10.1016/j.foodchem.2014.04.044.

Wu, T., C. Taylor, T. Nebl, K. Ng, and L. E. Bennett. 2017. “Effects of Chemical Composition and Baking on in Vitro Digestibility of Proteins in Breads Made from Selected Gluten-Containing and Gluten-Free Flours.” *Food Chemistry* 233 (October). Elsevier Ltd: 514–24. doi:10.1016/j.foodchem.2017.04.158.

Zhou, H., Y. Tan, and D. J. McClements. 2023. “Applications of the INFOGEST In Vitro Digestion Model to Foods: A Review.” *Annual Review of Food Science and Technology* 14: 135–56. doi:10.1146/annurev-food-060721-012235.

13 Protein Sources and Determining Ileal Amino Acid Digestibility

Suzanne M. Hodgkinson

ABBREVIATIONS

DIAAS	Digestible indispensable amino acid score
FAO	Food and Agriculture Organisation of the United Nations
N	Nitrogen
PDCAAS	Protein digestibility-corrected amino acid score
PEG	Polyethylene glycol

INTRODUCTION

Protein is an essential macronutrient in the diet, required for a plethora of metabolic processes. A typical diet contains a range of protein sources, which vary in terms of nutritional quality. When a protein-containing food or meal is consumed, not all of the amino acids will be digested and absorbed. The proportion of the different amino acids that are digested and absorbed varies depending on the protein source. The amount of amino acids that are digested and absorbed is determined based on the amount of each amino acid that was consumed and the amount excreted. This is termed protein digestibility, and calculating it on an individual amino acid basis determines amino acid digestibility. The nutritional quality of the protein in foods is based on two aspects: the amount of amino acids that are digested and absorbed (the amount digestible) and their relation to human amino acid requirements.

In countries such as the United States, the protein quality of foods and ingredients is evaluated based on protein digestibility-corrected amino acid score (PDCAAS), which was recommended by the Food and Agriculture Organisation and World Health Organisation as the preferred method in 1990 (FAO and WHO 1990). However, the more recently FAO-convened Expert Working Group on Protein Quality has endorsed digestible indispensable amino acid score (DIAAS) as the preferred method to evaluate protein quality (FAO 2013).

For both PDCAAS and DIAAS, the amino acid composition of the food or ingredient is first determined. A correction is then applied to adjust for amino acid availability (digestibility), and the resulting corrected composition is compared with reference values that reflect human requirements. Table 13.1 shows the reference values for the amount of digestible indispensable amino acids when DIAAS is determined for different age groups. A principal difference between PDCAAS and DIAAS is the manner in which the correction for amino acid availability is applied. The different aspects that differ between the determination of the correction factor for PDCAAS and DIAAS are discussed below.

DOI: 10.1201/b23328-13

TABLE 13.1
Recommended Scoring (Reference) Pattern (mg/g Protein) for Determination of Digestible Indispensable Amino Acid Score (DIAAS) for Different Age Groups[1]

Amino Acid	Infant (Birth to 6 Months)	Child (6 Months to 3 Years)	Older Child, Adolescent, Adult
Histidine	21	20	16
Isoleucine	55	32	30
Leucine	96	66	61
Lysine	69	57	48
Sulphur amino acids[2]	33	27	23
Aromatic amino acids[3]	94	52	41
Threonine	44	31	25
Tryptophan	17	8.5	6.6
Valine	55	43	40

[1] FAO, 2013
[2] Methionine + cysteine
[3] Phenylalanine + tyrosine

FAECAL VERSUS ILEAL DIGESTIBILITY

Traditionally, the proportion of a food's amino acids that are digested and absorbed was determined by measuring the difference between the amount consumed orally and the amount that was present in the faeces. This is the case when PDCAAS is determined for a food or ingredient. However, the large intestine contains many microbes that metabolise the amino acids passing from the small intestine. In pigs, around 80% of the amino acids present in the faeces are of microbial origin as opposed to dietary origin (Mason 1980). When a single dietary indispensable amino acid was infused into the large intestine in pigs that received a diet first-limiting in the same indispensable amino acid, there was no improvement in nitrogen balance (Darragh et al. 1994). This shows that if there is any absorption of amino acids from the large intestine, it appears to be of no nutritional significance (van der Wielen et al. 2017). From the protein metabolism perspective, digestion and absorption of amino acids conclude at the end of the small intestine, the terminal ileum. For these reasons, it is more accurate to determine the nutritional quality of proteins at the ileal level, thus giving ileal digestibility (Schaafsma 2012).

In Table 13.2, values for true nitrogen digestibility are presented at ileal and faecal levels from rats (Rutherfurd et al. 2015). While the difference between the true ileal protein digestibility determined at ileal and faecal levels is less notable for high-quality proteins such as WPI, with a difference of 2.7 percentage units, there are larger differences for lower quality proteins such as cooked rice (difference of 19 percentage units) and wheat bran (difference of 18 percentage units), highlighting the need to use ileal digestibility values. One important difference between PDCAAS and DIAAS is that PDCAAS is determined based on faecal digestibility, whereas for DIAAS, the digestibility of amino acids is determined at ileal level. The latter better represents the nutritional value of the amino acids.

APPARENT VERSUS TRUE DIGESTIBILITY

When amino acid digestibility is determined using the difference between the amount of amino acids consumed and those present in digesta collected from the terminal ileum, this is termed

TABLE 13.2
True Ileal and Faecal Nitrogen (N) Digestibility for Foods Determined in the Growing Rat[1]

Protein Source	True N digestibility, %		Overall SEM	P	Overestimation (%)
	Ileal	Faecal			
WPI	99.0	102	0.54	<0.001	2.7
WPC	95.3	99.8	0.90	<0.001	4.8
SPI	95.0	98.2	1.00	0.004	3.3
Roasted peanuts	90.9	98.4	1.71	<0.001	8.2
Cooked rolled oats	88.5	95.2	1.23	0.003	7.6
Cooked kidney beans	79.6	80.4	4.47	0.85	–
Cooked rice	72.8	86.6	3.60	0.003	19
Wheat bran	72.7	85.3	3.28	0.001	17

[1] Source of data: Rutherfurd et al. 2015

"apparent" digestibility. However, within the digesta there are amino acids of both dietary origin and endogenous origin. The endogenous amino acids are those of human (or animal, when an animal model is used) origin. They include amino acids originating from digestive enzymes, mucus that lines the digestive tract, and sloughed enterocytes lining the gut. Microbes, while not strictly endogenous, and serum albumin are also present in digesta. It has been estimated that around 70–80% of the endogenous secretions are digested and absorbed by the terminal ileum (Souffrant 1991). However, the remaining 20–30% are still present in the digesta. It is necessary to correct for the endogenous secretions when determining digestibility. When digestibility is corrected for the presence of endogenous secretions in the digesta, "true" digestibility is determined. Some research groups use the term "standardised digestibility" for true digestibility.

Several methods have been developed to measure endogenous secretions (for review, see Moughan et al. 2018). The currently preferred method (e.g., recommended by FAO Expert Working Groups; FAO 2013) uses a protein-free diet. All amino acids present in digesta collected following the consumption of a protein-free diet must, by definition, be of endogenous origin and can be used to correct apparent digestibility to true digestibility.

Both PDCAAS and DIAAS are calculated based on true digestibility. In both cases, the correction for endogenous secretions is determined by feeding a protein-free diet.

An important point when determining the amount of true ileal digestible amino acids is that these values can be used to combine foods or ingredients. When values are determined for single foods, they can then be combined with other foods to determine the amount of true ileal digestible amino acids in a mixture. For example, if the amounts of true ileal digestible amino acids are known for cornflakes and milk powder, they can then be combined to give values for the mixture (Fanelli et al. 2021). The predicted values were very close to the measured values.

CORRECTING FOR NITROGEN OR INDIVIDUAL AMINO ACID DIGESTIBILITY

When determining protein quality, the correction for availability (digestibility) can be made based on the digestibility of nitrogen (used in PDCAAS) or each amino acid can be individually corrected for its digestibility (used for DIAAS). The chemical analysis required when solely basing digestibility calculations on the digestibility of nitrogen (nitrogen analysis) is more economical compared with the amino acid analyses required to correct for individual amino acids.

Determining the amino acid composition for all indispensable amino acids in a sample requires three separate analyses. The first step for "standard" amino acid analysis requires hydrolysis of the sample to break the peptide bonds between the amino acids. The conditions required for this involve strong acid (6M HCl), 110°C for 22–24 hours in an oxygen-free environment, typically achieved by slowly applying a vacuum. While the peptide bonds are broken under these conditions, the amino acids tryptophan, cysteine, and partially methionine are destroyed. A second analysis involving a pre-hydrolysis reaction with performic acid is required for the accurate determination of methionine and cysteine. An alkali-based hydrolysis is required to determine tryptophan concentration. For a full review on the methods for amino acid analysis, see Rutherfurd and Moughan (2018).

Despite being more expensive, amino acid analyses offer greater accuracy than nitrogen-based corrections. This is because, firstly, not all of the nitrogen in a sample is amino nitrogen, so nitrogen digestibility does not only account for amino nitrogen digestibility. Secondly, there can be a notable variation in the digestibility of individual amino acids within the same food. This is particularly so for proteins that have lower digestibility (60–75%), such as many grains and legumes. In Table 13.3, the difference that can be observed is highlighted in the case of chickpeas (data from Project Proteos, unpublished). While the determined nitrogen digestibility for chickpeas is 91%, the digestibility of individual amino acids ranges from 68% for cysteine to 92% for reactive lysine. Calculating true ileal digestible amino acid content based on nitrogen digestibility yields different results compared to using individual amino acid digestibility. To show how these differences would affect protein quality determination, the values can be used to calculate DIAAS. It should be noted that the correct manner to calculate DIAAS involves individual amino acid digestibility figures, not nitrogen digestibility. In this case, if DIAAS was calculated based on nitrogen digestibility (the digestibility of each amino acid calculated by multiplying the amino acid composition by nitrogen digestibility), the value would be overestimated (DIAAS of 83) compared with the correctly calculated value of 76. The latter value is calculated by multiplying the amino acid composition by individual amino acid digestibility coefficients.

TABLE 13.3
Amino Acid Content, True Ileal Amino Acid Digestibility Coefficient (TIAAD) and Amount of Digestible Amino Acids Calculated Based on TIAAD and Nitrogen (N) Digestibility for Chickpeas

Amino Acid	Amino Acid Content mg/g DM	TIAAD	Amount Digestible AA mg/g DM	
			TIAAD	N Digestibility
Histidine	5.4	0.844	4.6	4.9
Isoleucine	9.3	0.834	7.8	8.5
Leucine	16.7	0.853	14.2	15.2
Rx Lysine	11.6	0.921	10.7	10.6
Methionine	3.3	0.869	2.9	3.0
Cysteine	2.5	0.683	1.7	2.3
Phenylalanine	12.9	0.884	11.4	11.7
Tyrosine	6.7	0.869	5.8	6.1
Threonine	7.7	0.761	5.9	7.0
Tryptophan	2.4	0.813	2.0	2.2
Valine	10.1	0.820	8.3	9.2

Source of data: Project Proteos, unpublished.

TABLE 13.4
Comparison between PDCAAS and DIAAS Determination

	PDCAAS	DIAAS
Officially recommended	FAO/WHO 1990	FAO 2013
Correction for digestibility	True faecal digestibility	True ileal digestibility
Method to correct for endogenous secretions	Feeding protein-free diet	Feeding protein-free diet
Digestibility correction	All amino acids, by nitrogen digestibility	Amino acids by individual amino acid digestibilities
Species for digestibility measurement	Rat	Human > pig > rat
Possible scores	0–1 >1 truncated to 1	No upper limit (no truncation)

Thus, to summarise the preceding discussion, protein quality should be determined based on the true ileal digestibility of the individual amino acids. The use of PDCAAS, which corrects for availability using the faecal digestibility of nitrogen, does not accurately reflect the nutritional quality of the protein. In contrast, DIAAS uses the true ileal digestibility of the individual amino acids, providing a more accurate representation of the nutritional quality of the food or ingredient (Mathai et al. 2017). Table 13.4 summarises the methodological differences between PDCAAS and DIAAS, as discussed above.

ILEAL DIGESTA COLLECTION – NASO-ILEAL INTUBATION

What can be considered to be the biggest complication for evaluating nutritional protein quality using DIAAS is that digestibility is determined at the ileal level, which necessitates collecting digesta from the terminal ileum (the end of the small intestine). This can be accomplished in humans using naso-ileal intubation or with the help of human ileostomates.

Naso-ileal intubation is performed in a hospital setting. It involves inserting a fine triple-lumen tube through the nose under local anaesthetic. The tube passes down the back of the throat into the oesophagus, through the stomach, and to the end of the small intestine. The tube is radio-opaque, and its correct positioning is confirmed via x-ray. The triple-lumen tube allows the infusion of a non-absorbable marker (e.g., polyethylene glycol, PEG) and the simultaneous collection of intestinal contents downstream from where the non-absorbable marker is infused. The third thin lumen is used to inflate a balloon on the end of the tube once it has passed through the stomach, facilitating the tube's movement through the small intestine via peristaltic movements (Figure 13.1).

Following an overnight fast and confirmation that the tube is in position, the participant consumes the test meal, which has the protein source being tested as the only source of protein (or a protein-free meal to measure endogenous secretions). For eight hours following the consumption of the test meal, ileal effluents are collected continuously through the tube via slight aspiration, providing a sample of ileal digesta. While ileal effluents are being collected, participants do not consume any food, although water consumption is allowed. Calvez et al. (2021) describe a typical protocol for determining true ileal amino acid digestibility using naso-ileal intubation.

While it is an expensive technique, the advantage of naso-ileal intubation is that it allows digesta to be collected from participants with a "normal" digestive tract. However, its primary limitation is that it is very invasive. Tolerance to tube insertion varies between participants, and many cannot tolerate it for the duration required to complete the study. Another important limitation is that due to the small lumen of the sampling tube, this method can only be used when the digesta does not contain particles, limiting the types of foods that can be tested. It is not appropriate for use with

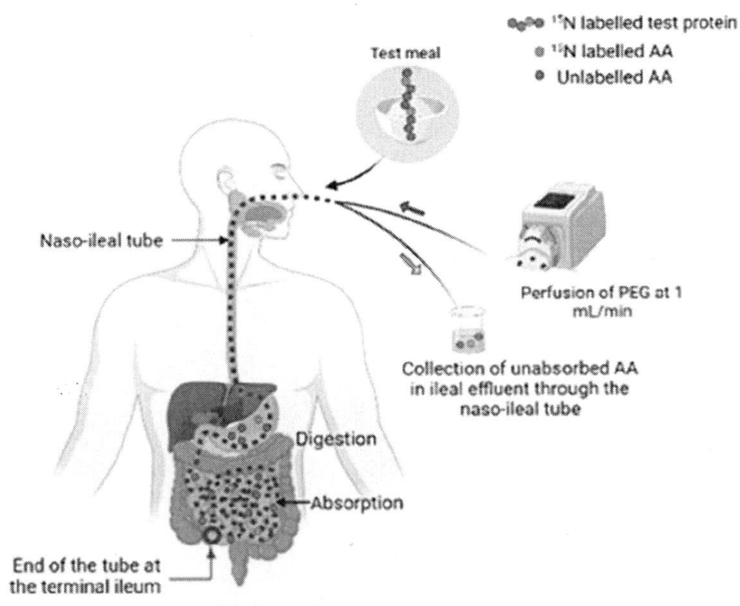

FIGURE 13.1 Naso-ileal intubation (diagram modified from Bandyopadhyay et al. 2022).

vulnerable groups, such as children. Only one test meal can be tested in each participant; the tube cannot remain in place for a longer period.

Concerns have been raised about the possible effect of the sampling tube's presence on digestive function, particularly the food's gastric and intestinal transit rate. Studies examining the effect of an intestinal tube on gastric emptying are not clear, with some studies reporting a delay in gastric emptying (Medhus et al. 1999; Longstreth et al. 1975; Read et al. 1983; Fone et al. 1991) while others report little or no effect on gastric emptying (Müller-Lissner et al. 1982). However, evidence indicates little, if any, effect of a food's transit rate on overall amino acid absorption and, thus, overall protein digestibility (Gaudichon et al. 1999). These results support the use of this method to collect ileal digesta from healthy individuals with intact digestive systems.

ILEAL DIGESTA COLLECTION – ILEOSTOMISED PARTICIPANTS

An alternative method for collecting digesta from humans involves the participation of adult ileostomates. These are people who, due to health issues with their large intestine, have had the end of the small intestine exteriorised via a stoma, allowing continuous collection of digesta into a stoma bag. This method enables a complete collection of digesta (Figure 13.2). To test protein quality in ileostomised participants, the procedure is as follows: following an overnight fast, participants consume a test meal that has the protein source being tested as the only source of protein (or a protein-free meal to measure endogenous secretions) and a fresh stoma bag is attached. All digesta entering the stoma bag over the following nine hours is collected. During this nine-hour period, the participants cannot eat but can drink sweetened drinks. Typically, each protein source/food is tested on two separate occasions with each ileostomate. Moughan et al. (2005) describe a commonly used protocol.

A significant advantage of working with ileostomised participants is that there are no limits to the types of foods that can be tested. Also, each participant can test numerous protein sources/foods (with a few days between test days). Endogenous secretions can be corrected on an individual basis; that is, the ileal digestibility of the food(s) is corrected for the endogenous secretions determined in the same individual.

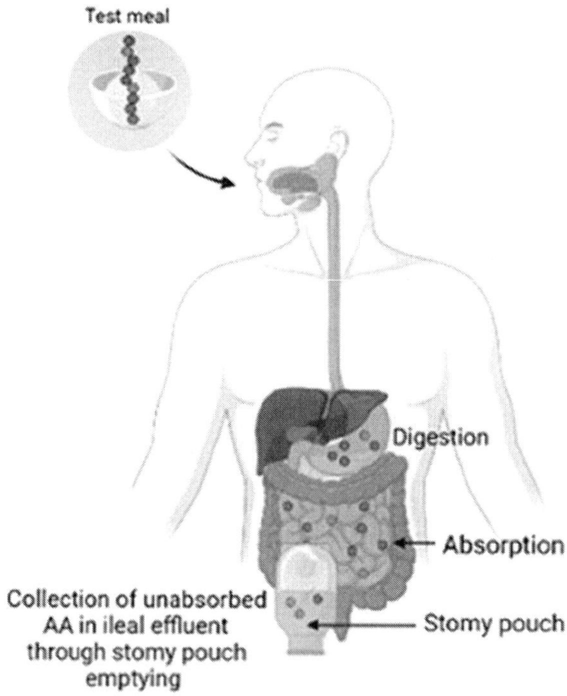

FIGURE 13.2 Ileostomy model (diagram modified from Bandyopadhyay et al. 2022).

The primary limitation of working with ileostomised participants is recruiting them; very few people have ileostomies. An additional complication is identifying participants who do not have other health conditions or take medications that could affect digestibility measurements. The situation is further complicated because, in many cases, once the medical condition involving the large intestine has healed, the ileostomy is reversed. Thus, for many ileostomates, there is only a short window of time between recovering sufficiently from the original surgery and having the ileostomy reversed. This makes the recruitment of a sufficient "n" of ileostomised participants very difficult. It should be noted that determining ileal amino acid digestibility is logistically demanding for ileostomised participants, as they must consume the test meal after an overnight fast but can only consume protein-free liquids for the following nine hours. It is important to recognise that they can easily reach study fatigue if they have too many study days within a short period.

Adult ileostomates have been used as a model for the "intact" human digestive system in numerous studies aimed at studying the digestion and absorption to the end of the small intestine (Sandberg et al. 1981, Kay et al. 1980; Englyst and Cummings 1985; Chapman et al. 1985; Cummings and Englyst 1991; Zhang et al. 1994; Sandberg et al. 1986; Muir and O'Dea 1993; Livesey et al. 1995; Rowan et al. 1994; Sandström et al. 1986).

It has been suggested that there may be an increased colonisation of the small intestine by microbes in ileostomates. However, no statistically significant differences have been found in the amounts of endogenous ileal amino acids (which include bacteria) in ileostomates before and after the administration of antibiotics (Fuller et al. 1994). Additionally, when polysaccharide digestion has been evaluated in adult ileostomates, including with the addition of metronidazole, which inhibits the metabolism of anaerobic flora in the gut, no significant degree of fermentation has been found (Englyst and Cummings 1985). After examining dietary fibre breakdown in adult ileostomates, Sandberg et al. (1981) also concluded that there was little, if any, microbial fermentation. Transit time has been shown to be the same in adult ileostomates and "intact" adults, with no difference between the two groups in the time taken for the "head" of the meal to travel from the mouth to the terminal ileum in ileostomates and mouth to the caecum in intact participants (Holgate and Read 1983). These studies support using adult ileostomates as a representative of the upper gastrointestinal tract in "intact" humans.

OTHER OPTIONS TO DETERMINE ILEAL AMINO ACID DIGESTIBILITY

The two options presented above that allow the collection of ileal digesta from humans are appropriate for use for some work, such as the validation of other methods to determine true ileal amino acid digestibility. However, neither method can be used on a routine basis for multiple tests of protein sources/foods. Other options have been developed, such as the use of an animal model.

The pig is the preferred animal model for determining true ileal amino acid digestibility for human foods (FAO 2013, 2018). The anatomy and physiology of the pig's digestive tract are very similar to those in humans, at least up to the terminal ileum (Miller and Ullrey 1987; Moughan and Rowan 1989; Deglaire and Moughan 2012). The true ileal amino acid digestibility of a range of protein-containing foods has been directly compared between the human ileostomate and the growing cannulated pig, and no statistically significant differences were reported (Hodgkinson et al. 2022). Figure 13.3 illustrates this comparison, showing the mean true ileal digestibility coefficients of the indispensable amino acids for different foods in humans and pigs. In Figure 13.3, each point represents a different food. The regression equation for the mean of the true ileal digestibility of the indispensable amino acids was found to be (y = human, x = pig) $y = 0.995x + 0.0004$, with $R^2 = 0.922$, indicating a strong correlation close to $y = x$. These values are used to calculate DIAAS based on the digestibility of the indispensable amino acids, supporting the use of pig-derived data for human nutrition studies. A protocol for determining true ileal amino acid digestibility in pigs is described by Hodgkinson et al. (2020).

Ideally, in vitro (laboratory-based) methods could estimate true ileal amino acid digestibility, eliminating the need for human or animal studies. In vitro methods such as INFOGEST (Brodkorb et al. 2019) have been developed, incorporating oral, gastric and intestinal phases. However, current methods cannot mimic the functions of the brush border, the layer lining the small intestine that contains further enzymes.

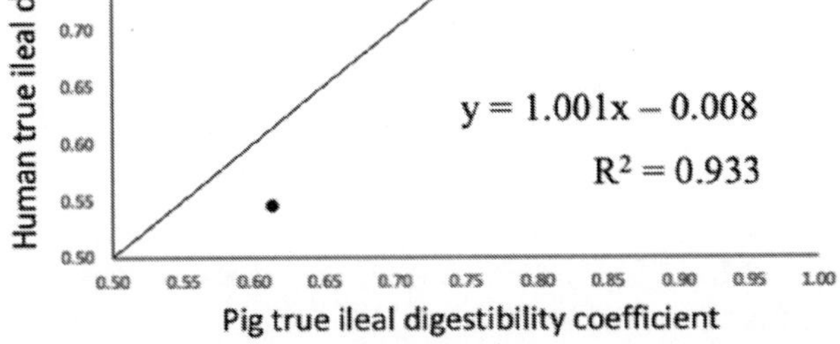

FIGURE 13.3 Mean true ileal digestibility of indispensable amino acids for protein sources, determined in the human and pig (Hodgkinson et al. 2022).

Compared with in vivo studies, in vitro methods are more economical and yield quicker results. However, while there is a lot of ongoing research in this area, no current in vitro method has undergone sufficient testing or validation to confidently predict values determined in humans or pigs. Current methods can, however, rank foods/ingredients by digestibility, which can be used in product development, and only the most promising would then require testing in pigs or humans.

A full description of in vitro models is beyond the scope of this chapter. For a detailed review, see Mackie et al. (2020).

SUMMARY

The nutritional quality of foods and ingredients is determined based on the amount of amino acids, particularly the indispensable amino acids (which must be supplied in the diet) that are digested and absorbed from food (digestibility), and the relationship between these amino acids and human requirements. This chapter discusses aspects related to amino acid digestibility.

Two methods are currently used to evaluate protein quality: the protein digestibility amino acid score (PDCAAS) and the digestible indispensable amino acid score (DIAAS). An important difference between these methods is how the correction for amino acid digestibility is applied, and different aspects related to this are presented in this chapter. DIAAS, which requires the determination of true (corrected for endogenous amino acids) ileal (end of the small intestine) digestibility of individual amino acids, is currently recommended by FAO-convened Expert Consultations to evaluate the nutritional quality of protein in foods and ingredients.

However, determining true ileal amino acid digestibility (e.g., for DIAAS calculation) requires the collection of digesta from the terminal ileum. Two methods can be used to collect ileal digesta in humans: naso-ileal intubation and the participation of ileostomised people. These methods are far from straightforward. Therefore, other methods have been developed to determine true ileal amino acid digestibility for human nutrition, such as animal models and in vitro (laboratory-based) methods.

MINI DICTIONARY OF TERMS

Digestible indispensable amino acid score (DIAAS): The method currently recommended by FAO Expert Consultations to evaluate dietary protein quality.

Dispensable amino acid: An amino acid that can be synthesised by the human body, and is therefore not strictly required in the diet.

Endogenous protein: Protein that originates from the human (or animal) body rather than the diet.

In vitro: An analysis method that is laboratory-based rather than animal-based.

Indispensable amino acid: An amino acid that cannot be synthesised by the human body, and must therefore be provided in the diet.

Protein digestibility-corrected amino acid score (PDCAAS): The method recommended by FAO and WHO groups in 1990 to evaluate protein quality, which is still used for regulatory purposes in countries such as the United States.

SUMMARY POINTS

- This chapter focuses on the determination of amino acid digestibility.
- Protein quality should be determined based on the true ileal digestibility of individual amino acids; the reasons for this are discussed.
- Determining true ileal amino acid digestibility requires collecting digesta from the end of the small intestine, the terminal ileum.

- Two methods are used to collect ileal digesta in humans: naso-ileal intubation and participation of ileostomates.
- Due to the complexity of ileal digesta collection in humans, alternative methods, such as animal models and in vitro (laboratory-based) methods, have been developed.

REFERENCES

Bandyopadhyay, S., S. Kashap, J. Calvez, S. Devi, D. Azzout-Marnish, and D. Tome. 2022. Evaluation of protein quality in humans and insights of stable isotope approaches to measure digestibility. *Advances in Nutrition* 13: 1130–1143.

Brodkorb, A., L. Egger, M. Alminger, P. Alvito, R. Assunção, S. Ballance, T. Bohn, C. Bourlieu-Lacanal, R. Boutrou, F. Carrière, A. Clemente, M. Corredig, D. Dupont, C. Dufour, C. Edwards, M. Golding, S. Karakaya, B. Kirkhus, S. Le Feunteun, U. Lesmes, A. Macierzanka, A. R. Mackie, C. Martins, S. Marze, D. J. McClements, O. Ménard, M. Minekus, R. Portmann, C. N. Santos, I. Souchon, R. P. Singh, G. E. Vegarud, M. S. J. Wickham, W. Weitschies, and I. Recio. 2019. INFOGEST static in vitro simulation of gastrointestinal food digestion. *Nature Protocols* 14: 991–1014.

Calvez, J., S. Benoit, J. Piedcoq, N. Khodorova, D. Azzout-Marniche, D. Tome, R. Benamouzig, G. Airinei, and C. Gaudichon. 2021. Very low ileal nitrogen and amino acid digestibility of zein compared to whey protein isolate in healthy volunteers. *American Journal of Clinical Nutrition* 113: 70–82.

Chapman, R. W., J. K. Sillery, M. M. Graham, and D. R. Saunders. 1985. Absorption of starch by healthy ileostomates: effect of transit time and of carbohydrate load. *American Journal of Clinical Nutrition* 41: 1244–1248.

Cummings, J. H., and H. N. Englyst. 1991. Measurement of starch fermentation in the human large intestine. *Canadian Journal of Physiology and Pharmacology* 69: 121–129.

Darragh, A. J., P. D. Cranwell, and P. J. Moughan. 1994. Absorption of lysine and methionine from the proximal colon of the piglet. *British Journal of Nutrition* 71: 739–752.

Deglaire, A., and P. J. Moughan. 2012. Animal models for determining amino acid digestibilty in humans - a review. *British Journal of Nutrition* 108: S273–S281.

Englyst, H. N., and J. H. Cummings. 1985. Digestion of the polysaccharides of some cereal foods in the human small intestine. *American Journal of Clinical Nutrition* 42: 778–787.

Fanelli, N. S., H. M. Bailey, L. V. Guardiola, and H. H. Stein. 2021. Values for digestible indispensable amino acid score (DIAAS) determined in pigs are greater for milk than for breakfast cereals, but DIAAS values for individual ingredients are additive in combined meals. *Journal of Nutrition* 151: 540–547.

Fone, D. R., M. Horowitz, R. Heddle, A. F. Maddox, P. J. Collins, N. W. Read, and J. Dent. 1991. Comparative effects of duodenal and ileal intubation on gastric emptying and postprandial antral, pyloric and duodenal motility. *Scandinavian Journal of Gastroenterology* 26: 16–22.

Food and Agriculture Organization of the United Nations (FAO). 2013. *Dietary Protein Quality Evaluation in Human Nutrition*. Report of an FAO Expert Consultation. FAO Food and Nutrition Paper 92. Rome: FAO.

Food and Agriculture Organization of the United Nations (FAO). 2018. *Protein Quality Assessment in Follow-up Formula for Young Children and Ready to Use Therapeutic Foods*. Report of an FAO Expert Working Group. Rome, Italy: FAO.

Food and Agriculture Organization of the United Nations, and World Health Organisation of the United Nations (WHO). 1990. *Report of a Joint FAO/WHO Expert Consultation on Protein Quality Evaluation*. Rome, Italy: Food and Agriculture Organization, United Nations.

Fuller, M. F., A. Milne, C. I. Harris, T. M. S. Reid, and R. Keenan. 1994. Amino acid losses in ileostomy fluid on a protein-free diet. *American Journal of Clinical Nutrition* 59: 70–73.

Gaudichon, C., S. Mahe, R. Benamouzig, C. Luengo, H Fouillet, S Dare, M Van Oycke, F Ferriere, J Rautureau, and D. Tome. 1999. Net postprandial utilization of [15N]-labeled milk protein nitrogen is influenced by diet composition in humans. *Journal of Nutrition* 129: 890–895.

Hodgkinson, S. M., H. H. Stein, S. de Vries, W. H. Hendriks, and P. J. Moughan. 2020. Determination of true ileal amino acid digestibility in the growing pig for calculation of Digestible Indispensable Amino Acid Score (DIAAS). *Journal of Nutrition* 150: 2621–2623.

Hodgkinson, S. M., N. Stroebinger, N. van der Wielen, M. Mensink, C. Montoya, W. H. Hendriks, S. de Vries, H. H. Stein, and P. J. Moughan. 2022. Comparison of true ileal amino acid digestibility between adult humans and growing pigs. *Journal of Nutrition* 152: 1635–1646.

Holgate, A. M., and N. W. Read. 1983. Relationship between small bowel transit time and absorption of a solid meal. Influence of metoclopramide, magnesium sulfate and lactulose. *Digestive Diseases and Sciences* 28: 812–819.

Kay, R. M., Z. Cohen, K. P. Siu, C. N. Petrunka, and S. M. Strasberg. 1980. Ileal excretion and bacterial modification of bile acids and cholesterol in patients with continent ileostomy. *Gut* 21: 128–132.

Livesey, G., J. A. Wilkinson, M. Roe, R. Faulks, S. Clark, J. C. Brown, H. Kennedy, and M. Elia. 1995. Influence of the physical form of barley grain on the digestion of its starch in the human small intestine and implications for health. *American Journal of Clinical Nutrition* 61: 75–81.

Longstreth, G. F., J. R. Malagelada, and V. L. W. Go. 1975. The gastric response to a transpyloric duodenal tube. *Gut* 16: 777–780.

Mackie, A., A.-I. Mulet-Cabero, and A. Torcello-Gomez. 2020. Simulating human digestion: developing our knowledge to create healthier and more sustainable foods. *Food and Function* 11: 9397–9431.

Mason, V. C. 1980. Role of the large intestine in the processes of digestion and absorption in the pig. In *Current Concepts of Digestion and Absorption in Pigs*, edited by A. G. Low and I. G. Partridge, 112–129. Scotland: National Institute for Research in Dairying, England and the Hannah Research Institute.

Mathai, J. K., Y. Liu, and H. H. Stein. 2017. Values for digestible indispensable amino acid scores (DIAAS) for some dairy and plant proteins may better describe protein quality than values calculated using the concept for protein digestibility-corrected amino acid scores (PDCAAS). *British Journal of Nutrition* 117: 490–499.

Medhus, A. W., O. Sandstad, J. Bredesen, and E. Husebye. 1999. Delay of gastric emptying by duodenal intubation: sensitive measurement of gastric emptying by the paracetamol absorption test. *Alimentary Pharmacology and Therapeutics* 13: 609–620.

Miller, E. R., and D. E. Ullrey. 1987. The pig as a model for human nutrition. *Annual Reviews of Nutrition* 7: 361–382.

Moughan, P. J., C. A. Butts, H. van Wijk, A. M. Rowan, and G. W. Reynolds. 2005. An acute ileal amino acid digestibility assay is a valid procedure for use in human ileostomates. *Journal of Nutrition* 135: 404–409.

Moughan, P. J., W. Miner-Williams, and R. Jha. 2018. The digestion of protein - amino acid digestibility. In *Feed Evaluation Science*, edited by P. J. Moughan and W. H. Hendriks, 173–217. Wageningen, The Netherlands: Wageningen Academic Publishers.

Moughan, P. J., and A. M. Rowan. 1989. The pig as a model animal for human nutrition research. *Proceedings of the Nutrition Society of New Zealand* 14: 116–123.

Muir, J. G., and K. O'Dea. 1993. Validation of an in vitro assay for predicting the amount of starch that escapes digestion in the small intestine of humans. *American Journal of Clinical Nutrition* 57: 540–546.

Müller-Lissner, S. A., C. J. Fimmel, N. Will, W. Müller-Duysing, F. Heinzel, and A. L. Blum. 1982. Effect of gastric and transpyloric tubes on gastric emptying and duodenogastric reflux. *Gastroenterology* 83: 1276–1279.

Read, N. W., M. N. Al Janabi, T. E. Bates, and D. C. Barber. 1983. Effect of gastrointestinal intubation on the passage of a solid meal through the stomach and small intestine in humans. *Gastroenterology* 84: 1568–1572.

Rowan, A. M., P. J. Moughan, M. N. Wilson, K. Maher, and C. Tasman-Jones. 1994. Comparison of the ileal and faecal digestibility of dietary amino acids in adult humans and evaluation of the pig as a model animal for digestion studies in man. *British Journal of Nutrition* 71: 29–42.

Rutherfurd, S. M., A. C. Fanning, B. J. Miller, and P. J. Moughan. 2015. Protein digestibility-corrected amino acid scores and digestible indispensable amino acid scores differently describe protein quality in growing male rats. *Journal of Nutrition* 145 (2): 372–379.

Rutherfurd, S. M., and P. J. Moughan. 2018. The chemical analysis of proteins and amino acids. In *Feed Evaluation Science*, edited by P. J. Moughan and W. H. Hendriks, 39–74. Wageningen, The Netherlands: Wageningen Academic Publishers.

Sandberg, A.-S., H. Andersson, B. Hallgren, K. Hasselblad, and B. Isaksson. 1981. Experimental model for in vivo determination of dietary fibre and its effect on the absorption of nutrients in the small intestine. *British Journal of Nutrition* 45: 283–294.

Sandberg, A. S., H. Andersson, and B. Sandström. 1986. Extrusion cooking of a high-fibre cereal product. 1. Effects on digestibility and absorption of protein, fat, starch, dietary fibre and phytate in the small intestine. *British Journal of Nutrition* 55: 245–254.

Sandström, B., H. Andersson, B. Kivistö, and A.-S. Sandberg. 1986. Apparent small intestinal absorption of nitrogen and minerals from soy and meat-protein-based diets. A study on human ileostomy subjects. *Journal of Nutrition* 116: 2209–2218.

Schaafsma, G. 2012. Advantages and limitations of the protein digestibility-corrected amino acid score (PDCAAS) as a method for evaluating protein quality in human diets. *British Journal of Nutrition* 108: S333–S336.

Souffrant, W. B. 1991. Endogenous nitrogen losses during digestion in pigs. In *Proceedings of the Vth International Symposium on Digestive Physiology in Pigs*, edited by M. W. A. Vestegen, J. Huisman, and L. A. den Hartog, 147–166. Wageningen, The Netherlands: Pudoc.

van der Wielen, N., P. J. Moughan, and M. Mensink. 2017. Amino acid absorption in the large intestine of humans and porcine models. *Journal of Nutrition* 147: 1493–1498.

Zhang, J. X., E. Lundin, G. Hallmans, H. Adlercreutz, H. Andersson, I. Bosaeus, P. Aman, R. Stenling, and S. Dahlgren. 1994. Effect of rye bran on excretion of bile acids, cholesterol, nitrogen, and fat in human subjects with ileostomies. *American Journal of Clinical Nutrition* 59: 389–394.

14 Kwashiorkor Malnutrition

Linking Sulfur Amino Acids to Signs of Kwashiorkor

Merry C. Fitzpatrick

ABBREVIATIONS

-CH_3	Chemical formulation for a methyl group
Cys	Cysteine/cystine
GAG	Glycosaminoglycan
GSH	Glutathione
Met	Methionine
MTHF	5-Methyltetrahydrofolate
SAA	Sulfur amino acid
SAH	S-adenosyl homocysteine
SAMe	S-adenosylmethionine
THF	Tetrahydrofolate

INTRODUCTION

Acute malnutrition is typically divided into two classifications, though a person may experience both types of malnutrition simultaneously. Wasting, when a person is measurably thin for their height, is the most commonly recognized form of acute malnutrition. Kwashiorkor malnutrition is much more complex, with a syndrome of signs. Indications include pale brittle hair, lethargy or irritability, symmetrical pitting edema in the face and feet and, in the most severe cases, dark, rough and fissured skin (Pham et al. 2021). Common metabolic alterations include low circulating albumin and sulfur amino acids (SAAs), electrolyte imbalance, fatty liver, and high oxidative stress, among others (Pham et al. 2021). Regardless of the variety of signs, kwashiorkor is diagnosed for treatment solely by the presence of bipedal pitting edema.

Although the exact etiology of kwashiorkor remains uncertain, it has long been considered a type of malnutrition because it is only found in populations with low protein intake and is cured through an improved diet (Gopalan 1955; Manary et al. 2009). Without treatment, severe kwashiorkor is often fatal. However, appropriate treatment – beginning with electrolyte rebalancing followed by low doses of high-quality protein combined with caloric sufficiency – often leads to rapid recovery (Schofield and Ashworth 1996; Odei Obeng-Amoako et al. 2022).

Numerous causal pathways for kwashiorkor have been proposed, but most have proven inconclusive. Over the decades, the theoretical focus has shifted from total protein insufficiency to the examination of specific amino acids (Phadke et al. 1995; Roediger and Waterlow 1995; Jahoor et al. 2005).

The most recent research indicates a central role for SAAs in the etiology of kwashiorkor (Fitzpatrick et al. 2021; Badaloo et al. 2012; Jahoor 2012). The latest understanding suggests that inadequate consumption of methionine, and to a lesser extent cysteine, likely contributes to the

 DOI: 10.1201/b23328-14

development of kwashiorkor's diverse symptoms through different but interconnected mechanisms. Additional trial evidence supporting this rising theory could significantly affect dietary recommendations related to protein intake to prevent kwashiorkor in humanitarian crises and low-income settings, as well as improve treatment approaches specific to kwashiorkor.

This chapter will begin with a discussion of current recommended SAA intakes and previous theories on kwashiorkor etiology. A more detailed discussion on the anomalies observed in the SAA metabolism of cases of kwashiorkor will describe how kwashiorkor manifests itself metabolically. With this foundation, a detailed description of the metabolic cycles of methionine and cysteine will be used to model how a dietary deficiency of methionine would affect these and related, interlinked cycles. The current array of signs will then be compared with these proposed metabolic changes to better understand the dynamics that potentially underlie and link the seemingly unconnected signs of kwashiorkor. Finally, the discussion will turn to the implications of this etiological model of kwashiorkor in the prevention and treatment of kwashiorkor. The chapter will end with a return to examining how lack of access to certain classes of protein-rich foods risks a higher incidence of kwashiorkor.

DIETARY REQUIREMENTS FOR SAAS IN HUMAN NUTRITION

While there are multiple amino acids that contain sulfur (see Table 14.1), the four most common SAAs in human nutrition are methionine (Met), cysteine (Cys), taurine (Tau) and homocysteine (Hcy). Multiple other SAAs either use these SAAs as substrates or are intermediaries, though each has its own metabolic uses. S-adenosylmethionine (SAMe) and S-adenosylhomocysteine (SAH) are active intermediaries between Met and Hcy. SAMe is formed in the body using Met as its primary substrate and is the primary 1-carbon or methyl group ($-CH_3$) donor. SAH is formed when SAMe loses its methyl group. SAH is the primary substrate for Hcy in a reversible reaction, so that as Hcy levels rise, Hcy can be reversibly transformed into SAH, facilitating Hcy homeostasis. Cys can be produced in the body, receiving sulfur from Hcy through an irriversible transulfuration pathway involving cystathionine as an intermediary.

Met is classified as an indispensable amino acid, meaning the human body cannot create Met and must obtain it through diet alone. Cys can be formed when Met donates its sulfur atom to serine through a series of reactions (transulfuration) to yield Cys. A crucial step in this process is not reversible, meaning that Met cannot be formed from Cys. Under certain conditions, such as in premature infants or when there is excessive oxidative stress, the body may not be able to create enough Cys from Met to meet nutritional needs, potentially rendering Cys conditionally essential.

Of all the sulfur-containing amino acids, only Met and Cys are incorporated into proteins and are necessary for the formation of every human protein. This chapter will focus exclusively on Met

TABLE 14.1
Classification of Sulfur Amino Acids

Essential	Conditionally Essential*	Non-Essential
Methionine	Cysteine	Homocysteine
		Taurine (and hypotaurine)
		S-adenosylmethionine
		S-adenosylhomocysteine
		Cystathionine

* Evidence regarding the conditionality of cysteine has not yet been well established.

TABLE 14.2
Sulfur Amino Acid Dietary Requirements

Age (years)	Protein Requirement[a] (g/kg/day)	SAA Combined Requirement[b] (mg/kg/day)	Met Requirement[c] (mg/kg/day)
0.5	1.12	31	22.2
1–2	0.86	22	15.8
3–14	0.73	17	12.2
15–18	0.70	16	11.5
>18	0.66	15	10.8

[a] Requirements for maintenance and growth combined (FAO 2013), grams protein/kilograms body weight/day
[b] FAO (2013)
[c] Calculated using the method described by Fitzpatrick (2021), using the ratio provided by WHO et al. (2007).

and Cys; therefore, throughout the rest of this chapter, the acronym SAA will refer only to these two amino acids.

As early as 1967, Finkelstein and Mudd noted that the addition of cysteine to a diet free of SAAs reduced the oxidation of Met in rats, essentially "sparing" Met (Finkelstein and Mudd 1967). Raguso et al. were able to confirm this in humans, estimating that cysteine can spare the amount of Met oxidized by around 50% in diets very low or devoid of Met (Raguso et al. 1997). Kurpad et al. further refined that relationship, showing that the sparing effect is reduced as the content of Met in the diet is increased (Kurpad et al. 2004). Calculations of Cys and Met requirements are complicated by this relationship, so the dietary requirements have often been combined into a single requirement for SAA (Met and Cys) (WHO et al. 2007; FAO 2013).

Based on the results of Kurpad et al., the WHO/FAO/UNU Expert Consultation in 2002 set the adult recommended requirement for Met at 10.4 mg/kg/day and for Cys at 4.1 mg/kg/day, resulting in a total SAA intake of 14.5 mg/kg/day, rounded up to 15 mg/kg/day for simplicity (WHO et al. 2007). The requirements for children presented in Table 14.2 were derived from the adult requirements, age-adjusted based on nitrogen balance for maintenance and protein deposition in tissue formation (FAO 2013; WHO et al. 2007).

Despite their close relationship and the practice of combining their requirements, each sulfur amino acid serves a unique purpose in both protein synthesis and metabolic function. For example, in protein synthesis, exposed Cys sulfur atoms in the Cys molecular structure link up to form S-S bonds (creating cystine), which determine the shape of the protein and stabilize it, critical to ensuring the proper functioning of the protein (Brosnan and Brosnan 2006). Met, on the other hand, is required as the first step in protein synthesis, linking with the t-RNA molecule to initiate transcription, and donating an entire methyl molecule to a series of substrates to create numerous other metabolites (Brosnan and Brosnan 2006).

THEORIES OF KWASHIORKOR ETIOLOGY

Multiple theories on the etiology of kwashiorkor have been proposed based on individual signs or anomalies associated with kwashiorkor. Most have been dismissed but have slowly led to the current thinking that includes a central role for SAAs.

The observation that children with kwashiorkor show high oxidative stress and low levels of circulating antioxidants such as glutathione led to the theory that diets insufficient in antioxidants contributed to the edema associated with kwashiorkor (Manary et al. 2000; Golden 1998). But antioxidant supplementation, including low doses of N-acetylcysteine, failed to reduce kwashiorkor incidence in populations, suggesting oxidative stress is secondary to other factors (Ciliberto et al. 2005; Fuchs 2005; Odigwe et al. 2010).

Following the observation that people with kwashiorkor have a unique gut microbiome, fecal matter from three children with kwashiorkor and three without was transplanted into germ-free mice (Smith et al. 2013). Mice receiving fecal matter from kwashiorkor cases paired with the typical Malawian diet lost weight, but fecal matter alone did not cause weight loss. The small sample size and interaction with the diet raised questions about causality and prompted a return to dietary deficiencies as a focus.

Because kwashiorkor is endemic only where the typical diet is very low in total protein and children with kwashiorkor consistently show low serum albumin, inadequate total protein intake was one of the earliest proposed causes (Williams 1933; Brock and Autret 1952). Studies estimating associations between protein intake and risk of kwashiorkor have been mixed, causing many to discount low protein as a causative factor. On closer inspection, though, diet studies failing to find a difference used food frequency surveys, which count food groups eaten per day but do not measure the amounts of foods eaten (Gopalan 1968; Sullivan et al. 2006). Studies that quantitatively measured total protein intake have generally found an inverse correlation between protein consumption and kwashiorkor incidence or prevalence (Gupte 1975; Gupte and Mehta 1971; Fitzpatrick et al. 2021).

Brock and Autret compared typical diets in different regions and sub-regions of Africa in the early 1950s, contrasting foods that appeared to be consumed more frequently in regions with high and low prevalence of kwashiorkor (Brock and Autret 1952). Although they did not survey diets, their analyses concluded that populations in high-prevalence areas consumed diets lacking in protein, pointing specifically to methionine, tryptophan and lysine as potentially missing amino acids.

In a detailed comparison of diets of populations with high or low prevalence of kwashiorkor, Fitzpatrick et al. (2021) found that children in high-prevalence populations consumed less SAAs than those in neighboring low-prevalence populations (32.4 [22.9–49.3] mg/kg versus 29.6 [18.1–44.3] mg/kg, $p < 0.05$), and a higher risk of inadequate intake of Met (35.1% versus 23.6%, $p < 0.05$). The limiting amino acid in the diets was Met.

SAA ANOMALIES IN KWASHIORKOR

Early analyses of circulating amino acid profiles consistently showed lower levels of essential amino acids relative to non-essential amino acids in people with kwashiorkor than in people without kwashiorkor (Arroyave et al. 1962; Holt et al. 1963; Dean and Whitehead 1964). Among these, Met levels were particularly low, even among children with the first mild signs of kwashiorkor. Additionally, deficiencies in Met were proportional to the severity of the condition and Met levels increased significantly after recovery (Whitehead and Dean 1964). Later research also included Cys and found it to be low in kwashiorkor cases (Jahoor et al. 2005, 2006b).

Jahoor and Badaloo demonstrated that children with kwashiokor not only had low circulating SAAs but also exhibited altered protein and SAA kinetics compared to malnourished children without edema or non-malnourished children (Jahoor 2012; Badaloo et al. 2012; Jahoor et al. 2005, 2006b). In one study, Badaloo found that children with kwashiorkor who were provided additional cysteine during treatment lost their edema in about 9 ± 1 days versus 14 ± 2 days among children in the control group ($p < 0.05$) (Badaloo et al. 2002). On the other hand, Ciliberto et al. found that supplementation of a population with an antioxidant cocktail that included 300 mg of N-acetylcysteine per day for 20 weeks did not reduce the incidence of kwashiorkor (Ciliberto et al. 2005).

Hypoalbuminemia is typical in kwashiorkor and has been attributed to low protein intake. Low serum albumin has been proposed as one of the causal factors leading to the edema characteristic of kwashiorkor. Most recently, Gonzales et al. found that while low albumin is nearly always present when there is nutritional edema, it is also occasionally present in wasted children without edema (Gonzales et al. 2022). On closer inspection, they also found that Syn1, a proteoglycan-bearing sulfated glycosaminoglycan (a water-binding component of the extracellular matrix), was not significantly associated with kwashiorkor compared to wasted cases. However, its concentration was negatively associated with edema severity. They concluded that while low albumin is necessary for the formation of nutritional edema, other factors must also be present.

Glutathione (GSH), a metabolite of Cys, is the most ubiquitous antioxidant in the body. Low levels of GSH have been linked to the high oxidative stress observed in kwashiorkor (Ciliberto et al. 2005). Reid and Badaloo showed that GSH levels associated with kwashiorkor improved with Cys supplementation, indicating that low levels of Cys were a limiting factor (Reid et al. 2000; Badaloo et al. 2002).

Other related anomalies include low lipid oxidation (Lenhartz et al. 1998), reduced production of glycosaminoglycans (Amadi et al. 2009), low excretion of urinary sulfates (Ittyerah 1969), low acylcarnitines and phosphatidylcholine (Di Giovanni et al. 2016), normal to high folate levels (Scott and Weir 1981), very high vitamin B12 concentrations (Adams et al. 1967), and normal glycine production (Jahoor et al. 2006a).

MET METABOLIC CYCLES, METABOLITES, AND ANOMALIES OF KWASHIORKOR

Met and Cys serve very different purposes in protein and non-protein processes. These amino acids are linked by transulfuration in a one-way reaction that carries sulfur from Met into Cys metabolic pathways. This section and the following two sections will review multiple intermediaries, metabolites and interlinking cycles that appear to be associated with kwashiorkor.

The two primary uses of Met are the initiation of protein synthesis for all human proteins and its role as a methyl (-CH_3) group donor. Met is the primary source of methyl groups in polyamine synthesis. Met is central to the transmethylation-remethylation cycle (Figure 14.1), leading to numerous metabolites used throughout the body. Most of these processes take place in the liver and kidneys, but some steps may also take place in specific tissues like the brain or muscle. Methionine donates its methyl group to create new metabolites through a multi-step process with multiple intermediaries.

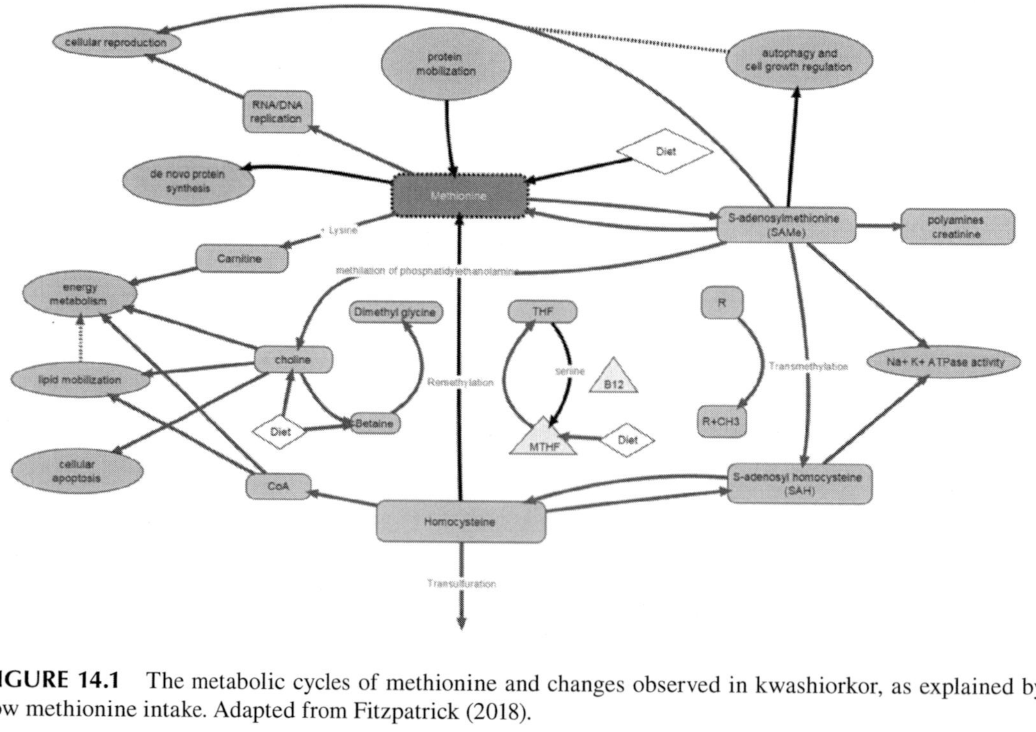

FIGURE 14.1 The metabolic cycles of methionine and changes observed in kwashiorkor, as explained by low methionine intake. Adapted from Fitzpatrick (2018).

First, Met is converted to S-adenosylmethionine (SAMe), which then donates the methyl group to multiple other substrates (transmethylation). With the loss of the methyl group, SAMe becomes S-adenosyl homocysteine (SAH) (Stipanuk and Ueki 2011). SAH then remains in flux with homocysteine (Hcy). At this critical juncture, Hcy can also be remethylated back to Met or transulfurated toward cystathionine and then on to Cys and the various pathways associated with Cys (Stipanuk and Ueki 2011).

The transmethylation of Hcy creates new metabolites via the remethylation process. It recycles and creates other metabolites by accepting a methyl group from them, often significantly reducing dietary demands for the original metabolite. A classic example is folate. Tetrahydrofolate (THF) accepts a methyl group from vitamin B12, facilitated by serine, forming 5-methyltetrahydrofolate (5-MTHF) (Scott and Weir 1981). 5-MTHF then remethylates Hcy to reform Met, relieving both B12 and folate of their methyl group and reverting them to a functional state once more rather than depend on a constant dietary supply of B12 and folate in their de-methylated state. In a similar process, betaine is relieved of its methyl group to form dimethylglycine.

Along the way, each intermediary has its own unique metabolic uses, some of which are noted in Figure 14.1. Examination of these pathways and how they might be affected by a deficiency of Met explains many of the anomalies seen in kwashiorkor.

SAMe is required for the methylation of phosphatidylethanolamine to form choline, an important component in lipid mobilization, energy metabolism, and regulation of cellular apoptosis, all mechanisms known to be deficient in kwashiorkor and potentially explaining the intense lethargy experienced by kwashiorkor patients. Both SAMe and SAH are factors in sodium-potassium ATPase activity, and whose shortage is potentially responsible for the electrolyte imbalance commonly observed in kwashiorkor. SAMe is also responsible for autophagy and cell growth regulation, potentially contributing to the macrocytic anemia seen in kwashiorkor. A lack of Met in this cycle would reduce the amount of Hcy available to be remethylated and possibly explain the very high concentrations of MTHF and vitamin B12 observed in kwashiorkor.

Beyond the transmethylation/remethylation cycle, Met is also responsible for initiating de novo protein synthesis, initiating cellular reproduction, and serves as a precursor for carnitine, another important player in energy metabolism (Roediger and Waterlow 1995; Borum 1983). Each of these processes has been observed to be inhibited in kwashiorkor.

TRANSULFURATION, LINKING CYS AND MET CYCLES

The transulfuration process is a one-way reaction that shifts the sulfur atom from Hcy to serine in the presence of vitamin B6 and glycine.

Although Met is necessary for good health, excess Met can be toxic but is not directly excreted from the body. Thus, its homeostasis is tightly controlled through the transulfuration process. High Met intake will result in high SAH, which will push the flux away from remethylation toward transulfuration and the eventual catabolism of Cys (Stipanuk 2020). Conversely, low SAH will shift the flux away from transulfuration to conserve more Hcy for remethylation (Courtney-Martin et al. 2012).

When lipid peroxide levels are high, it pushes the flux of Hcy toward transulfuration, allowing for the creation of more of the antioxidant glutathione (GSH) from Cys. However, it appears that the preference for flux toward remethylation when SAH is low may override this effect (Courtney-Martin et al. 2012; Fukagawa 2006). Green et al. found that, on average, children under treatment for kwashiorkor needed five days of Met supplementation before the flux shifted toward transulfuration (Green et al. 2014). Other studies have shown that cofactors and metabolites such as serine and glycine are present in normal to high levels in kwashiorkor, suggesting that low SAH is directing the flux away from transulfuration rather than a lack of cofactors inhibiting transulfuration reactions (Jahoor et al. 2006a; Whitehead and Dean 1964). The delay observed by Green et al. could be explained by the time needed for the supplements to fill unmet methyl group needs and to rebuild Met/SAH/Hcy pools.

While total Cys requirements may be met through transulfuration, it is an inefficient way to meet Cys requirements. As mentioned previously, acquiring some Cys through the diet lowers total SAA intake requirements.

CYS METABOLIC CYCLES, METABOLITES, AND ANOMALIES OF KWASHIORKOR

Cys can be sourced through diet, turnover of proteins in the body and the transulfuration of Hcy. The sulfur atom on Cys is an active site, with the strength of sulfur-sulfur bonds being the perfect strength for different Cys atoms in polyamine chains to attach to each other, creating the shapes necessary for proteins to function. In addition to this structural role, Cys serves myriad other uses through multiple potential pathways in which the sulfur atom is a central feature (Figure 14.2).

Cassava tubers and leaves are naturally high in cyanide and are a very common staple in many kwashiorkor-endemic regions (Kamalu 1993). Even with careful preparation, a certain amount of cyanide remains, especially in the leaves. The sulfur atom from Cys, in combination with rhodanese, is critical for the detoxification of cyanide, producing the less toxic product thiocyanate, which can be safely excreted in the urine (Zottola 2009). This process appears to take precedence over other possible uses for Cys, potentially drawing limited supplies of Cys away from other critical uses. Fitzpatrick et al. found that children in a high kwashiorkor prevalence population had higher levels of thiocyanate in their urine than those in a neighboring low prevalence population (Fitzpatrick et al. 2021).

Children with kwashiorkor have reduced production of sulfated glycosaminoglycans (GAGs) (Amadi et al. 2009). The sulfur from cysteine is also essential for the structural integrity of sulfated GAGs. GAGs are part of the glycocalyx, a gel in the interstitial space that binds a large amount of water. When GAGs lack sulfur, the glycocalyx can deteriorate, and water can move more rapidly and freely through the interstitial space. This is one of the mechanisms proposed for the edema characteristic of kwashiorkor (Manary et al. 2009). As noted above, Gonzales et al. found that lower levels of sulfated GAGs were associated with more severe edema in kwashiorkor (Gonzales et al. 2022).

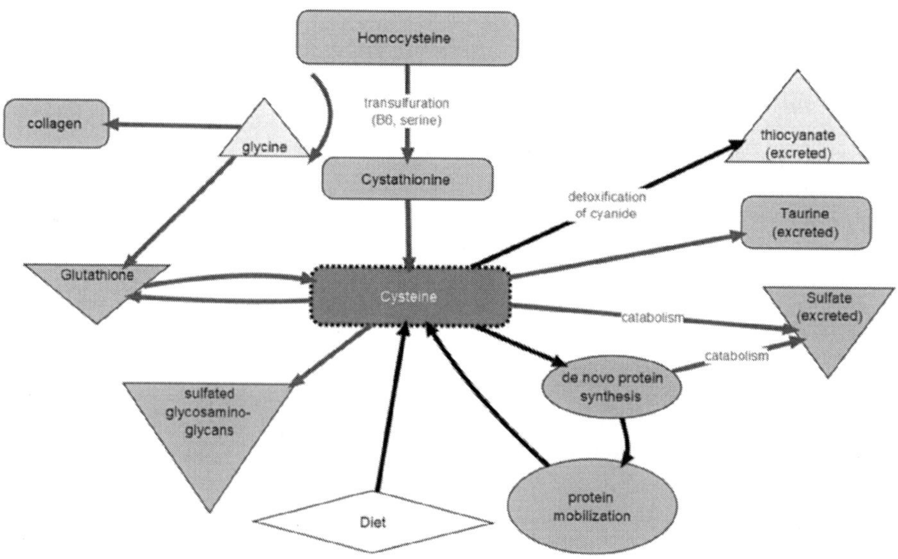

FIGURE 14.2 The metabolic cycles of cysteine and changes observed in kwashiorkor as explained by low cysteine availability. Adapted from Fitzpatrick (2018).

Glutathione (GSH) is an important antioxidant and is notably low in kwashiorkor, potentially responsible for the high levels of oxidative stress observed. GSH is derived from Cys and is reconverted to Cys when Cys levels are low, forming a sort of storage pool for Cys, but in a safer, less reactive form than Cys. Supplementation of Cys in kwashiorkor increases flux toward GSH and increases overall GSH levels in erythrocytes, indicating that Cys is likely the limiting factor in these cases (Badaloo et al. 2002).

When GSH levels are replete and excess Cys is available, it is catabolized into sulfate and excreted in the urine. The only way to dispose of excess Met is through transulfuration to Cys and then catabolism of Cys as sulfate. Sulfate in the urine, therefore, provides a general composite indication of Met and Cys sufficiency (Nakamura et al. 2002). Excretion of urinary sulfate has been observed to be low in cases of kwashiorkor, further indicating that SAA intakes are particularly low for this population (Ittyerah 1969).

IMPLICATIONS FOR A BETTER UNDERSTANDING OF THE ROLE OF SAAS AND PROTEIN QUALITY IN RISK OF KWASHIORKOR

The evidence presented suggests that diets low in sulfur amino acids, especially methionine, are central to the etiology of kwashiorkor. Nevertheless, more trials are necessary to test the causal relationship between low dietary SAAs and kwashiorkor and perhaps to better quantify deficiencies. To achieve this, surveys and studies need to reverse a current trend in both research and clinical practice of aggregating both kwashiorkor and wasting cases without subgroup analysis. It is clear that kwashiorkor is a unique and separate syndrome, with unique vulnerabilities and potential solutions.

The growing and deepening understanding of the relationship between insufficient dietary SAAs and kwashiorkor should help us to formulate better prevention strategies in low-resource settings where kwashiorkor is still endemic. Such strategies could include optimizing choices for feeding children as they are weaned and through toddlerhood to maximize SAA intake. Agricultural policies in populations prone to kwashiorkor need to reconsider the crops that are promoted, particularly those relating to the promotion of cassava as a staple. Where cassava is a staple, it is important to promote simple methods to reduce cyanide in the leaves as well as the tubers to reduce the loss of SAAs to the detoxification of cyanide present in these foods.

Formulations of products for supplemental feeding and treatment of kwashiorkor could be made more effective by incorporating dairy products that contain higher concentrations of methionine, such as casein instead of whey. Extending the duration of treatment with therapeutic foods beyond the loss of edema would provide the opportunity for patients to become fully replete in methyl groups, allowing metabolic anomalies to correct themselves.

Taking the full array of diverse signs of kwashiorkor together, the SAA metabolic pathways provide an explanation that ties them all together. SAA insufficiency alone, above all methionine insufficiency, may not explain all signs, and additional conditions may be necessary to trigger kwashiorkor. Nevertheless, SAA pathways provide a major framework upon which to build further research and prevention strategies.

THE EFFECTS OF OTHER PROTEIN SOURCES IN THE DIET

Animal-source foods are the richest sources of SAAs. Plants, on the other hand, generally have less total protein per gram of food and far less SAAs per gram of protein. For example, 100 g of boiled egg contains 12.6 g of protein and 0.68 g SAAs (USDA 2020). Maize has approximately 9.5 g of total protein and 0.42 g SAA per 100 g, about 61% of the amount in egg, and must be complemented with a pulse like pinto beans to provide sufficient lysine (Vasan et al. 2008). A 100 g portion of boiled pinto beans similarly contains 9.0 g of total protein, in which only 0.20 g is SAAs and Met is the limiting amino acid (USDA 2020).

Tubers have far less total protein per gram of food than grains or pulses. For example, 100 g of cassava contains only 0.25–1.6 g of total protein, of which only 0.04–0.07 g is SAA (Nassar et al. 2009). Cassava, therefore, can be considered virtually devoid of SAAs, while the natural cyanide may actually draw more sulfur out of the body than it provides. Tubers must therefore be complemented by an animal-source protein rather than pulses (in which Met is the limiting amino acid). This is especially important for young children as they have a much higher proportional need for SAAs per body weight (FAO 2013).

Although animal-source proteins are by far the best dietary sources of SAAs, these are also the most expensive sources of protein, inaccessible to many populations where kwashiorkor is endemic. Despite plant sources such as maize and beans individually containing less total protein and SAAs than a typical animal-source protein, when properly combined, they can provide enough to meet basic needs. For example, a two-year-old child weighing 13 kg could meet their SAA requirements of 0.40 g by eating, over the course of a day, 70 g of maize and 55 g of beans. However, the sheer bulk associated with this food can be a problem for small children who often eat only twice a day.

Cassava is a staple in many kwashiorkor-endemic areas and has been touted by agronomists as a solution to food insecurity because it grows in feeble soil and can tolerate drought well (Sengupta 2024). Considering the evidence presented here, policies promoting cassava need to consider if the benefits of the additional calories outweigh the reduced protein and SAA content, especially in the absence of animal-source proteins.

MINI DICTIONARY OF TERMS

1-carbon donor: Metabolites that donate a methyl group to another metabolite. Common examples include methionine, methylated-folate (5-CH_3-THF or also written as 5-MTHF), betaine and choline.

1-carbon metabolism: The metabolic pathways that create new metabolites by moving methyl groups from one molecule to another.

Bipedal pitting edema: A type of edema found symmetrically to equal degrees in both feet, which, when pressed for several seconds with a finger, leaves an impression in the edema.

Methyl group: One carbon atom with three hydrogen atoms attached, and the carbon atom attached to a root molecule. The chemical formulation is conventionally written as -CH_3.

Remethylation: The recycling of certain molecules (nutrients) that function by donating or receiving a methyl moiety.

Transmethylation: The molecular transfer of a single methyl moiety from one molecule to another.

Transulfuration: The process of moving a sulfur atom from one molecule to another.

SUMMARY POINTS

- This chapter describes the metabolic pathways of the sulfur amino acids methionine and cysteine through transmethylation, transulfuration and remethylation, using these pathways as a framework upon which to model clinical signs associated with insufficient intake of methionine and cysteine.
- These pathways are essential to numerous metabolic processes throughout the body.
- Disruption of these pathways by inadequate dietary intake of methionine or cysteine, particularly methionine, can result in a wide array of clinical signs.
- The signs one would expect to see from a disruption of the pathways closely match the signs associated with kwashiorkor malnutrition.
- Young children have higher proportional needs for methionine and cysteine, potentially explaining why children are more often affected by kwashiorkor.

- The framework of the SAA metabolic pathways provides an important tool to understand kwashiorkor and to design better treatments and preventive strategies.
- When framing policies on agricultural products to promote in kwashiorkor-endemic regions, the quality of the proteins must be considered.
- Kwashiorkor malnutrition is characterized by a unique collection of signs due to a unique causal pathway that distinguishes it from wasting, a different type of malnutrition caused primarily by insufficient energy intake.
- Surveys, research and treatment need to disaggregate data by type of acute malnutrition to improve understanding of the needs associated with each type of malnutrition, ensuring the best chances of survival for all.

REFERENCES

Adams, E. B., J. N. Scragg, B. T. Naidoo, S. K. Liljestrand, and V. I. Cockram. 1967. "Observations on the etiology and treatment of anaemia in kwashiorkor." *British Medical Journal* 3 (5563):451–454.

Amadi, Beatrice, Andrew O. Fagbemi, Paul Kelly, Mwiya Mwiya, Franco Torrente, Camilla Salvestrini, Richard Day, Michael H. Golden, Erik A. Eklund, Hudson H. Freeze, and Simon H. Murch. 2009. "Reduced production of sulfated glycosaminoglycans occurs in Zambian children with kwashiorkor but not marasmus." *The American Journal of Clinical Nutrition* 89 (2):592–600. doi: 10.3945/ajcn.2008.27092.

Arroyave, Guillermo, Dorothy Wilson, Carlota De Funes, and Moisés Béhar. 1962. "The free amino acids in blood plasma of children with kwashiorkor and marasmus." *The American Journal of Clinical Nutrition* 11 (5):517–524.

Badaloo, Asha, Jean W. Hsu, Carolyn Taylor-Bryan, Curtis Green, Marvin Reid, Terrence Forrester, and Farook Jahoor. 2012. "Dietary cysteine is used more efficiently by children with severe acute malnutrition with edema compared with those without edema." *The American Journal of Clinical Nutrition* 95 (1):84–90. doi: 10.3945/ajcn.111.024323.

Badaloo, Asha, Marvin Reid, Terrence Forrester, William C. Heird, and Farook Jahoor. 2002. "Cysteine supplementation improves the erythrocyte glutathione synthesis rate in children with severe edematous malnutrition." *The American Journal of Clinical Nutrition* 76 (3):646–652.

Borum, P. R. 1983. "Carnitine." *Annual Review of Nutrition* 3:233–259.

Brock, J. F., and M. Autret. 1952. "Kwashiorkor in Africa." *Bulletin of the World Health Organization* 5:1–71.

Brosnan, J. T., and M. E. Brosnan. 2006. "The sulfur-containing amino acids: an overview." *Journal of Nutrition* 136 (6 Suppl):1636S–1640S. doi: 10.1093/jn/136.6.1636S.

Ciliberto, Heather, Michael Ciliberto, Andreé Briend, Per Ashorn, Dennis Bier, and Mark Manary. 2005. "Antioxidant supplementation for the prevention of kwashiorkor in Malawian children: randomised, double blind, placebo controlled trial." *BMJ* 330 (7500):1109. doi: 10.1136/bmj.38427.404259.8F.

Courtney-Martin, Glenda, Ronald O. Ball, and Paul B. Pencharz. 2012. "Sulfur amino acid metabolism and requirements." *Nutrition Reviews* 70 (3):170–175. doi: 10.1111/j.1753-4887.2011.00466.x.

Dean, R. F. A., and R. G. Whitehead. 1964. "Plasma aminogram in kwashiorkor." *The Lancet* 284 (7350):98–99.

Di Giovanni, Valeria, Celine Bourdon, Dominic X. Wang, Swapna Seshadri, Edward Senga, Christian J. Versloot, Wieger Voskuijl, Richard D. Semba, Indi Trehan, Ruin Moaddel, M. Isabel Ordiz, Ling Zhang, John Parkinson, Mark J. Manary, and Robert H. J. Bandsma. 2016. "Metabolomic changes in serum of children with different clinical diagnoses of malnutrition." *The Journal of Nutrition*. doi: 10.3945/jn.116.239145.

FAO. 2013. "Dietary protein quality evaluation in human nutrition: report of an FAO expert consultation." In *FAO Food and Nutrition Paper 92*. Auckland, New Zealand: Food and Agricultural Organization of the United Nations.

Finkelstein, James D., and S. Harvey Mudd. 1967. "Trans-sulfuration in mammals." *Journal of Biological Chemistry* 242 (5):873–880. doi: 10.1016/s0021-9258(18)96205-8.

Fitzpatrick, Merry Colleen. 2018. "The role of sulfur amino acids in risk of kwashiorkor." Order No. 10743822, Tufts University, Gerald J. and Dorothy R. Friedman School of Nutrition Science and Policy. www-proquest-com.ezproxy.library.tufts.edu/dissertations-theses/role-sulfur-amino-acids-risk-kwashiorkor/docview/2015177153/se-2.

Fitzpatrick, Merry C., Anura V. Kurpad, Christopher P. Duggan, Shibani Ghosh, and Daniel G. Maxwell. 2021. "Dietary intake of sulfur amino acids and risk of kwashiorkor malnutrition in eastern democratic republic of the Congo." *The American Journal of Clinical Nutrition* 114 (3):925–933. doi: 10.1093/ajcn/nqab136.

Fuchs, George J. 2005. "Antioxidants for children with kwashiorkor: oxidative stress may not explain this deadly disease." *BMJ: British Medical Journal* 330 (7500):1095–1096. doi: 10.2307/25459644.

Fukagawa, Naomi K. 2006. "Sparing of methionine requirements: evaluation of human data takes sulfur amino acids beyond protein." *The Journal of Nutrition* 136 (6):1676S–1681S.

Golden, Michael. 1998. "Edematous malnutrition." *British Medical Bulletin* 54 (2):433–444.

Gonzales, G. B., J. M. Njunge, B. M. Gichuki, B. Wen, M. Ngari, I. Potani, J. Thitiri, D. Laukens, W. Voskuijl, R. Bandsma, J. Vanmassenhove, and J. A. Berkley. 2022. "The role of albumin and the extracellular matrix on the pathophysiology of edema formation in severe malnutrition." *EBioMedicine* 79:103991. doi: 10.1016/j.ebiom.2022.103991.

Gopalan, C. 1955. "Kwashiorkor in India." *Indian Journal of Medical Research* 43 (4):751–773.

Gopalan, C. 1968. "Kwashiorkor and marasmus: evolution and distinguishing features." In *Calorie Deficiencies and Protein Deficiencies: Proceedings of a Colloquium held in Cambridge April 1967*, edited by R. A. McCance and Elsie M. Widdowson, 49–58. Boston: Little, Brown and Company.

Green, C. O., A. V. Badaloo, J. W. Hsu, C. Taylor-Bryan, M. Reid, T. Forrester, and F. Jahoor. 2014. "Effects of randomized supplementation of methionine or alanine on cysteine and glutathione production during the early phase of treatment of children with edematous malnutrition." *American Journal of Clinical Nutrition* 99 (5):1052–1058. doi: 10.3945/ajcn.113.062729.

Gupte, Suraj. 1975. "Marasmus and kwashiorkor." *Pediatrics* 56 (1):152–152.

Gupte, Suraj, and S. Mehta. 1971. "Advanced protein-calorie malnutrition; clinical observations on North Indian children." *Pediatric Clinics of India* 6 (2):91–100.

Holt, L. Emmett, Selma Snyderman, Patricia Norton, Ellen Roitman, and Janet Finch. 1963. "The Plasma aminogram in kwashiorkor." *Lancet* 282 (7322):1343–1348.

Ittyerah, Roy Thomas. 1969. "Urinary excretion of sulfate in kwashiorkor." *Clinica Chimica Acta* 25 (3):365–369. doi: http://dx.doi.org/10.1016/0009-8981(69)90194-6.

Jahoor, Farook. 2012. "Effects of decreased availability of sulfur amino acids in severe childhood undernutrition." *Nutrition Reviews* 70 (3):176–187. doi: 10.1111/j.1753-4887.2011.00462.x.

Jahoor, Farook, Asha Badaloo, Marvin Reid, and Terrence Forrester. 2005. "Protein kinetic differences between children with edematous and nonedematous severe childhood undernutrition in the fed and postabsorptive states." *The American Journal of Clinical Nutrition* 82 (4):792–800.

Jahoor, Farook, Asha Badaloo, Marvin Reid, and Terrence Forrester. 2006a. "Glycine production in severe childhood undernutrition." *The American Journal of Clinical Nutrition* 84 (1):143–149.

Jahoor, Farook, Asha Badaloo, Marvin Reid, and Terrence Forrester. 2006b. "Sulfur amino acid metabolism in children with severe childhood undernutrition: cysteine kinetics." *The American Journal of Clinical Nutrition* 84 (6):1393–1399.

Kamalu, Beryl P. 1993. "Cassava (Manihot Esculenta Crantz) in the etiology of kwashiorkor." *Nutrition Research Reviews* 6 (1):121–135. doi:10.1079/NRR19930009.

Kurpad, Anura V., Meredith M. Regan, Sureka Varalakshmi, Justin Gnanou, Amarnath Lingappa, and Vernon R. Young. 2004. "Effect of cystine on the methionine requirement of healthy Indian men determined by using the 24-h indicator amino acid balance approach." *The American Journal of Clinical Nutrition* 80 (6):1526–1535.

Lenhartz, H., R. Ndasi, A. Anninos, D. Bötticher, E. Mayatepek, E. Tetanye, and M. Leichsenring. 1998. "The clinical manifestation of the kwashiorkor syndrome is related to increased lipid peroxidation." *The Journal of Pediatrics* 132 (5):879–881. http://dx.doi.org/10.1016/S0022-3476(98)70324-5.

Manary, Mark, Geert Tom Heikens, and Michael Golden. 2009. "Kwashiorkor: more hypothesis testing is needed to understand the etiology of edema." *Malawi Medical Journal* 21 (3):106–107.

Manary, Mark, Christiaan Leeuwenburgh, and Jay Heinecke. 2000. "Increased oxidative stress in kwashiorkor." *The Journal of Pediatrics* 137 (3):421–424. doi: 10.1067/mpd.2000.107512.

Nakamura, H., R. Kajikawa, and T. Ubuka. 2002. "A study on the estimation of sulfur-containing amino acid metabolism by the determination of urinary sulfate and taurine." *Amino Acids* 23 (4):427–431. doi: 10.1007/s00726-002-0208-9.

Nassar, M. A. Nagib, P. Osmindo Junior, V. Marcelo Sousa, and Rodomiro Ortiz. 2009. "Improving carotenoids and amino-acids in cassava." *Recent Patents on Food, Nutrition & Agriculture* 1 (1):32–38. doi: http://dx.doi.org/10.2174/2212798410901010032.

Odei Obeng-Amoako, G. A., H. Stobaugh, S. V. Wrottesley, T. Khara, P. Binns, I. Trehan, R. E. Black, P. Webb, M. Mwangome, J. Bailey, P. Bahwere, C. Dolan, E. Boyd, A. Briend, M. A. Myatt, and N. Lelijveld. 2022. "How do children with severe underweight and wasting respond to treatment? A pooled secondary data analysis to inform future intervention studies." *Matern Child Nutr*:e13434. doi: 10.1111/mcn.13434.

Odigwe, Chibuzo C., Geir Smedslund, Regina I. Ejemot-Nwadiaro, Chiedozie C. Anyanechi, and Michael B. Krawinke. 2010. "Supplementary vitamin E, selenium, cysteine and riboflavin for preventing kwashiorkor in preschool children in developing countries." *Cochrane Database of Systematic Reviews* (April 14)(4):CD008147. doi: 10.1002/14651858.CD008147.pub2.

Phadke, M. A., V. A. Khedkar, D. Pashankar, S. L. Kate, G. D. Mokashi, P. S. Gambhir, and S. M. Bhate. 1995. "Serum amino acids and genesis of protein energy malnutrition." *Indian Pediatrics* 32:301–306.

Pham, T. P., M. T. Alou, M. H. Golden, M. Million, and D. Raoult. 2021. "Difference between kwashiorkor and marasmus: Comparative meta-analysis of pathogenic characteristics and implications for treatment." *Microb Pathog* 150:104702. doi: 10.1016/j.micpath.2020.104702.

Raguso, C. A., A. M. Ajami, R. Gleason, and V. R. Young. 1997. "Effect of cystine intake on methionine kinetics and oxidation determined with oral tracers of methionine and cysteine in healthy adults." *The American Journal of Clinical Nutrition* 66 (2):283–292.

Reid, Marvin, Asha Badaloo, Terrence Forrester, John F. Morlese, Margaret Frazer, William C. Heird, and Farook Jahoor. 2000. "In vivo rates of erythrocyte glutathione synthesis in children with severe protein-energy malnutrition." *American Journal of Physiology - Endocrinology And Metabolism* 278 (3):E405.

Roediger, W. E. W., and J. Prof Waterlow. 1995. "New views on the pathogenesis of kwashiorkor: methionine and other amino acids." *Journal of Pediatric Gastroenterology & Nutrition* 21 (2):130–136.

Schofield, C., and A. Ashworth. 1996. "Why have mortality rates for severe malnutrition remained so high?" *Bulletin of the World Health Organization* 74 (2):223–229.

Scott, John M., and Donald G. Weir. 1981. "The methyl folate trap: A physiological response in man to prevent methyl group deficiency in kwashiorkor (methionine deficiency) and an explanation for folic-acid-induced exacerbation of subacute combined degeneration in pernicious anaemia." *The Lancet* 318 (8242):337–340. doi: https://doi.org/10.1016/S0140-6736(81)90650-4.

Sengupta, Somini. 2024. "A 'revolutionary' way to feed the world that's very old." *The New York Times*, January 22, 2024, Climate. https://www.nytimes.com/2024/01/22/climate/agriculture-africa-traditional-crops.html.

Smith, M. I., T. Yatsunenko, M. J. Manary, I. Trehan, R. Mkakosya, J. Cheng, A. L. Kau, S. S. Rich, P. Concannon, J. C. Mychaleckyj, J. Liu, E. Houpt, J. V. Li, E. Holmes, J. Nicholson, D. Knights, L. K. Ursell, R. Knight, and J. I. Gordon. 2013. "Gut microbiomes of malawian twin Pairs discordant for kwashiorkor." *Science* 339 (6119):548–554. doi: 10.1126/science.1229000.

Stipanuk, M. H. 2020. "Metabolism of sulfur-containing amino acids: how the body copes with excess methionine, cysteine, and sulfide." *Journal of Nutrition* 150 (Suppl 1):2494S–2505S. doi: 10.1093/jn/nxaa094.

Stipanuk, Martha H., and Iori Ueki. 2011. "Dealing with methionine/homocysteine sulfur: cysteine metabolism to taurine and inorganic sulfur." *Journal of Inherited Metabolic Disease* 34 (1):17–32. doi: 10.1007/s10545-009-9006-9.

Sullivan, Jesse, MacDonald Ndekha, Dawn Maker, Christine Hotz, and Mark J. Manary. 2006. "The quality of the diet in Malawian children with kwashiorkor and marasmus." *Maternal & Child Nutrition* 2 (2):114–122. doi: 10.1111/j.1740-8709.2006.00053.x.

USDA. 2020. "USDA food data central." https://fdc.nal.usda.gov/.

Vasan, P., A. B. Mandal, Narayan Dutta, S. K. Maiti, and K. Sharma. 2008. "Digestibility of amino acids of maize, low tannin sorghum, pearl millet and finger millet in caecectomized roosters." *Asian-Australasian Journal of Animal Sciences* 21 (5):701–706. doi: 10.5713/ajas.2008.70296.

Whitehead, R. G., and R. F. A. Dean. 1964. "Serum amino acids in kwashiorkor: I. Relationship to clinical condition." *The American Journal of Clinical Nutrition* 14 (6):313–319.

WHO, FAO, and UNU. 2007. *Protein and Amino acid Requirements in Human Nutrition: Report of a Joint FAO/WHO/UNU Expert Consultation*. WHO Technical Report Series; no. 935. Geneva: World Health Organization.

Williams, Cicely D. 1933. "A nutritional disease of childhood associated with a maize diet." *Archives of Disease in Childhood* 8 (48):423–433.

Zottola, Mark A. 2009. "A partial exploration of the potential energy surfaces of SCN and HSCN: implications for the enzyme-mediated detoxification of cyanide." *Journal of Molecular Graphics and Modelling* 28 (2):183–186. doi: http://dx.doi.org/10.1016/j.jmgm.2009.06.005.

15 Malnutrition and Refeeding Syndrome

Isil Ozer and Rajkumar Rajendram

ABBREVIATIONS

AIDS	Acquired immunodeficiency syndrome
ATP	Adenosine triphosphate
BMI	Body mass index
IGF-1	Insulin growth factor 1
IQ	Intelligence quotient
LIFDC	Low-income food-deficit countries
LNS	Lipid nutrient supplements
MAM	Moderate acute malnutrition
MUAC	Mid-upper arm circumference
PEM	Protein-energy deficiency
RDA	Recommended dietary allowance
RUSF	Ready-to-use supplementary food
RUTF	Ready-to-use therapeutic food
SAM	Severe acute malnutrition
TSFT	Triceps skinfold thickness
WHO	World Health Organization
WHZ	Weight-for-height Z-score

INTRODUCTION

"Malnutrition" is a very broad term. It encompasses the spectrum of states that occur when there is an imbalance between the nutrients an organism ingests and its needs. Malnutrition develops when an organism's dietary intake does not provide the nutrients required for optimal health, development, and growth. Malnutrition is dangerous. It negatively affects the intelligence quotient (IQ) of individuals and the population, increases the risk of infections, exacerbates chronic diseases, and ultimately may be fatal (World Health Organization (WHO) 2019).

Malnutrition usually manifests as undernutrition (i.e., inadequate intake of essential nutrients) or macronutrient deficiency. However, somewhat counterintuitively, overnutrition (i.e., excessive intake of some nutrients) may also cause a form of malnutrition that may manifest as obesity. Micronutrient deficiencies (inadequate intake of specific vitamins and minerals) may occur in both overnutrition and undernutrition.

Protein is particularly important for homeostasis, growth, development, maintenance of organs, and immune function. Protein malnutrition is a subtype of undernutrition. It occurs when the body does not get enough protein in the diet. Malnutrition can manifest in developed countries as well as underdeveloped nations, and the socioeconomic determinants of insufficient nutritional intake remain incompletely understood. Yet, the WHO has designated low-income food-deficit countries (LIFDCs) as the primary focus for global efforts to thwart malnutrition (WHO 2019).

DOI: 10.1201/b23328-15

The risk factors that predispose individuals to protein malnutrition (i.e., pandemics, natural disasters, wars, migration, misuse of alcohol and other substances, loss of homes and jobs, and the inability to buy sufficient food to eat every day) are certainly more common in LIFDCs than in developed countries. The prevalence of infectious diseases that may induce or exacerbate protein malnutrition in LIFDCs is also high. These diseases include the human immunodeficiency virus (HIV), tuberculosis, malaria, pneumonia, and acute gastroenteritis.

To reduce the prevalence and impact of malnutrition, the relevant stakeholders in LIFDCs (e.g., the WHO and governments) have generally focused on projects aimed at socioeconomic improvement and maternal education. Yet, malnutrition remains a major problem in the 21st century. Thus, the effectiveness of the current strategies must be questioned.

Malnutrition occurs when nutrient intake is abnormal and/or not balanced. The longer an organism is affected, the more it relies on its homeostatic adaptations to survive. However, these adaptations may have negative consequences if the organism is subsequently exposed to adequate nutrients. Nutritional therapies that ignore the physiological changes caused by malnutrition may trigger a process known as the "refeeding syndrome" (Reber et al. 2019).

Refeeding syndrome is a potentially life-threatening condition that may occur when nutrients are provided to severely malnourished organisms. It is characterized by complex metabolic and electrolyte disturbances such as hypokalemia (low plasma potassium concentration), hypomagnesemia (low plasma magnesium concentration), hypophosphatemia (low plasma phosphate concentration), and edema (fluid retention) (Reber et al. 2019). Refeeding syndrome is most likely to occur when carbohydrates and calories are given without sufficient electrolytes and micronutrients (Reber et al. 2019). This occurs because refeeding may induce rapid changes in plasma electrolytes in the setting of malnutrition.

Sadly, in the 21st century, malnutrition and refeeding syndrome remain common. More effective solutions to correct nutritional deficits are needed. This requires a reanalysis of current approaches to the prevention and treatment of malnutrition and refeeding syndrome. In this context, it is important to consider malnutrition and inadequate protein intake in more detail. Refeeding syndrome is a common, life-threatening problem that may occur in any malnourished individual. However, this chapter focuses on refeeding syndrome in the setting of protein-energy malnutrition (PEM).

CLASSIFICATION OF MALNUTRITION

The understanding of severe acute malnutrition (SAM) has changed over time (Pham et al. 2021). The initial descriptions were based on clinical observations, but there are now many ways to classify malnutrition. After distinguishing undernutrition (when nutritional intake is inadequate) from overnutrition, the duration of inappropriate nutrient intake must also be considered.

Acute malnutrition occurs when there is a rapid, severe decline in nutritional status over a relatively short period of time. In children, this may cause wasting (significant weight loss) or low weight-for-height (underweight). Chronic malnutrition, on the other hand, occurs when there is prolonged undernutrition. This may result in micronutrient deficiencies and/or low height-for-age (stunted) or underweight children.

The precise nature of malnutrition should also be considered. Insufficient intake of vitamins and minerals results in micronutrient deficiencies. "Hidden hunger" occurs when an individual's diet lacks essential micronutrients, despite adequate calorie intake. Thus, even obese patients can have hidden hunger (i.e., micronutrient deficiency). Protein-energy malnutrition (PEM) occurs when protein intake is insufficient.

Marasmus and kwashiorkor are both severe subtypes of PEM. Marasmus usually occurs after prolonged deficiency of both protein and calories. It typically manifests as cachexia (significant weight loss) with muscle wasting (Pham et al. 2021). Kwashiorkor typically occurs with severe protein deficiency despite adequate or only slightly reduced calorie intake. It is characterized by generalized edema but relatively preserved muscles (Pham et al. 2021). Biochemical and histopathological studies, nutritional analyses, and immunological tests suggest that patients with kwashiorkor are at a higher risk of death than those with marasmus (Pham et al. 2021). Insulin growth factor 1 (IGF-1) and leptin are involved in the pathogenesis of both PEM and refeeding syndrome (Pham et al. 2021). In fact, IGF-1 deficiency may trigger refeeding syndrome (Goyale et al. 2015). The efficacy of antibiotics is decreased in patients with PEM due to the immunological changes caused by malnutrition (Oshikoya et al. 2010).

AETIOPATHOLOGY OF MALNUTRITION

Insufficient nutrient intake precludes the rapid cellular regeneration required to maintain the function of the respiratory and gastrointestinal systems. In malnourished patients, atrophy of the intestinal villi reduces their ability to absorb dietary nutrients (Montoro-Huguet, Belloc, and Domínguez-Cajal 2021). The reduction in bile acids also causes malabsorption of fats and fat-soluble vitamins (Montoro-Huguet, Belloc, and Domínguez-Cajal 2021). Insufficiency of Vitamin A (a fat-soluble vitamin) further impairs tissue regeneration (especially that of mucous membranes) (Montoro-Huguet, Belloc, and Domínguez-Cajal 2021).

Protein intake is particularly important for cellular regeneration. The processes described above are involved in the pathogenesis of kwashiorkor (Manary, Heikens, and Golden 2009). Thus, the patient with PEM enters a vicious cycle of exogenous nutrient deficiency, malabsorption, insufficient endogenous nutrient production, and urinary losses. Vitamin C deficiency can illustrate this.

Vitamin C acts as a cofactor for two enzymes involved in carnitine synthesis (Furusawa et al. 2008). Hence, Vitamin C deficiency affects mitochondrial utilization of the already insufficient free fatty acids (Furusawa et al. 2008). The resulting lack of energy production leads to fatigue.

Insufficient protein consumption also limits the synthesis of the apolipoproteins required to transport fatty acids in the blood (Manary, Heikens, and Golden 2009). Moreover, low consumption of cholesterol-rich foods and the insufficient endogenous synthesis of cholesterol *in vivo* further reduce the concentration of cholesterol in the blood (Soliman 2018).

Cholesterol is necessary for the synthesis of bile acids, steroid hormones, and myelin (Clayton 1998). Thus, inadequate cholesterol intake and disorders of cholesterol synthesis can result in severe consequences, including intellectual disability due to hypomyelination (Clayton 1998). Neurodevelopmental disorders further decrease steroid synthesis and impair fat absorption. Deficiency of bile synthesis causes malabsorption of the fat-soluble vitamins (Montoro-Huguet, Belloc, and Domínguez-Cajal 2021). When very low-density lipoproteins are deficient, fatty acids cannot be transported to other tissues (De Silva and Nightingale 2019). This results in hepatic steatosis (De Silva and Nightingale 2019).

Iodine deficiency reduces the synthesis of thyroid hormones, slowing the basal metabolic rate (Lisco et al. 2023). When metabolism is slow, the exogenous administration of large amounts of carbohydrate significantly increases insulin production. In malnourished patients, this anabolic hormone rapidly directs the already insufficient extracellular phosphorus, magnesium, and potassium in the fluid into cells (De Silva and Nightingale 2019). This results in electrolyte imbalances, including hypophosphatemia, hypomagnesemia, and hypokalemia (De Silva and Nightingale 2019).

Of the patients with inadequate nutrient intake, it is difficult to identify those at risk of refeeding syndrome (De Silva and Nightingale 2019). Even fasting for as little as 5–7 days may result in thiamine and phosphorus deficiency (i.e., refeeding syndrome) upon resumption of nutritional intake (De Silva and Nightingale 2019). Refeeding syndrome can impair adenosine triphosphate (ATP) synthesis in two ways:

1. Thiamine deficiency (thiamine is a cofactor for mitochondrial enzymes)
2. Phosphorus deficiency (phosphate is the precursor of ATP).

When ATP is low, the Na/K ATPase cannot function. Sodium accumulates within cells, resulting in edema. Insulin-induced hypokalemia also impairs the electrical conduction in the heart (De Silva and Nightingale 2019), potentially causing heart failure and arrhythmias, which can cause morbidity and mortality (De Silva and Nightingale 2019). Phosphorus deficiency also reduces plasma 2,3-diphosphoglycerate concentrations (De Silva and Nightingale 2019). This compromises oxygen transport from erythrocytes to tissues, leading to hypoxia, which damages tissues.

The deficiency of antioxidants such as selenium, vitamins A, E, and C, and glutamine exacerbates damage to cells (Odigwe et al. 2010). Prophylaxis with a combination of vitamin E, selenium, cysteine, and riboflavin is beneficial (Odigwe et al. 2010). The clinical and pathological features of malnutrition are summarized in Table 15.1, and the pathogenesis of refeeding syndrome is summarized in Figure 15.1.

RECOMMENDATIONS FOR PREVENTING AND TREATING MALNUTRITION AND REFEEDING SYNDROME

The WHO aims to prevent permanent nutritional deficiencies for geographical reasons by providing food supplements or single/multivitamin and mineral supplements (WHO Regional Office for Africa 2017). Clean water and modifying local dietary habits according to needs are also essential (WHO 2019). The US government has also issued recommendations for nutritionally balanced food intake (Hopkins and Gunther 2015; US Department of Agriculture and US Department of Health and Human Services 2020).

Early detection of failure to thrive and abnormalities in the weight-to-height ratio is important for the early diagnosis and successful treatment of malnutrition in children (WHO 2019). At-risk children should receive frequent monitoring and treatment as early as possible (WHO 2019). Mothers with children in this group should be educated about healthy nutrition using local foods. The WHO also recommends follow-up of children at risk of malnutrition using mid-upper arm circumference (MUAC) measurements (WHO 2019).

To achieve its objectives, the WHO regularly updates its guidance for preventing and treating malnutrition worldwide. For example, in 2017, the WHO objectives to improve health in Africa included recommendations to monitor the nutritional status of expectant mothers and provide them with balanced nutrition (WHO Regional Office for Africa 2017).

Various other international bodies have developed strategies to reduce malnutrition in Africa by 2025 (WHO Regional Office for Africa 2017). The main targets include reducing the incidence of stunting, moderate acute malnutrition (MAM), and wasting (WHO Regional Office for Africa 2017). Preventing childhood obesity was also considered to be very important (WHO Regional Office for Africa 2017). Secondary targets include preventing anemia, reducing the incidence of low birth weights, and encouraging breastfeeding for at least six months (WHO Regional Office for Africa 2017). These strategies aim to reduce stunting (a sign of chronic malnutrition) to below 40% (WHO Regional Office for Africa, 2017). Screening for a height-for-age ratio below –2 standard deviations (SD) from the mean has been recommended for both diagnosis and follow-up (WHO Regional Office for Africa 2017).

However, recent data show that malnutrition remains problematic worldwide and suggest that the WHO's 2025 targets cannot be achieved with the current treatment protocols (Aloa et al. 2021). Acknowledging this, the WHO updated its guidance on preventing and managing acute malnutrition in 2023 (WHO 2019).

For high-risk groups, preventing, screening for, and treating thiamine deficiency is crucial (Hiffler et al. 2016). A meta-ethnographic systematic review of studies on hidden hunger and wasting in South Asia showed that iron, zinc, iodine, and vitamin deficiencies are very common (Wali et al. 2022). However, the literature reviewed by Wali et al. (2022) did not describe deficiencies in phosphorus and thiamine, which are important for refeeding syndrome.

TABLE 15.1
Clinical and Laboratory Characteristics of Various Types of Malnutrition

Characteristics		Marasmus	Kwashiorkor	Hidden Hunger	Obesity
Clinical characteristics	Stunting	+	+	+/–	+/–
	Reduced subcutaneous adipose tissue	+	+	–/+	–
	Edema	–	+	–	–/+*
	Hair shedding	+	+	+/–	–
	Dermatitis	–	+	–/+**	–
	Rickets	–	+	+/–***	+/–****
	Pulmonary infections	+	+	+/–	+/– ¥
	Gastroenteritis	+	+	–/+	–/+
	Chronic infections	+/–	+/–	+/–	–/+
	Learning disability	+	+	+	–
	Hepatomegaly	–	+	–/+	+
	Waist circumference	↓	↑(edema)	↑+/–	↑
Biochemical characteristic	Albumin	↓	↓	Unstable	Unstable
	Vitamin D	↓	↓	Unstable	Unstable
	Lactate	↑	↑	Unstable	Unstable
	Calcium	↓	↓	Unstable	Unstable
	Phosphorus	↓	↓	Unstable	Unstable
	Magnesium	↓	↓	Unstable	Unstable
	Urea	↑	↑	Unstable	Unstable
	Creatinine	↑	↑	Unstable	Unstable
	Liver enzymes	↑	↑	Unstable	Unstable
	Glucose	Unstable	Unstable	Unstable	Unstable
	Fat-soluble vitamins	↓	↓	Unstable	Unstable
	Water-soluble vaitamins	↓	↓	Unstable	Unstable
	Ketone bodies	Unstable	Unstable	Unstable	Unstable
	Immunoglobulins	↓	↓	Unstable	Unstable
	Lymphocytes	↓	↓	Unstable	Unstable
Histopathological features	Fatty liver	–	+	–/+	–/+
	Intestinal villous atrophy	+	+	–/+	–
	Delay in myelination	+	+	–/+	–
	Renal tubular dysfunction	+	+	–/+	–/+
	Osteoporosis §	+/–	+	+/–	+/–

This table presents the characteristic clinical and laboratory findings in patients with select types of malnutrition. *Cardiovascular diseases caused by obesity can cause edema. **If there is iron deficiency, hair loss may occur. ***If there is vitamin D deficiency, there may be signs of rickets. ****Obain legs, femoral head lesions may be confused with rickets in overweight children. ¥ Due to gastroesophageal reflux. § May be due to vitamin D and/or vitamin C deficiency.

Relatively few studies have investigated the prevalence of micronutrient deficiencies. However, studies in Western societies have demonstrated that vitamin C, D, folic acid, calcium, selenium, and iodine deficiencies are prominent in both adults (Viñas et al. 2011) and children (Kaganov et al. 2015). Studies on immigrants in Europe have identified thiamine deficiency (Matsangos et al. 2022). Homeless individuals are at particularly high risk of thiamine deficiency (Ijaz et al. 2017).

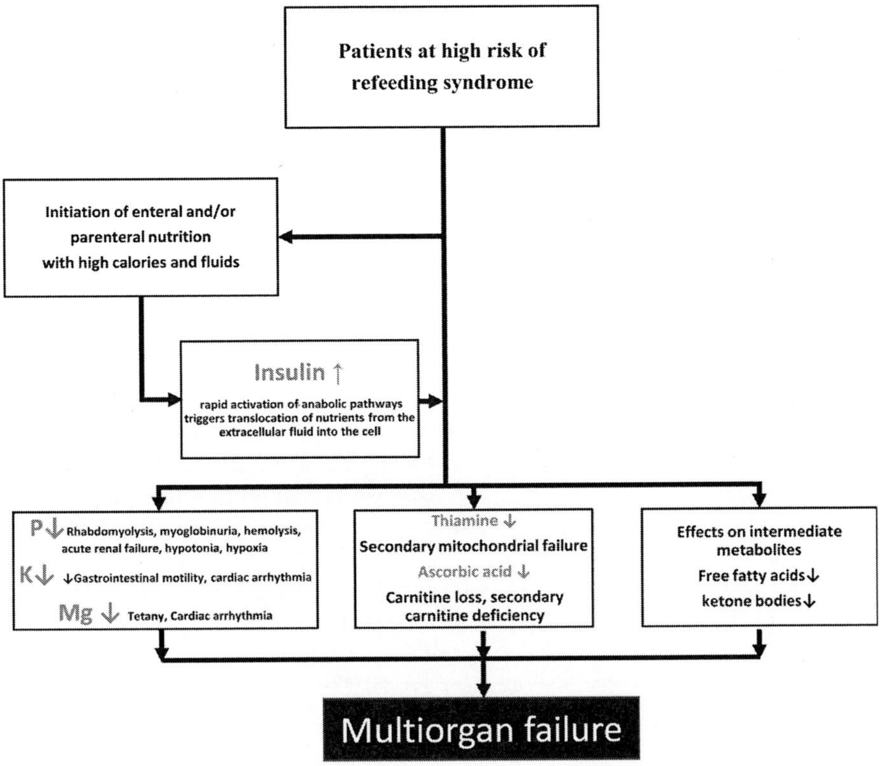

FIGURE 15.1 Pathophysiological mechanisms of refeeding syndrome. This figure summarizes the pathogenesis of the refeeding syndrome.

A report by the European Society of Clinical Nutrition and Metabolism identified thiamine deficiency as one of the most important risk factors for refeeding syndrome in the perioperative period and in hospitalized patients with chronic diseases (Lobo et al. 2020; Ponzo et al. 2021). Yet, healthcare providers often overlook the nutritional significance of thiamine. For example, when Fedele et al. (2021) reviewed studies of the nutritional status of hospitalized obese and low-body weight patients during the COVID-19 pandemic, they described deficiencies in vitamins A, D, C, and E but did not provide any information about thiamine (Fedele et al. 2021).

Healthcare providers in areas at high risk of malnutrition are advised to follow WHO protocols. These protocols include the use of ready-to-use therapeutic food (RUTF), ready-to-use supplementary food (RUSF), prophylactic antibiotics, and vitamin A supplementation (WHO 2013, 2019). The effectiveness of these protocols has been investigated in children under the age of five diagnosed with SAM and MAM (Das et al. 2020).

In view of the WHO recommendations and studies of their effectiveness, high-calorie foods can be used to provide 130–170 kcal/kg/day to treat children with SAM and MAM (Schoonees et al. 2019). These foods may include peanut oil, oleic acid, omega-3 fatty acids, corn, sorghum, soy milk, or milk powder (Schoonees et al. 2019). Healthcare providers can further improve malnourished patients' general condition in relation to edema, weight gain, and infection by administering high doses of vitamin A, iron, zinc, and prophylactic antibiotics (penicillin or trimethoprim-sulfamethoxazole). However, some high-risk, chronically malnourished patients with very low BMI may require parenteral nutrition. This is because malnutrition may be caused or exacerbated by atrophy of the intestinal villi.

Patients with SAM often develop refeeding syndrome when started on a high-carbohydrate diet. Suggestions for nutritional therapy to reduce the risk of refeeding syndrome in various types of malnutrition are presented in Table 15.2.

A meta-analysis compared the safety and effectiveness of RUTF and RUSF against lipid-based food supplements containing essential fatty acids, cheap animal proteins (skimmed milk powder), legumes, and micronutrients in treating MAM in patients under the age of five years (Gera et al. 2017). The lipid-based supplements were more effective. One of the nine studies included in this meta-analysis started with an appropriate amount of calories (40 kcal/kg/day) to prevent the development of refeeding syndrome. However, the studies reviewed by Gera et al. (2017) did not provide data on thiamine and phosphorus, B12, folate, and vitamin C.

CONDITIONS AND GROUPS AT INCREASED RISK OF MALNUTRITION

It is easy to recognize that children with chronic diseases, individuals with eating disorders (e.g., anorexia nervosa), and patients on restricted diets are at risk of malnutrition. It is less obvious that those who cannot gain weight due to genetic diseases associated with neurological, cardiac, renal, and/or gastrointestinal dysfunction also face this risk (Prasadajudio et al. 2022). Gastrointestinal conditions that require surgical treatment (e.g., pancreatitis, biliary tract diseases, and intestinal anomalies) also predispose individuals to malnutrition by inducing malabsorption. Iatrogenic malnutrition may also occur in underfed, critically ill patients (De Waele et al. 2021).

The COVID-19 pandemic increased the risk of malnutrition worldwide (UNICEF/WHO/World Bank Group Joint Child Malnutrition Forecasts 2021). This risk varies between countries and regions. People with low socioeconomic status often aspire to migrate to countries that offer better economic opportunities. In countries where the risk of malnutrition is high, common comorbidities in malnourished patients include tuberculosis, anemia, and pneumonia (Moate et al. 2022). In these countries, malnutrition is associated with a fatal outcome in 10–15% of cases (Moate et al. 2022). Table 15.3 describes the geographical differences in characteristics of patient groups at high risk of malnutrition.

THE ROLE OF REFEEDING SYNDROME IN MALNUTRITION TREATMENT FAILURES

A meta-analysis of patients hospitalized for SAM treatment over two decades revealed a relationship between mortality and low Z-score for height (WHZ), HIV infection, diarrhea, pneumonia, shock, anorexia, and enteral nutritional requirements (Karunaratne et al. 2020). Loss of consciousness and increased mortality were observed with refeeding syndrome in patients with loss of appetite and enteral tube feeding (Karunaratne et al. 2020). Thiamine deficiency is a significant risk factor for refeeding syndrome in African children (Hiffler et al. 2016) and hospitalized adult patients with chronic diseases (Stanga et al. 2008). Patients with refeeding syndrome who develop multiorgan failure also have decreased levels of thiamine, phosphorus, and magnesium (Boateng et al. 2010). While these deficiencies contribute to organ dysfunction, infection may precipitate multiorgan failure in patients with refeeding syndrome (Williams and Berkley 2018). Malnourished children may not show signs of clinical infection (Jones and Berkley 2014). Thus, it is important to administer broad-spectrum antibiotic therapy targeted to the affected organs in this cohort of patients, even if a causative organism has not been identified (Jones and Berkley 2014; Williams and Berkley 2018).

TABLE 15.2
Approach to Treating Malnutrition that Reduces the Risk of Inducing Refeeding Syndrome

Contents		Marasmus	Kwashiorkor	Hidden Hunger	Obesity
	Enteral	**Intravenous Fluids, Enteral Tube Feeding, Complete or Partially Hydrolyzed Nutritional Therapy***		**Balanced Diet**	**Balanced Diet and Lifestyle Change**
Nutritional Therapy	**Parenteral** **5–25 kcal/kg/day** **20–35 ml/kg/day**	**Fluid, Start with a Calorie-Restricted Diet, Slow, Controlled Increase in Administered Calories****			
Minerals	Sodium <1 mmol/kg/day	Personalized treatment based on heart and kidney function			
	Potassium 1–1.5 mmol/kg/day	Rapid treatment of hypokalemia, close follow-up			
	Phosphorus 0.3–0.6 mmol/kg/day	Rapid treatment of hypophosphatemia, close follow-up			
	Magnesium 0.2–0.4 mmol/kg/day	Rapid treatment of hypomagnesemia, close follow-up			
	Bicarbonate	Effective treatment of metabolic acidosis but must pay attention to the serum sodium concentration			
	Zinc***	For replication of the cells in the skin, hair, nails, and gastrointestinal tract			For replication of cells in the skin, hair, and nails
	Trace elements***	Especially for the antioxidant effect of copper and selenium			−/+
Vitamins	D***	For the treatment of osteoporosis or to enhance the function of the immune system			
	K***	If there is liver failure, fat malabsorption, or the intestinal flora are disturbed			+/−
	A***	If there is liver failure or fat malabsorption			
	E***	If there is liver failure or fat malabsorption or for antioxidant effect			
	C***	If there is carnitine deficiency, hemorrhages within tissues, or osteoporosis			
	B1200–300 mg/d	If the patient has encephalopathy, or lactic acidosis			
	B12***	If there is anemia, neuropathy, high homocysteine, hypotonia, or atrophy of the tongue or gastrointestinal mucosa			
	Folate 1–2 mg/day	If there is high homocysteine, or a history of thrombo-embolism			

(*Continued*)

TABLE 15.2 (CONTINUED)
Approach to Treating Malnutrition that Reduces the Risk of Inducing Refeeding Syndrome

	Contents	Marasmus	Kwashiorkor	Hidden Hunger	Obesity
	Enteral	**Intravenous Fluids, Enteral Tube Feeding, Complete or Partially Hydrolyzed Nutritional Therapy***		**Balanced Diet**	**Balanced Diet and Lifestyle Change**
Nutritional Therapy	**Parenteral 5–25 kcal/kg/day 20–35 ml/kg/day**	**Fluid, Start with a Calorie-Restricted Diet, Slow, Controlled Increase in Administered Calories****			
Aminoacids and other nutrients	Branched-chain amino acids	If there is muscle tissue loss			+/–
	Glutamine	If there is intestinal villous atrophy			–/+
	Carnitine 50–100 mg/kg/day	For primary or secondary deficiency			
	Ubiqinon (CoQ10) 30–200 mg/day	For primary or secondary deficiency			
	Omega 3 1,000 mg fish oil	Hypertriglyceridemia, fatty liver, delayed myelination		Hypertriglyceridemia, fatty liver	
	medium-chain triglycerides	If the patient has hypertriglyceridemia, fatty liver, lymphatic system problems			
	Prebiotics	For intestinal villous atrophy			+/–
	Probiotics	Support for mucosal immunity			+/–
Antimicrobial therapy	For tuberculosis, malaria, parasites, and viruses	In high-risk regions and groups at risk			
Other	Immunoglobulins	If there is primary or secondary immunodeficiency			+/–

This table presents strategies for the treatment of malnutrition that minimize the risk of precipitating the refeeding syndrome. *Enteral nutrition at home if needed after discharge. Ready-to-use liquid foods can reduce the risk of infection and minimize the risk of errors in the preparation or administration of food supplements. **If prolonged parenteral nutrition is required, enteral nutrition should be started as soon as possible, even if at a low dose, to prevent atrophy of the intestinal villi. ***Amount in accordance with the RDA. RDA: Recommended dietary allowances, EPA: eicosapentaenoic acid, DHA: docosahexaenoic acid

TABLE 15.3
Conditions that Increase the Risk of Malnutrition, Patient Groups, Geographical Differences

Conditions		Underdeveloped Countries	Developing Countries	Immigrants in Developed Countries	Developed Countries' Residents
		Southeast Asia, Africa	South America South Asia	Mediterranean, Balkans, Northern Europe	Western Europe, America, Canada, Australia
Nutritional status	Low income	+	+	+	+/–
	Unbalanced diet	+	+	+	+
	Insufficient, dirty water	+	+/–	–	–
Social problems	Uneducated, inexperienced mother	+	+/–	+/–	+/–
	Large family	+	+	+/–	–
	Separation from parent	–	–	+	+/–
	Lack of social security	+	+	+	+/–
	Eating behavior disorders	+	+	+	+
Secondary to chronic diseases	Malabsorption	+	+	+	+
	Diseases treated with diet	+	+	+	+
	Catabolic disease (Infection, tumor)	+	+	+	+
	Neurological and cardiac diseases	+	+	+	+

This table presents strategies for the treatment of malnutrition that minimize the risk of precipitating the refeeding syndrome.

ANTHROPOMETRIC MEASUREMENTS AND BIOMARKERS IN THE DIAGNOSIS OF MALNUTRITION AND THE STRATIFICATION OF THE RISK OF REFEEDING SYNDROME

Studies on the indicators for the early diagnosis of malnutrition using assessments of food consumption have revealed that the current paradigm of assessments and follow-up is insufficient to reduce malnutrition, and no consensus or standard plan has been established (Klanjsek et al. 2019). Anthropometric measurements used to monitor nutritional status include weight-to-height ratio, body mass index (BMI), MUAC, and triceps skinfold thickness (TSFT) (Frison et al. 2016). While patients aged 6–59 months are monitored using MUAC <115 mm or weight-for-height/length less than –3 Z-scores, adults can be monitored using BMI (Padilla, Ferreyro, and Arnold 2021).

Although anthropometric measurements cannot provide an early diagnosis of malnutrition, BMI is clearly relevant in the diagnosis and treatment of malnutrition and refeeding syndrome in adults (Krutkyte et al. 2022). A lower BMI correlates with a longer period of tolerable fasting (Krutkyte et al. 2022). Chronically malnourished patients with BMI $\leq$14 kg/m^2 may develop refeeding syndrome if nutrition is provided after 15 days of fasting. However, patients with a BMI between 16 kg/m^2 and 18.5 kg/m^2 with no other major risk factors may develop refeeding syndrome on refeeding after only five days of fasting (Krutkyte et al. 2022).

Thus, in Western societies, patients considered to be at minimal risk of malnutrition may develop symptoms of refeeding syndrome after a relatively short period of reduced nutrient intake. Similar findings were observed in adults treated with hemodialysis, chemotherapy, and gastrointestinal surgery (Reber et al. 2019).

Insulin-like growth factor (IGF1) and leptin levels are biomarkers that can be used in the diagnosis of refeeding syndrome (Goyale et al. 2015). However, for a parameter to be effective as a screening tool, it must be measured routinely, be inexpensive, and be rapidly available. A worsening base deficit and increasing lactate, both detectable using a blood gas analyzer, may indicate low thiamine levels. Thus, these widely available tests can be used to screen for malnutrition and therefore indicate the risk of refeeding syndrome.

DISCHARGE AND FOLLOW-UP AT HOME

The post-discharge prognosis of patients with SAM is poor. The complete recovery rate is less than 10% (Noble et al. 2022) and the readmission rate is high. Acute gastroenteritis was the main reason for readmission in these patients. Some studies recommend post-discharge treatment with RUTF, trimethoprim-sulfamethoxazole, vitamin A, zinc, pancreatic enzymes, and probiotics (Prendergast 2021).

Additionally, recommendations to improve the nutritional status of childbearing mothers living with the same nutritional conditions to regain mental and physical health were not emphasized. No information was provided about the rate of refeeding syndrome occurrence. In a similar study, discharged patients were not followed up, and neuropsychiatric evaluation was recommended in addition to anthropometric measurements (O'Sullivan et al. 2018). This study highlighted intestinal parasites and reported that acute gastroenteritis and respiratory tract infections are common causes of readmissions. Furthermore, excess sugar added to the RUTF content increased the risk of relapse, refeeding syndrome, and obesity in survivors. The effect of RUTF, which is recommended for patients aged six months to five years in the rehabilitation phase of SAM treatment, is beneficial only for weight gain but is controversial regarding neurocognitive parameters (Schoonees et al. 2019) and does not prevent blindness due to vitamin A deficiency (Bassey et al. 2022).

Up to 37% of relapses of SAM within six months of discharge are due to early discharge, noncompliance with recommendations, and intercurrent infections (Stobaugh et al. 2019). Long-term studies have demonstrated that patients surviving an episode of SAM have an increased risk of cardiometabolic complications in old age (Grey et al. 2021). Studies on children and adolescents who have experienced famine show a greater risk of developing hypertension, type 2 diabetes mellitus, coronary artery calcification, and myocardial infarction in the future (Hidayat et al. 2020). However, findings on the development of dyslipidemia were controversial. Additionally, few patients developed fatty liver, kidney failure, and thyroid dysfunction, and increased IGF1 levels were an important stimulus in high-risk patients. Given the deleterious effects of the COVID-19 pandemic, more studies are needed to explore the effect of hunger on health. Adults over 60 years of age and children are included in the group at risk of malnutrition (Ewunie et al. 2022). Thus, protocols are also recommended for the follow-up of malnourished elderly people.

ATTEMPTS TO ERADICATE MALNUTRITION

During the two World Wars and the Great Depression, malnutrition was a serious problem in the United States. During World War I, nutritional support was provided at school summer camps. After World War II, support was provided to children in schools in the form of ready-made breakfast cereals. However, only the most economical ready-to-eat foods were provided. The micronutrient content of ready-to-eat foods is generally poor. The food products provided to children during this period are responsible for the twin pandemics of obesity and hidden hunger that plague our time.

Nutritional research on the daily requirements for macronutrients and micronutrients was conducted to correct these problems. Recommended Dietary Allowance rules and tables were developed. The amount of salt and cholesterol in meals was reduced to prevent obesity. In 1946, milk, legumes, and eggs were added to school lunch menus. Fresh fruit, vegetables, and meat were included after 1994. From 2004, lunch meals were supplemented with seafood and probiotic yogurt. After 2010, trans fat was removed from the diet, and fruits and vegetables were included in separate meals (Hopkins and Gunther 2015). These changes were evaluated in terms of calorie, protein, fat, iron, sodium, and vitamins A and C content, but levels of thiamine and phosphorus (indicators of refeeding syndrome) were not measured.

In LIFDCs, the WHO recommended that the ready-to-eat cereals given to American children 100 years ago be called RUTF. While RUSF and legumes were added to the eggs given to American children in 1946, saturated fats and trans fats continue to be included. However, it is not given to school-age children as in the USA, but to children between six months and five years old.

Many meta-analyses have shown that the implementation of the recommendations the WHO made over two decades ago falls far short of their targets (Lenters et al. 2013; Scott et al. 2020). In fact, the problem has become global, affecting all age groups with obesity and hidden hunger symptoms exacerbated after the pandemic (Lazzerini et al. 2013; Ghodsi et al. 2021). Children under the age of five years who experience serious nutritional deficits may lose weight and have stunted growth, resulting in short stature (Kirolos et al. 2022). The damage to these children's IQ and neuropsychiatric function may be permanent. Moreover, survivors are prone to developing chronic diseases related to the cardiovascular and endocrine systems in later life (Grey et al. 2021).

Due to changing socioeconomic factors worldwide, an increasing number of regions are considered to be at high risk of malnutrition. Poor economic conditions in Eastern Mediterranean countries (Ghodsi et al. 2021) and nutritional imbalance in Europe (Kaganov et al. 2015) have increased hidden hunger and micronutrient deficiencies. The number of obese adolescents at risk of refeeding syndrome increased during the COVID-19 pandemic (Fedele et al. 2021).

Although researchers emphasize that obesity and nutritional imbalance are real issues, it is necessary to draw attention to another important aspect of malnutrition. Patients with eating disorders and phosphate deficiency are inevitably at increased risk of refeeding syndrome.

The WHO is developing and using online tools to track the current status of the fight against malnutrition (Global Database on the Implementation of Nutrition Action (GINA) 2019). Margolies et al. examined the concept of "not how much we reach the food, but how much the food we reach provides appropriate support." This included factors such as access to high nutritional quality and healthy foods for pregnant women, mothers, and babies and eradicating malnutrition through enabling countries and individuals to achieve socioeconomic self-sufficiency (Margolies et al.

2022). Although efforts to find solutions on an international scale for the follow-up and treatment of malnutrition have been made for many years, refeeding syndrome remains understudied. The early diagnosis approach in malnutrition risk groups is summarized in Figure 15.2.

Historically, pediatricians and intensivists have conducted the most research on malnutrition, made the best recommendations for its management, and had the greatest interest in the diagnosis and treatment of refeeding syndrome (Preiser et al. 2021). However, endocrinologists, neurologists, gastroenterologists, oncologists (Preiser et al. 2015), and surgeons (Nasir et al. 2018) are increasingly joining the search for remedies for malnutrition and refeeding syndrome. As the demographics

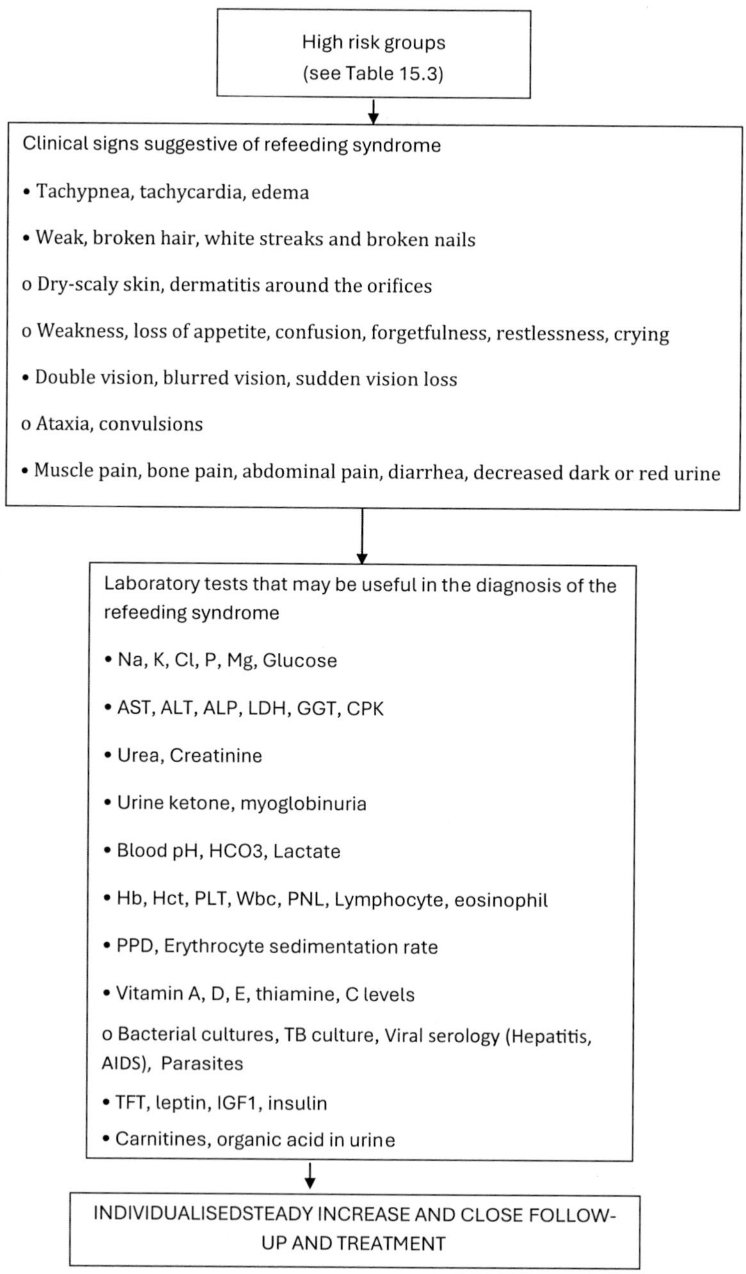

FIGURE 15.2 An evidence-based approach to the prevention and early diagnosis of refeeding syndrome in high-risk groups. This figure illustrates select strategies to restrict the development of the refeeding syndrome.

of the affected patients are explored, the economic dimension of the problem has also attracted more attention (Pradelli et al. 2023). Thus, all physicians should be aware of the risks of malnutrition, particularly during prolonged hospitalization. Healthcare systems should, therefore, monitor inpatients' length of stay and, when feasible, aim for less than 30 days of hospitalization to reduce the financial burden and death due to malnutrition.

The weight and neurocognitive development of children improve when they are supported by experts trained in nutrition therapy (Zhang et al. 2021). Given the high prevalence of malnutrition, this research clearly demonstrates the need for more staff to be trained to provide nutritional therapy.

CONCLUSION: RECOMMENDATIONS FOR MONITORING NUTRITIONAL STATUS TO PREVENT MALNUTRITION AND REFEEDING SYNDROME

The United Nations established the Sustainable Development Goals in 2015, one of which was to end malnutrition by 2030 (International Food Policy Research Institute 2016). Current data demonstrate that the international community is far from achieving this target (Sassi and Thakare 2022). Yet, the WHO does not seem to share the United Nations' goal to improve nutritional status worldwide by 2030. Awareness of refeeding syndrome, a significant cause of death, remains low in LIFDCs (see Table 15.4 for old facts and new concepts of malnutrition and refeeding syndrome).

The metabolic impact of malnutrition and refeeding syndrome differs significantly. The clinical manifestations of malnutrition are associated with secondary mitochondrial cytopathies. Thus, the

TABLE 15.4
Changes in the Understanding of Malnutrition and Refeeding Syndrome

	Obsolete, Outdated Concepts	Current Concepts
About malnutrition	Inadequate protein and energy	Inadequate and/or unbalanced nutrition
	Specific high-risk geographic locations (low-income countries)	The whole world, regardless of income level
	Protein, calorie deficiency	Deficiency of protein, calories, vitamins, minerals
	Wasting, stunting	Dermatitis, osteoporosis, reduced IQ, pallor, hypotonia
	Marasmus, kwashiorkor, marasmic kwashiorkor	Hidden hunger and obesity in addition to old terms
	Eradication is not a World Health Organization goal	Eradication is possible (nutritional support for pregnant and lactating mothers, self-sufficiency in agriculture and animal husbandry in high-risk countries)
About refeeding	The malnutrition risk group	Unbalanced diet and low nutrient reserves
	Electrolyte deficiency: phosphorous, magnesium, potassium decreased	Trace element deficiency, e.g., selenium and copper
	Vitamins: B1 deficiency	Deficiency of vitamin C, folate, fat-soluble vitamins
	Lactate high (due to thiamine low)	Carnitine low (vitamin C low)
	Tissue damage (vitamin mineral imbalance); phosphorus, magnesium, potassium, and thiamine deficiency	Deficiency of trace elements (selenium, copper, etc.), vitamin C, folate, and carnitine, alongside vitamin mineral imbalance
	Prediction markers: IGF1, leptin	Prediction markers: lactate, blood gases, liver enzymes, glucose

This table explores the changing concepts of malnutrition and the refeeding syndrome.

etiopathogenesis of SAM involves free radical damage due to the lack of trace elements and inefficient energy production due to the lack of vitamins (thiamine, ascorbic acid) and minerals (phosphorus, magnesium, potassium).

The mitochondrial dysfunction associated with malnutrition can be subtle and delayed. However, uncontrolled correction of nutritional deficiencies (i.e., inappropriate refeeding) may precipitate severe acute metabolic and physiological derangements. Thus, the metabolic derangements that occur in refeeding syndrome may be considered an acute form of SAM. Effective treatment must include therapies to support mitochondrial function and suppress free radical production. Regardless, for families to have enough food, the cohort of current and potential "breadwinners" must have the nutrients required to maintain their own health and allow physical and mental development. Healthy parents who provide sufficient food to their families will usually have well-nourished children.

MINI DICTIONARY OF TERMS

Body mass index (BMI): Weight (kg) / height (m^2). Normal range: 18.5–24.9

Mid-upper arm circumference (MUAC): Arm circumference measured at the midpoint between the shoulder joint and elbow. Used to monitor muscle tissue loss in the arm and assess the severity of malnutrition due to protein deficiency.

Moderate acute malnutrition (MAM): Weight-for-height Z-score (WHZ) between −2 and −3 SD, weight-for-height equal to 70–80% of the National Center for Health Statistics (NCHS) or WHO reference median, or MUAC of 115–125 mm.

Obesity: Condition associated with excess body weight and adverse health effects, BMI ≥30 kg/m^2.

Severe acute malnutrition (SAM): WHZ <–3 SD, WFH <70% WHO reference, MUAC<115 mm or edema.

Stunting: Height-for-age Z-score (HAZ) <–2 SD due to chronic nutritional deficiency, resulting in growth cessation and shorter stature compared to peers.

Triceps skinfold thickness (TSFT): Measurement of skinfold thickness over the triceps muscle, used to predict body nutrient reserves by monitoring changes in subcutaneous adipose tissue.

Undernutrition: BMI <18.5 kg/m^2 or WHZ <–2 SD.

Wasting: Weak body appearance with WHZ<–2 SD, developing as an acute symptom of essential nutrient insufficiency.

SUMMARY POINTS

- Malnutrition results from inadequate access to high-quality food.
- Refeeding syndrome is a metabolic response to the provision of food in individuals with inadequate body nutrient reserves.
- The final stage of refeeding syndrome is mitochondrial failure. Its treatment should include metabolic resuscitation.
- In high-risk societies, simple precautions can prevent malnutrition, and cheap, easily accessible dietary treatments can be provided before nutritional reserves reach levels that could result in refeeding syndrome.
- Eradicating malnutrition requires a social structure with self-sufficient income and access to adequate amounts of high-quality food.

REFERENCES

Alao, R., Nur, H., Fivian, E., Shankar, B., Kadiyala, S., & Harris-Fry, H. (2021). Economic inequality in malnutrition: A global systematic review and meta-analysis. *BMJ Global Health*, 6(12), e006906.

Bassey, C., Crooks, H., Paterson, K., Ball, R., Howell, K., Humphries-Cuff, I., Gaffigan, K., et al. (2022). Impact of home food production on nutritional blindness, stunting, wasting, underweight and mortality in children: A systematic review and meta-analysis of controlled trials. *Critical Reviews in Food Science and Nutrition*, 62(7), 1856–1869.

Boateng, A. A., Sriram, K., Meguid, M. M., & Crook, M. (2010). Refeeding syndrome: Treatment considerations based on collective analysis of literature case reports. *Nutrition* (Burbank, Los Angeles County, Calif.), 26(2), 156–167.

Clayton P. T. (1998). Disorders of cholesterol biosynthesis. *Archives of Disease in Childhood*, 78(2), 185–189.

Das, J. K., Salam, R. A., Saeed, M., Kazmi, F. A., & Bhutta, Z. A. (2020). Effectiveness of interventions for managing acute malnutrition in children under five years of age in low-income and middle-income countries: A systematic review and meta-analysis. *Nutrients*, 12(1), 116.

De Silva, A., & Nightingale, J. M. D. (2019). Refeeding syndrome: Physiological background and practical management. *Frontline Gastroenterology*, 11(5), 404–409.

De Waele, E., Demol, J., Verhelst, C., Geers, J., Loodts, I., & Jonckheer, J. (2021). Iatrogenic malnutrition is more prevalent in critically ill covid-19 patients than in non-covid-19 patients, especially when discharged to the ward. *Clinical Nutrition Espen*, 46, S643.

Ewunie, T. M., Hareru, H. E., Dejene, T. M., & Abate, S. M. (2022). Malnutrition among the aged population in Africa: A systematic review, meta-analysis, and meta-regression of studies over the past 20 years. *PLoS One*, 17(12), e0278904.

FAO, IFAD, UNICEF, WFP and WHO. 2019. *The state of food security and nutrition in the world 2019: Safeguarding against economic slowdowns and downturns*. Rome: FAO.

Fedele, D., De Francesco, A., Riso, S., & Collo, A. (2021). Obesity, malnutrition, and trace element deficiency in the coronavirus disease (COVID-19) pandemic: An overview. Nutrition (Burbank, Los Angeles County, Calif.), 81, 111016.

Frison, S., Kerac, M., Checchi, F. & Prudhon, C. (2016). Anthropometric indices and measures to assess change in the nutritional status of a population: A systematic literature review. *BMC Nutrition*, 2, 76.

Furusawa, H., Sato, Y., Tanaka, Y., Inai, Y., Amano, A., Iwama, M., Kondo, Y., et al. (2008). Vitamin C is not essential for carnitine biosynthesis in vivo: verification in vitamin C-depleted senescence marker protein-30/gluconolactonase knockout mice. *Biological & Pharmaceutical Bulletin*, 31(9), 1673–1679.

Gera, T., Pena-Rosas, J. P., Boy-Mena, E., & Sachdev, H. S. (2017). Lipid based nutrient supplements (LNS) for treatment of children (6 months to 59 months) with moderate acute malnutrition (MAM): A systematic review. *PLoS One*, 12(9), e0182096.

Ghodsi, D., Omidvar, N., Nikooyeh, B., Roustaee, R., Shakibazadeh, E., & Al-Jawaldeh, A. (2021). Effectiveness of community nutrition-specific interventions on improving malnutrition of children under 5 years of age in the Eastern mediterranean region: A systematic review and meta-analysis. *International Journal of Environmental Research and Public Health*, 18(15), 7844.

Global database on the Implementation of Nutrition Action (GINA). (2019). *Results of a user survey*. Geneva: World Health Organization. Licence: CC BY-NC-SA 3.0 IGO. Cataloguing-in-Publication (CIP) data. CIP data are available at http://apps.who.int/iris.

Goyale, A., Ashley, S. L., Taylor, D. R., Elnenaei, M. O., Alaghband-Zadeh, J., Sherwood, R. A., le Roux, C. W., et al. (2015). Predicting refeeding hypophosphataemia: insulin growth factor 1 (IGF-1) as a diagnostic biochemical marker for clinical practice. *Annals of Clinical Biochemistry*, 52(Pt 1), 82–87.

Grey, K., Gonzales, G. B., Abera, M., Lelijveld, N., Thompson, D., Berhane, M., Abdissa, A., et al. (2021). Severe malnutrition or famine exposure in childhood and cardiometabolic non-communicable disease later in life: A systematic review. *BMJ Global Health*, 6(3), e003161.

Hidayat, K., Du, X., Shi, B. M., & Qin, L. Q. (2020). Foetal and childhood exposure to famine and the risks of cardiometabolic conditions in adulthood: A systematic review and meta-analysis of observational studies. *Obesity Reviews: An Official Journal of the International Association for the Study of Obesity*, 21(5), e12981. https://doi.org/10.1111/obr.12981

Hiffler, L., Rakotoambinina, B., Lafferty, N., & Martinez Garcia, D. (2016). Thiamine deficiency in tropical pediatrics: New insights into a neglected but vital metabolic challenge. *Frontiers in Nutrition*, 3, 16.

Hopkins, L. C., & Gunther, C. (2015). A historical review of changes in nutrition standards of USDA child meal programs relative to research findings on the nutritional adequacy of program meals and the diet and nutritional health of participants: Implications for future research and the summer food service program. *Nutrients*, 7(12), 10145–10167.

Ijaz, S., Jackson, J., Thorley, H., Porter, K., Fleming, C., Richards, A., Bonner, A., et al. (2017). Nutritional deficiencies in homeless persons with problematic drinking: A systematic review. *International Journal for Equity in Health*, 16(1), 71.

International Food Policy Research Institute. 2016. *Global nutrition report 2016: From promise to impact: Ending malnutrition by 2030.* Washington, DC: International Food Policy Research Institute.

Jones, K. D., & Berkley, J. A. (2014). Severe acute malnutrition and infection. *Paediatrics and International Child Health*, 34(Suppl 1), S1–S29.

Kaganov, B., Caroli, M., Mazur, A., Singhal, A., & Vania, A. (2015). Suboptimal micronutrient intake among children in Europe. *Nutrients*, 7(5), 3524–3535.

Karunaratne, R., Sturgeon, J. P., Patel, R., & Prendergast, A. J. (2020). Predictors of inpatient mortality among children hospitalized for severe acute malnutrition: A systematic review and meta-analysis. *The American Journal of Clinical Nutrition*, 112(4), 1069–1079.

Kirolos, A., Goyheneix, M., Kalmus Eliasz, M., Chisala, M., Lissauer, S., Gladstone, M., & Kerac, M. (2022). Neurodevelopmental, cognitive, behavioural and mental health impairments following childhood malnutrition: A systematic review. *BMJ Global Health*, 7(7), e009330.

Klanjsek, P., Pajnkihar, M., Marcun Varda, N., & PovalejBrzan, P. (2019). Screening and assessment tools for early detection of malnutrition in hospitalised children: A systematic review of validation studies. *BMJ Open*, 9(5), e025444.

Krutkyte, G., Wenk, L., Odermatt, J., Schuetz, P., Stanga, Z., & Friedli, N. (2022). Refeeding syndrome: A critical reality in patients with chronic disease. *Nutrients*, 14(14), 2859.

Lazzerini, M., Rubert, L., & Pani, P. (2013). Specially formulated foods for treating children with moderate acute malnutrition in low- and middle-income countries. *The Cochrane Database of Systematic Reviews*, 6, CD009584.

Lenters, L. M., Wazny, K., Webb, P., Ahmed, T., & Bhutta, Z. A. (2013). Treatment of severe and moderate acute malnutrition in low- and middle-income settings: A systematic review, meta-analysis and Delphi process. *BMC Public Health*, 13(Suppl 3), S23.

Lisco, G., De Tullio, A., Triggiani, D., Zupo, R., Giagulli, V. A., De Pergola, G., Piazzolla, G., et al. (2023). Iodine deficiency and iodine prophylaxis: An overview and update. *Nutrients*, 15(4), 1004.

Lobo, D. N., Gianotti, L., Adiamah, A., Barazzoni, R., Deutz, N. E. P., Dhatariya, K., Greenhaff, P. L., et al. (2020). Perioperative nutrition: Recommendations from the ESPEN expert group. *Clinical Nutrition* (Edinburgh, Scotland), 39(11), 3211–3227.

Manary, M. J., Heikens, G. T., & Golden, M. (2009). Kwashiorkor: More hypothesis testing is needed to understand the aetiology of oedema. *Malawi Medical Journal: The Journal of Medical Association of Malawi*, 21(3), 106–107.

Margolies, A., Kemp, C. G., Choo, E. M., Levin, C., Olney, D., Kumar, N., Go, A., et al. (2022). Nutrition-sensitive agriculture programs increase dietary diversity in children under 5 years: A review and meta-analysis. *Journal of Global Health*, 12, 08001.

Matsangos, M., Ziaka, L., Exadaktylos, A. K., Klukowska-Rötzler, J., & Ziaka, M. (2022). Health status of afghan refugees in Europe: Policy and practice implications for an optimised healthcare. *International Journal of Environmental Research and Public Health*, 19(15), 9157.

Moate, T., Rabie, T., Minnie, C., & Mäenpää, A. (2022). Comorbidities of child malnutrition in low- and medium-income countries: A systematic review. *Journal of Pediatric Gastroenterology and Nutrition*, 75(4), 400–410.

Montoro-Huguet, M. A., Belloc, B., & Domínguez-Cajal, M. (2021). Small and large intestine (I): Malabsorption of nutrients. *Nutrients*, 13(4), 1254.

Nasir, M., Zaman, B. S., & Kaleem, A. (2018). What a trainee surgeon should know about refeeding syndrome: A literature review. *Cureus*, 10(3), e2388.

Noble, C. C. A., Sturgeon, J. P., Bwakura-Dangarembizi, M., Kelly, P., Amadi, B., Prasadajudio, M., Devaera, Y., et al. (2022). Disease-related malnutrition in pediatric patients with chronic disease: A developing country perspective. *Current Developments in Nutrition*, 7(4), 100021.

O'Sullivan, N. P., Lelijveld, N., Rutishauser-Perera, A., Kerac, M., & James, P. (2018). Follow-up between 6 and 24 months after discharge from treatment for severe acute malnutrition in children aged 6-59 months: A systematic review. *PLoS One*, 13(8), e0202053.

Odigwe, C. C., Smedslund, G., Ejemot-Nwadiaro, R. I., Anyanechi, C. C., & Krawinkel, M. B. (2010). Supplementary vitamin E, selenium, cysteine and riboflavin for preventing kwashiorkor in preschool children in developing countries. *The Cochrane Database of Systematic Reviews*, 2010(4), CD008147.

Oshikoya, K. A., Sammons, H. M., & Choonara, I. (2010). A systematic review of pharmacokinetics studies in children with protein-energy malnutrition. *European Journal of Clinical Pharmacology*, 66(10), 1025–1035.

Padilla, C. J., Ferreyro, F. A., & Arnold, W. D. (2021). Anthropometry as a readily accessible health assessment of older adults. *Experimental Gerontology*, 153, 111464.

Pham, T. P., Alou, M. T., Golden, M. H., Million, M., & Raoult, D. (2021). Difference between kwashiorkor and marasmus: Comparative meta-analysis of pathogenic characteristics and implications for treatment. *Microbial Pathogenesis*, 150, 104702.

Ponzo, V., Pellegrini, M., Cioffi, I., Scaglione, L., & Bo, S. (2021). The refeeding syndrome: A neglected but potentially serious condition for inpatients: A narrative review. *Internal and Emergency Medicine*, 16(1), 49–60.

Pradelli, L., Zaniolo, O., Sanfilippo, A., Lezo, A., Riso, S., & Zanetti, M. (2023). Prevalence and economic cost of malnutrition in Italy: A systematic review and metanalysis from the Italian Society of Artificial Nutrition and Metabolism (SINPE). *Nutrition* (Burbank, Los Angeles County, Calif.), 108, 111943.

Preiser, J. C., Arabi, Y. M., Berger, M. M., Casaer, M., McClave, S., Montejo-González, J. C., Peake, S., et al. (2021). A guide to enteral nutrition in intensive care units: 10 expert tips for the daily practice. *Critical Care* (London, England), 25(1), 424.

Preiser, J. C., van Zanten, A. R., Berger, M. M., Biolo, G., Casaer, M. P., Doig, G. S., Griffiths, R. D., et al. (2015). Metabolic and nutritional support of critically ill patients: Consensus and controversies. *Critical Care* (London, England), 19(1), 35.

Prendergast, A. J. (2021). Postdischarge interventions for children hospitalized with severe acute malnutrition: A systematic review and meta-analysis. *The American Journal of Clinical Nutrition*, 113(3), 574–585.

Reber, E., Friedli, N., Vasiloglou, M. F., Schuetz, P., & Stanga, Z. (2019). Management of refeeding syndrome in medical inpatients. *Journal of Clinical Medicine*, 8(12), 2202.

Roman Viñas, B., Ribas Barba, L., Ngo, J., Gurinovic, M., Novakovic, R., Cavelaars, A., de Groot, L. C., et al. (2011). Projected prevalence of inadequate nutrient intakes in Europe. *Annals of Nutrition & Metabolism*, 59(2–4), 84–95.

Sassi, M., & Thakare, H. (2022). Conflict and child malnutrition: A systematic review of the emerging quantitative literature. *Current Nutrition Reports*, 11(1), 1–8.

Schoonees, A., Lombard, M. J., Musekiwa, A., Nel, E., & Volmink, J. (2019). Ready-to-use therapeutic food (RUTF) for home-based nutritional rehabilitation of severe acute malnutrition in children from six months to five years of age. *The Cochrane Database of Systematic Reviews*, 5(5), CD009000.

Scott, N., Delport, D., Hainsworth, S., Pearson, R., Morgan, C., Huang, S., Akuoku, J. K., et al. (2020). Ending malnutrition in all its forms requires scaling up proven nutrition interventions and much more: A 129-country analysis. *BMC Medicine*, 18(1), 356.

Soliman G. A. (2018). Dietary cholesterol and the lack of evidence in cardiovascular disease. *Nutrients*, 10(6), 780.

Stanga, Z., Brunner, A., Leuenberger, M., Grimble, R. F., Shenkin, A., Allison, S. P., & Lobo, D. N. (2008). Nutrition in clinical practice-the refeeding syndrome: Illustrative cases and guidelines for prevention and treatment. *European Journal of Clinical Nutrition*, 62(6), 687–694.

Stobaugh, H. C., Mayberry, A., McGrath, M., Bahwere, P., Zagre, N. M., Manary, M. J., Black, R., et al. (2019). Relapse after severe acute malnutrition: A systematic literature review and secondary data analysis. *Maternal & Child Nutrition*, 15(2), e12702.

United Nations Children's Fund (UNICEF); World Health Organization; International Bank for Reconstruction and Development/The World Bank. (2021). *Levels and trends in child malnutrition: Key findings of the 2021 edition of the joint child malnutrition estimates*. Geneva, Switzerland: World Health Organization.

U.S. Department of Agriculture and U.S. Department of Health and Human Services. (2020, December). *Dietary guidelines for Americans, 2020–2025* (9th ed.). Available at DietaryGuidelines.gov. Accessed 08/10/2023

Wali, N., Agho, K., & Renzaho, A. M. (2022). Hidden hunger and child undernutrition in South Asia: A meta-ethnographic systematic review. *Asia Pacific Journal of Clinical Nutrition*, 31(4), 713–739.

Williams, P. C. M., & Berkley, J. A. (2018). Guidelines for the treatment of severe acute malnutrition: A systematic review of the evidence for antimicrobial therapy. *Paediatrics and International Child Health*, 38(Suppl 1), S32–S49.

World Health Organization Regional office of Africa. (2017). *Nutrition in the WHO African region*. Available at https://www.afro.who.int/sites/default/files/2017-11/Nutrition in the WHO African Region 2017.pdf Accessed 07/10/23

World Health Organization. (2019). *Nutrition Landscape Information System (NLiS) country profile indicators: Interpretation guide* (2nd ed.). ISBN 978-92-4-151695-2. Available at https://iris.who.int/bitstream/handle/10665/332223/9789241516952-eng.pdf?sequence=1 Accessed 07/10/23

World Health Organization (WHO). (2013). Guideline: Updates on the Management of Severe Acute Malnutrition in Infants and Children. World Health Organization. Available at https://www.who.int/publications/i/item/9789241506328

Zhang, Z., Li, F., Hannon, B. A., Hustead, D. S., Aw, M. M., Liu, Z., Chuah, K. A., Low, Y. L., et al (2021). Effect of oral nutritional supplementation on growth in children with undernutrition: A systematic review and meta-analysis. *Nutrients*, 13(9), 3036.

16 Postoperative Protein Malnutrition

The Example of Bariatric Surgery

Laís Lima de Castro Abreu, Jussilene Alves Amorim, Neusa Camila Cavalcante Andrade Oliveira, and Maria do Carmo de Carvalho e Martins

ABBREVIATIONS

AA	Amino acids
RYGB	Roux-en-Y gastric bypass
CKK	Cholecystokinin
GLP1	Glucagon-like peptide 1
GIP	Glucose-dependent insulinotropic polypeptide
RDA	Recommended dietary allowance
SBCBM	Brazilian Society of Bariatric and Metabolic Surgery
PYY	Peptide YY

INTRODUCTION

Bariatric surgery is a treatment indicated for patients with morbid obesity who have unsatisfactory weight loss and poor adherence to the usual therapeutic measures, which include lifestyle changes related to diet and physical exercise, as well as pharmacological and/or psychological treatments. In these cases, surgical treatments can provide greater and faster body weight loss (Arterburn et al. 2020).

As it is a major surgical procedure and is usually performed on patients who are at increased risk of complications, the indication should be made considering the risk-benefit ratio of the intervention (Lopes et al. 2022; Andrade et al. 2021). The surgical procedure provides a structural readaptation of the gastrointestinal tract which, depending on the surgical technique used, can influence the digestion and absorption of nutrients and promote metabolic changes (Marathe et al. 2013; Barros et al. 2015).

One of the main side effects following bariatric surgery is related to nutrient deficiency, as different types of surgery affect nutrient absorption (Gasmi et al. 2022). Therefore, a clear understanding of the consequences of bariatric surgical procedures on the digestion and absorption of proteins is essential to clarify the postoperative effects of these surgeries. This chapter seeks to describe the mechanisms involved in the onset of protein malnutrition in the postoperative period of bariatric surgery.

DOI: 10.1201/b23328-16

TYPES OF BARIATRIC SURGERY AND REPERCUSSIONS ON GASTROINTESTINAL FUNCTION

Bariatric procedures are selected based on the most appropriate technique for each patient. They can be performed via laparoscopy or endoscopy and, depending on their function, are categorized into three main types: restrictive, malabsorptive and combined/mixed (restrictive and malabsorptive). Bariatric surgeries, regardless of whether they are restrictive or malabsorptive, cause weight loss primarily due to malabsorption of macronutrients (up to 25% protein) and micronutrients. In addition, the reduction in gastrointestinal transit time after surgery results in secondary malabsorption of these nutrients. The following are aspects of each type of surgical procedure (SBCBM 2017).

Gastric bypass or gastroplasty with Roux-en-Y gastric bypass (RYGB) (Figure 16.1A) is a surgical procedure in which part of the stomach is sequentially stapled (gastroplasty), the gastric pouch is anastomosed to the intestine (gastroenteroanastomosis) and, finally, the distal intestinal portions are connected (enteroanastomosis). This method reduces the space for the passage of the food bolus in the stomach and, associated with this, creates a detour in the intestine that increases the production of peptide YY (PYY), a hormone associated with satiety. Consequently, RYGB is classified as a mixed technique (SBCBM 2017).

The gastroplasty component of RYGB results in decreased secretion of hydrochloric acid, leading to elevated gastric pH and subsequent alterations in the pH of the entire gastrointestinal tract from the stomach onwards. Another common gastrointestinal repercussion is that the reduced gastric capacity following RYGB causes the entry of the food bolus into the stomach to disproportionately stimulate the mechanoreceptors in the gastric mucosa. This results in faster gastric emptying into the small intestine in order to protect the stomach from excessive pressure, which impairs digestion and absorption of nutrients (Sinclair, Brennan, and Le Roux 2018).

Another surgical technique that can be adopted is vertical gastrectomy (Figure 16.1B), also known as sleeve gastrectomy, a restrictive surgical procedure involving gastric stapling (gastroplasty) and removal of the great gastric curvature (gastrectomy), thereby reducing the capacity of the organ. In this procedure, the stomach is transformed into a kind of tube, with a reduced capacity for accommodating and passing the food bolus (SBCBM 2017).

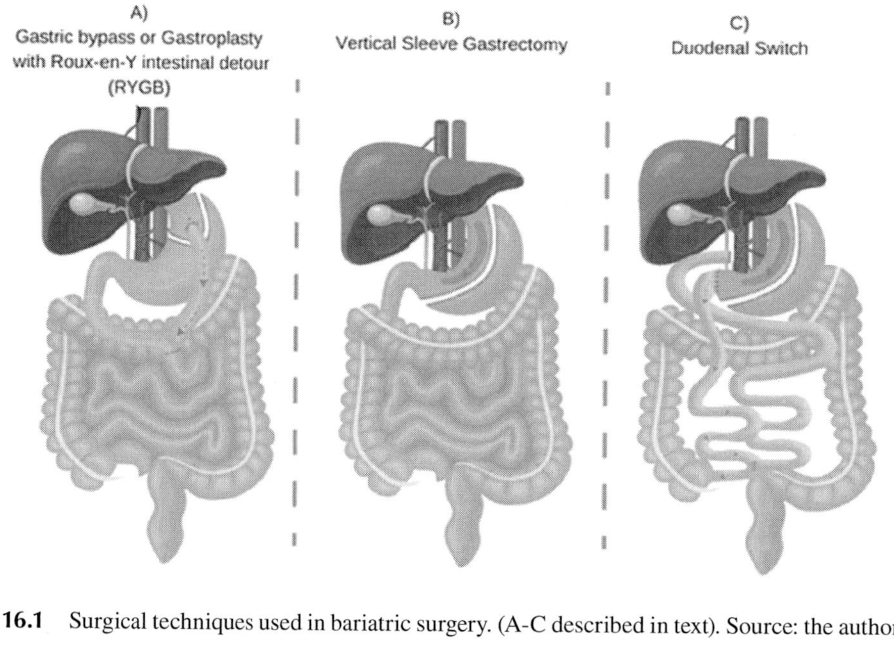

FIGURE 16.1 Surgical techniques used in bariatric surgery. (A-C described in text). Source: the authors, 2023.

After sleeve gastrectomy, it is common for food intake to decrease due to the reduced gastric capacity. In addition, there may be a reduction in hydrochloric acid secretion and intrinsic factor, as occurs in the RYGB technique. There is also an increase in the speed of gastric emptying and intestinal transit. The main consequence of these repercussions can be vitamin and mineral deficiencies (Rufino, Oliveira, and Rasera 2023).

Duodenal switch gastroplasty (Figure 16.1C), a less commonly used restrictive procedure, involves sequential acts of gastric stapling (gastroplasty) followed by removal of the great gastric curvature (gastrectomy), combined with one or more anastomoses in the intestine. In addition to transforming the stomach into a kind of tube, the surgeon performs an intestinal bypass. This bypass, when exposed to luminal contents, enables signaling to the central nervous system via humoral and neural mechanisms, promoting satiety (SBCBM 2017).

The adjustable gastric band procedure is another restrictive technique consisting of installing an inflatable and adjustable silicone ring around the stomach, enabling adjustment of the gastric capacity and emptying. The prosthetic ring, however, can lead to complications due to its presence in the abdominal cavity (SBCBM 2017).

Given that nutrient absorption primarily occurs in the small intestine, some nutritional deficiencies may arise after bariatric surgery, depending on the technique used and the individual's eating behavior. These deficiencies require prevention or compensation by intervening with a new nutritional approach (Costa et al. 2021).

REPERCUSSIONS OF BARIATRIC SURGERY ON NUTRITIONAL STATUS

Of the nutritional deficiencies associated with bariatric surgery, in addition to protein malnutrition, deficiencies in minerals such as calcium, iron and zinc, as well as by vitamins B12, B1, D and folic acid, are frequently observed. These deficiencies can lead to several clinical conditions, including megaloblastic anemia, pernicious anemia, iron deficiency anemia, hypocalcemia, secondary hyperparathyroidism, beriberi, Wernicke's encephalopathy and Korsakoff's syndrome (Ferraz et al. 2018; Leiro and Melendez-Araújo 2014; Torezan 2013; Rocha 2012).

Many factors are involved in causing these deficiencies, and their incidence varies depending on the specific procedure performed, so not all individuals experience the same vitamin and mineral deficits. This variability necessitates careful nutritional assessment and monitoring, especially when recommending supplements (Bordalo et al. 2011; Rocha 2012). Furthermore, even before any surgical intervention, obesity may be associated with subclinical nutritional deficiencies. These deficiencies can be aggravated by the anatomical and functional changes to the gastrointestinal tract following surgery, which will determine the degree of deficiency according to the absorption site (Bordalo et al. 2011). These absorption sites are illustrated in Figure 16.2.

A study conducted by Leiro and Melendez-Araújo (2014) at a nutrition outpatient clinic examined 36 women who underwent RYGB more than one year prior. The researchers observed that despite consuming vitamin B12 according to the recommended dietary allowance (RDA), these patients exhibited impaired absorption of the nutrient and consequently developed pernicious anemia.

Individuals with obesity have a high tendency to develop vitamin D deficiency due to several factors, including sedentary lifestyles, limited sun exposure, inadequate dietary intake of vitamin D-rich foods and uptake of significant amounts of this fat-soluble vitamin in excess adipose tissue present. Low postoperative vitamin D concentration can worsen after the patient has undergone bariatric surgery (Jesus 2016).

Vitamin D deficiency after bariatric surgery can result from lipid malabsorption, particularly following RYGB, which impairs the absorption of this micronutrient (Bordalo et al. 2011). Additionally, restrictive procedures that bypass the duodenum and proximal jejunum, where vitamin D absorption occurs, contribute to this deficiency (Leiro and Melendez-Araújo 2014).

Iron deficiency is a common complication in patients undergoing RYGB, affecting up to 47% of patients within two years post-surgery (Torezan 2013). Iron deficiency occurs due to reduced food

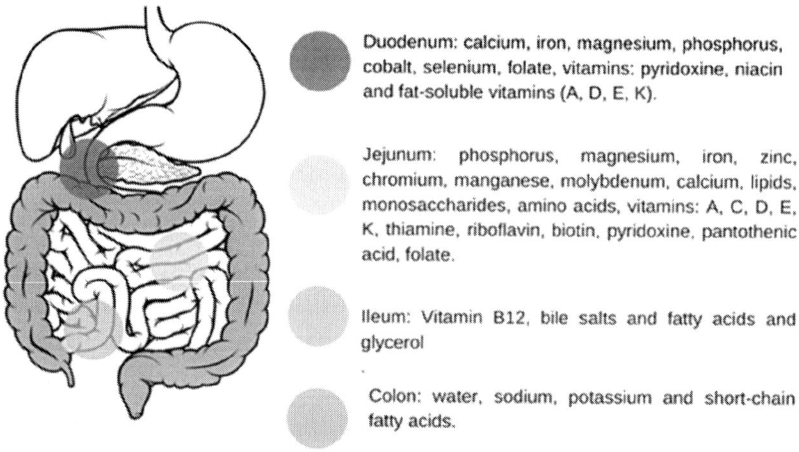

FIGURE 16.2 Site of nutrient absorption in the intestine. Source: the authors, 2023.

intake, reduced gastric secretion of hydrochloric acid and anatomical changes (exclusion) of the duodenum and proximal jejunum, where this mineral is absorbed.

Preoperatively, obese patients often have lower serum zinc levels compared to the general population. In the postoperative period, the prevalence of zinc deficiency can increase due to the metabolic changes that occur as a result of surgery. Zinc deficiency can negatively affect pancreatic secretion and the action of growth hormone (GH), further impairing the metabolism of patients undergoing bariatric surgery (Ferraz et al. 2018).

Additionally, studies have shown that dumping syndrome is one of the complications that most affect individuals after bariatric surgery, particularly following RYGB, with an incidence of up to 75.9%. This syndrome results from rapid gastric emptying, excessive food ingestions relative to the post-surgical gastric capacity or consumption of foods rich in simple carbohydrates with high glycemic indices, rich in fat or poorly tolerated by the patient (Rolim et al. 2018).

It is worth noting that the complications of bariatric surgery are most frequently observed in patients who do not follow the recommended postoperative lifestyle and dietary changes. This underscores the need for continuous postoperative monitoring, since the chronicity of the disease is associated with the side effects of surgery if they are not controlled (Pellegrini et al. 2021).

BARIATRIC SURGERY AND REPERCUSSIONS ON PROTEIN DIGESTION AND ABSORPTION

The processes of digestion and absorption are essential for maintaining the health and proper functioning of the body, influenced by factors such as the quality of the diet, the state of general health and the presence of medical conditions affecting the digestive system. After bariatric surgery, particularly RYGB, metabolic changes occur in the digestion and absorption of proteins. RYGB as shown in Figure 16.3 involves extensive rearrangement of the upper gastrointestinal tract, creating a gastric pouch that drains directly into the lower jejunum, excluding the remainder of the stomach, duodenum, and upper jejunum from dietary exposure and leading to alterations in the gastropancreatic enzyme mix, which may impact protein digestion and amino acid absorption (Moustarah and Hould 2015; Bojsen-Moller et al. 2015).

Protein digestion begins in the stomach, where proteins and polypeptides are denatured by the action of hydrochloric acid (HCl) and hydrolyzed by pepsin. However, digestion in the stomach accounts for only 10 to 20% of total protein digestion. The majority occurs in the lumen of the

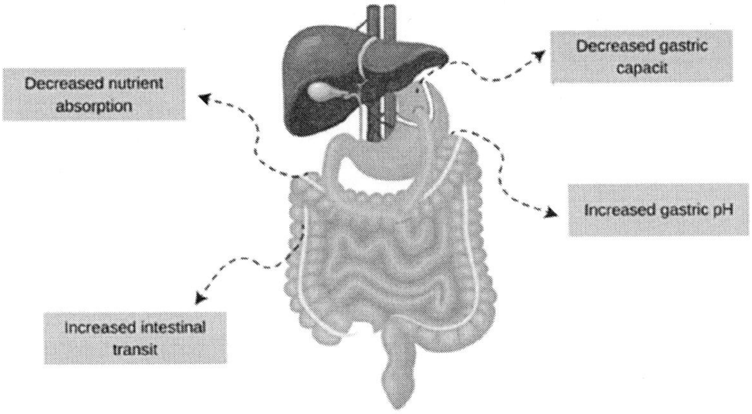

FIGURE 16.3 Most common bariatric surgery method (RYGB) and its repercussions. Source: the authors, 2023.

duodenum and jejunum under the influence of pancreatic juice, with final processing in the terminal ileum. In the small intestine, enteropeptidase, at neutral pH, activates trypsinogen to trypsin, which then promotes the activation of the other pro-peptidases in the pancreatic juice. The luminal hydrolysis of proteins and polypeptides then takes place, producing free amino acids (AA) and small peptides (2–6 AA). These products are further hydrolyzed by brush border peptidases into AA and di- and tripeptides, which are absorbed mainly in the proximal jejunum (Frenhani and Burini 1999; Aires 2019).

According to Steenackers, Gesquiere and Matthys (2018), both RYGB and sleeve gastrectomy can influence nutrient ingestion, digestion and absorption through reduced gastric capacity, reduced secretion of hydrochloric acid and digestive enzymes and rearrangement of the architecture of the small intestine.

A study by Bojsen-Moller et al. (2015) involving nine RYGB patients reported that eating a caseinate meal containing only essential amino acids as the protein source, compared to the presurgery diet, resulted in increased caseinate digestion and amino acid absorption. This led to a faster and more transient postprandial rise in plasma amino acid concentration, demonstrating that accelerated protein digestion three months post-RYGB, attributed to extensive gastrointestinal changes.

Svane et al. (2019) conducted a study with 36 individuals: 12 underwent duodenal switch gastroplasty, 12 underwent RYGB and 12 underwent no procedure, serving as controls. Plasma amino acid concentrations and the kinetics of proteins throughout the body (decomposition, synthesis, oxidation and net balance) were assessed using the total rate of appearance of metabolites. Results indicated accelerated digestion, absorption and synthesis of proteins throughout the body post-RYGB. In contrast, protein metabolism after duodenal switch gastroplasty was similar to that of control subjects. Both procedures altered the secretion of gastrointestinal hormones, with the secretion of glucagon-like peptide-1 (GLP-1), PYY and cholecystokinin (CCK) increasing after RYGB. In addition, a reduction in ghrelin secretion was observed and glucose-dependent insulinotropic polypeptide (GIP) responses were more prolonged after duodenal switch gastroplasty (Svane et al. 2019).

It is worth noting that bypass procedures RYGB and duodenal switch gastroplasty greatly reduce the surface area of the intestinal mucosa available for absorption by shortening the digestive tract. It is unclear whether absorption differs between individuals undergoing proximal small bowel bypass (gastric bypass) and those with distal small bowel bypass (jejunoileal bypass). There is not enough evidence to elucidate the exact dynamics of changes in protein digestion and absorption in these situations, so further research is needed (Padwal, Brocks, and Sharma 2010).

MECHANISMS AND CONSEQUENCES OF PROTEIN MALNUTRITION AFTER BARIATRIC SURGERY

Before bariatric surgery, some patients are at high risk of protein malnutrition because they have comorbidities associated with obesity that affect nutritional status, such as diabetes, gastroesophageal reflux disease, sleep apnea, hypertension, hyperlipidemia, hepatic steatosis, coronary artery disease and chronic kidney disease. It should be noted that these diseases may be diagnosed until patients are assessed for the surgical procedure (Schinkel et al. 2006).

Furthermore, excessive intake of foods rich in lipids and simple carbohydrates can lead to protein malnutrition, which is often underreported in patients with obesity. Protein malnutrition has been identified as a risk factor in gastrointestinal surgery, associated with greater morbidity and longer hospital stays. It can also hinder the healing process of the surgical wound and exacerbate comorbidities associated with obesity (Blume et al. 2012). For this reason, the combination of obesity and protein malnutrition is thought to increase surgical complications in patients undergoing bariatric surgery compared to normal weight and/or normoalbuminemic patients (Major et al. 2018).

In the immediate postoperative period, patients experience increased metabolic stress, heightened catabolism with increased protein requirements and anorexia. These conditions can trigger mechanical complications and increase susceptibility to infections. In addition, post-bariatric surgery protein malnutrition is linked to reduced protein intake resulting from decreased gastric capacity, coupled with impaired digestion due to reduced digestive enzyme secretion and decreased intestinal amino acids absorption (Llanos et al. 2015).

In this context, Parmar et al. (2020) reported that post-bariatric surgery intestinal lengths of between 150 and 250 cm increase patients' chances of developing protein malnutrition by 20%, and recommend a minimum length of 400 cm as essential to avoid its occurrence. Furthermore, clinical observations also indicate that a longer post-bariatric intestinal length is necessary to prevent metabolic disorders, since these disorders can reduce the rate of protein digestion and absorption, which seems to correlate with intestinal length (Chen et al. 2019).

Given these findings, protein malnutrition is a serious health problem requiring immediate treatment due to its numerous consequences, including muscle weakness, anemia, alopecia and edema (Llanos et al. 2015; Blume et al. 2012). According to Deitel (1998), the diagnosis of protein malnutrition requires the combination of hypoalbuminemia, anemia, edema, asthenia and alopecia. Kassir et al. (2016) emphasize that the presence of edema and a low serum albumin level (<3.5 g/dL) would be sufficient for the diagnosis of protein malnutrition.

During the immediate postoperative phase, most patients are placed on a liquid diet, as they cannot tolerate large volumes of solid food. This intolerance can be exacerbated in some cases by prolonged vomiting. These factors result in insufficient protein intake, which increases the risk of developing protein malnutrition (Mahopatra, Gangadharan, and Pitchumoni 2020).

Regarding the types of bariatric surgery, Lupoli et al. (2017) reported that 7 to 21% of patients undergoing duodenal switch gastroplasty experienced protein malnutrition as a result of poor protein digestion and absorption secondary to altered biliopancreatic function. The authors also noted that protein malnutrition occurred in 13% of patients two years post-RYGB, attributed to reduced gastric capacity and protein malabsorption. In contrast, sleeve gastrectomy patients may develop protein malnutrition due to non-adaptive eating behaviors after surgery, such as avoidance of protein food sources like red meat and dairy products.

To determine the prevalence of nutritional deficiencies in morbidly obese patients before surgery and compare them with the postoperative results of sleeve gastrectomy and RYGB, Gehrer et al. (2010) conducted a study. They reported that patients undergoing sleeve gastrectomy (group 1, n = 50) or RYGB (group 2, n = 86) had hypoalbuminemia in the late postoperative period (4% versus 8%, respectively). The authors attributed the presence of hypoalbuminemia to the small volume of meals consumed and the reduction in the production of digestive enzymes, causing impaired digestion of nutrients, including protein.

Following the same line of reasoning, Llanos et al. (2015) compared the incidence of protein malnutrition between RYGB and duodenal switch gastroplasty, using serum prealbumin as a predictor of protein malnutrition in 84 patients undergoing RYGB and 106 undergoing duodenal switch gastroplasty. Serum prealbumin levels before surgery were significantly higher in patients assigned to duodenal switch gastroplasty. However, the incidence of hypoalbuminemia was significantly higher during the first postoperative year of duodenal switch gastroplasty (16.5% versus 5.4% after RYGB). This suggests that protein malnutrition is more frequent in patients undergoing duodenal switch gastroplasty.

A possible explanation for these findings could be the longer hospitalization time of patients undergoing duodenal switch gastroplasty, which can lead to greater protein catabolism (Topart, Becoaurn, and Ritz 2012). Protein malnutrition associated with procedures such as RYGB and duodenal switch gastroplasty leads to an annual hospitalization rate of 1% per year, causing significant morbidity and undesirable postoperative outcomes (Kassir et al. 2016). Although bariatric surgery is considered an effective therapeutic procedure for patients with morbid obesity, it is important to note that it can result in nutritional deficiencies that can have serious consequences for the patient's recovery, requiring early diagnosis and treatment (Martins et al. 2015).

POSTOPERATIVE DIETARY MANAGEMENT TO PREVENT PROTEIN MALNUTRITION

Nutritional surveillance is an essential component in managing bariatric patients for the following reasons: (1) it increases patients' adherence to healthy eating habits and adequate supplementation regimes; (2) it mitigates the risk of weight regain; (3) it facilitates the detection of potential nutritional deficiencies that may develop despite medical therapy; and (4) it contributes to maintaining a good quality of life (Lupoli et al. 2017).

Individualized nutritional monitoring for bariatric patients is crucial, as it aims to assess the patient's dietary evolution in phases, including the types of diet and their consistency. Practioners should encourage long-term weight loss after bariatric surgery through realistic expectations. Furthermore, individuals undergoing postoperative follow-up require regular treatment and support from qualified health professionals. Adherence to follow-up is associated with a lower frequency of postoperative adverse events, greater loss of excess body weight and reduced occurrence of comorbidities (Dagan et al. 2017).

Therefore, postoperative food consumption should be monitored and accompanied by recommendations to consume more water, fruits and vegetables, proteins of high biological value and to limit the intake of carbohydrates and fats. Moreover, it is essential that all daily meals are eaten in small volumes and in larger portions. Patients should strictly adhere to healthy behavioral lifestyle practices that include conscious food-purchasing behaviors and regular medical and nutritional assessment (Afshan et al. 2019). Additionally, significant emphasis is placed on regular training and aerobic exercise as essential components for preserving lean mass, especially muscle mass (Lupoli et al. 2017).

According to the *Brazilian Guide to Nutrition in Bariatric and Metabolic Surgery* (2023), the first postoperative food intake should only be in liquid form with approximately 600 calories per day, progressing after the third day to a complete liquid diet with around 1000 to 1200 calories and moving on to a paste-like consistency after the 15th day, with approximately 1200 to 1500 calories and a supply of 60–80 g of protein/day. A solid or normal diet is indicated after the 22nd day post-surgery, maintaining a balanced diet in which the patient should be instructed to include foods of higher nutritional quality in their menu. At this stage, food selection should be more qualitative than quantitative, as shown in Table 16.1.

Bariatric surgery presents a subsequent risk of nutritional deficiencies. At the end of one month post-surgery, the most common nutritional alterations are anemia, with a reduction in hemoglobin concentration, as well as reductions in total proteins, iron, ferritin and serum calcium. Follow-up

TABLE 16.1
Main Nutritional Recommendations after Bariatric Surgery

Phase	Observations
First 3 days of surgery	Clear liquid diet with approximately 600 calories per day
3 to 14 days after surgery	Complete liquid diet with around 1000 to 1200 calories per day
15 to 21 days after surgery	Pasty diet with approximately 1200 to 1500 calories and 60 to 80 g of protein/day
22 days after surgery	Normal/solid diet with a maximum intake of up to 2.1 g/Kg of protein per ideal body weight

Source: the authors, 2023.

reports one year after surgery reveal the presence of low concentrations of vitamin B12, calcium and proteins. For this reason, most patients undergoing malabsorptive surgery receive multivitamin complexes, iron, calcium, vitamin D and vitamin B12 (Antoniewicz et al. 2019; Padwal, Brocks, and Sharma 2010).

The susceptibility to nutritional deficiency, especially protein deficiency, implies the need for long-term nutritional monitoring and intervention in post-bariatric surgery patients. Thus, it is essential to ensure dietary intake meets nutritional needs, which may require the use of supplements if intake is not sufficient to meet nutritional demands (Kuin et al. 2019).

Furthermore, according to the American Association of Clinical Endocrinologists, the Obesity Society and the American Society for Metabolic and Bariatric Surgery, preventing protein malnutrition requires individualized recommendations, assessed and guided by a nutritionist and considering gender, age and body weight. The average daily protein intake should be 60 to 120 g (1.1 g/Kg of ideal body weight). This recommendation can be increased to 1.5 g/Kg of ideal body weight, with a maximum intake of up to 2.1 g/Kg of ideal body weight (Mechanick et al. 2013). Additionally, patients with severe protein malnutrition should be treated with protein module supplements rich in branched-chain amino acids (Steenackers, Gesquiere, and Matthys 2018).

Finally, it is important to emphasize that the management of postoperative nutrition in bariatric surgery begins in the preoperative period with a thorough assessment of nutritional status, a strong educational program and follow-up to reinforce important principles associated with maintaining long-term weight loss. The practitioner should not only review the standard assessment components (i.e. medical comorbidities, weight history, laboratory tests and dietary intake) but also assess other issues that may affect nutritional status, including readiness for change, realistic goals, environment and general nutrition knowledge, as well as behavioral, cultural, psychosocial and economic issues (Aills et al. 2008).

FINAL CONSIDERATIONS

Bariatric surgery is an effective strategy for weight loss, but it can lead to a series of nutritional deficiencies in the postoperative period due to low food intake, impaired digestion and/or absorption, inadequate supplementation and insufficient nutritional monitoring. The severity of these deficiencies also varies depending on the type of surgical procedure performed, as each surgical technique causes different changes in food intake, digestion and nutrient absorption. Changes in nutritional status can range from alterations in plasma protein concentrations, such as albumin, to vitamin, mineral and electrolyte deficiencies. The most common deficiencies, apart from protein malnutrition, include iron, vitamin B12, calcium, vitamin D and folic acid deficiencies.

Protein malnutrition is a serious complication that can affect around 13% of patients after bariatric surgery. It can have severe consequences, such as compromising immunity, delaying the healing of surgical wounds and even contributing to the development of fistulas. Pre-existing hypoalbuminemia can be further aggravated by inadequate protein intake and lead to loss of lean body mass, reduced metabolic rate and functional damage, which can negatively impact the success of obesity treatment.

Considering the above, it is clear that bariatric surgery can negatively affect not only dietary protein intake, but also its digestion and absorption. Impaired protein intake can trigger a state of malnutrition which, in turn, leads to multiple complications, including the worsening or onset of liver damage, which can culminate in liver failure in more serious cases.

Therefore, it is crucial to emphasize the importance of the nutritionist's role and that ofthe rest of the multidisciplinary team in ensuring adequate protein intake and preventing, mitigating or addressing complications arising from surgery that affect nutritional status.

MINI DICTIONARY OF TERMS

Adjustable gastric band: Installation of an inflatable and adjustable silicone ring around the stomach, which allows adjustment of gastric capacity and emptying speed of the organ to be adjusted.

Bariatric surgery or gastroplasty: Surgical treatment indicated for morbidly obese patients with unsatisfactory weight loss.

Gastrectomy: Surgical technique in which part or all of the stomach is removed.

Gastroplasty duodenal switch: Restrictive procedure performed with gastric stapling, removal of the great gastric curvature and one or more anastomoses in the intestine.

Protein malnutrition: Protein deficiency resulting from intense catabolism or poor ingestion, digestion and/or absorption.

Roux-en-Y gastric bypass with intestinal detour: Surgical procedure involving the stapling of part of the stomach, anastomosis of the gastric pouch with the intestine and connection between the final intestinal portions.

Sleeve gastrectomy: Surgical procedure in which the stomach is stapled and the great curvature of the stomach is removed, reducing gastric capacity.

SUMMARY POINTS

- Bariatric surgery is an effective strategy for weight loss, but it can lead to a series of nutritional deficiencies in the post-surgical period.
- Protein malnutrition is a serious metabolic complication and is more prevalent in the postoperative period of malabsorptive bariatric surgery.
- Protein malnutrition is characterized by the presence of hypoalbuminemia, anemia, edema, asthenia and alopecia.
- When protein malnutrition is diagnosed, protein supplementation should be routine, and the nutritionist should encourage the intake of protein of high biological value.
- Key factors that should be monitored by a professional nutritionist to minimize the occurrence of protein malnutrition post-bariatric surgery include: loss of excess weight, assessment of body composition, the presence or absence of clinical signs associated with specific nutritional deficiencies, biochemical analyses and dietary intake.

REFERENCES

Afshan, M., L. Alsheddi, L. Alfayadh, B. Bukhari, R. Elawad, and A. A. Alfadda. 2019. Dietary and Lifestyle Factors Serve as Predictors of Successful Weight Loss Maintenance Postbariatric Surgery. *Journal of Obesity* 2019 (February): 1–6. https://doi.org/10.1155/2019/7295978.

Aills, L., J. Blankenship, C. Buffington, M. Furtado, and J. Parrott. 2008. ASMBS Allied Health Nutritional Guidelines for the Surgical Weight Loss Patient. *Surgery for Obesity and Related Diseases* 4 (5): S73–108. https://doi.org/10.1016/j.soard.2008.03.002.

Aires, Margarida de Mello. 2019. *Physiology*. 5th ed. Rio de Janeiro: Guanabara Koogan, 1376 p.

Andrade, V. B. P. K., C. V. Andreoli, B. Ejnisman, A. Pochini, and E. P. Biruel. 2021. Evidence on Clinical and Laboratory Parameters After Bariatric Surgery in Diabetics: Systematic Integrative Review / Evidence on Clinical and Laboratory Parameters After Bariatric Surgery in Diabetics: Systematic Integrative Review. *Brazilian Journal of Health Review* 4 (2): 9515–26. https://doi.org/10.34119/bjhrv4n2-444.

Antoniewicz, A., P. Kalinowski, K. J. Kotulecka, P. Kocoń, R. Paluszkiewicz, P. Remiszewski, and K. Zieniewicz. 2019. Nutritional Deficiencies in Patients after Roux-En-Y Gastric Bypass and Sleeve Gastrectomy during 12-Month Follow-Up. *Obesity Surgery* 29 (10): 3277–84. https://doi.org/10.1007/s11695-019-03985-3.

Arterburn, D. E., D. A. Telem, R. F. Kushner, and A. P. Courcoulas. 2020. Benefits and Risks of Bariatric Surgery in Adults: A Review. *JAMA: The Journal of the American Medical Association* 324 (9): 879–87. https://doi.org/10.1001/jama.2020.12567.

Barros, L. M., R. A. N. Moreira, N. M. Frota, T. M. Araújo, and J. Á. Caetano. 2015. Quality of Life Among Morbidly Obese People and Patients Undergoing Bariatric Surgery. *Revista Eletrônica de Enfermagem* 17 (2). https://doi.org/10.5216/ree.v17i2.27367.

Blume, C. A., C. C. Boni, D. S. Casagrande, J. R., A. V. Padoin, and C. C. Mottin. 2012. Nutritional Profile of Patients before and after Roux-En-Y Gastric Bypass: 3-Year Follow-Up. *Obesity Surgery* 22 (11): 1676–85. https://doi.org/10.1007/s11695-012-0696-y.

Bojsen-Møller, K. N., S. H. Jacobsen, C. Dirksen, N. B. Jørgensen, S. Reitelseder, J. E. B. Jensen, V. B. Kristiansen, J. J. Holst, G. V. Hall, and S. Madsbad. 2015. Accelerated Protein Digestion and Amino Acid Absorption after Roux-En-Y Gastric Bypass. *The American Journal of Clinical Nutrition* 102 (3): 600–7. https://doi.org/10.3945/ajcn.115.109298.

Bordalo, L. A., T. F. S. Teixeira, J. Bressan, and D. M. Mourão. 2011. Bariatric Surgery: How and Why to Supplement. *Revista da Associacao Medica Brasileira (1992)* 57 (1): 113–20. https://doi.org/10.1590/s0104-42302011000100025.

Chen, J. C., C. Y. Shen, W. J. Lee, P. L. Tsai, and Y. C. Lee. 2019. Protein Deficiency after Gastric Bypass: The Role of Common Limb Length in Revision Surgery. *Surgery for Obesity and Related Diseases: Official Journal of the American Society for Bariatric Surgery* 15 (3): 441–46. https://doi.org/10.1016/j.soard.2018.12.037.

Costa, G. C. 2021. *Bariatric Surgery: A Treatment for Everyone, the Surgery for Everyone.* Editora Senac Sao Paulo.

Dagan, S. S., A. Goldenshluger, I. Globus, C. Schweiger, Y. Kessler, G. K. Sandbank, T. B. Porat, and T. Sinai. 2017. Nutritional Recommendations for Adult Bariatric Surgery Patients: Clinical Practice. *Advances in Nutrition* (Bethesda, Md.) 8 (2): 382–94. https://doi.org/10.3945/an.116.014258.

Deitel, M. 1998. *Surgery for the Morbidly Obese Patient.* Toronto: Fd-Communications.

Ferraz, Á. A. M. R. C. Bandeira, L. T. Carvalho, F. S. C. Siqueira, and J. M. Campos. 2018. Micronutrient Deficiencies after Bariatric Surgery: Comparative Analysis Between Vertical Gastrectomy and Roux-en-Y Gastric Bypass. *Revista do Colegio Brasileiro de Cirurgioes* 45 (6): e2016. https://doi.org/10.1590/0100-6991e-20182016.

Frenhani, P. B., Burini, R. C. 1999. Mechanisms of Absorption of Amino Acids and Oligopeptides. Control and Implications in Human Dietotherapy. *Arquivos De Gastroenterologia*, 36(4), 227–237. https://doi.org/10.1590/S0004-28031999000400011

Gasmi, A., G. Bjørklund, P. K. Mujawdiya, Y. Semenova, M. Peana, A. Dosa, S. Piscopo, A. G. Benahmed, and D. O. Costea. 2022. Micronutrients Deficiences in Patients after Bariatric Surgery. *European Journal of Nutrition* 61 (1): 55–67. https://doi.org/10.1007/s00394-021-02619-8.

Gehrer, S., B. Kern, T. Peters, C. C. Courtin, and R. Peterli. 2010. Fewer Nutrient Deficiencies after Laparoscopic Sleeve Gastrectomy (LSG) than after Laparoscopic Roux-Y-Gastric Bypass (LRYGB)-a Prospective Study. *Obesity Surgery* 20 (4): 447–53. https://doi.org/10.1007/s11695-009-0068-4.

Jesus, A. 2016. Vitamin D levels after Gastric Bypass: Implications and Recommendations. *Acta portuguesa de nutrição* 06: 42–45. https://doi.org/10.21011/apn.2016.0608.

Kassir, R., T. Debs, P. Blanc, J. Gugenheim, I. B. Amor, C. Boutet, and O. Tiffet. 2016. Complications of Bariatric Surgery: Presentation and Emergency Management. *International Journal of Surgery* (London, England) 27: 77–81. https://doi.org/10.1016/j.ijsu.2016.01.067.

Kuin, C., F. den Ouden, H. Brandts, L. Deden, E. Hazebroek, M. Borren, and H. Boer. 2019. Treatment of Severe Protein Malnutrition after Bariatric Surgery. *Obesity Surgery* 29 (10): 3095–102. https://doi.org/10.1007/s11695-019-04035-8.

Leiro, L. S., and M. S. M. Araújo. 2014. Diet Micronutrient Adequacy of Women after 1 Year of Gastric Bypass. ABCD. *Arquivos Brasileiros de Cirurgia Digestiva (São Paulo)* 27: 21–25. https://doi.org/10.1590/S0102-67202014000S100006.

Llanos, J. P. S., M. F. Ferrer, L. A. S. Walther, B. G. Bray, L. M. González, I. B. Lesmes, and B. M. Esteban. 2015. Protein Malnutrition Incidence Comparison after Gastric Bypass versus Biliopancreatic Diversion. *Nutricion Hospitalaria: Organo Oficial de La Sociedad Espanola de Nutricion Parenteral y Enteral* 32 (1): 80–86. https://doi.org/10.3305/nh.2015.32.1.8963.

Lopes, V. S., M. T. N. Filho, L. G. Tosti, E. F. di Nardo, and S. A. Gabriel. 2022. Review of Current Indications and Surgical Techniques for Bariatric Surgery. *Revista Corpus Hippocraticum* 2 (1). http://189.112.117.16/index.php/revista-medicina/article/view/404.

Lupoli, R., E. Lembo, G. Saldalamacchia, C. K. Avola, L. Angrisani, and B. Capaldo. 2017. Bariatric Surgery and Long-Term Nutritional Issues. *World Journal of Diabetes* 8 (11): 464. https://doi.org/10.4239/wjd.v8.i11.464.

Major, P., P. Małczak, M. Wysocki, G. Torbicz, N. Gajewska, M. Pędziwiatr, and A. Budzyński. 2018. Bariatric Patients Nutritional Status as a Risk Factor for Postoperative Complications, Prolonged Length of Hospital Stay and Hospital Readmission: A Retrospective Cohort Study. *International Journal of Surgery* (London, England) 56: 210–14. https://doi.org/10.1016/j.ijsu.2018.06.022.

Marathe, C. S., C. K. Rayner, K. L. Jones, and M. Horowitz. 2013. Relationships between Gastric Emptying, Postprandial Glycemia, and Incretin Hormones. *Diabetes Care* 36 (5): 1396–1405. https://doi.org/10.2337/dc12-1609.

Martins, T. C. P., T. C. Duarte, E. R. T. Mosca, C. F. Pinheiro, M. A. Marçola, and D. A. Souza. 2015. Severe Protein Malnutrition in a Morbidly Obese Patient after Bariatric Surgery. *Nutrition* (Burbank, Los Angeles County, Calif.) 31 (3): 535–38. https://doi.org/10.1016/j.nut.2014.10.011.

Mechanick, J. I., A. Youdim, D. B. Jones, W. T. Garvey, D. L. Hurley, M. M. McMahon, and L. J. Heinberg, 2013. Clinical Practice Guidelines for the Perioperative Nutritional, Metabolic, and Nonsurgical Support of the Bariatric Surgery Patient-2013 Update: Cosponsored by American Association of Clinical Endocrinologists, the Obesity Society, and American Society Fo. *Obesity* 21 (S1): S1–27. https://doi.org/10.1002/oby.20461.

Mohapatra, S., K. Gangadharan, and C. S. Pitchumoni. 2020. Malnutrition in Obesity before and after Bariatric Surgery. *Disease-a-Month: DM* 66 (2): 100866. https://doi.org/10.1016/j.disamonth.2019.06.008.

Moustarah, F., and F. S. Hould. 2015. 35 Laparoscopic Malabsorption Procedures: Management of Nutritional Complications after Biliopancreatic Diversion. In *Em Minimally Invasive Bariatric Surgery*, 323–33. New York, NY: Springer New York.

Padwal, R., D. Brocks, and A. M. Sharma. 2010. A Systematic Review of Drug Absorption Following Bariatric Surgery and Its Theoretical Implications. *Obesity Reviews: An Official Journal of the International Association for the Study of Obesity* 11 (1): 41–50. https://doi.org/10.1111/j.1467-789X.2009.00614.x.

Parmar, C. D., J. Gan, C. Stier, Z. Dong, S. Chiappetta, L. El-Kadre, M. M. Bashah, C. Wang, and N. Sakran. 2020. One Anastomosis/Mini Gastric Bypass (OAGB-MGB) as Revisional Bariatric Surgery after Failed Primary Adjustable Gastric Band (LAGB) and Sleeve Gastrectomy (SG): A Systematic Review of 1075 Patients. *International Journal of Surgery* (London, England) 81: 32–38. https://doi.org/10.1016/j.ijsu.2020.07.007.

Pereira, S. E., C. Rossoni, M. P. C. Cambi et al. 2023. Brazilian Guide to Nutrition in Bariatric and Metabolic Surgery. *Langenbeck's Archives of Surgery* 408 (1):143. Published 2023 Apr 11. https://doi:10.1007/s00423-023-02868-7.

Pellegrini, M., F. Rahimi, S. Boschetti, A. Devecchi, A. Francesco, M. V. Mancino, and M. Toppino. 2021. Pre-Operative Micronutrient Deficiencies in Patients with Severe Obesity Candidates for Bariatric Surgery. *Journal of Endocrinological Investigation* 44 (7): 1413–23. https://doi.org/10.1007/s40618-020-01439-7.

Rocha, J. C. G. 2012. Vitamin B12 deficiency in the postoperative period of bariatric surgery. *International Journal of Nutrology* 05 (02): 082–089. https://doi.org/10.1055/s-0040-1703967.

Rolim, F. F. Araújo, F. S. Cruz, J. M. Campos, and Á. A. B. Ferraz. 2018. Long-Term Repercussions of Roux-en-Y Gastric Bypass in a Low-Income Population: Evaluation After Ten Years of Surgery. *Journal of the Brazilian College of Surgeons* 45 (4). https://doi.org/10.1590/0100-6991e-20181916.

Rufino, I. I., D. N. Oliveira, and I. R. Júnior. 2023. Obesity and Bariatric Surgery: An Analysis of the Bariatric Techniques Gastric Bypass and Sleeve Gastrectomy and Their Impact on the Health of Obese Patients. *Research, Society and Development* 12 (1): e16112139465. https://doi.org/10.33448/rsd-v12i1.39465.

Schinkel, E. R., S. M. Pettine, E. Adams, and M. Harris. 2006. Impact of Varying Levels of Protein Intake on Protein Status Indicators after Gastric Bypass in Patients with Multiple Complications Requiring Nutritional Support. *Obesity Surgery* 16 (1): 24–30. https://doi.org/10.1381/096089206775222168.

Sinclair, P., D. J. Brennan, and C. W. Roux. 2018. Gut Adaptation after Metabolic Surgery and Its Influences on the Brain, Liver and Cancer. *Nature Reviews. Gastroenterology & Hepatology* 15 (10): 606–24. https://doi.org/10.1038/s41575-018-0057-y.

Steenackers, N., I. Gesquiere, and C. Matthys. 2018. The Relevance of Dietary Protein after Bariatric Surgery: What Do We Know?. *Current Opinion in Clinical Nutrition and Metabolic Care* 21 (1): 58–63. https://doi.org/10.1097/MCO.0000000000000437.

Svane, M. S., K. N. Bojsen-Møller, C. Martinussen, C. Dirksen, J. L. Madsen, S. Reitelseder, and L. Holm, et al. 2019. Postprandial Nutrient Handling and Gastrointestinal Hormone Secretion after Roux-En-Y Gastric Bypass vs Sleeve Gastrectomy. *Gastroenterology* 156 (6): 1627–41.e1. https://doi.org/10.1053/j.gastro.2019.01.262.

Topart, P., G. Becouarn, and P. Ritz. 2012. Comparative Early Outcomes of Three Laparoscopic Bariatric Procedures: Sleeve Gastrectomy, Roux-En-Y Gastric Bypass, and Biliopancreatic Diversion with Duodenal Switch. *Surgery for Obesity and Related Diseases: Official Journal of the American Society for Bariatric Surgery* 8 (3): 250–54. https://doi.org/10.1016/j.soard.2011.05.012.

Torezan, E. F. G. 2013. Review of the Main Micronutrient Deficiencies in the Postoperative Period of Roux-en-Y Gastric Bypass. *International Journal of Nutrology* 06 (01): 037–042. https://doi.org/10.1055/s-0040-1705668.

17 The Paradox of Dietary Protein and Bone Physiology

Raiany Rosa Bergamo

ABBREVIATIONS

BMC	Bone mineral content
BMD	Bone mineral density
CSA	Cross-sectional area
CSMI	Cross-sectional moment of inertia
DF	Distal femur
FGF23	Fibroblast growth factor 23
IGF-1	Insulin-like growth factor 1
L1–L4	Lumbar spine
OB	Osteoblast
OC	Osteoclast
PF	Proximal femur
PMB	Peak bone mass
PTH	Parathyroid hormone
RDA	Recommended dietary allowance
TB	Total body
TBLH	Total body less head
Z	Section modulus

INTRODUCTION

This chapter explores the connection between dietary protein and bone health, highlighting their complex relationship. Various factors, such as age, health status, dietary intake, and habitual diet of the population, influence the impact of protein on bone health. While some studies suggest that excessive protein intake can negatively impact bone health by increasing urinary calcium excretion, other studies demonstrate that higher protein consumption can improve bone variables only when calcium intake meets the recommended guidelines, and can even enhance the positive effects of physical activity on bone mineral density (BMD) and bone mineral content (BMC). As bone is a complex tissue, it is not impacted solely by protein. Thus, the relationship between dietary protein and bone health is not straightforward and depends on multiple factors.

BONE PHYSIOLOGY

The process of bone modeling initiates during fetal growth and persists until the completion of the second decade of life (Gordon et al. 2017). This dynamic process begins during childhood and adolescence when bones respond to various factors like genetics, hormones, mechanical loading, and nutrition (Figure 17.1). Bone modeling involves the formation of new bone tissue by osteoblasts and the removal of existing bone by osteoclasts, and it aids in the adaptation of bone structure to meet diverse demands and contributes to the maintenance of overall bone health. (Weaver et al. 2016).

DOI: 10.1201/b23328-17

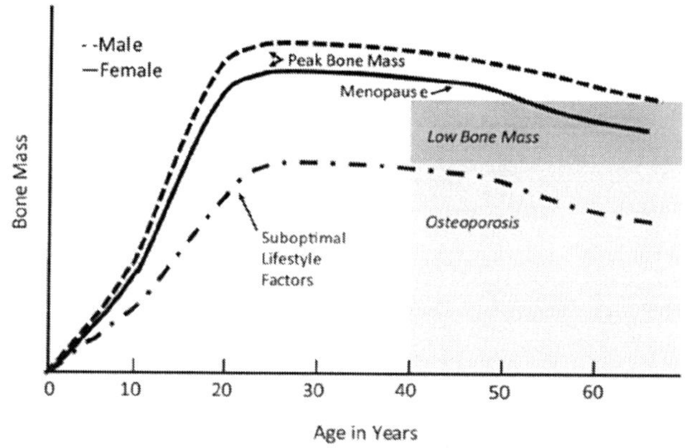

FIGURE 17.1 Bone mass across the lifespan with optimal and suboptimal lifestyle choices. Reproduced unmodified from Weaver et al. (2016). Creative Commons Attribution – Non-Commercial 4.0 International License (http://creativecommons.org/licenses/by-nc/4.0/).

Bone formation takes place in areas where the bone requires reinforcement or expansion, typically in response to growth or increased stress. This process is carried out by osteoblasts, which deposit collagen and mineralize it with calcium and phosphate ions, forming a new bone matrix. Bone resorption, the process of removing existing bone, plays a vital role in eliminating old or damaged bone and reshaping bones during growth. Osteoclasts are the key players in this process, responsible for breaking down and resorbing bone tissue.

Peak bone mass (PBM) is the quantity of bone achieved when accrual either halts or reaches a plateau, typically occurring sometime after the conclusion of the growth and development phase (Stagi et al. 2013). Inadequate nutrition and a sedentary lifestyle, considered suboptimal lifestyle factors, can lead to an initial reduction in PBM, thereby increasing an individual's vulnerability to osteoporosis (Redondo-Cuevas, Sanchis-Chordà, and Codoñer-Franch 2018). With advancing age, the rate of bone resorption tends to surpass that of bone formation, resulting in a gradual decline in bone density. This decline can lead to conditions such as osteoporosis (Robling, Castillo, and Turner 2006).

NUTRIENTS AND BONE MASS

Bone is a complex tissue, comprising minerals, organic matrix, water, and cells. Dietary protein and calcium serve as substrates for bone formation. Hydroxyapatite, a primary component of bone mineral, consists of calcium and phosphate ions along with hydroxide ions. It is crucial for imparting the hardness and rigidity that bones require to maintain their strength and density. One-third of the total bone mass comprises protein, with type 1 collagen fibers making up the majority (Bilezikian et al. 2018). Nutrition plays a central role in the accumulation of bone mass (Weaver et al. 2016). Calcium and phosphorus are key nutrients for bone formation, constituting 80% to 90% of BMC (René Rizzoli, Biver, and Brennan-Speranza 2021). Moreover, several nutrients, such as vitamin D, phosphorus, and magnesium, are pivotal in preserving bone integrity (Figure 17.2). These nutrients promote collagen synthesis and influence bone metabolism by enhancing the absorption of calcium and phosphorus, regulating parathyroid hormone levels, affecting vitamin D metabolism, and modulating the activities of osteoblasts and osteoclasts (Phillips, Chevalier, and Leidy 2016).

The internationally recommended protein intake guideline (Trumbo et al. 2002) for adults is generally 0.8 g/kg/d. However, some specialists recommend higher values for the elderly due to their unique needs. Protein requirements for infants, children, and adolescents differ across countries but tend to be higher than those for adults, primarily due to the increased demand for protein during periods of growth and development (Trumbo et al. 2002).

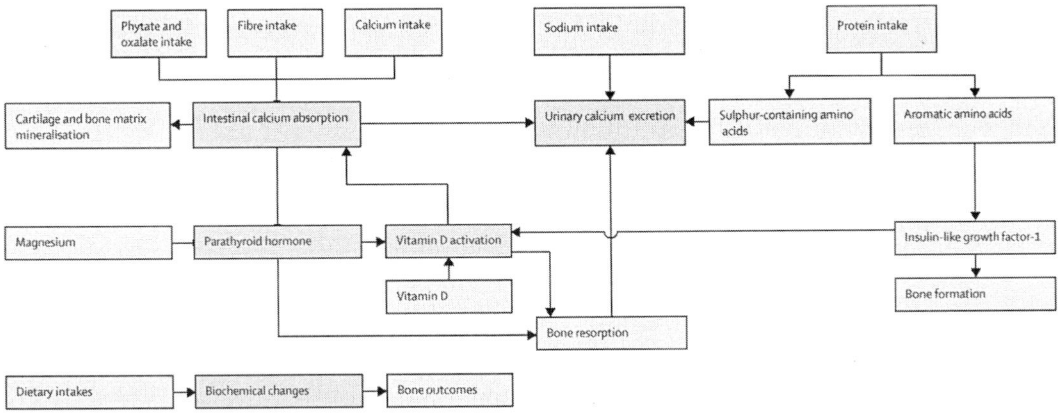

FIGURE 17.2 Nutrient intake, relevant physiological responses, and bone health. Arrows indicate stimulation; flat arrowheads indicate inhibition. Reprinted from Rizzoli, Biver, and Brennan-Speranza (2021), with permission from Elsevier.

Over a span of ten years, dietary choices can result in a substantial 15% reduction in bone calcium when they contribute to the mitigation of mild metabolic acidosis (Wachman and Bernstein 1968).

Protein's impact on bone health varies, with both positive and negative effects determined by several factors. These factors include the specific type of protein in one's diet, the source of protein, calcium intake, demographic factors such as age and gender, weight loss efforts, and the acid-base balance of the diet (Heaney and Layman 2008).

THE PROTEIN PARADOX

Protein has garnered significant attention and concern over the years, as evidenced by studies dating back to 1920 (Sherman, Rose, and Rose 1920), 1950 (Pullman et al. 1954), and 1970 (Johnson, Alcantara, and Linkswiler 1970). This interest stems from a combination of biochemical and nutritional factors. These studies have consistently revealed that elevated protein intake strongly influences urinary calcium loss, potentially disrupting calcium equilibrium and contributing to bone loss. In 1968, Wachman and Bernstein discussed the connection between daily dietary habits with the onset of osteoporosis. They hypothesized that the elevated occurrence of osteoporosis in older individuals might, in part, result from a lifelong diet rich in acid precursors, leading to higher systemic acidity and, consequently, continuous utilization of the bone's basic salt buffering capacity to counterbalance fluctuations in pH homeostasis. Furthermore, compelling evidence (Frassetto and Sebastian 1996) indicates that as individuals age, their renal function progressively declines, causing an increase in acidity. Consequently, humans tend to become slightly more acidic as they grow older.

Excessive protein intake could potentially lead to an unfavorable calcium balance, resulting in resorption and a decrease in BMD (Figure 17.3). This rise in urinary calcium levels may depend on both the quantity and nature of protein ingested, as dietary acid load can be influenced by the ratio of animal to plant protein intake (Lea, Crawford, and Worsley 2006).

On the other hand, dietary protein plays a significant role in promoting bone health by impacting insulin-like growth factor-1 (IGF-1) production, a critical hormone for bone growth. IGF-1 not only affects bone but also regulates phosphate reabsorption in the kidneys and stimulates the active absorption of calcium and phosphate in the intestines, promoting the synthesis of 1,25-dihydroxyvitamin D3 (calcitriol) in the kidneys (René Rizzoli et al. 2001). Reduced protein intake correlates with decreased IGF-1 levels; consequently, lower IGF-1 levels are linked to reduced bone mass,

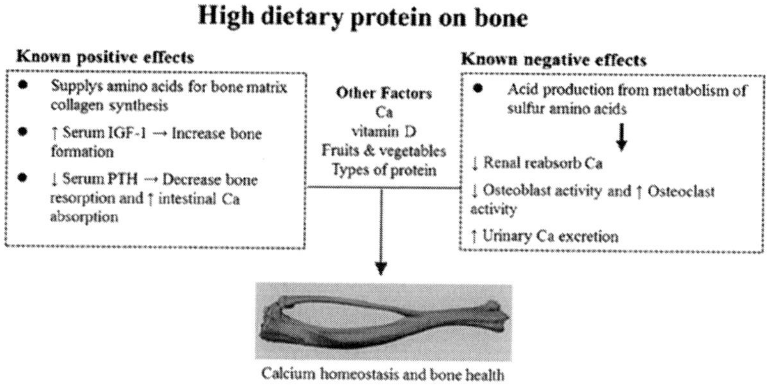

FIGURE 17.3 Known positive and negative effects of high dietary protein intake on bone, and the ultimate outcome may be influenced by other dietary factors. From Cao (2017), reproduced with permission from SNCSC.

potentially increasing the risk of fractures, as studies have shown a negative correlation between IGF-1 concentration and predicted fracture risk (René Rizzoli et al. 2001) (Figure 17.3).

Protein intake below 0.8 g/kg/d is typically regarded as inadequate for maintaining bone health (Burckhardt, Dawson-Hughes, and Weaver 2010), with increased bone tissue resorption linked to an increased risk of fractures, and it represents the minimum intake required to maintain nitrogen balance in elderly individuals (Kurpad and Vaz 2000).

However, the precise impact of protein type on bone health remains uncertain. While excessive protein intake may pose a potential risk to bone health and lead to an unfavorable calcium balance, there is also evidence indicating that inadequate protein consumption can also play a role in these health concerns, as it has been linked to an increased risk of fractures.

PROTEIN INTAKE ON PEAK BONE MASS DURING CHILDHOOD

The prepubertal phase provides an excellent window of opportunity to improve bone health in the long run. Adequate nutrient composition can maximize PBM and promote lifelong bone health (Jean Philippe Bonjour et al. 2011). Consequently, childhood nutrition may have far-reaching, lifelong implications. Knowledge is limited regarding the short- and long-term impacts of protein intake and supplementation on bone turnover and development. The prepubertal years are crucial in shaping bone growth, playing a vital role in achieving higher PBM and enhancing bone strength (Chevalley et al. 2017). Calcium and protein directly affect bone metabolism as essential components of the mineral and organic structure of bone tissue. Additionally, they indirectly impact by affecting endocrine and paracrine pathways, contributing to the overall accumulation of the skeleton during the growth phase (Jean Philippe Bonjour et al. 2011).

A three-year study of prepubertal Chinese girls (Q. Zhang et al. 2010) revealed negative associations between protein intake (average 55.9–61.0 g/d or 1.36 g/kg/d) and BMC accrual at total body (TB), proximal femur (PF), and distal femur (DF), especially with inadequate calcium intake (average 454 mg/d, less than half the recommended amount). Dietary animal protein intake seemed to negatively impact on bone mass accrual. A predictive model suggested that theoretically doubling protein intake could lead to a 1.92% reduction in BMC accrual for the TB. In healthy children and adolescents aged 6–18 years, long-term protein intake (Alexy et al. 2005) more strongly influences the acquisition of bone mass and strength than calcium intake. Specifically, high protein intake enhanced the positive impact of physical activity on BMC in eight-year-old prepubertal boys (Chevalley et al. 2008, 2017).

A 15-year prospective study of healthy boys aged approximately 7.4 and 15.2 years (Chevalley et al. 2017) demonstrated that the positive interaction between protein intake (an average of 47.3 and 63.7 g/d, respectively) and physical activity, rather than calcium intake (752 and 1031 mg/d, respectively) (Chevalley et al. 2008, 2014) plays a pivotal role in shaping bone health from prepuberty through mid-late adolescence.

An eight-year study tracking bone structure and strength development in healthy boys from prepuberty (7.4 ±0.4 years old) to mid-late adolescence (15.2 ±0.5 years old) (Chevalley et al. 2014) emphasized the combined impact of protein intake (average of 1.78 and 1.08 g/kg/d, respectively) and physical activity on the width of the femoral neck, an important attribute that predicts mechanical resistance (Robling, Castillo, and Turner 2006), thereby potentially lowering the risk of fragility fractures in later life (Bouxsein 2008).

PROTEIN INTAKE AND PUBERTY DURING ADOLESCENCE

The onset of puberty marks an accelerated somatic growth phase, and by the age of 18, 90% of PBM is achieved and remains stable throughout adulthood (Malina, Bouchard, and Bar-Or 2020). During this developmental stage, maintaining body composition and ensuring sufficient nutrient intake is crucial to support the acquisition of bone minerals, as adolescents are nutritionally susceptible due to increased growth demands and unhealthy dietary behaviors (Berg 2019). The potential advantages of increased protein consumption on bone-related factors, particularly in adolescents, remain uncertain.

Adolescent Brazilian male soccer players (Bergamo et al. 2023) showed a negative correlation between protein intake (average 2.04 g/kg/d) and BMC of total body less head (TBLH), as well as BMD of lumbar spine (L1–L4) and bone geometry variables such as section modulus (Z), cross-sectional moment of inertia (CSMI), and cross-sectional area (CSA). Calcium intake was 56% below the daily recommendation of 1300 mg, which may explain the negative association between high protein intake and bone parameters. It is of utmost importance to conduct more research focused on adolescent athletes since physically active individuals are more susceptible to stress fractures resulting from prolonged or high-intensity exercise.

The favorable impact of protein on bone mass in peri-adolescent and early adult females (Vatanparast et al. 2007), with protein intake above the RDA, might be influenced by calcium intake during adolescence (exceeding 1000 mg per day). In this scenario, higher protein intake positively predicted BMC and BMD for the total body. In a short-term randomized controlled study involving young adults (Cao et al. 2014), a high protein intake at three times the RDA with adequate calcium intake, consisting of 1000 mg of dietary calcium along with 162 mg of supplementation, did not affect calcium balance or impair skeletal integrity.

INFLUENCE OF PROTEIN INTAKE AND EXERCISE ON BONE MASS

There is growing acknowledgment of how diet and physical activity influence bone mineral density, bone mineral content, and remodeling. Engaging in sports provides a potent osteogenic stimulus, increasing bone mineral accrual, preserving and enhancing BMD (Hervás et al. 2018), and positively influencing bone strength (Zhu and Zheng 2021). Higher levels of both physical activity and protein intake are linked to increased size of weight-bearing bones (Bouxsein 2005). As with skeletal muscle development, adequate nutrition likely supports the anabolic impact of mechanical loading on bone tissue (Holick and Nieves 2015).

Malnutrition can have a multifaceted impact on health, affecting IGF-I serum levels and bone metabolism, while exerting a catabolic influence on muscle mass, which indirectly influences bone health (Bass, Eser, and Daly 2005). However, the role of protein intake in modifying the bone's response to exercise during growth remains unclear (Specker and Vukovich 2007).

As mentioned earlier, protein and calcium intake can amplify the effects of physical activity on bone development, as evidenced in children (Jean Philippe Bonjour et al. 2011; Chevalley et al.

2008, 2017) and adolescents (Jean Philippe Bonjour et al. 2015; J P Bonjour et al. 2010), and on bone maintenance in adults (Antonio et al. 2020) and elderly (Isanejad et al. 2017).

Research conducted with the elderly population (68.1 ±1.9 years old) indicated a potentially adverse relationship between protein intake (average of 0.96 g/kg/d) and BMD and BMC. The researchers suggested that physical activity might counteract these negative effects of high protein intake on bone health (Isanejad et al. 2017).

Throughout the growth phase and into adulthood, lifestyle factors such as diet and consistent physical activity are pivotal external factors determining bone mass development. This influence endures, actively contributing to the formation of a strong and resilient adult skeleton (Amato et al. 2021). Despite the significant interest in nutrition, many active individuals lack knowledge on this subject (Berg 2019; Smith, Holmes, and McAllister 2015). Hence, proper nutritional guidance is vital, especially for young athletes, as it plays a central role not only in improving athletic performance but also in ensuring proper growth, development, and overall health.

PROTEIN INTAKE AND OSTEOPOROSIS

After age 50, approximately 50% of women and 20% of men will experience an osteoporotic fracture (Teede et al. 2012). Ensuring nutritional adequacy in older adults is paramount, given their heightened susceptibility to falls, fractures, and potential malnutrition. Current research has raised concerns regarding osteoporosis and its connection to dietary factors, particularly the intake of protein (both animal and plant sources), calcium, phosphorus, magnesium, and essential vitamins. These concerns arise because age-related bone tissue loss primarily results from hormonal shifts, diminished absorption in the intestinal epithelium, and decreased secretion of digestive enzymes as individuals grow older (L. T. Ho-Pham et al. 2012). Consequently, bone resorption rates nearly double in individuals aged over 70 compared to those aged 60 to 69. While recent studies remain inconclusive regarding the impact of protein intake on bone health, several physiological mechanisms suggest increased protein consumption may help maintain or even enhance BMD.

Studies involving interventions with adults aged over 65 at risk of malnutrition, pre-frailty, or frailty yielded several significant findings that emerged from the research (van Dongen et al. 2020; Grootswagers et al. 2021). Both total and animal protein intake correlated with higher BMD in the TB and spine. Conversely, plant-based protein consumption was linked to lower BMD values in the same areas (Groenendijk et al. 2022). Total protein intake showed a stronger link with spine BMD in females, while plant protein intake had a more substantial association with spine BMD in males than in females. Combined analyses considering both protein intake (total and animal) and adequate calcium intake, along with sufficient serum 25(OH)D concentration, revealed stronger associations with TB and spine BMD. Furthermore, a high ratio of animal to plant protein intake was associated with higher BMD in the TB and spine (Groenendijk et al. 2022).

A study of middle-aged and older US adults (Y. W. Zhang et al. 2022) suggests a potential link between reduced dietary protein intake and increased risk of osteoporosis. ROC (Receiver Operating Characteristic) curve analysis determined that a daily dietary protein intake of 61.2 g serves as the optimal threshold for assessing the risk of osteoporosis (OP), exhibiting a specificity of 59.9% and a sensitivity of 56.6% (Figure 17.4).

Conversely, higher protein intake might reduce the odds of developing osteoporosis. A recent study (Kirk, Prokopidis, and Duque 2021) found that consuming 1.2–1.5 g of protein per kilogram of body weight per day could help mitigate osteosarcopenia, improving bone density and reducing fracture risk among individuals aged over 65. However, these benefits were observed only with adequate amounts of vitamin D (800 IU/day) and calcium (1000–1200 mg/day) intake. Evidence for benefits from protein supplementation beyond the recommended daily allowance (RDA) remains inconclusive.

A five-year cohort study (Meng et al. 2009) found that a protein intake exceeding 87 g/d (or 1.6 g/kg/d) was associated with advantageous outcomes in preserving lean body mass and BMC among

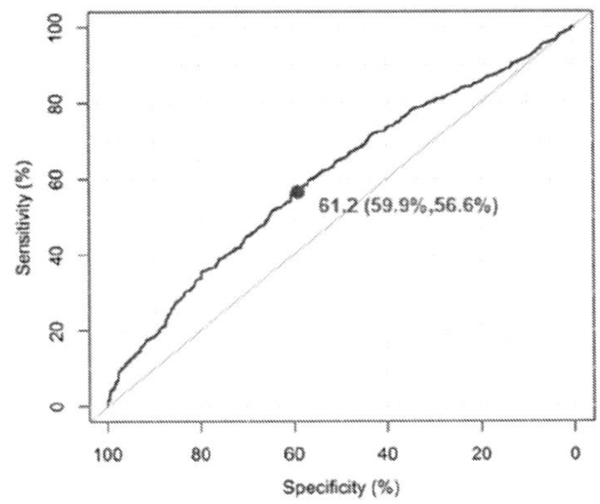

FIGURE 17.4 Cut-off value of protein intake for predicting the risk of osteoporosis through ROC curve analysis. The optimal cut-off value is a daily dietary protein intake of 61.2 g, with a specificity of 59.9% and a sensitivity of 56.6%. From Zhang et al. (2022), reproduced with permission from SNCSC.

women over 70 years of age. These findings align with recent reviews of nitrogen balance studies, suggesting that maintaining nitrogen balance in healthy elderly individuals necessitates a protein intake of 1.0–1.3 g/kg/d (Morais, Chevalier, and Gougeon 2006). Another review proposed that a safe protein intake for the elderly range from 1.0 to 1.25 g/kg/d of high-quality protein (Campbell et al. 1994). It is worth noting that the positive impact of protein on BMC may be mediated through its influence on muscle tissue.

PROTEIN INTAKE AND BONE STRENGTH: FRACTURE RISK

An expert consensus paper endorsed by the European Society for Clinical and Economical Aspects of Osteoporosis, Osteoarthritis, and Musculoskeletal Diseases, and the International Osteoporosis Foundation (R. Rizzoli et al. 2018) on the benefits and safety of dietary protein for bone health found evidence suggesting that higher dietary protein intake may slightly reduce the risk of hip fractures, as long as calcium intake meets recommended levels.

Regarding bone geometry variables and their relation to bone strength, evidence from healthy postmenopausal women (Durosier-Izart et al. 2017), with average dietary calcium intake (1156 mg/d) and protein intake (1.11 g/kg/d) at or slightly above the recommended amounts for this age group, supports a dose-dependent positive correlation between the anticipated failure load and the consumption of total, animal, and dairy protein consumption. This correlation is mainly related to changes in trabecular microstructure. The positive effects of dietary proteins on bone health appear to depend on having sufficient calcium intake.

Increased protein intake, whether from animal or plant sources, does not seem to contribute to osteoporosis development or elevated fracture risk. As long as calcium intake meets recommended levels, protein intake positively impacts mitigating bone loss and reducing fracture risk, particularly at the hip, even when exceeding the current RDA.

PLANT PROTEIN AND BONE HEALTH: IS THE ALKALIZING EFFECT SUFFICIENT?

The impact of vegetarian diets on bone homeostasis has various implications, with differences in reports and findings stemming from factors like population size, study design, and conclusions. Numerous studies suggest that higher dietary protein intake can increase urinary calcium excretion. Therefore, vegetarian diets, which primarily feature plant-based proteins with an alkalizing nature and generally lower overall protein intake, are expected to result in reduced urinary calcium losses, thus requiring less calcium (Falchetti et al. 2022). Recently, several studies have revealed that the interplay between calcium balance and protein consumption is considerably more intricate (Thorpe and Evans 2011).

A systematic review and meta-analysis from the National Osteoporosis Foundation (Shams-White et al. 2018) comparing the impact of animal versus plant protein sources on bone health in healthy postmenopausal women found no significant disparities in bone turnover markers or BMD between the two groups. Furthermore, most epidemiological data suggest that individuals following a plant-based diet experience no detriment to bone mineral density or association with osteoporosis when they consume adequate calcium and vitamin D.

A Bayesian meta-analysis (Lan T. Ho-Pham, Nguyen, and Nguyen 2009) verified that vegetarians, compared to omnivores, exhibited a 4% lower BMD at both the lumbar spine and femoral neck. Furthermore, vegans, compared to non-vegetarians, displayed a 6% lower BMD at the femoral neck and lumbar spine. However, these differences were not considered clinically significant for fracture risk. A cross-sectional study (Menzel et al. 2021) examining connections between veganism and BMD measured through calcaneal quantitative ultrasound (QUS), found that vegans had lower QUS parameter levels than omnivores. This disparity was attributed to decreased levels of various nutrients, including protein, zinc, vitamins A, D, B2, B12, selenium, iron, n-3 fatty acids, and calcium (Figure 17.5). These findings suggest compromised bone homeostasis in vegans compared to omnivores, implying a potential association between various nutrition-related biomarkers and bone health. With sufficient calcium intake and appropriate protein levels, there is no notable difference in fracture risk. Additionally, vitamin B12 deficiency, common in vegetarians and especially in vegans, has been linked to decreased BMD and increased risk of fractures (Falchetti et al. 2022).

While vegetarianism and veganism offer numerous benefits, they also have drawbacks, including an elevated risk of osteoporosis, as highlighted in a recent review article on this topic. Both vegetarian and vegan diets exert a notable alkalizing effect, which slows bone tissue resorption and

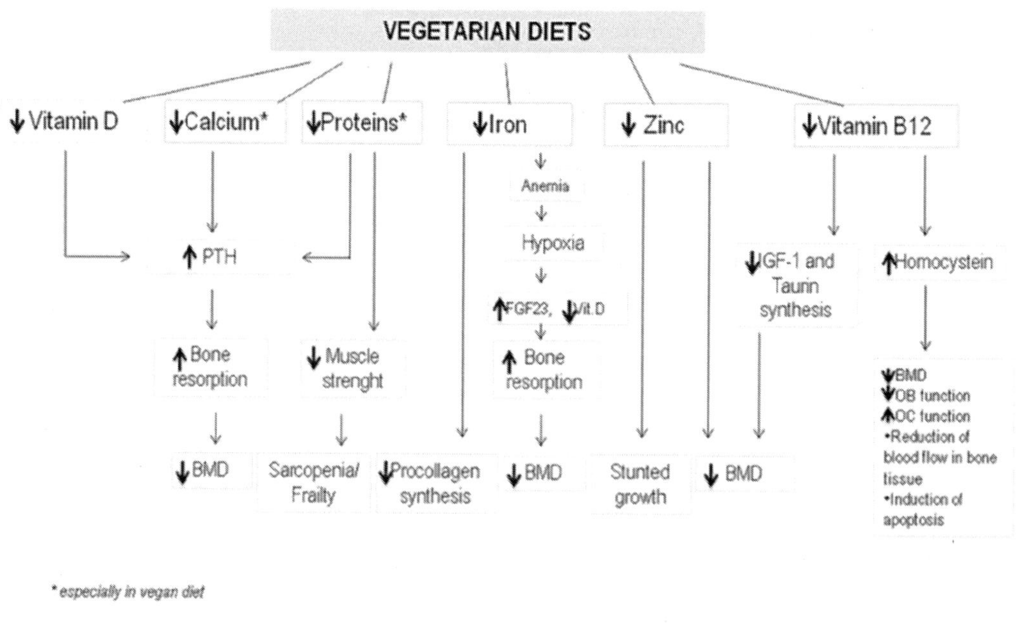

FIGURE 17.5 Impact of nutrient variability in vegetarian and vegan diets on bone health. This figure explores the diverse impact of nutrient variability in vegetarian and vegan diets on bone health, with a focus on PTH (parathyroid hormone), BMD (bone mineral density), FGF23 (fibroblast growth factor 23), IGF-1 (insulin growth factor-1), OB (osteoblast), and OC (osteoclast). Reproduced unmodified from Falchetti et al. (2022). Creative Commons Attribution – Non-Commercial 4.0 International License (http://creativecommons.org/licenses/by-nc/4.0/).

stimulates bone synthesis. Nevertheless, this protective impact is compromised by significantly deficient protein, calcium, vitamin D, or vitamin B12 intake (Galchenko, Gapparova, and Sidorova 2023). The crucial factor for bone health is not the source of protein but rather the variations in the quantitative composition of other nutrients in the diet (L. T. Ho-Pham et al. 2012).

The American Dietetic Association (Craig and Mangels 2009) recommends that vegetarians include foods providing ample amounts of calcium, vitamin D, vitamin K, potassium, and magnesium in their diets. A balanced protein intake is important, avoiding excessive amounts. A study investigating the connection between urinary potential renal acid load (UPRAL) in vegans versus omnivores and its impact on bone health (Penczynski et al. 2022) revealed contrasting acid-base profiles between the two groups. Vegans exhibited a pronounced alkaline surplus, while omnivores maintained a balanced to low acid load despite moderate protein intake. However, within this spectrum from alkaline to low acid load, no significant correlation with bone health was established.

Despite the common perception of protein intake as an acid-producing factor that may increase the risk of bone loss, when protein consumption is combined with adequate amounts of calcium and other essential minerals, it can actually benefit overall skeletal metabolism (Jean Philippe Bonjour 2005).

MINI DICTIONARY OF TERMS

Bone mineral content: The total amount of minerals in the bone, such as calcium and phosphorus, evaluated in relation to bone health and conditions like osteoporosis.

Bone mineral density: The amount of mineral in a specific area of bone, reflecting bone strength.

Bone modeling: Process by which bones change in shape and size, involving both the formation of new bone tissue and the removal of existing bone.

IGF-1: A hormone that promotes the growth and development of various tissues and organs.

Osteoporosis: A medical condition characterized by loss of bone density and weakening of the bones.

Peak bone mass: The highest amount of bone tissue accumulated in a lifetime.

SUMMARY POINTS

- Protein's impact on bone health varies based on protein type, source, calcium intake, age, and dietary acid-base balance, resulting in diverse effects.
- Insufficient nutrition and a sedentary lifestyle can diminish peak bone mass, increasing the risk of osteoporosis later in life.
- For older adults, a protein intake of at least 0.8 g/kg/d is essential to maintain bone health and nitrogen balance.
- The effect of specific protein types on bone health remains unclear. Both excessive and inadequate protein intake may pose risks, affecting calcium balance and fracture risk.
- When combined with sufficient calcium and essential minerals, protein intake, often considered an acid-producing factor linked to bone loss, can positively affect skeletal metabolism.
- Higher protein intake, even above current recommendations, from either animal or plant sources, does not appear to increase osteoporosis or fracture risk. Instead, it may help preserve bone health.

REFERENCES

Alexy, Ute, Thomas Remer, Friedrich Manz, Christina M. Neu, and Eckhard Schoenau. 2005. "Long-Term Protein Intake and Dietary Potential Renal Acid Load Are Associated with Bone Modeling and Remodeling at the Proximal Radius in Healthy Children." *American Journal of Clinical Nutrition* 82 (5). https://doi.org/10.1093/ajcn/82.5.1107.

Amato, Alessandra, Patrizia Proia, Gaetano Felice Caldara, Angelina Alongi, Vincenzo Ferrantelli, and Sara Baldassano. 2021. "Analysis of Body Perception, Preworkout Meal Habits and Bone Resorption in Child Gymnasts." *International Journal of Environmental Research and Public Health* 18 (4). https://doi.org/10.3390/ijerph18042184.

Antonio, Jose, Darren G. Candow, Scott C. Forbes, Michael J. Ormsbee, Patrick G. Saracino, and Justin Roberts. 2020. "Effects of Dietary Protein on Body Composition in Exercising Individuals." *Nutrients*. https://doi.org/10.3390/nu12061890.

Bass, Shona L., P. Eser, and R. Daly. 2005. "The Effect of Exercise and Nutrition on the Mechanostat." *Journal of Musculoskeletal Neuronal Interactions* 5 (3): 239–54.

Berg, Erin K. 2019. "Performance Nutrition for the Adolescent Athlete: A Realistic Approach." *Clinical Journal of Sport Medicine*. https://doi.org/10.1097/JSM.0000000000000744.

Bergamo, Raiany Rosa, Mauro Alexandre Páscoa, Jefferson Eduardo Hespanhol, Anderson Marques de Moraes, and Gil Guerra-Júnior. 2023. "Positive Association of Lean Mass and Negative Association of Protein Intake on Bone Mass and Bone Geometry of Adolescent Soccer Players." *Nutrition (Burbank, Los Angeles County, Calif.)* 105 (January). https://doi.org/10.1016/J.NUT.2022.111857.

Bilezikian, John P., Roger Bouillon, Thomas Clemens, Juliet Compston, Douglas C. Bauer, Peter R. Ebeling, Klaus Engelke, et al. 2018. "Primer on the Metabolic Bone Diseases and Disorders of Mineral Metabolism." https://doi.org/10.1002/9781119266594.

Bonjour, Jean Philippe. 2005. "Dietary Protein: An Essential Nutrient for Bone Health." *Journal of the American College of Nutrition* 24 (6 Suppl): 526S–36S. https://doi.org/10.1080/07315724.2005.10719501.

Bonjour, Jean Philippe, Thierry Chevalley, Patrick Amman, and René Rizzoli. 2015. "Protein Intake and Bone Health." *Nutrition and Bone Health* (January): 301–17. https://doi.org/10.1007/978-1-4939-2001-3_20/FIGURES/10.

Bonjour, Jean Philippe, T. Chevalley, S. Ferrari, and R. Rizzoli. 2010. "Importance of Nutrition for Optimal Bone Development during Childhood and Adolescence." Osteoporosis International Conference (var. pagings).

Bonjour, Jean Philippe, Thierry Chevalley, Serge Ferrari, and Rene Rizzoli. 2011. "Peak Bone Mass and Its Regulation." In *Pediatric Bone: Biology and Diseases*. https://doi.org/10.1016/B978-0-12-382040-2.10009-7.

Bouxsein, Mary L. 2005. "Determinants of Skeletal Fragility." *Best Practice and Research: Clinical Rheumatology*. https://doi.org/10.1016/j.berh.2005.07.004.

Bouxsein, Mary L. 2008. "Biomechanics of Age-Related Fractures." In *Osteoporosis, Two-Volume Set*. https://doi.org/10.1016/B978-012370544-0.50025-2.

Burckhardt, Peter, Bess Dawson-Hughes and Connie M. Weaver. 2010. "Nutritional Influences on Bone Health." https://doi.org/10.1007/978-1-84882-978-7.

Campbell, Wayne W., Marilyn C. Crim, Gerard E. Dallai, Vernon R. Young, and William J. Evans. 1994. "Increased Protein Requirements in Elderly People: New Data and Retrospective Reassessments." *American Journal of Clinical Nutrition* 60 (4). https://doi.org/10.1093/ajcn/60.4.501.

Cao, Jay J. 2017. "High Dietary Protein Intake and Protein-Related Acid Load on Bone Health." *Current Osteoporosis Reports* 15 (6): 571–76. https://doi.org/10.1007/S11914-017-0408-6.

Cao, Jay J., Stefan M. Pasiakos, Lee M. Margolis, Edward R. Sauter, Leah D. Whigham, James P. McClung, Andrew J. Young, and Gerald F. Combs. 2014. "Calcium Homeostasis and Bone Metabolic Responses to High-Protein Diets during Energy Deficit in Healthy Young Adults: A Randomized Controlled Trial1-4." *American Journal of Clinical Nutrition* 99 (2). https://doi.org/10.3945/ajcn.113.073809.

Chevalley, Thierry, Jean Philippe Bonjour, Marie Claude Audet, Fanny Merminod, Bert Van Rietbergen, Rene Rizzoli, and Serge Ferrari. 2017. "Prepubertal Impact of Protein Intake and Physical Activity on Weight-Bearing Peak Bone Mass and Strength in Males." *Journal of Clinical Endocrinology and Metabolism* 102 (1). https://doi.org/10.1210/jc.2016-2449.

Chevalley, Thierry, Jean Philippe Bonjour, Serge Ferrari, and René Rizzoli. 2008. "High-Protein Intake Enhances the Positive Impact of Physical Activity on BMC in Prepubertal Boys." *Journal of Bone and Mineral Research* 23 (1). https://doi.org/10.1359/jbmr.070907.

Chevalley, Thierry, Jean Philippe Bonjour, Bert Van Rietbergen, Serge Ferrari, and René Rizzoli. 2014. "Tracking of Environmental Determinants of Bone Structure and Strength Development in Healthy Boys: An Eight-Year Follow up Study on the Positive Interaction between Physical Activity and Protein Intake from Prepuberty to Mid-Late Adolescence." *Journal of Bone and Mineral Research* 29 (10). https://doi.org/10.1002/jbmr.2247.

Craig, Winston J., and Ann Reed Mangels. 2009. "Position of the American Dietetic Association: Vegetarian Diets." *Journal of the American Dietetic Association* 109 (7). https://doi.org/10.1016/s0002-8223(21)01979-9.

Dongen, Ellen J. I. van, Annemien Haveman-Nies, Esmée L. Doets, Berber G. Dorhout, and Lisette C. P. G. M. de Groot. 2020. "Effectiveness of a Diet and Resistance Exercise Intervention on Muscle Health in Older Adults: ProMuscle in Practice." *Journal of the American Medical Directors Association* 21 (8). https://doi.org/10.1016/j.jamda.2019.11.026.

Durosier-Izart, Claire, Emmanuel Biver, Fanny Merminod, Bert Van Rietbergen, Thierry Chevalley, François R. Herrmann, Serge L. Ferrari, and Reñe Rizzoli. 2017. "Peripheral Skeleton Bone Strength Is Positively Correlated with Total and Dairy Protein Intakes in Healthy Postmenopausal Women." *American Journal of Clinical Nutrition* 105 (2). https://doi.org/10.3945/ajcn.116.134676.

Falchetti, Alberto, Guido Cavati, Roberto Valenti, Christian Mingiano, Roberta Cosso, Luigi Gennari, Iacopo Chiodini, and Daniela Merlotti. 2022. "The Effects of Vegetarian Diets on Bone Health: A Literature Review." *Frontiers in Endocrinology* 13 (August). https://doi.org/10.3389/FENDO.2022.899375.

Frassetto, Lynda, and Anthony Sebastian. 1996. "Age and Systemic Acid-Base Equilibrium: Analysis of Published Data." *Journals of Gerontology - Series A Biological Sciences and Medical Sciences* 51 (1). https://doi.org/10.1093/gerona/51A.1.B91.

Galchenko, Alexey, K. Gapparova, and E. Sidorova. 2023. "The Influence of Vegetarian and Vegan Diets on the State of Bone Mineral Density in Humans." *Critical Reviews in Food Science and Nutrition* 63 (7): 845–61. https://doi.org/10.1080/10408398.2021.1996330.

Gordon, Catherine M., Babette S. Zemel, Tishya A.L. Wren, Mary B. Leonard, Laura K. Bachrach, Frank Rauch, Vicente Gilsanz, Clifford J. Rosen, and Karen K. Winer. 2017. "The Determinants of Peak Bone Mass." *Journal of Pediatrics*. https://doi.org/10.1016/j.jpeds.2016.09.056.

Groenendijk, Inge, Pol Grootswagers, Aurelia Santoro, Claudio Franceschi, Alberto Bazzocchi, Nathalie Meunier, Aurélie Caille, et al. 2022. "Protein Intake and Bone Mineral Density: Cross-Sectional Relationship and Longitudinal Effects in Older Adults." https://doi.org/10.1002/jcsm.13111.

Grootswagers, Pol, Ellen Smeets, Antwi Boasiako Oteng, and Lisette de Groot. 2021. "A Novel Oral Nutritional Supplement Improves Gait Speed and Mitochondrial Functioning Compared to Standard Care in Older Adults with (or at Risk of) Undernutrition: Results from a Randomized Controlled Trial." *Aging* 13 (7). https://doi.org/10.18632/aging.202912.

Heaney, Robert P., and Donald K. Layman. 2008. "Amount and Type of Protein Influences Bone Health." *American Journal of Clinical Nutrition* 87. https://doi.org/10.1093/ajcn/87.5.1567s.

Hervás, Gotzone, Fátima Ruiz-Litago, Jon Irazusta, Ainhoa Fernández-Atutxa, Ana Belen Fraile-Bermúdez, and Idoia Zarrazquin. 2018. "Physical Activity, Physical Fitness, Body Composition, and Nutrition Are Associated with Bone Status in University Students." *Nutrients* 10 (1). https://doi.org/10.3390/nu10010061.

Holick, Michael F., and Jeri W. Nieves. 2015. *Nutrition and Bone Health. Nutrition and Bone Health*. New York: Springer. https://doi.org/10.1007/978-1-4939-2001-3.

Ho-Pham, Lan T., Nguyen D. Nguyen, and Tuan V. Nguyen. 2009. "Effect of Vegetarian Diets on Bone Mineral Density: A Bayesian Meta-Analysis." *American Journal of Clinical Nutrition* 90 (4). https://doi.org/10.3945/ajcn.2009.27521.

Ho-Pham, L. T., B. Q. Vu, T. Q. Lai, N. D. Nguyen, and T. V. Nguyen. 2012. "Vegetarianism, Bone Loss, Fracture and Vitamin D: A Longitudinal Study in Asian Vegans and Non-Vegans." *European Journal of Clinical Nutrition* 66 (1). https://doi.org/10.1038/ejcn.2011.131.

Isanejad, Masoud, J. Sirola, J. Mursu, H. Kröger, M. Tuppurainen, and A. T. Erkkilä. 2017. "Association of Protein Intake with Bone Mineral Density and Bone Mineral Content among Elderly Women: The OSTPRE Fracture Prevention Study." *Journal of Nutrition, Health and Aging* 21 (6). https://doi.org/10.1007/s12603-016-0800-4.

Johnson, N. E., E. N. Alcantara, and H. Linkswiler. 1970. "Effect of Level of Protein Intake on Urinary and Fecal Calcium and Calcium Retention of Young Adult Males." *The Journal of Nutrition* 100 (12). https://doi.org/10.1093/jn/100.12.1425.

Kirk, Ben, Konstantinos Prokopidis, and Gustavo Duque. 2021. "Nutrients to Mitigate Osteosarcopenia: The Role of Protein, Vitamin D and Calcium." *Current Opinion in Clinical Nutrition and Metabolic Care* 24 (1): 25–32. https://doi.org/10.1097/MCO.0000000000000711.

Kurpad, A. V., and M. Vaz. 2000. "Protein and Amino Acid Requirements in the Elderly." *European Journal of Clinical Nutrition* 54. https://doi.org/10.1038/sj.ejcn.1601035.

Lea, Emma J., D. Crawford, and A. Worsley. 2006. "Public Views of the Benefits and Barriers to the Consumption of a Plant-Based Diet." *European Journal of Clinical Nutrition* 60 (7): 828–37. https://doi.org/10.1038/sj.ejcn.1602387.

Malina, Robert M, Claude Bouchard, and Oded Bar-Or. 2020. "Bone Tissue In Skeletal Growth And Body Composition." In *Growth, Maturation, and Physical Activity.* https://doi.org/10.5040/9781492596837.ch-006.

Meng, Xingqiong, Kun Zhu, Amanda Devine, Deborah A. Kerr, Colin W. Binns, and Richard L. Prince. 2009. "A 5-Year Cohort Study of the Effects of High Protein Intake on Lean Mass and BMC in Elderly Postmenopausal Women." *Journal of Bone and Mineral Research* 24 (11): 1827–34. https://doi.org/10.1359/JBMR.090513.

Menzel, Juliane, Klaus Abraham, Gabriele I. Stangl, Per Magne Ueland, Rima Obeid, Matthias B. Schulze, Isabelle Herter-Aeberli, Tanja Schwerdtle, and Cornelia Weikert. 2021. "Vegan Diet and Bone Health-Results from the Cross-Sectional RBVD Study." *Nutrients* 13 (2): 1–16. https://doi.org/10.3390/NU13020685.

Morais, José A., S. Chevalier, and R. Gougeon. 2006. "Protein Turnover and Requirements in the Healthy and Frail Elderly." *Journal of Nutrition, Health and Aging* 10 (4).

Penczynski, Katharina J., Thomas Remer, Juliane Menzel, Klaus Abraham, and Cornelia Weikert. 2022. "Urinary Potential Renal Acid Load (UPRAL) among Vegans Versus Omnivores and Its Association with Bone Health in the Cross-Sectional Risks and Benefits of a Vegan Diet Study." *Nutrients* 14 (21). https://doi.org/10.3390/nu14214468.

Phillips, Stuart M., Stéphanie Chevalier, and Heather J. Leidy. 2016. "Protein 'Requirements' beyond the RDA: Implications for Optimizing Health." In *Applied Physiology, Nutrition, and Metabolism = Physiologie Appliquee, Nutrition et Metabolisme.* https://doi.org/10.1139/apnm-2015-0550.

Pullman, Theodore N., Alf S. Alving, Raymond J. Dern, and Milton Landowne. 1954. "The Influence of Dietary Protein Intake on Specific Renal Functions in Normal Man." *The Journal of Laboratory and Clinical Medicine* 44 (2): 320–32.

Redondo-Cuevas, Lucía, Jesús Sanchis-Chordà, and Pilar Codoñer-Franch. 2018. "Child Nutrition and Bone Health." *Journal of Child Science.* Georg Thieme Verlag. https://doi.org/10.1055/s-0038-1669402.

Rizzoli, René, Patrick Ammann, Thierry Chevalley, and Jean Philippe Bonjour. 2001. "Protein Intake and Bone Disorders in the Elderly." *Joint Bone Spine* 68 (5). https://doi.org/10.1016/S1297-319X(01)00295-0.

Rizzoli, René, Emmanuel Biver, and Tara C. Brennan-Speranza. 2021. "Nutritional Intake and Bone Health." *Www.Thelancet.Com/Diabetes-Endocrinology* 9: 606–27. https://doi.org/10.1016/S2213-8587(21)00119-4.

Rizzoli, R., E. Biver, J. P. Bonjour, V. Coxam, D. Goltzman, J. A. Kanis, J. Lappe, et al. 2018. "Benefits and Safety of Dietary Protein for Bone Health—an Expert Consensus Paper Endorsed by the European Society for Clinical and Economical Aspects of Osteopororosis, Osteoarthritis, and Musculoskeletal Diseases and by the International Osteoporosis Foundation." *Osteoporosis International* 29 (9): 1933–48. https://doi.org/10.1007/S00198-018-4534-5.

Robling, Alexander G., Alesha B. Castillo, and Charles H. Turner. 2006. "Biomechanical and Molecular Regulation of Bone Remodeling." *Annual Review of Biomedical Engineering.* https://doi.org/10.1146/annurev.bioeng.8.061505.095721.

Shams-White, Marissa M., Mei Chung, Zhuxuan Fu, Karl L. Insogna, Micaela C. Karlsen, Meryl S. LeBoff, Sue A. Shapses, et al. 2018. "Animal versus Plant Protein and Adult Bone Health: A Systematic Review and Meta-Analysis from the National Osteoporosis Foundation." *PLoS One.* https://doi.org/10.1371/journal.pone.0192459.

Sherman, H. C., A. R. Rose, and M. S. Rose. 1920. "Calcium Requirement of Maintenance in Man." *Journal of Biological Chemistry* 44 (1). https://doi.org/10.1016/s0021-9258(18)86268-8.

Smith, JohnEric W., Megan E. Holmes, and Matthew J. McAllister. 2015. "Nutritional Considerations for Performance in Young Athletes." *Journal of Sports Medicine.* https://doi.org/10.1155/2015/734649.

Specker, B., and M. Vukovich. 2007. "Evidence for an Interaction between Exercise and Nutrition for Improved Bone Health during Growth." *Medicine and Sport Science.* https://doi.org/10.1159/000103004.

Stagi, Stefano, Loredana Cavalli, Chiara Iurato, Salvatore Seminara, Maria Luisa Brandi, and Maurizio De Martino. 2013. "Bone Metabolism in Children and Adolescents: Main Characteristics of the Determinants of Peak Bone Mass." *Clinical Cases in Mineral and Bone Metabolism* 10 (3): 172–79.

Teede, H. J., D. A. Renouf, I. A. Jayasuriya, and D. Barrie. 2012. "STOP Fracture Study: Southern Health Osteoporotic Fracture Screening Project." *Internal Medicine Journal* 42 (5). https://doi.org/10.1111/j.1445-5994.2010.02258.x.

Thorpe, Matthew P., and Ellen M. Evans. 2011. "Dietary Protein and Bone Health: Harmonizing Conflicting Theories." *Nutrition Reviews* 69 (4): 215–30. https://doi.org/10.1111/J.1753-4887.2011.00379.X.

Trumbo, Paula, Sandra Schlicker, Allison A. Yates, and Mary Poos. 2002. "Dietary Reference Intakes for Energy, Carbohydrate, Fiber, Fat, Fatty Acids, Cholesterol, Protein and Amino Acids." *Journal of the American Dietetic Association* 102 (11). https://doi.org/10.1016/S0002-8223(02)90346-9.

Vatanparast, Hassanali, Donald A. Bailey, Adam D. G. Baxter-Jones, and Susan J. Whiting. 2007. "The Effects of Dietary Protein on Bone Mineral Mass in Young Adults May Be Modulated by Adolescent Calcium Intake." *Journal of Nutrition* 137 (12). https://doi.org/10.1093/jn/137.12.2674.

Wachman, A., and D. S. Bernstein. 1968. "Diet and Osteoporosis." *Lancet* 1 (7549). https://doi.org/10.1016/s0140-6736(68)90908-2.

Weaver, C. M., C. M. Gordon, K. F. Janz, H. J. Kalkwarf, J. M. Lappe, R. Lewis, M. O'Karma, T. C. Wallace, and B. S. Zemel. 2016. "The National Osteoporosis Foundation's Position Statement on Peak Bone Mass Development and Lifestyle Factors: A Systematic Review and Implementation Recommendations." *Osteoporosis International* 27 (4). https://doi.org/10.1007/s00198-015-3440-3.

Zhang, Qian, Guansheng Ma, Heather Greenfield, Kun Zhu, Xueqin Du, Leng Huat Foo, Xiaoqi Hu, and David R. Fraser. 2010. "The Association between Dietary Protein Intake and Bone Mass Accretion in Pubertal Girls with Low Calcium Intakes." *British Journal of Nutrition* 103 (5). https://doi.org/10.1017/S0007114509992303.

Zhang, Y. W., M. M. Cao, Y. J. Li, G. C. Dai, P. P. Lu, M. Zhang, L. Y. Bai, et al. 2022. "Dietary Protein Intake in Relation to the Risk of Osteoporosis in Middle-Aged and Older Individuals: A Cross-Sectional Study." *Journal of Nutrition, Health and Aging* 26 (3): 252–58. https://doi.org/10.1007/S12603-022-1748-1/TABLES/4.

Zhu, Xiaowei, and Houfeng Zheng. 2021. "Factors Influencing Peak Bone Mass Gain." *Frontiers of Medicine*. https://doi.org/10.1007/s11684-020-0748-y.

18 Addressing Protein-Energy Wasting in Hemodialysis Patients

The Case for High-Flux and Highly-Efficient Convective-Based Therapies

Bernard Canaud, Marion Morena-Carrere, Hélène Leray-Moragues, and Jean-Paul Cristol

ABBREVIATIONS

BIS	BioImpedance spectroscopy
CI	Creatinine index
CKD	Chronic kidney disease
HCO	High cut-off
HDF	Hemodiafiltration
HF-HD	High-flux hemodialysis
IDPN	Intradialytic parenteral nutrition
MCO	Mid cut-off
MIA	Malnutrition inflammation atherosclerosis
MICS	Malnutrition inflammation complex syndrome
nPNA	Normalized protein nitrogen appearance
ONS	Oral nutrition supplements
PEM	Protein-energy malnutrition
PEW	Protein-energy wasting
RKF	Residual kidney function
TPN	Total parenteral nutrition

INTRODUCTION

Malnutrition presents a significant and multifaceted challenge for end-stage kidney disease patients undergoing dialysis. Firstly, its prevalence is substantial, affecting approximately 30 to 60% of patients, depending on the diagnostic criteria used (Locatelli et al. 2002). Secondly, severe malnutrition can lead to suboptimal health conditions and, in some cases, worsen existing illnesses, leading to poor outcomes. Thirdly, malnutrition is frequently associated with poor patient outcomes. Thus, addressing and rectifying malnutrition in dialysis patients is a primary but complex task.

Hemodialysis patients are particularly susceptible to malnutrition due to factors including the chronic nature of kidney disease, the presence of other coexisting medical conditions, advanced

DOI: 10.1201/b23328-18

age, metabolic effects of uremia, and the nature of renal replacement therapy (Dukkipati and Kopple 2009; Fouque et al. 2011).

Renal replacement therapy modalities are designed to improve uremic disorders and prevent further deterioration of nutritional status. However, inadequate dialysis treatment may inadvertently increase the risk of malnutrition. This chapter will explore the impact of high-flux hemodialysis (HF-HD) and high-volume hemodiafiltration (HDF) through a case study.

The primary objectives of this chapter are to define protein-energy malnutrition (PEM), elucidate the fundamental aspects of protein-energy imbalance in hemodialysis patients, discuss the nutritional management strategies to mitigate this risk, and conclude with key takeaways.

DEFINITION OF MALNUTRITION

Defining malnutrition in dialysis patients is challenging due to varying approaches and definitions (Collins and Yates 2003). The World Health Organization defines malnutrition as a clinical condition encompassing deficiencies, excesses, or imbalances in energy and nutrient intake, including undernutrition, micronutrient-related malnutrition, and overweight/obesity. However, this definition oversimplifies malnutrition in dialysis patients, as dietary changes alone do not address the issue.

An expert panel led by the International Society of Renal Nutrition and Metabolism proposed a definition that considers the overlapping syndromes of malnutrition, inflammation, cachexia, and wasting in chronic kidney disease (CKD) patients. They define severe protein-energy wasting (PEW) as decreased body stores of protein and energy fuels leading to cachexia (Ikizler et al. 2013). Diagnostic criteria for PEW include biochemical indicators, low body weight or weight loss, decreased muscle mass, and low protein or energy intake. However, this approach has limitations: it is not easily applicable clinically and does not address abnormal micronutrient or excess protein/calorie intake. Overnutrition, characterized by overweight and obesity, is increasingly prevalent in the dialysis population and may mask protein wasting in some cases, such as peritoneal dialysis patients receiving high amounts of dextrose. Furthermore, the high prevalence of diabetic patients undergoing dialysis complicates PEW evaluation. The American Society of Parenteral and Enteral Nutrition introduced another definition of malnutrition, encompassing under- or overnutrition with or without inflammatory activity that affects body composition and function (Soeters and Schols 2009). However, this definition does not specifically address PEW in dialysis patients.

Given these limitations, the concept of PEM should be used as an overarching standard for defining malnutrition in dialysis patients (Kopple 1997). This approach acknowledges that malnutrition can manifest in different profiles, such as protein malnutrition, energy deficiency, or a mixed malnutrition state (cachexia, frailty). Frailty, characterized by muscle wasting and reduced physical activity, represents an extreme form of PEM. These malnutrition states stem from complex pathophysiologic processes encompassing PEW, energy imbalances, disruptions in micronutrient levels, inadequate physical activity, and undesired or dialysis-related side effects (Kim, Kalantar-Zadeh, and Kopple 2013).

NUTRITIONAL EQUILIBRIUM IN DIALYSIS PATIENTS

In dialysis patients, nutritional status reflects the equilibrium of a complex interaction of factors in four categories: patient-related, uremic-related, dialysis- or technique-related, and care-related. This is briefly schematized in Figure 18.1.

PATIENT-RELATED FACTORS

It is well documented in the medical literature that older age represents a major risk for PEM. The aging process results in considerable physiological body composition changes, including lean body mass loss and fat gain. In healthy individuals, lean tissue and muscle mass decrease by half between

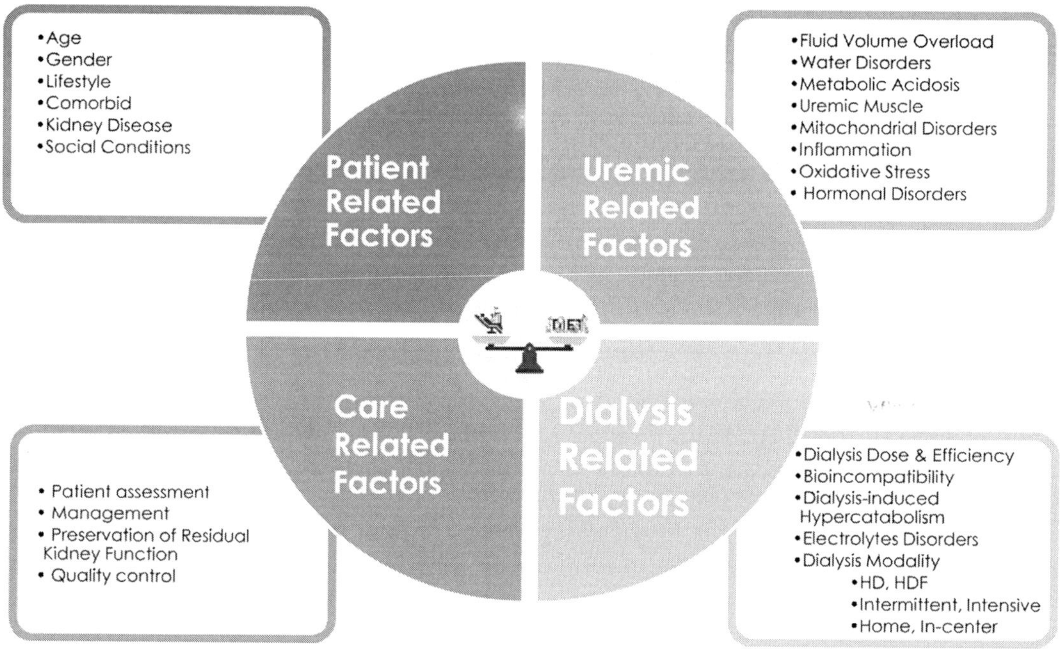

FIGURE 18.1 Factors related to nutritional status in hemodialysis patients. The nutritional status reflects the equilibrium of complex interactions between several factors that may be schematically clustered in four categories: patient-related; uremic-related; dialysis or technique-related; care management (HD, hemodialysis; HDF, hemodiafiltration).

20 and 80 years of age. Several factors may restrict caloric protein intake, including spontaneous loss of appetite, mastication difficulties, subtle micronutrient deficiencies, and limited physical activity. Long-standing CKD and associated comorbid conditions (diabetes mellitus, hypertension, heart failure, limb arteriopathy, depression, etc.) negatively impact the nutritional behavior of uremic patients (Goodkin et al. 2003). Poor environment and lack of family or social support significantly affect the eating habits of the elderly. Precarious socioeconomic conditions also commonly cause diet restrictions. Furthermore, some of the many medications prescribed to these patients may affect dietary intake or protein catabolism (e.g., steroids, non-steroidal anti-inflammatory drugs), contributing to nutritional imbalance. Intercurrent illnesses (e.g., infection, vascular access revision, cardiovascular events) are common in dialysis patients and may lead to hospitalization, representing a major factor for sudden nutritional decompensation (Hakim and Levin 1993).

Consequently, dialysis patients are considered a vulnerable population, marked by an unstable nutritional balance. Their nutritional status can deteriorate rapidly during concurrent events like hospitalization. Early implementation of appropriate measures (dry weight revision) and nutritional support should be considered for these high-risk situations.

Uremic-Related Factors

The most common cause of PEM associated with uremia is inadequate protein and/or calorie intake, primarily due to anorexia resulting from the accumulation of nitrogen-based uremic toxins, dysregulation of appetite mediators (e.g., leptin), and loss of hypothalamic amino acid sensing capacity. Insufficient dietary intake can also be influenced by past diet restrictions, gut malabsorption disorders, medication interactions, depression, or socioeconomic issues (Ekramzadeh et al. 2014).

Fluid volume overload and water-related issues such as hyponatremia have recently emerged as significant contributors to malnutrition (Canaud, Morena-Carrere et al. 2022). They are closely linked to inflammation, either as a direct cause or a resultant factor, along with other oxidative mechanisms that hinder the nutritional health of dialysis patients (Dekker et al. 2018). Given this, effective management of dialysis patients and accurate restoration of fluid and volume balance are of utmost significance. This crucial aspect has been extensively explored in recent studies and reviews (Dekker et al. 2016).

Muscle metabolism and mitochondrial disorders significantly contribute to PEW in uremic patients. Muscle loss and atrophy are common in aging individuals but are also prevalent among dialysis patients. Reduced muscle mass and reduced muscle function in CKD patients also contribute to increased frailty. Uremic myopathy in CKD patients can be attributed to various factors, including inflammation, exposure to high levels of reactive oxygen species, malnutrition, and metabolic acidosis, potentially exacerbated by hemodialysis treatment itself. Mitochondria, essential for muscle cells, play a central role in regulating muscle metabolism, energy supply, energy-sensitive signaling pathways, reactive oxygen species production, calcium balance, and apoptosis. Patients with kidney disease often exhibit lower mitochondrial content in their muscles, along with altered morphology and impaired functionality, which can worsen mitochondrial disorders. Furthermore, muscle mass serves as a crucial reservoir of amino acids and nitrogen reserves for the body. This reserve provides amino acids for protein synthesis when needed, making hemodialysis patients particularly vulnerable due to their compromised muscle mass and function. In this context, food may be utilized as a medicine to alleviate muscle and mitochondrial dysfunction in dialysis patients (Mafra et al. 2021).

Metabolic acidosis has detrimental effects on the nutritional status of uremic patients (Kalantar-Zadeh et al. 2004; Mitch and Maroni 1999). It enhances protein breakdown, reduces protein synthesis, promotes insulin resistance, decreases serum leptin levels, and induces inflammation (Kalantar-Zadeh et al. 2004). Studies suggest that metabolic acidosis upregulates the ATP-dependent ubiquitin-proteasome system and the branched-chain ketoacid dehydrogenase complex, both implicated in skeletal muscle metabolism. The ubiquitin-proteasome system primarily degrades protein in skeletal muscle, while the branched-chain ketoacid dehydrogenase complex regulates valine metabolism, an essential amino acid for muscle protein synthesis during starvation.

Chronic inflammation and oxidative stress (Morena, Cristol, and Canaud 2000), observed in 50 to 60% of dialysis patients, can lead to progressive muscle and visceral protein loss, vascular injury, and altered serum albumin production. Muscle and visceral proteins undergo catabolism, releasing amino acids that the liver utilizes for energy supply or as substrates for acute-phase protein production, such as albumin and fibrinogen, crucial for the body's defense mechanisms. Consequently, reduced serum albumin production by the liver leads to hypoalbuminemia. Additionally, alterations in the vascular endothelium facilitate the access of immune-competent cells to the interstitial space, further supporting the immune response. This microinflammation state is associated with overlapping syndromes known as malnutrition inflammation atherosclerosis (MIA) syndrome or malnutrition inflammation complex syndrome (MICS), reflecting the increased risk of cardiovascular events.

Endocrine and hormonal disorders are prevalent in CKD. Uremia affects hormone production, excretion, regulation, and receptor mechanisms. Hormonal imbalances impact carbohydrate metabolism, thyroid hormone excretion, the hypothalamic-pituitary axis, and the regulation of nutritional and sexual hormones. From a nutritional perspective, hormonal disorders can be categorized into three main groups based on their metabolic impact: those stimulating protein breakdown (insulin resistance, altered adiponectin and resistin metabolism, and increased glucocorticoid activity), those impairing protein synthesis (resistance to growth hormone and insulin-like growth factor-1, testosterone deficiency, altered thyroxine and triiodothyronine metabolism, and tissue sodium accumulation (Deger et al. 2017), and those affecting appetite regulation (leptin, ghrelin).

In summary, uremic-related factors contributing to malnutrition in dialysis patients include inadequate protein and calorie intake, fluid overload, water imbalance, metabolic acidosis, chronic inflammation and oxidative stress, muscle and mitochondrial disorders, and hormone disorders affecting protein metabolism, muscle metabolism abnormalities, synthesis, and appetite regulation. These factors collectively play a significant role in the nutritional challenges faced by these patients.

Dialysis or Technique-Related Factors

Extracorporeal renal replacement modalities aim to improve the nutritional status of dialysis patients by addressing key factors involved in PEM, such as reducing uremic toxin levels, correcting fluid overload, treating metabolic acidosis, and restoring electrolyte balance. The hemodialyzer lies at the core of the extracorporeal treatment system and is a pivotal component within the realm of bioexchangers. Its primary function is to facilitate the transfer of solutes via a dialysate-to-plasma gradient. However, the hemodialyzer is not inert; it exhibits bioreactive properties that activate diverse blood cells and protein cascades. The interaction between blood and the membrane of the hemodialyzer triggers the release of a multitude of mediators. These mediators are a direct result of the activation of cascades involving both cells and proteins. This process includes contact phase and complement activation, cytokine release, and crosstalk between pathways, leading to what is commonly recognized as bioincompatibility reactions, which are key components of inflammation. It's worth noting that this process is not without its consequences (Grooteman and Nubé 2004). The repetitive release of these mediators three times a week due to ongoing blood-membrane interaction has significant and persistent adverse effects on various target organs (Grooteman and Nubé 2004). Dialysis fluids are used to exchange electrolytes and remove uremic toxins. Nevertheless, these modalities have unintended consequences that may contribute to nutritional disorders. These include bioincompatibility (such as acetate), inflammation (due to microbial contaminants), hemodynamic stress, thermal exchange, nutrient loss (including amino acids, small peptides, and water-soluble vitamins), and albumin depletion. These factors collectively exert a negative influence on the nutritional status of dialysis patients (Lim and Kopple 2000).

The question arises whether more efficient renal replacement modalities, such as HF-HD and high-volume HDF, offer benefits or pose additional risks for nutritional management.

Dialysis adequacy is crucial for preserving or restoring nutritional status and improving outcomes for dialysis patients (Combe et al. 2001). Dialysis dose, commonly assessed using Kt/V, provides insight into the clearance of small uremic solutes but not larger toxins. Recent recommendations propose using indicators that reflect the clearances of intermediate and larger uremic toxins (Rosner et al. 2021). For instance, in convective modalities like HDF, this could involve assessing the total ultrafiltered volume per session. Alternatively, biomarkers such as β2-microglobulin that closely mirror the clearance of middle or higher molecular weight accumulated uremic compounds could also be considered (Liabeuf et al. 2012).

Previous studies have demonstrated the impact of diffusive dialysis dose on nutritional outcomes. A parabolic dose-relationship exists between dialysis dose (Kt/V) and normalized protein nitrogen appearance (nPNA) as a surrogate for dietary protein intake, with a minimum dose of 1.2 to 1.5 g/kg/24h required for thrice-weekly treatment. Determining the role of convective dose on nutritional status is more challenging to ascertain due to the influence of diffusive dialysis dose. Recent studies have demonstrated that enhancing convective clearance, either through HF-HD (MPO study) or high-volume HDF in both adults (Maduell et al. 2016; Molina et al. 2018) and children, has shown positive effects on nutritional outcomes. While establishing a definitive explanation for these findings is challenging, research shows that convective-based therapies, namely HDF, may have direct and indirect positive effects on the nutritional status of HDF-treated patients (Canaud, Blankestijn et al. 2022). Convective-based therapies tend to improve leptin clearance, leading to a more efficient reduction of circulating levels of appetite-inhibiting mediators. They also reduce inflammation and oxidative stress, reverse resistance to growth hormone and insulin action, facilitate correction of anemia, and promote better maintenance of physical activity (Canaud, Blankestijn et al. 2022; Pecoits-Filho et al. 2021). Together, these factors mitigate the muscle catabolic effects associated with dialysis while enhancing the muscle anabolic effects.

The bioincompatibility of the hemodialysis system, which triggers acute-phase reactions, significantly affects the nutritional outcomes of dialysis patients. Biocompatibility should be considered as a comprehensive system encompassing the dialyzer membrane, tubing line, and the microbial quality of dialysis fluid. Reactions at the blood-membrane interface during dialysis sessions are triggered by membrane bioincompatibility but are exacerbated by impurities or contaminants in the dialysis fluid. Both the type of dialyzer membrane (synthetic and biocompatible) and the purity of the dialysis fluid play integral roles in improving nutritional outcomes and long-term results. Complement-activating membranes, such as cellulosic-based membranes, can activate pro-inflammatory cells and proteins, negatively impacting the nutritional status of dialysis patients. Contaminated dialysis fluid further amplifies these deleterious effects, inducing cytokine activation and release. Micro-inflammation resulting from complement activation via membrane contact or contaminated dialysis fluid predisposes patients to PEM, cardiovascular disease, β2-microglobulin amyloidosis, and other serious conditions.

Hypercatabolism induced during the hemodialysis session is a well-documented phenomenon (Ikizler et al. 1996). It involves several mechanisms, including hemodynamic stress (resulting in the release of vasoactive peptides and acetate load), hormone derangement (involving glucocorticoid release), intradialytic nutrient losses (such as glucose, amino acids, small peptides, and micronutrients), as well as complement and inflammatory mechanisms, including cytokine activation (due to bioincompatibility) (Lim and Kopple 2000). Amino acid and protein losses during dialysis, coupled with low nutrient intake during the interdialytic period, result in reduced nutrient availability for muscle synthesis. High-protein diets have been shown to have a positive impact on nitrogen balance, indicating the importance of dietary protein intake. In a study comparing high (1.4 g/kg/day) versus low (0.5 g/kg/day) protein diets, it was found that a negative nitrogen balance occurred on the days with a low-protein diet, while a positive nitrogen balance was observed on the days with high protein intake.

Hemodialysis exerts pronounced and deleterious effects on whole-body and skeletal muscle protein homeostasis, with increased net skeletal muscle protein breakdown being a common feature, which persists several hours after the completion of dialysis. Factors contributing to protein breakdown include the absolute decline in amino acid levels due to dialysis losses, imbalances in amino acid levels, and activation of the inflammatory cascade. Interleukin-6 release plays a central role in regulating protein turnover during hemodialysis. Amino acids released as a result of muscle protein breakdown are absorbed by the liver, where they contribute to the creation of acute-phase proteins and serve as an energy source during HD. The amount of amino acid loss in various renal replacement modalities (low-flux HD, HF-HD, and HDF), calculated from seven studies, are summarized in Table 18.1.

TABLE 18.1
Amino Acid Losses According to Renal Replacement Modality

AA Loss (g/session)	Low-Flux HD	High-Flux HD	HDF
Total AA	7.260 [5.587–8.063]	7.300 [6.125–8.875]	5.370 [4.275–5.400]
EAA	2.380 [2.322–2.962]	2.900 [1.955–3.350]	1.630 [1.260–1.645]
NEAA	4.965 [3.275–5.388]	4.440 [3.875–5.760]	3.730 [3.010–3.765]
BCAA	0.900 [0.785–0.930]	0.790 [0.345–0.810]	0.250 [0.245–0.355]

Amount of amino acid loss in various renal replacement modalities (low-flux HD, HF-HD, and HDF) calculated from seven studies.

HD, hemodialysis; HF-HD, high-flux hemodialysis; HDF, hemodiafiltration; AA, amino acid; EAA, essential amino acid; NEAA, non-essential amino acid; BCAA, branched-chain amino acid.

As shown, more efficient modalities such as HF-HD and HDF do not significantly increase amino acid losses compared to low-flux HD, with a mean value of amino acid losses close to 7 g per session. This counterintuitive finding is consistent with the fact that the mean molecular weight of amino acids is 150 daltons, reflecting the clearances of small molecular weight compounds. Additionally, amino acid losses correlate linearly with their plasma concentration. In other words, supplementing amino acids during dialysis significantly increases losses, but the overall net balance is improved. From that perspective, providing amino acids through the dialysate makes more sense to prevent such losses, as shown in some studies (Deleaval et al. 2020). Research demonstrates that these amino acid losses can potentially be counterbalanced through appropriate oral dietary supplementation, as an easily implementable approach (Ikizler et al. 2013). Although intradialytic nutritional supplementation, whether administered orally, intravenously, or through dialysis fluids, enhances protein synthesis, it does not mitigate muscle protein breakdown (Pupim et al. 2006). Notably, no study has yet investigated the immediate effects of ultrapure dialysis fluid on overall and muscle-specific protein breakdown. However, over an extended duration, ultrapure dialysis fluid has demonstrated considerable benefits for nutritional outcomes (Schiffl et al. 2001).

Electrolyte prescription may impact nutritional outcomes in dialysis patients, though systematic studies in this area are limited. Recent evidence highlights the importance of matching dialysate sodium levels to the patient's plasma sodium concentration. This alignment affects not only sodium mass balance and hemodynamic stability but also plasma tonicity, and potentially improves thirst, appetite, and interdialytic weight gain (Kuhlmann et al. 2019; Petitclerc and Mercadal 2023). Case reports show positive effects of correcting electrolyte disorders, such as acidosis and hypercalcemia, on nutritional status (de Brito-Ashurst et al. 2009). Clinicians adjust dialysate calcium concentration to manage calcium balance in dialysis patients.

Renal replacement modalities are evolving to include enhanced techniques, such as HF-HD, convective-based methods like high-volume HDF (external HDF), more permeable membranes (high cut-off, HCO, or medium cut-off, MCO) (internal HDF), and extended dialysis treatment duration (ranging from more frequent sessions to nocturnal or intensive dialysis) (Canaud et al. 2021). These highly efficient modalities aim to increase solute mass removal. However, they may also increase the clearance of essential nutritional components like amino acids, small peptides, hormones (retinol binding protein), micronutrients, and water-soluble vitamins. Certain nutrients, particularly folate, pyridoxine, and thiamine, deserve special attention due to their cardiovascular impact. This potential risk needs individual evaluation. Nephrologists must consider how these approaches affect effective nutritional management and the risk of PEM, especially regarding albumin losses. Different treatment schedules (daily, alternate days, nocturnal sessions), specific modalities (MCO membrane, HCO membrane, protein-leaking membrane), or combinations thereof (nocturnal HDF, MCO, or expanded dialysis) could enable more precise assessment of these assumptions for individual patients (Bévier et al. 2022; Morena et al. 2002). When choosing dialyzers and treatment schedules (i.e., modality, frequency, treatment time), clinicians should consider both efficacy in clearing middle and larger molecules and nutritional safety. Individual assessments are crucial in estimating the risk/benefit ratio and determining appropriate treatment options and nutrient supplementation. Nevertheless, certain membrane types may significantly increase albumin losses, as recently highlighted.

In summary, dialysis or technique-related factors influence the nutritional outcomes of dialysis patients. Efficient solute removal can result in the loss of essential nutritional compounds such as amino acids, micronutrients, and water-soluble vitamins. The biocompatibility of the hemodialysis system, hypercatabolism during dialysis, electrolyte prescription, and choice of renal replacement modalities all critically influence nutritional management. Clinicians must consider patient-specific nutritional status and balance the risks and benefits when selecting appropriate treatment options. Preserving residual kidney function (RKF) may significantly benefit dialysis patients, including improving nutritional status and overall outcomes.

Care Management-Related Factors

Nutritional assessment and monitoring are crucial for managing dialysis patients (Combe et al. 2001; Cano et al. 2002; Chauveau et al. 2001). Guidelines recommend regular evaluation of nutritional status, with quarterly assessments at a minimum and more frequent monitoring if necessary. While laboratory tests and technology-based methods can provide precise quantification, the diagnosis of nutritional status should primarily be based on clinical context, nutritional history, and physical examination (Marcelli, Wabel et al. 2015). Figure 18.2 provides a comprehensive overview of the tools needed to explore the various facets of nutritional assessment in dialysis patients. Notably, within these approaches, a specialized renal dietitian can effectively assist in analyzing dietary intake and history, including consideration of unconventional diets (such as a very low-salt diet) that could potentially impact appetite. Table 18.2 presents current dietary intake recommendations for patients with Stage-5 CKD undergoing dialysis.

Anthropometric assessment, focusing on trends in body weight, skinfold thickness, mid-arm peripheral circumference, and muscle strength measures (e.g., handgrip test), is easily performed and relevant for routine nutritional monitoring. Laboratory tests, such as albumin, transthyretin, retinol-binding protein, and C-reactive protein (CRP), are essential for assessing malnutrition risk and guiding nutritional support. The nPNA is an objective tool for assessing dietary protein intake and predicting outcomes. The creatinine index (CI), calculated using a simplified bedside formula, can monitor muscle mass behavior, PEW risk, and outcomes (Canaud et al. 2014, 2020). Body composition monitoring

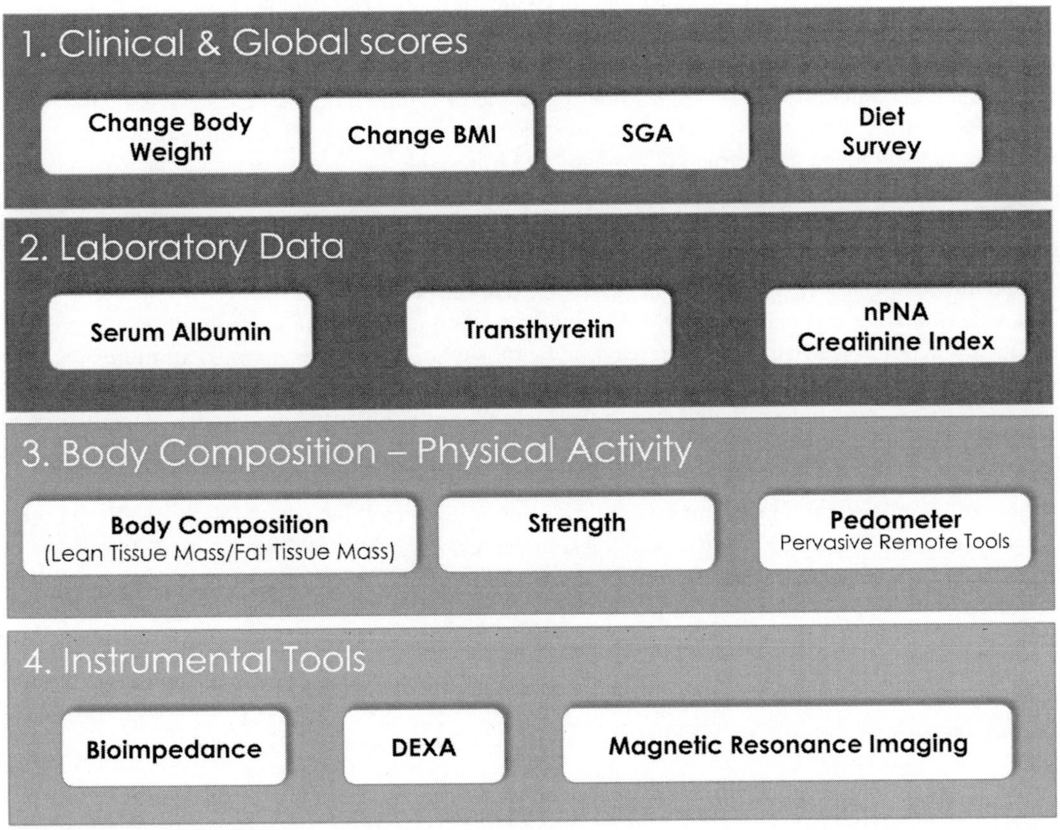

FIGURE 18.2 Tools for assessing nutritional status in hemodialysis patients. This figure presents a comprehensive overview of the tools needed to explore the various facets involved in the nutritional assessment of a dialysis patient (BMI, body mass index; SGA, subjective global assessment; nPNA, normalized protein nitrogen appearance).

TABLE 18.2
Diet Intake Recommendations in Dialysis Patients

Nutritional Parameter CKD5-HD	ERBP	KDOQI
Calories (Kcal/kg IBW/day)	30–40	35 (<60 yrs)
	Adjusted to age, gender, and physical activity	30–35 (>60 yrs)
Protein (g/kg IBW/day)	>1.1	1.0–1.2
Carbohydrates (% of calories)	50–60	50–60
Fat (% of calories)	<30 and <10% saturated fat	<30 and <10% saturated fat
Sodium (mg/day)	2000	2000
Potassium (mg/day)	2000–3000	2000–3000
Calcium (mg/day)	<2000	<2000
Phosphorus (mg/day)	800–1000	800–1000
Fluid (mL/day)	1000 + Urine	1000 + Urine

ERBP: European renal best practice; KDOQI: Kidney Disease Outcomes Quality Initiative.

using bioimpedance spectroscopy (BIS) is valuable for assessing fluid overload and changes in body composition, including lean tissue mass and fat tissue mass, in CKD patients (Marcelli, Usvyat et al. 2015). More sophisticated imaging tools, such as DEXA, tomography, and isotopic methods, including total potassium, have been proposed to assess body composition. However, these remain within the domain of clinical research and are not applicable in routine clinical practice (Marcelli, Wabel et al. 2015). BIS offers a cost-effective and reliable non-invasive tool for guiding clinical decisions, including achieving the appropriate dry weight and improving outcomes (Broers et al. 2020). When assessing the nutritional status of elderly dialysis patients, clinicians should consider two additional factors: dental status (severely affected in about two-thirds of cases) and physical activity levels (significantly reduced compared to healthy populations, indicating a sedentary lifestyle). These factors should inform the nutritional management plan for dialysis patients.

Preserving RKF can play a crucial role in safeguarding dialysis patients from PEM and enhancing their outcomes. RKF presents an intriguing factor supporting the development of more adaptable dialysis treatment schedules, such as incremental dialysis. This approach has been suggested for regions facing heavy dialysis demand or to facilitate the initiation of dialysis under specific circumstances (Caton et al. 2023). Studies have shown that even low residual glomerular filtration rates correlate with better outcomes. Although the exact mechanisms by which RKF benefits dialysis patients are not fully understood, factors such as improved uremic control, reduced fluid overload, decreased dialysis-induced stress, and better nutritional status likely contribute to these positive effects.

Maintaining continuous quality control and meeting dialysis adequacy targets is essential for better outcomes in hemodialysis patients, as demonstrated in a recent large cohort study. In this context, preserving nutritional parameters is paramount, as discussed earlier.

In summary, regular nutritional assessment and monitoring, including dietary evaluation, anthropometric assessment, laboratory tests, body composition monitoring, and consideration of dental status and physical activity, are essential for effectively managing dialysis patients' care.

NUTRITIONAL MANAGEMENT OF DIALYSIS PATIENTS

Assessing Nutritional Status Risk

Various tools and scoring methods have been used to assess PEM in dialysis patients. However, these tools are often clinically based, lack precision, or are not user-friendly. To effectively manage nutritional status, it is important to estimate the severity of PEM through a stepwise quantitative scoring system that combines multiple methods. The assessment should be simple, reliable, and dynamic for clinical use.

Clinicians should categorize the severity of PEM to determine appropriate nutritional management. They can consider four categories: normal nutritional status, minor, moderate, and severe PEM. It is crucial to recognize that dialysis patients are metabolically unstable and more vulnerable to factors such as intercurrent illness, surgical interventions, and hospitalization, which heighten their risk of PEM compared to healthier individuals.

Optimizing Renal Replacement Therapy

This section does not cover the indicators of renal replacement adequacy or dialysis efficiency. Interested readers are referred to textbooks or documents that cover this aspect.

However, it is crucial to emphasize that inadequate dialysis significantly contributes to PEM in dialysis patients. PEM is part of a constellation of symptoms reflecting the uremic state, including anorexia, gastrointestinal disorders, pruritus, and peripheral neuropathy. Therefore, when diagnosing PEM, the first step is to assess dialysis efficiency and identify potential causes of failure. Clinicians should adopt a stepwise approach, checking for vascular access dysfunction, inadequate dialysis prescription or treatment delivery, poor patient compliance, or a discrepancy between the delivered dialysis dose and the patient's metabolic needs. In these situations, the focus should be on swiftly correcting the cause of dialysis efficacy failure. In most cases, nutritional intervention is not required if prompt corrective action is taken.

Preserving RKF

Retrospective studies have consistently shown that preserving residual kidney function in dialysis patients correlates with better outcomes. It is linked to reduced mortality and improved intermediate outcomes, including lower interdialytic weight gain, preserved nutritional status, and fewer instances of fluid overload and electrolyte disorders. Therefore, strategies aiming to preserve residual kidney function are appealing, such as reducing hemodynamic stress (e.g., low ultrafiltration rate, controlled volume), implementing incremental dialysis programs, and using more biocompatible dialysis systems (e.g., high-flux synthetic membrane, ultrapure dialysis fluid, HDF). However, further research is needed to determine the specific impact of preserving residual kidney function on dialysis patients' nutritional status.

Preventing PEM

Opinions suggest that stable HD patients generally require higher daily protein and caloric intake compared to the general population, although individual needs may vary. Scientific societies have issued recommendations for daily dietary protein and caloric requirements for dialysis patients (Ikizler et al. 2013; Fouque et al. 2007).

Treating protein and/or calorie undernutrition involves providing patients with sufficient protein and/or calories to meet their daily energy and protein expenditure, while also addressing the underlying causes of PEM, such as depression or dysphagia. Regular oral diet supplementation during or after dialysis sessions has been shown to mitigate dialysis-induced hypercatabolism, and clinicians should encourage this practical measure to prevent PEM (Pupim et al. 2006). Practitioners can use dialysis fluid to prevent nutrient losses or provide supplemental glucose or amino acids as dialytic nutrition (Deleaval et al. 2020). Patients with dietary restrictions on potassium, sodium, and phosphorus require special attention, as standard oral supplements may not be suitable. Clinicians should consider specific renal supplements, taking into account individual needs and concerns, such as avoiding high-potassium supplements in patients with hyperkalemia. Apart from prescriptions primarily related to renal replacement therapy, healthcare providers should encourage and educate patients about increasing physical activity through any means to prevent deconditioning and maintain their muscle mass and cardiac health.

Correcting PEM

Once PEM is identified, a treatment plan can be formulated based on the severity of the condition (Ikizler et al. 2013). There are three major degrees of PEM: minor, moderate, and severe. Figure 18.3 illustrates a stepwise approach according to the severity level.

Minor PEM

For minor cases of PEM, partial or intermittent oral supplementation is suitable, with a focus on increasing dietary energy and protein intake. This can be achieved by recommending high-protein and caloric foods and using oral nutrition supplements (ONS) such as fortified beverages or powders (Clarkson et al. 2019). ONS provide additional proteins, calories, vitamins, minerals, and trace elements to meet the patient's needs. Typically, an ONS provides approximately 250–350 calories, along with 9–13 g of protein, and a rich array of 20 or more essential vitamins and minerals each day. It's important to bear in mind that oral nutritional supplements are intended to complement, not replace, regular meals. Patients can consume them as an extra snack between meals or even during a dialysis session. ONS have demonstrated significant improvements in outcomes for dialysis patients. Temporary supplementation of vitamins, micronutrients, and trace elements may also help address subclinical deficiencies that affect taste, appetite, and physical exercise ability (Wright and Weiner 2012). Healthcare providers should incorporate physical activity and aerobic exercise into the nutritional and health rehabilitation program, including regular intradialytic exercise tailored to the patient's abilities to enhance muscle strength and overall well-being.

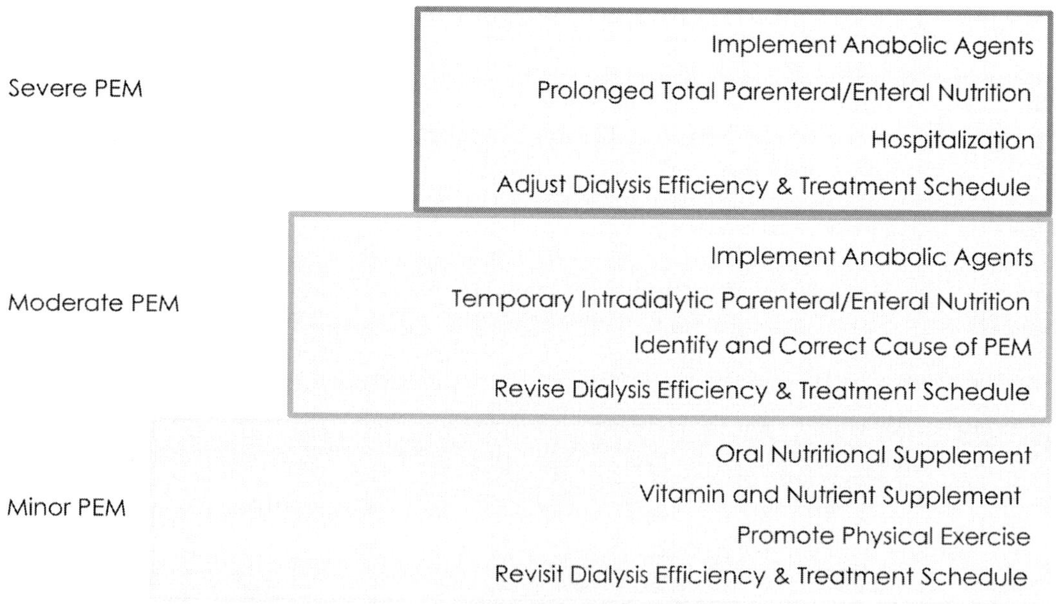

FIGURE 18.3 Stepwise approach for addressing protein-energy malnutrition in dialysis patients. Once protein energy malnutrition is identified, a treatment plan can be formulated based on the severity of the condition with three major degrees, minor, moderate, and severe (PEM, protein-energy malnutrition).

Moderate PEM

For moderate PEM or when initial nutritional interventions prove ineffective, intradialytic parenteral nutrition (IDPN) can be considered (Pupim et al. 2006). IDPN is administered intravenously during dialysis sessions, typically through a central venous route or dialysis shunt. The appropriate prescription needs to be calculated. The use of IDPN as supplemental therapy for malnourished dialysis patients remains controversial regarding its impact on overall outcomes. While some studies have shown that IDPN improves protein synthesis rates, others have found no additional benefit in mortality reduction compared to oral supplements alone. A comprehensive, randomized study in France involving hemodialysis patients with PEM found that nutritional support—via either IDPN or oral supplements—could elevate serum albumin and transthyretin levels. However, it's noteworthy to highlight that IDPN showed no advantage over oral supplements alone in reducing mortality. Conversely, a randomized German multicenter trial demonstrated that IDPN improved subjective global assessment, nutritional biomarkers, and patient outcomes. IDPN should be considered when oral or enteral feedings are not feasible or have not been successful.

Severe PEM

In cases of severe PEM or inadequate response to previous interventions, patients require continuous nutritional support. This can be provided through ambulatory home therapy or hospitalization-based methods. For patients who cannot tolerate enteral feeding, total parenteral nutrition (TPN) is an option. TPN is administered via a central venous route during the interdialytic period or via an AV fistula during dialysis (Martin-Alemañy et al. 2016). Calculating the appropriate prescription and coordinating with dialysis and fluid volume adjustments requires the assistance of a nutritional consultant or trained dietitian. TPN should be considered when intradialytic parenteral nutrition (IDPN) and other measures have failed to improve nutritional status. A risk-benefit assessment is necessary when deciding on the route of nutrition administration.

Anabolic agents, such as recombinant human growth hormone, can stimulate protein and energy anabolism and help reverse PEM in end-stage kidney disease patients. However, the clinical evidence supporting their efficacy is limited to short-term uncontrolled studies. Researchers have studied the use of nandrolone decanoate, an anabolic steroid, in pre-dialysis patients for its effect on body composition, but long-term safety and efficacy data are lacking.

Appetite stimulation is another approach to reversing PEM. Correcting micronutrient deficiencies through aggressive supplementation has shown promise in improving appetite. Dialysis patients may have concerns about micronutrient deficiencies, particularly water-soluble vitamins lost during hemodialysis. Supplementing with daily dialysis-specific vitamins and trace element compounds is common, although research on the benefits of routine supplementation is limited. Supplementation with single micronutrients or mixtures of trace elements (e.g., zinc, cobalt) has been associated with appetite improvement and taste recovery. The optimal intake of micronutrients in dialysis patients remains undefined, and various prescription dialysis-specific vitamin supplements are available.

Other strategies to stimulate appetite, such as megestrol acetate and ghrelin, are still in the early research stages and lack sufficient evidence to support their routine use (Cano et al. 2009). L-carnitine, an essential co-factor in fatty acid metabolism, has been linked to certain dialysis-related functional deficiencies. However, the use of L-carnitine supplements for treating these disorders is controversial, and current recommendations suggest limited trial periods, discontinuing if there is no response.

Promoting Physical Exercise

Physical activity is significantly impaired in dialysis patients. Observational studies using monitoring devices have shown that daily activity is reduced by almost 50% compared to a healthy population during the interdialytic period and by close to 75% on dialysis days. Encouraging physical activity through individualized wellness coaching programs, similar to those used for the elderly population, can prevent muscle wasting and increase dietary protein and energy intake. Preliminary short-term studies promoting cycling activity during dialysis have shown promising results in improving physical fitness and nutritional status, including changes in body composition.

CONCLUSION

PEM is a significant concern in dialysis patients and requires proper management to improve patient outcomes. While PEM is the most common form of malnutrition in these patients, it is not the only one. It arises from a combination of factors including the chronic nature of kidney disease, underlying comorbidities, advanced age, and the metabolic effects of uremia.

Renal replacement therapy modalities aim to address uremic disorders and prevent the worsening of PEM. However, if hemodialysis treatment is not appropriately dosed or performed, it can contribute to the aggravation or onset of PEM. It is crucial for physicians to prioritize nutritional monitoring in dialysis patients and provide timely and adequate nutritional support when needed. By doing so, the damaging consequences in this high-risk population can be prevented.

MINI DICTIONARY OF TERMS

Care management: A method for managing and treating end-stage kidney disease patients.

End-stage kidney disease: Stage 5 chronic kidney disease (CKD) patient requiring maintenance dialysis.

Intercurrent illnesses: Diseases associated with CKD.

Malnutrition: Protein-energy malnutrition, protein malnutrition, energy malnutrition.

Physical exercise: Methods for reconditioning dialysis patients through exercise.

Renal replacement therapy: Extracorporeal therapy relying on hemodialysis or hemodiafiltration.

Residual kidney function: Kidney function maintained while the patient is on maintenance dialysis.

SUMMARY POINTS

- Malnutrition affects an estimated 30 to 60% of dialysis patients, depending on the diagnostic criteria.
- Severe malnutrition can compromise health conditions, potentially worsening existing illnesses and leading to poor outcomes for end-stage kidney disease (ESKD) patients.
- Malnutrition consistently correlates with adverse patient outcomes, underscoring the critical need to address this complex issue within the context of dialysis care.
- Protein-energy malnutrition emerges as a comprehensive standard, encompassing different profiles such as protein malnutrition, energy deficiency, or mixed malnutrition states, reflecting complex pathophysiologic processes.
- Nutritional status in hemodialysis patients results from a delicate equilibrium influenced by patient-related, uremic-related, dialysis- or technique-related, and care management factors.
- Older age, comorbidities, disabilities, and intercurrent illness contribute to malnutrition in dialysis patients.
- Key contributors to malnutrition in dialysis patients include: inadequate protein and calorie intake, fluid volume overload and hyponatremia, muscle metabolism and mitochondrial disorders, metabolic acidosis, chronic inflammation, oxidative stress, and endocrine and hormonal disorders.
- Factors impacting nutritional status include: dialyzer type, dialysis fluid purity, dialysis adequacy, bioincompatibility, hemodialysis-induced hypercatabolism, electrolyte

prescription, and renal replacement modalities (high-flux hemodialysis, high-volume hemodiafiltration).

- Crucial aspects of care management include: nutritional assessment and monitoring, maintaining residual kidney function, continuous quality control, and achieving dialysis adequacy targets.
- Nutritional management strategies should stratify nutritional status risk, provide optimal renal replacement therapy, preserve residual kidney function, and prevent and correct protein-energy malnutrition (PEM).
- Clinicians should develop a treatment plan tailored to the severity of PEM, categorized as minor, moderate, or severe.
- Treatment plans should incorporate physical exercise interventions to mitigate and/or correct PEM.

REFERENCES

Bévier, A., E. Novel-Catin, E. Blond, S. Pelletier, F. Parant, L. Koppe, and D. Fouque. 2022. "Water-soluble vitamins and trace elements losses during on-line hemodiafiltration." *Nutrients* 14 (17). doi: 10.3390/nu14173454.

Broers, N. J. H., B. Canaud, M. J. E. Dekker, F. M. van der Sande, S. Stuard, P. Wabel, and J. P. Kooman. 2020. "Three compartment bioimpedance spectroscopy in the nutritional assessment and the outcome of patients with advanced or end stage kidney disease: What have we learned so far?" *Hemodial Int* 24 (2):148–61. doi: 10.1111/hdi.12812.

Canaud, B., P. J. Blankestijn, M. P. C. Grooteman, and A. Davenport. 2022. "Why and how high volume hemodiafiltration may reduce cardiovascular mortality in stage 5 chronic kidney disease dialysis patients? A comprehensive literature review on mechanisms involved." *Semin Dial* 35 (2):117–28. doi: 10.1111/sdi.13039.

Canaud, B., A. Granger Vallee, N. Molinari, L. Chenine, H. Leray-Moragues, A. Rodriguez, L. Chalabi, M. Morena, and J. P. Cristol. 2014. "Creatinine index as a surrogate of lean body mass derived from urea Kt/V, pre-dialysis serum levels and anthropometric characteristics of haemodialysis patients." *PLoS One* 9 (3):e93286.

Canaud, B., M. Morena-Carrere, H. Leray-Moragues, and J. P. Cristol. 2022. "Fluid overload and tissue sodium accumulation as main drivers of protein energy malnutrition in dialysis patients." *Nutrients* 14 (21). doi: 10.3390/nu14214489.

Canaud, B., S. Stuard, F. Laukhuf, G. Yan, M. I. G. Canabal, P. S. Lim, and M. A. Kraus. 2021. "Choices in hemodialysis therapies: variants, personalized therapy and application of evidence-based medicine." *Clin Kidney J* 14 (Suppl 4):i45–i58. doi: 10.1093/ckj/sfab198.

Canaud, B., X. Ye, L. Usvyat, J. Kooman, F. van der Sande, J. Raimann, Y. Wang, and P. Kotanko. 2020. "Clinical and predictive value of simplified creatinine index used as muscle mass surrogate in end-stage kidney disease haemodialysis patients-results from the international monitoring dialysis outcome initiative." *Nephrol Dial Transplant* 35 (12):2161–71. doi: 10.1093/ndt/gfaa098.

Cano, N. J., M. Aparicio, G. Brunori, J. J. Carrero, B. Cianciaruso, E. Fiaccadori, B. Lindholm, V. Teplan, D. Fouque, and G. Guarnieri. 2009. "Espen guidelines on parenteral nutrition: adult renal failure." *Clin Nutr* 28 (4):401–14. doi: 10.1016/j.clnu.2009.05.016.

Cano, N. J., H. Roth, M. Aparicio, R. Azar, B. Canaud, P. Chauveau, C. Combe, D. Fouque, M. Laville, and X. M. Leverve. 2002. "Malnutrition in hemodialysis diabetic patients: evaluation and prognostic influence." *Kidney Int* 62 (2):593–601.

Caton, E., S. Sharma, E. Vilar, and K. Farrington. 2023. "Impact of incremental initiation of haemodialysis on mortality: a systematic review and meta-analysis." *Nephrol Dial Transplant* 38 (2):435–46. doi: 10.1093/ndt/gfac274.

Chauveau, P., C. Combe, M. Laville, D. Fouque, R. Azar, N. Cano, B. Canaud, H. Roth, X. Leverve, and M. Aparicio. 2001. "Factors influencing survival in hemodialysis patients aged older than 75 years: 2.5-year outcome study." *Am J Kidney Dis* 37 (5):997–1003. doi: 10.1016/s0272-6386(05)80016-2.

Clarkson, M. J., P. N. Bennett, S. F. Fraser, and S. A. Warmington. 2019. "Exercise interventions for improving objective physical function in patients with end-stage kidney disease on dialysis: a systematic review and meta-analysis." *Am J Physiol Renal Physiol* 316 (5):F856–72. doi: 10.1152/ajprenal.00317.2018.

Collins, S., and R. Yates. 2003. "The need to update the classification of acute malnutrition." *Lancet* 362 (9379):249. doi: 10.1016/s0140-6736(03)13926-8.

Combe, C., P. Chauveau, M. Laville, D. Fouque, R. Azar, N. Cano, B. Canaud, H. Roth, X. Leverve, and M. Aparicio. 2001. "Influence of nutritional factors and hemodialysis adequacy on the survival of 1,610 French patients." *Am J Kidney Dis* 37 (1 Suppl 2):S81–8.

de Brito-Ashurst, I., M. Varagunam, M. J. Raftery, and M. M. Yaqoob. 2009. "Bicarbonate supplementation slows progression of CKD and improves nutritional status." *J Am Soc Nephrol* 20 (9):2075–84. doi: 10.1681/asn.2008111205.

Deger, S. M., P. Wang, R. Fissell, C. D. Ellis, C. Booker, F. Sha, J. L. Morse, T. G. Stewart, J. C. Gore, E. D. Siew, J. Titze, and T. A. Ikizler. 2017. "Tissue sodium accumulation and peripheral insulin sensitivity in maintenance hemodialysis patients." *J Cachexia Sarcopenia Muscle* 8 (3):500–7. doi: 10.1002/jcsm.12179.

Dekker, M. J. E., C. Konings, B. Canaud, F. M. van der Sande, S. Stuard, J. G. Raimann, E. Öztürk, L. Usvyat, P. Kotanko, and J. P. Kooman. 2018. "Interactions between malnutrition, inflammation, and fluid overload and their associations with survival in prevalent hemodialysis patients." *J Ren Nutr* 28 (6):435–44. doi: 10.1053/j.jrn.2018.06.005.

Dekker, M. J. E., D. Marcelli, B. Canaud, C. J. Konings, K. M. Leunissen, N. W. Levin, P. Carioni, V. Maheshwari, J. G. Raimann, F. M. van der Sande, L. A. Usvyat, P. Kotanko, and J. P. Kooman. 2016. "Unraveling the relationship between mortality, hyponatremia, inflammation and malnutrition in hemodialysis patients: results from the international MONDO initiative." *Eur J Clin Nutr* 70 (7):779–84. doi: 10.1038/ejcn.2016.49.

Deleaval, P., B. Luaire, P. Laffay, D. Jambut-Cadon, M. Stauss-Grabo, B. Canaud, and C. Chazot. 2020. "Short-term effects of branched-chain amino acids-enriched dialysis fluid on branched-chain amino acids plasma level and mass balance: a randomized cross-over study." *J Ren Nutr* 30 (1):61–8. doi: 10.1053/j.jrn.2019.03.079.

Dukkipati, R., and J. D. Kopple. 2009. "Causes and prevention of protein-energy wasting in chronic kidney failure." *Semin Nephrol* 29 (1):39–49. doi: 10.1016/j.semnephrol.2008.10.006.

Ekramzadeh, M., Z. Mazloom, P. Jafari, M. Ayatollahi, and M. M. Sagheb. 2014. "Major barriers responsible for malnutrition in hemodialysis patients: challenges to optimal nutrition." *Nephrourol Mon* 6 (6):e23158. doi: 10.5812/numonthly.23158.

Fouque, D., S. Pelletier, D. Mafra, and P. Chauveau. 2011. "Nutrition and chronic kidney disease." *Kidney Int* 80 (4):348–57. doi: 10.1038/ki.2011.118.

Fouque, D., M. Vennegoor, P. ter Wee, C. Wanner, A. Basci, B. Canaud, P. Haage, K. Konner, J. Kooman, A. Martin-Malo, L. Pedrini, F. Pizzarelli, J. Tattersall, J. Tordoir, and R. Vanholder. 2007. "EBPG guideline on nutrition." *Nephrol Dial Transplant* 22 Suppl 2:ii45–87.

Goodkin, D. A., J. L. Bragg-Gresham, K. G. Koenig, R. A. Wolfe, T. Akiba, V. E. Andreucci, A. Saito, H. C. Rayner, K. Kurokawa, F. K. Port, P. J. Held, and E. W. Young. 2003. "Association of comorbid conditions and mortality in hemodialysis patients in Europe, Japan, and the United States: the Dialysis Outcomes and Practice Patterns Study (DOPPS)." *J Am Soc Nephrol* 14 (12):3270–7. doi: 10.1097/01.asn.0000100127.54107.57.

Grooteman, M. P., and M. J. Nubé. 2004. "Impact of the type of dialyser on the clinical outcome in chronic haemodialysis patients: does it really matter?" *Nephrol Dial Transplant* 19 (12):2965–70. doi: 10.1093/ndt/gfh502.

Hakim, R. M., and N. Levin. 1993. "Malnutrition in hemodialysis patients." *Am J Kidney Dis* 21 (2):125–37. doi: 10.1016/s0272-6386(12)81083-3.

Ikizler, T. A., N. J. Cano, H. Franch, D. Fouque, J. Himmelfarb, K. Kalantar-Zadeh, M. K. Kuhlmann, P. Stenvinkel, P. TerWee, D. Teta, A. Y. Wang, and C. Wanner. 2013. "Prevention and treatment of protein energy wasting in chronic kidney disease patients: a consensus statement by the international society of renal nutrition and metabolism." *Kidney Int* 84 (6):1096–107. doi: 10.1038/ki.2013.147.

Ikizler, T. A., R. L. Wingard, M. Sun, J. Harvell, R. A. Parker, and R. M. Hakim. 1996. "Increased energy expenditure in hemodialysis patients." *J Am Soc Nephrol* 7 (12):2646–53. doi: 10.1681/asn.v7122646.

Kalantar-Zadeh, K., R. Mehrotra, D. Fouque, and J. D. Kopple. 2004. "Metabolic acidosis and malnutrition-inflammation complex syndrome in chronic renal failure." *Semin Dial* 17 (6):455–65.

Kim, J. C., K. Kalantar-Zadeh, and J. D. Kopple. 2013. "Frailty and protein-energy wasting in elderly patients with end stage kidney disease." *J Am Soc Nephrol* 24 (3):337–51. doi: 10.1681/asn.2012010047.

Kopple, J. D. 1997. "McCollum Award Lecture, 1996: protein-energy malnutrition in maintenance dialysis patients." *Am J Clin Nutr* 65 (5):1544–57. doi: 10.1093/ajcn/65.5.1544.

Kuhlmann, U., A. Maierhofer, B. Canaud, J. Hoyer, and M. Gross. 2019. "Zero diffusive sodium balance in hemodialysis provided by an algorithm-based electrolyte balancing controller: a proof of principle clinical study." *Artif Organs* 43 (2):150–8.

Liabeuf, S., A. Lenglet, L. Desjardins, N. Neirynck, G. Glorieux, H. D. Lemke, R. Vanholder, M. Diouf, G. Choukroun, and Z. A. Massy. 2012. "Plasma beta-2 microglobulin is associated with cardiovascular disease in uremic patients." *Kidney Int* 82 (12):1297–303. doi: 10.1038/ki.2012.301.

Lim, V. S., and J. D. Kopple. 2000. "Protein metabolism in patients with chronic renal failure: role of uremia and dialysis." *Kidney Int* 58 (1):1–10. doi: 10.1046/j.1523-1755.2000.00135.x.

Locatelli, F., D. Fouque, O. Heimburger, T. B. Drüeke, J. B. Cannata-Andía, W. H. Hörl, and E. Ritz. 2002. "Nutritional status in dialysis patients: a European consensus." *Nephrol Dial Transplant* 17 (4):563–72. doi: 10.1093/ndt/17.4.563.

Maduell, F., R. Ojeda, M. Arias-Guillen, F. Rossi, N. Fontseré, M. Vera, N. Rico, L. N. Gonzalez, G. Piñeiro, M. Jiménez-Hernández, L. Rodas, and J. L. Bedini. 2016. "Eight-year experience with nocturnal, every-other-day, online haemodiafiltration." *Nephron* 133 (2):98–110. doi: 10.1159/000446970.

Mafra, D., N. A. Borges, B. Lindholm, P. G. Shiels, P. Evenepoel, and P. Stenvinkel. 2021. "Food as medicine: targeting the uraemic phenotype in chronic kidney disease." *Nat Rev Nephrol* 17 (3):153–71. doi: 10.1038/s41581-020-00345-8.

Marcelli, D., L. A. Usvyat, P. Kotanko, I. Bayh, B. Canaud, M. Etter, E. Gatti, A. Grassmann, Y. Wang, C. Marelli, L. Scatizzi, A. Stopper, F. M. van der Sande, and J. Kooman. 2015. "Body composition and survival in dialysis patients: results from an international cohort study." *Clin J Am Soc Nephrol* 10 (7):1192–200. doi: 10.2215/cjn.08550814.

Marcelli, D., P. Wabel, S. Wieskotten, A. Ciotola, A. Grassmann, A. Di Benedetto, and B. Canaud. 2015. "Physical methods for evaluating the nutrition status of hemodialysis patients." *J Nephrol* 28 (5):523–30. doi: 10.1007/s40620-015-0189-x.

Martin-Alemañy, G., R. Valdez-Ortiz, G. Olvera-Soto, I. Gomez-Guerrero, G. Aguire-Esquivel, G. Cantu-Quintanilla, J. C. Lopez-Alvarenga, P. Miranda-Alatriste, and A. Espinosa-Cuevas. 2016. "The effects of resistance exercise and oral nutritional supplementation during hemodialysis on indicators of nutritional status and quality of life." *Nephrol Dial Transplant* 31 (10):1712–20. doi: 10.1093/ndt/gfw297.

Mitch, W. E., and B. J. Maroni. 1999. "Factors causing malnutrition in patients with chronic uremia." *Am J Kidney Dis* 33 (1):176–9. doi: 10.1016/s0272-6386(99)70279-9.

Molina, P., B. Vizcaíno, M. D. Molina, S. Beltrán, M. González-Moya, A. Mora, C. Castro-Alonso, J. Kanter, A. I. Ávila, J. L. Górriz, N. Estañ, L. M. Pallardó, D. Fouque, and J. J. Carrero. 2018. "The effect of high-volume online haemodiafiltration on nutritional status and body composition: the ProtEin stores prEservaTion (PESET) study." *Nephrol Dial Transplant* 33 (7):1223–35. doi: 10.1093/ndt/gfx342.

Morena, M., J. P. Cristol, J. Y. Bosc, C. Tetta, G. Forret, C. L. Leger, C. Delcourt, L. Papoz, B. Descomps, and B. Canaud. 2002. "Convective and diffusive losses of vitamin C during haemodiafiltration session: a contributive factor to oxidative stress in haemodialysis patients." *Nephrol Dial Transplant* 17 (3):422–7. doi: 10.1093/ndt/17.3.422.

Morena, M., J. P. Cristol, and B. Canaud. 2000. "Why hemodialysis patients are in a prooxidant state? What could be done to correct the pro/antioxidant imbalance." *Blood Purif* 18 (3):191–9. doi: 10.1159/000014418.

Pecoits-Filho, R., J. Larkin, C. E. Poli-de-Figueiredo, A. L. Cuvello-Neto, A. B. L. Barra, P. B. Gonçalves, S. Sheth, M. Guedes, M. Han, V. Calice-Silva, M. C. M. de Castro, P. Kotanko, T. P. de Moraes, J. G. Raimann, and M. E. F. Canziani. 2021. "Effect of hemodiafiltration on measured physical activity: primary results of the HDFIT randomized controlled trial." *Nephrol Dial Transplant* 36 (6):1057–70. doi: 10.1093/ndt/gfaa173.

Petitclerc, T., and L. Mercadal. 2023. "Automated adjustment of dialysate sodium by the hemodialysis monitor: rationale, implementation, and clinical benefits." *Semin Dial* 36 (3):184–92. doi: 10.1111/sdi.13132.

Pupim, L. B., K. M. Majchrzak, P. J. Flakoll, and T. A. Ikizler. 2006. "Intradialytic oral nutrition improves protein homeostasis in chronic hemodialysis patients with deranged nutritional status." *J Am Soc Nephrol* 17 (11):3149–57. doi: 10.1681/asn.2006040413.

Rosner, M. H., T. Reis, F. Husain-Syed, R. Vanholder, C. Hutchison, P. Stenvinkel, P. J. Blankestijn, M. Cozzolino, L. Juillard, K. Kashani, M. Kaushik, H. Kawanishi, Z. Massy, T. L. Sirich, L. Zuo, and C. Ronco. 2021. "Classification of uremic toxins and their role in kidney failure." *Clin J Am Soc Nephrol* 16 (12):1918–28. doi: 10.2215/cjn.02660221.

Schiffl, H., S. M. Lang, D. Stratakis, and R. Fischer. 2001. "Effects of ultrapure dialysis fluid on nutritional status and inflammatory parameters." *Nephrol Dial Transplant* 16 (9):1863–9. doi: 10.1093/ndt/16.9.1863.

Soeters, P. B., and A. M. Schols. 2009. "Advances in understanding and assessing malnutrition." *Curr Opin Clin Nutr Metab Care* 12 (5):487–94. doi: 10.1097/MCO.0b013e32832da243.

Wright, S., and D. E. Weiner. 2012. "Oral nutritional supplement use in dialysis patients: full speed ahead?" *Am J Kidney Dis* 60 (4):507–9. doi: 10.1053/j.ajkd.2012.07.004.

19 The Importance of Protein Intake in Relation to Oral Health

A New Narrative

Thilini Jayasinghe, Sharon Erdrich, Gangani Dharmarathne, and Axel Spahr

ABBREVIATIONS

AMDR	Acceptable macronutrient distribution range
BCAA	Branched-chain amino acids
IgA	Immunoglobulin A
IGF-1	Insulin growth factor-1
IL-8	Interleukin-8
MBP	Milk basic protein
MMP-8	Matrix metalloproteinase-8
MSUD	Maple syrup urine disease
PEM	Protein-energy malnutrition
PKU	Phenylketonuria
TNF-α	Tumour necrosis factor-α
VCAM-1	Vascular cell adhesion molecule-1

INTRODUCTION

Proper nutrition is essential for maintaining optimal oral health. While many nutrients contribute to oral well-being, protein intake plays a crucial role in supporting oral health. Proteins are macronutrients composed of amino acids, which are the building blocks for various structures and processes in the body. In the context of oral health, protein intake is particularly important due to its numerous benefits for the teeth, tooth-supportive tissues, gums, and overall integrity of the oral cavity. These benefits include enhancing tooth development, stability, and strength, supporting tissue repair, regeneration, and maintenance, and promoting wound healing and supporting immune function.

Dietary proteins are integral to oral health, influencing tooth development and strength (Price 2009). They function as vital building blocks, contributing to the formation of bone and the periodontium (Kotronia et al. 2021). Collagen, an important structural protein, contains amino acids such as glycine, proline, and hydroxyproline and plays a crucial role in supporting oral health. It contributes to the formation of dental elements and also provides essential support to various periodontal tissues (Binlateh et al. 2022).

Adequate protein intake is crucial for providing essential amino acids that have multifaceted roles in maintaining and repairing oral tissues. Amino acids are the constituents of proteins, fortifying tooth structures, significantly enhancing their strength, and making them more resistant to the

DOI: 10.1201/b23328-19

detrimental effects of decay and erosion (Tungare and Paranjpe 2022). This is particularly crucial for teeth, where proteins play a vital role in ensuring the resilience and structural integrity of their mineralised components like enamel, dentine, and cementum (Woelber et al. 2016). However, inadequate protein intake may result in a short supply of key amino acids required for maintenance, healing, and adaptive processes, disrupting the synthesis of new tissue (Demling 2009). Research indicates that inadequate protein intake increases susceptibility to infections, hinders the pace of wound healing, and leads to the degeneration of periodontal connective tissues (Adegboye et al. 2016). One example of the impact of protein deficiency is lower levels of insulin growth factor-1 (IGF-1) in the bloodstream. IGF-1 plays a pivotal role in promoting cell proliferation and tissue repair; thus, any inadequacy can hinder overall regenerative processes in the body (Kaigler, Cirelli, and Giannobile 2006).

Proteins are actively involved in various stages of the wound-healing process, including inflammation, cell proliferation, and tissue remodelling (Dodington et al. 2021). By providing an adequate supply of essential amino acids, dietary protein supports the timely and effective healing of oral wounds. This not only reduces the risk of complications but also promotes better oral health outcomes. For instance, both animal and human intervention studies have demonstrated the benefits of collagen peptides rich in hydroxyproline, glycine, and proline in enhancing wound healing (Li et al. 2007; Wu et al. 2011)

Protein intake is instrumental in bolstering the immune system's capacity to combat harmful bacteria and mitigate the risk of oral infections (Adegboye et al. 2012). Studies suggest that branched-chain amino acids (BCAAs), especially leucine, can reduce levels of pro-inflammatory cytokines and mediators in macrophages (Lee et al. 2017). Additionally, amino acids are essential for the formation of antibodies such as salivary immunoglobulin A (IgA), which help to resist oral infections (Cruzat et al. 2018). Research also indicates that dietary protein intake can influence the concentration of specific salivary proteins, including amylase and proline-rich proteins (Louro et al. 2021).

Incorporating protein-rich foods into the diet enables individuals to provide their bodies with the necessary amino acids to support oral health and enjoy the benefits of a healthy and functional oral cavity as shown in Figure 19.1. Conversely, compromised oral health can alter food choices and negatively impact food intake, leading to suboptimal nutritional status, which can lead to chronic systemic diseases. Recognising and treating oral health and nutrition problems are important in improving overall health and quality of life.

IMPORTANCE OF OPTIMAL DIETARY PROTEIN INTAKE IN VARIOUS LIFE STAGES

Tooth development starts well before birth. Within the first six weeks of gestation, the tooth germ begins forming in the developing foetus (Rathee and Jain 2024). By the eighth week, the dental lamina starts forming, providing the foundation for subsequent enamelisation. This highlights the

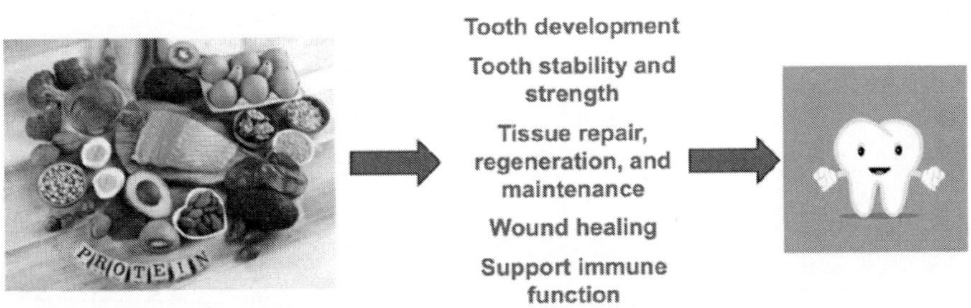

FIGURE 19.1 Importance of protein to oral health.

critical importance of nutrition for women during the preconception phase and throughout pregnancy to supply the necessary nutrients for foetal teeth development. The development of the secondary (permanent) teeth commences in the second trimester. This involves tooth mineralisation, the formation of the alveolar bone (the bone surrounding the tooth socket), and cementum (a calcified tissue covering the tooth root), with the maturation of these tissues continuing after birth and into childhood and adolescence. Protein, along with calcium, phosphorus, vitamin D, vitamin K, boron, and magnesium, is essential for this process (Urrutia and Thorp 2012). Conversely, inadequate nutrient intake during pregnancy can result in oral malformations in the child (Wallenstein et al. 2013). Thus, proper nutrition during pregnancy is vital for ensuring healthy tooth development and improving oral health outcomes in later years.

Infant oral health is closely tied to protein intake, with preclinical research revealing that both maternal protein restriction during gestation and lactation and neonatal protein malnutrition can have adverse effects, including reduced bone density, delayed tooth development, reduced vascularity in dental pulp, decreased matrix mineralisation, and increased susceptibility to dental caries (Calsa et al. 2022; Menaker and Navia 1974). Due to ethical constraints, there are no experimental results available for humans, but it is evident that adequate protein intake is essential during these critical stages of life to promote the proper development of the infant's oral structures, including the teeth (Elango and Ball 2016; Yenen and Ataçağ 2019).

Tooth development continues until approximately three years of age, encompassing the formation of tooth root structures. Children need protein for tooth development because this macronutrient is integral to the formation and maintenance of tooth structures, such as enamel, dentine, root cementum, and tooth-supportive structures like the periodontal ligament and alveolar bone. As outlined above, protein is an important contributor to gum health, supporting wound healing in the oral cavity and helping maintain a strong immune system.

A lack of protein in animals has also been associated with zinc deficiency, which may cause issues with salivary gland function, leading to the development of dental cavities (Johnson, Lopez, and Navia 1995), breakdown of periodontal ligaments, degeneration of gingival tissues, and resorption of the alveolar bone (Stahl, Sandler, and Cahn 1955). Therefore, during infancy and childhood, a protein-rich diet is crucial for fostering the proper development of teeth and gums and other oral tissues (Naidoo and Myburgh 2007; Nutten 2016).

The role of dietary protein in oral health remains important throughout childhood and into later stages of life. As adolescents undergo rapid growth, proteins aid in tissue repair and the development of oral structures like the gums and jawbones. Additionally, proteins contribute to the formation and mineralisation of teeth, supporting overall dental health (Garcia-Iborra et al. 2023; Sbricoli et al. 2022). At this life stage, the recommended daily protein intake varies depending on actual age, gender, activity level, and growth rate, but on average, adolescents typically need about 0.77 to 0.99 grams of protein per kilogram of body weight per day (Ministry of Health Australia 2014). However, an excessive intake of protein should be balanced with overall dietary considerations to prevent potential health issues such as reduced saliva production, contributing to dry mouth (Delimaris 2013). Conversely, a deficiency in protein intake during this period can have adverse effects on oral health, with problems such as weakened tooth structures and tooth-supportive tissues, compromised gum health, delayed wound healing, and reduced immune response. In turn, individuals may become even more susceptible to oral health problems (Russell et al. 2010). Therefore, a nutritionally balanced, protein-rich diet is essential for promoting optimal oral health during adolescence.

In young adulthood, dietary protein supports oral tissue development, preserves gum and tooth health, and aids in tissue repair. At this life stage, protein needs are quite varied, with lifestyle being a major determinant. In women, the reproductive years are particularly important due to the increased demands of pregnancy and lactation.

As individuals transition into later life stages, maintaining adequate protein intake becomes even more crucial (Jayasinghe et al. 2022). Proteins continue to support the healing process after oral surgery or tooth extraction. They enhance immune function, aid in preventing and combating oral

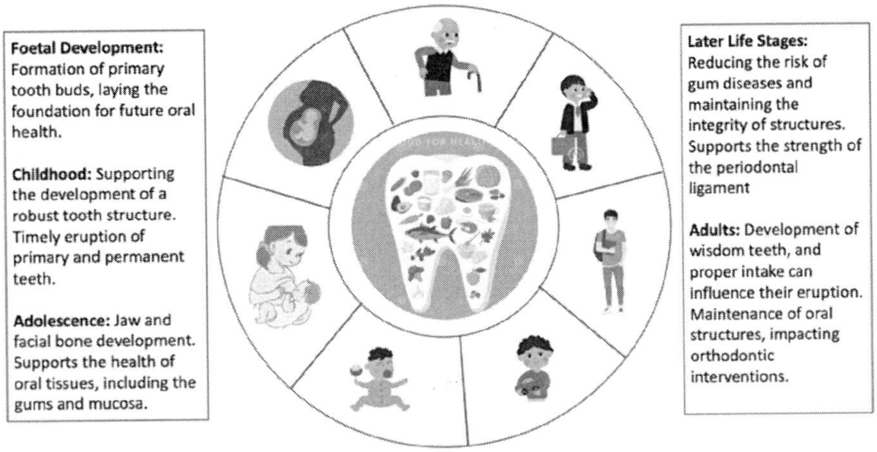

FIGURE 19.2 Importance of protein for oral health in various life stages. This figure illustrates the multifaceted impact of dietary protein on oral health throughout the lifecycle. Starting from prenatal development, through childhood and adolescence, into young adulthood, and spanning into later life stages, protein plays a crucial role in tooth development, maintenance of gum and tooth health, tissue repair, and immune function.

infections, and contribute to the preservation of gum and bone health, a crucial factor in the ageing population (Giridhar 2016). Sufficient protein intake can help mitigate common oral health issues that older adults often face, such as dry mouth, dental caries, periodontitis, edentulism, and oral cancers (Gil-Montoya et al. 2015; Jayasinghe et al. 2022). Figure 19.2 illustrates the importance of protein on oral health throughout the life stages.

In summary, the importance of optimal dietary protein intake is intricately linked to oral health throughout the various stages of life. Its contributions extend to supporting oral tissue development, maintaining gum and tooth health, aiding in tissue repair, and even assisting with the unique challenges faced from pregnancy to old age. A nutritionally balanced, protein-rich diet remains essential for promoting strong oral health throughout life's diverse stages.

EFFECT OF VARIOUS DIETARY PROTEIN SOURCES ON ORAL HEALTH

Whey Protein and Oral Health

Whey protein, derived from milk, has shown promising effects on oral health by enhancing the immune system's functionality and effectively inhibiting dental plaque formation (Krissansen 2007; Merritt, Qi, and Shi 2006). The commercialisation of whey proteins has gained momentum, positioning them as potential adjuncts to antioxidant and anti-inflammatory interventions, wound healing, and bone repair facilitators (Ebaid et al. 2011).

Immune dysfunction plays a pivotal role in numerous oral health issues, and there is potential for whey protein to enhance both inflammation control and glycaemic management (Patel 2015). For instance, whey protein, rich in precursor of the antioxidant glutathione, shows promise in mitigating the adverse effects of stressors such as oxidative stress, inflammation, and immune system suppression. Studies have demonstrated that whey protein can accelerate the release of bioactive peptides from food, increase intracellular glutathione levels, decrease circulating serum C-reactive protein levels, and reduce inflammatory cytokines, such as interleukin-8 (IL-8) (Lands et al. 2010; Piccolomini et al. 2012). Whey protein hydrolysate and BCAAs exhibit anti-inflammatory properties by diminishing the expression of inflammatory markers like tumour necrosis factor-α (TNF-α) and vascular cell adhesion molecule-1 (VCAM-1) (Patel 2015; Da Silva et al. 2017). These anti-inflammatory effects may have broader implications for systemic inflammation, which can have direct and indirect impacts on oral health.

Additionally, whey protein, particularly its core fraction known as milk basic protein (MBP), plays a role in stimulating bone formation via the proliferation of osteoblasts while suppressing bone resorption in adults (Aoe et al. 2001; Toba et al. 2001). An acidic whey protein fraction has demonstrated potential in restoring bone, as observed in studies involving ovariectomised rats (Toba et al. 2000; Kruger et al. 2005). This suggests that MBP may contribute to maintaining healthy bone mineralisation and density – essential for the integrity of the tooth-supporting structures that anchor teeth in the jawbone. An intricate relationship exists between protein concentration, acidity (pH), and oral physiology in the context of whey protein beverages (Beecher et al. 2008). Higher protein concentrations lead to increased saliva flow rates, demonstrating another potential benefit of whey protein for oral well-being.

Lactoferrin and Oral Health

Lactoferrin, a multifunctional glycoprotein found in milk, plays a significant role in promoting oral health by supporting bone health, boosting immunity, and providing antimicrobial properties (Siqueiros-Cendón et al. 2014). In relation to bone health, lactoferrin promotes osteoblast growth, inhibits osteoclast formation, and reduces osteoblast apoptosis (Cornish et al. 2006; Gao et al. 2018). Its immunomodulatory functions include downregulating the immune response to mitigate inflammatory diseases (Krissansen 2007). Lactoferrin's antimicrobial properties, including its ability to inhibit the growth of *Streptococcus mutans*, a significant oral pathogen implicated in dental caries (tooth decay), contribute to reduced dental plaque formation (Rosa et al. 2021; Naot et al. 2005). These multifaceted effects indicate that lactoferrin may contribute to maintaining tooth integrity as well as bone mineralisation and density, crucial for the supporting structures that anchor teeth in the jaw.

Plant Proteins

Proteins from plant sources, such as legumes, nuts, and seeds, have been shown to be more beneficial for oral health than a high carbohydrate, meat-containing diet (Eberhard et al. 2022). Overall, lower consumption of red and/or processed meat and higher intakes of whole grains and fruits have also been associated with a reduced probability of developing severe periodontitis (Salazar et al. 2018).

Soya beans contain specific compounds called isoflavonoids, which have been shown in animal and *in vitro* studies to have a range of beneficial effects on oral health. These include improved salivary flow rates, stimulation of bone formation, inhibition of bacterial pathogens, reduction of plaque accumulation, and accelerated healing of gingival and periodontal tissues (Valizadeh et al. 2021). As yet, in-depth exploration in human clinical models is lacking.

OPTIMAL PROTEIN INTAKE FOR ORAL HEALTH

As discussed in the previous section, maintaining optimal protein intake is vital for supporting oral health across the lifespan.

Although there are no specific guidelines exclusively addressing protein intake for oral health, ensuring that the diet meets optimal protein requirements plays a fundamental role in supporting various bodily functions, including the health of bones, oral structures, and teeth. Following general protein guidelines is crucial for overall good health and optimal nutrition throughout the various life stages, from infancy through childhood, young adulthood, and into the senior years (Ministry of Health Australia 2014).

Inadequate protein intake, falling below 10 percent of daily caloric intake, can result in protein malnutrition. Conversely, excessive protein consumption, exceeding 35 percent, may elevate the risk of chronic diseases (Delimaris 2013). Therefore, a nutritionally balanced diet, providing the right amount of protein, typically within the Acceptable Macronutrient Distribution Range (AMDR) of 15 to 25 percent of energy intake for adults, is essential for overall health (Wolfe et al. 2017). Overall diet quality is also important; ideally, the diet should include a variety of nutrient-rich foods from different food groups.

A further consideration is the influence of dietary patterns on the composition of saliva. For example, liquid diets have the potential to alter salivary protein and electrolyte concentrations, while those rich in vegetables, fruits (providing a good source of polyphenols), and high-protein foods promote an enriched salivary proteome (Louro et al. 2021). Salivary proteins include mucin, amylase, lysozyme, lactoferrin, and immunoglobulins, all of which have important roles in oral health and digestion (Loo et al. 2010). Additionally, emerging research is shedding light on the role of salivary proteins in taste modulation and feeding behaviour, showing potential implications for orosensory and post-ingestive feedback (Martin, Gutierrez, and Torregrossa 2023).

Salivary gland function, crucial for oral protection, is also influenced by protein deficiency and dietary consistency as shown in Figure 19.3. The health of glandular tissue influences salivary protein synthesis and salivary secretion. One study exploring the effects of different dietary protein content on parotid gland growth and function demonstrated the significant role of diet composition on parotid gland weight, parotid saliva protein concentration, and liver zinc levels (Johnson, Lopez, and Navia 1995).

Saliva, in addition to essential functions such as lubrication and enzymatic pre-digestion, is an important first line of defence against the potentially damaging effects of food and microorganisms. Saliva's antioxidant properties are critical in countering reactive oxygen species, which pose a risk to oral health. While food contributes antioxidant compounds to this balance, research on the impact of food on saliva's antioxidant capacity has yielded conflicting results (Manfredo Vieira et al. 2018). Together, these studies demonstrate the intricate relationship between protein intake, saliva, oral health, and taste perception (Torregrossa et al. 2014).

Protein's crucial role in immune function and wound healing also prompted the investigation of its association with periodontal healing (McClave et al. 2009). A daily protein intake of at least 1 gram per kilogram of body weight was associated with reduced periodontal disease burden, especially in individuals over 65 years of age undergoing non-surgical periodontal treatment (Dodington et al. 2021). Given the scarcity of data on protein intake and periodontal healing, this study focused on protein as the primary nutrient of interest. Periodontitis, often prevalent in adults over 65 years, is linked to age-related conditions. Noteworthy is that older adults require adequate dietary protein for maintaining muscle mass and nutrient absorption, both of which directly affect oral health. Dietary intakes of 1.0 to 1.5 g/kg body weight/day of protein intake for older adults ($\geq$65 years) are recommended to combat sarcopenia, the age-related decline in muscle mass and function (Baum, Kim, and Wolfe 2016; Phillips 2017).

DIETARY PROTEIN DEFICIENCY AND ITS IMPACT ON ORAL HEALTH

It is important to highlight that specific rare protein deficiency diseases, including Maple Syrup Urine Disease (MSUD), phenylketonuria (PKU), homocystinuria, cystinuria, argininaemia, and tyrosinaemia, have the potential to affect oral health (Ballikaya et al. 2021; Abola et al. 2022). These conditions may have diverse effects on aspects such as tooth development, enamel health, gum condition, motor skills, coordination, oral hygiene practices, and even liver function, thus ultimately influencing oral well-being (Ballikaya et al. 2021; Abola et al. 2022). Additionally, the requirement of diets with specific amino acid restrictions for these diseases can affect the development of teeth and enamel, potentially leading to dental problems such as enamel hypoplasia, causing teeth to be more susceptible to caries or decay and sensitivity (Table 19.1). Some examples are outlined herewith:

TABLE 19.1
Impact of Selected Amino Acid Imbalances on Oral Health

Amino Acid	Effect on Oral Health	Reference
Alanine	Alanine and histidine form citrulline. A higher concentration of citrulline in saliva is correlated with periodontitis.	(Balci et al. 2021)
Arginine	Higher concentration in saliva in Stage III Grade C generalised periodontitis.	(Balci et al. 2021)
L-Arginine	L-arginine monohydrochloride in saliva inhibits bacterial coaggregation in the oral cavity by decreasing the viscosity of extracellular polymeric substances produced by bacteria and altering cellular metabolism, resulting in biofilm dispersion and reduced antibiotic tolerance.	(Kolderman et al. 2015)
Glutamic acid	Higher in Stage III Grade C generalised periodontitis.	(Balci et al. 2021)
Glycine	Lower salivary glycine has been associated with an increased risk of dental caries and periodontitis through increased collagen degradation and collagenase activity, leading to higher inflammation in gingiva.	(Fonteles et al. 2009) (Syrjänen, Piironen, and Markkanen 1987)
Histidine†	Lack of histidine and its derivatives in saliva results in chelation, i.e., formation of metal complexes with amino acids, leading to the initial lesion and, subsequently, to destruction of the organic matrix by the action of proteolytic bacteria.	(Vranić, Granić, and Rajić 1991)
Isoleucine†	Found in carious dentine.	(Lin 1989)
Proline	Salivary proline-rich glycoprotein regulates oral calcium homeostasis by controlling the supersaturated state of saliva with respect to calcium phosphate salts, countering the plaque acidity, forming the dental pellicle, and influencing the composition of plaque.	(Pateel et al. 2021) (Zakhary et al. 2007)
Tryptophan†	Higher saliva tryptophan levels were observed in Stage III Grade B generalised periodontitis.	(Balci et al. 2021)
Tyrosine	Potential biomarker of oral lichen planus (lower levels).	(Darczuk et al. 2019)
Valine†	Detected in sound dentine compared to carious dentine.	(Lin 1989)
Homocysteine‡	Associated with a high narrow palate, mandibular prognathia (protruding lower jaw), crowding and early eruption of teeth, and short dental roots.	(Björksved and Arnrup 2012)

Amended Table 1 from (Jayasinghe et al. 2022).

- Homocystinuria can lead to a detrimental accumulation of homocysteine, which can compromise the integrity of connective tissues in the body (Chapman et al. 2023). This may negatively impact oral health by contributing to an increased susceptibility to gum disease (Chapman et al. 2023).
- Argininaemia occurs due to a deficiency of arginase I, resulting in defective enzymatic metabolisation of arginine to urea and ornithine. Argininaemia is managed by severe restriction of dietary protein intake to decrease arginine intake (Scaglia and Lee 2006). Arginine deficiency alone can impact the nervous system and influence oral health by affecting motor skills, coordination, and, consequently, the ability to practice adequate oral hygiene (Kelsey and Lamster 2008).

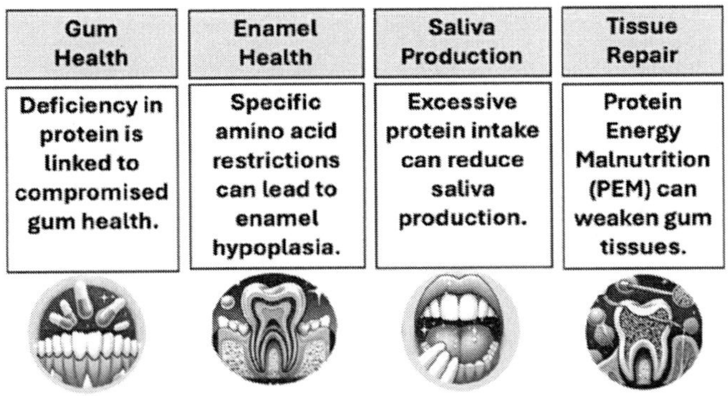

FIGURE 19.3 Connection between protein intake and oral health.

- Tyrosinaemia occurs through genetic defects in breaking down tyrosine. Type 1 tyrosinaemia can lead to liver dysfunction, impacting the metabolisation of nutrients, which can have detrimental effects on oral health (Chinsky et al. 2017).

Any condition that requires protein restriction poses a risk for deficiency-related sequelae. These include increased susceptibility to various oral health issues, such as dental caries and gum disease or periodontitis, which can compromise oral defence mechanisms (Vieira et al. 2020).

Protein-energy malnutrition (PEM) can have profound implications for oral health, as the body's nutritional status plays a critical role in maintaining the integrity and function of the oral cavity (Rahman and Walls 2020; Sheetal 2013). PEM can impact oral health via dental development and enamel health, gum health, salivary gland function, immune function, and wound healing as outlined in Figure 19.3 (Sheetal 2013). Severe malnutrition during childhood, with insufficient protein and energy intake, can lead to delayed tooth eruption, malformation of teeth, and enamel hypoplasia, a condition characterised by underdeveloped tooth enamel (Setiawan et al. 2022). In individuals with PEM, impaired wound healing may lead to prolonged recovery times following oral surgeries, tooth extractions, or the management of dental conditions (Mahmoodpoor et al. 2018). Furthermore, PEM often coexists with other nutrient deficiencies, and these deficiencies can further exacerbate oral health problems, as vitamins and minerals are essential for the development and maintenance of oral tissues. In addition to weakening gum tissues and compromising the body's ability to defend against infections, malnutrition undermines the immune response, rendering individuals more susceptible to inflammatory gum diseases like gingivitis and periodontitis (Azzolino et al. 2019). Table 19.1 provides a summary of the impact of selected amino acid imbalances, including both deficiencies and excesses, from oral health-based human studies.

EXCESSIVE PROTEIN INTAKE AND ORAL HEALTH

The potential impact of excessive protein intake on oral health is a subject of growing concern. While sufficient protein intake is generally acknowledged for its positive impact on overall health, an excess, especially when derived predominantly from animal sources, may present notable challenges to oral health, as supported by various research findings (Pesta and Samuel 2014). High-protein diets, often with more than 30 percent of calories from protein, have gained popularity due to claims of aiding weight loss (Carreiro et al. 2016). Ongoing research aims to ascertain potential long-term health consequences of high-protein diets (Brehm and D'Alessio 2008). While observational studies have linked diets high in animal protein, particularly red meat, to an increased risk of kidney stones, kidney disease, liver dysfunction, and colorectal cancer, it is worth noting that these diets also tend to be high in saturated fats and cholesterol, complicating direct attributions to protein

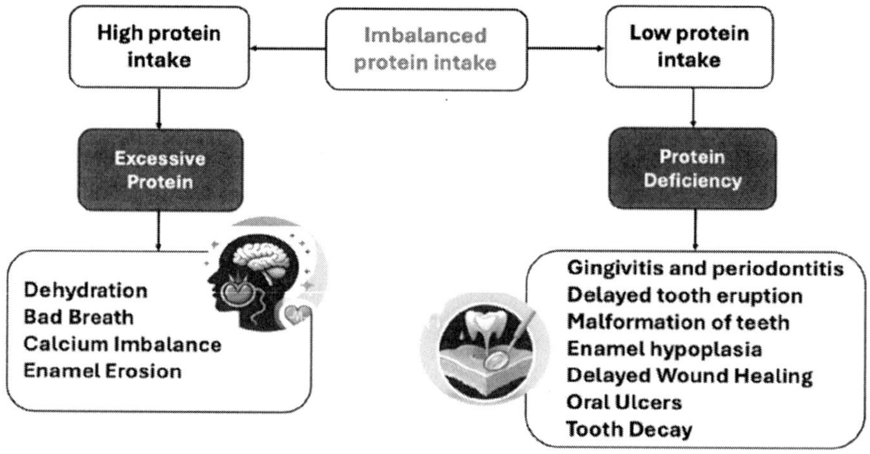

FIGURE 19.4 Effects of imbalanced protein intake on dental health and long-term health risks.

content (Rodriguez et al. 2019). While such diets might exacerbate kidney and liver issues in those predisposed, the connection between high protein intake and oral health remains contested. Worthy of consideration is the possibility that excessive protein intake can lead to the neglect of other essential nutrients, as high-protein diets often involve the restriction of a diverse range of nutrient-rich, fibre-rich foods that normally provide a variety of vitamins, minerals, starches, antioxidants, and polyphenols.

Excessively high-protein diets have been associated with a decreased production of saliva (Figure 19.3), potentially leading to dry mouth, a condition that compromises the oral environment's natural defence mechanisms (Müller et al. 2023). Moreover, the breakdown of protein in the mouth can contribute to the production of sulphur compounds, contributing to halitosis (Zabokova Bilbilova 2021). Research has also indicated that diets rich in protein, especially when combined with frequent consumption of acidic and sugary foods, may elevate the risk of cavities due to increased oral acidity and enamel erosion (Tungare and Paranjpe 2022). Additionally, certain animal proteins such as aspartate, serine, and cysteine have been shown to influence the composition of oral bacteria, potentially favouring the growth of bacteria associated with periodontal diseases (Sedghi et al. 2021). To mitigate these risks, adopting a balanced diet that includes a variety of nutrient-rich foods, ensuring adequate hydration, incorporating plant-based proteins, and maintaining diligent oral hygiene is advised.

Consulting with qualified healthcare professionals with oral health and nutritional knowledge is crucial for personalised guidance, considering individual health needs and considerations. It is essential to emphasise the importance of a balanced and diverse diet to ensure not only sufficient protein intake but also the inclusion of other nutrients that support overall oral health. Striking the right balance in protein consumption as demonstrated in Figure 19.4, coupled with a well-rounded diet, remains key to promoting oral well-being and preventing potential adverse effects associated with excessive protein intake.

SUMMARY

The relationship between protein intake and oral health is multifaceted and holds significant implications for overall well-being. Proteins, composed of amino acids, serve as fundamental building blocks for tissues and processes in the body, making them essential for maintaining optimal oral health. Proteins have crucial roles in tissue repair, enhancing tooth strength, supporting immune function, and contributing to wound healing. They also contribute to the structural integrity of teeth and various periodontal tissues, ensuring the overall health and function of the oral environment.

As shown in Figure 19.3, maintaining optimal protein intake is critical, *as both deficiencies and excesses can have negative consequences.* Ensuring optimal levels of protein intake is crucial, as both deficiencies and excesses can have negative consequences. While protein deficiency can increase susceptibility to oral health problems such as dental caries and compromised oral defence mechanisms, excessive dietary protein intake has raised concerns about potential long-term health risks, including kidney issues and certain cancers.

The literature highlights the potential benefits of well-balanced protein intake for periodontal healing, bone health, and immune function. Whey protein, in particular, has shown promise in enhancing the immune system, inhibiting dental plaque formation, and supporting bone repair.

As research on the relationship between protein intake and oral health continues, it is essential for individuals to prioritise moderation and dietary variety. Adequate protein intake is essential, but it should be part of a balanced diet that includes other essential nutrients to maintain overall health. With additional research, a better understanding of the complexities of protein's influence on oral health will foster the development of targeted recommendations for different life stages and specific oral health conditions.

MINI DICTIONARY OF TERMS

Inflammatory cytokines: Signalling molecules that regulate immune responses and contribute to inflammation in the body.

Salivary proteome: The complete set of proteins present in saliva, encompassing a diverse array of molecules that play various roles in oral health, immunity, and disease detection.

SUMMARY POINTS

- Proteins, comprising amino acids, are vital for oral health, contributing to tissue repair, tooth strength, and immune support.
- Optimal protein levels are crucial, as deficiencies can lead to oral health issues, while excesses may pose long-term risks such as kidney problems and certain cancers.
- Research suggests well-balanced protein intake promotes periodontal healing, bone health, and immune function. Whey protein, in particular, shows promise in enhancing the immune system and supporting oral health.
- Individuals should prioritise moderation and dietary variety, incorporating protein into a balanced diet with essential nutrients for overall health.

REFERENCES

Abola, Iveta, Darta Elizabete Emulina, Ingus Skadins, Anda Brinkmane, Linda Gailite, and Madara Auzenbaha. 2022. "Dental Status and Periodontal Health of Patients with Phenylketonuria in Latvia." *Acta Stomatologica Croatica* 56 (2): 109–19. https://doi.org/10.15644/asc56/2/2.

Adegboye, A. R., Barbara J. Boucher, Johanne Kongstad, Nils-Erik Fiehn, Lisa B. Christensen, and Berit L. Heitmann. 2016. "Calcium, Vitamin D, Casein and Whey Protein Intakes and Periodontitis among Danish Adults." *Public Health Nutrition* 19 (3): 503–10. https://doi.org/10.1017/S1368980015001202.

Adegboye, A. R., Lisa B. Christensen, Poul Holm-Pedersen, Kirsten Avlund, Barbara J. Boucher, and Berit L. Heitmann. 2012. "Intake of Dairy Products in Relation to Periodontitis in Older Danish Adults." *Nutrients* 4 (9): 1219–29. https://doi.org/10.3390/nu4091219.

Aoe, Seichiro, Yasuhiro Toba, Jun-ichi Yamamura, Hiroshi Kawakami, Masatoshi Yahiro, Masayoshi Kumegawa, Akira Itabashi, and Yukihiro Takada. 2001. "Controlled Trial of the Effects of Milk Basic Protein (MBP) Supplementation on Bone Metabolism in Healthy Adult Women." *Bioscience, Biotechnology, and Biochemistry* 65 (4): 913–18. https://doi.org/10.1271/bbb.65.913.

Azzolino, Domenico, Pier Carmine Passarelli, Paolo De Angelis, Giovan Battista Piccirillo, Antonio D'Addona, and Matteo Cesari. 2019. "Poor Oral Health as a Determinant of Malnutrition and Sarcopenia." *Nutrients* 11 (12): 2898. https://doi.org/10.3390/nu11122898.

Balci, Nur, Şivge Kurgan, Ali Çekici, Tülin Çakır, and Muhittin A. Serdar. 2021. "Free Amino Acid Composition of Saliva in Patients with Healthy Periodontium and Periodontitis." *Clinical Oral Investigations* 25 (6): 4175–83. https://doi.org/10.1007/s00784-021-03977-7.

Ballikaya, Elif, Yılmaz Yildiz, Nagihan Koç, Ayşegül Tokatli, Meryem Uzamis Tekcicek, and Hatice Serap Sivri. 2021. "Oral Health Status of Children and Young Adults with Maple Syrup Urine Disease in Turkey." *BMC Oral Health* 21 (1): 8. https://doi.org/10.1186/s12903-020-01376-0.

Baum, Jamie, Il-Young Kim, and Robert Wolfe. 2016. "Protein Consumption and the Elderly: What Is the Optimal Level of Intake?" *Nutrients* 8 (6): 359. https://doi.org/10.3390/nu8060359.

Beecher, J. W., M. A. Drake, P. J. Luck, and E. A. Foegeding. 2008. "Factors Regulating Astringency of Whey Protein Beverages." *Journal of Dairy Science* 91 (7): 2553–60. https://doi.org/10.3168/jds.2008-1083.

Binlateh, Thunwa, Peungchaleoy Thammanichanon, Pawornwan Rittipakorn, Natthapol Thinsathid, and Paiboon Jitprasertwong. 2022. "Collagen-Based Biomaterials in Periodontal Regeneration: Current Applications and Future Perspectives of Plant-Based Collagen." *Biomimetics* 7 (2): 34. https://doi.org/10.3390/biomimetics7020034.

Björksved, Margitha, and Kristina Arnrup. 2012. "Homocystinuria and Oral Health. A Report of 14 Cases." *Swedish Dental Journal* 36 (2): 101–8. http://www.ncbi.nlm.nih.gov/pubmed/22876397.

Brehm, Bonnie J., and David A. D'Alessio. 2008. "Benefits of High-Protein Weight Loss Diets: Enough Evidence for Practice?" *Current Opinion in Endocrinology, Diabetes & Obesity* 15 (5): 416–21. https://doi.org/10.1097/MED.0b013e328308dc13.

Calsa, Bruno, Tainá Jaqueline Bortolança, Beatriz Calloni Masiero, Marcelo Augusto Marretto Esquisatto, Camila Andrea de Oliveira, Rosana Catisti, and Milton Santamaria-Jr. 2022. "Maxillary and Dental Development in the Offspring of Protein-restricted Female Rats." *European Journal of Oral Sciences* 130 (6). https://doi.org/10.1111/eos.12895.

Carreiro, Alicia L., Jaapna Dhillon, Susannah Gordon, Kelly A. Higgins, Ashley G. Jacobs, Breanna M. McArthur, Benjamin W. Redan, Rebecca L. Rivera, Leigh R. Schmidt, and Richard D. Mattes. 2016. "The Macronutrients, Appetite, and Energy Intake." *Annual Review of Nutrition* 36 (1): 73–103. https://doi.org/10.1146/annurev-nutr-121415-112624.

Chapman, Kimberly A., Danae Bartke, Vanessa Vogel-Farley, Mary Cobb, and Mary Chapman. 2023. "Dental Complications in Homocystinurias." *Molecular Genetics and Metabolism Reports* 36 (September): 100999. https://doi.org/10.1016/j.ymgmr.2023.100999.

Chinsky, Jeffrey M., Rani Singh, Can Ficicioglu, Clara D. M. van Karnebeek, Markus Grompe, Grant Mitchell, Susan E. Waisbren, et al. 2017. "Diagnosis and Treatment of Tyrosinemia Type I: A US and Canadian Consensus Group Review and Recommendations." *Genetics in Medicine* 19 (12): 1380–95. https://doi.org/10.1038/gim.2017.101.

Cornish, Jillian, K. Palmano, K. E. Callon, M. Watson, J. M. Lin, P. Valenti, D. Naot, A. B. Grey, and I. R. Reid. 2006. "Lactoferrin and Bone; Structure–Activity Relationships." *Biochemistry and Cell Biology* 84 (3): 297–302. https://doi.org/10.1139/o06-057.

Cruzat, Vinicius, Marcelo Macedo Rogero, Kevin Noel Keane, Rui Curi, and Philip Newsholme. 2018. "Glutamine: Metabolism and Immune Function, Supplementation and Clinical Translation." *Nutrients* 10 (11): 1564. https://doi.org/10.3390/nu10111564.

Darczuk, Dagmara, Wirginia Krzyściak, Beata Bystrowska, Barbara Kęsek, Dorota Kościelniak, Maria Chomyszyn-Gajewska, and Tomasz Kaczmarzyk. 2019. "The Relationship between the Concentration of Salivary Tyrosine and Antioxidants in Patients with Oral Lichen Planus." *Oxidative Medicine and Cellular Longevity* 2019 (November): 1–11. https://doi.org/10.1155/2019/5801570.

Delimaris, Ioannis. 2013. "Adverse Effects Associated with Protein Intake above the Recommended Dietary Allowance for Adults." *ISRN Nutrition* 2013 (July): 1–6. https://doi.org/10.5402/2013/126929.

Demling, Robert H. 2009. "Nutrition, Anabolism, and the Wound Healing Process: An Overview." *Eplasty* 9: e9. http://www.ncbi.nlm.nih.gov/pubmed/19274069.

Dodington, David W., Hannah E. Young, Jennifer R. Beaudette, Peter C. Fritz, and Wendy E. Ward. 2021. "Improved Healing after Non-Surgical Periodontal Therapy Is Associated with Higher Protein Intake in Patients Who Are Non-Smokers." *Nutrients* 13 (11): 3722. https://doi.org/10.3390/nu13113722.

Ebaid, Hossam, Amir Salem, Abdalla Sayed, and Ali Metwalli. 2011. "Whey Protein Enhances Normal Inflammatory Responses during Cutaneous Wound Healing in Diabetic Rats." *Lipids in Health and Disease* 10 (1): 235. https://doi.org/10.1186/1476-511X-10-235.

Eberhard, Joerg, Kate Ruiz, Jian Tan, Thilini N. Jayasinghe, Shahrukh Khan, Elif Eroglu, Christina Adler, et al. 2022. "A Randomized Clinical Trial to Investigate the Effect of Dietary Protein Sources on Periodontal Health." *Journal of Clinical Periodontology* 49 (4): 388–400. https://doi.org/10.1111/jcpe.13587.

Elango, Rajavel, and Ronald O. Ball. 2016. "Protein and Amino Acid Requirements during Pregnancy." *Advances in Nutrition* 7 (4): 839S–844S. https://doi.org/10.3945/an.115.011817.

Fonteles, Cristiane S. R., Milene H. Guerra, Thyciana R. Ribeiro, Danielle N. Mendonça, Cibele B. M. de Carvalho, André J. Monteiro, Daniela O. Toyama, Marcos H. Toyama, and Manasses C. Fonteles. 2009. "Association of Free Amino Acids with Caries Experience and Mutans Streptococci Levels in Whole Saliva of Children with Early Childhood Caries." *Archives of Oral Biology* 54 (1): 80–85. https://doi.org/10.1016/j.archoralbio.2008.07.011.

Gao, Ryan, Maureen Watson, Karen E. Callon, Donna Tuari, Michael Dray, Dorit Naot, Satya Amirapu, Jacob T. Munro, Jillian Cornish, and David S. Musson. 2018. "Local Application of Lactoferrin Promotes Bone Regeneration in a Rat Critical-Sized Calvarial Defect Model as Demonstrated by Micro-CT and Histological Analysis." *Journal of Tissue Engineering and Regenerative Medicine* 12 (1): e620–26. https://doi.org/10.1002/term.2348.

Garcia-Iborra, Maria, Esther Castanys-Munoz, Elena Oliveros, and Maria Ramirez. 2023. "Optimal Protein Intake in Healthy Children and Adolescents: Evaluating Current Evidence." *Nutrients* 15 (7): 1683. https://doi.org/10.3390/nu15071683.

Gil-Montoya, José, Ana Lucia Ferreira de Mello, Rocio Barrios, Miguel Angel Gonzalez-Moles, and Manuel Bravo. 2015. "Oral Health in the Elderly Patient and Its Impact on General Well-Being: A Nonsystematic Review." *Clinical Interventions in Aging* (February): 461. https://doi.org/10.2147/CIA.S54630.

Giridhar, VUsha. 2016. "Role of Nutrition in Oral and Maxillofacial Surgery Patients." *National Journal of Maxillofacial Surgery* 7 (1): 3. https://doi.org/10.4103/0975-5950.196146.

Jayasinghe, Thilini N., Sanaa Harrass, Sharon Erdrich, Shalinie King, and Joerg Eberhard. 2022. "Protein Intake and Oral Health in Older Adults—A Narrative Review." *Nutrients* 14 (21): 4478. https://doi.org/10.3390/nu14214478.

Johnson, D. A., H. Lopez, and J. M. Navia. 1995. "Effects of Protein Deficiency and Diet Consistency on the Parotid Gland and Parotid Saliva of Rats." *Journal of Dental Research* 74 (8): 1444–52. https://doi.org/10.1177/00220345950740080301.

Kaigler, Darnell, Joni A. Cirelli, and William V. Giannobile. 2006. "Growth Factor Delivery for Oral and Periodontal Tissue Engineering." *Expert Opinion on Drug Delivery* 3 (5): 647–62. https://doi.org/10.1517/17425247.3.5.647.

Kelsey, Jennifer L., and Ira B. Lamster. 2008. "Influence of Musculoskeletal Conditions on Oral Health Among Older Adults." *American Journal of Public Health* 98 (7): 1177–83. https://doi.org/10.2105/AJPH.2007.129429.

Kolderman, Ethan, Deepti Bettampadi, Derek Samarian, Scot E. Dowd, Betsy Foxman, Nicholas S. Jakubovics, and Alexander H. Rickard. 2015. "L-Arginine Destabilizes Oral Multi-Species Biofilm Communities Developed in Human Saliva." Edited by Ali Al-Ahmad. *PLoS One* 10 (5): e0121835. https://doi.org/10.1371/journal.pone.0121835.

Kotronia, Eftychia, Heather Brown, A. Olia Papacosta, Lucy T. Lennon, Robert J. Weyant, Peter H. Whincup, S. Goya Wannamethee, and Sheena E. Ramsay. 2021. "Poor Oral Health and the Association with Diet Quality and Intake in Older People in Two Studies in the UK and USA." *British Journal of Nutrition* 126 (1): 118–30. https://doi.org/10.1017/S0007114521000180.

Krissansen, Geoffrey W. 2007. "Emerging Health Properties of Whey Proteins and Their Clinical Implications." *Journal of the American College of Nutrition* 26 (6): 713S–723S. https://doi.org/10.1080/07315724.2007.10719652.

Kruger, Marlena C., Gabrielle G. Plimmer, Linda M. Schollum, Neill Haggarty, Satyendra Ram, and Kate Palmano. 2005. "The Effect of Whey Acidic Protein Fractions on Bone Loss in the Ovariectomised Rat." *British Journal of Nutrition* 94 (2): 244–52. https://doi.org/10.1079/BJN20051454.

Lands, L. C., M. Iskandar, N. Beaudoin, B. Meehan, N. Dauletbaev, and Y. Berthiuame. 2010. "Dietary Supplementation with Pressurized Whey in Patients with Cystic Fibrosis." *Journal of Medicinal Food* 13 (1): 77–82. https://doi.org/10.1089/jmf.2008.0326.

Lee, Jae Hoon, Eunju Park, Hyue Ju Jin, Yunjeong Lee, Seung Jun Choi, Gyu Whan Lee, Pahn-Shick Chang, and Hyun-Dong Paik. 2017. "Anti-Inflammatory and Anti-Genotoxic Activity of Branched Chain Amino Acids (BCAA) in Lipopolysaccharide (LPS) Stimulated RAW 264.7 Macrophages." *Food Science and Biotechnology* 26 (5): 1371–77. https://doi.org/10.1007/s10068-017-0165-4.

Li Peng, Yu-Long Yin, Defa Li, Sung Woo Kim, and Guoyao Wu. 2007. "Amino Acids and Immune Function." *British Journal of Nutrition* 98 (2): 237–52. https://doi.org/10.1017/S000711450769936X.

Lin, Y. H. 1989. "Changes of Dentin Matrices during Carious Process." *Tsurumi Shigaku. Tsurumi University Dental Journal* 15 (1): 249–66. http://www.ncbi.nlm.nih.gov/pubmed/2626796.

Loo, J. A., W. Yan, P. Ramachandran, and D. T. Wong. 2010. "Comparative Human Salivary and Plasma Proteomes." *Journal of Dental Research* 89 (10): 1016–23. https://doi.org/10.1177/0022034510380414.

Louro, Teresa, Carla Simões, Wilmara Lima, Laura Carreira, Paula Midori Castelo, Henrique Luis, Pedro Moreira, and Elsa Lamy. 2021. "Salivary Protein Profile and Food Intake: A Dietary Pattern Analysis." Edited by Roxana Valdés-Ramos. *Journal of Nutrition and Metabolism* 2021 (April): 1–10. https://doi.org/10.1155/2021/6629951.

Mahmoodpoor, Ata, Kamran Shadvar, Seied Saghaleini, Kasra Dehghan, and Zohreh Ostadi. 2018. "Pressure Ulcer and Nutrition." *Indian Journal of Critical Care Medicine* 22 (4): 283–89. https://doi.org/10.4103/ijccm.IJCCM_277_17.

Manfredo Vieira, S., M. Hiltensperger, V. Kumar, D. Zegarra-Ruiz, C. Dehner, N. Khan, F. R. C. Costa, et al. 2018. "Translocation of a Gut Pathobiont Drives Autoimmunity in Mice and Humans." *Science* 359 (6380): 1156–61. https://doi.org/10.1126/science.aar7201.

Martin, Laura E., Verenice Ascencio Gutierrez, and Ann-Marie Torregrossa. 2023. "The Role of Saliva in Taste and Food Intake." *Physiology & Behavior* 262 (April): 114109. https://doi.org/10.1016/j.physbeh.2023.114109.

McClave, Stephen A., Robert G. Martindale, Vincent W. Vanek, Mary McCarthy, Pamela Roberts, Beth Taylor, Juan B. Ochoa, Lena Napolitano, and Gail Cresci. 2009. "Guidelines for the Provision and Assessment of Nutrition Support Therapy in the Adult Critically Ill Patient." *Journal of Parenteral and Enteral Nutrition* 33 (3): 277–316. https://doi.org/10.1177/0148607109335234.

Menaker, Lewis, and Juan M. Navia. 1974. "Effect of Undernutrition During the Perinatal Period on Caries Development in the Rat: V. Changes in Whole Saliva Volume and Protein Content." *Journal of Dental Research* 53 (3): 592–97. https://doi.org/10.1177/00220345740530031201.

Merritt, Justin, Fengxia Qi, and Wenyuan Shi. 2006. "Milk Helps Build Strong Teeth and Promotes Oral Health." *Journal of the California Dental Association* 34 (5): 361–66. https://doi.org/10.1080/19424396.2006.12222204.

Ministry of Health Australia. 2014. "Nutrient Reference Values for Australia and New Zealand." https://www.nrv.gov.au/nutrients/protein.

Müller, Frauke, Najla Chebib, Sabrina Maniewicz, and Laurence Genton. 2023. "The Impact of Xerostomia on Food Choices—A Review with Clinical Recommendations." *Journal of Clinical Medicine* 12 (14): 4592. https://doi.org/10.3390/jcm12144592.

Naidoo, Sudeshni, and Neil Myburgh. 2007. "Nutrition, Oral Health and the Young Child." *Maternal & Child Nutrition* 3 (4): 312–21. https://doi.org/10.1111/j.1740-8709.2007.00115.x.

Naot, D., A. Grey, I. R. Reid, and J. Cornish. 2005. "Lactoferrin - A Novel Bone Growth Factor." *Clinical Medicine & Research* 3 (2): 93–101. https://doi.org/10.3121/cmr.3.2.93.

Nutten, Sophie. 2016. *Proteins, Peptides and Amino Acids: Role in Infant Nutrition*. Edited by J. Bhatia R. Shamir, and Y. Vandenplas, 1–10. Nestlé Nutrition Institute Workshop Series. Karger. https://doi.org/10.1159/000442697.

Pateel, Deepak Gowda Sadashivappa, Shilpa Gunjal, Liew Fong Fong, and Nur Sulwana Mohd Hanapi. 2021. "Association of Salivary Statherin, Calcium, and Proline-Rich Proteins on Oral Hygiene: A Cross-Sectional Study." Edited by Sandrine Bittencourt Berger. *International Journal of Dentistry* 2021 (February): 1–6. https://doi.org/10.1155/2021/1982083.

Patel, Seema. 2015. "Emerging Trends in Nutraceutical Applications of Whey Protein and Its Derivatives." *Journal of Food Science and Technology* 52 (11): 6847–58. https://doi.org/10.1007/s13197-015-1894-0.

Pesta, Dominik H., and Varman T. Samuel. 2014. "A High-Protein Diet for Reducing Body Fat: Mechanisms and Possible Caveats." *Nutrition & Metabolism* 11 (1): 53. https://doi.org/10.1186/1743-7075-11-53.

Phillips, Stuart M. 2017. "Determining the Protein Needs of 'Older' Persons One Meal at a Time." *The American Journal of Clinical Nutrition* 105 (2): 291–92. https://doi.org/10.3945/ajcn.116.150839.

Piccolomini, André F., Michèle M. Iskandar, Larry C. Lands, and Stan Kubow. 2012. "High Hydrostatic Pressure Pre-Treatment of Whey Proteins Enhances Whey Protein Hydrolysate Inhibition of Oxidative Stress and IL-8 Secretion in Intestinal Epithelial Cells." *Food & Nutrition Research* 56 (1): 17549. https://doi.org/10.3402/fnr.v56i0.17549.

Price, W. A. 2009. *Nutrition and Physical Degeneration*. Edited by Price-Pottenger Nutrition Foundation. 8th ed. Houston, Texas: Price Pottenger Nutrition.

Rahman, Naomi, and Angus Walls. 2020. *Chapter 12: Nutrient Deficiencies and Oral Health*. Edited by Naomi Rahman and Angus Walls, 114–24. Karger. https://doi.org/10.1159/000455379.

Rathee, Manu, and Prachi Jain. 2024. *Embryology, Teeth*. StatPearls. http://www.ncbi.nlm.nih.gov/pubmed /29745448.

Rodriguez, A., G. C. Curhan, E. N. Taylor, G. Gambaro, and P. M. Ferraro. 2019. "Mediterranean Diet Adherence and the Risk of Kidney Stones." *European Urology Supplements* 18 (7): e2878–79. https:// doi.org/10.1016/S1569-9056(19)33057-X.

Rosa, Luigi, Maria Stefania Lepanto, Antimo Cutone, Giusi Ianiro, Stefania Pernarella, Riccardo Sangermano, Giovanni Musci, Livia Ottolenghi, and Piera Valenti. 2021. "Lactoferrin and Oral Pathologies: A Therapeutic Treatment." *Biochemistry and Cell Biology* 99 (1): 81–90. https://doi.org/10.1139/bcb-2020 -0052.

Russell, Stefanie L., Walter J. Psoter, Germain Jean-charles, Samuel Prophte, and Bette Gebrian. 2010. "Protein-energy Malnutrition during Early Childhood and Periodontal Disease in the Permanent Dentition of Haitian Adolescents Aged 12–19 Years: A Retrospective Cohort Study." *International Journal of Paediatric Dentistry* 20 (3): 222–29. https://doi.org/10.1111/j.1365-263X.2010.01031.x.

Salazar, Christian R., Nadia Laniado, Yasmin Mossavar-Rahmani, Luisa N. Borrell, Qibin Qi, Daniela Sotres-Alvarez, Douglas E. Morse, et al. 2018. "Better-quality Diet Is Associated with Lower Odds of Severe Periodontitis in US Hispanics/Latinos." *Journal of Clinical Periodontology* 45 (7): 780–90. https://doi.org/10.1111/jcpe.12926.

Sbricoli, Luca, Lia Bernardi, Fatima Ezeddine, Christian Bacci, and Adolfo Di Fiore. 2022. "Oral Hygiene in Adolescence: A Questionnaire-Based Study." *International Journal of Environmental Research and Public Health* 19 (12): 7381. https://doi.org/10.3390/ijerph19127381.

Scaglia, Fernando, and Brendan Lee. 2006. "Clinical, Biochemical, and Molecular Spectrum of Hyperargininemia Due to Arginase I Deficiency." *American Journal of Medical Genetics Part C: Seminars in Medical Genetics* 142C (2): 113–20. https://doi.org/10.1002/ajmg.c.30091.

Sedghi, Lea, Vincent DiMassa, Anthony Harrington, Susan V. Lynch, and Yvonne L. Kapila. 2021. "The Oral Microbiome: Role of Key Organisms and Complex Networks in Oral Health and Disease." *Periodontology 2000* 87 (1): 107–31. https://doi.org/10.1111/prd.12393.

Setiawan, Arlette Suzy, Ratna Indriyanti, Netty Suryanti, Laili Rahayuwati, and Neti Juniarti. 2022. "Neonatal Stunting and Early Childhood Caries: A Mini-Review." *Frontiers in Pediatrics* 10 (July). https://doi.org /10.3389/fped.2022.871862.

Sheetal, Aparna. 2013. "Malnutrition and Its Oral Outcome – A Review." *Journal of Clinical and Diagnostic Research*. https://doi.org/10.7860/JCDR/2012/5104.2702.

Silva, Marine S. Da, Cyril Bigo, Olivier Barbier, and Iwona Rudkowska. 2017. "Whey Protein Hydrolysate and Branched-Chain Amino Acids Downregulate Inflammation-Related Genes in Vascular Endothelial Cells." *Nutrition Research* 38 (February): 43–51. https://doi.org/10.1016/j.nutres.2017.01.005.

Siqueiros-Cendón, Tania, Sigifredo Arévalo-Gallegos, Blanca Flor Iglesias-Figueroa, Isui Abril García-Montoya, José Salazar-Martínez, and Quintín Rascón-Cruz. 2014. "Immunomodulatory Effects of Lactoferrin." *Acta Pharmacologica Sinica* 35 (5): 557–66. https://doi.org/10.1038/aps.2013.200.

Stahl, S.Sigmund, Henry C. Sandler, and Lester Cahn. 1955. "The Effects of Protein Deprivation upon the Oral Tissues of the Rat and Particularly upon Periodontal Structures under Irritation." *Oral Surgery, Oral Medicine, Oral Pathology* 8 (7): 760–68. https://doi.org/10.1016/0030-4220(55)90040-2.

Syrjänen, S., P. Piironen, and H. Markkanen. 1987. "Free Amino-Acid Content of Wax-Stimulated Human Whole Saliva as Related to Periodontal Disease." *Archives of Oral Biology* 32 (9): 607–10. https://doi .org/10.1016/0003-9969(87)90032-X.

Toba, Yasuhiro, Yukihiro Takada, Yasuhiro Matsuoka, Yoshikazu Morita, Mutsumi Motouri, Tomoko Hirai, Toshiaki SUGURI, et al. 2001. "Milk Basic Protein Promotes Bone Formation and Suppresses Bone Resorption in Healthy Adult Men." *Bioscience, Biotechnology, and Biochemistry* 65 (6): 1353–57. https://doi.org/10.1271/bbb.65.1353.

Toba, Yasuhiro, Yukihiro Takada, J. Yamamura, M. Tanaka, Y. Matsuoka, H. Kawakami, A Itabashi, S. Aoe, and M. Kumegawa. 2000. "Milk Basic Protein: A Novel Protective Function of Milk against Osteoporosis." *Bone* 27 (3): 403–8. https://doi.org/10.1016/s8756-3282(00)00332-x.

Torregrossa, Ann-Marie, Larissa Nikonova, Michelle B. Bales, Maria Villalobos Leal, James C. Smith, Robert J. Contreras, and Lisa A. Eckel. 2014. "Induction of Salivary Proteins Modifies Measures of Both Orosensory and Postingestive Feedback during Exposure to a Tannic Acid Diet." Edited by John I. Glendinning. *PLoS One* 9 (8): e105232. https://doi.org/10.1371/journal.pone.0105232.

Tungare, Sujata, and Arati G. Paranjpe. 2022. *Diet and Nutrition To Prevent Dental Problems*. StatPearls. http://www.ncbi.nlm.nih.gov/pubmed/30480981.

Urrutia, Rachel P., and John M. Thorp. 2012. "Vitamin D in Pregnancy." *Current Opinion in Obstetrics & Gynecology* 24 (2): 57–64. https://doi.org/10.1097/GCO.0b013e3283505ab3.

Valizadeh, Maryam, Farnoosh Alimohammadi, Ali Azarm, Zeynab Pourtaghi, Mohammad moein Derakhshan barjoei, Hamoun Sabri, Aryan Jafari, et al. 2021. "Uses of Soybean Isoflavonoids in Dentistry: A Literature Review." *Journal of Dental Sciences* (December). https://doi.org/10.1016/j.jds.2021.11.020.

Vieira, Karlla A., Laerto S. Rosa-Júnior, Maria A. V. Souza, Natanael B. Santos, Telma M. M. T. Florêncio, and Sandra K. Bussadori. 2020. "Chronic Malnutrition and Oral Health Status in Children Aged 1 to 5 Years." *Medicine* 99 (18): e19595. https://doi.org/10.1097/MD.0000000000019595.

Vranić, L., P. Granić, and Z. Rajić. 1991. "Basic Amino Acid in the Pathogenesis of Caries." *Acta Stomatologica Croatica* 25 (2): 71–76. https://doi.org/1819935.

Wallenstein, Matthew B., Gary M. Shaw, Wei Yang, and Suzan L. Carmichael. 2013. "Periconceptional Nutrient Intakes and Risks of Orofacial Clefts in California." *Pediatric Research* 74 (4): 457–65. https://doi.org/10.1038/pr.2013.115.

Woelber, J. P., K. Bremer, K. Vach, D. König, E. Hellwig, P. Ratka-Krüger, A. Al-Ahmad, and C. Tennert. 2016. "Erratum to: An Oral Health Optimized Diet Can Reduce Gingival and Periodontal Inflammation in Humans - a Randomized Controlled Pilot Study." *BMC Oral Health* 16 (1): 109. https://doi.org/10.1186/s12903-016-0304-y.

Wolfe, Robert R., Amy M. Cifelli, Georgia Kostas, and Il-Young Kim. 2017. "Optimizing Protein Intake in Adults: Interpretation and Application of the Recommended Dietary Allowance Compared with the Acceptable Macronutrient Distribution Range." *Advances in Nutrition* 8 (2): 266–75. https://doi.org/10.3945/an.116.013821.

Wu, Guoyao, Fuller W. Bazer, Robert C. Burghardt, Gregory A. Johnson, Sung Woo Kim, Darrell A. Knabe, Peng Li, et al. 2011. "Proline and Hydroxyproline Metabolism: Implications for Animal and Human Nutrition." *Amino Acids* 40 (4): 1053–63. https://doi.org/10.1007/s00726-010-0715-z.

Yenen, Zeynep, and Tijen Ataçağ. 2019. "Oral Care in Pregnancy." *Journal of the Turkish-German Gynecological Association* 20 (4): 264–68. https://doi.org/10.4274/jtgga.galenos.2018.2018.0139.

Zabokova Bilbilova, Efka. 2021. "Dietary Factors, Salivary Parameters, and Dental Caries." In *Dental Caries*. IntechOpen. https://doi.org/10.5772/intechopen.92392.

Zakhary, G. M., R. M. Clark, S. I. Bidichandani, W. L. Owen, R. L. Slayton, and M. Levine. 2007. "Acidic Proline-Rich Protein Db and Caries in Young Children." *Journal of Dental Research* 86 (12): 1176–80. https://doi.org/10.1177/154405910708601207.

20 Protein in the Diet *Impact on Food Intake and Energy Regulation*

Jamie I. Baum, Chetanjot Kaur Romana, and Samuel Selorm Attu

ABBREVIATIONS

AEE	Activity energy expenditure
BCAA	Branched-chain amino acids
BMI	Body mass index
CCK	Cholecystokinin
CVD	Cardiovascular disease
DEE	Daily energy expenditure
EE	Energy expenditure
FFM	Fat-free mass
GLP-1	Glucagon-like peptide-1
mTOR	Mammalian target of rapamycin
PYY	Peptide YY
REE	Resting energy expenditure
RMR	Resting metabolic rate
T2D	Type 2 diabetes mellitus
TEF	Thermic effect of food

INTRODUCTION

Body weight that exceeds what is considered healthy for a given height is defined as overweight or obesity (CDC 2022). Obesity has become a major cause of mortality and morbidity worldwide, with more deaths attributed to excess weight than underweight (WHO 2021). In 2022, approximately 890 million people – one in eight globally – were living with obesity. An additional 2.5 billion adults aged 18 years and older were classified as overweight (WHO 2024). This widespread health problem may have a negative impact on cardiometabolic health (Smith, Mittendorfer, and Klein 2019) and increases the risk of chronic diseases such as cardiovascular disease, type 2 diabetes (T2D), stroke, and certain cancers (Munsters and Saris 2014). The high prevalence of obesity in both developed and developing countries has made it a major public health concern.

The development of obesity is multifactorial, arising from interactions between genetic predisposition and environmental risk factors. These factors include physical inactivity, excessive caloric intake, adverse intrauterine conditions, medications, socioeconomic factors, insufficient sleep, endocrine disruption, and alterations in the gut microbiome (Goodarzi 2018; Chait and den Hartigh 2020). Therefore, it is essential to identify effective treatment and preventive approaches for weight loss and weight maintenance to reduce the prevalence of obesity and its associated chronic diseases. Behavioural changes, including dietary modifications (e.g., regulation of food intake), are readily

DOI: 10.1201/b23328-20

accessible and have been shown to be effective in managing obesity, potentially reducing the need for more expensive medications and surgeries (Bluher 2019).

Food intake and energy expenditure are key regulators of body weight (Cifuentes and Acosta 2022). This chapter focuses on dietary protein's role in the regulation of food intake and energy expenditure for the treatment and prevention of overweight and obesity.

REGULATION OF FOOD INTAKE

Food intake is highly regulated by internal physiological information and external environmental cues. The central nervous system (CNS) tightly controls food intake, integrating several inputs and operating on at least three levels: behaviour (eating, physical activity, and sleep), the autonomic nervous system response (through regulating energy expenditure and metabolic processes), and neuroendocrine signals (regulating food intake and energy expenditure) (Chamorro et al. 2023). The hypothalamus and brain stem integrate neuronal and hormonal signals from the periphery and other brain areas to stimulate feeding behaviours (e.g., homeostatic control of food intake and energy balance).

However, humans eat not only in response to homeostatic signals but also for psychological and sociocultural reasons (Cifuentes and Acosta 2022). This integration of information determines what, when, and how much we eat. Homeostatic regulation depends on peripheral metabolic signals such as nutrients (e.g., amino acids, glucose, fatty acids) and neuroendocrine signals including hormones, peptides, and visceral information from the gastrointestinal tract. These include hormones signalling orexigenic (e.g., ghrelin) and anorexigenic (e.g., gut hormones such as peptide YY (PYY), glucagon-like peptide-1 (GLP-1), and cholecystokinin (CCK)) effects (Chamorro et al. 2023). These neuroendocrine signals are secreted in response to macronutrients (e.g., dietary protein) entering the gut lumen (Campos, Port, and Acosta 2022; Hirschberg 1998) with the ultimate goal of preserving the stability of energy stores (Chamorro et al. 2023). Table 20.1 provides a detailed description of these gut hormones' functions.

Food intake is divided into hunger, satiation, and satiety (Blundell and Halford 1994), coordinated by the gastrointestinal, endocrine, and nervous systems to maintain energy homeostasis (Strominger and Brobeck 1953). Hunger is the desire to eat, characterized by an uncomfortable emptiness in the abdomen; satiation refers to the sensation of fullness that results in meal termination; and satiety describes the postprandial (i.e., post-meal) events that determine the timing for the next meal (Cifuentes and Acosta 2022). Although it may seem that food intake is a step-by-step system, these are dynamic processes that are controlled by both homeostatic (i.e., metabolic) and hedonic (i.e., reward) components of food intake regulation (Campos, Port, and Acosta 2022). Collectively, these processes are termed the satiety cascade (Hopkins and Blundell 2016)

Macronutrient composition can influence the expression of the satiety cascade and appetite-related processes, with dietary macronutrients exerting a hierarchical effect on satiety, food reward, and short-term food intake (Hopkins and Blundell 2016). Protein exerts the strongest effect on satiety, while fat exerts the weakest effect (Blundell et al. 1993).

ENERGY BALANCE, BODY COMPOSITION, AND APPETITE

Energy balance (Figure 20.1) in humans involves a complex interplay between energy intake (e.g., food and beverage intake) and energy expenditure (e.g., calories burned) for metabolism and physical activity (Hopkins and Blundell 2016). The basic components of energy balance include energy intake, energy expenditure, and energy storage (Casanova et al. 2019). The relationship between energy intake and energy expenditure determines the stability of body weight over time (Hill, Wyatt, and Peters 2012, 2013). Obesity occurs when there is an imbalance in energy homeostasis and energy intake exceeds energy expenditure, leading to excessive energy storage (Cifuentes and Acosta 2022; Hopkins and Blundell 2016; Casanova et al. 2019). Both sides of the energy balance

TABLE 20.1
Summary of Gut Hormones Regulating Food Intake

Appetite Hormone	Description
Peptide YY (PYY)	PYY is secreted by enteroendocrine L-cells of the ileum and colon in response to nutrients, mainly fat. Plasma levels increase within 15 minutes post-intake, peak around 60 minutes, and remain elevated for up to six hours. PYY plays a role in satiety control.
Glucagon-like Peptide-1 (GLP-1)	GLP-1 is synthesized as pre-pro-glucagon in the L-cells of the ileum and colon, and in the pancreas by brain pre-pro-glucagon neurons. It is released in response to intestinal luminal nutrients and/or bile acids. Concentrations are lowest after overnight fasting and increase upon meal ingestion in a biphasic pattern. The first peak is 10–15 minutes after food intake, followed by a second peak at 30–60 minutes. GLP-1 plays a role in satiety.
Cholecystokinin (CCK)	CCK is released from intestinal I-cells in the duodenal and jejunal mucosa in response to fat, protein, and, to a lesser extent, carbohydrates. Plasma levels increase 10–15 minutes after oral intake and remain elevated for up to five hours postprandial. CCK is mediated through the CCK-1 receptor (CCK1R). CCK plays a role in satiation.
Ghrelin	Ghrelin is produced by X/A-like cells in the gastric oxyntic mucosa and the small intestine. These cells express G-protein-coupled receptors. Upon activation by nutrients, these receptors suppress ghrelin's release. Plasma ghrelin peaks before a meal and decreases postprandially in response to nutrient ingestion. Protein is the primary suppressant of ghrelin. GLP-1 plays a role in hunger.

The table summarizes the role of homeostatic gastrointestinal in response to macronutrient (e.g., dietary protein) components of food intake (Cifuentes and Acosta 2022; Hopkins and Blundell 2016).

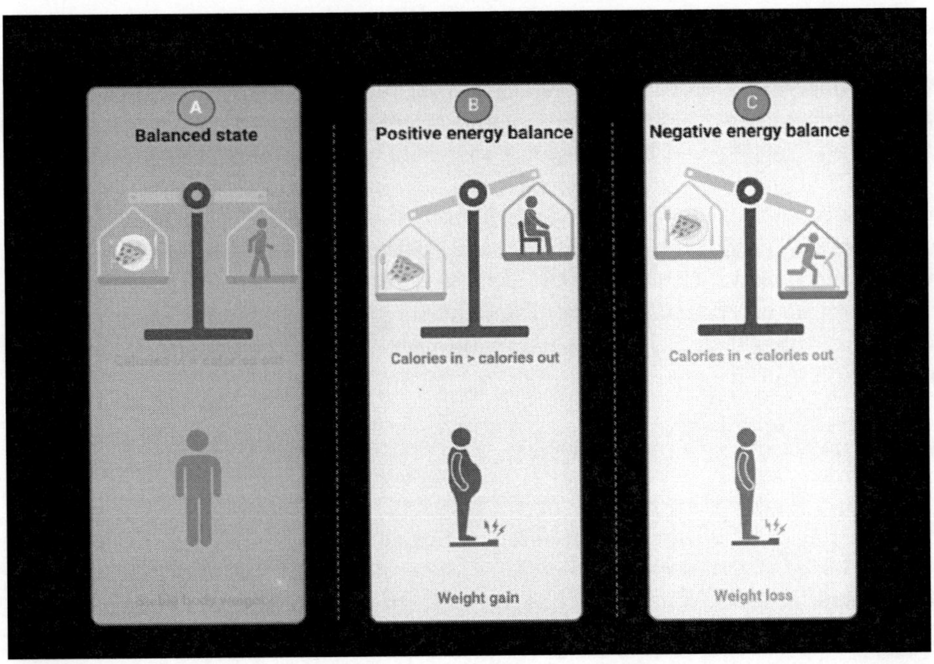

FIGURE 20.1 The role of energy balance in body weight gain. A. When energy intake (e.g., food intake/calories in) equals energy out (e.g., energy expenditure/calories out), there is a balanced energy state, and a stable weight is maintained. B. When calories in exceeds calories out, this leads to positive energy balance and may lead to weight gain over time. C. When calories in are less than calories out, this leads to negative energy balance and may lead to weight loss over time.

equation, energy intake and energy expenditure, are influenced by a complex interaction of behavioural, physiological, and environmental variables (Figure 20.1).

Energy expenditure (EE) occurs through resting metabolic rate (RMR) (also referred to as resting energy expenditure (REE)), the thermic effect of food (TEF), and physical activity. RMR, accounting for 60–75% of daily EE (DEE), represents the energy necessary to maintain normal bodily processes and homeostasis, and is proportional to body mass, specifically fat-free mass (FFM). TEF, responsible for 10–15% of DEE, is the amount of energy required to absorb, digest, and metabolize the food consumed. Physical activity, including exercise and non-exercise activity, contributes 10–30% to DEE (Casanova et al. 2019). Energy intake is influenced by factors such as body composition, sex, appetite signals, and nutrient composition (Casanova et al. 2019).

Individual energy requirements vary with age, sex, body size, and physiological status. REE measurements form the basis of estimating total EE and, consequently, overall energy requirements (Soares and Muller 2018). FFM (e.g., skeletal muscle) is the primary determinant of RMR, explaining 64–80% of its inter-individual variability (Soares and Muller 2018; Cunningham 1980).

Recent research suggests that FFM also influence appetite regulation through an energy-sensing mechanism that translates FFM energy requirements into energy intake signals (Blundell et al. 2012b; Hopkins and Blundell 2016; Blundell et al. 2012a; Blundell et al. 2015; Caudwell et al. 2013). In the 1950s, Edholm et al. (Edholm et al. 1955) first observed that the difference in food intake between individuals was related to differences in energy intake. They argued that energy intake was driven by energy expenditure and this regulates energy balance (Edholm et al. 1955). In addition, since FFM is a strong determinant of energy expenditure, it could be driving appetite response and energy intake (Edholm et al. 1955). This idea was further supported by an early investigation into the relationships between body composition and *ad libitum* food intake in obese and normal-weight women, which showed that energy intake for the maintenance of body weight was not correlated with FM but was positively correlated with FFM (Lissner et al. 1989). More recently, Blundell et al. (Blundell et al. 2012a) demonstrated that FFM, but not FM, was positively associated with self-selected meal size and total energy intake in overweight and obese individuals. This work has since been further supported in adults (Caudwell et al. 2013; Weise et al. 2014; Piaggi et al. 2015) and in adolescents (Cameron et al. 2016). Cameron et al. (Cameron et al. 2016) demonstrated that FFM and skeletal muscle mass predicted energy intake in 304 post-pubertal male and female adolescents. The effect of FFM on energy intake in pre-pubertal children still needs to be explored.

In the work described above, not only is the observation that FFM is positively correlated with appetite important, but so is the observation that RMR, which is largely determined by FFM, correlates with energy intake. Caudwell et al. (Caudwell et al. 2013) demonstrated that RMR is not only correlated with energy intake but also strongly correlated with hunger and meal size in overweight and obese individuals. The mechanisms through which RMR influences energy intake are not known.

DIETARY PROTEIN AND REGULATION OF FOOD INTAKE

There is substantial evidence supporting higher protein dietary patterns (1.2–1.6 g protein/kg body weight/day) as a successful strategy in treating or preventing obesity (Lasker, Evans, and Layman 2008; Leidy et al. 2015). Individuals following a higher protein dietary pattern have reductions in body weight and fat mass while preserving FFM (Clifton, Bastiaans, and Keogh 2009; Clifton, Condo, and Keogh 2014; Dong et al. 2013). The effectiveness of these diets may be due to the regulation of energy metabolism and appetite signalling leading to reduced energy intake (Leidy et al. 2015).

Protein-induced satiety has been shown to reduce energy intake following acute, single meals (e.g., meals given one time) with protein composition ranging from 25–81% of energy (i.e., calories) (M. Veldhorst et al. 2008). Standard protein intake is typically defined as 10–15% of energy, while higher protein intake ranges from 20–30% of energy, based on the acceptable macronutrient distribution range (ADMR) (Baum and Wolfe 2015). Higher protein intake mediates satiety through both neural and endocrine signals originating from the gastrointestinal tract and portal circulation, increasing levels of GLP-1, PYY, CCK, and glucagon, while reducing ghrelin (Khan et al. 2021).

Randomized controlled trials have shown that healthy, normal-weight individuals fed a high-protein lunch have significantly higher postprandial satiety and TEF than after a normal-protein lunch (Lejeune et al. 2006; Smeets et al. 2008). While acute studies show no differences in ghrelin and PYY responses, GLP-1 response is typically lower following high-protein lunches. Longer-term studies, ranging from 24 hours to five days, show consistently higher satiety and fullness throughout the day with high-protein diets compared to energy-matched, normal-protein diets. These effects are accompanied by reduced hunger, appetite, desire to eat, and perceived food intake (Westerterp-Plantenga et al. 1999; Lejeune et al. 2006).

DIETARY PROTEIN AND REGULATION OF ENERGY EXPENDITURE

Higher-protein dietary patterns increase energy expenditure through increased TEF (also known as postprandial thermogenesis) and elevated resting metabolism. Protein metabolism requires 20–30% of its usable energy for processing or storage, significantly higher than carbohydrates (5–10%) or fats (0–3%) (Westerterp-Plantenga et al. 2009; Leidy et al. 2015). Several studies have confirmed that dietary protein elicits a higher TEF than carbohydrates and fat (Halton and Hu 2004; Eisenstein et al. 2002). A meta-analysis examining TEF and appetite in adults (Ravn et al. 2013) revealed a positive correlation between protein intake and TEF, even after adjusting for variables such as sex and dinner intake. Moreover, during weight loss, high-protein dietary patterns also prevent a decline in REE, possibly by preserving FFM (Wycherley et al. 2012, 2013).

TEF, accounting for approximately 10% of DEE, is a potential target for obesity prevention due to its responsiveness to dietary macronutrient composition (Morgan et al. 1982). Researchers have explored TEF manipulation as a strategy to increase total DEE to assist with weight control. Altering the macronutrient ratios in meals directly impacts TEF, with protein demonstrating the most significant effect. High-protein meals can elevate TEF by up to 20% compared to low-protein meals (Halton and Hu 2004), a phenomenon attributed to the higher energetic cost of protein metabolism compared to other macronutrients (Halton and Hu 2004).

THE ROLE OF BRANCHED-CHAIN AMINO ACIDS IN FOOD INTAKE AND ENERGY REGULATION

Dietary protein serves as a source of essential amino acids, including the branched-chain amino acids (BCAAs): leucine, isoleucine, and valine. Several studies have suggested that BCAAs, particularly leucine, may act as direct signals of protein intake in the brain (Khan et al. 2021). Studies in rodents have found that when leucine is administered into the brain, food intake is suppressed by impacting BCAA metabolism, mammalian target of rapamycin (mTOR) signalling, and AMP-activated protein kinase (AMPK) pathways (Cota et al. 2006; Tsang et al. 2020; Laeger et al. 2014). Leucine acts in both the hypothalamus and the brain stem to suppress food intake, an effect that is not observed with other amino acids (Laeger et al. 2014). It has also been shown that brain leucine signalling is required for the anorectic effects of high-protein diets (Tsang et al. 2020).

THE EFFECTS OF OTHER PROTEINS

While this chapter primarily reviews the general effect of dietary protein on food intake and energy regulation, it is crucial to recognize that both the quantity and source of protein significantly affect appetite response (M. Veldhorst et al. 2008). Veldhorst et al. conducted a comparative study of casein, soy, and whey proteins at two different energy levels (10% and 25%). They discovered that satiety outcomes differed based on protein type, quantity, or a combination of both factors. At lower protein levels (10% of energy), the source of protein made a difference in satiety response, with whey being more satiating than casein and soy. However, at higher levels (25% of energy), the protein source had no effect on appetite response. This suggests that appetite modulation by different protein types may depend on specific amino acid threshold values (M.A. Veldhorst et al. 2009).

A retrospective analysis of several studies in which 350 kcal meals with varying protein content (15–30 g) identified a potential protein-satiety threshold. Only meals containing 30 g elicited a more substantial and sustained increase in postprandial fullness compared with the lower protein quantities. No differences were observed between the lower protein quantities (Phillips, Chevalier, and Leidy 2016; Paddon-Jones and Leidy 2014). This finding suggests that a minimum protein quantity may be necessary to trigger significant satiety responses. In another study comparing the acute effects of equivalent quantities (24 g) of high-quality animal (beef) and plant (soy) proteins, there was no difference in postprandial appetite, satiety, or food intake. This suggests that both protein quality and quantity influence postprandial appetite response (Douglas, Lasley, and Leidy 2015).

MINI DICTIONARY OF TERMS

Appetite: The desire or inclination to eat, driven by physiological and psychological factors such as hunger, taste, and emotional cues.

Dietary protein: Complex molecules composed of chains of amino acids, essential for the body's growth, repair, and tissue maintenance. Obtained from food sources.

Energy balance: The relationship between the calories consumed through diet and the calories expended through metabolic processes and physical activity.

Energy expenditure: The total energy output of the body, encompassing basal metabolic rate (BMR), physical activity, and the thermic effect of food (TEF). A key factor in determining overall energy balance and weight management.

Food intake: The consumption of food and beverages, providing essential nutrients and energy for various physiological functions.

Protein source: Any food or substance containing significant amounts of dietary protein.

Satiety: The feeling of fullness and satisfaction that occurs after eating, signalling the body to stop consuming food. Involves both physiological and psychological processes that regulate food intake.

SUMMARY POINTS

- This chapter examines how dietary protein intake influences food intake and energy regulation, with implications for the treatment and prevention of obesity and associated chronic diseases.
- Obesity is primarily caused by chronic positive energy balance, where caloric intake exceeds total energy expenditure over time, leading to excess adiposity.
- Energy balance is influenced by a complex interplay of behavioural, physiological, and environmental factors, including food intake and energy regulation.
- Dietary protein intake influences food intake through its satiating effects and regulation of energy expenditure via the thermic effect of feeding.
- Both animal and plant-based protein sources contribute to a balanced diet, with animal proteins being highly satiating and plant proteins offering additional benefits such as fibre content. Balancing protein intake within the diet is crucial for effective appetite control and overall energy balance.

REFERENCES

Baum, J. I., and R. R. Wolfe. 2015. "The link between dietary protein intake, skeletal muscle function and health in older adults." *Healthcare (Basel)* 3 (3): 529–43. https://doi.org/10.3390/healthcare3030529. https://www.ncbi.nlm.nih.gov/pubmed/27417778.

Bluher, M. 2019. "Obesity: global epidemiology and pathogenesis." *Nat Rev Endocrinol* 15 (5): 288–98. https://doi.org/10.1038/s41574-019-0176-8. https://www.ncbi.nlm.nih.gov/pubmed/30814686.

Blundell, J. E., V. J. Burley, J. R. Cotton, and C. L. Lawton. 1993. "Dietary fat and the control of energy intake: evaluating the effects of fat on meal size and postmeal satiety." *Am J Clin Nutr* 57 (5 Suppl): 772S–7S; discussion 777S–778S. https://doi.org/10.1093/ajcn/57.5.772S. https://www.ncbi.nlm.nih.gov/pubmed/8475895.

Blundell, J. E., P. Caudwell, C. Gibbons, M. Hopkins, E. Naslund, N. A. King, and G. Finlayson. 2012a. "Body composition and appetite: fat-free mass (but not fat mass or BMI) is positively associated with self-determined meal size and daily energy intake in humans." *Br J Nutr* 107 (3): 445–9. https://doi.org/10.1017/S0007114511003138. https://www.ncbi.nlm.nih.gov/pubmed/21733267.

Blundell, J. E., P. Caudwell, C. Gibbons, M. Hopkins, E. Naslund, N. A. King, and G. Finlayson. 2012b. "Role of resting metabolic rate and energy expenditure in hunger and appetite control: a new formulation." *Dis Model Mech* 5 (5): 608–13. https://doi.org/10.1242/dmm.009837. https://www.ncbi.nlm.nih.gov/pubmed/22915022.

Blundell, J. E., G. Finlayson, C. Gibbons, P. Caudwell, and M. Hopkins. 2015. "The biology of appetite control: Do resting metabolic rate and fat-free mass drive energy intake?" *Physiol Behav* 152 (Pt B): 473–8. https://doi.org/10.1016/j.physbeh.2015.05.031. https://www.ncbi.nlm.nih.gov/pubmed/26037633.

Blundell, J. E., and J. C. Halford. 1994. "Regulation of nutrient supply: the brain and appetite control." *Proc Nutr Soc* 53 (2): 407–18. https://doi.org/10.1079/pns19940046. https://www.ncbi.nlm.nih.gov/pubmed/7972155.

Cameron, J. D., R. J. Sigal, G. P. Kenny, A. S. Alberga, D. Prud'homme, P. Phillips, S. Doucette, and G. Goldfield. 2016. "Body composition and energy intake - skeletal muscle mass is the strongest predictor of food intake in obese adolescents: The HEARTY trial." *Appl Physiol Nutr Metab* 41 (6): 611–17. https://doi.org/10.1139/apnm-2015-0479. https://www.ncbi.nlm.nih.gov/pubmed/27111402.

Campos, A., J. D. Port, and A. Acosta. 2022. "Integrative hedonic and homeostatic food intake regulation by the central nervous system: insights from neuroimaging." *Brain Sci* 12 (4). https://doi.org/10.3390/brainsci12040431. https://www.ncbi.nlm.nih.gov/pubmed/35447963.

Casanova, N., K. Beaulieu, G. Finlayson, and M. Hopkins. 2019. "Metabolic adaptations during negative energy balance and their potential impact on appetite and food intake." *Proc Nutr Soc* 78 (3): 279–89. https://doi.org/10.1017/S0029665118002811. https://www.ncbi.nlm.nih.gov/pubmed/30777142.

Caudwell, P., G. Finlayson, C. Gibbons, M. Hopkins, N. King, E. Naslund, and J. E. Blundell. 2013. "Resting metabolic rate is associated with hunger, self-determined meal size, and daily energy intake and may represent a marker for appetite." *Am J Clin Nutr* 97 (1): 7–14. https://doi.org/10.3945/ajcn.111.029975. https://www.ncbi.nlm.nih.gov/pubmed/23193010.

CDC. 2022. *Defining Adult Overweight & Obesity*. Overweight & Obesity. Centers for Disease Control and Prevention. Accessed April 14. https://www.cdc.gov/obesity/basics/adult-defining.html.

Chait, Alan, and Laura J. den Hartigh. 2020. "Adipose tissue distribution, inflammation and its metabolic consequences, including diabetes and cardiovascular disease." *Frontiers in Cardiovascular Medicine* 7: 22. https://doi.org/10.3389/fcvm.2020.00022.

Chamorro, R., C. Jouffe, H. Oster, N. H. Uhlenhaut, and S. M. Meyhofer. 2023. "When should I eat: A circadian view on food intake and metabolic regulation." *Acta Physiol (Oxf)* 237 (3): e13936. https://doi.org/10.1111/apha.13936. https://www.ncbi.nlm.nih.gov/pubmed/36645134.

Cifuentes, L., and A. Acosta. 2022. "Homeostatic regulation of food intake." *Clin Res Hepatol Gastroenterol* 46 (2): 101794. https://doi.org/10.1016/j.clinre.2021.101794. https://www.ncbi.nlm.nih.gov/pubmed/34481092.

Clifton, P. M., K. Bastiaans, and J. B. Keogh. 2009. "High protein diets decrease total and abdominal fat and improve CVD risk profile in overweight and obese men and women with elevated triacylglycerol." *Nutr Metab Cardiovasc Dis* 19 (8): 548–54. https://doi.org/10.1016/j.numecd.2008.10.006. https://www.ncbi.nlm.nih.gov/pubmed/19179060.

Clifton, P. M., D. Condo, and J. B. Keogh. 2014. "Long term weight maintenance after advice to consume low carbohydrate, higher protein diets--a systematic review and meta analysis." *Nutr Metab Cardiovasc Dis* 24 (3): 224–35. https://doi.org/10.1016/j.numecd.2013.11.006. https://www.ncbi.nlm.nih.gov/pubmed/24472635.

Cota, D., K. Proulx, K. A. Smith, S. C. Kozma, G. Thomas, S. C. Woods, and R. J. Seeley. 2006. "Hypothalamic mTOR signaling regulates food intake." *Science* 312 (5775): 927–30. https://doi.org/10.1126/science.1124147. https://www.ncbi.nlm.nih.gov/pubmed/16690869.

Cunningham, J. J. 1980. "A reanalysis of the factors influencing basal metabolic rate in normal adults." *Am J Clin Nutr* 33 (11): 2372–4. https://doi.org/10.1093/ajcn/33.11.2372. https://www.ncbi.nlm.nih.gov/pubmed/7435418.

Dong, J. Y., Z. L. Zhang, P. Y. Wang, and L. Q. Qin. 2013. "Effects of high-protein diets on body weight, glycaemic control, blood lipids and blood pressure in type 2 diabetes: meta-analysis of randomised controlled trials." *Br J Nutr* 110 (5): 781–9. https://doi.org/10.1017/S0007114513002055. https://www.ncbi.nlm.nih.gov/pubmed/23829939.

Douglas, S. M., T. R. Lasley, and H. J. Leidy. 2015. "Consuming beef vs. Soy protein has little effect on appetite, satiety, and food intake in healthy adults." *J Nutr* 145 (5): 1010–16. https://doi.org/10.3945/jn.114.206987. https://www.ncbi.nlm.nih.gov/pubmed/25809680.

Edholm, O. G., J. G. Fletcher, E. M. Widdowson, and R. A. McCance. 1955. "The energy expenditure and food intake of individual men." *Br J Nutr* 9 (3): 286–300. https://www.ncbi.nlm.nih.gov/pubmed/13250128.

Eisenstein, J., S. B. Roberts, G. Dallal, and E. Saltzman. 2002. "High-protein weight-loss diets: are they safe and do they work? A review of the experimental and epidemiologic data." *Nutr Rev* 60 (7 Pt 1): 189–200. https://doi.org/10.1301/00296640260184264. https://www.ncbi.nlm.nih.gov/pubmed/12144197.

Goodarzi, M. O. 2018. "Genetics of obesity: what genetic association studies have taught us about the biology of obesity and its complications." *Lancet Diabetes Endocrinol* 6 (3): 223–36. https://doi.org/10.1016/S2213-8587(17)30200-0. https://www.ncbi.nlm.nih.gov/pubmed/28919064.

Halton, T. L., and F. B. Hu. 2004. "The effects of high protein diets on thermogenesis, satiety and weight loss: a critical review." *J Am Coll Nutr* 23 (5): 373–85. http://www.ncbi.nlm.nih.gov/pubmed/15466943.

Hill, James O., Holly R. Wyatt, and John C. Peters. 2012. "Energy Balance and Obesity." *Circulation (New York, N.Y.)* 126 (1): 126–32. https://doi.org/10.1161/CIRCULATIONAHA.111.087213.

Hill, James O., Holly R. Wyatt, and John C. Peters. 2013. "The Importance of Energy Balance." *European Endocrinology* 9 (2): 111–15. https://doi.org/10.17925/ee.2013.09.02.111.

Hirschberg, A. L. 1998. "Hormonal regulation of appetite and food intake." *Ann Med* 30 (1): 7–20. https://doi.org/10.3109/07853899808999380. https://www.ncbi.nlm.nih.gov/pubmed/9556085.

Hopkins, M., and J. E. Blundell. 2016. "Energy balance, body composition, sedentariness and appetite regulation: pathways to obesity." *Clin Sci (Lond)* 130 (18): 1615–28. https://doi.org/10.1042/CS20160006. https://www.ncbi.nlm.nih.gov/pubmed/27503946.

Khan, M. S., R. A. Spann, H. Munzberg, S. Yu, V. L. Albaugh, Y. He, H. R. Berthoud, and C. D. Morrison. 2021. "Protein appetite at the interface between nutrient sensing and physiological homeostasis." *Nutrients* 13 (11). https://doi.org/10.3390/nu13114103. https://www.ncbi.nlm.nih.gov/pubmed/34836357.

Laeger, T., S. D. Reed, T. M. Henagan, D. H. Fernandez, M. Taghavi, A. Addington, H. Munzberg, R. J. Martin, S. M. Hutson, and C. D. Morrison. 2014. "Leucine acts in the brain to suppress food intake but does not function as a physiological signal of low dietary protein." *Am J Physiol Regul Integr Comp Physiol* 307 (3): R310–20. https://doi.org/10.1152/ajpregu.00116.2014. https://www.ncbi.nlm.nih.gov/pubmed/24898843.

Lasker, D. A., E. M. Evans, and D. K. Layman. 2008. "Moderate carbohydrate, moderate protein weight loss diet reduces cardiovascular disease risk compared to high carbohydrate, low protein diet in obese adults: A randomized clinical trial." *Nutr Metab (Lond)* 5: 30. https://doi.org/10.1186/1743-7075-5-30. https://www.ncbi.nlm.nih.gov/pubmed/18990242.

Leidy, H. J., P. M. Clifton, A. Astrup, T. P. Wycherley, M. S. Westerterp-Plantenga, N. D. Luscombe-Marsh, S. C. Woods, and R. D. Mattes. 2015. "The role of protein in weight loss and maintenance." *Am J Clin Nutr* 101 (6): 1320S–9S. https://doi.org/10.3945/ajcn.114.084038. https://www.ncbi.nlm.nih.gov/pubmed/25926512.

Lejeune, M. P., K. R. Westerterp, T. C. Adam, N. D. Luscombe-Marsh, and M. S. Westerterp-Plantenga. 2006. "Ghrelin and glucagon-like peptide 1 concentrations, 24-h satiety, and energy and substrate metabolism during a high-protein diet and measured in a respiration chamber." *Am J Clin Nutr* 83 (1): 89–94. https://doi.org/10.1093/ajcn/83.1.89. https://www.ncbi.nlm.nih.gov/pubmed/16400055.

Lissner, L., J. P. Habicht, B. J. Strupp, D. A. Levitsky, J. D. Haas, and D. A. Roe. 1989. "Body composition and energy intake: do overweight women overeat and underreport?" *Am J Clin Nutr* 49 (2): 320–5. https://www.ncbi.nlm.nih.gov/pubmed/2916451.

Morgan, J. B., D. A. York, A. Wasilewska, and J. Portman. 1982. "A study of the thermic responses to a meal and to a sympathomimetic drug (ephedrine) in relation to energy balance in man." *Br J Nutr* 47 (1): 21–32. http://www.ncbi.nlm.nih.gov/pubmed/7037048.

Munsters, M. J., and W. H. Saris. 2014. "Body weight regulation and obesity: dietary strategies to improve the metabolic profile." *Annu Rev Food Sci Technol* 5: 39–51. https://doi.org/10.1146/annurev-food-030212-182557. https://www.ncbi.nlm.nih.gov/pubmed/24580072.

Paddon-Jones, D., and H. Leidy. 2014. "Dietary protein and muscle in older persons." *Curr Opin Clin Nutr Metab Care* 17 (1): 5–11. https://doi.org/10.1097/MCO.0000000000000011. https://www.ncbi.nlm.nih.gov/pubmed/24310053.

Phillips, S. M., S. Chevalier, and H. J. Leidy. 2016. "Protein "requirements" beyond the RDA: implications for optimizing health." *Appl Physiol Nutr Metab* 41 (5): 565–72. https://doi.org/10.1139/apnm-2015-0550. https://www.ncbi.nlm.nih.gov/pubmed/26960445.

Piaggi, P., M. S. Thearle, J. Krakoff, and S. B. Votruba. 2015. "Higher daily energy expenditure and respiratory quotient, rather than fat-free mass, independently determine greater ad libitum overeating." *J Clin Endocrinol Metab* 100 (8): 3011–20. https://doi.org/10.1210/jc.2015-2164. https://www.ncbi.nlm.nih.gov/pubmed/26086330.

Ravn, A. M., N. T. Gregersen, R. Christensen, L. G. Rasmussen, O. Hels, A. Belza, A. Raben, T. M. Larsen, S. Toubro, and A. Astrup. 2013. "Thermic effect of a meal and appetite in adults: an individual participant data meta-analysis of meal-test trials." *Food Nutr Res* 57. https://doi.org/10.3402/fnr.v57i0.19676. https://www.ncbi.nlm.nih.gov/pubmed/24376394.

Smeets, A. J., S. Soenen, N. D. Luscombe-Marsh, O. Ueland, and M. S. Westerterp-Plantenga. 2008. "Energy expenditure, satiety, and plasma ghrelin, glucagon-like peptide 1, and peptide tyrosine-tyrosine concentrations following a single high-protein lunch." *J Nutr* 138 (4): 698–702. https://doi.org/10.1093/jn/138.4.698. https://www.ncbi.nlm.nih.gov/pubmed/18356323.

Smith, G. I., B. Mittendorfer, and S. Klein. 2019. "Metabolically healthy obesity: facts and fantasies." *J Clin Invest* 129 (10): 3978–89. https://doi.org/10.1172/JCI129186. https://www.ncbi.nlm.nih.gov/pubmed/31524630.

Soares, M. J., and M. J. Muller. 2018. "Resting energy expenditure and body composition: critical aspects for clinical nutrition." *Eur J Clin Nutr* 72 (9): 1208–14. https://doi.org/10.1038/s41430-018-0220-0. https://www.ncbi.nlm.nih.gov/pubmed/30185843.

Strominger, J. L., and J. R. Brobeck. 1953. "A mechanism of regulation of food intake." *Yale J Biol Med* 25 (5): 383–90. https://www.ncbi.nlm.nih.gov/pubmed/13057269.

Tsang, A. H., D. Nuzzaci, T. Darwish, H. Samudrala, and C. Blouet. 2020. "Nutrient sensing in the nucleus of the solitary tract mediates non-aversive suppression of feeding via inhibition of AgRP neurons." *Mol Metab* 42: 101070. https://doi.org/10.1016/j.molmet.2020.101070. https://www.ncbi.nlm.nih.gov/pubmed/32898712.

Veldhorst, M. A., A. G. Nieuwenhuizen, A. Hochstenbach-Waelen, A. J. van Vught, K. R. Westerterp, M. P. Engelen, R. J. Brummer, N. E. Deutz, and M. S. Westerterp-Plantenga. 2009. "Dose-dependent satiating effect of whey relative to casein or soy." *Physiol Behav* 96 (4–5): 675–82. https://doi.org/10.1016/j.physbeh.2009.01.004. https://www.ncbi.nlm.nih.gov/pubmed/19385022.

Veldhorst, M. A., A. Smeets, S. Soenen, A. Hochstenbach-Waelen, R. Hursel, K. Diepvens, M. Lejeune, N. Luscombe-Marsh, and M. Westerterp-Plantenga. 2008. "Protein-induced satiety: effects and mechanisms of different proteins." *Physiol Behav* 94 (2): 300–7. https://doi.org/10.1016/j.physbeh.2008.01.003. https://www.ncbi.nlm.nih.gov/pubmed/18282589.

Weise, C. M., M. G. Hohenadel, J. Krakoff, and S. B. Votruba. 2014. "Body composition and energy expenditure predict ad-libitum food and macronutrient intake in humans." *Int J Obes (Lond)* 38 (2): 243–51. https://doi.org/10.1038/ijo.2013.85. https://www.ncbi.nlm.nih.gov/pubmed/23736368.

Westerterp-Plantenga, M. S., A. Nieuwenhuizen, D. Tome, S. Soenen, and K. R. Westerterp. 2009. "Dietary protein, weight loss, and weight maintenance." *Annu Rev Nutr* 29: 21–41. https://doi.org/10.1146/annurev-nutr-080508-141056. https://www.ncbi.nlm.nih.gov/pubmed/19400750.

Westerterp-Plantenga, M. S., V. Rolland, S. A. Wilson, and K. R. Westerterp. 1999. "Satiety related to 24 h diet-induced thermogenesis during high protein/carbohydrate vs high fat diets measured in a respiration chamber." *Eur J Clin Nutr* 53 (6): 495–502. https://doi.org/10.1038/sj.ejcn.1600782. https://www.ncbi.nlm.nih.gov/pubmed/10403587.

WHO, 2021. *Obesity*. World Health Organization. Accessd March 6, 2025. https://www.who.int/news-room/facts-in-pictures/detail/6-facts-on-obesity.

WHO. 2024. *Obesity and Overweight.* World Health Organization. Last Modified March 1, 2024. Accessed April 14. https://www.who.int/news-room/fact-sheets/detail/obesity-and-overweight.

Wycherley, T. P., J. D. Buckley, M. Noakes, P. M. Clifton, and G. D. Brinkworth. 2013. "Comparison of the effects of weight loss from a high-protein versus standard-protein energy-restricted diet on strength and aerobic capacity in overweight and obese men." *Eur J Nutr* 52 (1): 317–25. https://doi.org/10.1007/s00394-012-0338-0. https://www.ncbi.nlm.nih.gov/pubmed/22406907.

Wycherley, T. P., L. J. Moran, P. M. Clifton, M. Noakes, and G. D. Brinkworth. 2012. "Effects of energy-restricted high-protein, low-fat compared with standard-protein, low-fat diets: a meta-analysis of randomized controlled trials." *Am J Clin Nutr* 96 (6): 1281–98. https://doi.org/10.3945/ajcn.112.044321. https://www.ncbi.nlm.nih.gov/pubmed/23097268.

21 Protein-Calorie Malnutrition in Cirrhosis

A Narrative

Jessica Faccioli, Silvia Nardelli, Stefania Gioia, and Lorenzo Ridola

ABBREVIATIONS

AAA	Aromatic amino acids
ADH	Antidiuretic hormone
AMMI	Appendicular muscle mass index
BCAA	Branched-chain amino acids
BCM	Body cell mass
BIA	Bioimpedance analysis
BMI	Body mass index
DEXA	Dual-energy X-ray absorptiometry
EASL	European Association for the Study of the Liver
ESPEN	European Society of Parenteral and Enteral Nutrition
HBV	Hepatitis B virus
HCC	Hepatocellular carcinoma
HCV	Hepatitis C virus
HE	Hepatic encephalopathy
HU	Hounsfield unit
MAMC	Mid-arm muscle circumference
MELD	Model for end-stage liver disease
NAFLD	Non-alcoholic fatty liver disease
PA	Phase angle (angolo di fase)
PCM	Protein-calorie malnutrition
PS	Performance status
REE	Resting energy expenditure
SGA	Subjective global assessment
SIBO	Small intestinal bacterial overgrowth
SMI	Skeletal muscle index
TACE	Transarterial chemoembolization
TC	Computed tomography
TSF	Triceps skinfold thickness

INTRODUCTION: THE LIVER AND LIVER CIRRHOSIS

Liver cirrhosis is the late and irreversible stage of liver fibrosis and is a significant cause of morbidity and mortality among patients with chronic liver disease. Although often considered a liver-specific

DOI: 10.1201/b23328-21

pathology, cirrhosis is better defined as a systemic disease due to its wide-ranging consequences affecting various organs.

The liver, the second-largest organ in the body, is the primary site for metabolizing carbohydrates, fats, and proteins. It plays a crucial role in distributing, storing, and detoxifying nutrients from the gastrointestinal tract.

Among the liver's main functions are those related to metabolism. It contributes to maintaining glucose homeostasis by storing glycogen when blood glucose levels are high and releasing it during fasting periods. The liver also participates in emulsifying and digesting fats through bile acid production, cholesterol synthesis, and beta-oxidation of fatty acids. It is involved in storing fat-soluble vitamins, such as A, D, E, and K, and minerals, such as iron and copper (Yao et al. 2018). The liver produces about 80% of plasma proteins, including coagulation factors, transport proteins, and albumin. It also contributes to protein metabolism through amino acid conversion and protein deamidation, a fundamental process for energy synthesis (Charlton et al. 1996).

Given the liver's central role in protein metabolism, alterations due to liver disease can lead to serious complications, including protein-calorie malnutrition (PCM).

PROTEIN-CALORIE MALNUTRITION

Malnutrition is defined as a measurable change in physical and mental function resulting from altered body composition and cell mass, resulting in an impaired quality of life and prognosis.

PCM is frequently reported in cirrhotic patients, with a prevalence ranging from 5 to 92%. It is considered one of the most common complications of cirrhosis (Traub et al. 2021). However, it is often underdiagnosed due to complications such as edema and ascites masking weight changes.

In patients with liver cirrhosis, PCM typically affects both lean and fat mass, with potential gender differences: female patients more frequently develop reduced fat stores, while males lose muscle tissue more rapidly (Italian Multicentre Cooperative Project on Nutrition in Liver Cirrhosis 1994).

The prevalence of malnutrition is higher among patients with severe cirrhosis. However, it can also affect those with compensated cirrhosis and can worsen their prognosis even with low Model for End-Stage Liver Disease (MELD) values (Atiemo et al. 2017).

PCM increases the risk of morbidity and mortality. Therefore, early screening and management of PCM are critical for improving patient outcomes.

PATHOPHYSIOLOGICAL MECHANISMS: REDUCED INTAKE AND MALABSORPTION

The pathogenesis of PCM in liver cirrhosis is multifactorial and complex, with reduced caloric intake and nutrient malabsorption playing major roles. (Figure 21.1).

Portal hypertension leads to impaired gastric motility, tissue congestion, and increased intestinal permeability. These changes trigger an increase in pro-inflammatory cytokines, causing anorexia, hypermetabolism, malabsorption, altered microbiota, and gut barrier dysfunction (Palmer et al. 2019).

Frequently, deficiencies of vitamins and micronutrients such as zinc and vitamin A can cause olfactory changes and dysgeusia, which can reduce appetite and nutrient intake (Aqel et al. 2005).

In addition, diets for cirrhotic patients are often poorly palatable and low in sodium and proteins. Without adequate nutritional counseling, low-sodium diets can lead to a reduction in daily caloric intake by up to 20% (Morando et al. 2014).

Fat malabsorption and deficiency of fat-soluble vitamins are frequent in cirrhotic patients. Bile salts, which are essential for fat digestion, may be unavailable because of reduced synthesis or cholestasis, and this deficiency may cause steatorrhea resulting from malabsorption of fat-soluble vitamins and long-chain fatty acids.

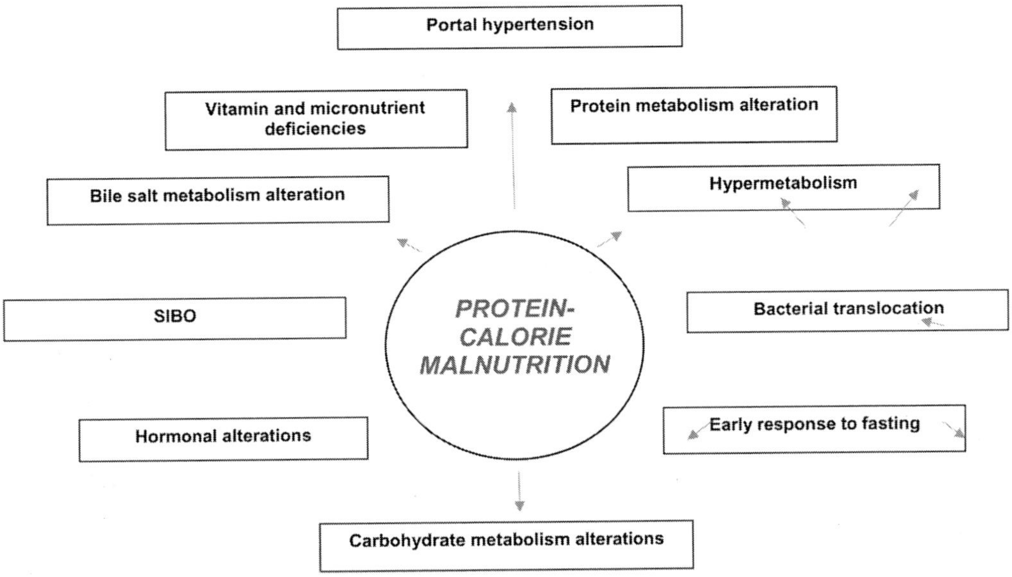

FIGURE 21.1 Physiopathology of PCM.

Fat malabsorption can also be due to small intestine bacterial overgrowth (SIBO), which is found in up to 60% of cirrhotic patients. In this condition, colon bacteria colonize the small intestine and impair microvilli function, digestive enzyme production, and intestinal barrier function, causing malabsorption and altered metabolism (Fukui and Wiest 2016).

Hormonal imbalances further contribute to malnutrition. Lower ghrelin concentration and elevated leptin levels are observed in cirrhotic patients. Ghrelin, produced by gastric endocrine cells, is the only circulating gastrointestinal hormone that has orexigenic effects. Its levels increase before meals, stimulating appetite and promoting fat deposition (Gardiner et al. 2008). Conversely, leptin, secreted by adipose tissue, has anorectic functions, suppressing appetite and promoting energy expenditure. These hormonal alterations have been related to anorexia (Cheung et al. 2012).

PATHOPHYSIOLOGICAL MECHANISMS: METABOLIC ALTERATIONS

PCM may also be caused by metabolic alterations. Cirrhotic patients are frequently hypermetabolic, a condition attributed to hyperdynamic circulation and increased activation of the orthosympathetic nervous system (Müller et al. 1999). Studies have shown that up to 34% of cirrhotic patients have resting energy expenditure (REE) above the normal value. This represents a predisposing factor for malnutrition as greater energy requirements are needed to maintain balance.

Alteration of macronutrient metabolism also represents a fundamental element underlying malnutrition in cirrhotic patients. Alterations in carbohydrate metabolism result in insulin resistance and type II diabetes mellitus in 70% and 14–46% of cases, respectively (Silva et al. 2015). The pathogenesis of insulin resistance is not completely clear but may be related to altered membrane transporters or insulin receptors (Changani et al. 2001).

In cirrhosis, the ability of hepatocytes to synthesize, store, and release glycogen is lost. Consequently, proteins become the main source for gluconeogenesis. This mechanism, activated during fasting, promotes muscle proteolysis and sarcopenia (Anand et al. 2017). While in healthy subjects this response is observed after 2–3 days of deprivation, in cirrhotic patients these changes occur as early as overnight, during which the catabolic activity is very high.

The main source of proteins is skeletal muscle, which provides both branched-chain amino acids (BCAAs) and aromatic amino acids (AAAs). However, tissues can only use BCAAs, and this generates a plasma level imbalance between these two classes of amino acids, with aromatic ones predominating (Dasarathy et al. 2016). When the BCAA:AAA ratio is <3, AAAs cross the blood-brain barrier and act as substrates for the synthesis of false neurotransmitters that displace major neurotransmitters (Fallahzadeh et al. 2019). A low BCAA:AAA ratio has been shown to be associated with a worse prognosis and risk of developing hepatic encephalopathy (HE) in cirrhotic patients.

MUSCLE ALTERATIONS: SARCOPENIA AND MYOSTEATOSIS

Generally, skeletal muscle is maintained by a balance between the synthesis and degradation of its proteins.

Sarcopenia, defined as a loss of muscle mass and strength leading to reduced physical function, is present in up to 70% of cirrhotic patients and indicates an advanced state of malnutrition. In the most severe forms, muscle mass, muscle strength, and muscle function are impaired (Cruz-Jentoft et al. 2019). The risk of developing this complication increases with the presence of ascites, advanced stages of the disease, and in men.

Sarcopenia results from chronic muscle breakdown in which proteins are mobilized to provide amino acids as alternative substrates for gluconeogenesis.

In patients awaiting liver transplantation, this condition increases the risk of waitlist mortality, preoperative risk, postoperative complications, and length of stay (Hsu et al. 2018). Regardless of liver function, it is a marker of pre- and post-transplant mortality and increases the risk of infectious complications, sepsis, HE (Bhanjii et al. 2018). and medical costs (van Vugt et al. 2017). In patients with hepatocarcinoma (HCC), sarcopenia may worsen performance status (PS) and influence the type of treatment. For patients undergoing liver resection or trans-catheter chemoembolization (TACE), sarcopenia is predictive of complications and lower survival (Voron et al. 2015; Dodson et al. 2013).

The relationship between sarcopenia and HE is related to muscle's role as an alternative site for ammonia detoxification. During muscle catabolism, glutamine is released, which can be converted to ammonia. Consequently, cirrhotic patients with sarcopenia have higher ammonium levels, associated with an increased risk of HE (Merli et al. 2012).

Sarcopenia may exacerbate the severity of liver disease. In fact, the presence of sarcopenia appears equivalent to adding ten points to the MELD value (Montano-Loza et al. 2015). Incorporating this factor could improve the predictive power of the MELD score, especially in cases with low values.

In cirrhotic patients, a qualitative alteration of muscle, known as myosteatosis, can also occur. This condition involves fat accumulation in muscle fibers or skeletal muscle fascia, is independent of dietary lipid introduction, and is associated with insulin resistance/type 2 diabetes mellitus. Myosteatosis is found in up to 50% of patients (Zoico et al. 2013).

Patients with myosteatosis may or may not be sarcopenic or obese. The underlying pathogenetic mechanism is not completely known, but it could be related to high muscle uptake of ammonia causing mitochondrial dysfunction.

Myosteatosis further reduces muscle strength and function, is associated with physical disability, and increases mortality by 1.5–2 times in cirrhotic patients (Eslamparast et al. 2018). In transplanted patients, it reduces post-transplant survival, mainly due to septic and respiratory complications (Czigany et al. 2021). Myosteatosis also reduces five-year survival in HCC patients undergoing hepatectomy surgery compared with patients with normal fat deposits (Kaibori et al. 2015).

Sarcopenia and myosteatosis may coexist in 17% of cirrhotic patients (Ebadi et al. 2002), and this coexistence may further worsen a patient's prognosis by increasing mortality (Montano-Loza et al. 2016) and the risk of HE (Bhanjii et al. 2018) (Figure 21.2).

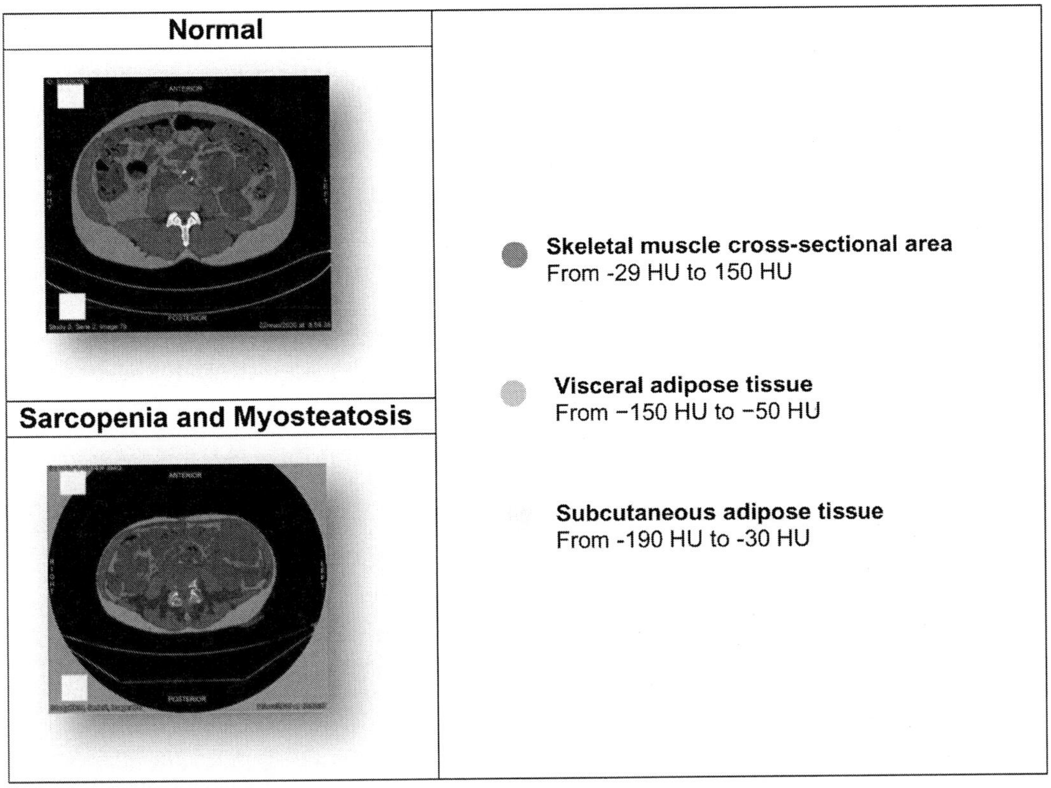

FIGURE 21.2 CT image of sarcopenia.

SARCOPENIC OBESITY

The increasing prevalence of obesity in industrialized countries has revealed the new condition of sarcopenic obesity. Obesity does not exclude malnutrition and is associated with metabolic dysregulation, visceral fat accumulation, and insulin resistance. Although sarcopenia is readily apparent in malnourished patients, it may be difficult to detect in obese patients because obesity may mask this condition.

It is difficult to define a cause-and-effect relationship between sarcopenia and non-alcoholic fatty liver disease (NAFLD) as both may influence each other: NAFLD, through myostatin, may promote the occurrence of sarcopenia, while sarcopenia may have deleterious effects on the liver by activating the myostatin receptor on stellate cells and promoting fibrosis (Bhanji et al. 2017).

The longitudinal study by Sinn et al. showed that patients with NAFLD experience more accelerated muscle mass loss than those without NAFLD over a five-year period, suggesting that NAFLD may be a risk factor for sarcopenia (Sinn et al. 2022).

Other possible causes of sarcopenic obesity include a sedentary lifestyle, which is prevalent among cirrhotic patients, increased inflammatory markers, and decreased testosterone and growth hormone levels (Schrager et al. 2007).

The definition of sarcopenic obesity is based on the concurrent assessment of skeletal muscle index (SMI), measured by computed tomography (CT), and body mass index (BMI) $\geq$30 kg/m^2 in Caucasian populations.

Sarcopenic obesity is found in 2–46% of cirrhotic patients, particularly in those with NAFLD-cirrhosis, and its prevalence is expected to increase in the coming years (Montano-Loza et al. 2016).

It may worsen the stage of cirrhosis, but data regarding the impact on mortality are inconsistent, both due to the failure to correct BMI for ascites and the "obesity paradox" phenomenon. This paradox refers to the observation of reduced mortality among obese patients with liver cirrhosis, possibly because BMI does not accurately assess adipose tissue distribution; excessive visceral adiposity can be present even when BMI is normal (Karagozian et al. 2016).

The simultaneous presence of obesity and sarcopenia is associated with higher rates of mortality and a greater impact on physical function and metabolic profile compared to either condition alone.

DIAGNOSIS

Given the possible complications and prognostic implications associated with malnutrition, early diagnosis is crucial.

In 2019, new guidelines on nutrition for patients with chronic liver disease were published by the European Association for the Study of the Liver (EASL) to clarify aspects of identifying patients at increased risk of malnutrition, available diagnostic tools, and recommended therapeutic strategies (Merli et al. 2019).

The first step in diagnosing PCM and muscle changes is to define the patient's nutritional status.

Unfortunately, the scores used to define the stage of liver cirrhosis, such as Child-Pugh and MELD, do not consider these issues, so they risk underestimating a patient's prognosis. However, they are useful in identifying patients at higher risk of PCM, along with BMI.

Patients at higher risk of PCM are those with advanced disease, identified by Child-Pugh class C, and/or those who are underweight (BMI <18.5 kg/m^2) (Amodio et al. 2013).

However, in cirrhotic patients, the definition of nutritional status can be complicated for several reasons.

Firstly, hydrosaline retention can influence anthropometric measurements and BMI. Secondly, there are differences in body composition between men and women, even after weight loss: men tend to lose predominantly lean mass, while women lose predominantly fat mass (Riggio et al. 1997).

CLINICO-BIOCHEMICAL DIAGNOSIS

Assessment of nutritional status in cirrhotic patients can be performed using several tools:

- Food questionnaire: The Subjective Global Assessment (SGA) is the most widely used questionnaire in clinical practice, validated for cirrhotic patients and providing prognostic information. This 15-minute questionnaire includes both clinical-anamnestic and objective examination information and classifies patients according to the degree of malnutrition. However, it is not very sensitive in detecting the earliest stages of malnutrition (Silva et al. 2015).
- Anthropometric measures: Mid-arm muscle circumference (MAMC) and triceps skin fold thickness (TSF) are quickly executable and correlate well with both muscle mass and fat mass, respectively. However, values may be influenced by hydrosaline retention and obesity. In addition, there are gender differences: in men, there is greater impairment of MAMC, while in women of TSF (Romeiro et al. 2015).
- BMI: This is calculated from the patient's dry weight divided by height squared. In the presence of hydrosaline retention, BMI should be corrected by estimating the patient's dry weight, commonly derived from post-paracentesis weight or weight prior to fluid accumulation. Alternatively, a percentage of weight can be subtracted based on the severity of ascites (Merli et al. 2019). However, its accuracy is low because it does not take into account the effect of sarcopenia on weight reduction, especially in cases of sarcopenic obesity.
- Plasma protein assay: Levels of albumin, pre-albumin, and retinol-binding protein, which are serum markers of nutritional-protein status, are unreliable in patients with cirrhosis because their deficiency is more indicative of advanced liver disease than of malnutrition (Silva et al. 2015).

TABLE 21.1
Clinical and Biochemical Methods for Diagnosing PCM

Method	Description	Advantages	Limitations
SGA	Questionnaire considering nutritional characteristics and objective findings	Rapid, easy to perform; noninvasive; inexpensive	Insensitive to early-stage malnutrition
Anthropometric measures (MAMC and TSF)	Physical measurements of muscle and fat mass	Easy to perform, noninvasive, inexpensive	Gender variability; results can be affected by hydrosaline retention
Plasma protein assay	Measurement of albumin, retinol-binding protein, and pre-albumin levels	–Objective biochemical markers, inexpensive	Values influenced by the severity of cirrhosis
Micronutrients and plasma vitamins assays	Measurement of sodium, vitamin D, vitamin B12, folic acid, etc.	–Objective biochemical markers, inexpensive, identify specific deficiencies	May not reflect overall nutritional status

- Plasma micronutrient dosage: Cirrhotic patients frequently experience hyponatremia, both hypervolemic and hypovolemic. The former is related to splanchnic vasodilation and effective hypovolemia, while the latter is caused by negative sodium balance due to excessive diuretic therapy or sodium restriction. Zinc deficiency is also common in patients with cirrhosis and is often due to diuretic use. These conditions can cause neurological changes (Silva et al. 2015).
- Plasma vitamin dosage: Cirrhotic can lead to various vitamin deficiencies. Hypovitaminosis D can be caused by impaired absorption or reduced plasma-binding proteins due to diminished hepatic synthesis. In patients with alcoholic etiology, thiamine deficiency can cause Wernicke's encephalopathy and chronic cerebellar damage. Vitamin B12 and folic acid deficiency can be due to loss of liver reserve function and can be responsible for anemia and neuropsychiatric symptoms. Furthermore, all fat-soluble vitamins are deficient in patients with cholestasis, with vitamin A deficiency potentially causing visual disturbances and increasing fibrotic damage (caused by stellate cells releasing collagen when they lack vitamin A) (Pazgan-Simon et al. 2015) (Table 21.1).

INSTRUMENTAL DIAGNOSIS

Assessment of nutritional status in cirrhosis may employ the following techniques:

- Dual-energy X-ray absorptiometry (DEXA): Provides precise differentiation between lean and fat mass, with high reproducibility. The appendicular muscle mass index (AMMI) offers an accurate measure of muscle mass unaffected by ascites. However, DEXA exposes patients to ionizing radiation (Romeiro et al. 2015).
- Bioimpedance analysis (BIA): A simple, noninvasive method to estimate body composition, including lean and fat mass, based on the specific electrical impedance of different tissue types. The results are translated into clinical data considering age, sex, and anthropometric data. Key parameters include:
  - Phase angle (PA): Indicates muscle strength and mass, decreasing in advanced stages of liver cirrhosis. A study by Román et al. showed that PA is related to hospitalizations, falls, and mortality (Román et al 2021).

- Body cell mass (BCM): Represents the metabolically active component of fat mass, normally 40% of body weight. BCM is considered the best indicator of a subject's nutritional status (Talluri et al. 2003).
- SMI: Used for sarcopenia diagnosis (<7.0 cm^2/m^2 in males and <5.7 cm^2/m^2 in females). The values may be affected by the presence of edema (Puri et al. 2021).
- Indirect calorimetry: Calculates REE based on oxygen consumption and carbon dioxide production. Values >10–20% above baseline are considered pathological and a sign of hypermetabolism. Limitations include cost, availability, and interference from hydrosaline retention (Silva et al. 2015).
- Abdominal CT: Analyzes muscle mass at the L3–L4 intervertebral level using Hounsfield unit (HU) cut-offs to distinguish tissue types. The 35 HU cut-off distinguishes between muscle and adipose tissue, while 150 HU differentiates muscle tissue from bone. Sarcopenia is present when SMI is <39 cm^2/m^2 in women and 50 cm^2/m^2 in men (Romeiro et al. 2015). While accurate, CT is expensive and exposes patients to ionizing radiation.
- Ultrasound of the quadriceps muscle: This radiation-free, cost-effective technique assesses thigh muscle thickness at one-third and half the distance between the patella and iliac crest with two degrees of pressure. Tandon et al. demonstrated that combining BMI with ultrasound-assessed muscle thickness can identify sarcopenia comparably to cross-sectional imaging techniques (Tandon et al. 2016).
- Handgrip strength: This dynamometer-based assessment offers prognostic value, predicting major complications in malnourished cirrhotic patients (Álvares-da-Silva et al. 2005). Merli et al. found a higher prevalence of minimal and overt HE in patients with impaired handgrip strength. However, the strength-muscle mass correlation is robust only in men, limiting its utility in women (Amodio et al. 2013) (Table 21.2).

TABLE 21.2
Instrumental Methods for Diagnosing PCM

Method	Description	Advantages	Limitations
DEXA	X-ray attenuation-based body composition analysis	Reproducible measures of lean and fat mass; correlates with anthropometric measurements	Not always available; hydrosaline retention affects results; ionizing radiation exposure
BIA	Tissue electrical conductivity measurement	Measures muscle strength, mass (PA), and body cell mass (BCM)	Hydrosaline retention affects results
Indirect calorimetry	Based on measurement of oxygen consumption and carbon dioxide production	Measure of REE	High cost, low availability, and results affected by hydrosaline retention
CT	Differentiation of bone, muscle, and fat using HU thresholds	Measure of SMI for the diagnosis of sarcopenia; not influenced by hydrosaline retention	High cost; exposure to ionizing radiations
Handgrip strength	Dynamometer-based assessment	Provides information on muscle strength	Gender-specific variability
Muscle ultrasound	Quadriceps femoris thickness measurement	Muscle mass assessment; radiation-free	Operator-dependent variability

THERAPY

Early recognition and timely intervention for malnutrition are crucial for improving survival and quality of life in cirrhotic patients. The best results are achieved when the approach is multidisciplinary and includes dietitian involvement, identifying and addressing nutrition-related problems.

Nutritional interventions should aim to improve PCM by ensuring adequate nutrient intake and avoiding loss of muscle mass. Obese patients require specific counseling to prevent sarcopenia while following low-calorie diets.

Nutritional counseling in cirrhotic patients should assess the types of foods consumed, the number of daily meals, potential obstacles to eating, and the level of daily physical activity.

This information enables accurate, functional guidance for individual patients' needs.

PROTEIN-CALORIE REQUIREMENTS

The patient's daily caloric requirement must take into account REE, daily physical activity, and food-related thermogenesis.

The daily requirement for a malnourished cirrhotic patient should be at least 35 kcal/kg/day to achieve an anabolic condition. However, caloric excess should be avoided because of its negative effect on lipogenesis and liver function (Fallahzadeh et al. 2019). Obese patients require a modified approach, with 20–25 kcal/kg/day prescribed to balance weight loss and adequate protein intake (Amodio et al. 2014).

Protein intake must maintain a positive nitrogen balance to prevent sarcopenia. Protein intake has beneficial effects on cognitive functioning, and protein restriction, even if for a short period, can have a deleterious effect in cirrhotic patients with HE (Merli et al. 2008). The European Society of Parenteral and Enteral Nutrition (ESPEN) guidelines recommend 1.2–1.5 g/kg/day of protein for liver disease should patients, adjusted for degree of disease severity and renal function (Plauth et al. 2006).

Some studies have reported that plant-based proteins (from fruits, vegetables, legumes, and cereals) are preferable to animal sources due to their high fermentable fiber content, which increases nitrogen clearance and reduces HE risk, and their potential to modulate gut microbiota, favoring beneficial species (such as *Lachnospiraceae*) over pathogenic ones (such as *Enterobacteriaceae* and *Streptococcaceae*) (Amodio et al. 2001; Bajaj et al. 2012).

Given the rapid catabolic response to fasting in cirrhotic patients, it is important to fractionate calorie intake into multiple daily meals. Thus, the consumption of 4–6 meals per day, including snacks at mid-morning, mid-afternoon, and before bedtime, should be encouraged to avoid hepatic gluconeogenesis from muscle proteins (Iwasa et al. 2013) (Table 21.3).

TABLE 21.3
Key Points for Initial Nutritional Assessment

Assessment Area	Factors to Consider
PCM risk factors	Severity of cirrhosis (Child-Pugh class)
	BMI value
Daily dietary intake	Food questionnaire results
	Presence of barriers to nutrition
Weight	Three-month weight change
	Presence of hydrosaline retention
	Obesity status
Muscle mass	Anthropometric measurements (MAMC)
	Handgrip strength (if available)
Physical activity	Type
	Frequency and duration

MACRONUTRIENTS, MICRONUTRIENTS, FIBER, AND VITAMINS

Regarding carbohydrate metabolism, the main concern is the risk of developing diabetes mellitus. A diet rich in complex carbohydrates with a low glycemic index provides adequate energy intake for cirrhotic patients without increasing this risk (Yao et al. 2018). In general, carbohydrates should account for 50–60% of daily caloric intake, comparable to recommendations for non-cirrhotic patients. This approach is preferable to a hypoglicemic diet with 20% carbohydrate intake, as it does not alter blood glucose or insulin levels (Schulte-Frohlinde et al. 2007).

For fat intake, there are no particular recommendations except for obese patients, who should limit consumption.

Cirrhotic patients should be encouraged to consume fruits and vegetables for their prebiotic and laxative effects. The daily fiber intake should be 25–45 g/day, introduced gradually and in line with the patient's tolerance level (Amodio et al. 2014).

All vitamin deficiencies should be corrected in cirrhotic patients. Hypovitaminosis D should be corrected when levels are <30 ng/mL by supplementing with 5000 IU/day or 50000 IU/week of vitamin D3 for at least three months (Fallahzadeh et al. 2020).

Hyponatremia, the most common electrolyte disturbance in cirrhosis, requires treatment based on its underlying cause. In hypovolemic hyponatremia, ongoing diuretic therapy must be discontinued, and fluids administered to restore circulatory volume. For hypervolemic hyponatremia, hydrosaline restriction should be prescribed to achieve a negative water balance. In cirrhotic patients with severe hyponatremia (<110 mEq/L), who are symptomatic or unresponsive to water restriction, hypertonic saline solutions should be administered slowly.

In patients with ascites, daily sodium intake should never be less than 60 mmol (1.38 g), as lower dosages are associated with malnutrition (Amodio et al. 2013) (Table 21.4).

PHYSICAL EXERCISE

Exercise recommendations for cirrhotic patients are derived from those for patients with chronic diseases and disabilities. A professional evaluation, preferably by a physiatrist, is essential before initiating an exercise program to ensure it is appropriate and safe.

Physical activity offers numerous benefits for cirrhotic patients, including improvements in muscle mass and function, quality of life, and overall prognosis. A prospective study by Zenith et al. showed that eight weeks of controlled aerobic exercise in patients with Child-Pugh class A and B increased muscle mass and reduced fatigue without causing adverse events (Zenith et al. 2014).

Exercise plays a significant role in managing malnutrition through several mechanisms: increasing appetite, maintaining muscle mass, and improving metabolism. The primary goals of physical exercise in cirrhosis are to improve muscle strength, endurance, and balance. Aerobic exercises such as walking or swimming, anaerobic exercises, balance exercises, and stretching exercises are

TABLE 21.4
Nutritional Recommendations in Patients with Liver Cirrhosis

Parameter	Normal	Malnourished	Obese
Caloric intake (kcal/day)	35–40 kcal/day		20–35 kcal/day
Carbohydrate intake (%)	50–60% of daily caloric requirement		
Protein intake (g/day)	1.2–1.5 g/day		
Type of protein	Dairy products and vegetables		
Fiber (g/day)	25–45 g/day		
Number of meals/day	4–6 meals/day, ideally with an evening snack		

indicated for these patients. They should be performed at increasing intensity until reaching an individually tolerated level (Tandon et al. 2018).

CONCLUSIONS

The management of PCM and sarcopenia is a key aspect of the treatment of patients with liver cirrhosis. A timely and comprehensive assessment of nutritional status, involving a multidisciplinary team of nutrition specialists and hepatologists, is essential. This assessment should utilize both non-instrumental methods, such as food diary and anthropometric measurements, and instrumental methods such as BIA, DEXA, CT, and muscle ultrasound.

Dietary recommendations should be individualized, aiming to provide adequate caloric and protein intake, preferably of plant origin, to prevent sarcopenia and optimize muscle function. In addition, any micronutrient and vitamin deficiencies should be corrected, as these can adversely affect both liver and neurological function.

Physical activity is another key element in managing cirrhosis. Exercise, both aerobic and anaerobic, can contribute to the maintenance and enhancement of muscle mass, promoting ammonium detoxification and reducing the risk of HE.

Overall, the management of malnutrition and sarcopenia in patients with cirrhosis requires a comprehensive and integrated approach, based on scientific evidence and supported by a multidisciplinary team. The hope is to improve patients' quality of life, reduce the burden of complications, and prolong survival, thus contributing to better clinical management of this condition.

MINI DICTIONARY OF TERMS

Gluconeogenesis: The process of producing glucose from non-glucose sources.

Hypermetabolism: A state of elevated metabolic rate, where the body consumes more energy than normal.

Myosteatosis: Accumulation of fat in muscle.

Protein-calorie malnutrition: A condition in which insufficient dietary intake of protein and calories causes impaired physical and mental function.

Protein-calorie requirements: The specific amounts of calories and proteins needed to maintain an adequate nutritional balance.

Sarcopenia: Progressive and generalized loss of skeletal muscle mass, strength, and function, leading to increased disability, reduced quality of life and higher mortality risk

Sarcopenic obesity: Loss of muscle mass and strength in obese patients.

SUMMARY POINTS

- Protein-calorie malnutrition is frequently found in cirrhotic patients, with sarcopenia being its most prominent clinical manifestation.
- Multiple factors contribute to rprotein-calorie malnutrition in patients with liver cirrhosis, including reduced caloric intake, malabsorption, inflammation, hormonal alterations, and metabolic changes.
- Risk factors for protein-calorie malnutrition include underweight (BMI <18.5) and severe disease (Child-Pugh class C).
- Nutritional status assessment should be performed by a multidisciplinary team using both non-instrumental methods, such as food diaries and anthropometric measurements, and instrumental methods, such as bioimpedance, dual-energy X-ray absorptiometry, computed tomography, and muscle ultrasound.
- Recommended daily caloric for cirrhotic patients is 35–40–kcal/kg/day with a protein intake of 1–1.5 g/kg/day, ideally divided into 4–6 meals.

- Physical activity, both aerobic and anaerobic, should be encouraged to maintain or increase muscle mass.

REFERENCES

Álvares-da-Silva, M.ário Reis, and Themis Reverbel da Silveira. 2005. "Comparison between Handgrip Strength, Subjective Global Assessment, and Prognostic Nutritional Index in Assessing Malnutrition and Predicting Clinical Outcome in Cirrhotic Outpatients." *Nutrition* 21 (2): 113–17. https://doi.org/10.1016/j.nut.2004.02.002.

Amodio, Piero, Fabio Canesso, and Sara Montagnese. 2014. "Dietary Management of Hepatic Encephalopathy Revisited." *Current Opinion in Clinical Nutrition and Metabolic Care* 17 (5): 448–52. https://doi.org/10.1097/mco.0000000000000084.

Amodio, P., L. Caregaro, E. Pettenó, M. Marcon, F. Del Piccolo, and A. Gatta. 2001. "Vegetarian Diets in Hepatic Encephalopathy: Facts or Fantasies?" *Digestive and Liver Disease* 33 (6): 492–500. https://doi.org/10.1016/s1590-8658(01)80028-1.

Amodio, Piero, Chantal Bemeur, Roger Butterworth, Juan Cordoba, Akinobu Kato, Sara Montagnese, Misael Uribe, Hendrik Vilstrup, and Marsha Y. Morgan. 2013. "The Nutritional Management of Hepatic Encephalopathy in Patients with Cirrhosis: International Society for Hepatic Encephalopathy and Nitrogen Metabolism Consensus." *Hepatology* 58 (1): 325–36. https://doi.org/10.1002/hep.26370.

Anand, Anil C. 2017. "Nutrition and Muscle in Cirrhosis." *Journal of Clinical and Experimental Hepatology* 7 (4): 340–57. https://doi.org/10.1016/j.jceh.2017.11.001.

Aqel, Bashar A., James S. Scolapio, Rolland C. Dickson, Duane D. Burton, and Ernest P. Bouras. 2005. "Contribution of Ascites to Impaired Gastric Function and Nutritional Intake in Patients with Cirrhosis and Ascites." *Clinical Gastroenterology and Hepatology* 3 (11): 1095–1100. https://doi.org/10.1016/s1542-3565(05)00531-8.

Bajaj, Jasmohan S., Jason M. Ridlon, Phillip B. Hylemon, Leroy R. Thacker, Douglas M. Heuman, Sean Smith, Masoumeh Sikaroodi, and Patrick M. Gillevet. 2012. "Linkage of Gut Microbiome with Cognition in Hepatic Encephalopathy." *American Journal of Physiology-Gastrointestinal and Liver Physiology* 302 (1): G168–75. https://doi.org/10.1152/ajpgi.00190.2011.

Bhanji, Rahima A., Carlos Moctezuma-Velazquez, Andres Duarte-Rojo, Maryam Ebadi, Sunita Ghosh, Christopher Rose, and Aldo J. Montano-Loza. 2018. "Myosteatosis and Sarcopenia Are Associated with Hepatic Encephalopathy in Patients with Cirrhosis." *Hepatology International* 12 (4): 377–86. https://doi.org/10.1007/s12072-018-9875-9.

Bhanji, Rahima A., Praveena Narayanan, Alina M. Allen, Harmeet Malhi, and Kymberly D. Watt. 2017. "Sarcopenia in Hiding: The Risk and Consequence of Underestimating Muscle Dysfunction in Nonalcoholic Steatohepatitis." *Hepatology* 66 (6): 2055–65. https://doi.org/10.1002/hep.29420.

Changani, K. K. 2001. "Evidence for Altered Hepatic Gluconeogenesis in Patients with Cirrhosis Using in Vivo 31-Phosphorus Magnetic Resonance Spectroscopy." *Gut* 49 (4): 557–64. https://doi.org/10.1136/gut.49.4.557.

Charlton, Michael R. 1996. "Protein Metabolism and Liver Disease." *Baillière's Clinical Endocrinology and Metabolism* 10 (4): 617–35. https://doi.org/10.1016/s0950-351x(96)80771-3.

Cheung, Kally, Samuel S. Lee, and Maitreyi Raman. 2012. "Prevalence and Mechanisms of Malnutrition in Patients with Advanced Liver Disease, and Nutrition Management Strategies." *Clinical Gastroenterology and Hepatology* 10 (2): 117–25. https://doi.org/10.1016/j.cgh.2011.08.016.

Córdoba, Juan, López-HellínJuan, Mercé Planas, SabínPilar, Francesc Sanpedro, Francisco Castro, Rafael Esteban, and Jaume Guardia. 2004. "Normal Protein Diet for Episodic Hepatic Encephalopathy: Results of a Randomized Study." *Journal of Hepatology* 41 (1): 38–43. https://doi.org/10.1016/j.jhep.2004.03.023.

Cruz-Jentoft, Alfonso J., Gülistan Bahat, Jürgen Bauer, Yves Boirie, Olivier Bruyère, Tommy Cederholm, Cyrus Cooper, et al. 2019. "Sarcopenia: Revised European Consensus on Definition and Diagnosis." *Age and Ageing* 48 (4): 601–1. https://doi.org/10.1093/ageing/afz046.

Dasarathy, Srinivasan, and Manuela Merli. 2016. "Sarcopenia from Mechanism to Diagnosis and Treatment in Liver Disease." *Journal of Hepatology* 65 (6): 1232–44. https://doi.org/10.1016/j.jhep.2016.07.040.

Dodson, Rebecca M., Amin Firoozmand, Omar Hyder, Vania Tacher, David P. Cosgrove, Nikhil Bhagat, Joseph M. Herman, et al. 2013. "Impact of Sarcopenia on Outcomes Following Intra-Arterial Therapy of Hepatic Malignancies." *Journal of Gastrointestinal Surgery* 17 (12): 2123–32. https://doi.org/10.1007/s11605-013-2348-5.

Dong Hyun Sinn, Danbee Kang, Mira Kang, Eliseo Guallar, Yun-Chul Hong, Kyung-Hyun Lee, Ji-Yeon Park, Juhee Cho, and Geum-Youn Gwak. 2022. "Nonalcoholic Fatty Liver Disease and Accelerated Loss of Skeletal Muscle Mass: A Longitudinal Cohort Study." *Hepatology* 76 (6): 1746–54. https://doi.org/10.1002/hep.32578.

E. Schulte-Frohlinde, Stefan Wagenpfeil, Johanna Willis, C. Lersch, Florian Eckel, Roland M. Schmid, and V. Schusdziarra. 2007. "Role of Meal Carbohydrate Content for the Imbalance of Plasma Amino Acids in Patients with Liver Cirrhosis." *Journal of Gastroenterology and Hepatology* 22 (8): 1241–48. https://doi.org/10.1111/j.1440-1746.2006.04620.x.

Ebadi, Maryam, Cynthia Tsien, Rahima A. Bhanji, Abha R. Dunichand-Hoedl, Elora Rider, Maryam Motamedrad, Vera C. Mazurak, Vickie E Baracos, and Aldo J. Montano-Loza. 2022. "Skeletal Muscle Pathological Fat Infiltration (Myosteatosis) Is Associated with Higher Mortality in Patients with Cirrhosis." *Cells* 11 (8): 1345–45. https://doi.org/10.3390/cells11081345.

Eslamparast, Tannaz, Aldo J. Montano-Loza, Maitreyi Raman, and Puneeta Tandon. 2018. "Sarcopenic Obesity in Cirrhosis-the Confluence of 2 Prognostic Titans." *Liver International* 38 (10): 1706–17. https://doi.org/10.1111/liv.13876.

Fallahzadeh, Mohammad Amin, and Robert S. Rahimi. 2019. "Hepatic Encephalopathy and Nutrition Influences: A Narrative Review." *Nutrition in Clinical Practice* 35 (1): 36–48. https://doi.org/10.1002/ncp.10458.

Fukui, Hiroshi, and Reiner Wiest. "Changes of Intestinal Functions in Liver Cirrhosis." *Inflammatory Intestinal Diseases* 1, no. 1 (2016): 24–40. https://doi.org/10.1159/000444436.

Gardiner, J. V., C. N. Jayasena, and S. R. Bloom. 2008. "Gut Hormones: A Weight off Your Mind." *Journal of Neuroendocrinology* 20 (6): 834–41. https://doi.org/10.1111/j.1365-2826.2008.01729.x.

Hsu, Ching-Sheng, and Jia-Horng Kao. 2018. "Sarcopenia and Chronic Liver Diseases." *Expert Review of Gastroenterology & Hepatology* 12 (12): 1229–44. https://doi.org/10.1080/17474124.2018.1534586.

Italian Multicentre Cooperative Project on Nutrition in Liver Cirrhosis. 1994. "Nutritional Status in Cirrhosis." *Journal of Hepatology* 21 (3): 317–25. https://doi.org/10.1016/s0168-8278(05)80308-3.

Iwasa, Motoh, Kazuko Iwata, Nagisa Hara, Ayana Hattori, Masumi Ishidome, Noriko Sekoguchi-Fujikawa, Rumi Mifuji-Moroka, et al. 2013. "Nutrition Therapy Using a Multidisciplinary Team Improves Survival Rates in Patients with Liver Cirrhosis." *Nutrition* 29 (11–12): 1418–21. https://doi.org/10.1016/j.nut.2013.05.016.

Karagozian, Raffi, Gaurav Bhardwaj, Dorothy B. Wakefield, and Gyorgy Baffy. 2016. "Obesity Paradox in Advanced Liver Disease: Obesity Is Associated with Lower Mortality in Hospitalized Patients with Cirrhosis." *Liver International* 36 (10): 1450–56. https://doi.org/10.1111/liv.13137.

Kofi Atiemo, Anton I. Skaro, Haripriya Maddur, L. Zhao, Samantha Montag, Lisa B VanWagner, Swati Goel, et al. 2017. "Mortality Risk Factors among Patients with Cirrhosis and a Low Model for End-Stage Liver Disease Sodium Score (≤15): An Analysis of Liver Transplant Allocation Policy Using Aggregated Electronic Health Record Data." *American Journal of Transplantation* 17 (9): 2410–19. https://doi.org/10.1111/ajt.14239.

Masaki Kaibori, Morihiko Ishizaki, Hiroya Iida, Kosuke Matsui, Tatsuma Sakaguchi, Kentaro Inoue, Toshihiko Mizuta, et al. 2015. "Effect of Intramuscular Adipose Tissue Content on Prognosis in Patients Undergoing Hepatocellular Carcinoma Resection." *Journal of Gastrointestinal Surgery* 19 (7): 1315–23. https://doi.org/10.1007/s11605-015-2838-8.

Merli, Manuela, Annalisa Berzigotti, Shira Zelber-Sagi, Srinivasan Dasarathy, Sara Montagnese, Laurence Genton, Mathias Plauth, and Albert Parés. 2019. "EASL Clinical Practice Guidelines on Nutrition in Chronic Liver Disease." *Journal of Hepatology* 70 (1): 172–93. https://doi.org/10.1016/j.jhep.2018.06.024.

Merli, Manuela, Michela Giusto, Cristina Lucidi, Valerio Giannelli, Ilaria Pentassuglio, Vincenza Di Gregorio, Barbara Lattanzi, and Oliviero Riggio. 2012. "Muscle Depletion Increases the Risk of Overt and Minimal Hepatic Encephalopathy: Results of a Prospective Study." *Metabolic Brain Disease* 28 (2): 281–84. https://doi.org/10.1007/s11011-012-9365-z.

Merli, Manuela, and Oliviero Riggio. 2008. "Dietary and Nutritional Indications in Hepatic Encephalopathy." *Metabolic Brain Disease* 24 (1): 211–21. https://doi.org/10.1007/s11011-008-9127-0.

Montano-Loza, Aldo J., Paul Angulo, Judith Meza-Junco, Carla M. M. Prado, Michael B. Sawyer, Crystal Beaumont, Nina Esfandiari, Mang Ma, and Vickie E. Baracos. 2015. "Sarcopenic Obesity and Myosteatosis Are Associated with Higher Mortality in Patients with Cirrhosis." *Journal of Cachexia, Sarcopenia and Muscle* 7 (2): 126–35. https://doi.org/10.1002/jcsm.12039.

Montano-Loza, Aldo J., Andres Duarte-Rojo, Judith Meza-Junco, Vickie E. Baracos, Michael B. Sawyer, Jack X. Q. Pang, Crystal Beaumont, Nina Esfandiari, and Robert P. Myers. 2015. "Inclusion of Sarcopenia within MELD (MELD-Sarcopenia) and the Prediction of Mortality in Patients with Cirrhosis." *Clinical and Translational Gastroenterology* 6 (7): e102. https://doi.org/10.1038/ctg.2015.31.

Morando, Filippo, Silvia Rosi, Elisabetta Gola, Mariateresa Nardi, Salvatore Piano, Silvano Fasolato, Marialuisa Stanco, et al. 2014. "Adherence to a Moderate Sodium Restriction Diet in Outpatients with Cirrhosis and Ascites: A Real-Life Cross-Sectional Study." *Liver International* 35 (5): 1508–15. https://doi.org/10.1111/liv.12583.

Müller, Manfred J., Joachim Böttcher, Oliver Selberg, Stefanie Weselmann, Klaus H. W. Böker, Mechthild Schwarze, Alexander von zur Mühlen, and Michael P. Manns. 1999. "Hypermetabolism in Clinically Stable Patients with Liver Cirrhosis." *The American Journal of Clinical Nutrition* 69 (6): 1194–201. https://doi.org/10.1093/ajcn/69.6.1194.

Palmer, Lena B., Gabriela Kuftinec, Michelle Pearlman, and Caitlin Homberger Green. 2019. "Nutrition in Cirrhosis." *Current Gastroenterology Reports* 21 (8). https://doi.org/10.1007/s11894-019-0706-5.

Pazgan-Simon, Monika, Jolanta Zuwała-Jagiełło, Sylwia Serafińska, and Krzysztof Simon. 2015. "Nutrition Principles and Recommendations in Different Types of Hepatic Encephalopathy." *Clinical and Experimental Hepatology* 4: 121–26. https://doi.org/10.5114/ceh.2016.57759.

Plauth, M., E. Cabré, O. Riggio, M. Assis-Camilo, M. Pirlich, J. Kondrup, P. Ferenci, and E. Holm. 2006. "ESPEN Guidelines on Enteral Nutrition: Liver Disease $." *Clinical Nutrition* 25 (Clin Nutr): 285–94. https://doi.org/10.1016/j.clnu.2006.01.018.

Puri, Pankaj, Radha K. Dhiman, Sunil Taneja, Puneeta Tandon, Manuela Merli, Anil C. Anand, Anil Arora, et al. 2021. "Nutrition in Chronic Liver Disease: Consensus Statement of the Indian National Association for Study of the Liver." *Journal of Clinical and Experimental Hepatology* 11 (1): 97–143. https://doi.org/10.1016/j.jceh.2020.09.003.

Riggio, O., A. Andreoli, F. Diana, P. Fiore, P. Meddi, R. Lionetti, F. Montagnese, M. Merli, L. Capocaccia, and A. De Lorenzo. 1997. "Whole Body and Regional Body Composition Analysis by Dual-Energy X-Ray Absorptiometry in Cirrhotic Patients." *European Journal of Clinical Nutrition* 51 (12): 810–14. https://doi.org/10.1038/sj.ejcn.1600490.

Román, Eva, Maria Poca, Gerard Amorós-Figueras, Javier Rosell-Ferrer, Cristina Gely, Juan C. Nieto, Silvia Vidal, et al. 2021. "Phase Angle by Electrical Bioimpedance Is a Predictive Factor of Hospitalisation, Falls and Mortality in Patients with Cirrhosis." *Scientific Reports* 11 (1): 20415. https://doi.org/10.1038/s41598-021-99199-8.

Romeiro, Fernando Gomes. 2015. "Nutritional Assessment in Cirrhotic Patients with Hepatic Encephalopathy." *World Journal of Hepatology* 7 (30): 2940. https://doi.org/10.4254/wjh.v7.i30.2940.

Silva, Marco, Sara Gomes, Armando Peixoto, Paulo Torres-Ramalho, Hélder Cardoso, Rosa Azevedo, Carla Cunha, and Guilherme Macedo. "Nutrition in Chronic Liver Disease." GE Portuguese Journal of Gastroenterology 22, no. 6 (November 2015): 268–76. https://doi.org/10.1016/j.jpge.2015.06.004

Schrager, Matthew A., E. Jeffrey Metter, Eleanor Simonsick, Alessandro Ble, Stefania Bandinelli, Fulvio Lauretani, and Luigi Ferrucci. 2007. "Sarcopenic Obesity and Inflammation in the InCHIANTI Study." *Journal of Applied Physiology* 102 (3): 919–25. https://doi.org/10.1152/japplphysiol.00627.2006.

Talluri, Antonio, Reinhard Liedtke, Ehab I. Mohamed, C. Maiolo, R. Martinoli, and Antonino De Lorenzo. 2003. "The Application of Body Cell Mass Index for Studying Muscle Mass Changes in Health and Disease Conditions." *Acta Diabetologica* 40 (S1): s286–89. https://doi.org/10.1007/s00592-003-0088-9.

Tandon, Puneeta, Kathleen Patricia Ismond, Kenneth Riess, Andres Duarte-Rojo, Bandar Al-Judaibi, Michael Andrew Dunn, Jennifer Holman, et al. 2018. "Exercise in Cirrhosis: Translating Evidence and Experience to Practice." *Journal of Hepatology* 69 (5): 1164–77. https://doi.org/10.1016/j.jhep.2018.06.017.

Tandon, Puneeta, Gavin Low, Marina Mourtzakis, Laura Zenith, Robert P. Myers, Juan G. Abraldes, Abdel Aziz M. Shaheen, et al. 2016. "A Model to Identify Sarcopenia in Patients with Cirrhosis." *Clinical Gastroenterology and Hepatology* 14 (10): 1473–1480.e3. https://doi.org/10.1016/j.cgh.2016.04.040.

Traub, Julia, Lisa Reiss, Benard Aliwa, and Vanessa Stadlbauer. 2021. "Malnutrition in Patients with Liver Cirrhosis." *Nutrients* 13 (2): 540. https://doi.org/10.3390/nu13020540.

Voron, Thibault, Lambros Tselikas, Daniel Pietrasz, Frederic Pigneur, Alexis Laurent, Philippe Compagnon, Chady Salloum, Alain Luciani, and Daniel Azoulay. 2015. "Sarcopenia Impacts on Short- and Long-Term Results of Hepatectomy for Hepatocellular Carcinoma." *Annals of Surgery* 261 (6): 1173–83. https://doi.org/10.1097/sla.0000000000000743.

Vugt, Jeroen L. A. van, Stefan Buettner, Louise J. M. Alferink, Niek Bossche, Ron W. F. de Bruin, Sarwa Darwish Murad, Wojciech G. Polak, Herold J. Metselaar, and Jan N. M. IJzermans. 2017. "Low Skeletal Muscle Mass Is Associated with Increased Hospital Costs in Patients with Cirrhosis Listed for Liver Transplantation-a Retrospective Study." *Transplant International* 31 (2): 165–74. https://doi.org/10.1111/tri.13048.

Yao, Chu Kion, James Fung, Natural Hoi Sing Chu, and Victoria Ping Yi Tan. 2018. "Dietary Interventions in Liver Cirrhosis." *Journal of Clinical Gastroenterology* 52 (8): 663–73. https://doi.org/10.1097/MCG.0000000000001071.

Zenith, Laura, Neha Meena, Ailar Ramadi, Milad Yavari, Andrea Harvey, Michelle Carbonneau, Mang Ma, et al. 2014. "Eight Weeks of Exercise Training Increases Aerobic Capacity and Muscle Mass and Reduces Fatigue in Patients with Cirrhosis." *Clinical Gastroenterology and Hepatology: The Official Clinical Practice Journal of the American Gastroenterological Association* 12 (11): 1920–26.e2. https://doi.org/10.1016/j.cgh.2014.04.016.

Zoico, Elena, Francesca Corzato, Clara Bambace, Andrea P. Rossi, Rocco Micciolo, Saverio Cinti, Tamara B. Harris, and Mauro Zamboni. 2013. "Myosteatosis and Myofibrosis: Relationship with Aging, Inflammation and Insulin Resistance." *Archives of Gerontology and Geriatrics* 57 (3): 411–16. https://doi.org/10.1016/j.archger.2013.06.001.

22 Dietary Protein and Cancer

Lucas Jurado-Fasoli and Carmen Rodríguez-García

ABBREVIATIONS

AICR	American Institute for Cancer Research
DNA	Deoxyribonucleic acid
GCN2	General control nonderepressible 2
GH	Growth hormone
HMB	β-hydroxy β-methylbutyrate
IARC	International Agency for Research on Cancer
IGF	Insulin-like growth factor
mTOR	Mammalian target of rapamycin
WCRF	World Cancer Research Fund

INTRODUCTION

Cancer is the second-leading cause of mortality worldwide, and the global cancer burden will continue growing for the coming decades (Foreman et al. 2018; Kocarnik et al. 2022). Although some cancers cannot be prevented, reduced exposure to cancer risk factors can decrease the risk of different cancer types (Tran et al. 2022). Indeed, 44% of cancer deaths and 42% of disability-adjusted life years are attributable to risk factors (Tran et al. 2022). In light of this, lifestyle factors such as physical activity, tobacco and alcohol cessation, and diet have been shown to protect against cancer and help with disease progression during medical treatment (Tran et al. 2022; Muscaritoli et al. 2021; International 2018). Therefore, understanding the influence of modifiable risk factors on cancer prevention and treatment is urgently needed.

Among modifiable risk factors, a healthy diet helps maintain adequate body weight and nutritional status and decreases exposure to dietary carcinogens, reducing cancer risk (International 2018; Afshin et al. 2019). Once the disease has appeared, the impact of the disease and the side effects of medical treatments make cancer patients a nutritionally vulnerable population with a high risk of adverse outcomes, comorbidities, muscle loss, and malnutrition (Ford, Orsso, et al. 2022). Nutritional interventions during cancer treatment can improve/maintain nutritional intake, decrease the risk of malnutrition and sarcopenia, mitigate metabolic consequences, improve quality of life, decrease systemic toxicity, and potentiate medical treatments (Muscaritoli et al. 2021; Ravasco 2019). Thus, dietary recommendations for cancer patients under medical treatment cannot be extrapolated from nutritional recommendations for cancer prevention.

Among dietary factors, dietary protein quantity and sources (e.g., red meat and processed meats) have been hot topics of scientific and clinical debate. Therefore, this chapter aims to review the current evidence regarding the influence of dietary protein on cancer risk and cancer treatment.

DIETARY PROTEIN AND CANCER RISK

When addressing cancer risk, it is important to realize that a single nutrient or dietary factor will not be the sole driver for developing the disease. Cancer is a multifactorial disease that can be triggered by exposure to unhealthy lifestyle habits, such as unhealthy dietary patterns. Therefore, clinical

DOI: 10.1201/b23328-22

recommendations should encourage the inclusion of healthy foods and discourage the consumption of unhealthy foods, with the aim of maintaining adequate body weight and nutritional status and avoiding the inclusion of carcinogens through diet (International 2018). Within this context, among nutrients, dietary protein has become the cornerstone of the scientific debate.

LOW VS. HIGH DIETARY PROTEIN INTAKE AND CANCER RISK

Evidence regarding levels of protein intake and cancer risk remains controversial. Mechanistically, protein or amino acid restriction might affect metabolic pathways that have been associated with healthspan extension (Mirzaei, Suarez, and Longo 2014). It can inhibit the mTOR pathway and activate the GCN2 pathway, and it can affect the production of acetyl-CoA, playing a role in the autophagy response and decreasing cancer risk (Mirzaei, Suarez, and Longo 2014). Moreover, protein or amino acid restriction can modulate the GH-IGF-I axis, abolishing the pro-anabolic effects and decreasing cancer risk (Levine et al. 2014; Mirzaei, Suarez, and Longo 2014). Other studies have also demonstrated an increase in immunosurveillance through macrophage reprogramming and the activation of the IRE1α pathway (Martínez-Garay and Djouder 2023). In addition, different pre-clinical studies restricting specific dietary amino acids have shown positive effects in cancer prevention. Specifically, restriction of dietary methionine, serine, cysteine, glycine, leucine, or asparagine has been demonstrated to decrease tumor formation and burden and to increase immunity in pre-clinical studies (Martínez-Garay and Djouder 2023). However, other studies in mice have shown that high-protein diets can delay cancer onset and tumor growth (Ho et al. 2011; Dooley et al. 2020). Thus, evidence remains controversial, and further studies are needed.

Epidemiological evidence in humans is equally controversial, since some studies have suggested that dietary protein intake is not directly related to cancer risk (Pang and Wang 2018; Al-Zahrani et al. 2022) and cancer mortality (Naghshi et al. 2020), whereas others have found a positive or negative relationship, specifically between certain dietary protein sources and cancer risk (Cross et al. 2007; Al-Zahrani et al. 2022; Fan et al. 2022). A previous human study observed that individuals (50–65 years old) with high-protein intake experienced a four-fold increase in cancer death risk during a follow-up of 18 years (Levine et al. 2014). However, this association was not observed when only plant-derived protein sources were taken into account (Levine et al. 2014). Although future studies are needed to elucidate the potential role of dietary protein intake on cancer risk in humans, it seems that dietary protein sources are more important than total protein intake for preventing cancer.

DIETARY PROTEIN SOURCES AND CANCER RISK

Strong evidence suggests prioritizing plant-based dietary protein sources over animal dietary protein sources for cancer prevention (Wu et al. 2016; Fan et al. 2022; Cross et al. 2007; International 2018) (Figure 22.1). The potential protective effects of plant-based dietary protein sources may stem from their fiber and phytochemical content (International 2018; Fan et al. 2022). Dietary fiber reduces the contact of dietary carcinogens with the mucosal barrier of the gastrointestinal tract, potentially mitigating the carcinogenic effect of different foods (International 2018). Additionally, dietary fiber has a satiating effect, which could decrease energy intake and help to maintain a healthy body weight (International 2018). Dietary fiber also offers positive metabolic effects, such as helping to maintain blood glucose and lipids within a normal range (International 2018). Furthermore, phytochemicals in vegetables may protect cells and DNA from damage that could lead to cancer (International 2018) and have antioxidant and anti-inflammatory effects, contributing to an adequate health status of the host. Notably, a high intake of soy has been associated with a lower risk of cancer, primarily attributed to its isoflavones content (the main phytochemical in soy) (Fan et al. 2022). Consequently, dietary protein sources such as legumes, whole-grain proteins, soy, and soy-derived products have shown protective effects against cancer (International 2018; Wu et al. 2016; Fan et al. 2022).

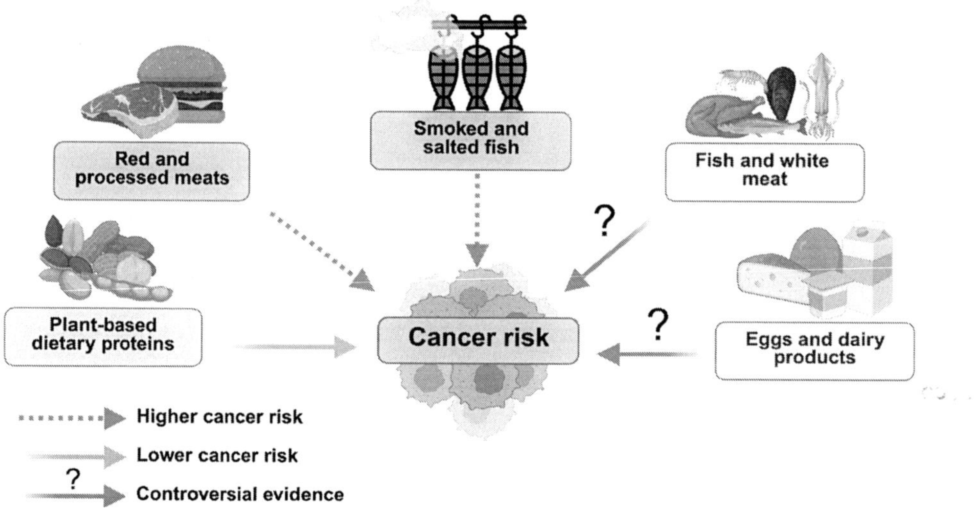

FIGURE 22.1 Dietary protein sources and cancer risk. Solid arrows indicate sources associated with a higher or lower cancer risk. Dashed arrows represent sources linked to a higher cancer risk. Arrows with a question mark represent sources for which the evidence is controversial or inconclusive.

Conversely, total animal protein has been linked to an increased risk of cancer. However, different studies have demonstrated a differential effect based on the animal protein source. There is strong evidence that red and processed meat increase the risk of cancer, especially colorectal cancer (International 2018). Indeed, the IARC has classified red meat as probably carcinogenic (Group 2A), whereas processed meat has been classified as carcinogenic to humans (Group 1) (IARC Working Group on the Evaluation of Carcinogenic Risks to Humans 2018). The mechanisms behind the potential carcinogenic effect of red and processed meat include genotoxic and epigenetic effects, increased oxidative stress, altered cell proliferation, cell death, and nutrient supply (IARC Working Group on the Evaluation of Carcinogenic Risks to Humans 2018). Moreover, cooking red and processed meats at high temperatures or in direct contact with flames or hot surfaces (i.e., barbecuing, frying) can increase the production of carcinogens (IARC Working Group on the Evaluation of Carcinogenic Risks to Humans 2018).

In contrast, white meat has lower fat and iron content than red meat and contains no nitrites, unlike processed meat. However, evidence regarding its relationship with cancer risk remains controversial. Some studies have found a protective effect of white meat consumption, whereas others have found no relationship between its consumption and cancer risk (Kim et al. 2019; Yun et al. 2023). Moreover, the WCRF/AICR report has not classified white meat as positive or negative in terms of cancer prevention (International 2018). Therefore, further studies are needed to elucidate the potential effects of white meat consumption on cancer risk.

Evidence regarding the relationship between fish consumption and cancer risk is mixed. Some systematic reviews have found a protective effect of fish consumption, whereas others have found no relationship between its consumption and cancer risk (Lee et al. 2020; International 2018). Indeed, the WCRF/AICR report does not consider fish consumption to increase or decrease cancer risk (International 2018). However, it is important to differentiate between salted fish and fresh fish; salted fish is related to an increased risk of cancer mainly due to its preservation and processing (International 2018). Additionally, smoked fish may be related to an increased cancer risk since the smoking process produces polycyclic aromatic hydrocarbons and heterocyclic amines, which are classified as carcinogenic to humans (Group 1; IARC). Thus, when recommending high fish consumption, it is important to promote healthy fish sources to prevent cancer.

Regarding dairy products, evidence is controversial since some studies have observed that dairy products increase the risk of prostate cancer (Al-Zahrani et al. 2022), whereas expert reports have suggested strong evidence that dairy products decrease colorectal cancer risk (International 2018). However, it is important to note that some studies have shown a differential effect of milk and fermented dairy products on cancer prevention (Kaluza et al. 2021). Fermented dairy products (i.e., yogurt or kefir) might exert protective effects against cancer, which might be explained by their probiotic content, nutrient content, and bioactive molecules (K. Zhang et al. 2019).

Evidence regarding egg consumption and cancer risk is controversial. Several studies have observed a positive association between egg consumption and cancer risk in different locations (Tse and Eslick 2014; Aminianfar et al. 2019), whereas other studies have observed a protective effect of egg consumption (Fang et al. 2012). Therefore, since the evidence is still not strong enough, further studies are needed to elucidate the potential role of egg consumption on cancer risk.

DIETARY PROTEIN DURING CANCER TREATMENT

The quantity and type of dietary protein during cancer treatment might influence the side effects derived from the disease and treatment. The metabolic disturbances, tumor obstruction, and cancer symptoms, together with the physical inactivity that usually characterizes this population, increase the risk of malnutrition and muscle mass loss. Malnutrition can decrease overall survival by approximately 49% (Zhang et al. 2021), and sarcopenia can increase mortality by about 35% in cancer patients (Ryan and Sullivan 2021). Therefore, dietary protein is an essential nutrient to monitor during cancer treatment.

LEVELS OF DIETARY PROTEIN INTAKE DURING CANCER TREATMENT

Muscle protein synthesis can be normal in cancer patients with cachexia (MacDonald et al. 2015), including those with insulin resistance (Winter, MacAdams, and Chevalier 2012). Consequently, protein intake is a critical factor during cancer treatment, requiring close monitoring to prevent or delay muscle mass loss. While protein recommendations for cancer patients during treatment are higher than for healthy populations, the precise quantity has not been defined. Most nutrition oncology guidelines suggest a minimum intake of 1 g/kg/day and a range of 1–1.5 g/kg/day (Ford, Arends, et al. 2022; Muscaritoli et al. 2021; Arends et al. 2017). Notably, protein intakes exceeding 1.5 g/kg/day can maintain or improve muscle mass in cancer patients with tumor cachexia/sarcopenia (Op Den Kamp et al. 2009). Protein intakes of approximately 2 g/kg/day can maintain protein balance in cancer patients and positively affect protein metabolism in cancer patients with severe malnutrition (Bozzetti and Bozzetti 2013; Arends et al. 2017).

Thus, protein recommendations ranging from 1 to 2g/kg/day may be appropriate for cancer patients during treatment (Figure 22.2). It is important to be aware that these recommendations should be individualized based on each patient's nutritional status, secondary symptoms, and adverse events (e.g., malnutrition, sarcopenia). Moreover, protein requirements could change depending on the medical treatment, with critical phases such as surgery or chemotherapy treatments potentially exacerbating muscle loss.

Although not directly demonstrated in cancer patients, their heightened risk of malnutrition and muscle loss suggests that certain nutritional recommendations from sports nutrition might be applicable to promote muscle hypertrophy. Specifically, dietary protein doses can be divided into different meals (every 3–5h) across the day to optimize muscle protein synthesis (Stokes et al. 2018) (Figure 22.2). A single dose of protein (~0.4g/kg) can also maximize muscle protein synthesis after a long period of fasting or after exercise (particularly resistance exercise) (Figure 22.2) (Stokes et

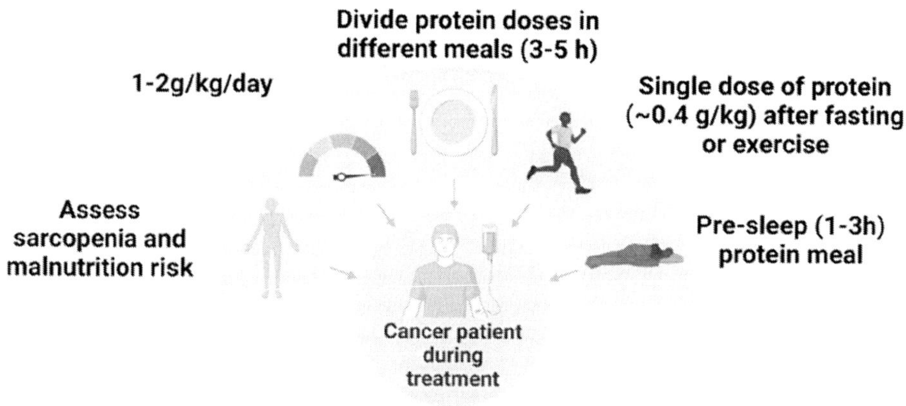

FIGURE 22.2 Key points of dietary protein intake during cancer treatment. Key points of the dietary protein intake during cancer treatment, specifically focused on preventing malnutrition and sarcopenia. Figure was compiled by the authors and created with BioRender.

al. 2018). Moreover, including a pre-sleep (1–3h) protein meal can help attenuate the reduction in muscle protein synthesis during overnight fasting (Stokes et al. 2018) (Figure 22.2).

DIETARY PROTEIN SOURCES DURING CANCER TREATMENT

The optimal dietary protein source remains a topic of debate among patients, dietitians, and nutritionists, considering nutritional values, anabolic potential, and, recently, environmental impact. When choosing the source of protein for the cancer patient during treatment, there are several things to consider: i) the presence or risk of muscle loss or malnutrition; ii) patients with overweight or obesity or early satiation; iii) the presence or risk of anemia or iron deficiency (Knight, Wade, and Balducci 2004; Steegmann et al. 2013); iv) gastrointestinal symptoms derived from the medical treatment; and v) individual preferences of the cancer patient.

Misconceptions persist regarding the anabolic potential of plant-based proteins. While no studies have specifically addressed this in cancer patients, research in healthy adults indicates that a vegan diet induces comparable muscle protein synthesis at rest and after exercise compared to an omnivorous diet (Monteyne et al. 2023). Therefore, ensuring adequate total protein intake (>1.2g/kg/day) may be more critical than the specific dietary protein source. Recent expert guidelines (Ford, Arends, et al. 2022) also suggest promoting the ingestion of more than 65% of protein sources from animal protein foods. However, this recommendation has not been verified in any clinical trial conducted in humans. Furthermore, cancer patients can increase the amount of plant protein, use plant-based protein supplements, or choose smart plant-based protein sources with adequate amounts of essential amino acids and high dietary protein density (i.e., soy) to promote anabolism (Pinckaers et al. 2021).

Whey protein stands out as one of the most valuable protein sources based on its essential amino acid content, specifically leucine, and its digestibility. Whey protein supplementation may improve muscle mass and decrease cancer cell proliferation in cancer patients with cachexia (Thampy et al. 2022). Animal studies have demonstrated its positive effects on tumor incidence, survival, and growth (Teixeira et al. 2019). In humans, whey protein supplementation has been shown to increase functional capacity, nutritional status, strength, lean mass, and quality of life in cancer patients (Teixeira et al. 2019). Therefore, whey protein supplementation can be a practical and feasible strategy to optimize cancer cachexia/sarcopenia syndrome.

For vegan cancer patients, soy protein isolate supplements offer a promising and effective alternative for meeting daily protein requirements. Soy protein boasts a high protein digestibility-corrected amino acid score and may promote satiation (Ford, Arends, et al. 2022). Studies have demonstrated its ability to increase muscle protein synthesis in healthy adults (Tang et al. 2009) and elderly men (with lower muscle protein synthesis rates) (Yang et al. 2012), and to induce similar muscle mass and strength gains in response to resistance training in healthy adults (Messina et al. 2018). However, as these effects have not been specifically demonstrated in cancer patients, future research should investigate the effects of soy protein isolate supplements on cancer patients with cachexia or sarcopenia.

The satiation potential of plant-based proteins merits consideration. Plant-based sources typically contain higher fiber and lower protein density, which can directly influence satiation (Ford, Arends, et al. 2022) (Figure 22.3). For patients experiencing early satiation as a symptom of medical treatments, animal proteins may be preferable to plant-based proteins, given their higher energy content, protein density, essential amino acids, and digestibility (Ford, Arends, et al. 2022) (Figure 22.3). Conversely, for overweight or obese cancer patients, increased consumption of plant-based proteins, which contain lower energy and higher dietary fiber than animal proteins (Ford, Arends, et al. 2022) may facilitate decreased energy intake and improved body weight management (an essential intervention in this patient population) (Pérez-Segura et al. 2017) (Figure 22.3).

The high prevalence of anemia (30–90%) and iron deficiency (60%) in cancer patients necessitates vigilant monitoring (Knight, Wade, and Balducci 2004; Steegmann et al. 2013). Animal

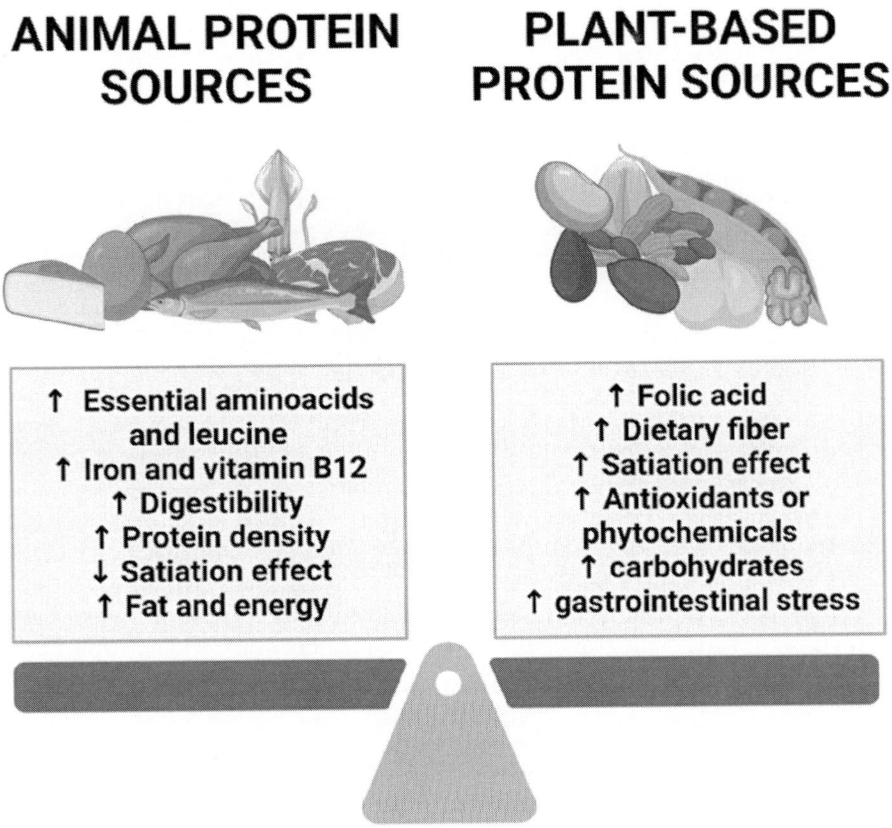

FIGURE 22.3 Differences between animal and plant-based protein sources. Figure summarizing the main nutritional differences between animal and plant-based protein sources. Figure was compiled by the authors and created with BioRender.

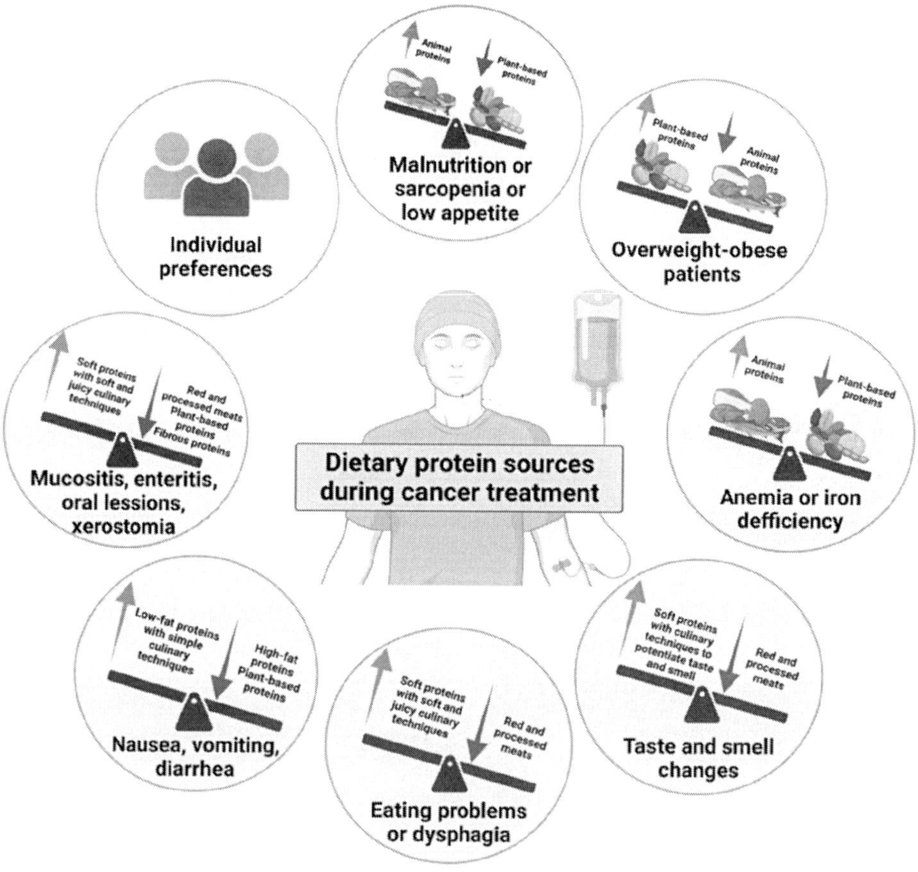

FIGURE 22.4 Dietary protein sources during cancer treatment. Main points to choose different protein sources in the cancer patient during treatment depending on the secondary symptoms. Figure was compiled by the authors and created with BioRender.

proteins provide higher levels of iron and vitamin B12, whereas vegetables are rich in folic acid (critical nutrients in anemia) (Ford, Arends, et al. 2022) (Figure 22.3). Incorporating animal protein sources may aid in preventing and/or treating anemia or iron deficiency in cancer patients, ideally combined with adequate amounts of folic acid from vegetables.

Cancer patients can develop gastrointestinal symptoms that may directly impact food consumption and nutrient intake (Ravasco 2019). Notably, protein may decrease after a cancer diagnosis, potentially increasing the risk of malnutrition and muscle loss (Ford, Orsso, et al. 2022). These symptoms can include deterioration in taste and smell, eating problems (teeth, chewing), dysphagia, nausea and vomiting, diarrhea, oral lesions, xerostomia, mucositis, and enteritis, among others (Ravasco 2019) (Figure 22.4):

i) Taste and smell changes: Patients should prioritize soft dietary protein sources such as white meat, fish, dairy products, and eggs, and reduce red and processed meats. It is also recommended that different culinary strategies are used to enhance taste and smell: marinate dietary proteins or combine them with sauces, spices, and herbs (Mahan, Escott-Stump, and Raymond 2013; Leser 2013) (Figure 22.4).
ii) Eating problems or dysphagia: Choose soft dietary protein sources that are easy to chew and swallow (i.e., fish instead of red meats). Use culinary preparations with a soft and juicy

consistency. Lubricate drier foods with juices, broths, sauces, milk, etc. (Mahan, Escott-Stump, and Raymond 2013; Leser 2013) (Figure 22.4).

iii) Nausea, vomiting, and diarrhea: Avoid high-fat dietary protein sources and choose low-fat dietary proteins (i.e., white meat, white fish, low-fat dairy products). Also avoid plant-based dietary proteins which can induce a flatulent effect (i.e., legumes). Choose simple culinary techniques: steam, boil, oven, griddle. Avoid cooking them with high quantities of fats (e.g., frying) (Mahan, Escott-Stump, and Raymond 2013; Leser 2013) (Figure 22.4).

iv) Oral lesions, xerostomia, mucositis, and enteritis: Use culinary preparations with a soft and juicy consistency. Lubricate drier foods with juices, broths, sauces, milk, etc. Avoid hard foods or foods that can break into small pieces or leave fibrous residues (i.e., legumes can remain in the mouth). Plant-based proteins can induce a flatulent effect (i.e., legumes) with a negative effect on enteritis. Choose simple culinary techniques and prioritize liquid preparations over solid preparations (Mahan, Escott-Stump, and Raymond 2013; Leser 2013) (Figure 22.4).

When recommending a source of protein, it is crucial to consider the individual preferences of the cancer patient. During medical treatment, cancer patients can experience alterations in food preferences, food avoidance, or food aversion, which might affect dietary protein sources. It is necessary to adapt the dietary protein sources according to these alterations to avoid a decrease in total protein intake in cancer patients (Ford, Orsso, et al. 2022). Additionally, if a cancer patient decides to follow a completely vegan dietary pattern, efforts should be made to achieve adequate amounts of protein, considering the secondary symptoms derived from the medical treatment and optimizing protein intake while minimizing the negative consequences for the patient.

DIETARY PROTEIN METABOLITES DURING CANCER TREATMENT

In addition to dietary protein sources, different dietary protein metabolites such as glutamine, leucine, β-hydroxy β-methylbutyrate (HMB), and creatine have emerged as potential therapeutic tools in cancer patients undergoing treatment.

Glutamine is an amino acid used in the synthesis of proteins (Figure 22.5). A recent meta-analysis of randomized controlled trials showed that supplementation with low (<10g) and medium (10–<20g) doses of glutamine reduces the risk of moderate-severe oral mucositis during chemotherapy or radiotherapy (Peng et al. 2021). In addition, glutamine may decrease peripheral neuropathy in cancer patients (Van Der Hulst et al. 1993) and help preserve intestinal integrity, which could be affected by medical treatments (Wang et al. 2007). Notably, glutamine supplementation can also be a useful strategy in increasing lean mass and improving the selectivity of different anti-cancer drugs (Prado, Purcell, and Laviano 2020).

Leucine, recognized as the most potent amino acid in stimulating the mTOR anabolic pathway, plays a key role in regulating muscle protein synthesis (Mitchell et al. 2016) (Figure 22.5). It may also influence the immune system and inflammatory processes (Beaudry and Law 2022). While some studies have found a potential effect on muscle mass maintenance in cancer patients, these studies typically combined leucine supplementation with other functional nutrients (i.e., fish oil or other amino acids). Therefore, further research is needed to elucidate the specific utility of leucine supplementation in cancer patients.

HMB, a product of leucine metabolism, has a role in both protein synthesis and degradation (Figure 22.5). A recent systematic review of randomized controlled trials in cancer patients showed that HMB supplementation has a positive effect on muscle mass and muscle function (Prado et al. 2022). Some studies have also suggested potential effects on hospitalization, cancer treatment-related toxicity, tumor response, or survival (Prado et al. 2022). However, additional research is

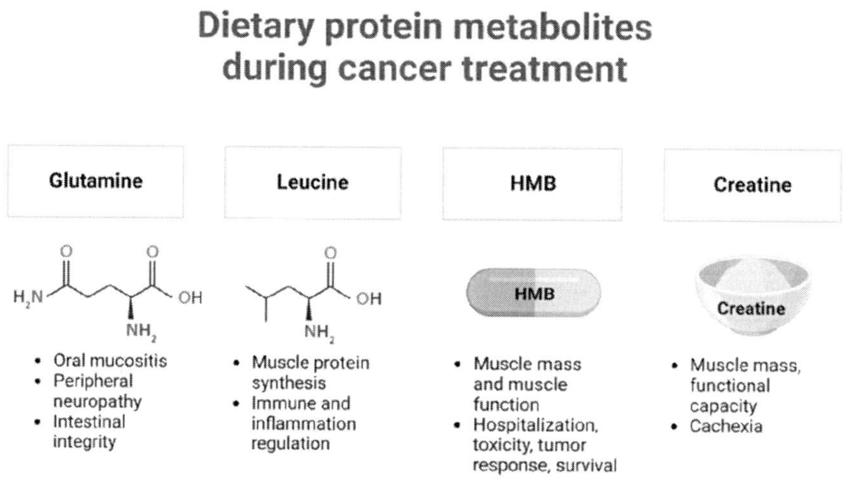

FIGURE 22.5 Dietary protein metabolites during cancer treatment. Summary of the effects of dietary protein metabolites during cancer treatment. Abbreviation: HMB, β-hydroxy β-methylbutyrate. Figure was compiled by the authors and created with BioRender.

needed to confirm the effects of HMB supplementation on these outcomes in cancer patients during treatment.

Creatine is a tripeptide commonly used in sports nutrition for its ergogenic properties (Figure 22.5). Notably, serum creatine levels are reduced in patients with cancer cachexia compared to those without cachexia (Wei et al. 2022). When combined with resistance exercise, creatine supplementation has demonstrated positive effects on muscle mass, fat mass, and functional capacity in cancer patients (Lønbro et al. 2013). Given its physiological functions, cancer patients at increased risk of muscle loss may benefit the most from creatine supplementation (Fairman et al. 2019; Prado, Purcell, and Laviano 2020). However, further randomized controlled trials are needed to confirm this hypothesis.

DIETARY PROTEIN IN PALLIATIVE CANCER PATIENTS

In palliative care scenarios, nutritional support remains important, as it can enhance quality of life and potentially prolong survival (Arends et al. 2017; Muscaritoli et al. 2021) (Figure 22.6). During palliative stages, a low protein intake can be driven by secondary symptoms that decrease oral food intake and induce functional impairments, fatigue, and overall wasting (Cotogni et al. 2021). Given that life expectancy for palliative cancer patients can range from months to several years, protein intake should be closely monitored to alleviate secondary symptoms and consequences of the disease, ultimately aiming to improve patients' quality of life and survival (Arends et al. 2017; Muscaritoli et al. 2021; Cotogni et al. 2021). Protein requirements, protein sources, and strategies can be extrapolated from the recommendations for cancer patients during treatment (see above). However, we need to keep in mind that for dying patients, nutritional recommendations should prioritize comfort, with an emphasis on individualized approaches, since they will not have functional or metabolic benefits (Arends et al. 2017; Muscaritoli et al. 2021).

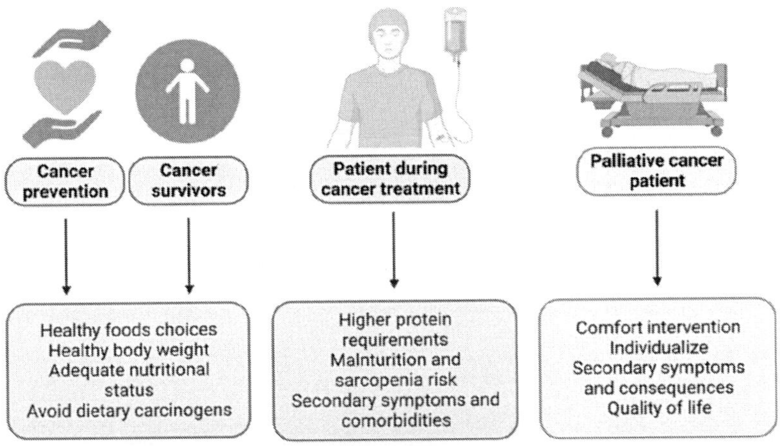

FIGURE 22.6 Overview of the main objectives of dietary protein intake during the cancer continuum. Main objectives that should be achieved through an adequate protein intake and sources for cancer prevention, cancer survivors, patients during cancer treatment, and palliative patients. Figure was compiled by the authors and created with BioRender.

DIETARY PROTEIN IN CANCER SURVIVORS

Once cancer treatment has finished, nutritional recommendations for cancer survivors should align with those for cancer prevention (Arends et al. 2017; Muscaritoli et al. 2021) (Figure 22.6). Specifically, nutrition strategies for cancer survivors should: i) help maintain a healthy body weight; ii) support an optimal nutritional status; and iii) avoid the incorporation of carcinogens through diet (e.g., alcohol). Regarding dietary protein, cancer survivors should prioritize plant-based sources and reduce the consumption of animal proteins. Additionally, cancer survivors should avoid consuming red and processed meats and smoked and salty fish. Unfortunately, there are no specific scientific recommendations for fish, egg, or dairy product consumption. Therefore, cancer survivors are advised to moderate their intake of these protein sources.

CONCLUSION

This chapter has demonstrated the complex role of dietary protein in modulating cancer risk and progression. Different protein sources can either increase or decrease cancer risk, while the relationship between overall dietary protein quantity and cancer risk remains controversial. Notably, dietary protein quantity is a key factor in nutritional interventions for cancer patients undergoing medical treatment. However, protein sources can be differentially managed depending on the secondary symptoms arising from medical treatments. Altogether, this chapter underscores the importance of protein across all stages of the disease.

MINI DICTIONARY OF TERMS

Cachexia: A complex wasting syndrome characterized by weight loss, muscle atrophy, appetite loss, and fatigue.

Malnutrition: A nutritional disorder resulting from an inadequate nutritional status which leads to significant weight loss, depletion of body resources, and impaired quality of life and disease prognosis.

Mucositis: Inflammation and potential ulceration of the mucous membranes lining the gastrointestinal tract.

Sarcopenia: A syndrome characterized by the loss of skeletal muscle mass, strength, and function.

Xerostomia: A condition characterized by reduced or insufficient production of saliva resulting in dryness of the mouth which can dampen the capacity of ingesting foods, nutritional status, quality of life, and comfort of cancer patients.

SUMMARY POINTS

- Evidence suggests that protein sources play a more significant role than total protein quantity in cancer prevention.
- Plant-based dietary proteins should be prioritized over animal protein sources for cancer prevention.
- The relationship between different dietary protein sources (i.e., eggs, white meat, dairy products) and cancer risk remains inconclusive.
- During cancer treatment, patients should increase their protein intake by around 1–2 g/kg/day.
- Protein source recommendations for cancer patients undergoing treatment should be individualized, based on secondary symptoms and comorbidities.
- Certain dietary protein metabolites (i.e., glutamine, leucine, β-hydroxy β-methylbutyrate, and creatine) may offer therapeutic benefits for patients during cancer treatment.
- Nutritional recommendations for palliative cancer patients should prioritize comfort and be tailored to individual needs, considering secondary symptoms.
- Both active cancer patients and survivors should follow protein intake recommendations similar to those for preventing cancer.

REFERENCES

Afshin, Ashkan, Patrick John Sur, Kairsten A. Fay, Leslie Cornaby, Giannina Ferrara, Joseph S. Salama, Erin C. Mullany, Kalkidan Hassen Abate, Cristiana Abbafati, and Zegeye Abebe. 2019. "Health Effects of Dietary Risks in 195 Countries, 1990–2017: A Systematic Analysis for the Global Burden of Disease Study 2017." *The Lancet* 393 (10184). Elsevier: 1958–72.

Al-Zahrani, Meshari A., Mohammad Ahmad, Mohammad Alkhamees, Ahmed Aljuhayman, Saleh Binsaleh, Raed Almannie, and Rahul Tiwari. 2022. "Dietary Protein Intake and Prostate Cancer Risk in Adults: A Systematic Review and Dose-Response Meta-Analysis of Prospective Cohort Studies." *Complementary Therapies in Medicine*. Elsevier 70: 102851.

Aminianfar, Azadeh, Roohallah Fallah-Moshkani, Asma Salari-Moghaddam, Parvane Saneei, Bagher Larijani, and Ahmad Esmaillzadeh. 2019. "Egg Consumption and Risk of Upper Aero-Digestive Tract Cancers: A Systematic Review and Meta-Analysis of Observational Studies." *Advances in Nutrition* 10 (4). Oxford University Press: 660–72.

Arends, Jann, Patrick Bachmann, Vickie Baracos, Nicole Barthelemy, Hartmut Bertz, Federico Bozzetti, Ken Fearon, et al. 2017. "ESPEN Guidelines on Nutrition in Cancer Patients." *Clinical Nutrition* 36 (1). Elsevier Ltd: 11–48. doi:10.1016/j.clnu.2016.07.015.

Beaudry, Anna G., and Michelle L. Law. 2022. "Leucine Supplementation in Cancer Cachexia: Mechanisms and a Review of the Pre-Clinical Literature." *Nutrients* 14 (14). MDPI: 2824.

Bozzetti, Federico, and Valentina Bozzetti. 2013. "Is the Intravenous Supplementation of Amino Acid to Cancer Patients Adequate? A Critical Appraisal of Literature." *Clinical Nutrition* 32 (1). Elsevier: 142–46.

Cotogni, Paolo, Silvia Stragliotto, Marta Ossola, Alessandro Collo, Sergio Riso, and Intersociety Italian Working Group for Nutritional Support in Cancer. 2021. "The Role of Nutritional Support for Cancer Patients in Palliative Care." *Nutrients* 13 (2). MDPI: 306.

Cross, Amanda J., Michael F. Leitzmann, Mitchell H. Gail, Albert R. Hollenbeck, Arthur Schatzkin, and Rashmi Sinha. 2007. "A Prospective Study of Red and Processed Meat Intake in Relation to Cancer Risk." *PLoS Medicine* 4 (12). Public Library of Science San Francisco, USA: e325.

Dooley, James, Vasiliki Lagou, Jermaine Goveia, Anna Ulrich, Katerina Rohlenova, Nathalie Heirman, Tobias Karakach, Yulia Lampi, Shawez Khan, and Jun Wang. 2020. "Heterogeneous Effects of Calorie Content and Nutritional Components Underlie Dietary Influence on Pancreatic Cancer Susceptibility." *Cell Reports* 32 (2). Elsevier: 107880.

Fairman, Ciaran M., Kristina L. Kendall, Nicolas H. Hart, Dennis R. Taaffe, Daniel A. Galvao, and Robert U. Newton. 2019. "The Potential Therapeutic Effects of Creatine Supplementation on Body Composition and Muscle Function in Cancer." *Critical Reviews in Oncology/Hematology* 133. Elsevier: 46–57.

Fan, Yahui, Mingxu Wang, Zhaofang Li, Hong Jiang, Jia Shi, Xin Shi, Sijiao Liu, Jinping Zhao, Liyun Kong, and Wei Zhang. 2022. "Intake of Soy, Soy Isoflavones and Soy Protein and Risk of Cancer Incidence and Mortality." *Frontiers in Nutrition* 9. Frontiers Media SA: 847421.

Fang, Danbo, Fuqing Tan, Chaojun Wang, Xuanwen Zhu, and Liping Xie. 2012. "Egg Intake and Bladder Cancer Risk: A Meta-Analysis." *Experimental and Therapeutic Medicine* 4 (5). Spandidos Publications: 906–12.

Ford, Katherine L., Jann Arends, Philip J. Atherton, Mariëlle P. K. J. Engelen, Thiago J. M. Gonçalves, Alessandro Laviano, Dileep N. Lobo, Stuart M. Phillips, Paula Ravasco, and Nicolaas E. P. Deutz. 2022. "The Importance of Protein Sources to Support Muscle Anabolism in Cancer: An Expert Group Opinion." *Clinical Nutrition* 41 (1). Elsevier: 192–201.

Ford, Katherine L., Camila E. Orsso, Nicole Kiss, Skyler B. Johnson, Sarah A. Purcell, Adele Gagnon, Alessandro Laviano, and Carla M. Prado. 2022. "Dietary Choices Following a Cancer Diagnosis: A Narrative Review." *Nutrition*. Elsevier103: 111838.

Foreman, Kyle J., Neal Marquez, Andrew Dolgert, Kai Fukutaki, Nancy Fullman, Madeline McGaughey, Martin A. Pletcher, Amanda E. Smith, Kendrick Tang, and Chun-Wei Yuan. 2018. "Forecasting Life Expectancy, Years of Life Lost, and All-Cause and Cause-Specific Mortality for 250 Causes of Death: Reference and Alternative Scenarios for 2016–40 for 195 Countries and Territories." *The Lancet* 392 (10159). Elsevier: 2052–90.

Ho, Victor W., Kelvin Leung, Anderson Hsu, Beryl Luk, June Lai, Sung Yuan Shen, Andrew I. Minchinton, Dawn Waterhouse, Marcel B. Bally, and Wendy Lin. 2011. "A Low Carbohydrate, High Protein Diet Slows Tumor Growth and Prevents Cancer InitiationLow Carbohydrate, High Protein Diets Slow Tumor Growth." *Cancer Research* 71 (13). AACR: 4484–93.

Hulst, Rene, R. W. J. Van Der, M. F. Von Meyenfeldt, N. E. P. Deutz, P. B. Soeters, R. J. M. Brummer, B. K. von Kreel, and J. W. Arends. 1993. "Glutamine and the Preservation of Gut Integrity." *The Lancet* 341 (8857). Elsevier: 1363–65.

IARC Working Group on the Evaluation of Carcinogenic Risks to Humans. 2018. "Red Meat and Processed Meat." *International Agency for Research on Cancer* 114. International Agency for Research on Cancer.

International, World Cancer Research Fund. 2018. *Diet, Nutrition, Physical Activity and Cancer: A Global Perspective: A Summary of the Third Expert Report*. World Cancer Research Fund International.

Kaluza, Joanna, Shoko Komatsu, Mara Lauriola, Holly R. Harris, Leif Bergkvist, Karl Michaëlsson, and Alicja Wolk. 2021. "Long-Term Consumption of Non-Fermented and Fermented Dairy Products and Risk of Breast Cancer by Estrogen Receptor Status–Population-Based Prospective Cohort Study." *Clinical Nutrition* 40 (4). Elsevier: 1966–73.

Kim, Seong Rae, Kyuwoong Kim, Sang Ah Lee, Sung Ok Kwon, Jong-Koo Lee, NaNa Keum, and Sang Min Park. 2019. "Effect of Red, Processed, and White Meat Consumption on the Risk of Gastric Cancer: An Overall and Dose–Response Meta-Analysis." *Nutrients* 11 (4). MDPI: 826.

Knight, Kevin, Sally Wade, and Lodovico Balducci. 2004. "Prevalence and Outcomes of Anemia in Cancer: A Systematic Review of the Literature." *The American Journal of Medicine* 116 (7). Elsevier: 11–26.

Kocarnik, Jonathan M., Kelly Compton, Frances E. Dean, Weijia Fu, Brian L. Gaw, James D. Harvey, Hannah Jacqueline Henrikson, Dan Lu, Alyssa Pennini, and Rixing Xu. 2022. "Cancer Incidence, Mortality, Years of Life Lost, Years Lived with Disability, and Disability-Adjusted Life Years for 29 Cancer Groups from 2010 to 2019: A Systematic Analysis for the Global Burden of Disease Study 2019." *JAMA Oncology* 8 (3). American Medical Association: 420–44.

Lee, Keum Hwa, Hyo Jin Seong, Gaeun Kim, Gwang Hun Jeong, Jong Yeob Kim, Hyunbong Park, Eunyoung Jung, Andreas Kronbichler, Michael Eisenhut, and Brendon Stubbs. 2020. "Consumption of Fish and ω-3 Fatty Acids and Cancer Risk: An Umbrella Review of Meta-Analyses of Observational Studies." *Advances in Nutrition* 11 (5). Oxford Academic: 1134–49.

Leser, Maureen. 2013. *Oncology Nutrition for Clinical Practice*. Oncology Nutrition Dietetic Practice Group.

Levine, Morgan E., Jorge A. Suarez, Sebastian Brandhorst, Priya Balasubramanian, Chia-Wei Cheng, Federica Madia, Luigi Fontana, Mario G Mirisola, Jaime Guevara-Aguirre, and Junxiang Wan. 2014. "Low Protein Intake Is Associated with a Major Reduction in IGF-1, Cancer, and Overall Mortality in the 65 and Younger but Not Older Population." *Cell Metabolism* 19 (3). Elsevier: 407–17.

Lønbro, Simon, Ulrik Dalgas, Hanne Primdahl, Jens Overgaard, and Kristian Overgaard. 2013. "Feasibility and Efficacy of Progressive Resistance Training and Dietary Supplements in Radiotherapy Treated Head and Neck Cancer Patients–the DAHANCA 25A Study." *Acta Oncologica* 52 (2). Taylor & Francis: 310–18.

MacDonald, Alisdair J., Neil Johns, Nathan Stephens, Carolyn Greig, James A. Ross, Alexandra C. Small, Holger Husi, Kenneth C. H. Fearon, and Tom Preston. 2015. "Habitual Myofibrillar Protein Synthesis Is Normal in Patients with Upper GI Cancer Cachexia." *Clinical Cancer Research* 21 (7). AACR: 1734–40.

Mahan, L. Kathleen, Sylvia Escott-Stump, and Janice L. Raymond. 2013. *Krause Dietoterapia*. Elsevier Amsterdam.

Martínez-Garay, Carlos, and Nabil Djouder. 2023. "Dietary Interventions and Precision Nutrition in Cancer Therapy." *Trends in Molecular Medicine*. Elsevier 29(7): 489-511.

Messina, Mark, Heidi Lynch, Jared M. Dickinson, and Katharine E. Reed. 2018. "No Difference between the Effects of Supplementing with Soy Protein versus Animal Protein on Gains in Muscle Mass and Strength in Response to Resistance Exercise." *International Journal of Sport Nutrition and Exercise Metabolism* 28 (6). Human Kinetics: 674–85.

Mirzaei, Hamed, Jorge A. Suarez, and Valter D. Longo. 2014. "Protein and Amino Acid Restriction, Aging and Disease: From Yeast to Humans." *Trends in Endocrinology & Metabolism* 25 (11). Elsevier: 558–66.

Mitchell, W. Kyle, Daniel J. Wilkinson, Bethan E. Phillips, Jonathan N. Lund, Kenneth Smith, and Philip J. Atherton. 2016. "Human Skeletal Muscle Protein Metabolism Responses to Amino Acid Nutrition." *Advances in Nutrition* 7 (4). Oxford University Press: 828S–38S.

Monteyne, Alistair J., Mariana O. C. Coelho, Andrew J. Murton, Doaa R. Abdelrahman, Jamie R. Blackwell, Christopher P. Koscien, Karen M. Knapp, Jonathan Fulford, Tim J. A. Finnigan, and Marlou L. Dirks. 2023. "Vegan and Omnivorous High Protein Diets Support Comparable Daily Myofibrillar Protein Synthesis Rates and Skeletal Muscle Hypertrophy in Young Adults." *The Journal of Nutrition*. Elsevier 153(6):1680–1695.

Muscaritoli, Maurizio, Jann Arends, Patrick Bachmann, Vickie Baracos, Nicole Barthelemy, Hartmut Bertz, Federico Bozzetti, et al. 2021. "ESPEN Practical Guideline: Clinical Nutrition in Cancer." *Clinical Nutrition* (March). Elsevier Ltd: 1–16. doi:10.1016/j.clnu.2021.02.005.

Naghshi, Sina, Omid Sadeghi, Walter C. Willett, and Ahmad Esmaillzadeh. 2020. "Dietary Intake of Total, Animal, and Plant Proteins and Risk of All Cause, Cardiovascular, and Cancer Mortality: Systematic Review and Dose-Response Meta-Analysis of Prospective Cohort Studies." *BMJ* 370:1680–1695. British Medical Journal Publishing Group.

Op Den Kamp, Céline M., Ramon C. Langen, Astrid Haegens, and Annemie M. Schols. 2009. "Muscle Atrophy in Cachexia: Can Dietary Protein Tip the Balance?" *Current Opinion in Clinical Nutrition and Metabolic Care*. doi:10.1097/MCO.0b013e3283319399.

Pang, Yanyang, and Wu Wang. 2018. "Dietary Protein Intake and Risk of Ovarian Cancer: Evidence from a Meta-Analysis of Observational Studies." *Bioscience Reports* 38 (6): BSR20181857. Portland Press.

Peng, Tzu-Rong, Hung-Hong Lin, Li-Jou Yang, and Ta-Wei Wu. 2021. "Effectiveness of Glutamine in the Management of Oral Mucositis in Cancer Patients: A Meta-Analysis of Randomized Controlled Trials." *Supportive Care in Cancer* 29. Springer: 4885–92.

Pérez-Segura, P., J. E. Palacio, L. Vázquez, S. Monereo, R. de Las Peñas, P. Martínez de Icaya, C. Grávalos, A. Lecube, A. Blasco, and J. M. García-Almeida. 2017. "Peculiarities of the Obese Patient with Cancer: A National Consensus Statement by the Spanish Society for the Study of Obesity and the Spanish Society of Medical Oncology." *Clinical and Translational Oncology* 19. Springer: 682–94.

Pinckaers, Philippe J. M., Jorn Trommelen, Tim Snijders, and Luc J. C. van Loon. 2021. "The Anabolic Response to Plant-Based Protein Ingestion." *Sports Medicine* 51 (Suppl 1). Springer: 59–74.

Prado, Carla M., Camila E. Orsso, Suzette L. Pereira, Philip J. Atherton, and Nicolaas E. P. Deutz. 2022. "Effects of Β -hydroxy Β -methylbutyrate (HMB) Supplementation on Muscle Mass, Function, and Other Outcomes in Patients with Cancer: A Systematic Review." *Journal of Cachexia, Sarcopenia and Muscle* 13 (3). Wiley Online Library: 1623–41.

Prado, Carla M., Sarah A. Purcell, and Alessandro Laviano. 2020. "Nutrition Interventions to Treat Low Muscle Mass in Cancer." *Journal of Cachexia, Sarcopenia and Muscle* 11 (2). Wiley Online Library: 366–80.

Ravasco, Paula. 2019. "Nutrition in Cancer Patients." *Journal of Clinical Medicine* 8 (8). MDPI: 1211.

Ryan, Aoife M., and Erin S. Sullivan. 2021. "Impact of Musculoskeletal Degradation on Cancer Outcomes and Strategies for Management in Clinical Practice." *Proceedings of the Nutrition Society* 80 (1). Cambridge University Press: 73–91.

Steegmann, J. L., J. M. Sanchez Torres, R. Colomer, A. Vaz, J. Lopez, I. Jalon, M. Provencio, A. González-Martín, and M. Pérez. 2013. "Prevalence and Management of Anaemia in Patients with Non-Myeloid Cancer Undergoing Systemic Therapy: A Spanish Survey." *Clinical and Translational Oncology* 15. Springer: 477–83.

Stokes, Tanner, Amy J. Hector, Robert W. Morton, Chris McGlory, and Stuart M. Phillips. 2018. "Recent Perspectives Regarding the Role of Dietary Protein for the Promotion of Muscle Hypertrophy with Resistance Exercise Training." *Nutrients* 10 (2). MDPI: 180.

Tang, Jason E., Daniel R. Moore, Gregory W. Kujbida, Mark A. Tarnopolsky, and Stuart M. Phillips. 2009. "Ingestion of Whey Hydrolysate, Casein, or Soy Protein Isolate: Effects on Mixed Muscle Protein Synthesis at Rest and Following Resistance Exercise in Young Men." *Journal of Applied Physiology*. American Physiological Society 107(3):987–92.

Teixeira, Filipe J., Heitor O. Santos, Scott L. Howell, and Gustavo D. Pimentel. 2019. "Whey Protein in Cancer Therapy: A Narrative Review." *Pharmacological Research* 144. Elsevier: 245–56.

Thampy, Anjana, Meena Kumari Palani Kumar, Muthukumar Serva Peddha, and Madhavi Reddy. 2022. "The Effectiveness of Whey Proteins in Prevention and Treatment of Cancer: A Review." *Critical Reviews in Food Science and Nutrition*. Taylor & Francis, 64(8):2088–2104.

Tran, Khanh Bao, Justin J. Lang, Kelly Compton, Rixing Xu, Alistair R. Acheson, Hannah Jacqueline Henrikson, Jonathan M. Kocarnik, Louise Penberthy, Amirali Aali, and Qamar Abbas. 2022. "The Global Burden of Cancer Attributable to Risk Factors, 2010–19: A Systematic Analysis for the Global Burden of Disease Study 2019." *The Lancet* 400 (10352). Elsevier: 563–91.

Tse, Genevieve, and Guy D. Eslick. 2014. "Egg Consumption and Risk of GI Neoplasms: Dose–Response Meta-Analysis and Systematic Review." *European Journal of Nutrition* 53. Springer: 1581–90.

Wang, Wei-Shu, Jen-Kou Lin, Tzu-Chen Lin, Wei-Shone Chen, Jeng-Kae Jiang, Huann-Sheng Wang, Tzeon-Jye Chiou, Jin-Hwang Liu, Chueh-Chuan Yen, and Po-Min Chen. 2007. "Oral Glutamine Is Effective for Preventing Oxaliplatin-Induced Neuropathy in Colorectal Cancer Patients." *The Oncologist* 12 (3). Oxford University Press: 312–19.

Wei, Lulu, Ranran Wang, Kai Lin, Xiaolu Jin, Li Li, Junaid Wazir, Wenyuan Pu, Panpan Lian, Renwei Lu, and Shiyu Song. 2022. "Creatine Modulates Cellular Energy Metabolism and Protects against Cancer Cachexia-Associated Muscle Wasting." *Frontiers in Pharmacology* 13. Frontiers: 5212.

Winter, Aaron, Jacqueline MacAdams, and Stéphanie Chevalier. 2012. "Normal Protein Anabolic Response to Hyperaminoacidemia in Insulin-Resistant Patients with Lung Cancer Cachexia." *Clinical Nutrition* 31 (5). Elsevier: 765–73.

Wu, Jing, Rong Zeng, Junpeng Huang, Xufeng Li, Jiren Zhang, James Chung-Man Ho, and Yanfang Zheng. 2016. "Dietary Protein Sources and Incidence of Breast Cancer: A Dose-Response Meta-Analysis of Prospective Studies." *Nutrients* 8 (11). MDPI: 730.

Yang, Yifan, Tyler A. Churchward-Venne, Nicholas A. Burd, Leigh Breen, Mark A. Tarnopolsky, and Stuart M. Phillips. 2012. "Myofibrillar Protein Synthesis Following Ingestion of Soy Protein Isolate at Rest and after Resistance Exercise in Elderly Men." *Nutrition & Metabolism* 9 (1). BioMed Central: 1–9.

Yun, Zhangjun, Mengdie Nan, Xiao Li, Zhu Liu, Jing Xu, Xiaofeng Du, Qing Dong, and Li Hou. 2023. "Processed Meat, Red Meat, White Meat, and Digestive Tract Cancers: A Two-Sample Mendelian Randomization Study." *Frontiers in Nutrition* 10. Frontiers Media SA: 1078963.

Zhang, Kui, Hao Dai, Weibo Liang, Lin Zhang, and Zhenhua Deng. 2019. "Fermented Dairy Foods Intake and Risk of Cancer." *International Journal of Cancer* 144 (9). Wiley Online Library: 2099–108.

Zhang, Xiaotao, Linda Pang, Shreela V. Sharma, Ruosha Li, Alan G. Nyitray, and Beatrice J. Edwards. 2021. "Malnutrition and Overall Survival in Older Patients with Cancer." *Clinical Nutrition* 40 (3). Elsevier: 966–77.

23 Dietary Protein and Gut Microbiota

Carmen Rodríguez-García and Lucas Jurado-Fasoli

ABBREVIATIONS

BCFA	Branched-chain fatty acids
CRP	C-reactive protein
H_2S	Hydrogen sulfide
LPS	Lipopolysaccharide
NAFLD	Non-alcoholic fatty liver disease
SCFA	Short-chain fatty acids
T2DM	Type 2 diabetes mellitus
TMAO	Trimethylamine-N-oxide
TNFα	Tumour necrosis factor-alpha

INTRODUCTION

The human gut microbiota is a dynamic ecosystem composed of a large number of host-specific microorganisms, where bacteria represent over 99% of the total composition (Sender, Fuchs, and Milo 2016). A "healthy microbiota" community usually demonstrates ecological stability (resilience and resistance to changes), high microbial gene richness, and high taxonomic alpha diversity (the average species diversity within a habitat) (Hou et al. 2022). The gut microbiota is a critical component of human physiology, performing a range of essential functions such as fermentation of non-digestible nutrients, maintenance of intestinal-mucosal barrier integrity, and regulation of the host's immune response (M.a. Zhao et al. 2023). Prolonged disruption of this ecosystem, known as dysbiosis, can lead to local and systemic inflammation, increasing the risk of diseases such as gastrointestinal and metabolic disorders, cardiovascular diseases, respiratory issues, and certain types of cancer (Carding et al. 2015).

The relative distribution of microbes in the gut is unique to each individual. Its composition is determined by genetic and environmental factors, such as sex, age, use of antibiotics, and gastrointestinal transit time (Jardon et al. 2022). Among these factors, diet plays an important role in defining and shaping the gut microbiota. However, evidence about the effects of dietary protein on the gut microbiota is limited (Bartlett and Kleiner 2022), although microbial metabolites derived from protein fermentation can lead to several diseases (Cai et al. 2022). Unfortunately, it is difficult to establish a direct link between dietary protein, gut microbiota, and human health due to inconsistent published evidence on the effects of these metabolites (Hou et al. 2022). Therefore, this chapter aims to review the evidence for the impact of dietary protein on the composition and function of the gut microbiota and its implications for human health.

DIETARY PROTEIN AND GUT MICROBIOME INTERACTION

Before exploring the relationship between gut microbiota and dietary protein in health, it is essential to understand protein digestion and fermentation processes.

DOI: 10.1201/b23328-23

DIETARY PROTEIN FERMENTATION IN GUT MICROBIOTA AND END PRODUCTS

Once ingested, protein reaches the stomach where, due to the acidic pH, proteins unfold, allowing easier access for proteolytic enzymes. In the small intestine, trypsin and other peptidases hydrolyse dietary protein into amino acids and small peptides, which are mainly absorbed into the bloodstream. However, a small fraction of the dietary protein escapes digestion and reaches the large intestine, where it is minimally absorbed. The undigested proteins are then fermented by the gut microbiota in the distal colon (Bartlett and Kleiner 2022). Proteolytic fermentation produces a variety of metabolites (Figure 23.1) that can be utilised by colonocytes or excreted in the faecal matter:

- Short-chain fatty acids (SCFA) (butyrate, propionate, and acetate) are usually produced by saccharolytic fermentation but can also result from proteolytic fermentation by some species of *Bifidobacterium* and *Lactobacillus*. Among others, butyrate is produced from lysine and glutamate; propionate results from the fermentation of methionine, alanine, and aspartate; and acetate from the fermentation of alanine, aspartate, glutamate, and serine (Rodríguez-Romero et al. 2022). Acetate and propionate positively impact glucose and lipid metabolism, while colonocytes largely use butyrate as their major energy source.
- Branched-chain fatty acids (BCFA) are produced by some *Clostridium* species from branched-chain amino acids. Leucine, isoleucine, and valine are the precursors of the BCFA isovalerate, 2-methylbutyrate, and isobutyrate, respectively (Rodríguez-Romero et al. 2022). Their circulating levels have been inversely related to serum C-reactive protein (CRP), insulin resistance, and cardiometabolic risk in humans (Yehia et al. 2023).
- Hydrogen sulfide (H_2S) is produced by sulfate-reducing bacteria such as *Bacillus*, *Enterobacter*, *Salmonella*, and *E. coli* through the metabolism of sulfur-containing amino acids such as cysteine (Linden 2013). High levels of H_2S are associated with inflammation and ulcerative colitis risk (Teigen et al. 2022).
- Ammonia is produced by acidic amino acid fermentation (aspartate and glutamate), and it is considered a harmful microbial metabolite because of its inflammatory activity on the colonic epithelium and its effects on the liver-brain axis in liver encephalopathy (Rodríguez-Romero et al. 2022).
- Polyamines, such as tryptamine, tyramine, and phenylethylamine, originate from the aromatic amino acids tryptophan, tyrosine, and phenylalanine, respectively, by some species of *Staphylococcus*, *Ruminococcus*, and *Morganella*, and from basic amino acids (Liu et al. 2020). They act as neuromodulators and vasoactive agents in the human body; for example, tryptamine stimulates gastrointestinal motility (Rodríguez-Romero et al. 2022; Liu et al. 2020).
- Indolic and phenolic compounds result from the metabolism of aromatic amino acids. Tryptophan may be metabolised by *Bacteroides*, *Bifidobacterium*, and *Clostridium* genera to produce indole and its derivatives (which play an important role in host defence), tyrosine metabolism produces phenol derivatives such as p-cresol (which is genotoxic), and phenylalanine leads to benzoic acid, whose effects on health are still unclear (Rodríguez-Romero et al. 2022; Liu et al. 2020).

The production of metabolites from amino acids depends mainly on the type of bacteria involved in their fermentation. In dysbiosis, the absence of beneficial bacteria such as *Bifidobacterium* or *Lactobacillus* leads to a decrease in metabolites involved in the immunometabolic regulation of the host and an increase in harmful metabolites.

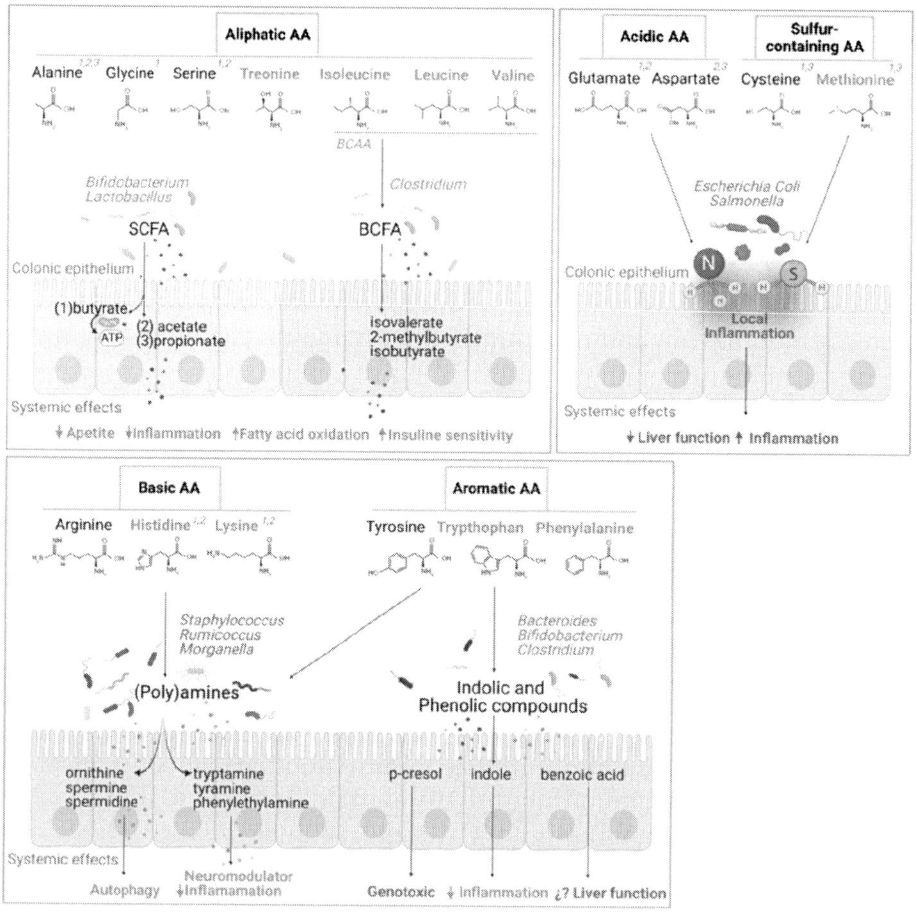

FIGURE 23.1 Gut microbiota fermentation of amino acids. The main end products resulting from gut microbiota fermentation of amino acids. The superscript numbers 1, 2, and 3 in aliphatic, acidic, sulfur-containing, and basic amino acids refer to butyrate, acetate, and propionate production, respectively. BCAA: branched-chain amino acids; BCFA: branched-chain fatty acids; H_2S: hydrogen sulfur; NH_3: ammonia; SCFA: short-chain fatty acids. This figure was created by the authors with BioRender.com.

RELATIONSHIP BETWEEN DIETARY PROTEIN AND GUT MICROBIOME ACROSS THE LIFESPAN

The colonisation of the gastrointestinal tract begins early in life and evolves with the host during growth, adulthood, and ageing. Its modulation can promote the development of some diseases, affect infant growth, and increase age-associated comorbidities (Di Profio et al. 2022).

PREGNANCY

The first factor that shapes the human gut microbiota comes from the mother during foetal development. The foetus is exposed through the placenta to metabolites produced by the maternal microbiota that shape its immune system. For example, acetate has been shown to affect epigenetic imprinting in the foetus, which is associated with the generation of regulatory T cells in adulthood, and has been linked to protection against the development of asthma later in life (Martino et al.

2022). Therefore, the modulation of gut metabolite production can directly impact the health of the foetus throughout its life. Although there are not many clinical trials or epidemiologic studies evaluating the effect of dietary protein on the gut microbiota during pregnancy, some of them have observed that a higher protein intake in healthy pregnant women was negatively associated with *Proteobacteria* (a phylum known to encompass multiple pathogens), while animal protein intake was negatively correlated with the *Actinobacteria* phylum (butyrate-producing bacteria) (Mandal et al. 2016; Selma-Royo et al. 2021). In obese pregnant women, total protein dietary intake appears to have beneficial effects by increasing the abundance of *Faecalibacterium*, a butyrate-producing bacterium with anti-inflammatory properties that promote gut health (Figure 23.2) (Ferrocino et al. 2018). Despite the limited evidence, it appears that dietary protein plays an important role in gut microbiota homeostasis during pregnancy, but more studies are necessary to elucidate the effect of dietary protein on gut microbiota metabolites and its impact on the foetal microbiota.

NEONATES

The infant microbiota is less stable and diverse than the adult microbiota, and it is dependent on factors such as milk consumption and mode of delivery. The neonatal diet consists of exclusive milk feeding until 4–6 months of age, followed by the introduction of complementary foods. Breastfeeding is recommended as the primary nutrition during infancy to support growth and neurodevelopment because it provides all necessary nutrients (lactose, lipids, proteins, and human milk oligosaccharides) (Di Profio et al. 2022). Moreover, oligosaccharides serve as metabolic substrates for some *Bifidobacterium* species, producing metabolites that have beneficial effects on host health and suppress the growth of opportunistic bacteria such as *Staphylococcaceae*, *Enterobacteriaceae*, and *Clostridiaceae* spp. (Laursen 2021). Conversely, exclusive formula feeding, characterised by higher protein content (~2.2 g/100 kcal vs. ~1.5 g/100 kcal in breast milk) and a lack of human milk oligosaccharides, leads to a higher abundance of *Enterobacteriaceae* and *Clostridiaceae* and lower *Bifidobacterium* spp., whose reduction has been associated with autoimmune diseases and asthma (Figure 23.2) (Ma et al. 2022). In infants with cow milk protein allergy, the use of soy-based formula reduces the abundance of *Bifidobacterium* spp. (Brink et al. 2020), with the above-mentioned negative effects on infant health. Interestingly, evidence suggests that when infant formulas are supplemented with human milk oligosaccharides, infants have a microbiota similar to that of breastfed infants, regardless of protein source (Berger et al. 2020; Pärnänen et al. 2022).

Complementary feeding is characterised by a shift from exclusive milk to a mixed diet that includes solid food. Due to the introduction of new sources of protein (such as legumes, dairy

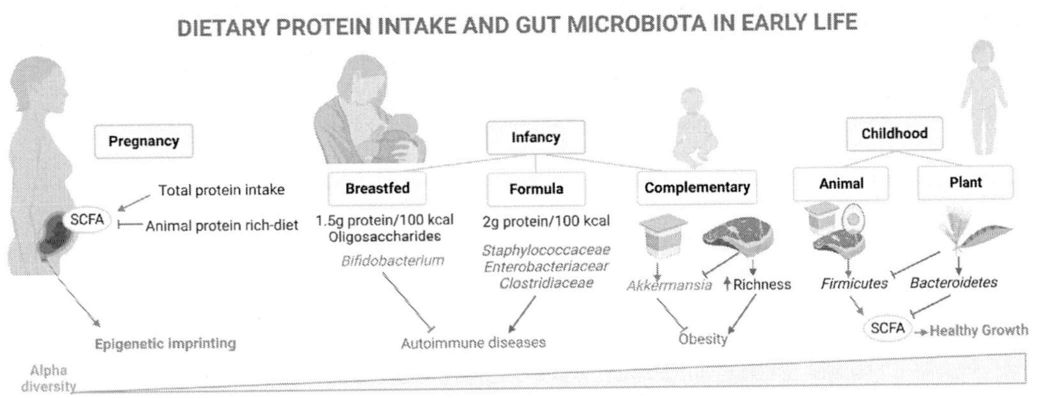

FIGURE 23.2 Dietary protein intake and gut microbiota in early life. Figure summarising the modulation of gut microbiota with dietary protein during pregnancy, infancy, and childhood and its impact on health. SCFA: short-chain fatty acids. Figure created with BioRender.com.

products, and meat) and fibre (from grains, vegetables, and fruits), the alpha diversity and abundance of *Ruminococcaceae*, *Lachnospiraceae*, and *Bacteroidaceae* species increase during this phase, and *Bifidobacterium* species decrease (Ma et al. 2022). Complementary food choices can affect gut microbiota and affect infant growth. A randomised controlled trial comparing dairy or meat-based complementary diets for infants aged 5–12 months observed that, in comparison to dairy, meat increased microbial community richness and decreased levels of *Akkermansia* spp. (Tang et al. 2023), which have been related to an increased risk of obesity (Reyna et al. 2022). These effects could be related to differences in the amino acid composition of each protein source or even to the presence of probiotics in dairy products.

FROM CHILDHOOD TO ADOLESCENCE

Data on the effect of dietary protein intake on the gut microbiota in children and adolescents are limited. During childhood, alpha diversity increases, as does ecological stability. Alterations in microbiota that colonise the gut during childhood may impair the host-microbiome interaction, leading to a higher predisposition to non-communicable diseases throughout life. The source of protein in this stage determines the composition of gut microbiota. Dairy consumption (especially yoghurt) is positively associated with *Firmicutes* abundance, while vegetable proteins such as nuts, legumes, and soy are negatively associated but positively associated with *Bacteroidetes* abundance. Contrary to this, animal protein is negatively associated with *Bacteroidetes* abundance (Smith-Brown et al. 2016; Herman et al. 2020). A study on egg supplementation (10 eggs/week) in 8–14-year-olds found that egg intake was associated with lower levels of atherogenic lipoproteins, higher levels of high-density lipoprotein cholesterol, and higher abundance of *Bifidobacterium*, improving gut microbiota, nutritional biomarkers, and growth without adverse events (Suta et al. 2023). Altogether, a healthy and varied diet appears to positively modulate gut microbiota during growth (Figure 23.2). However, during childhood, in the absence of contraindications such as allergies, consuming high-quality proteins from animal sources is essential for gut microbiota colonisation.

ADULTHOOD

Under healthy conditions, the gut microbiota is ecologically stable and predominantly composed of six bacterial phyla: *Firmicutes* and *Bacteroidetes*, which represent about 90%, along with *Fusobacteria*, *Proteobacteria*, *Actinobacteria*, and *Verrucomicrobia*. The *Firmicutes* phylum includes the *Enterococcus*, *Clostridium*, *Bacillus*, *Lactobacillus*, and *Ruminococcus* genera. *Bacteroidetes* is represented by *Prevotella* and *Bacteroides*, while the *Actinobacteria* phylum predominantly consists of the *Bifidobacterium* genus (Jardon et al. 2022).

The modulation of adult gut microbiota by dietary protein was first described in 1977. The ratio of anaerobic (*Bacteroides*) to facultative and aerobic organisms was 34:1 on a meat-rich diet, compared to 15:1 on a meat-free diet, suggesting that the source of protein was critical to the composition of the gut microbiota (Hentges et al. 1977). Moreover, the type of meat consumed can differentially affect gut microbiota. Compared with a beef and pork-based diet, a chicken-based diet reduced the abundance of the *Bacteroides* genus, increased the concentration of SCFA and BCFA, and decreased inflammation-related blood indexes (D. Zhao, Shan, et al. 2022). When lean red meat is incorporated into a healthy dietary pattern, whether processed or unprocessed, it may induce modifications in gut microbiota composition that are related to positive modulation of cardiometabolic risk factors such as reductions in serum total cholesterol and low-density lipoprotein cholesterol concentrations, but without changes in blood pressure, serum triglycerides, and faecal SCFA or BCFA (Y. Wang et al. 2023).

Regarding non-meat protein sources, such as pea and whey protein, they have been associated with a lower abundance of *Clostridium perfringens* and *Bacteroides fragilis* and a higher abundance of commensal *Lactobacillus* and *Bifidobacterium* (Mohammadkhah et al. 2018). Similarly,

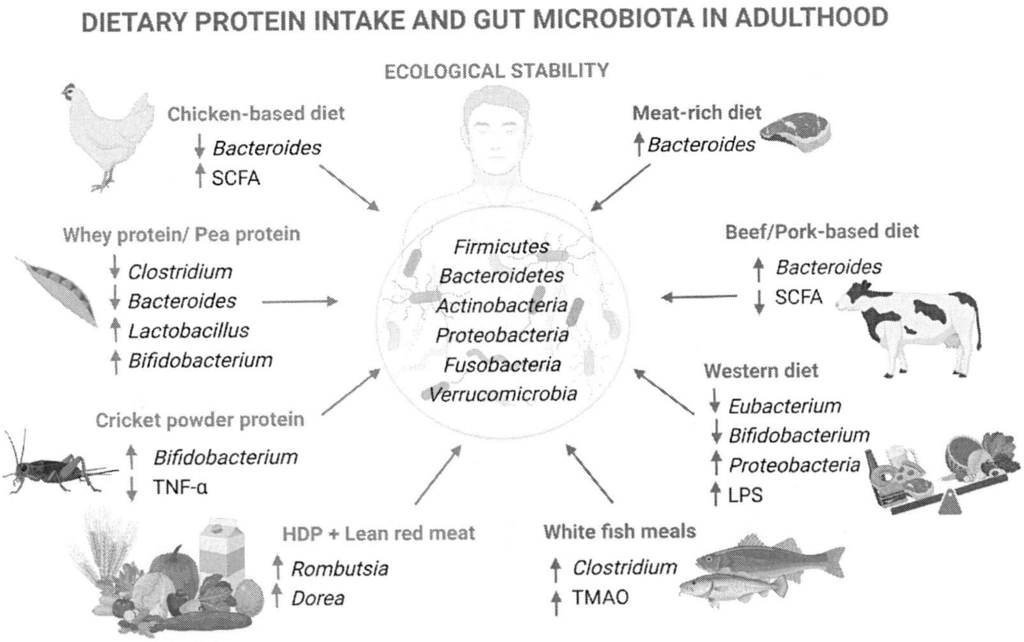

FIGURE 23.3 Dietary protein intake and gut microbiota in adulthood. Summary of the effects of different sources of dietary proteins and dietary patterns impacting ecological stability. Dietary protein left: beneficial. Dietary protein right: harmful. HDP: healthy dietary pattern; LPS: lipopolysaccharide; SCFA: short-chain fatty acids; TMAO: trimethylamine-N-oxide; TNFα: tumour necrosis factor-alpha. Figure created with BioRender.com.

the intake of 25 g/day of cricket powder reduced plasma tumour necrosis factor-alpha and increased *Bifidobacterium* abundance, suggesting that insect-based protein may reduce systemic inflammation and improve gut health (Stull et al. 2018). In adulthood, it appears that plant protein sources may have a more positive effect on gut health-related outcomes.

Some dietary patterns, such as the Western diet (low in plant-derived fibre and high in animal protein), decrease *Eubacterium* and *Bifidobacterium* genera and increase the abundance of *Proteobacteria*. These changes have been associated with systemic inflammation due to increasing levels of lipopolysaccharide (LPS), and cardiometabolic diseases (Wu et al. 2011). However, the effects of animal protein on gut microbiota and health remain contradictory. A comparison of high-protein meals containing trimethylamine-N-oxide (TMAO) such as fish, and its dietary precursors, such as carnitine (beef) or choline (eggs), revealed that the consumption of fish increased TMAO levels (which have been related to cardiovascular risk), and *Clostridium* abundance more than the other sources (Cho et al. 2017; Schmedes et al. 2019). However, this effect cannot be generalised to all fish sources since white fish, such as seabass and cod, tend to have higher concentrations of TMAO than blue fish. Although the evidence remains limited and controversial, in adults, diets high in protein and low in plant-based foods may negatively impact the gut microbiota and have adverse effects on health. In addition, the high intake of red meat may negatively modify the gut microbiota, but the intake of lean red meat in a healthy dietary pattern could be beneficial to host health (Figure 23.3).

AGEING

Due to the accumulation of damage throughout life and biological programming, ageing impacts gut microbiota composition, resulting in a reduction in ecological stability, gut microbiota alpha

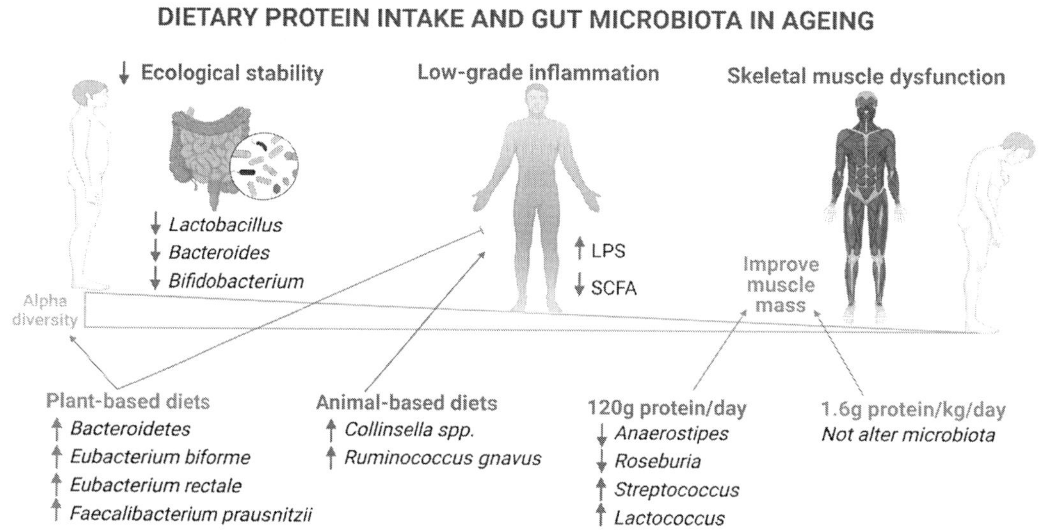

FIGURE 23.4 Dietary protein intake and gut microbiota in ageing. Effects of dietary proteins on age-related alterations in gut microbiota-muscle axis. LPS: lipopolysaccharide; SCFA: short-chain fatty acids. Figure created with BioRender.com.

diversity, lower *Lactobacillus*, *Bacteroides,* and *Bifidobacterium* abundance, and SCFA production, which promotes low-grade inflammation (Martino et al. 2022). In addition, sarcopenic patients exhibit increased LPS biosynthesis and a significant reduction in *Roseburia inulinivorans* and *Faecalibacterium prausnitzii*, species with high metabolic capacity in the production of SCFA, promoting systemic inflammation and supporting a gut-muscle axis that could explain skeletal muscle dysfunction during ageing (Prokopidis et al. 2021). High-protein diets are often recommended for older people. Research on the microbiota of older women showed relative stability to the high-protein diet, with the suppression of butyrate-producing genera, *Anaerostipes* and *Roseburia*, an enhancement of *Streptococcus* and *Lactococcus*, and improvement in muscle mass (Ford et al. 2020). Another clinical trial conducted by Mitchell et al. observed that a diet containing twice the recommended dietary intake (1.6 g/kg/day) provided via whole foods with appropriate dietary fibre composition did not alter gut microbiota or proteolytic metabolites but had beneficial effects on lean body mass (Mitchell et al. 2020). Regarding dietary patterns, plant-based diets have been positively correlated with alpha diversity, the *Bacteroidetes* phylum, and some anti-inflammatory species, such as *Eubacterium biforme*, *Eubacterium rectale,* and *Faecalibacterium prausnitzii* (Y. Wang et al. 2023). Conversely, animal-based diets have been associated with pro-inflammatory species (*Collinsella* spp. and *Ruminococcus gnavus)* (van Soest et al. 2020). Furthermore, higher intakes of processed meat have been associated with higher atherosclerotic cardiovascular disease risk partly mediated by microbiota-derived metabolites of L-carnitine (TMAO) (M. Wang et al. 2022).

Dietary modulation of the gut microbiota is a potential target for improving healthy ageing and age-related comorbidities. A high-protein diet (1.6 g/kg/day) in the elderly, combined with an appropriate amount of plant-based foods, could help restore the gut microbiota, increasing alpha diversity and SCFA-producing bacteria, along with improvements in muscle mass (Figure 23.4).

RELATIONSHIP BETWEEN DIETARY PROTEIN, GUT MICROBIOME, AND HEALTH

Gut microbial metabolites such as SCFA have important metabolic functions. SCFAs could prevent obesity and type 2 diabetes mellitus (T2DM) by improving appetite regulation, increasing anorexic hormone production, and increasing energy expenditure (Canfora et al. 2019). However, the microbial metabolites derived exclusively from protein fermentation are often considered detrimental to metabolic health and gut integrity. Although the evidence in humans is limited, some studies have evaluated the effect of dietary protein on health.

OVERWEIGHT AND OBESITY

Short-term high-protein diets are usually recommended for weight loss. In overweight individuals, three weeks of isocaloric supplementation with soy protein or casein in high-protein diets did not modify the microbiota composition (Beaumont et al. 2017). However, a high-protein diet induced a shift in bacterial metabolism toward amino acid degradation with different metabolite profiles depending on the protein source (Beaumont et al. 2017). Casein raised faecal concentrations of BCFA, while soy protein increased the relative concentration of tyramine, phenylacetate, and valerate (Beaumont et al. 2017). An intervention with a high-protein/low-carbohydrate diet in obese individuals increased concentrations of N-nitroso compounds, phenylacetic acid, and BCFA, and decreased the abundance of *Roseburia* and *Eubacterium*, which was concomitant with the reduction in the proportion of butyrate (Russell et al. 2011). Thus, long-term adherence to such diets may increase the risk of colonic disease due to the reduction of SCFA.

On the other hand, after 12 weeks of intervention with whey protein isocaloric snack bars, obese individuals showed a significant reduction in fat mass, lower appetite, and no changes in the microbiota (Reimer et al. 2017). Similarly, after an eight-week intervention in obese women with an energy-restricted diet combined with 20 g/day of whey protein or whey protein hydrolysate, there were no significant changes in microbiota despite significant weight loss (Sun et al. 2022). On the contrary, a calorie-restricted high-protein diet (30% of calorie intake) for eight weeks showed weight loss accompanied by high microbial diversity, depletion of *Prevotella* spp., and enrichment of *Bifidobacterium* spp. and *Akkermansia* spp. (Dong et al. 2020). Similarly, a very low-calorie, high-protein diet for 45 days, incorporating animal, vegetal, or whey protein, in patients with obesity and insulin resistance, produced a decrease in waist circumference, blood pressure, and body weight. Moreover, diets based on vegetable or whey protein have a healthier microbiota composition while animal-based protein worsened the markers of renal function (Figure 23.5) (Basciani et al. 2020). Therefore, in obesity, a calorie-restricted high-protein diet may be more effective than a standard high-protein diet in contributing to weight loss and positively modulating the composition of the gut microbiota.

TYPE 2 DIABETES MELLITUS

Evidence about protein impact on gut microbiota during T2DM is still understudied. In two independent cohorts, it has been observed that the intake of animal but not plant protein was related to T2DM (Bel Lassen et al. 2021). Although there was no significant association between alpha diversity and protein intake, the abundance of *Roseburia hominis*, one of the major butyrate producers associated with healthy metabolic status, was negatively associated with animal protein intake in T2DM. Some metabolites from proteolytic fermentation, such as indole, are involved in the secretion of glucagon-like peptide-1, a neuromodulatory molecule that promotes insulin secretion (Rodríguez-Romero et al. 2022). Similarly, indole propionic acid produced through tryptophan deamination has been associated with insulin secretion (due to its direct effect on β-cell function), negatively with CRP concentrations, and inversely related to the incidence of T2DM (Figure 23.5) (de Mello et al. 2017; Tuomainen et al. 2018). Overall, this suggests that some metabolites of proteolytic fermentation may have positive effects on insulin sensitivity, protecting against T2DM; conversely, high-protein diets are associated with increased risk.

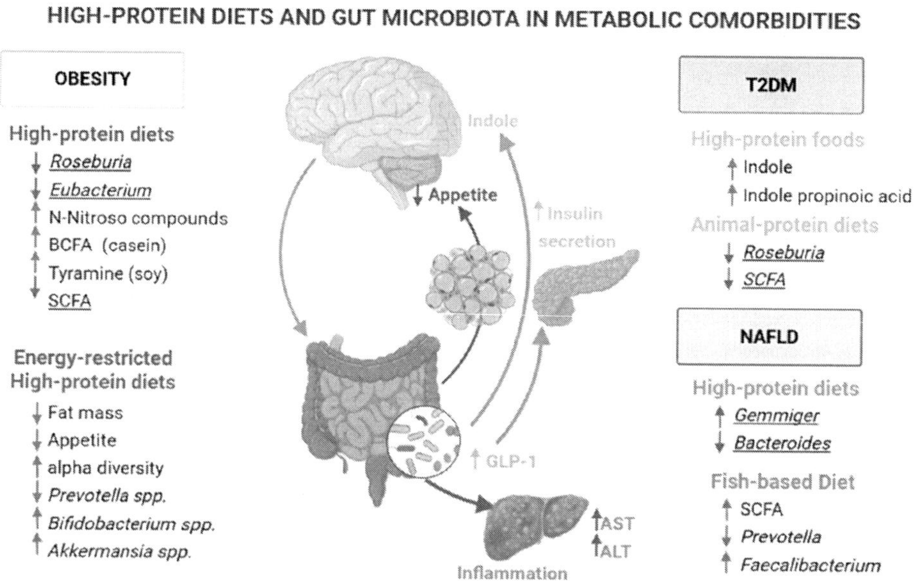

FIGURE 23.5 High-protein diets and gut microbiota in metabolic comorbidities. Summary of the main effects of high-protein diets on gut microbiota in obesity, type 2 diabetes mellitus, and non-alcoholic fatty liver disease. Underlined: harmful. AST: aspartate aminotransferase; ALT: alanine aminotransferase; BCFA: branched-chain fatty acids; GLP-1: glucagon-like peptide-1; NAFLD: non-alcoholic fatty liver disease; SCFA: short-chain fatty acids; T2DM: type 2 diabetes mellitus. Figure created with BioRender.com.

NON-ALCOHOLIC FATTY LIVER DISEASE

In non-alcoholic fatty liver disease (NAFLD) patients, the intake of high-protein diets has been positively associated with the abundance of the *Gemmiger* genus and negatively correlated with the abundance of *Bacteroides* genus. These changes in gut microbiota have been associated with increased levels of aspartate aminotransferase and alanine aminotransferase, as well as abnormal liver function (Lang et al. 2020). In these patients, an isoenergetic diet based on fish intake compared to red meat intake improves clinical outcomes associated with NAFLD, increases levels of unconjugated bile acids and SCFA, decreases the *Prevotella,* and increases the abundance of *Faecalibacterium* spp. (Figure 23.5) (He et al. 2022). Therefore, in NAFLD patients, a higher intake of fish may regulate gut microbiota and metabolites involved in liver function.

COLONIC INFLAMMATION

Gut microbiota contribute to epithelial integrity and the maintenance of intestinal homeostasis. Although dietary changes are not sufficient to produce cancer, it has been suggested that high-protein diets cause dysbiosis and inflammation, increasing the risk of inflammatory bowel disease and colorectal cancer (Jantchou et al. 2010). Some studies observed that a high-protein diet produces dysbiosis similar to that in Crohn's disease, which has been positively associated with *E. coli* abundance, elevated LPS concentration, high tryptophan metabolism from *Helicobacter*, and reduced *Lachnospiraceae* (a butyrate-producing bacteria) (Y. Zhao, Chen et al. 2022). Metabolites such as amines react with nitric oxide, resulting in nitrite compounds that could lead to chronic

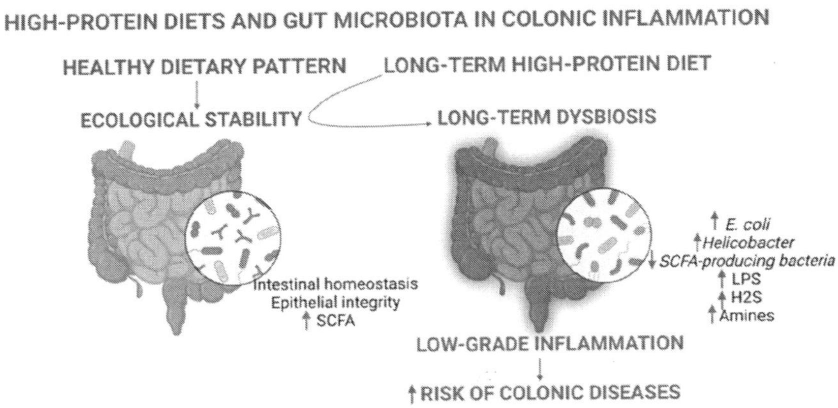

FIGURE 23.6 High-protein diets and gut microbiota in colonic inflammation. Summary of the impact of a long-term high-protein diet on ecological stability and its relationship with colonic disease risk. H_2S: hydrogen sulfur; LPS: lipopolysaccharide; SCFA: short-chain fatty acids. Figure created with BioRender.com.

inflammation and increase the risk of gastrointestinal cancer (Zhu et al. 2014). Furthermore, colorectal cancer patients showed high levels of phenylalanine, H_2S, and p-cresol, which could be predictors of genotoxicity against colonocytes (Cai et al. 2022). Therefore, prolonged adherence to high-protein diets should be avoided along with foods rich in cysteine, aspartate, and glutamate, because they increase concentrations of metabolites such as amines and H_2S, which are directly linked to colon inflammation and an increased risk of colorectal cancer (Figure 23.6).

CONCLUSION

Despite limited evidence on the effects of dietary protein on the gut microbiota and its consequences for health, it seems clear that both the source and the amount of protein can induce changes in the composition and function of the gut microbiota. However, these effects vary across the lifespan and in certain diseases; thus dietary recommendations should be accordingly adjusted. In addition, when high-protein diets are designed to assess their effects on the microbiota, they result in low-fat or low-carbohydrate diets depending on the protein sources, producing side effects due to the depletion of these nutrients.

MINI DICTIONARY OF TERMS

Alpha diversity: The average species diversity of the microbiota.

Dysbiosis: A pathological state characterised by changes in the microbiota involving an increase in harmful bacteria and a decrease in bacteria beneficial to the health of the host.

Microbiota: Living microorganisms inhabiting an environment, such as the gut microbiota.

Microbiome: The collection of the genomes of all microorganisms in the environment, including the environmental conditions, the metabolites, and the structural elements of the microorganisms.

Resilience: The capacity of a system to maintain its status and recover easily from disturbances.

Resistance: The ability of a system to remain unchanged in the presence of different stimuli.

SUMMARY POINTS

- Dietary protein can modulate gut microbiota composition and function.
- Metabolites from undigested protein fermentation by gut microbiota have beneficial and detrimental effects on host health.
- During infancy, breastfeeding modulates the gut microbiota to promote healthy growth, whereas feeding formula with a higher protein content has detrimental effects.
- Complementary feeding based on dairy products is associated with a higher abundance of *Akkermansia* spp.and a lower risk of obesity than meat.
- During infancy, healthy animal protein sources (dairy, eggs, and meat) have more beneficial effects on healthy growth by increasing short-chain fatty acid concentrations than plant proteins.
- To maintain the gut microbiota's ecological stability during adulthood, high-protein diets, meat-rich diets, and red meat-based diets should be avoided.
- In ageing, animal protein diets can increase the circulating lipopolysaccharide and worsen age-related comorbidities, while plant-based diets can help restore ecological stability.
- In obese individuals, energy-restricted high-protein diets may contribute to weight loss, reduce appetite, increase alpha diversity, and increase the abundance of *Bifidobacterium* and *Akkermansia* spp.
- Long-term high-protein diets are associated with a higher abundance of lipopolysaccharide-producing bacteria, which increase the risk of developing non-alcoholic fatty liver disease and colonic diseases such as colorectal cancer.

REFERENCES

Bartlett, Alexandria, and Manuel Kleiner. 2022. "Dietary protein and the intestinal microbiota: An understudied relationship." *iScience* 25 (11). Elsevier: 105313.

Basciani, S., E. Camajani, S. Contini, A. Persichetti, R. Risi, L. Bertoldi, L. Strigari, G. Prossomariti, M. Watanabe, S. Mariani, C. Lubrano, A. Genco, G. Spera, and L. Gnessi. 2020. "Very-low-calorie ketogenic diets with whey, vegetable, or animal protein in patients with obesity: A randomized pilot study." *The Journal of clinical endocrinology and metabolism* 105 (9). Oxford Academic: 2939–2949.

Beaumont, Martin, Kevin Joseph Portune, Nils Steuer, Annaïg Lan, Victor Cerrudo, Marc Audebert, Florent Dumont, Giulia Mancano, Nadezda Khodorova, Mireille Andriamihaja, Gheorghe Airinei, Daniel Tomé, Robert Benamouzig, Anne-Marie Davila, Sandrine Paule Claus, Yolanda Sanz, and François Blachier. 2017. "Quantity and source of dietary protein influence metabolite production by gut microbiota and rectal mucosa gene expression: A randomized, parallel, double-blind trial in overweight humans1." *The American Journal of Clinical Nutrition* 106 (4). Elsevier: 1005–19.

Bel Lassen, P., I. Attaye, S. Adriouch, M. Nicolaou, J. Aron-Wisnewsky, T. Nielsen, R. Chakaroun, E. Le Chatelier, S. Forslund, E. Belda, P. Bork, F. Bäckhed, M. Stumvoll, O. Pedersen, H. Herrema, A. K. Groen, S. J. Pinto-Sietsma, A. H. Zwinderman, M. Nieuwdorp, K. Clement, and Consortium on Behalf of Metacardis. 2021. "Protein intake, metabolic status and the gut microbiota in different ethnicities: Results from two independent cohorts." *Nutrients* 13 (9). MDPI: 3159.

Berger, B., N. Porta, F. Foata, D. Grathwohl, M. Delley, D. Moine, A. Charpagne, L. Siegwald, P. Descombes, P. Alliet, G. Puccio, P. Steenhout, A. Mercenier, and N. Sprenger. 2020. "Linking human milk oligosaccharides, infant fecal community types, and later risk to require antibiotics." *mBio* 11 (2). ASM Journals: e03196-19.

Brink, Lauren R., Kelly E. Mercer, Brian D. Piccolo, Sree V. Chintapalli, Ahmed Elolimy, Anne K. Bowlin, Katelin S. Matazel, Lindsay Pack, Sean H. Adams, Kartik Shankar, Thomas M. Badger, Aline Andres, and Laxmi Yeruva. 2020. "Neonatal diet alters fecal microbiota and metabolome profiles at different ages in infants fed breast milk or formula." *The American Journal of Clinical Nutrition* 111 (6). Elsevier: 1190–1202.

Cai, J., Z. Chen, W. Wu, Q. Lin, and Y. Liang. 2022. "High animal protein diet and gut microbiota in human health." *Critical Reviews in Food Science and Nutrition* 62 (22). Taylor & Francis: 6225–37.

Canfora, Emanuel E., Ruth C. R. Meex, Koen Venema, and Ellen E. Blaak. 2019. "Gut microbial metabolites in obesity, NAFLD and T2DM." *Nature Reviews Endocrinology* 15 (5). Springer Nature: 261–73.

Carding, S., K. Verbeke, D. T. Vipond, B. M. Corfe, and L. J. Owen. 2015. "Dysbiosis of the gut microbiota in disease." *Microbial Ecology in Health and Disease* 26. Taylor & Francis: 26191.

Cho, C. E., S. Taesuwan, O. V. Malysheva, E. Bender, N. F. Tulchinsky, J. Yan, J. L. Sutter, and M. A. Caudill. 2017. "Trimethylamine-N-oxide (TMAO) response to animal source foods varies among healthy young men and is influenced by their gut microbiota composition: A randomized controlled trial." *Molecular Nutrition & Food Research* 61 (1). Wiley Online Library: 1600324.

de Mello, Vanessa D., Jussi Paananen, Jaana Lindström, Maria A. Lankinen, Lin Shi, Johanna Kuusisto, Jussi Pihlajamäki, Seppo Auriola, Marko Lehtonen, Olov Rolandsson, Ingvar A. Bergdahl, Elise Nordin, Pirjo Ilanne-Parikka, Sirkka Keinänen-Kiukaanniemi, Rikard Landberg, Johan G. Eriksson, Jaakko Tuomilehto, Kati Hanhineva, and Matti Uusitupa. 2017. "Indolepropionic acid and novel lipid metabolites are associated with a lower risk of type 2 diabetes in the finnish diabetes prevention study." *Scientific Reports* 7 (1). Springer Nature: 46337.

Di Profio, E., V. C. Magenes, G. Fiore, M. Agostinelli, A. La Mendola, M. Acunzo, R. Francavilla, F. Indrio, A. Bosetti, E. D'Auria, E. Borghi, G. Zuccotti, and E. Verduci. 2022. "Special diets in infants and children and impact on gut microbioma." *Nutrients* 14 (15). MDPI: 3198.

Dong, T. S., K. Luu, V. Lagishetty, F. Sedighian, S. L. Woo, B. W. Dreskin, W. Katzka, C. Chang, Y. Zhou, N. Arias-Jayo, J. Yang, A. Ahdoot, Z. Li, J. R. Pisegna, and J. P. Jacobs. 2020. "A high protein calorie restriction diet alters the gut microbiome in obesity." *Nutrients* 12 (10). MDPI: 3221.

Ferrocino, I., V. Ponzo, R. Gambino, A. Zarovska, F. Leone, C. Monzeglio, I. Goitre, R. Rosato, A. Romano, G. Grassi, F. Broglio, M. Cassader, L. Cocolin, and S. Bo. 2018. "Changes in the gut microbiota composition during pregnancy in patients with gestational diabetes mellitus (GDM)." *Scientific Reports* 8 (1). Springer Nature: 12216.

Ford, Amanda L., Varuni Nagulesapillai, Amanda Piano, Jérémie Auger, Stephanie-Anne Girard, Mary Christman, Thomas A. Tompkins, and Wendy J. Dahl. 2020. "Microbiota stability and gastrointestinal tolerance in response to a high-protein diet with and without a prebiotic, probiotic, and synbiotic: A randomized, double-blind, placebo-controlled trial in older women." *Journal of the Academy of Nutrition and Dietetics* 120 (4). Elsevier: 500–16.e10.

He, K., L. L. Guo, H. Tang, X. Peng, J. Li, S. Feng, C. Bie, W. Chen, Y. Li, M. Wang, and S. Tang. 2022. "A freshwater fish-based diet alleviates liver steatosis by modulating gut microbiota and metabolites: A clinical randomized controlled trial in chinese participants with nonalcoholic fatty liver disease." *The American Journal of Gastroenterology* 117 (10). Wolters Kluwer: 1621–1631.

Hentges, D. J., B. R. Maier, G. C. Burton, M. A. Flynn, and R. K. Tsutakawa. 1977. "Effect of a high-beef diet on the fecal bacterial flora of humans." *Cancer Research* 37 (2). AACR Journals: 568–71.

Herman, D. R., N. Rhoades, J. Mercado, P. Argueta, U. Lopez, and G. E. Flores. 2020. "Dietary habits of 2- to 9-year-old American children are associated with gut microbiome composition." *Journal of the Academy of Nutrition and Dietetics* 120 (4). Elsevier: 517–34.

Hou, K., Z. X. Wu, X. Y. Chen, J. Q. Wang, D. Zhang, C. Xiao, D. Zhu, J. B. Koya, L. Wei, J. Li, and Z. S. Chen. 2022. "Microbiota in health and diseases." *Signal Transduction and Targeted Therapy* 7 (1). Springer Nature: 135.

Jantchou, P., S. Morois, F. Clavel-Chapelon, M. C. Boutron-Ruault, and F. Carbonnel. 2010. "Animal protein intake and risk of inflammatory bowel disease: The E3N prospective study." *The American Journal of Gastroenterology* 105 (10). Wolters Kluwer: 2195–201.

Jardon, K. M., E. E. Canfora, G. H. Goossens, and E. E. Blaak. 2022. "Dietary macronutrients and the gut microbiome: A precision nutrition approach to improve cardiometabolic health." *Gut* 71 (6). British Medical Journal Publishing Group: 1214–26.

Lang, S., A. Martin, F. Farowski, H. Wisplinghoff, Mjgt Vehreschild, J. Liu, M. Krawczyk, A. Nowag, A. Kretzschmar, J. Herweg, B. Schnabl, X. M. Tu, F. Lammert, T. Goeser, F. Tacke, K. Heinzer, P. Kasper, H. M. Steffen, and M. Demir. 2020. "High protein intake is associated with histological disease activity in patients with NAFLD." *Hepatoly Communications* 4 (5). Wiley Online Library: 681–95.

Laursen, M. F. 2021. "Gut microbiota development: influence of diet from infancy to toddlerhood." *Annals of Nutrition and Metabolism* 77 (Suppl.3). *Karger*: 21–34..

Linden, David R. 2013. "Hydrogen sulfide signaling in the gastrointestinal tract." *Antioxidants & Redox Signaling* 20 (5). Mary Ann Liebert: 818–30.

Liu, Yali, Yuanlong Hou, Guangji Wang, Xiao Zheng, and Haiping Hao. 2020. "Gut microbial metabolites of aromatic amino acids as signals in host–microbe interplay." *Trends in Endocrinology & Metabolism* 31 (11). Cell Press: 818–34.

Ma, J., Z. Li, W. Zhang, C. Zhang, Y. Zhang, H. Mei, N. Zhuo, H. Wang, and D. Wu. 2022. "Comparison of the gut microbiota in healthy infants with different delivery modes and feeding types: a cohort study." *Frontiers in Microbiology* 13. Frontiers Media SA: 868227.

Mandal, S., K. M. Godfrey, D. McDonald, W. V. Treuren, J. V. Bjørnholt, T. Midtvedt, B. Moen, K. Rudi, R. Knight, A. L. Brantsæter, S. D. Peddada, and M. Eggesbø. 2016. "Fat and vitamin intakes during pregnancy have stronger relations with a pro-inflammatory maternal microbiota than does carbohydrate intake." *Microbiome* 4 (1). BMC: 55.

Martino, Cameron, Amanda Hazel Dilmore, Zachary M. Burcham, Jessica L. Metcalf, Dilip Jeste, and Rob Knight. 2022. "Microbiota succession throughout life from the cradle to the grave." *Nature Reviews Microbiology* 20 (12). Springer Nature: 707–20.

Mitchell, S. M., E. J. McKenzie, C. J. Mitchell, A. M. Milan, N. Zeng, R. F. D'Souza, F. Ramzan, P. Sharma, E. Rettedal, S. O. Knowles, N. C. Roy, A. Sjödin, K. H. Wagner, J. M. O'Sullivan, and D. Cameron-Smith. 2020. "A period of 10 weeks of increased protein consumption does not alter faecal microbiota or volatile metabolites in healthy older men: A randomised controlled trial." *Journal of Nutritional Science* 9. Cambridge University Press: e25.

Mohammadkhah, A. I., E. B. Simpson, S. G. Patterson, and J. F. Ferguson. 2018. "Development of the gut microbiome in children, and lifetime implications for obesity and cardiometabolic disease." *Children (Basel)* 5 (12). MDPI: 160.

Pärnänen, K. M. M., J. Hultman, M. Markkanen, R. Satokari, S. Rautava, R. Lamendella, J. Wright, C. J. McLimans, S. L. Kelleher, and M. P. Virta. 2022. "Early-life formula feeding is associated with infant gut microbiota alterations and an increased antibiotic resistance load." *The American Journal of Clinical Nutrition* 115 (2). Elsevier: 407–21.

Prokopidis, Konstantinos, Edward Chambers, Mary Ni Lochlainn, and Oliver C. Witard. 2021. "Mechanisms linking the gut-muscle axis with muscle protein metabolism and anabolic resistance: Implications for older adults at risk of sarcopenia." *Frontiers in Physiology* 12. Frontiers Media SA: 770455.

Reimer, Raylene A., Holly J. Willis, Jasmine M. Tunnicliffe, Heekuk Park, Karen L. Madsen, and Adriana Soto-Vaca. 2017. "Inulin-type fructans and whey protein both modulate appetite but only fructans alter gut microbiota in adults with overweight/obesity: A randomized controlled trial." *Molecular Nutrition & Food Research* 61 (11). Wiley Online Library: 1700484.

Reyna, M. E., C. Petersen, D. L. Y. Dai, R. Dai, A. B. Becker, M. B. Azad, K. Miliku, D. L. Lefebvre, T. J. Moraes, P. J. Mandhane, R. C. T. Boutin, B. B. Finlay, E. Simons, A. L. Kozyrskyj, W. Lou, S. E. Turvey, and P. Subbarao. 2022. "Longitudinal body mass index trajectories at preschool age: children with rapid growth have differential composition of the gut microbiota in the first year of life." *International Journal of Obesity (Lond)* 46 (7). Springer Nature: 1351–58.

Rodríguez-Romero, José de Jesús, Alba Cecilia Durán-Castañeda, Alicia Paulina Cárdenas-Castro, Jorge Alberto Sánchez-Burgos, Victor Manuel Zamora-Gasga, and Sonia Guadalupe Sáyago-Ayerdi. 2022. "What we know about protein gut metabolites: Implications and insights for human health and diseases." *Food Chemistry: X* 13. Elsevier: 100195.

Russell, Wendy R., Silvia W. Gratz, Sylvia H. Duncan, Grietje Holtrop, Jennifer Ince, Lorraine Scobbie, Garry Duncan, Alexandra M. Johnstone, Gerald E. Lobley, R. John Wallace, Garry G. Duthie, and Harry J. Flint. 2011. "High-protein, reduced-carbohydrate weight-loss diets promote metabolite profiles likely to be detrimental to colonic health." *The American Journal of Clinical Nutrition* 93 (5). Elsevier: 1062–72.

Schmedes, M., A. D. Brejnrod, E. K. Aadland, P. Kiilerich, K. Kristiansen, H. Jacques, C. Lavigne, I. E. Graff, Ø Eng, A. Holthe, G. Mellgren, J. F. Young, U. K. Sundekilde, B. Liaset, and H. C. Bertram. 2019. "The effect of lean-seafood and non-seafood diets on fecal metabolites and gut microbiome: Results from a randomized crossover intervention study." *Molecular Nutrition & Food Research* 63 (1). Wiley Online Library: e1700976.

Selma-Royo, M., I. García-Mantrana, M. Calatayud, A. Parra-Llorca, C. Martínez-Costa, and M. C. Collado. 2021. "Maternal diet during pregnancy and intestinal markers are associated with early gut microbiota." *European Journal of Nutrition* 60 (3). Springer Nature: 1429–42.

Sender, R., S. Fuchs, and R. Milo. 2016. "Are we really vastly outnumbered? Revisiting the ratio of bacterial to host cells in humans." *Cell* 164 (3): 337–40.

Smith-Brown, P., M. Morrison, L. Krause, and P. S. Davies. 2016. "Dairy and plant based food intakes are associated with altered faecal microbiota in 2 to 3 year old Australian children." *Scientific Reports* 6. Springer Nature: 32385.

Stull, Valerie J., Elijah Finer, Rachel S. Bergmans, Hallie P. Febvre, Colin Longhurst, Daniel K. Manter, Jonathan A. Patz, and Tiffany L. Weir. 2018. "Impact of edible cricket consumption on gut microbiota in healthy adults, a double-blind, randomized crossover trial." *Scientific Reports* 8 (1). Springer Nature: 10762.

Sun, Y., C. Ling, L. Liu, J. Zhang, J. Wang, X. Tong, K. Hidayat, M. Chen, X. Chen, H. Zhou, J. Xu, L. Qin, W. Zhu, and J. Yang. 2022. "Effects of whey protein or its hydrolysate supplements combined with an energy-restricted diet on weight loss: A randomized controlled trial in older women." *Nutrients* 14 (21). MDPI: 4540.

Suta, S., A. Surawit, P. Mongkolsucharitkul, B. Pinsawas, T. Manosan, S. Ophakas, T. Pongkunakorn, S. Pumeiam, K. Sranacharoenpong, S. Sutheeworapong, P. Poungsombat, S. Khoomrung, P. Akarasereenont, I. Thaipisuttikul, B. Suktitipat, and K. Mayurasakorn. 2023. "Prolonged egg supplement advances growing child's growth and gut microbiota." *Nutrients* 15 (5). MDPI: 1143.

Tang, Minghua, Cheng Ma, Eileen M. Weinheimer-Haus, Charles E. Robertson, Jennifer M. Kofonow, Lillian M. Berman, Akbar Waljee, Ji Zhu, Daniel N. Frank, and Nancy F. Krebs. 2023. "Different gut microbiota in U.S. formula-fed infants consuming a meat vs. dairy-based complementary foods: A randomized controlled trial." *Frontiers in Nutrition* 9. Frontiers Media SA: 1063518.

Teigen, Levi, Prince P. Mathai, Sharon Lopez, Michael Matson, Baila Elkin, Daria Kozysa, Amanda J. Kabage, Matthew Hamilton, Byron P. Vaughn, and Michael J. Sadowsky. 2022. "Differential hydrogen sulfide production by a human cohort in response to animal-and plant-based diet interventions." *Clinical Nutrition* 41 (6). Elsevier: 1153–162.

Tuomainen, Marjo, Jaana Lindström, Marko Lehtonen, Seppo Auriola, Jussi Pihlajamäki, Markku Peltonen, Jaakko Tuomilehto, Matti Uusitupa, Vanessa D. de Mello, and Kati Hanhineva. 2018. "Associations of serum indolepropionic acid, a gut microbiota metabolite, with type 2 diabetes and low-grade inflammation in high-risk individuals." *Nutrition & Diabetes* 8 (1). Springer Nature: 35.

van Soest, A. P. M., G. D. A. Hermes, A. A. M. Berendsen, O. van de Rest, E. G. Zoetendal, S. Fuentes, A. Santoro, C. Franceschi, Lcpgm de Groot, and W. M. de Vos. 2020. "Associations between pro- and anti-inflammatory gastro-intestinal microbiota, diet, and cognitive functioning in dutch healthy older adults: The NU-AGE study." *Nutrients* 12 (11). MDPI: 3471.

Wang, M., Z. Wang, Y. Lee, H. T. M. Lai, M. C. de Oliveira Otto, R. N. Lemaitre, A. Fretts, N. Sotoodehnia, M. Budoff, J. A. DiDonato, B. McKnight, W. H. W. Tang, B. M. Psaty, D. S. Siscovick, S. L. Hazen, and D. Mozaffarian. 2022. "Dietary meat, trimethylamine N-oxide-related metabolites, and incident cardiovascular disease among older adults: The cardiovascular health study." *Arteriosclerosis, Thrombosis, and Vascular Biology* 42 (9): e273–88.

Wang, Y., S. R. Lindemann, T. L. Cross, M. Tang, C. M. Clark, and W. W. Campbell. 2023. "Effects of adding lean red meat to a u.s.-style healthy vegetarian dietary pattern on gut microbiota and cardiovascular risk factors in young adults: A crossover randomized controlled trial." *The Journal of Nutrition* 153 (5). Elsevier: 1439–52.

Wu, G. D., J. Chen, C. Hoffmann, K. Bittinger, Y. Y. Chen, S. A. Keilbaugh, M. Bewtra, D. Knights, W. A. Walters, R. Knight, R. Sinha, E. Gilroy, K. Gupta, R. Baldassano, L. Nessel, H. Li, F. D. Bushman, and J. D. Lewis. 2011. "Linking long-term dietary patterns with gut microbial enterotypes." *Science* 334 (6052): 105–8.

Yehia, N. A., K. Z. H. Lai, Z. Semnani-Azad, S. Blanco Mejia, R. P. Bazinet, J. L. Beaudry, and A. J. Hanley. 2023. "Association of branched chain fatty acids with cardiometabolic disorders in humans: A systematic review." *Nutrition Reviews* 81 (2). Oxford Academic: 180–90.

Zhao, D., K. Shan, Y. Xie, G. Zhang, Q. An, X. Yu, G. Zhou, and C. Li. 2022. "Body weight index indicates the responses of the fecal microbiota, metabolome and proteome to beef/chicken-based diet alterations in Chinese volunteers." *NPJ Biofilms and Microbiomes* 8 (1). Springer Nature: 56.

Zhao, Min'an, Jiayi Chu, Shiyao Feng, Chuanhao Guo, Baigong Xue, Kan He, and Lisha Li. 2023. "Immunological mechanisms of inflammatory diseases caused by gut microbiota dysbiosis: A review." *Biomedicine & Pharmacotherapy* 164. Elservier: 114985.

Zhao, Yiming, Lulu Chen, Liyu Chen, Jing Huang, Shuijiao Chen, and Zheng Yu. 2022. "Exploration of the potential relationship between gut microbiota remodeling under the influence of high-protein diet and crohn's disease." *Frontiers in Microbiology* 13. Frontiers Media: 831176.

Zhu, Y., P. P. Wang, J. Zhao, R. Green, Z. Sun, B. Roebothan, J. Squires, S. Buehler, E. Dicks, J. Zhao, M. Cotterchio, P. T. Campbell, M. Jain, P. S. Parfrey, and J. R. McLaughlin. 2014. "Dietary N-nitroso compounds and risk of colorectal cancer: a case-control study in Newfoundland and Labrador and Ontario, Canada." *British Journal of Nutrition* 111 (6). Cambridge University Press: 1109–17.

24 How Protein Deficiency Impacts Immunity and Infection Response

The Example of Helminth Parasite Infection

Cecilia C. Vila, María P. Saracino, Anabel N. Pallaro, and Pablo C. Baldi

ABBREVIATIONS

ADCC	Antibody-dependent cellular cytotoxicity
AW	Adult worms
IL	Interleukin
LT and LB	T and B lymphocytes
ML	Muscle larvae
NBL	Newborn larvae
PD	Protein deficiency
PEM	Protein-energy malnutrition
RELM-β	Resistin-like molecule-beta
Th2	T helper 2
WHO	World Health Organization

INTRODUCTION

Helminth infections, which include species from *Nematoda*, *Cestodes*, *Trematodes*, and *Acanthocephala* phyla, affect 1.5 billion people worldwide (Wright et al. 2018). These parasitic helminths typically have complex life cycles, with different stages and hosts. In most cases, these infections are chronic in the mammalian host, with a type 2 immune response being the dominant immune response generated (Allen and Maizels 2011).

Malnutrition is a serious public health problem affecting millions of people worldwide, but it is observed most frequently in non-industrialized countries among children under five years old (Ibrahim et al. 2017). According to the World Health Organization (WHO), malnutrition refers to deficiencies, excesses, or imbalances in a person's intake of energy and/or nutrients (WHO 2021). Alarmingly, not only did 735 million people in the world face hunger in 2022, but 2.4 billion people did not have access to sufficient food and a healthy diet. Children under five years of age continue to suffer from stunting (148 million), wasting (45 million), and overweight conditions (37 million). Child wasting is a life-threatening condition caused by insufficient nutrient intake, poor nutrient absorption, and/or frequent or prolonged illness. Consequently, affected children suffer from weakened immunity and a higher risk of mortality (FAO et al. 2023).

 DOI: 10.1201/b23328-24

Protein malnutrition not only increases the morbidity of helminth-infected individuals with higher parasitic load and prolonged infections but might also contribute to mortality in these patients (Bhutta et al. 2017). Research has shown that malnutrition alters cellular and humoral immunity at both the systemic and mucosal levels; meanwhile, episodes of infection enhance malnutrition through anorexia, reduced nutrient absorption, nutrient losses, and nutrient utilization for the inflammatory response and tissue repair (Jones et al. 2014).

Chronic malnutrition is associated with an increased risk of infectious disease morbidity and mortality (Caulfield et al. 2004). Both malnutrition and gastrointestinal helminth infections are associated with chronic diseases (Pelletier 1995; Solomons 1993). Given these considerations, this chapter analyzes the effects of protein malnutrition on the anti-parasitic immune response during helminth infections.

MALNUTRITION AND IMMUNE RESPONSE

Several studies have shown that the relationship between infection and malnutrition is bidirectional, with one enhancing the other (Brown 2003).

The initiation of both innate and adaptive immune responses involves the activation and proliferation of immune cells, as well as the synthesis of different molecules, all of which consume anabolic energy. Consequently, the host's nutritional status influences the magnitude of the immune response against infections (Schaible and Kaufmann 2007). Malnutrition is commonly associated with deficiencies in the immune response, particularly in cellular immunity, phagocyte function, cytokine production, antibody response, and the complement system (Ibrahim et al. 2017; Chandra 1991; 1990). The integrity of the gastrointestinal mucosa is commonly impaired in malnutrition, leading to an increased susceptibility to pathogens (Martinsen et al. 2005; Peterson and Artis 2014).

Protein Malnutrition

Experimental animal models have contributed to understanding the effects of malnutrition on the immune system. Studies have shown that rats with protein deficiency (PD) induced by a protein-deficient diet experienced bone marrow atrophy with gelatinous degeneration, expansion of the extracellular matrix, a loss in markers of cell proliferation, and suppression of the cell cycle progression of hematopoietic progenitor cells (Nakajima et al. 2014; Xavier et al. 2007). Administration of a PD diet to Wistar rats from weaning resulted in lower proliferation and cellular maturation in the thymus. Protein deficiency generated a modification of the histoarchitecture of the intestinal villi with a decrease in the subpopulations of T and B lymphocytes (LT and LB, respectively) in the intestine (Vidueiros et al. 2008). Another study showed that PD mice were anemic and leukopenic, with reduced numbers of neutrophils, lymphocytes, and monocytes (Cunha et al. 2013). In protein-malnourished individuals, the innate immune response is affected due to a reduced number of macrophages and a modified production of proinflammatory cytokines (Santos et al. 2017).

Protein-energy malnutrition (PEM) generates various alterations in the immune system, including atrophy of lymphoid organs, a reduced number of T cells, altered proportions of subsets of T cells, decreased activity of NK cells, and lower cytokine production (Rodríguez et al. 2011; Scrimshaw and SanGiovanni 1997). PEM significantly compromises mucosal epithelial barriers (Rodríguez et al. 2011). In the gastrointestinal tract, flattened atrophied microvilli, reduced lymphocyte counts in Peyer's patches, and reduced IgA secretion are observed (Lima et al. 2005). The ability of lymphocytes to respond adequately to cytokines is impaired in severely malnourished children (Keusch 2003; Rodríguez et al. 2005).

IMMUNE RESPONSE AGAINST HELMINTHS

In general terms, the immune response against helminth infections has a type 2 profile, characterized by innate immune cells such as innate lymphoid cells, basophils, eosinophils, and/or mast cells, which may be effector cells and/or important sources of type 2 cytokines (Gause et al. 2020). The main interleukins (IL) involved in this immune response are IL-4, IL-5, and IL-13, and the adaptive immune cells include CD4$^+$ T helper 2 (Th2) cells and B cells (Allen and Maizels 2011) (Figure 24.1). Although most helminth parasites induce type 2 responses, the specific effector pathway mediating protection varies between different parasites, life cycle stages, and the site of infection.

When intestinal helminths invade and migrate through tissues, they cause considerable damage. The death of host cells is associated with the release of damage-associated molecular patterns, which trigger signaling pathways that contribute to the initiation of the type 2 response. IL-4 and IL-13 induce the proliferation of goblet cells, which secrete mucus and resistin-like molecule-beta (RELM-β). IL-4 induces IgE production; then, IgE-antigen immune complexes bind to high-affinity IgE receptors on basophils and mast cells, causing the degranulation and release of several proinflammatory mediators (Allen and Maizels 2011). IL-5 is responsible for the activation and recruitment of eosinophils, and IL-9 causes mast cell activation. Collectively, immune cells and the secreted cytokines coordinate parasite expulsion by increasing fluid and mucus production, encapsulation and barrier formation, epithelial cell turnover, smooth muscle cell contraction, and production of anthelmintic effector molecules, such as RELM-β (Allen and Maizels 2011). Type 2 cytokines also stimulate the repair of damaged tissues and have anti-inflammatory actions (Allen and Maizels 2011).

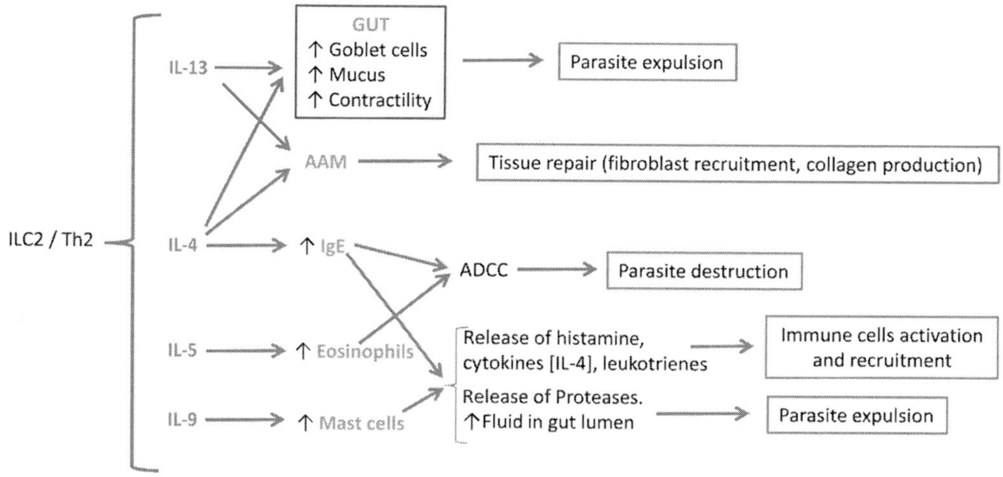

FIGURE 24.1 Main effectors of type 2 responses against helminths. Type 2 cytokines are produced early in the infection by type 2 innate lymphoid cells (ILC2) and later by T helper 2 lymphocytes (Th2). They mediate several processes that either limit helminth infection or repair the tissue damage produced by these pathogens. IL-4 and IL-13 stimulate gut cells to produce several responses that aid in parasite expulsion. They also induce the differentiation of macrophages to the M2 or alternative activated macrophages (AAM) profile that mediates tissue repair. In addition, IL-4 induces the production of IgE by antigen-specific B cells, and these antibodies bind to specific receptors on the surface of eosinophils and mast cells. Helminth recognition by these antibodies induces the release of the granule content of cells to mediate either antibody-dependent cell-mediated cytotoxicity (ADCC, eosinophils) or the release of proinflammatory mediators (mast cells). In addition, mast cells also secrete proteases that increase the permeability of gut epithelium and the discharge of fluids into the intestinal lumen, thus contributing to parasite expulsion.

PROTEIN MALNUTRITION AND HELMINTH INFECTIONS

In this chapter, we analyze the effects of protein malnutrition on the following parasitic infections: *Ascaris* spp., *Echinococcus* spp., hookworms, *Schistosoma* spp., *Trichinella* spp., and *Trichuris* spp. The main findings of the relevant studies are summarized in Table 24.1.

TABLE 24.1
Effects of Different Types of Malnutrition during Helminth Infections

Helminth	Type of Malnutrition	Immune/Hematological Parameters	Parasitic Disease Parameters	References
	Protein deficiency		↓ worm weight	Forsum et al. 1981
Ascaris spp.	Malnutrition in humans	↑ serum IL-4 ↑ total IgE ↓ specific IgE ↓ CD3⁺ and CD4⁺ cells ↓ IL-2 receptor (Th) ↓ memory cells	↑ parasite load	Hagel et al. 1995 Hagel et al. 2003 Thein-Hlaing et al. 1991
Echinococcus spp.	Malnutrition in humans	↓ hemoglobin	↑ lesion size	Wang et al. 2020 Liang et al. 2020
Hookworms	Protein deficiency		↑ worm burden ↓ egg production ↓ adult worm expulsion	Pacanaro et al. 2014 Bolin et al. 1977
Schistosoma spp.	Protein deficiency	↓ specific sera antibodies ↓ liver fibrosis ↓ cytokines	↓ size of worms ↓ parasite fecundity ↓ schistosome eggs ↓ size of granulomas	Coutinho 2004 Maciel et al. 2021
	Protein-energy malnutrition		↓ size of worms ↓ schistosome eggs ↓ pathologic consequences	Knauft and Warren 1969 Oliveira et al. 2003
Trichinella spp.	Protein deficiency	↓ eosinophilia ↓ total and specific antibodies ↓ gut and lung mast and goblet cells ↓ lung neutrophils and eosinophils	↑ adult worms and muscle larvae ↑ adult worm fecundity index ↓ NBL mortality ↑ host mortality	Gbakima 1993 Vila et al. 2019 Vila et al. 2021
Trichuris spp.	Protein deficiency	↑ specific antibodies ↓ eosinophil counts	↑ adult worms and eggs ↑ adult worms ↓ adult worm length	Michael and Bundy 1992 Pedersen et al. 2002
	Protein-energy malnutrition		↓ adult worm length	Pedersen et al. 2002

Ascaris Infection

Hosts contract *Ascaris* spp. infection via ingestion of embryonated eggs. After a complex cycle, the released larvae reach the small intestine, where they mature into adult worms (Figure 24.2). *Ascaris* spp. infection affects the nutrition of the host. Research has shown that ascariasis can cause erosion in intestinal villi, leading to malabsorption of different nutrients (Ojha et al. 2014). This leads to growth retardation, undernutrition, impaired cognitive functions, and low educational achievements in children (Dickson et al. 2000).

Several studies have evaluated the immune response in malnourished persons with ascariasis. Hagel et al. found that malnourished children (below the 10th percentile of height and weight for age) had higher serum levels of IL-4 and total IgE but lower levels of specific anti-*Ascaris* IgE

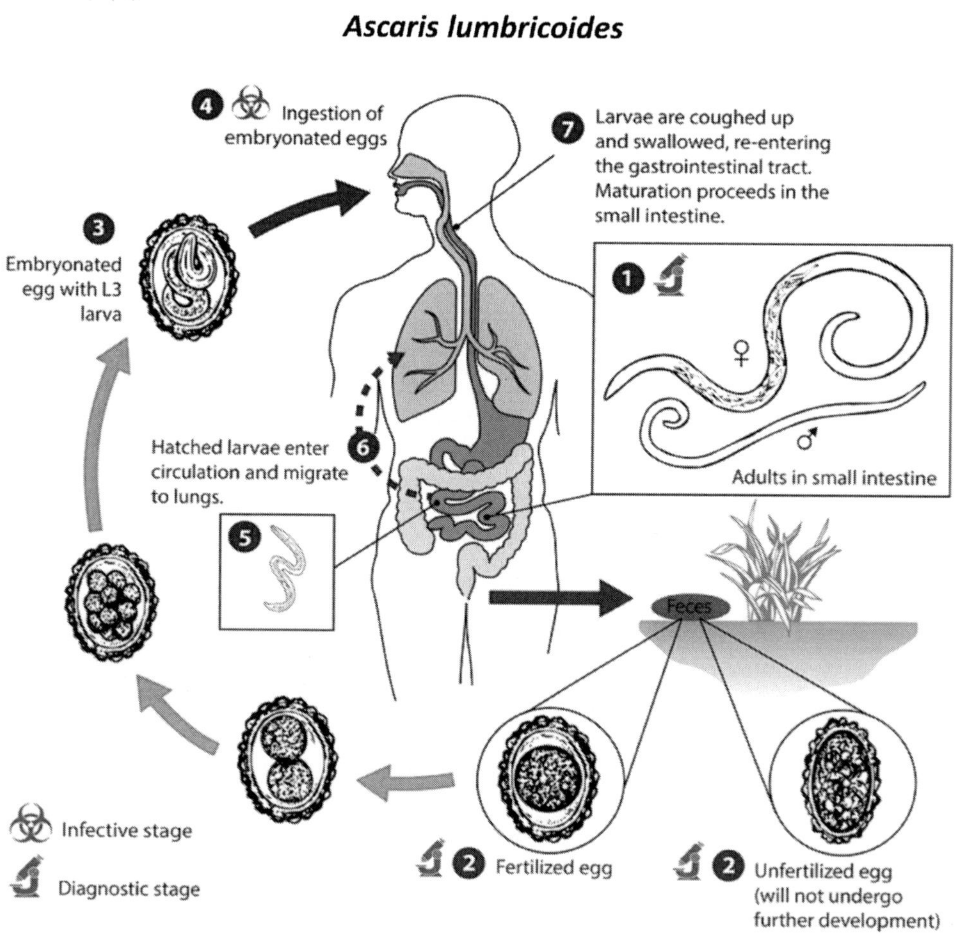

FIGURE 24.2 Life cycle of *Ascaris lumbricoides*. Unembryonated eggs released with the feces of infected patients mature in the environment to contain infective L3 larvae. After accidental ingestion of the eggs by a new host, the larvae hatch, invade the intestinal mucosa, and reach the systemic circulation. Mature pulmonary larvae breach the alveolar walls, reach the throat through the bronchial tree, and are swallowed. Adult worms develop in the small intestine. This image is reproduced from the Centers for Disease Control and Prevention (CDC) webpage (https://www.cdc.gov/parasites/). Use of this image does not imply endorsement by CDC to the contents of this chapter.

antibodies compared to well-nourished children (Hagel et al. 1995). Specific IgE is critical for the immune response against helminths, but the polyclonal IgE response stimulated by these parasites may reduce its effectiveness, an effect that seems to be exacerbated by malnutrition (Ayelign et al. 2020). In another study on malnourished children, a lower proportion of $CD3^+$ and $CD4^+$ cells, diminished expression of the IL-2 receptor by helper T cells, and lower proportion of memory T cells ($CD4^+CD45RO^+$) were also found (Hagel et al. 2003). These works show the negative impact of malnutrition on the cellular and humoral immune response to *Ascaris* spp.

After 12 months of regular anthelmintic treatment (oxantel/pyrantel 20 mg/kg), the levels of total IgE and IL-4 decreased in well-nourished children (Hagel et al. 1995). Specific anti-*Ascaris* IgE levels increased after anthelmintic treatment in well-nourished children, but no change was detected in the malnourished group. Researchers also observed significantly inverse correlations between nutritional status (Z-score: height for age and weight for age) and IgE or IL-4 levels.

Another study on children with ascariasis found that well-nourished children and those with mild malnutrition were more frequent among those with the lowest mean worm burden but constituted a minority among children with the highest mean worm burden. The opposite effect was found in children with a severe degree of malnutrition (Thein-Hlaing et al. 1991).

Studies on the effects of malnutrition on *Ascaris* infection have also been performed in animal models. Forsum et al. analyzed the effects of *A. suum* infection in young pigs receiving low-protein diets (7.8% vs. 17% protein). For both diets, *Ascaris*-infected pigs showed a reduction in growth rate. Surprisingly, the weight of worms was lower in pigs with the higher parasite burden in the PD group, showing that protein malnutrition also had a detrimental impact on helminth development (Forsum et al. 1981).

Overall, the research about *Ascaris* spp. and malnutrition in humans indicates that malnutrition has a negative effect on the specific cellular and humoral response. Treatments with anthelmintics generate benefits not only by reducing parasite burden but also at the nutritional level. The better results obtained in the mild-malnutrition group (with the lowest mean worm burden) also indicate that treatment should be accompanied by a nutritional recovery diet to enhance its effects.

Echinococcus Infection

Human echinococcosis is a parasitic disease caused by cestodes of the genus *Echinococcus*, affecting over a million people. The two most important forms in humans are cystic echinococcosis (hydatidosis) and alveolar echinococcosis. Humans become infected via ingestion of eggs, which hatch and develop into metacestodes (fluid-filled cysts, or hydatids) in tissues, particularly the liver and lungs. Among patients hospitalized with echinococcosis, Wang et al. found lower values of weight, body mass index, albumin, and hemoglobin and higher lesion size in malnourished patients compared to well-nourished ones (Wang et al. 2020). Liang et al. studied the preoperative nutritional status of patients (well-nourished or malnourished) with hepatic alveolar echinococcosis. They found that the European Nutrition Risk Screening 2002 (NRS 2002; a score higher than 3 indicates malnutrition risk) was negatively correlated with body weight, body mass index, and albumin and positively correlated with the diameter of the lesion. In the well-nourished group, hospitalization time was shorter, with fewer postoperative complications (Liang et al. 2020).

Hookworm Infection

Hookworms (*Ancylostoma duodenale*, *Necator americanus*) are soil-transmitted nematode parasites that can reside for many years in the small intestine of their human hosts. These helminths feed on blood and can cause iron deficiency anemia (Loukas et al. 2016).

Nippostrongylus brasiliensis is a rat nematode with a similar life cycle, frequently used as a model of human hookworm infections.

Pacanaro et al. studied the consequences of PD and hookworm infection in hamsters. The researchers fed hamsters a standard diet or an isocaloric PD diet for 50 days and then infected them with hookworm larvae. Animals fed the PD diet presented a more severe infection, with a higher number of worms recovered in the intestine. Nonetheless, egg production was reduced in malnourished hamsters, suggesting that protein restriction was a determinant in egg formation (Pacanaro et al. 2014).

Bolin et al. analyzed the effect of iron deficiency and PD (10% vs. 30% protein) in rats infected with *N. brasiliensis* six weeks after diet treatment. They found a delay in the expulsion of *N. brasiliensis* from the small intestines of animals with iron deficiency, PD, or both (Bolin et al. 1977). In another study, Duncombe et al. analyzed the effect of iron deficiency and PD (the same model described previously) on the efficacy of benzimidazole anthelmintics in rats infected with *N. brasiliensis*. They found that the anthelmintics mebendazole and fenbendazole were less effective in eradicating parasites when animals had iron deficiency and PD (Duncombe et al. 1977). These studies showed how iron deficiency and PD delay the expulsion of worms during *N. brasiliensis* infection. The lower effectiveness of anthelmintics found in animals with iron deficiency and PD suggests that a protein-supplemented diet would be recommended in combination with the anthelmintic treatment.

Schistosoma Infection

Schistosoma spp. are trematodes. Infection in mammalian hosts occurs through skin penetration by cercariae, which transform into schistosomula that reach the liver, where adult worms develop (Figure 24.3). Worms finally lodge within mesenteric vessels or vesical veins, from where the gravid females release eggs. Many bloodborne eggs become lodged within vascularized tissues and organs, which is the main cause of pathology during schistosomiasis (Burke et al. 2009).

During the first week of murine *S. mansoni* infection, a Th1 immune response is observed. With the onset of egg deposition, a strong Th2 immune response emerges. Finally, when the infection becomes chronic and egg production continues, a general down-modulation of immune reactivity takes place, leading to a diminished Th2 response together with a smaller size of newly formed granulomas (Pearce and MacDonald 2002). The eggs induce a granulomatous host immune response largely characterized by lymphocytes (which mainly produce Th2 cytokines) eosinophils, and alternatively activated macrophages (Pearce and MacDonald 2002; Fairfax et al. 2012). This response can lead to severe hepatic fibrosis and portal hypertension.

There is evidence that host-parasite responses in schistosomiasis may be modified by the nutritional status of the host. Coutinho et al. evaluated the immune response in mice with PD during schistosomiasis infection and reinfections. One of the studies analyzed two different diets – a PD diet (7–8% protein) and control diet (22% protein) – and evaluated the effect of shifting from a deficient (16 weeks) to a balanced diet (8 weeks) and vice versa in mice during infection with *S. mansoni*. They found that in the well-nourished controls and "late undernourished" mice, the volume density and the mean size of periovular granulomas were higher in comparison with the other groups. When repeated infections were evaluated, liver and spleen/body weight ratios from undernourished mice were always lower than those from well-nourished reinfected controls. No differences were found between the diet groups regarding the lymphoproliferative response *in vitro* (spleen cells stimulated with concanavalin-A or *Schistosoma* soluble egg antigen (SEA)) and production of IFN-γ and IL-5. However, serum levels of SEA-specific antibodies in PD mice were lower than those detected in control animals (Coutinho 2004). These results showed how the immune response against schistosomiasis is affected by PD at the cellular and humoral levels.

Maciel et al. evaluated the effect of protein and fat deficiency during schistosomiasis. They divided mice into control (14% protein and 10% lipids), PD (3% or 8% protein), and low-fat groups. The 3% PD diet was the only one capable of establishing malnutrition in mice. Mice fed this diet showed a reduction in body weight and serum albumin levels before infection, decreased levels of

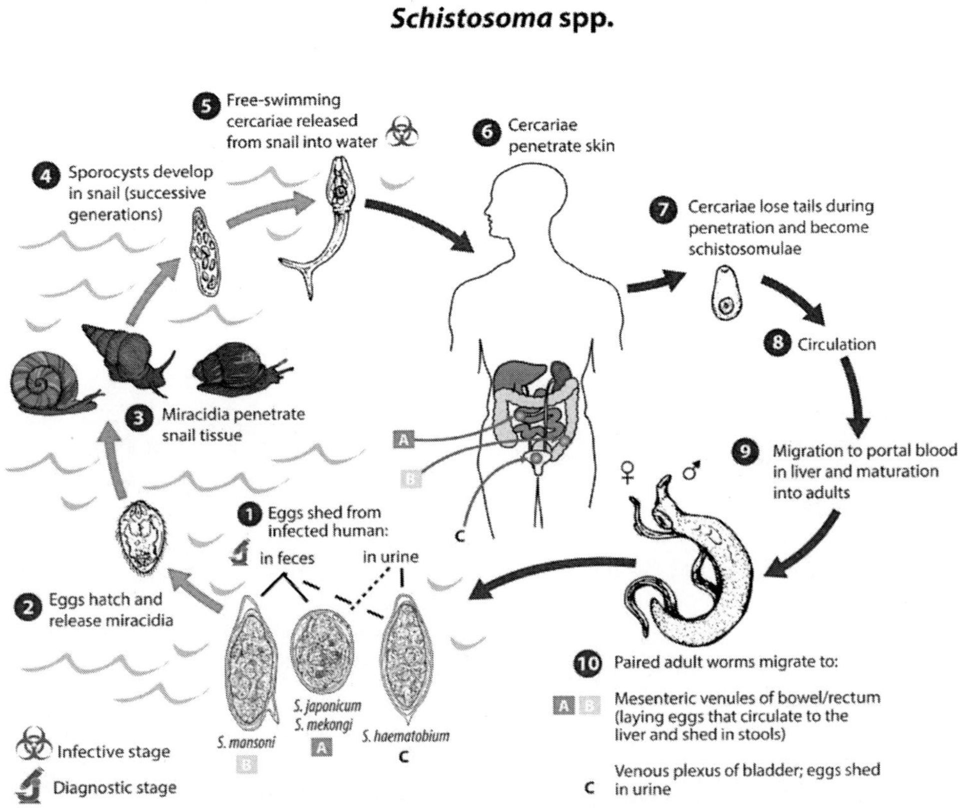

FIGURE 24.3 Life cycle of *Schistosoma* spp. Schistosoma eggs eliminated with feces or urine hatch and release miracidia, which penetrate specific snail hosts. Infective cercariae released from the snail penetrate the skin of the human host and transform into schistosomulae. The latter mature in the liver and exit via the portal vein system. Adult worms copulate and reside in the mesenteric veins (S. japonicum, S. mansoni) or the venous plexus of the bladder (S. haematobium). The eggs deposited by the females reach the lumen of the intestine and of the bladder and ureters, and are eliminated with feces or urine, respectively. Reproduced free of charge from the Centers for Disease Control and Prevention (CDC) webpage (https://www.cdc.gov/parasites/). Use of this image does not imply endorsement by CDC to the contents of this chapter.

all biochemical parameters, decreased numbers of schistosome eggs trapped in intestines, impaired parasite fecundity, a delay in granuloma development with a smaller granuloma area, and reduced levels of IL-4 and IFN-γ in the liver (Maciel et al. 2021).

Knauft et al. analyzed the effect of protein (20%, 12%, 8%, and 4% casein) and energy restriction in Swiss albino female mice during *Schistosoma* infection. They observed a decrease in the size of worms from the mice fed the most energy-restricted and PD diets. Additionally, they found a decrease in the number of eggs per worm pair in the liver in all the animals on deficient diets (Knauft and Warren 1969). Another study analyzed morphological parameters in male and female adult worms recovered from undernourished albino mice compared to parasites recovered from well-fed infected mice. Undernourished animals were fed a multi-deficient, essentially PD diet (similar to that consumed in northeast Brazil) and compared to well-fed control mice. Significant differences were detected in body length and width of parasites, parameters of suckers, uterine egg width, ovary length, and area of testicular lobes, with lower values for specimens from undernourished mice.

These findings suggest that the nutritional status of the host negatively influences *S. mansoni* adult worms, likely due to the unavailability of essential nutrients for the parasites (Oliveira et al. 2003).

In summary, PD during schistosomiasis results in a decrease in the number of schistosome eggs trapped in intestines, impaired parasite fecundity, delayed granuloma development resulting in smaller granuloma areas, and alterations in immune responses at both cellular and humoral levels.

Trichinellosis

The entire life cycle of the nematode *Trichinella spiralis* occurs within one host (Figure 24.4). Infection begins when raw skeletal muscle containing muscle larvae (ML) is ingested. In the stomach, ML move into the small intestine, where they invade epithelial cells. ML molt four times, becoming adult worms (AW), which then mate. Female worms release newborn larvae (NBL), which enter the lymphatics and eventually reach the bloodstream. NBL migrate to skeletal muscle cells, where they develop into ML. Clearance of AW from the intestine is mediated by a potent type 2 immune response (Saracino et al. 2020). NBL are killed by antibody-dependent cellular cytotoxicity (ADCC) at both systemic and tissue levels (Venturiello et al., 1993; Falduto et al. 2015).

Several studies have evaluated the effect of PD in rats during trichinellosis. Gbakima examined how changes in host dietary protein concentration affect the propagation and expulsion of *T. spiralis* adult worms as well as larva-induced inflammatory responses in tongue muscle tissue. Five-week-old CD1 female mice were fed PD (4%) or normal (26%) protein diets for six weeks before being infected with 200 *T. spiralis* larvae. The PD diet led to diminished expulsion of *T. spiralis* from the intestinal tract. Counts of adult worms (days 10 and 14 post-infection (p.i.)) and larvae (day 35 p.i.) were higher in mice fed the PD diet compared to those on normal protein diets. The PD diet also induced leucopenia, suppressed eosinophilia and its related allergic response, reduced food intake during infection, delayed initiation of inflammation, weakened inflammatory responses overall, and prolonged the chronic stages of infection in muscle tissue. In contrast, well-nourished animals exhibited severe yet effective inflammatory responses (Gbakima 1993).

A study by Vila et al. also examined how PD affects mucosal and systemic immunity during *T. spiralis* infection. Two groups of weaning Wistar rats received a PD diet (6.5% casein), while two other groups received a control diet (20% casein). After ten days, one group of each diet was infected with 900 ML. Food and protein intakes were lower in infected PD rats compared to controls; body weight was also lower in infected PD rats than in non-infected PD rats in their dietary group. Blood eosinophil counts, as well as total and specific antibodies, were reduced in the infected PD rats compared to controls. Additionally, gut and lung cell populations showed fewer mast cells and goblet cells in infected PD rats compared to infected controls. The persistence of AW in the intestines and migrant larvae in the lungs lasted longer in PD rats than in controls; AW fecundity index scores were higher among PD animals than in controls. Muscular parasite burdens were also elevated among PD rats compared to those on control diets (Vila et al. 2019).

These findings show the effect of PD on mortality rates, parasite burden levels, and immune responses in *T. spiralis*-infected animals. In particular, the lower immune response against AW in the intestine (mast cells, goblet cells, and antibodies) and the higher AW fecundity index may help explain the higher parasite burden in muscle.

Vila et al. later focused on the immune response to NBL. They conducted *in vitro* studies using lung cells and soluble extracts from the four groups of rats mentioned above (infected or not, PD diet or control), and evaluated ADCC against NBL. The researchers found that the infected PD group had lower titers of anti-NBL antibodies in lung tissue extract than the control group. In ADCC assays using non-infected control cells, NBL mortality percentage was lower with lung tissue extract from the infected PD group than from the control group In assays using control cytotoxic sera, ADCC was exerted by lung cell suspensions from the control group at all days p.i., but only by lung cell suspension from 13 days p.i. in the PD group. ADCC assays combining lung tissue extract and lung cell suspension from the same group showed a lower response for the PD

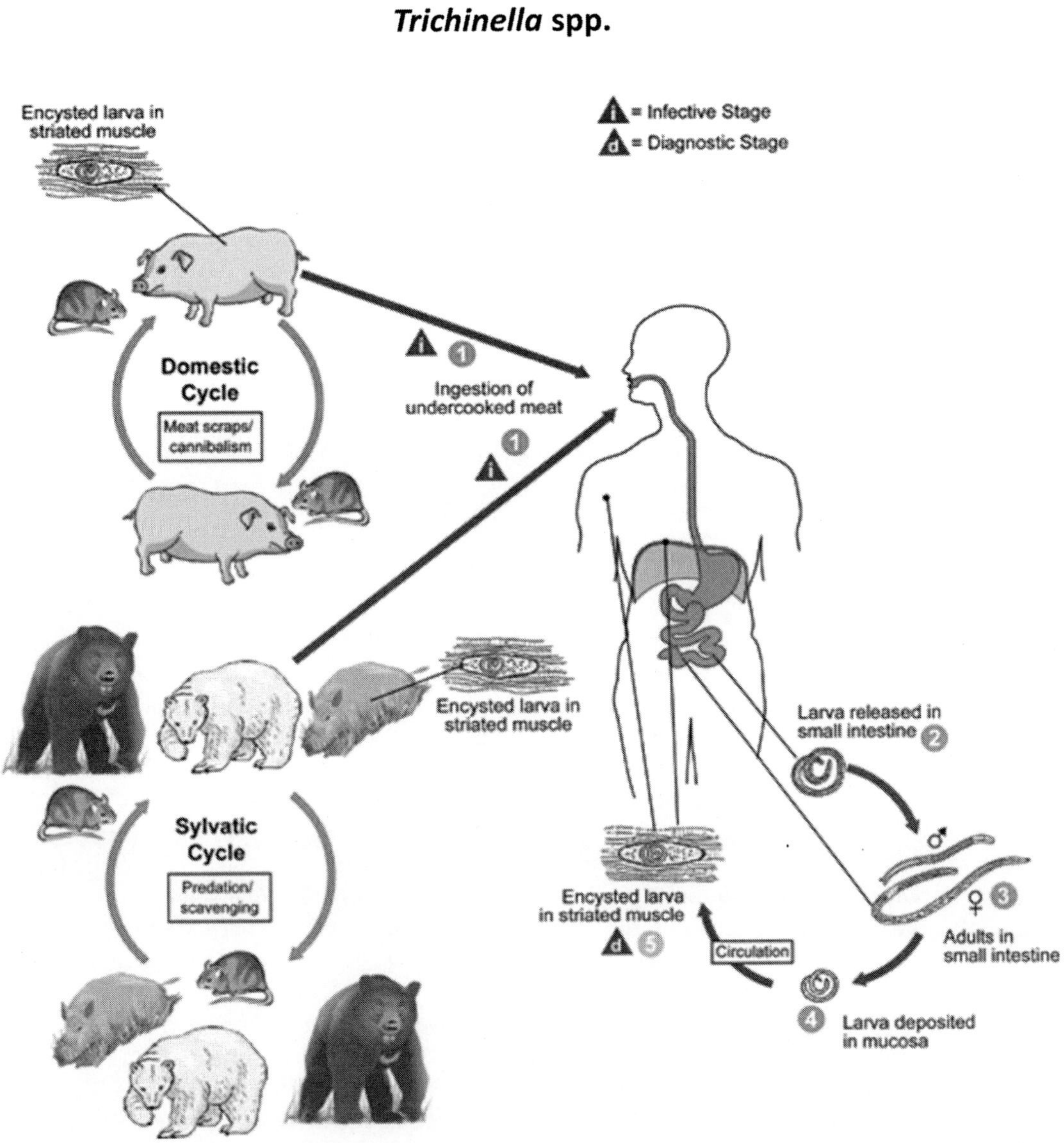

FIGURE 24.4 Life cycle of *Trichinella* spp. The entire life cycle of the nematode *Trichinella spiralis* occurs in a single host. After ingestion of raw skeletal muscle containing muscle larvae, the latter are released in the stomach and move into the small intestine to invade epithelial cells. Larvae transform into adult worms, which mate. Female worms release newborn larvae, which reach the bloodstream and migrate to skeletal muscle cells to become muscle larvae. Reproduced free of charge from the Centers for Disease Control and Prevention (CDC) webpage (https://www.cdc.gov/parasites/). Use of this image does not imply endorsement by CDC to the contents of this chapter.

group compared to the control group. Analysis of lung cell suspensions revealed that the PD group contained fewer neutrophils, eosinophils, and cells expressing the high-affinity IgE receptor than the control infected group (Vila et al. 2021). These findings demonstrate a lower immune response against NBL in the PD group, which, combined with the presence of AW in the gut at later p.i. times (which continue to release NBL), may also contribute to the higher parasite burden (ML) found in the muscle of the PD group.

Overall, studies on *Trichinella* spp. infection and protein malnutrition indicate that PD generates a higher burden of AW (with a higher fecundity index) and ML. The PD diet also diminished the expulsion of *T. spiralis* from the intestinal tract. Furthermore, PD generated lower eosinophil counts, reduced total and specific antibodies, and fewer mast and goblet cells in the intestine and lungs compared to well-nourished conditions. Moreover, the PD group exhibited weaker immune responses against NBL (via the ADCC mechanism). This diet resulted in the death of some animals under certain conditions.

TRICHURIASIS

The life cycle of the nematode *Trichuris* spp. begins when a host ingests water or food contaminated with embryonated eggs. These eggs hatch in the intestine to release infective L1 larvae, which then migrate to the caecum and colon and undergo four molts to become adult worms. Trichuriasis, particularly in cases of heavy infections, contributes to PEM in children, though growth rates improve following treatment (Gilman et al. 1976; Gilman et al. 1983; Bowie et al. 1978).

Several studies have analyzed the effects of protein malnutrition during *Trichuris* spp. infection in animal models. Michael et al. studied the influence of host dietary protein (2%, 4%, and 16% casein) on acquired immunity and intestinal helminth population dynamics during repeated *T. muris* infection. They found that CBA/Ca mice fed a 2% protein diet *ad libitum* maintained body weight during the experiment, while those fed 4% or 16% protein diets gained weight. Mice fed 2% or 4% protein diets accumulated adult worms in proportion to infection dose when repeatedly infected with either 5 or 50 eggs every 10 days, but those receiving the 16% protein diet showed a decrease in worm numbers over time. These results show that protein deficiency enhances larval establishment and adult worm survival. *T. muris* egg output per mouse increased with infection dose in mice fed the PD diet, but no parasite eggs were detected in hosts fed the 16% protein diet. The specific anti-adult worm antibody response was more intense in mice fed the PD diets (Michael and Bundy 1992). This suggests that the higher antibody response in these mice may be a function of their heavier worm burdens. Additionally, the fecal egg density of malnourished mice increased with the infection dose, presumably reflecting the long-term survival and dose-related accumulation of adult worms.

In a separate study, Pedersen et al. (2002) investigated the interaction between PD, PEM, and intestinal nematode infection in pigs. The researchers fed pigs either a PEM, PD, or normal protein-energy diet, and inoculated them with *T. suis* (4000 infective eggs) six weeks later. Subsequently, they inoculated the pigs with *Ascaris suum* (600 infective eggs). Their findings revealed higher numbers of worm-positive animals among *T. suis*-infected pigs in the PD group than in the control diet group at ten weeks p.i. However, like the observations of *Schistosoma* spp. infections, malnutrition seemed to affect the development of *T. suis* in the host. *T. suis* worms in the PEM and PD groups were decreased in length, and the pre-patency period was extended in these groups (reflected by lower fecal egg output). More *A. suum* worms were recovered from the PD group compared with the control group, accompanied by higher fecal egg output. The PD diets resulted in lower pig body weight gains, serum albumin, hemoglobin, and packed cell volume levels, as well as diminished peripheral eosinophil counts. The number of pigs with *T. suis* and *A. suum* worms at ten weeks p.i. was higher in PD pigs compared to control pigs, whereas the PEM group had a higher number of *A. suum*-positive pigs compared with the control group. The researchers detected a higher total *A. suum* egg excretion in the PD group compared to PEM and control pigs (Pedersen et al. 2002). These results demonstrate that PEM and PD affect worm growth rates and the host's ability to resolve infections, consistent with findings in other helminth infections.

CONCLUSIONS

Considering all the presented data, we may conclude that, in general, protein malnutrition has a severe effect on the immune response of the host against helminth infections, aggravating the

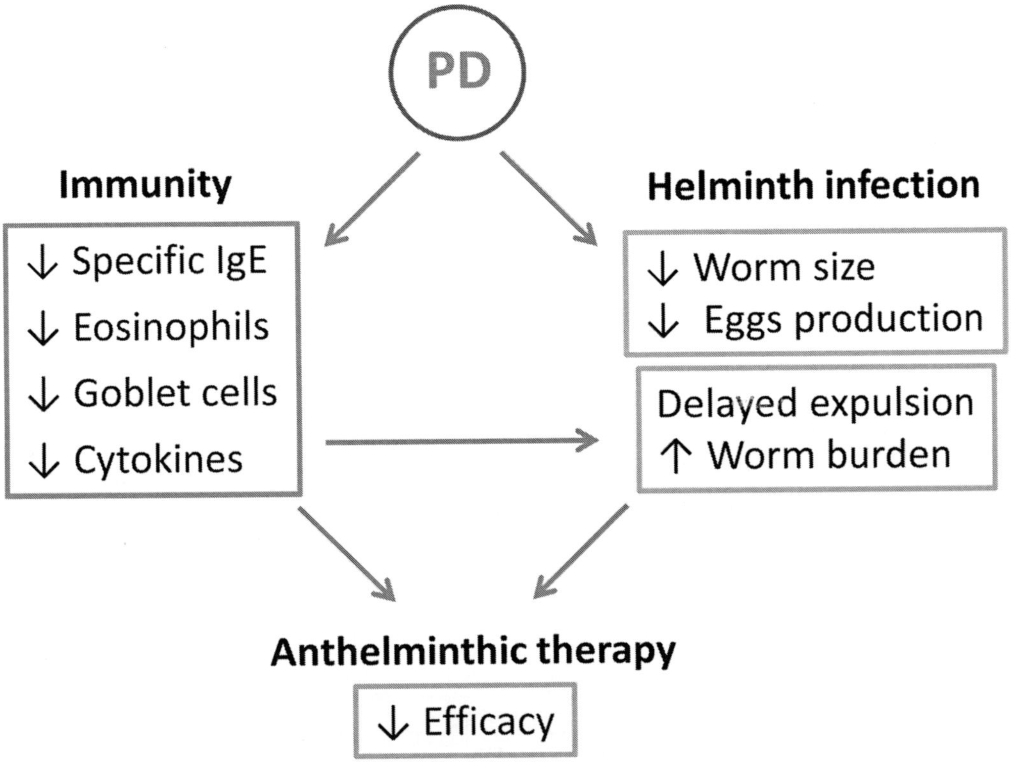

FIGURE 24.5 Consequences of protein deficiency (PD) in the context of helminth infections. PD impairs several cellular and humoral mediators of the immune response against parasitic helminths, although it may also affect the development of different stages of these parasites. The impairment of the immune response correlates with higher burdens and prolonged presence of parasites in the host tissues. Both factors in turn contribute to a reduced efficacy of the anthelmintic therapy in the presence of PD.

situation for the host and even leading to its death in some cases. There are also negative effects of malnutrition for some helminth species, affecting their growth (i.e., *Ascaris* spp. and *Schistosoma* spp.) (Figure 24.5).

Regarding the immune response, protein malnutrition reduced effective type 2 responses in different helminth models, which correlated with a higher parasite burden, a higher presence of helminth stages over time, and an incomplete resolution of these infections. There is a need to continue studying the immune mechanisms altered. In particular, the role of malnutrition during the initiation of the immune response against helminths requires further analysis.

MINI DICTIONARY OF TERMS

Antibody-dependent cellular cytotoxicity (ADCC): A mechanism whereby specific antibodies bound to the surface of a phagocyte recognize a parasite and mediate phagocyte activation to attack the pathogen.

Cercariae: Motile larval stages of trematodes found in water sources.

Cestodes: Segmented flat hermaphrodite helminths.

Cytokines: Soluble substances secreted by immune cells to mediate responses in a paracrine or endocrine manner.

Helminths: Free-living or parasitic invertebrate animals, commonly known as worms.

Helper T cells (Th): CD4+ T lymphocytes that interact directly with other immune cells to activate specific functions. Th cells are classified into different phenotypes (Th1, Th2, Th17, etc.) depending on the cytokines they produce.

Nematodes: Round, non-segmented helminths, mostly hermaphroditic.

Trematodes: Flat, non-segmented helminths, predominantly hermaphroditic (except for *Schistosoma* spp.).

SUMMARY POINTS

- Protein deficiency (PD) has a strong effect on the immune response during helminth infections.
- PD suppresses type 2 responses required to control helminthiasis.
- PD leads to increased parasite burden in tissues and prolonged presence of helminth stages.
- PD also negatively affects the development of some helminths.
- PD may reduce the efficacy of anthelmintics.

REFERENCES

Allen, J.E., and R.M. Maizels. 2011. "Diversity and Dialogue in Immunity to Helminths." *Nature Reviews Immunology* 11 (6): 375–88. https://doi.org/10.1038/nri2992.

Ayelign, B., Y. Akalu, B. Teferi, M.D. Molla, and T. Shibabaw. 2020. "Helminth Induced Immunoregulation and Novel Therapeutic Avenue of Allergy." *Journal of Asthma and Allergy* 13: 439–51. https://doi.org/10.2147/JAA.S273556.

Bhutta, Z.A., J.A. Berkley, R.H.J. Bandsma, M. Kerac, I. Trehan and A. Briend. 2017. "Severe Childhood Malnutrition". *Nature Reviews Disease Primers* 3:17067. https://doi.org/10.1038/nrdp.2017.67.

Bolin, T.D., A.E. Davis, A.G. Cummins, V.M. Duncombe, and J.D. Kelly. 1977. "Effect of Iron and Protein Deficiency on the Expulsion of *Nippostrongylus brasiliensis* from the Small Intestine of the Rat." *Gut* 18 (3): 182–86. https://doi.org/10.1136/gut.18.3.182.

Bowie, M.D., A. Morison, J.D. Ireland, and P.J. Duys. 1978. "Clubbing and Whipworm Infestation." *Archives of Disease in Childhood* 53 (5): 411–13. https://doi.org/10.1136/adc.53.5.411.

Brown, K.H. 2003. "Symposium: Nutrition and Infection, Prologue and Progress Since 1968." *The Journal Of Nutrition* 133 (1): 328–32.

Burke, M.L., M.K. Jones, G.N. Gobert, Y.S. Li, M.K. Ellis, and D.P. McManus. 2009. "Immunopathogenesis of Human Schistosomiasis." *Parasite Immunology* 31 (4): 163–76. https://doi.org/10.1111/j.1365-3024.2009.01098.x.

Caulfield, L.E., M. de Onis, M. Blössner, and R.E. Black. 2004. "Undernutrition as an Underlying Cause of Child Deaths Associated with Diarrhea, Pneumonia, Malaria, and Measles." *The American Journal of Clinical Nutrition* 80 (1): 193–98. https://doi.org/10.1093/ajcn/80.1.193.

Chandra, R.K. 1990. "Micronutrients and Immune Functions." *Annals of the New York Academy of Sciences* 587 (1): 9–16. https://doi.org/10.1111/j.1749-6632.1990.tb00128.x.

Chandra, R.K. 1992. . "Protein-Energy Malnutrition and Immunological Responses." *The Journal of Nutrition* 122 (3): 597–600. https://doi.org/10.1093/jn/122.suppl_3.597

Coutinho, E.M. 2004. "Malnutrition and Hepatic Fibrosis in Murine Schistosomiasis." *Memorias Do Instituto Oswaldo Cruz* 99 (5): 85–92. https://doi.org/10.1590/s0074-02762004000900015.

Cunha, M.C.R., F. da Silva Lima, M.A. Ramirez Vinolo, A. Hastreiter, R. Curi, P. Borelli, and R.A. Fock. 2013. "Protein Malnutrition Induces Bone Marrow Mesenchymal Stem Cells Commitment to Adipogenic Differentiation Leading to Hematopoietic Failure." *PLoS One* 8 (3): 1–12. https://doi.org/10.1371/journal.pone.0058872.

Dickson, R., S. Awasthi, P. Williamson, C. Demellweek, and P. Garner. 2000. "Effects of Treatment for Intestinal Helminth Infection on Growth and Cognitive Performance in Children: Systematic Review of Randomised Trials." *BMJ (Clinical Research Ed.)* 320 (7251): 1697–701. https://doi.org/10.1136/bmj.320.7251.1697.

Duncombe, V.M., T.D. Bolin, A.E. Davis, and J.D. Kelly. 1977. "*Nippostronglylus brasiliensis* Infection in the Rat: Effect of Iron and Protein Deficiency and Dexamethasone on the Efficacy of Benzimidazole Anthelmintics." *Gut* 18 (11): 892–96. https://doi.org/10.1136/gut.18.11.892.

Fairfax, K., M. Nascimento, S. Ching Cheng Huang, B. Everts, and E.J. Pearce. 2012. "Th2 Responses in Schistosomiasis." *Seminars in Immunopathology* 34 (6): 863–71. https://doi.org/10.1007/s00281-012-0354-4.

Falduto, G.H., C.C. Vila, M.P. Saracino, M.A. Calcagno, and S.M. Venturiello. 2015. "*Trichinella spiralis*: Killing of Newborn Larvae by Lung Cells." *Parasitology Research* 114 (2): 679–85. https://doi.org/10.1007/s00436-014-4233-x.

FAO; IFAD; UNICEF; WFP; WHO. 2023. *The State of Food Security and Nutrition in the World 2023.* FAO. https://doi.org/10.4060/cc3017en.

Forsum, E., M.C. Nesheim, and D.W.T. Crompton. 1981. "Nutritional Aspects of Ascaris Infection in Young Protein-Deficient Pigs." *Parasitology* 83 (3): 497–512. https://doi.org/10.1017/S0031182000080483.

Gause, W.C., C. Rothlin, and P. Loke. 2020. "Heterogeneity in the Initiation, Development and Function of Type 2 Immunity." *Nature Reviews Immunology* 20 (10): 603–14. https://doi.org/10.1038/s41577-020-0301-x.

Gbakima, A.A. 1993. "The Effect of Dietary Protein on *Trichinella spiralis* Infection and Inflammatory Reactions in the Tongue in CD1 Mice." *Nutrition Research* 13 (7): 787–800. https://doi.org/10.1016/S0271-5317(05)80803-6.

Gilman, R.H., Y.H. Chong, C. Davis, B. Greenberg, H.K. Virik, and H.B. Dixon. 1983. "The Adverse Consequences of Heavy *Trichuris* Infection." *Transactions of the Royal Society of Tropical Medicine and Hygiene* 77 (4): 432–38. https://doi.org/10.1016/0035-9203(83)90103-7.

Gilman, R.H., C. Davis, and F. Fitzgerald. 1976. "Heavy *Trichuris* Infection and Amoebic Dysentery in Orang Asli Children. A Comparison of the Two Diseases." *Transactions of the Royal Society of Tropical Medicine and Hygiene* 70 (4): 313–16. https://doi.org/10.1016/0035-9203(76)90085-7.

Hagel, I., N.R. Lynch, M.C. Di Prisco, J. Sanchez, and M. Pérez. 1995. "Nutritional Status and the IgE Response against *Ascaris lumbricoides* in Children from a Tropical Slum." *Transactions of the Royal Society of Tropical Medicine and Hygiene* 89 (5): 562–65. https://doi.org/10.1016/0035-9203(95)90109-4.

Hagel, I., N.R. Lynch, F. Puccio, O. Rodriguez, R. Luzondo, P. Rodríguez, P. Sánchez, M. Cabrera, and M.C. Di Prisco. 2003. "Defective Regulation of the Protective IgE Response against Intestinal Helminth *Ascaris lumbricoides* in Malhourished Children." *Journal of Tropical Pediatrics* 49 (3): 136–42. https://doi.org/10.1093/tropej/49.3.136.

Ibrahim, M.K., M. Zambruni, C.L. Melby, and P.C. Melby. 2017. "Impact of Childhood Malnutrition on Host Defense and Infection." *Clinical Microbiology Reviews* 30 (4): 919–71. https://doi.org/10.1128/CMR.00119-16.

Jones, K.D., J. Thitiri, M. Ngari, and J.A. Berkley. 2014. "Childhood Malnutrition: Toward an Understanding of Infections, Inflammation, and Antimicrobials." *Food and Nutrition Bulletin* 35 (2_suppl1): S64–70. https://doi.org/10.1177/15648265140352S110.

Keusch, G.T. 2003. "Symposium: Nutrition and Infection, Prologue and Progress Since 1968 The History of Nutrition: Malnutrition, Infection and Immunity." *Journal of Nutrition* 133 (June): 336–40.

Knauft, R.F., and K.S. Warren. 1969. "The Effect of Calorie and Protein Malnutrition on Both the Parasite and the Host in Acute Murine *Schistosomiasis mansoni*." *Journal of Infectious Diseases* 120 (5): 560–75. https://doi.org/10.1093/infdis/120.5.560.

Liang, X., W. Shu, Z. Linyong, L. Jianshui, G. Junqing, D. Enzhu, and X. Mingqing. 2020. "Preoperative Nutritional Evaluation of Patients with Hepatic Alveolar Echinococcosis." *PLoS One* 15 (2): 1–11. https://doi.org/10.1371/journal.pone.0229396.

Lima, A.A.M., L.F.B. Brito, H.B. Ribeiro, M.C.V. Martins, A.P. Lustosa, E.M. Rocha, N.L. Lima, C.M.G. Monte, and R.L. Guerrant. 2005. "Intestinal Barrier Function and Weight Gain in Malnourished Children Taking Glutamine Supplemented Enteral Formula." *Journal of Pediatric Gastroenterology and Nutrition* 40 (1): 28–35. https://doi.org/10.1097/00005176-200501000-00006.

Loukas, A., P.J. Hotez, D. Diemert, M. Yazdanbakhsh, J.S. McCarthy, R. Correa-Oliveira, J. Croese, and J.M. Bethony. 2016. "Hookworm Infection." *Nature Reviews Disease Primers* 2:16088. https://doi.org/10.1038/nrdp.2016.88.

Maciel, P.S., R. Gonçalves, L. Ribeiro do Valle Antonelli, and C. Toscano Fonseca. 2021. "*Schistosoma mansoni* Infection Is Impacted by Malnutrition." *Frontiers in Microbiology* 12 (March): 1–12. https://doi.org/10.3389/fmicb.2021.635843.

Martinsen, T.C., K. Bergh, and H.L. Waldum. 2005. "Gastric Juice: A Barrier against Infectious Diseases." *Basic and Clinical Pharmacology and Toxicology* 96 (2): 94–102. https://doi.org/10.1111/j.1742-7843.2005.pto960202.x.

Michael, E., and D.A.P. Bundy. 1992. "Protein Content of CBA/Ca Mouse Diet: Relationship with Host Antibody Responses and the Population Dynamics of *Trichuris muris* (Nematoda) in Repeated Infection." *Parasitology* 105 (1): 139–50. https://doi.org/10.1017/S0031182000073790.

Nakajima, K., A.R. Crisma, G.B. Silva, M.M. Rogero, R.A. Fock, and P. Borelli. 2014. "Malnutrition Suppresses Cell Cycle Progression of Hematopoietic Progenitor Cells in Mice via Cyclin D1 Down-Regulation." *Nutrition* 30 (1): 82–89. https://doi.org/10.1016/j.nut.2013.05.029.

Ojha, S.C., C. Jaide, N. Jinawath, P. Rotjanapan, and P. Baral. 2014. "Geohelminths: Public Health Significance." *Journal of Infection in Developing Countries* 8 (1): 5–16. https://doi.org/10.3855/jidc.3183.

Oliveira, S.A., A.A. Barbosa, D.C. Gomes, J.R. Machado-Silva, A.F. Barros, R.H. Neves, and E.M. Coutinho. 2003. "Morphometric Study of *Schistosoma mansoni* Adult Worms Recovered from Undernourished Infected Mice." *Memorias Do Instituto Oswaldo Cruz* 98 (5): 623–27. https://doi.org/10.1590/S0074-02762003000500007.

Pacanaro, C.P., S.R. Dias, L.R. Serafim, M.P. Costa, E. Aguilar, P.R. Paes, J.I. Alvarez-Leite, and É.M. Rabelo. 2014. "Evaluation of Biochemical, Hematological and Parasitological Parameters of Protein-Deficient Hamsters Infected with *Ancylostoma Ceylanicum*." *PLoS Neglected Tropical Diseases* 8 (9):e3184. https://doi.org/10.1371/journal.pntd.0003184.

Pearce, E.J., and A.S. MacDonald. 2002. "The Immunobiology of Schistosomiasis." *Nature Reviews Immunology* 2 (7): 499–511. https://doi.org/10.1038/nri843.

Pedersen, S., I. Saeed, K.F. Michaelsen, H. Friis, and K.D. Murrell. 2002. "Impact of Protein Energy Malnutrition on *Trichuris suis* Infection in Pigs Concomitantly Infected with *Ascaris suum*." *Parasitology* 124 (5): 561–68. https://doi.org/10.1017/S0031182002001592.

Pelletier, D.L. 1995. "Potentiating Effects of Malnutrition on Child Mortality: Epidemiologic Evidence and Policy Implications." *Food and Nutrition Bulletin* 16 (3): 206–13. https://doi.org/10.1177/156482659501600304.

Peterson, L.W., and D. Artis. 2014. "Intestinal Epithelial Cells: Regulators of Barrier Function and Immune Homeostasis." *Nature Reviews Immunology* 14 (3): 141–53. https://doi.org/10.1038/nri3608.

Rodríguez, L., E. Cervantes, and R. Ortiz. 2011. "Malnutrition and Gastrointestinal and Respiratory Infections in Children: A Public Health Problem." *International Journal of Environmental Research and Public Health* 8 (4): 1174–205. https://doi.org/10.3390/ijerph8041174.

Rodríguez, L., C. González, L. Flores, L. Jiménez-Zamudio, J. Graniel, and R. Ortiz. 2005. "Assessment by Flow Cytometry of Cytokine Production in Malnourished Children." *Clinical and Diagnostic Laboratory Immunology* 12 (4): 502–7. https://doi.org/10.1128/CDLI.12.4.502-507.2005.

Santos, E.W., D.C. Oliveira, G.B. Silva, M. Tsujita, J.O. Beltran, A. Hastreiter, R.A. Fock, and P. Borelli. 2017. "Hematological Alterations in Protein Malnutrition." *Nutrition Reviews* 75 (11): 909–19. https://doi.org/10.1093/nutrit/nux041.

Saracino, M.P., C.C. Vila, M. Cohen, M.V. Gentilini, G.H. Falduto, M.A. Calcagno, E. Roux, S.M. Venturiello, and E.L. Malchiodi. 2020. "Cellular and Molecular Changes and Immune Response in the Intestinal Mucosa during *Trichinella spiralis* Early Infection in Rats." *Parasites and Vectors* 13 (1): 1–19. https://doi.org/10.1186/s13071-020-04377-8.

Schaible, U.E., and S.H.E. Kaufmann. 2007. "Malnutrition and Infection: Complex Mechanisms and Global Impacts." *PLoS Medicine* 4 (5): 0806–12. https://doi.org/10.1371/journal.pmed.0040115.

Scrimshaw, N.S., and J.P. SanGiovanni. 1997. "Synergism of Nutrition, Infection, and Immunity: An Overview." *American Journal of Clinical Nutrition* 66 (2): 464–77. https://doi.org/10.1093/ajcn/66.2.464S.

Solomons, N.W. 1993. "Pathways to the Impairment of Human Nutritional Status by Gastrointestinal Pathogens." *Parasitology* 107 (S1): S19–35. https://doi.org/10.1017/S003118200007548X.

Thein-Hlaing, Thane-Toe, Than-Saw, Myat-Lay-Kyin, and Myint-Lwin. 1991. "A Controlled Chemotherapeutic Intervention Trial on the Relationship between *Ascaris lumbricoides* Infection and Malnutrition in Children." *Transactions of the Royal Society of Tropical Medicine and Hygiene* 85 (4): 523–28. https://doi.org/10.1016/0035-9203(91)90242-Q.

Venturiello, S.M., G.H. Giambartolomei, and S.N. Costantino. 1993. "Immune Killing of Newborn *Trichinella* Larvae by Human Leucocytes." *Parasite Immunology* 15 (10): 559–64. https://doi.org/10.1111/pim.1993.15.10.559.

Vidueiros, S.M., I. Fernandez, N. Slobodianik, M.E. Roux, and A. Pallaro. 2008. "Nutrition Disorder and Immunologic Parameters: Study of the Intestinal Villi in Growing Rats." *Nutrition* 24 (6): 575–81. https://doi.org/10.1016/j.nut.2008.01.055.

Vila, C.C., M.P. Saracino, G.H. Falduto, M.A. Calcagno, S.M. Venturiello, A.N. Pallaro, and P.C. Baldi. 2019. "Protein Malnutrition Impairs the Immune Control of *Trichinella spiralis* Infection." *Nutrition* 60: 161–69. https://doi.org/10.1016/j.nut.2018.10.024.

Vila, C.C., M.P. Saracino, T. Lombardo, G.H. Falduto, M. Díaz, M.A. Calcagno, A.N. Pallaro, and P.C. Baldi. 2021. "Protein Deficiency during *Trichinella spiralis* Infection Impairs Lung Immunity against Newborn Larvae." *Parasite Immunology* 43 (4): 1–10. https://doi.org/10.1111/pim.12820.

Wang, Z., J. Xu, G. Song, M. Pang, B. Guo, X. Xu, H. Wang, et al. 2020. "Nutritional Status and Screening Tools to Detect Nutritional Risk in Hospitalized Patients with Hepatic Echinococcosis." *Parasite* 27:74. https://doi.org/10.1051/parasite/2020071.

World Health Organization. 2021. "Malnutrition." June 9, 2021. https://www.who.int/news-room/fact-sheets/detail/malnutrition.

Wright, J.E., M. Werkman, J.C. Dunn, and R.M. Anderson. 2018. "Current Epidemiological Evidence for Predisposition to High or Low Intensity Human Helminth Infection: A Systematic Review." *Parasites and Vectors* 11 (1): 1–12. https://doi.org/10.1186/s13071-018-2656-4.

Xavier, J.G., M.E. Favero, M.A.R. Vinolo, M.M. Rogero, M.L.Z. Dagli, V.E. Arana-Chavez, R. Borojevic, and Primavera Borelli. 2007. "Protein-Energy Malnutrition Alters Histological and Ultrastructural Characteristics of the Bone Marrow and Decreases Haematopoiesis in Adult Mice." *Histology and Histopathology* 22 (4–6): 651–60. https://doi.org/10.14670/HH-22.651.

25 Linking Protein Intolerance, Immunity, and Disease
The Example of Celiac Disease

Mohammad Rostami-Nejad, Nastaran Asri, and Hamidreza Houri

ABBREVIATIONS

CD	Celiac disease
DH	Dermatitis herpetiformis
GA	Gluten ataxia
GFD	Gluten-free diet
Gln	Glutamine
GRDs	Gluten-related disorders
HLA	Human leukocyte antigens
NCGS	Non-celiac gluten sensitivity
Pro	Proline
WA	Wheat allergy

INTRODUCTION

Protein intolerance, also known as protein hypersensitivity or protein allergy, refers to an adverse reaction of the immune system to proteins found in certain foods (Walker-Smith 1988). The term "food protein intolerance" is often preferred over "food protein allergy" because it includes all adverse reactions to food proteins, regardless of the pathogenesis (Borschel et al. 2014; Zepeda-Ortega et al. 2021). This condition is characterized by a range of symptoms that can vary in severity from mild discomfort to life-threatening reactions (Yu, Freeland, and Nadeau 2016; Turner et al. 2016). Understanding the causes, symptoms, and mechanisms of protein intolerance is crucial for the effective diagnosis, management, and prevention of this condition (Abrams and Sicherer 2016; Vandenplas et al. 2007). Protein intolerance can manifest in numerous ways, affecting different organ systems (Guler et al. 2020). Gastrointestinal symptoms are commonly observed, including abdominal pain, bloating, diarrhea, and vomiting (Heine 2015; Reda 2009). Some individuals may experience skin reactions, including hives, eczema, or swelling (Dhar and Srinivas 2016; Katta and Schlichte 2014; Turner et al. 2022). Respiratory symptoms like wheezing, coughing, and difficulty breathing, and circulatory symptoms like a drop in blood pressure can also occur (di Palmo et al. 2019; Foong, du Toit, and Fox 2017). In severe cases, a potentially life-threatening allergic reaction called anaphylaxis may develop, resulting in a sudden drop in blood pressure, difficulty breathing, and loss of consciousness (Turner et al. 2019; Cardona et al. 2020). Diagnosing protein intolerance often requires a comprehensive evaluation, which may include a detailed medical history, physical examination, and specific diagnostic tests (Vandenplas et al. 2007; Walker-Smith 1988).

The immune system plays a crucial role in the development and management of protein intolerance. When individuals with protein intolerance consume specific proteins, their immune system

DOI: 10.1201/b23328-25

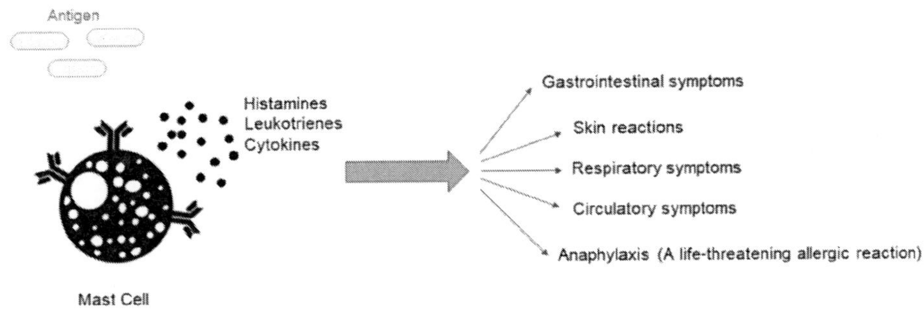

FIGURE 25.1 Various presentations of protein intolerance.

recognizes these proteins as foreign or harmful antigens. This triggers an immune response, aimed at protecting the body from potential threats (Walker-Smith 1988; Kopko et al. 2022). The immune system responds to protein antigens by activating various components (Chaplin 2010; Nicholson 2016). Upon exposure to the antigen, specialized immune cells, such as mast cells and T cells, recognize and interact with the antigen (El Ansari, Kanagaratham, and Oettgen 2020; Nakano and Kitaura 2022). Mast cells release mediators, including histamine, which contribute to the allergic symptoms associated with protein intolerance, such as inflammation, itching, and gastrointestinal distress (Ramsay et al. 2010; Gupta and Harvima 2018). Additionally, T cells play a crucial role in the immune response. They recognize specific protein fragments (peptides) derived from the ingested proteins presented on the surface of antigen-presenting cells (Satitsuksanoa et al. 2018; Berin et al. 2022). This recognition triggers the activation and proliferation of T cells, leading to the production of immune mediators (Satitsuksanoa et al. 2018; Berin et al. 2022; Yu, Freeland, and Nadeau 2016). These mediators can further contribute to the symptoms experienced by individuals with protein intolerance (Yu, Freeland, and Nadeau 2016) (Figure 25.1).

Understanding the interplay between the immune system and protein intolerance is crucial for developing effective diagnostic and therapeutic strategies (Ahmad et al. 2022). By deciphering the specific immune pathways involved in protein intolerance, researchers and clinicians can identify potential targets for intervention, leading to better management and improved quality of life for individuals affected by this condition (Chen et al. 2018). Figure 25.2 shows a flowchart of food protein intolerance management.

Protein intolerance encompasses various conditions, including milk protein intolerance, peanut protein intolerance, soy protein intolerance, egg protein intolerance, fish and shellfish protein intolerance, and gluten protein intolerance (Walker-Smith 1988). These examples highlight the diverse manifestations of protein intolerance and the importance of accurate diagnosis, management, and prevention strategies (Walker-Smith 1988). By understanding the underlying mechanisms and immune responses associated with these forms of protein intolerance, researchers and clinicians can work towards improving the quality of life for affected individuals through targeted interventions and therapies.

MILK PROTEIN INTOLERANCE

Milk protein intolerance, also known as cow's milk protein intolerance, is a condition where individuals experience adverse reactions to the proteins found in cow's milk, such as casein (Giannetti et al. 2021). Milk protein intolerance is an immune-mediated response to milk proteins. This condition is particularly common in infants and young children but can also affect adults (Al-Beltagi et al. 2022). When individuals with milk protein intolerance consume milk or dairy products, they may experience a range of gastrointestinal and systemic symptoms. Anaphylaxis, though rare, is another

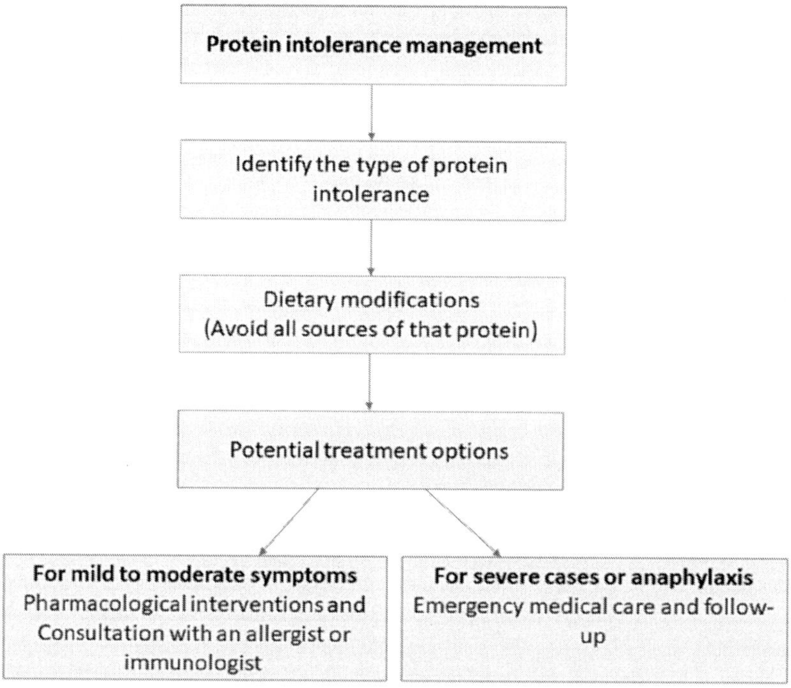

FIGURE 25.2 Protein intolerance management.

potential life-threatening reaction that may develop in individuals with milk protein intolerance (Santoro et al. 2019).

Management of milk protein intolerance involves the complete avoidance of cow's milk and dairy products containing milk proteins. Individuals can opt for alternative sources of nutrition, such as hypoallergenic formulas (e.g., extensively hydrolyzed formulas or amino acid-based formulas) for infants, or plant-based milk alternatives (e.g., almond milk, oat milk, or soy milk) for older children and adults. Furthermore, individuals with milk protein intolerance need to ensure they maintain an adequate intake of calcium and other nutrients found in dairy products. Calcium-rich alternatives, such as fortified plant-based milks or calcium supplements, can help meet their nutritional needs (D'Auria et al. 2021 Maryniak et al. 2022).

PEANUT PROTEIN INTOLERANCE

Peanut protein intolerance, also known as peanut allergy, is a severe form of protein intolerance characterized by an immune system reaction to proteins found in peanuts. It is one of the most common food allergies, particularly in children, and can cause significant health risks (Al-Muhsen, Clarke, and Kagan 2003). Peanut protein intolerance can cause a variety of symptoms, ranging from mild discomfort to life-threatening reactions. These symptoms can affect different parts of the body, including the skin, respiratory system, and gastrointestinal tract. In severe cases, peanut protein intolerance can lead to anaphylaxis, a rapid and severe allergic reaction that affects multiple systems in the body (Zhou et al. 2013).

The mainstay of peanut protein intolerance management is strict avoidance of peanuts and peanut products (Venter, Sicherer, and Greenhawt 2019). Individuals with a peanut allergy should meticulously read food labels, educate themselves about hidden sources of peanuts, and be cautious of cross-contamination. In cases of accidental exposure, prompt administration of epinephrine using an autoinjector is necessary to counteract the severe allergic reaction (Venter, Sicherer, and

Greenhawt 2019). Epinephrine helps alleviate symptoms such as breathing difficulties, low blood pressure, and hives (Coombs et al. 2011).

In recent years, there have been advancements in research aiming to develop potential treatments for peanut protein intolerance. These include oral immunotherapy, where individuals gradually consume small amounts of peanuts to desensitize their immune system, and sublingual immunotherapy, where peanut extracts are placed under the tongue. These treatments aim to train the immune system to tolerate peanuts and reduce the severity of allergic reactions. However, they are still considered experimental and are primarily conducted under the supervision of allergists or immunologists (Chinthrajah et al. 2020).

SOY PROTEIN INTOLERANCE

Soy protein intolerance refers to an adverse reaction to proteins found in soybeans. It is more common in infants and children, who typically outgrow the intolerance as they age. Symptoms can vary and may include gastrointestinal or skin reactions, respiratory symptoms, or systemic symptoms (Kattan, Cocco, and Järvinen 2011).

The primary treatment for soy protein intolerance is the strict avoidance of soy products and the use of alternative protein sources. Individuals with soy protein intolerance must carefully read food labels to identify hidden sources of soy and ensure that their diet is free from soy-based ingredients or derivatives. In cases where soy is a crucial component of one's diet, such as in vegetarian or vegan diets, consultation with a registered dietitian or nutritionist can help develop a balanced and nutritionally adequate eating plan that meets the individual's needs (Meinlschmidt et al. 2016).

EGG PROTEIN INTOLERANCE

Egg protein intolerance is an adverse reaction to proteins found in eggs, most commonly those found in egg whites. It is more prevalent in children but can also affect adults. Symptoms can range from mild to severe and may include skin reactions, gastrointestinal, respiratory, or systemic symptoms (Caubet and Wang 2011).

The primary treatment for egg protein intolerance is strict avoidance of eggs and egg-containing products. Individuals with egg protein intolerance need to be vigilant in reading food labels to identify potential hidden sources of eggs, as eggs can be present in various processed foods. Additionally, they may need to discuss alternative cooking methods and ingredient substitutions with a registered dietitian or nutritionist to ensure a nutritious diet that adequately meets their nutritional needs (Caubet and Wang 2011).

FISH AND SHELLFISH PROTEIN INTOLERANCE

Fish and shellfish protein intolerance refers to adverse reactions to proteins found in fish and shellfish, such as tuna, salmon, shrimp, lobster, and crab. It is considered one of the most common types of food allergies in adults and can manifest in various ways, ranging from mild discomfort to severe allergic reactions. Understanding the mechanisms and symptoms associated with fish and shellfish protein intolerance is vital for accurate diagnosis, effective management, and prevention of adverse reactions (Caubet and Wang 2011).

Symptoms of fish and shellfish protein intolerance can vary widely, ranging from mild skin reactions, such as itching or hives, to severe systemic manifestations, such as difficulty breathing, swelling of the lips or throat (angioedema), rapid heartbeat, and a drop in blood pressure. In severe cases, anaphylaxis, a potentially life-threatening condition, can occur. Prompt administration of epinephrine via an autoinjector is crucial in managing anaphylactic reactions (Caubet and Wang 2011).

Due to the potential severity of fish and shellfish protein intolerance, strict avoidance of fish and shellfish is the primary management strategy. Individuals with fish and shellfish protein intolerance should inform restaurant staff about their specific dietary restrictions to minimize the risk of accidental exposure. In cases where accidental ingestion does occur, immediate medical attention should be sought, especially if symptoms worsen or anaphylactic reactions are suspected (Davis et al. 2020).

It is important to note that fish and shellfish protein intolerance can differ among individuals. Some people may only be sensitive to specific types of fish or shellfish, while others may have cross-reactivity between different species (Fernandes et al. 2015).

In recent years, researchers have been investigating potential treatments for fish and shellfish protein intolerance. One promising approach is oral immunotherapy, which involves gradually increasing doses of fish or shellfish proteins under medical supervision to desensitize the immune system. This method aims to promote tolerance and reduce the severity of allergic reactions (Fernandes et al. 2015).

GLUTEN INTOLERANCE

Gluten protein intolerance refers to an abnormal immune response triggered by the ingestion of gluten proteins (Balakireva and Zamyatnin 2016; Scherf, Koehler, and Wieser 2016). It shares commonalities with other protein intolerances, necessitating its classification within the broader context of protein intolerance conditions (Balakireva and Zamyatnin 2016; Scherf, Koehler, and Wieser 2016). Gluten is composed of two main protein fractions, gliadins and glutenins (Barak, Mudgil, and Khatkar 2014). Gliadins are alcohol-soluble proteins, while glutenins are insoluble, providing the elastic properties necessary for dough formation (Urade, Sato, and Sugiyama 2018; Kłosok et al. 2021). The specific composition of gluten proteins contributes to their unique characteristics and their potential to act as allergens (Balakireva and Zamyatnin 2016).

Individuals with gluten protein intolerance experience an immune reaction upon gluten consumption involving both innate and adaptive immune pathways (Sharma et al. 2020; Lebwohl, Ludvigsson, and Green 2015). Certain gluten peptides can activate immune cells in the gut, leading to the release of pro-inflammatory cytokines and the recruitment of additional immune cells (Garrote et al. 2008; Cukrowska et al. 2017). Gluten proteins possess several features that contribute to their allergenic potential. These include a high content of proline (Pro) and glutamine (Gln) residues, which render them resistant to complete digestion in the gastrointestinal tract (Tian et al. 2015; Smith et al. 2015). Consequently, partially digested gluten peptides can interact with the immune system, triggering an immune response in susceptible individuals (Cebolla et al. 2018).

Gluten protein intolerance is associated with increased intestinal permeability, commonly referred to as "leaky gut" (Caio et al. 2020). Gluten peptides can disrupt the integrity of the intestinal lining, allowing the passage of gluten fragments into the underlying tissues (Cardoso-Silva et al. 2019). The presence of gluten peptides beyond the gut lumen activates immune cells in the gut-associated lymphoid tissue, leading to local inflammation and potential systemic effects (Voisine and Abadie 2021).

Gluten protein intolerance manifests through a range of symptoms affecting various organ systems (Cenni et al. 2023; Losurdo et al. 2018). Gastrointestinal symptoms often include abdominal pain, bloating, diarrhea, and constipation. Non-gastrointestinal symptoms such as fatigue, headaches, joint pain, skin rashes, and cognitive impairments can also occur. The diverse clinical presentation further supports the classification of gluten intolerance as a class of protein intolerance (Cenni et al. 2023; Losurdo et al. 2018).

A summary of common types of protein intolerance and management strategies is presented in Table 25.1.

TABLE 25.1
Summary of Common Types of Protein Intolerance and Management Strategies

Type of Protein Intolerance	Management Strategies
Milk protein intolerance	• Avoid cow's milk and dairy products containing milk proteins • Choose hypoallergenic or plant-based alternatives • Ensure adequate intake of essential nutrients
Peanut protein intolerance	• Strictly avoid peanuts and peanut products • Read food labels carefully for hidden sources of peanuts • Be cautious of cross-contamination • Administer epinephrine in case of accidental exposure • Consider oral or sublingual immunotherapy (experimental)
Soy protein intolerance	• Avoid soy products and read food labels for hidden sources of soy • Consult with a dietitian for alternative proteinsources and balanced diet planning
Egg protein intolerance	• Strictly avoid eggs and egg-containing products • Read food labels carefully for hidden sources of eggs • Consult with a dietitian for alternative cooking methods and ingredient substitutions
Fish and shellfish protein intolerance	• Strictly avoid fish and shellfish • Inform restaurant staff about dietary restrictions • Seek immediate medical attention in case of accidental ingestion or worsening symptoms • Consider oral immunotherapy (experimental)
Gluten intolerance	• Strictly avoid gluten-containing foods • Read food labels carefully for hidden sources of gluten • Consult with a dietitian for gluten-free alternatives • Manage symptoms associated with gluten intolerance and intestinal permeability • Consider additional medical interventions if necessary

GLUTEN-RELATED DISORDERS

Gluten-related disorders (GRDs) refer to a group of conditions caused by a reaction to gluten. GRDs, encompassing autoimmune, allergic, and non-autoimmune-allergic etiologies, have heterogeneous manifestations (Taraghikhah et al. 2020). Autoimmune conditions such as CD, dermatitis herpetiformis (DH), and gluten ataxia (GA) are well-established within this spectrum. Conversely, wheat allergy (WA) and non-celiac gluten sensitivity (NCGS) represent allergic and non-autoimmune-allergic disorders (Sapone et al. 2012) (Figure 25.3). Previously thought to predominantly affect European populations, GRDs now demonstrate a global prevalence of approximately 5% (Taraghikhah et al. 2020). Through the advancement of sensitive serological tests, the diagnosis of GRDs has risen significantly, shedding light on the considerable impact these conditions have on global health (Asri and Rostami-Nejad 2022).

The cultivation of ancient grasses, including the precursors to modern wheat and barley, originated in the Middle East's Fertile Crescent around 10,000 years ago. As cultivation spread across northern Europe, correlating with the rise of early civilizations, reports of symptoms akin to GRDs emerged (Leonard and Vasagar 2014; Gujral, Freeman, and Thomson 2012; Taraghikhah et al. 2020). Subsequently, the industrialization of agriculture, coupled with the utilization of pesticides, nitrogen-based fertilizers, and genetic modifications, has yielded copious amounts of high-gluten wheat now utilized in the global food industry. Such rapid changes in wheat production quantity and quality have potentially contributed to the escalating prevalence of GRDs worldwide (Tovoli et al. 2015; Shewry 2018). Among GRDs, CD has been extensively researched, with its pathogenesis firmly linked to gluten.

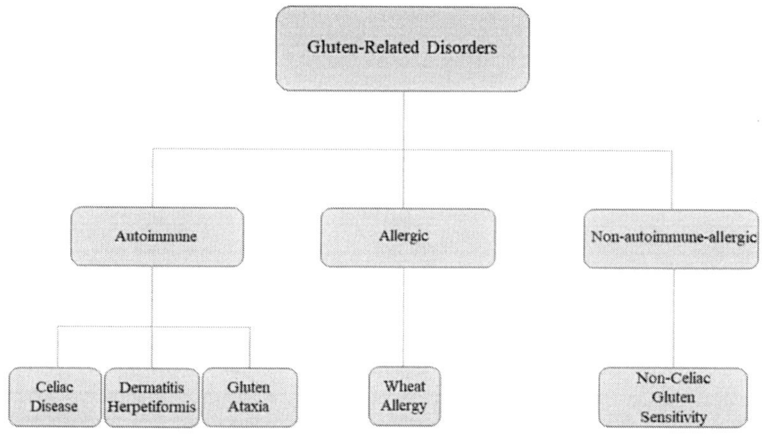

FIGURE 25.3 Classification of gluten-related disorders.

CELIAC DISEASE, PROTEIN INTOLERANCE, AND IMMUNITY

Celiac disease (CD), also known as celiac sprue, non-tropical sprue, and gluten-sensitive enteropathy, is classified as a type of protein intolerance, specifically an autoimmune disorder triggered by the ingestion of gluten proteins (Asri and Rostami-Nejad 2019; Taraghikhah et al. 2020). In individuals with CD, the immune system mounts an abnormal immune response against gluten that targets the small intestine, causing inflammation, tissue damage, and malabsorption of nutrients (Asri et al. 2021; Iversen and Sollid 2023). Understanding the intricate connections between protein intolerance, immunity, and CD pathogenesis is crucial for developing targeted interventions and improving the management of this debilitating autoimmune condition (Caio et al. 2019; Asri et al. 2020). This chapter delves deeper into the complex mechanisms underlying the interplay between protein intolerance, immune responses, and the development of CD.

CD arises from an inappropriate immune response to gluten peptides, triggering an autoimmune process (Caio et al. 2019). CD is a complex autoimmune disorder characterized by inflammation and damage in the small intestine, specifically targeting the villi, which are responsible for nutrient absorption (Taraghikhah et al. 2020). A genetic background, specifically human leukocyte antigen (HLA)-DQ2/DQ8 positivity and non-HLA genes, is a mandatory determinant for the development of the disease, which occurs with the contribution of environmental factors such as viral infections and dysbiosis of gut microbiota (Aboulaghras et al. 2022; Rossi et al. 2023). The prevalence of CD in the general population is approximately 1%, with a female predominance (El-Metwally et al. 2020). CD can present with a variety of symptoms and manifestations, leading to several phenotypes, including gastrointestinal, extraintestinal, and subclinical (Taraghikhah et al. 2020). The classic symptoms of the condition result from inflammation affecting the gastrointestinal tract, leading to diarrhea and poor absorption of nutrients (Caio et al. 2019). If left untreated, CD can lead to additional serious complications, including malnutrition, osteoporosis, infertility, and an increased risk of certain cancers (Barker and Liu 2008).

The pathogenesis of CD involves a complex interplay between protein intolerance and immune dysregulation (Kupfer and Jabri 2012). Upon gluten ingestion, individuals with CD mount an immune response, primarily mediated by gluten-specific T cells (Taraghikhah et al. 2020). These T cells recognize gluten peptides presented by HLA class II molecules on antigen-presenting cells, leading to the secretion of pro-inflammatory cytokines (Taraghikhah et al. 2020). The inflammatory process subsequently triggers the activation of innate immune cells, such as macrophages and dendritic cells, further perpetuating the immune cascade (Molaaghaee-Rouzbahani et al. 2022). Consequently, tissue damage occurs within the small intestine, primarily affecting the absorptive

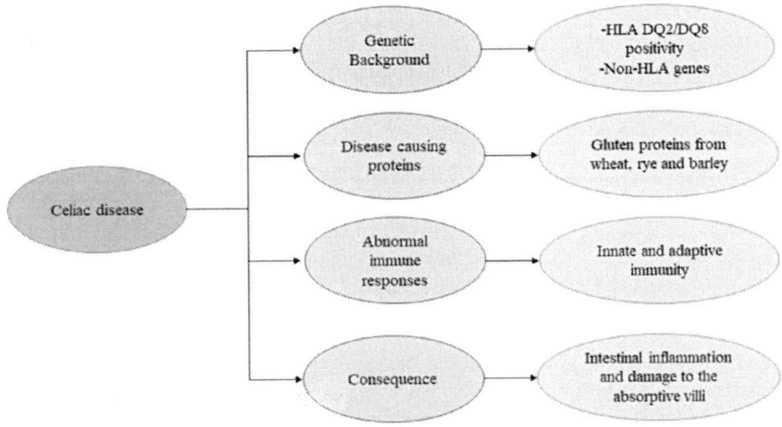

FIGURE 25.4 Various aspects of celiac disease pathogenesis.

villi (Asri and Rostami-Nejad 2019). The impairment of the intestinal barrier function promotes the translocation of gluten peptides and microbial antigens across the intestinal mucosa, leading to sustained immune activation and systemic inflammation (Stolfi et al. 2022). Moreover, the chronic intestinal inflammation in celiac disease may disrupt immune homeostasis, resulting in alterations in immune cell populations, cytokine profiles, and immune regulatory mechanisms (Asri and Rostami-Nejad 2019).

The only effective treatment for CD is a strict, lifelong gluten-free diet (GFD). By completely eliminating gluten from the diet, individuals with CD can manage their symptoms, allow the intestinal lining to heal, and prevent further damage to their small intestine. However, the scientific community is continuously exploring innovative approaches to address CD beyond dietary restrictions (Asri et al. 2020). Ongoing research endeavors focus on understanding and modulating the immune response associated with this protein intolerance disorder (Asri et al. 2020). The goal is to develop alternative treatment options that can provide relief and improve the quality of life for individuals living with this autoimmune disorder. Different aspects of CD pathogenesis are presented in Figure 25.4.

CONCLUSION

In conclusion, protein intolerance, including celiac disease, involves an abnormal immune response to specific food proteins. Understanding its causes and mechanisms is crucial for accurate diagnosis and effective management. The immune system plays a key role, triggering various symptoms and affecting different organ systems. By unraveling the complexities of the immune response, researchers aim to develop targeted interventions and improve treatment options. While a strict gluten-free diet is currently the primary treatment for celiac disease, ongoing research offers promising advancements for better understanding and management of protein intolerance and related disorders.

MINI DICTIONARY OF TERMS

Anaphylaxis: A severe and potentially life-threatening allergic reaction. It occurs rapidly and affects multiple systems of the body.

HLA genes: A group of genes that encode proteins called human leukocyte antigens. These antigens play a crucial role in the immune system by presenting foreign substances, such as antigens from bacteria or viruses, to immune cells called T lymphocytes.

Marsh classification: A system used to classify the severity and extent of damage to the small intestine in individuals with celiac disease.

Protein intolerance: A general term referring to various conditions or situations where an individual has difficulty digesting or metabolizing certain types of proteins.

SUMMARY POINTS

- The immune system plays a crucial role in the development and management of protein intolerance.
- Understanding the interplay between the immune system and protein intolerance is crucial for developing effective diagnostic and therapeutic strategies.
- Celiac disease is classified as a type of gluten protein intolerance triggered by the ingestion of gluten proteins.
- Individuals with CD experience an immune reaction upon gluten consumption involving both innate and adaptive immune pathways.
- Intestinal inflammation in CD results in alterations in immune cell populations, cytokine profiles, and immune regulatory mechanisms.
- This chapter focused on understanding the immune responses associated with this protein intolerance disorder.

REFERENCES

Aboulaghras, Sara, Daniela Piancatelli, Khadija Oumhani, Abdelaali Balahbib, Abdelhakim Bouyahya, and Khalid Taghzouti. 2022. Pathophysiology and immunogenetics of celiac disease. *Clinica Chimica Acta* 528:74–83. https://doi.org/10.1016/j.cca.2022.01.022.

Abrams, E. M., and S. H. Sicherer. 2016. Diagnosis and management of food allergy. *CMAJ* 188 (15):1087–93. https://doi.org/10.1503/cmaj.160124.

Ahmad, Hafiz Ishfaq, Abdul Jabbar, Nadia Mushtaq, Zainab Javed, Muhammad Umar Hayyat, Javaria Bashir, Iqra Naseeb, Zain Ul Abideen, Nisar Ahmad, and Jinping Chen. 2022. Immune tolerance vs. Immune resistance: The interaction between host and pathogens in infectious diseases. *Front Vet Sci* 9. https://doi.org/10.3389/fvets.2022.827407.

Al-Beltagi, M., N.K. Saeed, A.S. Bediwy, R. Elbeltagi. 2022. Cow's milk-induced gastrointestinal disorders: From infancy to adulthood. *World J Clin Pediatr* 11(6):437–454. doi:10.5409/wjcp.v11.i6.437.

Al-Muhsen, S., A. E. Clarke, and R. S. Kagan. 2003. Peanut allergy: an overview. *CMAJ* 168 (10):1279–85.

Asri, Nastaran, and Mohammad Rostami-Nejad. 2019. The facts of celiac disease; A comprehensive review. *Int J Celiac Dis* 7 (2):48–52. doi: 10.12691/ijcd-7-2-7.

Asri, N., M. Rostami-Nejad. 2022. Gluten-related disorders definition. In *Gluten-related disorders*, edited by M. Rostami-Nejad, 49-57. Academic Press. doi: 10.1016/B978-0-12-821846-4.00003-6

Asri, Nastaran, Mohammad Rostami-Nejad, Robert P. Anderson, and Kamran Rostami. 2021. The gluten gene: unlocking the understanding of gluten sensitivity and intolerance. *Appl Clin Genet* 14 (Null):37–50. doi: 10.2147/TACG.S276596.

Asri, N., M. Rostami-Nejad, M. Rezaei-Tavirani, M. Razzaghi, H. Asadzadeh-Aghdaei, and M. R. Zali. 2020. Novel therapeutic strategies for celiac disease. *Middle East J Dig Dis* 12 (4):229–37. doi: 10.34172/mejdd.2020.187.

Balakireva, A. V., and A. A. Zamyatnin. 2016. Properties of gluten intolerance: gluten structure, evolution, pathogenicity and detoxification capabilities. *Nutrients* 8 (10). doi: 10.3390/nu8100644.

Barak, Sheweta, Deepak Mudgil, and B. S. Khatkar. 2014. Influence of gliadin and glutenin fractions on rheological, pasting, and textural properties of dough. Int J *Food Prop* 17 (7):1428–38. doi: 10.1080/10942912.2012.717154.

Barker, J. M., and E. Liu. 2008. Celiac disease: pathophysiology, clinical manifestations, and associated autoimmune conditions. *Adv Pediatr* 55:349–65. doi: 10.1016/j.yapd.2008.07.001.

Berin, M. C., C. Agashe, A. W. Burks, D. Chiang, W. F. Davidson, P. Dawson, A. Grishin, A. K. Henning, S. M. Jones, E. H. Kim, D. Y. M. Leung, M. Masilamani, A. M. Scurlock, S. H. Sicherer, R. A. Wood, and H. A. Sampson. 2022. Allergen-specific T cells and clinical features of food allergy: lessons from CoFAR immunotherapy cohorts. *J Allergy Clin Immunol* 149 (4):1373–82.e12. doi: 10.1016/j.jaci.2021.09.029.

Borschel, Marlene, Dean Antonson, Nancy Murray, Maria Oliva-Hemker, Lynn Mattis, and Geraldine Baggs. 2014. Evaluation of a free amino acid-based formula in infants with presumptive food protein-induced proctocolitis. *SAGE Open Medicine* 2. doi: 10.1177/2050312114551857.

Caio, G., L. Lungaro, N. Segata, M. Guarino, G. Zoli, U. Volta, and R. De Giorgio. 2020. Effect of gluten-free diet on gut microbiota composition in patients with celiac disease and non-celiac gluten/wheat sensitivity. *Nutrients* 12 (6). doi: 10.3390/nu12061832.

Caio, G., U. Volta, A. Sapone, D. A. Leffler, R. De Giorgio, C. Catassi, and A. Fasano. 2019. Celiac disease: a comprehensive current review. *BMC Med* 17 (1):142. doi: 10.1186/s12916-019-1380-z.

Cardona, V., I. J. Ansotegui, M. Ebisawa, Y. El-Gamal, M. Fernandez Rivas, S. Fineman, M. Geller, A. Gonzalez-Estrada, P. A. Greenberger, M. Sanchez Borges, G. Senna, A. Sheikh, L. K. Tanno, B. Y. Thong, P. J. Turner, and M. Worm. 2020. World allergy organization anaphylaxis guidance 2020. *World Allergy Organ J* 13 (10):100472. doi: 10.1016/j.waojou.2020.100472.

Cardoso-Silva, D., D. Delbue, A. Itzlinger, R. Moerkens, S. Withoff, F. Branchi, and M. Schumann. 2019. Intestinal barrier function in gluten-related disorders. *Nutrients* 11 (10). doi: 10.3390/nu11102325.

Caubet, J. C., and J. Wang. 2011. Current understanding of egg allergy. *Pediatr Clin North Am* 58 (2):427–43, xi. doi: 10.1016/j.pcl.2011.02.014.

Cebolla, Á, M. L. Moreno, L. Coto, and C. Sousa. 2018. Gluten immunogenic peptides as standard for the evaluation of potential harmful prolamin content in food and human specimen. *Nutrients* 10 (12). doi: 10.3390/nu10121927.

Cenni, Sabrina, Veronica Sesenna, Giulia Boiardi, Marianna Casertano, Giuseppina Russo, Alfonso Reginelli, Susanna Esposito, and Caterina Strisciuglio. 2023. The role of gluten in gastrointestinal disorders: a review. *Nutrients* 15 (7):1615.

Chaplin, D. D. 2010. Overview of the immune response. *J Allergy Clin Immunol* 125 (2 Suppl 2):S3–23. doi: 10.1016/j.jaci.2009.12.980.

Chen, L., H. Deng, H. Cui, J. Fang, Z. Zuo, J. Deng, Y. Li, X. Wang, and L. Zhao. 2018. Inflammatory responses and inflammation-associated diseases in organs. *Oncotarget* 9 (6):7204–18. doi: 10.18632/oncotarget.23208.

Chinthrajah, R. S., S. Cao, T. Dunham, V. Sampath, S. Chandra, M. Chen, S. Sindher, and K. Nadeau. 2020. Oral immunotherapy for peanut allergy: The pro argument. *World Allergy Organ J* 13 (8):100455. doi: 10.1016/j.waojou.2020.100455.

Coombs, R., E. Simons, R. G. Foty, D. M. Stieb, and S. D. Dell. 2011. Socioeconomic factors and epinephrine prescription in children with peanut allergy. *Paediatr Child Health* 16 (6):341–44. doi: 10.1093/pch/16.6.341.

Cukrowska, B., A. Sowińska, J. B. Bierła, E. Czarnowska, A. Rybak, and U. Grzybowska-Chlebowczyk. 2017. Intestinal epithelium, intraepithelial lymphocytes and the gut microbiota - Key players in the pathogenesis of celiac disease. *World J Gastroenterol* 23 (42):7505–18. doi: 10.3748/wjg.v23.i42.7505.

D'Auria, E., S. Salvatore, M. Acunzo, D. Peroni, E. Pendezza, E. Di Profio, G. Fiore, G.V. Zuccotti, E. Verduci. 2021. Hydrolysed formulas in the management of cow's milk allergy: New insights, pitfalls and tips. *Nutrients* 13(8):2762. doi: 10.3390/nu13082762.

Davis, C. M., R. S. Gupta, O. N. Aktas, V. Diaz, S. D. Kamath, and A. L. Lopata. 2020. Clinical management of seafood allergy. *J Allergy Clin Immunol Pract* 8 (1):37–44. doi: 10.1016/j.jaip.2019.10.019.

Dhar, S., and S. M. Srinivas. 2016. Food allergy in atopic dermatitis. *Indian J Dermatol* 61 (6):645–48. doi: 10.4103/0019-5154.193673.

di Palmo, E., M. Gallucci, F. Cipriani, L. Bertelli, A. Giannetti, and G. Ricci. 2019. Asthma and food allergy: which risks? *Medicina* 55 (9). doi: 10.3390/medicina55090509.

El-Metwally, Ashraf, Paivi Toivola, Khalid AlAhmary, Salwa Bahkali, Ali AlKhathaami, Munira K. AlSaqabi, Shatha A. Al Ammar, Munazza Jawed, and Saleh M. Alosaimi. 2020. The epidemiology of celiac disease in the general population and high-risk groups in arab countries: a systematic review. *BioMed Res Int* 2020:6865917. doi: 10.1155/2020/6865917.

El Ansari, Y. S., C. Kanagaratham, and H. C. Oettgen. 2020. Mast cells as regulators of adaptive immune responses in food allergy. *Yale J Biol Med* 93 (5):711–18.

Fernandes, Telmo J. R., Joana Costa, M. Beatriz P. P. Oliveira, and Isabel Mafra. 2015. An overview on fish and shellfish allergens and current methods of detection. *Food Agric Immunol* 26 (6):848–69. doi: 10.1080/09540105.2015.1039497.

Foong, R. X., G. du Toit, and A. T. Fox. 2017. Asthma, food allergy, and how they relate to each other. *Front Pediatr* 5:89. doi: 10.3389/fped.2017.00089.

Garrote, J. A., E. Gómez-González, D. Bernardo, E. Arranz, and F. Chirdo. 2008. Celiac disease pathogenesis: the proinflammatory cytokine network. *J Pediatr Gastroenterol Nutr* 47 (Suppl 1):S27–32. doi: 10.1097/MPG.0b013e3181818fb9.

Giannetti, A., G. Toschi Vespasiani, G. Ricci, A. Miniaci, E. di Palmo, and A. Pession. 2021. Cow's milk protein allergy as a model of food allergies. *Nutrients* 13 (5). doi: 10.3390/nu13051525.

Gujral, N., H. J. Freeman, and A. B. Thomson. 2012. Celiac disease: prevalence, diagnosis, pathogenesis and treatment. *World J Gastroenterol* 18 (42):6036–59. doi: 10.3748/wjg.v18.i42.6036.

Guler, N., F. C. Cokugras, N. Sapan, A. Selimoglu, I. Turktas, H. Cokugras, M. Aydogan, and O. F. Beser. 2020. Diagnosis and management of cow's milk protein allergy in Turkey: Region-specific recommendations by an expert-panel. *Allergol Immunopathol (Madr)* 48 (2):202–10. doi: 10.1016/j.aller.2019.05.004.

Gupta, K., and I. T. Harvima. 2018. Mast cell-neural interactions contribute to pain and itch. *Immunol Rev* 282 (1):168–187. doi: 10.1111/imr.12622.

Heine, R. G. 2015. Gastrointestinal food allergies. *Chem Immunol Allergy* 101:171–80. doi: 10.1159/000371700.

Iversen, Rasmus, and Ludvig M. Sollid. 2023. The immunobiology and pathogenesis of celiac disease. *Annu Rev Pathol* 18 (1):47–70. doi: 10.1146/annurev-pathmechdis-031521-032634.

Katta, R., and M. Schlichte. 2014. Diet and dermatitis: food triggers. *J Clin Aesthet Dermatol* 7 (3):30–36.

Kattan, J. D., R. R. Cocco, and K. M. Järvinen. 2011. Milk and soy allergy. *Pediatr Clin North Am* 58 (2):407–26, x. doi: 10.1016/j.pcl.2011.02.005.

Kłosok, K., R. Welc, E. Fornal, and A. Nawrocka. 2021. Effects of physical and chemical factors on the structure of gluten, gliadins and glutenins as studied with spectroscopic methods. *Molecules* 26 (2). doi: 10.3390/molecules26020508.

Kopko, C., J. A. Garthoff, K. Zhou, L. Meunier, A. J. O'Sullivan, and V. Fattori. 2022. Are alternative proteins increasing food allergies? Trends, drivers and future perspectives. *Trends Food Sci Technol* 129:126–33. doi: https://doi.org/10.1016/j.tifs.2022.09.008.

Kupfer, S. S., and B. Jabri. 2012. Pathophysiology of celiac disease. *Gastrointest Endosc Clin N Am* 22 (4):639–60. doi: 10.1016/j.giec.2012.07.003.

Lebwohl, B., J. F. Ludvigsson, and P. H. Green. 2015. Celiac disease and non-celiac gluten sensitivity. *BMJ* 351:h4347. doi: 10.1136/bmj.h4347.

Leonard, M. M., and B. Vasagar. 2014. US perspective on gluten-related diseases. *Clin Exp Gastroenterol* 7:25–37. doi: 10.2147/ceg.S54567.

Losurdo, G., M. Principi, A. Iannone, A. Amoruso, E. Ierardi, A. Di Leo, and M. Barone. 2018. Extra-intestinal manifestations of non-celiac gluten sensitivity: An expanding paradigm. *World J Gastroenterol* 24 (14):1521–30. doi: 10.3748/wjg.v24.i14.1521.

Maryniak, N.Z., A.I. Sancho, E.B. Hansen, K.L. Bøgh. 2022. Alternatives to cow's milk-based infant formulas in the prevention and management of cow's milk allergy. *Foods* 11(7):926. doi: 10.3390/foods11070926.

Meinlschmidt, Pia, Daniela Sussmann, Ute Schweiggert-Weisz, and Peter Eisner. 2016. Enzymatic treatment of soy protein isolates: effects on the potential allergenicity, technofunctionality, and sensory properties. *Food Sci Nutr* 4 (1):11–23. doi: https://doi.org/10.1002/fsn3.253.

Molaaghaee-Rouzbahani, Sara, Nastaran Asri, Somayeh Jahani-Sherafat, Davar Amani, Andrea Masotti, Kaveh Baghaei, Abbas Yadegar, Hamed Mirjalali, and Mohammad Rostami-Nejad. 2022. The modulation of macrophage subsets in celiac disease pathogenesis. *Immun Inflamm Dis* 10 (12):e741. https://doi.org/10.1002/iid3.741.

Nakano, N., and J. Kitaura. 2022. Mucosal mast cells as key effector cells in food allergies. *Cells* 11 (3). doi: 10.3390/cells11030329.

Nicholson, L. B. 2016. The immune system. *Essays Biochem* 60 (3):275–301. doi: 10.1042/ebc20160017.

Ramsay, D. B., S. Stephen, M. Borum, L. Voltaggio, and D. B. Doman. 2010. Mast cells in gastrointestinal disease. Gastroenterol Hepatol *(N Y)* 6 (12):772–77.

Reda, Shereen M. 2009. Gastrointestinal manifestations of food allergy. *Pediatric Health* 3 (3):217–29. doi: 10.2217/phe.09.13.

Rossi, R. E., G. Dispinzieri, A. Elvevi, and S. Massironi. 2023. Interaction between gut microbiota and celiac disease: from pathogenesis to treatment. *Cells* 12 (6). doi: 10.3390/cells12060823.

Santoro, A., L. Andreozzi, G. Ricci, C. Mastrorilli, C. Caffarelli. 2019. Allergic reactions to cow's milk proteins in medications in childhood. *Acta Biomed* 90(3-S):91–93. doi: 10.23750/abm.v90i3-S.8169.

Sapone, A., J. C. Bai, C. Ciacci, J. Dolinsek, P. H. Green, M. Hadjivassiliou, K. Kaukinen, K. Rostami, D. S. Sanders, M. Schumann, R. Ullrich, D. Villalta, U. Volta, C. Catassi, and A. Fasano. 2012. Spectrum of gluten-related disorders: consensus on new nomenclature and classification. *BMC Med* 10:13. doi: 10.1186/1741-7015-10-13.

Satitsuksanoa, Pattraporn, Kirstin Jansen, Anna Głobińska, Willem van de Veen, and Mübeccel Akdis. 2018. Regulatory immune mechanisms in tolerance to food allergy. *Front Immunol* 9. doi: 10.3389/fimmu.2018.02939.

Scherf, Katharina Anne, Peter Koehler, and Herbert Wieser. 2016. Gluten and wheat sensitivities – An overview. *J Cereal Sci* 67:2–11. doi: https://doi.org/10.1016/j.jcs.2015.07.008.

Sharma, N., S. Bhatia, V. Chunduri, S. Kaur, S. Sharma, P. Kapoor, A. Kumari, and M. Garg. 2020. Pathogenesis of celiac disease and other gluten related disorders in wheat and strategies for mitigating them. *Front Nutr* 7:6. doi: 10.3389/fnut.2020.00006.

Shewry, Peter R. 2018. Do ancient types of wheat have health benefits compared with modern bread wheat? *J Cereal Sci* 79:469–76. doi: https://doi.org/10.1016/j.jcs.2017.11.010.

Smith, F., X. Pan, V. Bellido, G. A. Toole, F. K. Gates, M. S. Wickham, P. R. Shewry, S. Bakalis, P. Padfield, and E. N. Mills. 2015. Digestibility of gluten proteins is reduced by baking and enhanced by starch digestion. *Mol Nutr Food Res* 59 (10):2034–43. doi: 10.1002/mnfr.201500262.

Stolfi, C., C. Maresca, G. Monteleone, and F. Laudisi. 2022. Implication of intestinal barrier dysfunction in gut dysbiosis and diseases. *Biomedicines* 10 (2). doi: 10.3390/biomedicines10020289.

Taraghikhah, N., S. Ashtari, N. Asri, B. Shahbazkhani, D. Al-Dulaimi, M. Rostami-Nejad, M. Rezaei-Tavirani, M. R. Razzaghi, and M. R. Zali. 2020. An updated overview of spectrum of gluten-related disorders: clinical and diagnostic aspects. *BMC Gastroenterol* 20 (1):258. doi: 10.1186/s12876-020-01390-0.

Tian, N., D. A. Leffler, C. P. Kelly, J. Hansen, E. V. Marietta, J. A. Murray, D. Schuppan, and E. J. Helmerhorst. 2015. Despite sequence homologies to gluten, salivary proline-rich proteins do not elicit immune responses central to the pathogenesis of celiac disease. *Am J Physiol Gastrointest Liver Physiol* 309 (11):G910–7. doi: 10.1152/ajpgi.00157.2015.

Tovoli, F., C. Masi, E. Guidetti, G. Negrini, P. Paterini, and L. Bolondi. 2015. Clinical and diagnostic aspects of gluten related disorders. *World J Clin Cases* 3 (3):275–84. doi: 10.12998/wjcc.v3.i3.275.

Turner, Paul J., Stefania Arasi, Barbara Ballmer-Weber, Alessia Baseggio Conrado, Antoine Deschildre, Jennifer Gerdts, Susanne Halken, Antonella Muraro, Nandinee Patel, Ronald Van Ree, Debra de Silva, Margitta Worm, Torsten Zuberbier, Graham Roberts, and Asthma European Network Food Allergy Guideline Group the Global Allergy. 2022. Risk factors for severe reactions in food allergy: rapid evidence review with meta-analysis. *Allergy* 77 (9):2634–52. doi: https://doi.org/10.1111/all.15318.

Turner, P. J., J. L. Baumert, K. Beyer, R. J. Boyle, C. H. Chan, A. T. Clark, R. W. R. Crevel, A. DunnGalvin, M. Fernández-Rivas, M. H. Gowland, L. Grabenhenrich, S. Hardy, G. F. Houben, J. O'B Hourihane, A. Muraro, L. K. Poulsen, K. Pyrz, B. C. Remington, S. Schnadt, R. van Ree, C. Venter, M. Worm, E. N. C. Mills, G. Roberts, and B. K. Ballmer-Weber. 2016. Can we identify patients at risk of life-threatening allergic reactions to food? *Allergy* 71 (9):1241–55. doi: https://doi.org/10.1111/all.12924.

Turner, P. J., M. Worm, I. J. Ansotegui, Y. El-Gamal, M. F. Rivas, S. Fineman, M. Geller, A. Gonzalez-Estrada, P. A. Greenberger, L. K. Tanno, M. S. Borges, G. Senna, A. Sheikh, B. Y. Thong, M. Ebisawa, and V. Cardona. 2019. Time to revisit the definition and clinical criteria for anaphylaxis? *World Allergy Organ J* 12 (10):100066. doi: 10.1016/j.waojou.2019.100066.

Urade, R., N. Sato, and M. Sugiyama. 2018. Gliadins from wheat grain: an overview, from primary structure to nanostructures of aggregates. *Biophys Rev* 10 (2):435–43. doi: 10.1007/s12551-017-0367-2.

Vandenplas, Y., S. Koletzko, E. Isolauri, D. Hill, A. P. Oranje, M. Brueton, A. Staiano, and C. Dupont. 2007. Guidelines for the diagnosis and management of cow's milk protein allergy in infants. *Arch Dis Child* 92 (10):902–8. doi: 10.1136/adc.2006.110999.

Venter, Carina, Scott H. Sicherer, and Matthew Greenhawt. 2019. Management of peanut allergy. *J Allergy Clin Immunol Pract* 7 (2):345–55.e2. https://doi.org/10.1016/j.jaip.2018.10.043.

Voisine, Jordan, and Valérie Abadie. 2021. Interplay between gluten, HLA, innate and adaptive immunity orchestrates the development of coeliac disease. *Front Immunol* 12. doi: 10.3389/fimmu.2021.674313.

Walker-Smith, J. 1988. Dietary protein intolerance. In *Diseases of the Small Intestine in Childhood*, edited by J. Walker-Smith, 144–84. London: Btitterworth. doi: 10.1016/B978-0-407-01320-9.50011-X. Epub 2013 Oct 21.

Yu, W., D. M. H. Freeland, and K. C. Nadeau. 2016. Food allergy: immune mechanisms, diagnosis and immunotherapy. *Nat Rev Immunol* 16 (12):751–65. doi: 10.1038/nri.2016.111.

Zepeda-Ortega, Benjamin, Anne Goh, Paraskevi Xepapadaki, Aline Sprikkelman, Nicolaos Nicolaou, Rosa Hernandez, Amir Abdul Latiff, Miu Yat, Mohamed Diab, Bakr Hussaini, Budi Setiabudiawan, Urszula Kudla, Joost van Neerven, Leilani Muhardi, and John Warner. 2021. Strategies and future opportunities for the prevention, diagnosis, and management of cow milk allergy. *Front Immunol* 12:608372. doi: 10.3389/fimmu.2021.608372.

Zhou, Yang, Jin-shui Wang, Xiao-jia Yang, Dan-hua Lin, Yun-fang Gao, Yin-jie Su, Sen Yang, Yan-jie Zhang, and Jing-jing Zheng. 2013. Peanut allergy, allergen composition, and methods of reducing allergenicity: a review. *Int J Food Sci* 2013:909140. doi: 10.1155/2013/909140.

26 Modeling Protein Malnutrition in Rodents

Implications for Clinical Investigations

Murali Badanthadka, Madhura R. J., Varsha A., and Vinitha D'Souza

ABBREVIATIONS

ASD	Autism Spectrum Disorder
EMLc	Essential Medicines List for Children
LBM	Lean body mass
NFHS5	National Family Health Survey 5
PMN	Protein malnutrition
SOFI	State of Food Security and Nutrition in the World
UN	United Nations
WHO	World Health Organization

INTRODUCTION

Malnutrition occurs when the intake of nutrients does not meet the body's requirements. It encompasses two types: undernutrition and overnutrition, although the terms malnutrition/malnourished typically refer to undernutrition/undernourished. Controlling malnutrition in all its forms is one of the greatest health challenges globally, as it affects every country to some degree. Poverty magnifies the risk of malnutrition, with people from poor financial backgrounds being more likely to be affected. Malnutrition amplifies healthcare costs, decreases productivity, and slows down economic growth, thus perpetuating the cycle of poverty and ill health (Maleta, 2006; Siddiqui et al., 2020; WHO factsheets, 2021).

Protein malnutrition (PMN) is one of the main forms of undernutrition. Its causes are multifactorial and closely linked to other forms of malnutrition, such as inadequate micronutrient intake, calorie deficit, and an unhealthy diet, which collectively contribute to non-communicable disease-related deaths worldwide. Low-income families face increased vulnerability to poor nutrition due to inadequate household resources and under-resourced communities. Infants, children, adolescents, and women are at particularly high risk of malnutrition. Experts recommed both monitoring and maintaining nutrition for infants up to two years of age, and avoiding conception during this period, as the best possible strategy to prevent protein malnutrition development (Keats et al., 2021).

Age-related physical changes, which affect cognitive ability and individual psychology, may accelerate malnutrition development in old age. Additional social factors like depression, isolation, limited earnings, and dementia may further increase the risk of nutritional deficiency. Moreover, food insecurity, poverty, limited access to nutritious food, exposure to infectious diseases, compromised

DOI: 10.1201/b23328-26

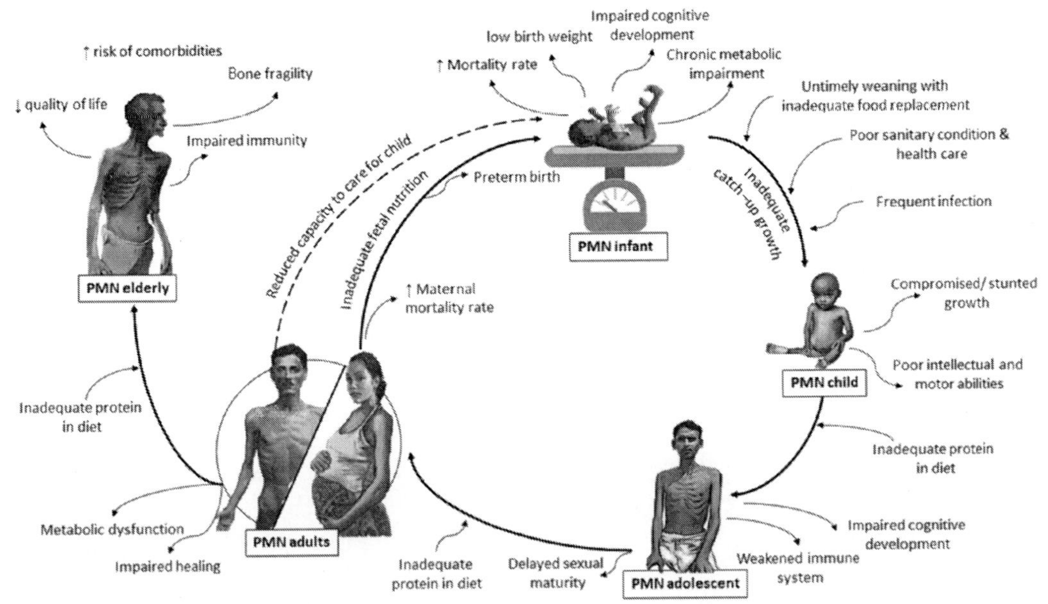

FIGURE 26.1 This schematic illustration presents the protein malnutrition cycle across various age groups and its associated consequences on health. The illustration depicts the protein malnutrition cycle throughout different life cycle stages: infancy, childhood, adolescence, adulthood, and old age, highlighting the impact of inadequate protein intake at various life stages and its associated consequences on health.

hygiene, and poor breastfeeding practices also contribute to PMN development (Maleta, 2006; Siddiqui et al., 2020; Panda et al., 2020; Keats et al., 2021; Chatterjee, 2021; Dent et al., 2023).

Developing protein malnutrition models in laboratory animals is vital for PMN research. These models help scientists overcome treatment barriers and differentiate the pharmacokinetics and pharmacodynamics mechanisms involved in interactions, including drug toxicity under PMN conditions. Rodent protein malnutrition models have proven valuable in research due to their pathological similarities to clinical conditions, ease of handling, and cost-effectiveness (Figure 26.1).

TYPES OF PROTEIN MALNUTRITION

Protein malnutrition, a form of undernutrition, occurs when dietary protein intake falls below the body's requirements. Protein malnutrition can be classified as mild, moderate, or severe based on its severity, using body mass index (BMI) calculations that consider an individual's body weight and height. In developing countries, PMN predominantly affects younger populations, while in developed countries, it primarily impacts older adults. PMN can be further categorized as primary or secondary, depending on its developmental context.

Primary protein malnutrition typically emerges in childhood due to insufficient protein intake via breastfeeding or dietary sources (often characterized by low nutritional value and additives) during critical developmental periods. Secondary protein malnutrition results from other diseases, infections, and critical illnesses (Dent et al., 2023).

Primary protein malnutrition is classified into three forms:

1) Kwashiorkor: This condition affects children and stems from short periods of breastfeeding, often due to consecutive pregnancies. It manifests as edema, causing a swollen head and abdomen, and can lead to acute illness. Kwashiorkor predominantly occurs in rural African regions, where staple foods are high in carbohydrates but low in proteins.

2) Marasmus: Characterized by low birth weight and undernutrition, marasmus leads to weight loss and depletion of fat and muscles in affected children. This form is more common than kwashiorkor.
3) Marasmic-kwashiorkor malnutrition: This type exhibits symptoms of both kwashiorkor and marasmus, including wasting and edema (Madhura et al., 2023; D'Souza et al., 2022; Victora et al., 2021).

EPIDEMIOLOGY OF PROTEIN MALNUTRITION

Protein malnutrition is a significant societal challenge, affecting individuals of all ages. Its prevalence is highest among children under five years of age. Inadequate protein intake during critical growth and development periods can lead to stunting, wasting, and other long-term health consequences. Older adults also face elevated risks of protein malnutrition due to poor appetite, metabolic changes, and associated medical conditions that compromise nutrient absorption and utilization. The global burden of malnutrition significantly impacts social, medical, and developmental domains, with serious and long-lasting consequences (Dent et al., 2023).

The State of Food Security and Nutrition in the World (SOFI) report indicates that in 2022, approximately 691–783 million people faced severe hunger, representing almost 122 million more cases than before the COVID-19 pandemic. Additionally, man-made complications such as the Ukraine war, climate changes, and natural calamities exacerbate the situation. Despite various government initiatives, global hunger numbers have plateaued, and food crises persist in some regions. The 2022 report highlights that Asia and Latin America made progress in hunger reduction compared to Western Asia, certain sub-regions of Africa, and the Caribbean, where hunger continued to increase. Throughout the African continent, one in five people experiences hunger, a rate twice the global average, making it the worst-affected region globally.

In 2022, moderate or severe food insecurity affected approximately 29.6% of the global population (~2.4 billion people), who lacked consistent access to food. Of these, 900 million people faced severe food insecurity (nutritional insufficiency) resulting in malnutrition. Compared to 2019, the capacity to afford healthy food deteriorated for about 134 million people post-COVID. Among children under five years of age, 148 million (22.3%) experienced stunting, and 45 million (6.8%) suffered from wasting (Unicef reports, 2023; Ma et al., 2022; UN reports, 2022; Chatterjee, 2021; Victora et al., 2021).

Overview of Developed Nations: United States

In 2021, hunger affected more than 9 million children (1 in 8) in the US. Approximately 3.4% of children under five years of age showed signs of stunting, while 0.1% experienced wasting, the lowest rate globally.

The US government implemented a food security policy for 40 million Americans, including 12 million children, who either skip meals or consume non-nutritional food. This situation arises because many families suffering from hunger and poverty reside in areas with limited access to fresh, unprocessed food or where such food is prohibitively expensive. Low food security affected 6.4% (8.4 million) of the population, while 3.8% (5.1 million) experienced very low food security due to low economic status, increased family size, and financial constraints resulting from the COVID-19 pandemic (Unicef reports, 2023; Seferidi et al., 2022; UN reports, 2022)

Overview of Underdeveloped African Countries

Approximately 60 million people face food insecurity in Djibouti, Ethiopia, Kenya, Somalia, South Sudan, Sudan, and Uganda across the greater Horn of Africa. In 2023, over 10 million children under five years suffered from acute malnutrition. According to a World Health Organization (WHO) report, Somalia, South Sudan, and parts of Kenya have reported the highest numbers of severely malnourished children seeking medical care since the COVID-19 pandemic (Kambale et al., 2022; Diallo et al., 2022; Victora et al., 2021).

Overview of Fast-Growing Economies: India

Despite its status as a developing/developed country, India remains the world's largest contributor to the global undernourished population. Protein malnutrition and other forms of malnutrition continue to pose significant public health challenges. The Government of India's National Family Health Survey 5 (NFHS5) reports that about 35.7% of children under five in India are stunted, indicating chronic malnutrition, which often includes protein deficiencies. The survey also found that 19% of children under five were affected by wasting and 32% were underweight. Protein malnutrition also affects pregnant women, lactating mothers, and the elderly. Regional disparities, with higher malnutrition rates in certain states, occur because of differences in educational background and socioeconomic status. Furthermore, children born to undernourished mothers perpetuate the cycle of malnutrition (Diallo et al., 2022; Panda et al., 2020; Chatterjee, 2021; Worldmeter, 2023).

PATHOPHYSIOLOGY OF PROTEIN MALNUTRITION (FIGURE 26.2)

Protein malnutrition commonly occurs during starvation, when protein intake falls below the body's minimum requirements. Unlike protein-energy malnutrition, which encompasses deficiencies in both protein and overall energy intake, protein malnutrition specifically arises when the protein content in the diet is insufficient, even when the overall energy supply remains adequate. Individuals may unconsciously prioritize other dietary components over protein to meet their energy needs, potentially leading to protein malnutrition even in cases of optimal overall energy

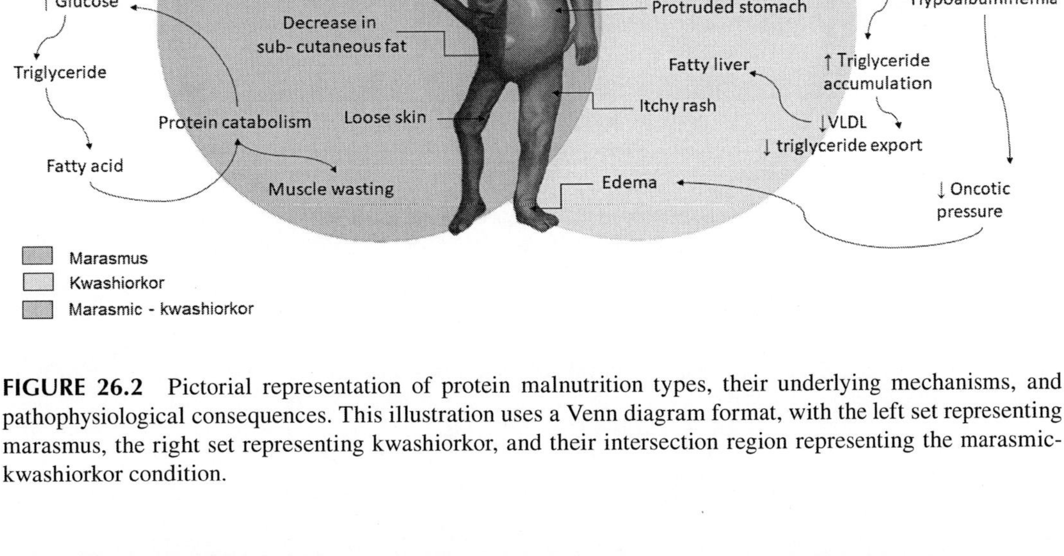

FIGURE 26.2 Pictorial representation of protein malnutrition types, their underlying mechanisms, and pathophysiological consequences. This illustration uses a Venn diagram format, with the left set representing marasmus, the right set representing kwashiorkor, and their intersection region representing the marasmic-kwashiorkor condition.

intake. Initial changes in protein malnutrition include strategic reduction of lean body mass (LBM) and decreased bioavailability of methionine-like essential amino acids (Wharton, 1984). The body's initial response is a decrease in metabolic activity, which slows or stops development (Dipasquale et al., 2020). This minimizes nutrient (protein) usage and maintains system function for a prolonged period. At first, adipose tissues are broken down to release energy until depleted. The process slows to conserve energy and maintain homeostasis (Hoffer, 2001). Once adipose tissues are depleted, the body may use proteins for energy production, resulting in muscle breakdown and a negative nitrogen balance (Victora et al., 2021).

As protein depletion progresses, the requirement for dietary proteins also decreases, leading to weight loss. Visceral organs, such as the liver, kidney, and heart, are relatively preserved up to a certain threshold level. Unless stored protein and energy are critically low, protein requirements decrease to maintain homeostasis. These adaptations result in reduced cardiac and respiratory capacity, impaired normal functioning, muscle weakness, and low protein reserves (Hoffer, 2001). Consequently, individuals experience growth retardation, impaired protein synthesis, and neurodegenerative effects. The resulting inflammation compounds existing complications and increases the risk of developing incurable neurodegeneration. Pathophysiological changes in protein malnutrition primarily stem from malabsorption, which is the result of inadequate protein intake or gastrointestinal disease that compromises digestion and absorption (Neale, 1968).

Protein malnutrition affects not only energy production and utilization pathways but also impairs hormonal mechanisms and immunity. It impacts growth and thyroid hormones, cellular glucose utilization mechanisms, and insulin secretion (Dipasquale et al., 2020; Divyaveer, 2019). Insulin release is hampered by functional and structural alterations in the endocrine pancreas, leading to insulin resistance and eventual hyperglycemia (Brooks et al., 1993) as a consequence of compromised glucose metabolism and inflammation associated with malnutrition (Ingenbleek, 2020).

Protein malnutrition manifests in various forms, each with distinct pathophysiological mechanisms:

A) Kwashiorkor: This severe form of protein malnutrition affects children exposed to a low-protein and high-carbohydrate diet after weaning. Characteristics include:
   - Impaired protein synthesis: Essential amino acid deficiency reduces protein synthesis, compromising the immune system, neurotransmission, tissue repair, and growth.
   - Hypoalbuminemia: Decreased blood albumin levels cause edema (fluid accumulation) in various tissues, as albumin is the major blood protein responsible for maintaining oncotic/osmotic pressure. Hypoalbuminemia influences fluid movement from blood vessels to interstitial spaces.
   - Immune system dysfunction: Both innate and adaptive immunity are impaired (Rytter et al., 2014). Impairments include reduced epithelial barrier function in the skin and gut, decreased granulocyte microbicidal activity, fewer circulating dendritic cells, and reduced complement proteins, all of which compromise innate immunity. Adaptive immunity is impaired due to reduced soluble IgA levels, diminished delayed-type hypersensitivity responses, fewer circulating B cells, lymphoid organ atrophy, hyporesponsiveness of lymphocytes to phytohemagglutinin, and a shift in cytokines from Th-1- to Th-2-mediated mechanisms (Bourke et al., 2016).
   - Liver dysfunction: This leads to a decreased capacity to synthesize proteins.

B) Marasmus: This condition results from an overall insufficient intake of protein and calories. Characteristics include:
   - Generalized wasting of body tissues.
   - Energy deficiency compensated by breakdown of body tissues, including fat and muscle, for energy production.

- Muscle wasting due to energy deficiency. Muscle proteins break down to release amino acids for energy production, reducing functional capacity and causing weakness.
- Organ dysfunction, including heart and immune system impairment, due to insufficient essential amino acids for metabolic function.
- Stunted growth in infants and children, who fail to reach developmental milestones due to inadequate nutritional intake.

C) Marasmic-kwashiorkor: Shows overlapping pathophysiology of both marasmus and kwashiorkor.

Protein malnutrition often leads to additional complications. Fatty liver and hepatomegaly are among the conditions that can develop as a result of PMN (Ampong et al., 2020; Solon-Biet et al., 2015).

RODENT MODELS OF PROTEIN MALNUTRITION

Protein malnutrition remains a global challenge. Researchers have conducted extensive studies through various means to minimize its complications. One approach involves the use of rodent models, which are commonly employed in scientific research. Scientists have developed multiple rodent models of protein malnutrition, utilizing approaches such as starvation, restricted feeding, low-protein diet, and no-protein diet. While some models fail to represent the consequences accurately, and others fail to mimic the clinical conditions, a few stable protein malnutrition rodent models effectively demonstrate the implications and simulate human clinical conditions.

a) Rationale for using protein malnutrition rodent models in research:
   - Physiological similarities: Rodents share physiological and genetic similarities with humans, making them suitable for understanding various biological processes and diseases/disorders, including malnutrition.
   - Study duration: Rodents have shorter life spans, considerably shorter reproduction cycles, and faster maturation. This enables researchers to conduct studies and obtain quick results. Additionally, results from these studies often show similarities to clinical conditions.
   - Study cost: Rodents' low body weight (compared to other mammals) allows researchers to use smaller quantities of experimental products/molecules/drugs, enhancing cost-effectiveness. Their shorter reproduction cycle, minimal space requirement for feeding and housing, and low maintenance costs further boost cost-effectiveness.
   - Ethical constraints: Ethical restrictions are fewer for rodent models than for higher mammals or humans. However, ethical clearance remains mandatory before conducting experiments.
   - Genetic manipulation: Rodents can be genetically modified to mimic specific clinical conditions. This allows researchers to extrapolate study results from a controlled biological environment to clinical medicine with specific outcomes, facilitating bench-to-bedside translation (Figure 26.3).

b) Limitations of protein malnutrition rodent models:
   - Species differences: Though rodents and humans are highly similar, distinct metabolic and physiological differences (albeit minor) may result in different responses in humans.
   - Isolated factors: Rodent studies often focus on specific/isolated aspects of malnutrition, with few exceptions. In clinical settings, malnutrition typically consists of multiple nutritional deficiencies.

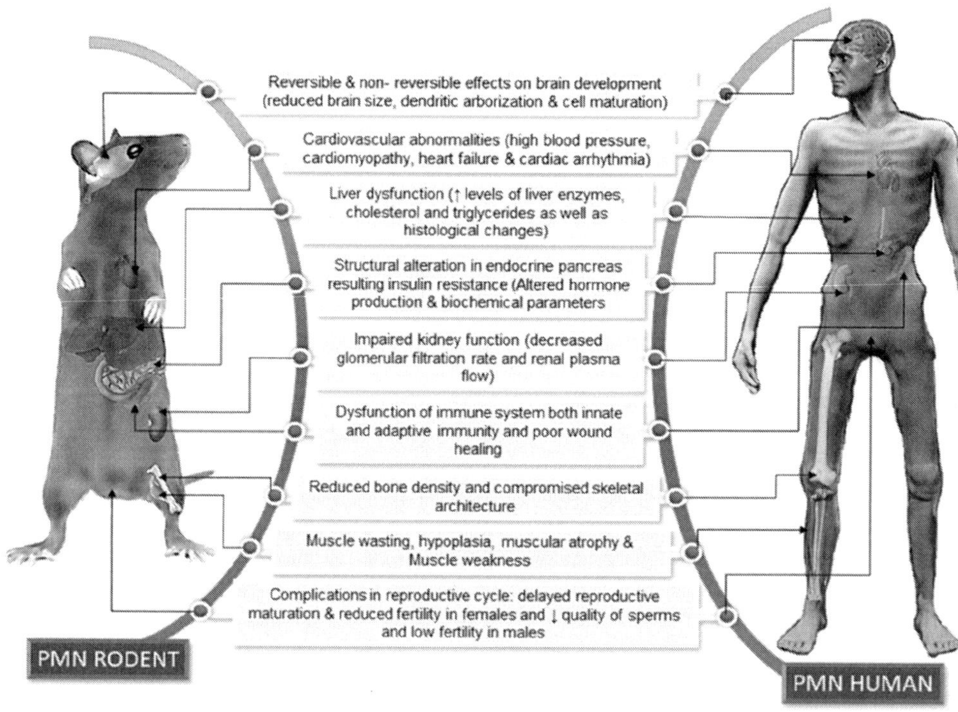

FIGURE 26.3 Comparative analysis of protein malnutrition in rodent and human models. This figure compares a protein-malnourished rodent model with corresponding pathophysiological conditions observed in humans. It highlights significant resemblances between the two, shedding light on key similarities in the manifestation of protein malnutrition.

- Diet complexity: Human dietary complexity cannot be precisely replicated in rodent diets. Researchers employ simplified dietary options to induce malnutrition in rodents, which differ from human dietary compositions.
- Ethical and moral considerations: Ethical constraints govern rodent use in experiments, varying slightly depending on the region and country where the investigation is conducted.
- Bench-to-bedside translation: Results obtained from rodent studies may not directly correlate to human populations. Despite similarities, biological differences between the two species may influence outcomes.
- Study duration dissimilarities: Malnutrition is a long-term process. Some rodent studies draw conclusions based on short-term observations. These study durations are often briefer than necessary to accurately reflect the human lifespan.

Any model developed for research must fulfill vital criteria and serve its intended purpose. Key considerations include:

- Stability of the model
- Similarities to the development of human protein malnutrition
- Suitability for short-term as well as long-term studies.
- Adherence to ethical practices by upholding animal welfare. For example, avoiding starvation and extreme diet restriction, which can lead to CNS-related complications such as anxiety, hyperactivity, and aggressiveness in animals (Figure 26.4).

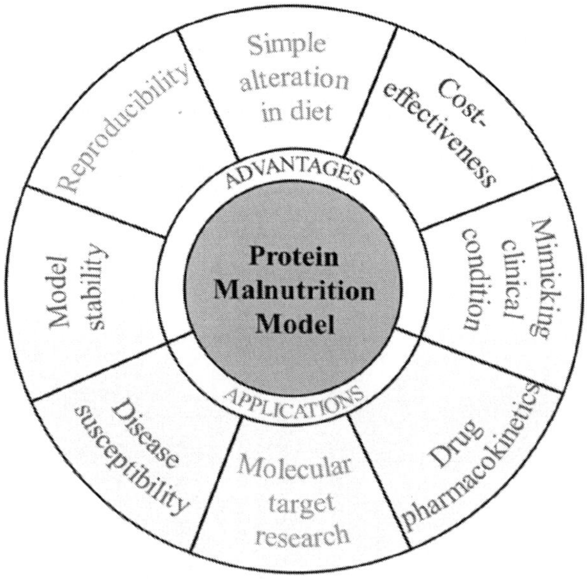

FIGURE 26.4 Advantages and applications of protein malnutrition rodent models in research. This figure outlines the advantages of protein malnutrition rodent models and their research applications, providing valuable information for developing new research protocols.

CONSEQUENCES OF PROTEIN MALNUTRITION IN PRECLINICAL ANIMAL MODELS

In rodent models, protein malnutrition at an early age has significant consequences in both the immediate and later stages of life. These consequences may be categorized as a) early consequences, b) delayed consequences, c) reversible and non-reversible conditions, d) susceptibility to disease, and e) susceptibility to drug-induced toxicity and environmental hazards (Figure 26.5).

Early Consequences of PMN in Rodent Models

Protein is necessary for optimal growth. Studies have demonstrated the effect of protein malnutrition during critical stages like pregnancy, lactation, and the post-weaning period. In rodents, PMN during these stages decreases growth rate, lowers body weight, and delays/impairs development, ultimately resulting in growth retardation (Naik et al., 2017). Muscle atrophy, which results in weakened muscles and reduced physical activity, also occurs due to protein deficiency. Malnutrition impairs brain development in both size and function. Malnutrition-induced abnormalities at the cell maturation levels have cumulative, long-lasting effects on cognitive and behavioral aspects. Hence, deficiency in protein at an early age causes cognitive deficits that impact learning and memory processes. PMN also affects organ functions, including vital organs, leading to potential metabolic impairment and cardiovascular problems (Naik et al., 2017). Altered levels of various biochemical parameters, neurotransmitters, and oxidative stress markers further illustrate these effects, eventually leading to developmental abnormalities in adulthood. A compromised immune system also

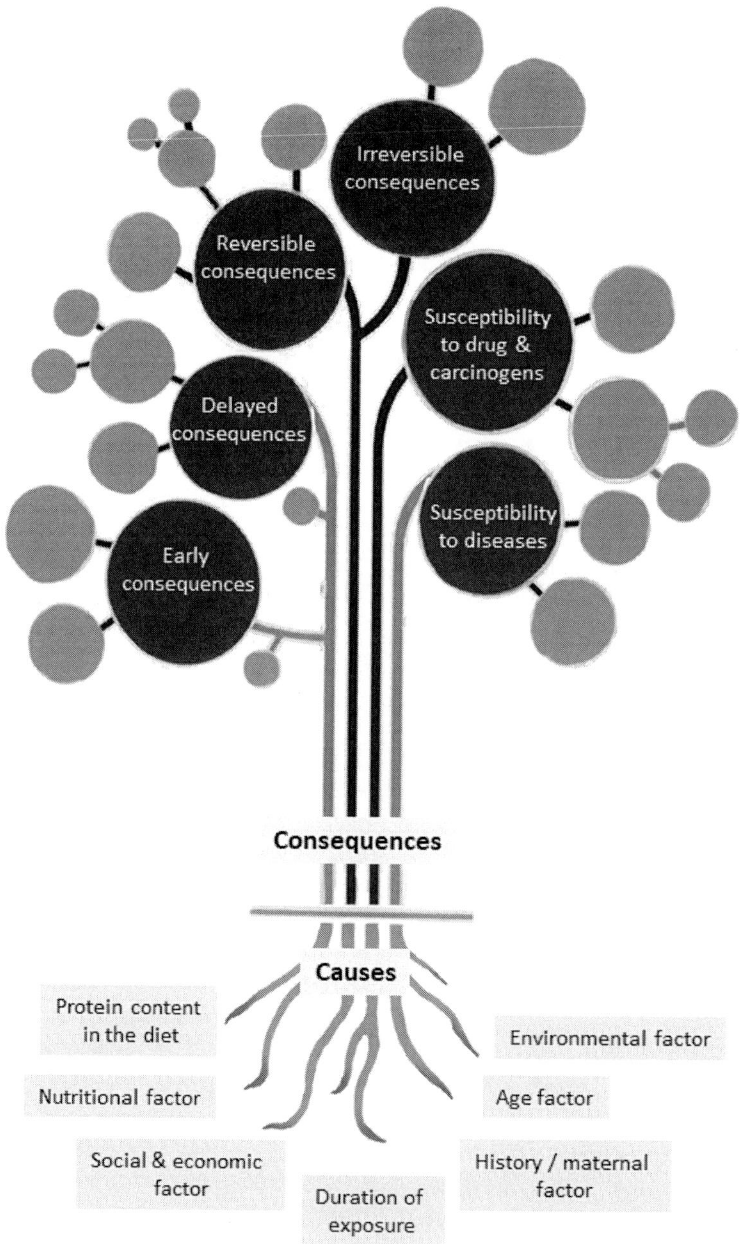

FIGURE 26.5 Tree diagram illustrating causes and consequences of protein malnutrition. The tree's root represents various factors contributing to protein malnutrition, while the branches and nodes illustrate the various consequences resulting from its onset.

makes rodents susceptible to infections and diseases (Naik et al., 2017; Chertoff, 2015; Sampaio et al., 2008; Del Carmen Miñana-Solis and Carolina, 2008; Malafaia et al., 2014; Chan et al., 1996). Short-term protein malnutrition can also cause changes in body weight, liver function, and mild histological changes. Paternal protein malnutrition in rats affects offspring as well, leading to delays in eye-opening, compromised weight gain, and impaired behavior in the early postnatal period (Galler et al., 1994). As in clinical scenarios, protein malnutrition in rodents results in multiple consequences; some are remarkable, while others overlap across disease, dysfunction, or symptom categories.

Delayed Consequences of PMN in Rodent Models

Physical strength depends on bone structure and mineral density. Chronic protein malnutrition leads to structural impairment due to decreased bone mineral density. This increases the risk of bone fractures, compromised architecture, and bone-related disorders. Bone repair mechanisms are also impaired, resulting in slow or poor recovery from fractures and delayed healing due to slower tissue regeneration. Brain structure and development may be compromised, leading to behavioral abnormalities (Borelli et al., 1995; Ozanne and C. Nicholas, 1999). Long-term protein malnutrition can lead to the development of chronic low-grade inflammation, which contributes to the development of metabolic and non-metabolic diseases. Vascular function impairment can result in cardiovascular complications and impaired regulation of blood pressure (Del Carmen Miñana-Solis and Carolina, 2008). These changes are influenced by metabolic adaptations that affect hormone production, glucose regulation, and energy utilization (Borelli et al., 1995). Such adaptive mechanisms can cause complications in the reproductive cycle, including delayed reproductive maturation, reduced fertility, and increased pregnancy complications in females, while diminishing sperm quality and lowering the fertility rate in males.

Reversible and Non-Reversible Conditions

Studies on protein malnutrition in rodent models show that the effects of dietary restriction in adulthood are reversible through dietary rehabilitation, demonstrating that adulthood is not a crucial period sensitive to metabolic reprogramming (Malta et al., 2016). On the other hand, low-protein diet exposure in the early/perinatal period leads to brain developmental abnormalities with irreversible behavioral consequences that persist into adulthood. Metabolic changes due to protein malnutrition in infancy also indicate irreversible effects (Chertoff, 2015; Borelli et al., 1995). These observations show that both age and duration of exposure to malnourished diets are critical determinants of reversibility or irreversibility in protein malnutrition rodent models.

Susceptibility to Disease

As in humans, protein malnutrition in rodents increases susceptibility to various diseases. Susceptibility increases due to the compromised immune system and impairments of developmental and metabolic systems. Protein malnutrition lowers the threshold for infection, increasing vulnerability to lower pathogen loads. Even minimal exposure to a low pathogen load can result in severe infections, e.g., tuberculosis. Several rodent studies have produced similar observations. Mice on a low-protein diet exposed to influenza infection experienced greater severity, possibly due to reduced influenza-specific CD8(+) T cells or diminished virus-specific antibody response compared to mice fed a normal protein diet (Taylor et al., 2013). Similarly, rats exhibited altered carbohydrate metabolism and lower glucose levels, indicating greater susceptibility to metabolic disorders (Lloyd and J. Collinge, 2005). In another study by Lago et al. (1993), rats experiencing early PMN showed enhanced susceptibility to seizures in adulthood, which was only partly reversed by restoring an optimal protein diet.

TABLE 26.1
Recent Studies Involving Low-Protein Diets in Rats and Mice

Sl. no.	Species	Model Details	Study	Reference
1.	Rats	Low-protein diet	DNA methylation and gene expression	Qu Zhi et al., 2022
		Sprague–Dawley: Low-protein diet (4%)	Ileal and colonic permeability	Eyzaguirre-Velásquez et al., 2017
		Wistar: Low-protein diet (6%)	Impaired recognition memory	Braga et al., 2014
		Sprague–Dawley: Low-protein diet (4.6%)	Craniofacial growth	Lobe et al., 2006
		Low-protein diet (9.5%)	Growth curve, hematology, macrophage function	Prestes-Carneiro et al., 2006
2.	Mice	BALB/c: Low-protein diet (10%)	Long-term PMN model	Madhura et al., 2023; D'Souza et al., 2022
		BALB/c: Low-protein diet (10%)	Pharmacokinetic study	Souza et al., 2022
		C57BL/6: Low-protein diet (5.8%), calorie restriction (50%)	Undernutrition with enteropathy	Salameh et al., 2020
		C57BL/6: Low-protein diet (2%)	Decreased immunity susceptibility to Influenza infection	Taylor et al., 2013
		HR-1 hairless mice: Low-protein diet (2%)	Effect on skin epidermis	Sugiyama et al., 2011

The table shows the details of various rodent models used in malnutrition research using different low-protein diets. It gives a snapshot understanding of the recent rodent models used for research and their protocols.

Susceptibility to Drug-Induced Toxicity and Environmental Hazards

Literature generally indicates that protein-malnourished animals are more susceptible to drug-induced liver injury compared to matched controls. Protein-malnourished rodents exposed to toxicants, carcinogens, tobacco, and environmental hazards tend to develop organ toxicity, organ dysfunction, and cancer more rapidly than rodents receiving adequate protein diets. However, outcomes also depend on factors such as the chemical properties of a molecule, dose, duration/route of exposure, age, gender, species, and strain of the animal. The liver, as a primary metabolic organ, shows both developmental and metabolic impairments in protein-malnourished animals. Consequently, nutritional status plays a vital role in the differential outcome of either survival or death from chemical-mediated toxicity (Table 26.1).

CLINICAL SCENARIO CONSEQUENCES OF PROTEIN MALNUTRITION

Globally, protein malnutrition remains one of the greatest barriers to developmental goals because health is the main driving force behind any development. Understanding the underlying complications of protein malnutrition is essential. While social complexities are addressed by routine surveys and reports, these do not address the complications within the protein-malnourished individual. Surveys or reports merely help create awareness, and programs only educate and prevent further decline in the malnutrition graph. However, they do not address how to overcome, treat, recover, and prevent a worsening of the condition in already-malnourished individuals. Preclinical studies play an important role in understanding these aspects and developing remedial strategies.

Correlation of PMN Rodent Model to Clinical Scenarios

Rodent models of protein malnutrition demonstrate a correlation to clinical PMN conditions. Malnutrition in rodent pups/infantile malnutrition develops metabolic changes (such as low blood glucose levels and altered carbohydrate metabolism) similar to those in human infants (Lago et al., 1993). Maternal protein malnutrition in rats causes behavioral changes resembling autism spectrum disorder (ASD) and associated symptoms corresponding to those in humans (Batista et al., 2019). Rapid weight loss, fat accumulation in the liver, and impaired synthesis of lipoproteins are significant clinical outcomes of protein malnutrition affecting liver function. Similar results have been reported in rats fed low-protein, high-carbohydrate diets (Wikramanayake, 1966).

Development of PMN in Infants

Breast milk provides a complete, healthy diet for infants because it contains all the essential nutrients, including carbohydrates, fats, minerals, proteins, and vitamins. Infants need to be exclusively breastfed for up to six months. Continuing breastfeeding with solid food up to two years is recommended. Compromised breastfeeding may occur due to poor maternal health, early subsequent pregnancy, suboptimal breastfeeding due to maternal occupation, maternal protein malnutrition, and poor lactation. The results can include marasmus and kwashiorkor. Multiple reports suggest that infants who do not receive colostrum during the first hour after birth, experience unhygienic feeding practices, and are weaned early are more likely to develop malnutrition (Ukarande and Ajit, 2015). Low protein content in breast milk can also lead to protein malnutrition in infants (Nakano et al., 2017).

The protein leverage theory states that when dietary protein content is low, individuals prioritize eating more to meet their body's total protein requirements. This contributes to overeating when dietary protein content is inadequate. However, this theory remains controversial. Recent research by Yingga Wu et al. reported that an increase in low-protein dietary intake was not observed in experimental mice. Instead, they found a significant decrease in body weight and glucose tolerance (Yingga Wu et al., 2021).

Development of PMN in Childhood

A nutritionally balanced, complete diet is essential for children after two years of age. Children who do not receive adequate nutrition early after weaning risk becoming malnourished, potentially leading to long-term complications. Protein malnutrition models developed using low-protein diets at weaning show similar conditions of long-term metabolic impairment, glucose intolerance, lower body weight, and altered biochemical parameters (Zheng et al., 2021; Branco et al., 2019; Madhura et al., 2023; D'Souza et al., 2022). These not only enhance our understanding of malnutrition physiology but also provide a platform for new findings/research questions, including drug and dose optimization for protein malnutrition therapy and drug discoveries.

TREATMENT STRATEGIES FOR PROTEIN MALNUTRITION

Treatment strategies are based on clinical expertise. The WHO has published guidelines for managing severe acute malnutrition in infants and children (Guideline: WHO, 2013). The United Nations (UN) designated 2016–2025 as the UN's decade of action for nutrition to achieve global nutrition goals related to non-communicable diseases. This initiative aims to eradicate hunger and transform

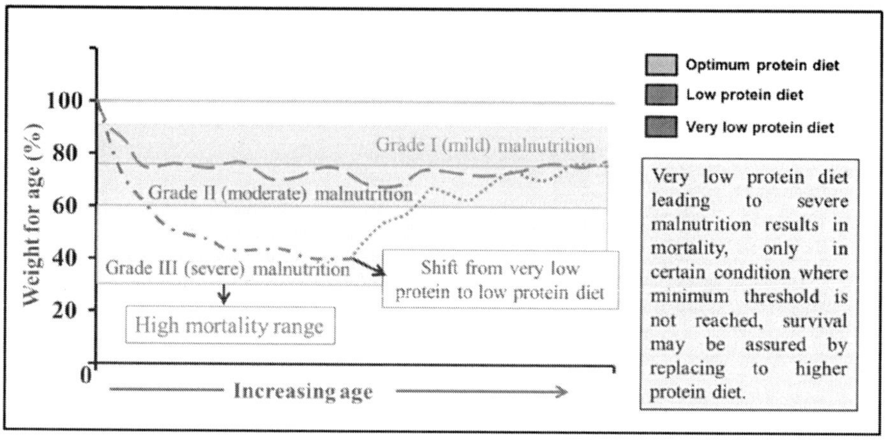

FIGURE 26.6 Graphical representation of weight-for-age in three different groups receiving different percentages of protein in their diets, resulting in various degrees of malnutrition. The graph illustrates body weight in relation to age for three different diet regimens. Optimum protein diet consumption maintains body weight within the normal range for a given age. A low-protein diet leads to mild and moderate malnutrition, while a very low-protein diet leads to severe malnutrition and potential mortality. Increasing dietary protein content can improve body weight and survival rates, but only in individuals who have not developed irreversible complications.

food systems to promote healthy practices by raising awareness and encouraging the sale of nutritional diets at affordable prices, rather than prioritizing profits in commercial settings. The WHO has also released a guide for training healthcare workers in managing severe acute malnutrition in inpatients to reduce disease burden (Training material: WHO, 2022). Furthermore, the WHO's Essential Medicines List for Children (EMLc) has recommended adding ready-to-use therapeutic food for treating children up to five years old with severe acute malnutrition (UN's decade of action on nutrition: WHO, 2023).

There is a pressing need to develop individualized protocols tailored to the nature of malnutrition, age, duration of exposure, and co-morbidities, such as secondary PMN. Treatment often incorporates various supplements, including vitamins, minerals, and modified diets to support the recovery of compromised biochemical pathways. However, the primary focus of therapy remains higher dietary protein intake, supplemented with additional nutrients as appropriate.

Addressing protein malnutrition requires a comprehensive approach encompassing improved access to nutritious food, promotion of breastfeeding practices, implementation of targeted supplementation and fortification programs, and strengthening of healthcare systems. Additionally, nutrition education and behavior change programs are crucial in raising awareness about proper dietary practices and promoting healthy eating habits. Governments worldwide have devised various strategies and programs and set goals to eradicate hunger and malnutrition (Figure 26.6).

MINI DICTIONARY OF TERMS

Bioavailability: The proportion of a substance readily available in a particular biological system.

Granulocyte: A type of white blood cell characterized by granules in the cytoplasm. The three main types are neutrophils, basophils, and eosinophils.

Homeostasis: The maintenance of a stable state between two different independent elements.

Malabsorption: A digestive condition characterized by difficulty absorbing nutrients.

Microbicidal: The ability of a substance or agent to destroy microbes.

Phytohemagglutinin: A protein that can bind to carbohydrates on T cell membranes and helps in metabolic activities and cell divisions.

SUMMARY POINTS

- Protein malnutrition is caused by inadequate nutrient intake to meet the body's requirements.
- It poses a global burden with long-lasting social, economic, medical, and developmental impacts.
- When considering population size, India has the highest number of undernourished individuals globally.
- Protein malnutrition affects elderly populations in high-income countries, while in low-income countries, it primarily impacts younger populations.
- Protein malnutrition highlights the critical importance of adequate protein intake, especially during the early stages of life, for maintaining proper growth, immune function, and metabolic health.
- Breastfeeding up to two years of age is critical in preventing protein malnutrition.
- The pathophysiology of protein malnutrition involves decreased body weight, stunted growth, impaired protein synthesis, compromised immune function, and edema, which can lead to marasmus and kwashiorkor-like conditions.
- Researchers have extensively studied protein malnutrition rodent models to understand the importance of nutrition.
- These rodent models demonstrate significant health implications that correlate with clinical conditions and pathophysiological outcomes, including metabolic changes similar to those in protein malnutrition infants. Studies have also observed autism-like behavior due to maternal malnutrition and liver function impairments.
- Protein malnutrition rodent models enable the study of disease susceptibility, drug-induced toxicity, and the effects of environmental hazards. Despite their limitations, these models provide valuable data on reversible and non-reversible conditions, offering insights for drug discovery research.
- The adverse effects of protein malnutrition, including complications like fatty liver and hepatomegaly, emphasize the critical importance of addressing protein deficiency to ensure overall health and well-being.
- Globally, researchers and policymakers have initiated multilevel interventions, including government policies, international cooperation, and community programs, to break the cycle of malnutrition.

REFERENCES

Ampong, Isaac, Adam Watkins, Jorge Gutierrez-Merino, John Ikwuobe, and Helen R. Griffiths. "Dietary protein insufficiency: an important consideration in fatty liver disease?." *British Journal of Nutrition* 123, no. 6 (2020): 601–609.

Batista, Tatiane Helena, AlexandreGiusti-Paiva, and Fabiana Cardoso Vilela. "Maternal protein malnutrition induces autism-like symptoms in rat offspring." *Nutritional Neuroscience* 22, no. 9 (2019): 655–663.

Borelli, Primavera, Mario Mariano, and Radovan Borojevic. "Protein malnutrition: effect on myeloid cell production and mobilization into inflammatory reactions in mice." *Nutrition Research* 15, no. 10 (1995): 1477–1485.

Bourke, Claire D., James A. Berkley, and Andrew J. Prendergast."Immune dysfunction as a cause and consequence of malnutrition." *Trends in Immunology* 37, no. 6 (2016): 386–398.

Braga, Natalia Nassiff, Marisa Tomoe Hebihara Fukuda, and Sebastião Sousa Almeida. "Early postnatal protein malnutrition impairs recognition memory in rats (Rattusnorvegicus)." *Psychology & Neuroscience* 7 (2014): 103–111.

Branco, Renato Chaves Souto, Rafael Ludemann Camargo, Thiago Martins Batista, Jean FranciescoVettorazzi, CamilaLubaczeuski, Lucas Henrique Montes Bomfim, Leonardo Reis Silveira, Antônio Carlos Boschero, Cláudio Cesar Zoppi, and EverardoMagalhãesCarneiro. "Protein malnutrition mitigates the effects of a high-fat diet on glucose homeostasis in mice." *Journal of Cellular Physiology* 234, no. 5 (2019): 6313–6323.

Brooks, Stanley E. H., Michael H. N. Golden, and Hazel M. Payne-Robinson. "Ultrastructure of islets of langerhans in protein-energy malnutrition." *West Indian Medical Journal* 42 (1993): 101–101.

Chan, John, Yu Tian, Kathryn E. Tanaka, Ming S. Tsang, Keming Yu, PadminiSalgame, Dinah Carroll, Yvonne Kress, Rachel Teitelbaum, and Barry R. Bloom. "Effects of protein calorie malnutrition on tuberculosis in mice." *Proceedings of the National Academy of Sciences* 93, no. 25 (1996): 14857–14861.

Chatterjee, Patralekha. "India's child malnutrition story worsens." *The Lancet Child & Adolescent Health* 5, no. 5 (2021): 319–320.

Chertoff, Mariela Sandra Juana. "Protein malnutrition and brain development." *Brain Disorders and Therapy* 4, no. 3 (2015): 1–6.

del Carmen Miñana-Solis, María, and Carolina Escobar. "Post-weaning protein malnutrition in the rat produces short and long term metabolic impairment, in contrast to earlier and later periods." *International Journal of Biological Sciences* 4, no. 6 (2008): 422.

Dent, Elsa, Olivia R. L. Wright, Jean Woo, and Emiel O. Hoogendijk. "Malnutrition in older adults." *The Lancet* 10380 (2023): 951–966.

Diallo, Abdoulaye Hama, Abu Sadat Mohammad Sayeem Bin Shahid, Ali Fazal Khan, Ali Faisal Saleem, Benson O. Singa, BlaiseSiezangaGnoumou, Caroline Tigoi et al. "Childhood mortality during and after acute illness in Africa and south Asia: a prospective cohort study." *The Lancet Global Health* 10, no. 5 (2022): e673–e684.

Dipasquale, Valeria, UgoCucinotta, and Claudio Romano. "Acute malnutrition in children: pathophysiology, clinical effects and treatment." *Nutrients* 12, no. 8 (2020): 2413.

Divyaveer, Smita. "Pathophysiology of protein–energy wasting." *Journal of Renal Nutrition and Metabolism* 5, no. 3 (2019): 64.

D'Souza, Vinitha, Madhura R. J., MeghashreeShetty, AnirbanChakraborthy, and Murali Badanthadka. "Protein malnutrition in BALB/c mice: an experimental model resembling clinical scenario." *Journal of Health and Allied Sciences NU* 12, no. 1 (2022): 1–6.

Eyzaguirre-Velásquez, Johana, Loreto Olavarría-Ramírez, Camila González-Arancibia, CamilaDíaz-Merino, RaúlAriz, SilvanaLópez, Waldo Quiroz, Caroll J. Beltrán, Javier A. Bravo, and Marcela Julio-Pieper. "Protein malnutrition during juvenile age increases ileal and colonic permeability in rats." *Journal of Pediatric Gastroenterology and Nutrition* 64, no. 5 (2017): 707–712.

Galler, Janina R., John Tonkiss, and Carmen S. Maldonado-Irizarry. "Prenatal protein malnutrition and home orientation in the rat." *Physiology & Behavior* 55, no. 6 (1994): 993–996.

Guideline: updates on the management of severe acute malnutrition in infants and children. WHO. 2013. https://www.who.int/publications/i/item/9789241506328

Hoffer, L. John. "Clinical nutrition: 1. Protein–energy malnutrition in the inpatient." *CMAJ* 165, no. 10 (2001): 1345–1349.

Ingenbleek, Yves. "Implications of protein malnutrition and inflammatory disorders in the pathophysiology of Alzheimer's disease." *Asia Pacific Journal of Clinical Nutrition* 29, no. 3 (2020): 450–461.

Kambale, Richard Mbusa, and Isia Nancy Francisca. "Optimising the management of acute malnutrition." *The Lancet Global Health* 10, no. 4 (2022): e453–e454.

Keats, Emily C., Jai K. Das, Rehana A. Salam, Zohra S. Lassi, AamerImdad, Robert E. Black, and Zulfiqar A. Bhutta. "Effective interventions to address maternal and child malnutrition: an update of the evidence." *The Lancet Child & Adolescent Health* 5, no. 5 (2021): 367–384.

Lago, E. S., N. R. Teodosio, C. R. Araujo, M. C. Azevedo, D. C. Pessoa, F. A. Campos, S. M. Zucas, and H. Flores. "Rat models of protein and protein-energy malnutrition. International Journal for Vitamin and Nutrition research. InternationaleZeitschrift fur Vitamin-und Ernahrungsforschung." *Journal International de Vitaminologieet de Nutrition* 63, no. 1 (1993): 52–56.

Lloyd, S. E., and J. Collinge. "Genetic susceptibility to prion diseases in humans and mice." *Current Genomics* 6, no. 1 (2005): 1–11.

Lobe, Shannon L., Marica C. Bernstein, and Rebecca Z. German. "Life-long protein malnutrition in the rat (Rattusnorvegicus) results in altered patterns of craniofacial growth and smaller individuals." *Journal of Anatomy* 208, no. 6 (2006): 795–812.

Ma, ZhengFeei, CheeWoon Wang, and YeongYeh Lee. "Malnutrition: A cause or a consequence of poverty?." *Frontiers in Public Health* 9 (2022): 796435.

Madhura, R. J., A. Varsha, Anirban Chakraborthy, Veena Shetty, and Murali Badanthadka. "Protein malnutrition in BALB/C mice: A model mimicking clinical scenario of marasmic-kwashiorkor malnutrition." *Journal of Pharmacological and Toxicological Methods* 119 (2023): 107231.

Malafaia, Guilherme, Abraão Tiago Batista Guimarães, CamilaGracyelle de CarvalhoLemes, and Dieferson da Costa Estrela. "Short-term malnutrition in Wistar rats." *Scientia Plena* 10, no. 7 (2014): 1–6.

Maleta, Ken. "Undernutrition." *Malawi Medical Journal: The Journal of Medical Association of Malawi* 18, no. 4 (2006): 189.

Malta, Ananda, Egberto Gaspar de Moura, TatianeAparecidaRibeiro, Laize Peron Tófolo, LatifaAbdennebi-Najar, Didier Vieau, Luiz Felipe Barella, Paulo Cezar de Freitas Mathias, Patrícia Cristina Lisboa, and JúlioCezar de Oliveira. "Protein-energy malnutrition at mid-adulthood does not imprint long-term metabolic consequences in male rats." *European Journal of Nutrition* 55 (2016): 1423–1433.

Naik, Aijaz Ahmad, NishaPatro, Pankaj Seth, and Ishan K. Patro. "Intra-generational protein malnutrition impairs temporal astrogenesis in rat brain." *Biology Open* 6, no. 7 (2017): 931–942.

Nakano, Yuya, Chizu Hirabayashi, Masahiko Murase, Katsumi Mizuno, and Kazuo Itabashi. "Malnutrition caused by unexpectedly low protein concentration in breast milk." *Pediatrics International: Official Journal of the Japan Pediatric Society* 59, no. 10 (2017): 1100–1101.

Neale, G. "Protein malnutrition in gastro-intestinal disease." *Postgraduate Medical Journal* 44, no. 514 (1968): 642.

Ozanne, Susan E., and C. Nicholas Hales."The long-term consequences of intra-uterine protein malnutrition for glucose metabolism." *Proceedings of the Nutrition Society* 58, no. 3 (1999): 615–619.

Panda, Basant Kumar, Sanjay K. Mohanty, ItishreeNayak, Vishal DevShastri, and S. V. Subramanian. "Malnutrition and poverty in India: does the use of public distribution system matter?." *BMC Nutrition* 6, no. 1 (2020): 1–14.

Prestes-Carneiro, LuizEuribel, Rodrigo DominguesLaraya, Paula Roberta Colacino Silva, Ricardo Alberto Moliterno, Ionice Felipe, and Paulo Cezar Mathias. "Long-term effect of early protein malnutrition on growth curve, hematological parameters and macrophage function of rats." *Journal of Nutritional Science and Vitaminology* 52, no. 6 (2006): 414–420.

Qu, Zhi, Liying Fu, Chenchen Wang, Suting Liu, and Bo Li. "Effects of DNA methylation and gene expression on rats with protein malnutrition in early life." *Gene & Protein in Disease* 1, no. 2 (2022): 169.

Rytter, Maren Johanne Heilskov, Lilian Kolte, André Briend, Henrik Friis, and Vibeke Brix Christensen. "The immune system in children with malnutrition—a systematic review." *PLoS One* 9, no. 8 (2014): e105017.

Salameh, Emmeline, Marine Jarbeau, Fanny B. Morel, MamaneZeilani, Moutaz Aziz, Pierre Déchelotte, and Rachel Marion-Letellier. "Modeling undernutrition with enteropathy in mice." *Scientific Reports* 10, no. 1 (2020): 15581.

Sampaio, Valdomiro de Freitas, Luiz Marcellino de Oliveira, Sebastião Sousa Almeida, Julio Sérgio Marchini, José Antunes-Rodrigues, and Lucila Leico Kagohara Elias. "The effects of early protein malnutrition and environmental stimulation on behavioral and biochemical parameters in rats." *Psychology & Neuroscience* 1 (2008): 109–119.

Seferidi, Paraskevi, Thomas Hone, Ana Clara Duran, Antonio Bernabe-Ortiz, and Christopher Millett. "Global inequalities in the double burden of malnutrition and associations with globalisation: a multilevel analysis of Demographic and Health Surveys from 55 low-income and middle-income countries, 1992–2018." *The Lancet Global Health* 10, no. 4 (2022): e482–e490.

Siddiqui, Faareha, Rehana A. Salam, Zohra S. Lassi, and Jai K. Das."The intertwined relationship between malnutrition and poverty." *Frontiers in Public Health* 8 (2020): 453.

Solon-Biet, Samantha M., Sarah J. Mitchell, Sean C. P. Coogan, Victoria C. Cogger, Rahul Gokarn, Aisling C. McMahon, David Raubenheimer, Rafael de Cabo, Stephen J. Simpson, and David G. Le Couteur. "Dietary protein to carbohydrate ratio and caloric restriction: comparing metabolic outcomes in mice." *Cell Reports* 11, no. 10 (2015): 1529–1534.

Souza, Vinitha D., Meghashree Shetty, Murali Badanthadka, B. S. Mamatha, and K. Vijayanarayana."The effect of nutritional status on the pharmacokinetic profile of acetaminophen." *Toxicology and Applied Pharmacology* 438 (2022): 115888.

Sugiyama, Akihiko, Yuka Fujita, Toshihiro Kobayashi, MizuyukiRyu, Yasushi Suzuki, Aino Masuda, Tairin Ochi, and Takashi Takeuchi. "Effect of protein malnutrition on the skin epidermis of hairless mice." *Journal of Veterinary Medical Science* 73, no. 6 (2011): 831–835.

Taylor, Andrew K., Weiping Cao, Keyur P. Vora, Juan De La Cruz, Wun-JuShieh, Sherif R. Zaki, Jacqueline M. Katz, SuryaprakashSambhara, and ShivaprakashGangappa. "Protein energy malnutrition decreases immunity and increases susceptibility to influenza infection in mice." *The Journal of Infectious Diseases* 207, no. 3 (2013): 501–510.

Training course on the inpatient management of severe acute malnutrition: Course director's guide. WHO. 2022. https://www.who.int/publications/i/item/9789240029781

Ukarande, Balaji V., and Ajit S. Nagaonkar. "A study of protein energy malnutrition with respect to breast feeding practices in less than six year children." *National Journal of Community Medicine* 6, no. 04 (2015): 626–628.

UN reports 2022, World is moving backwards on eliminating hunger and malnutrition. https://news.un.org/en/story/2022/07/1122032

Unicef reports 2023, 122 million more people pushed into hunger since 2019 due to multiple crises. https://www.unicef.org/press-releases/122-million-more-people-pushed-hunger–2019-due-multiple-crises-reveals-un-report

United Nation's decade of action on nutrition. https://www.un.org/nutrition/about

Victora, Cesar G., Parul Christian, Luis Paulo Vidaletti, Giovanna Gatica-Domínguez, PurnimaMenon, and Robert E. Black. "Revisiting maternal and child undernutrition in low-income and middle-income countries: variable progress towards an unfinished agenda." *The Lancet* 397, no. 10282 (2021): 1388–1399.

Wharton, Brian. "The pathophysiology of infantile malnutrition protein-energy malnutrition and failure to thrive." *Archives of Disease in Childhood* 59, no. 12 (1984): 1201.

WHO 2023, WHO endorses landmark public health decisions on Essential Medicines for Multiple Sclerosis. https://www.who.int/news/item/26-07-2023-who-endorses-landmark-public-health-decisions-on-essential-medicines-for-multiple-sclerosis

WHO factsheets and details 2021. Malnutrition. https://www.who.int/news-room/fact-sheets/detail/malnutrition

Wikramanayake, T. W. "Protein malnutrition in the rat induced by protein-calorie imbalance." *British Journal of Nutrition* 20, no. 3 (1966): 641–648.

World meter 2023, Undernourished people in the world.https://www.worldometers.info/undernourishment/

Wu, Yingga, Baoguo Li, Li Li, Sharon E. Mitchell, Cara L. Green, Giuseppe D'Agostino, Guanlin Wang et al. "Very-low-protein diets lead to reduced food intake and weight loss, linked to inhibition of hypothalamic mTOR signaling, in mice." *Cell Metabolism* 33, no. 5 (2021): 888–904.

Zheng, Jia, Ling Zhang, Jiayi Liu, Yanli Li, and Junqing Zhang. "Long-term effects of maternal low-protein diet and post-weaning high-fat feeding on glucose metabolism and hypothalamic POMC promoter methylation in offspring mice." *Frontiers in Nutrition* 8 (2021): 657848.

27 High-Protein Diets and Their Impact on Hypercholesterolemia

Maria do Carmo de Carvalho e Martins,
Maísa Guimarães Silva Primo,
Liriane Andressa Alves da Silva, and
Amanda Suellenn da Silva Santos Oliveira

ABBREVIATIONS

ABCA1	ATP-binding cassette transporter A1
CYP27A1	Cytochrome P450 family 27 subfamily A member 1
FDA	Food and Drug Administration
GIP	Gastric inhibitory peptide or glucose-dependent insulinotropic polypeptide
HDL-c	High-density lipoprotein cholesterol
HMGCR	3-hydroxy-3-methylglutaryl-CoA reductase
IIAEK	Peptide Ile-Ile-Ala-Glu-Lys, lactostatin
LCAT	Lecithin–cholesterol acyltransferase
LDL-c	Low-density lipoprotein cholesterol
LPYP	Peptide Leu-Pro-Tyr-Pro
SFA	Saturated Fatty Acids
TG	Triglycerides
VLDL-c	Very low-density lipoprotein cholesterol

INTRODUCTION

Hypercholesterolemia, also known as dyslipidemia, is a metabolic alteration resulting in increased serum concentrations of total cholesterol (TC), triglycerides (TG), and low-density lipoprotein cholesterol (LDL-c). It is often associated with reduced serum concentrations of high-density lipoprotein cholesterol (HDL-c) and increased triglycerides, especially in TG-rich lipoproteins such as chylomicrons and very low-density lipoprotein (VLDL) (Trautwein et al. 2020).

Increased serum LDL-c concentrations directly contribute to cardiovascular diseases (CVDs), making the reduction of LDL-c crucial for CVD prevention and treatment (Ference et al. 2017; Mach et al. 2020). It is important to emphasize that adopting a healthy diet and lifestyle should always be the basis for reducing LDL-c and TG levels.

There is evidence to support the idea that certain nutrients, food groups, or dietary patterns can have a positive impact on dyslipidemia and CVD prevention (Willett et al. 2019) (Figure 27.1). A plant-focused diet can serve as an effective therapeutic strategy in clinical settings for managing systemic arterial hypertension, hypercholesterolemia, and other cardiovascular risk factors, potentially reducing the need for medication (Najjar et al. 2018).

Plant-based diets prioritize the intake of complex carbohydrates and plant proteins, while reducing total lipid content, especially saturated and trans fats, and increasing unsaturated fatty acids.

DOI: 10.1201/b23328-27

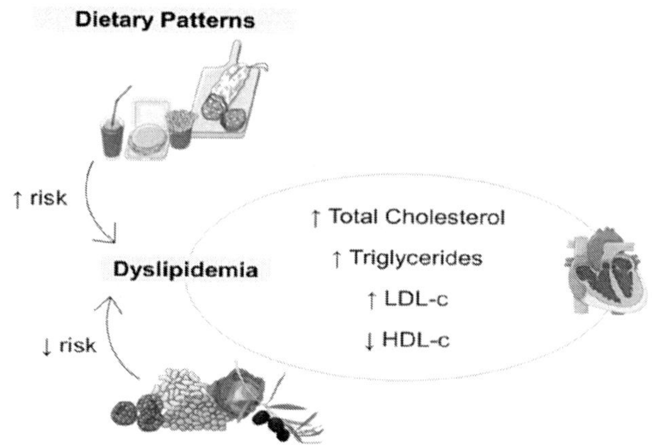

FIGURE 27.1 Dietary patterns and risk for dyslipidemia. LDL-c, low-density lipoprotein cholesterol; HDL-c, high-density lipoprotein cholesterol. A diet rich in saturated fat and animal protein may increase the risk of developing dyslipidemia and cardiovascular disease, while plant-based proteins may lower this risk (compiled by the authors; images from smart.servier.com.)

Plant foods are also rich in dietary fiber and include phytosterols and other health-beneficial natural compounds known as phytonutrients (Trautwein et al. 2020).

The quality of fat in foods is influenced by the types of fatty acids present. Animal-derived fats, such as those in fatty meat, butter, and whole dairy products, as well as tropical vegetable oils like coconut and palm, contain high levels of saturated fatty acids (SFA) and are the main SFA sources in Western diets (Eilander et al. 2015). In contrast, vegetable oils are generally rich in unsaturated fatty acids. This means that a high-protein diet can be harmful, depending on the type of protein consumed in excess (Eilander et al. 2015; Trautwein et al. 2020).

Studies investigating the effects of diets rich in animal and plant proteins, which are common in various regions of the world, have shown varying health impacts (Xu et al. 2020; Díaz-Rúa et al. 2017). High-protein diets improve HDL-c levels, and high intakes of plant and dairy proteins may protect against obesity-related problems and insulin resistance. However, excessive consumption of meat protein has been associated with excess weight gain and adverse metabolic effects (Fogelholm et al. 2012; Mozaffarian 2016).

Given this context, assessing the effects of high-protein diets on cardiovascular health and lipid profiles requires a comprehensive analysis of the various sources of dietary protein, both animal- and plant-based. Although there are promising indications regarding the benefits of plant proteins on dyslipidemia, the relationship between animal protein intake and circulating lipoproteins requires more detailed analysis. This chapter seeks to deepen the understanding of the effects of protein-rich diets on hypercholesterolemia, providing a holistic view of how different protein sources affect the lipid profile and associated conditions, thus contributing to a more complete understanding of dietary aspects and their effects on cardiovascular health.

EFFECTS OF HIGH-PROTEIN DIETS CONTAINING PROTEINS FROM DIFFERENT FOOD SOURCES ON HEALTH AND LIPID PROFILE

Protein intake and its health effects are priority topics in nutrition research and the prevention of cardiovascular and metabolic diseases (Lamberg-Allardt et al. 2023). In recent years, there has been an increase in the popularity of protein-rich diets, especially as a strategy for controlling body weight (Wang et al. 2022).

High-fat diets are associated with increased hepatic lipid content and the development of steatosis in rodents and humans (Eng et al. 2021). In contrast, protein-rich diets can help prevent the accumulation of lipids in the liver (Xu et al. 2020). However, some studies have also shown that high protein intake can increase morbidity and mortality from various causes and is associated with negative effects due to the metabolites produced during amino acid metabolism. (Díaz-Rúa et al. 2017). Furthermore, an association has been shown between high intakes of total protein and protein of animal origin and an increased risk of type 2 diabetes mellitus (Zhao et al. 2019).

Plant proteins have a protective effect against hypercholesterolemia (Shi et al. 2021; Tong et al. 2021). However, this effect should be considered in conjunction with other dietary measures, such as reducing consumption of saturated and trans fats, as well as limiting animal protein sources, which tend to have higher saturated fat and cholesterol content (Campbell 2017).

High-protein diets can increase VLDL levels and reduce triglyceride production due to their lower carbohydrate content (Velázquez et al. 2019). However, the effects of these diets on lipid profiles are still controversial, especially regarding animal protein intake, which may be associated with higher intake of cholesterol and saturated fatty acids, increasing the risk of CVD (Amini et al. 2016). Guasch-Ferré et al. (2019) found that replacing red meat with high-quality vegetable protein sources, but not fish or low-quality carbohydrates, resulted in favorable changes in blood lipid and lipoprotein concentrations. These findings highlight the importance of considering not only the amount of protein in a diet but also the origin of these proteins, as different sources can have varying effects on lipid profiles (Figure 27.2) and, consequently, on cardiovascular health.
The effects of high-protein, high-carbohydrate diets can be influenced by other dietary components, such as fiber and total fat content. Replacing a portion of carbohydrates with protein can have positive effects on the triglyceride/HDL-c ratio (Shi et al. 2021).

Proteins can affect the plasma concentration of cholesterol through various mechanisms, including: 1. interfering with intestinal absorption of bile acids, affecting the absorption of cholesterol, and the tissue storage, synthesis, and excretion of lipids; 2. influencing cholesterol metabolism by affecting hormones such as insulin, glucagon, thyroid hormones, and gastrointestinal hormones like gastric inhibitory peptide or glucose-dependent insulinotropic polypeptide (GIP) (Bhoite et al. 2021).

High-quality dietary proteins contain essential amino acids that regulate gene expression and influence lipid metabolism. Branched-chain amino acids such as leucine, isoleucine, and valine reduce lipogenesis, preventing excessive lipid accumulation in the liver (Schwarz et al. 2012).

The health effects of high-protein diets remain inconclusive, with protein sources being an important factor to consider. While these diets may benefit body weight and some cardiovascular risk factors, they also raise concerns about possible negative effects on blood lipid levels (Dong et al. 2013). The consumption of red meat, a common source of protein, has been associated with an increased prevalence of hyperlipidemia (Diarz et al. 2020). On the other hand, diets rich in low-fat

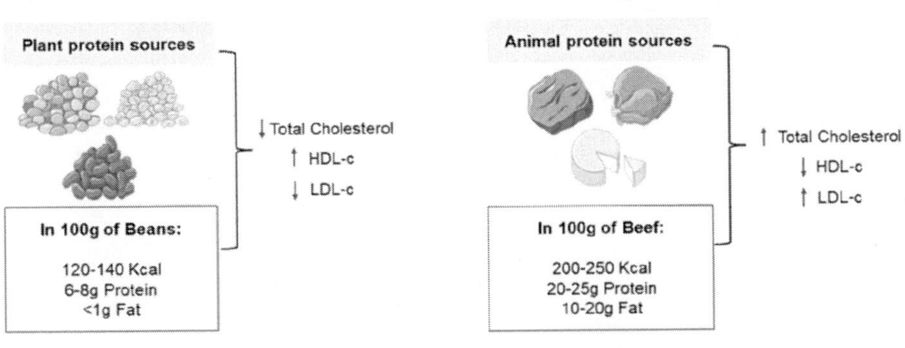

FIGURE 27.2 Effects of different protein sources on lipid profiles. LDL-c, low-density lipoprotein cholesterol; HDL-c, high-density lipoprotein cholesterol. The nutritional composition of portions may vary depending on the specific beef cut and bean type, as well as the cooking method used for both (compiled by the authors; images from smart.servier.com).

proteins, such as vegetable proteins, are linked to a reduced risk of hypercholesterolemia and CVD (Coelho-Júnior et al. 2023).

EFFECTS OF SOY PROTEIN ON DYSLIPIDEMIA

Dyslipidemia is a risk factor for CVD, and non-pharmacological treatment often involves dietary strategies known to improve lipid profiles (Wang et al. 2017). These interventions typically include reducing saturated fat intake and increasing fiber and protein consumption. However, there is limited guidance on the specific types of protein that should be consumed to improve lipid profiles (Figure 27.3) (Zhao et al. 2020).

With regard to the effects of proteins, vegetable proteins, particularly those from legumes such as soybeans, peas, and beans, are associated with reduced serum cholesterol concentrations. This effect may be attributed to the lower levels of saturated fat and the absence of cholesterol in legumes. Additionally, legumes have a high fiber content and are rich in iron, calcium, potassium, and B vitamins compared to other vegetables (Caponio et al. 2021).

Soy (*Glycine max*) is a legume considered a functional food, containing phytochemical substances such as carotenoids, flavonoids, lignans, and phytic acid, which have hypocholesterolemic, antioxidant, anti-obesity, and anticarcinogenic properties (Kim et al. 2021).

Soybeans comprise approximately 40% proteins, of which 85% are globulins (2S, 7S, 11S, and 15S) (Caponio et al. 2021). The main proteins in this legume are β-conglycinin (7S), which has three subunits (α', α, and β) and glycinin (11S), with subunits A1ab1b and A2b1a. Both proteins have a bile acid sequestering action, which decreases intestinal absorption of cholesterol, contributing to lower plasma cholesterol concentrations (Caponio et al. 2021) (Table 27.1).

A meta-analysis carried out by Mejia et al. (2019), which included 43 controlled clinical trials reviewed by the Food and Drug Administration (FDA), showed that ingesting at least 25 g/day of soy proteins for six weeks reduced plasma concentrations of LDL-c and total cholesterol in adults with hypercholesterolemia.

It is worth highlighting the effect of isoflavones in soy protein content. Sacks et al. (2006) reported that soy protein containing isoflavones was more effective in lowering LDL-c than soy protein with reduced amounts of this bioactive compound. Similarly, studies have shown that the consumption of soy proteins and isoflavones resulted in more benefits than harm to human health (Li et al. 2020).

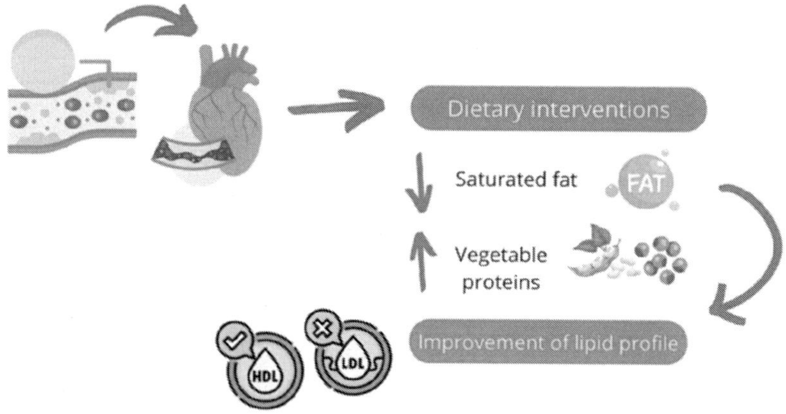

FIGURE 27.3 Impact of nutritional interventions on lipid profile. LDL-c, low-density lipoprotein cholesterol; HDL-c, high-density lipoprotein cholesterol. Dyslipidemias are risk factors for chronic diseases such as cardiovascular diseases, and changes in the dietary profile may modify lipid profiles (compiled by the authors; images from smart.servier.com).

TABLE 27.1
Effects of Vegetable Proteins Present in Legumes on Lipid Profile Parameters

Vegetable Protein	Effect on Lipid Profile
Soy	Reduction of plasma cholesterol concentrations by the sequestering action of proteins in bile acids.
Pea	Reduction in plasma cholesterol concentrations and triglyceride and VLDL-c content in the liver due to increased activity of the hepatic cholesterol-7a-hydroxylase enzyme.
Cowpea	Reduction of cholesterol, non-HDL-c cholesterol, and increase in HDL-c due to the consumption of beans with fiber, saponins, and resistant starches that can reduce intestinal cholesterol absorption.

Source: compiled by the authors. VLDL-c: very low-density lipoprotein; HDL-c: high-density lipoprotein cholesterol; non-HDL-c: cholesterol non-high-density lipoprotein. Dyslipidemias are risk factors for chronic diseases such as cardiovascular diseases, and changes in the dietary profile may be capable of modifying the lipid profile.

Soy protein can be hydrolyzed into peptides through enzymatic reactions and, thus, has gel-forming properties, as well as emulsifying and foaming functions (Kim et al. 2021).

Soy peptides such as Ile-Ile-Ala-Glu-Lys or lactostatin (IIAEK) can decrease circulating concentrations of total cholesterol, as well as LDL-C and fat accumulation, thus contributing to the prevention of atherosclerosis and CVD (Chatterjee et al. 2018). Another peptide, Leu-Pro-Tyr-Pro (LPYP), can inhibit 3-hydroxy-3-methylglutaryl-coenzyme A reductase (HMGR), an enzyme involved in cholesterol synthesis (Yoshikawa et al. 2000; Chatterjee et al. 2018).

EFFECTS OF PEA AND COWPEA PROTEIN ON DYSLIPIDEMIA

Peas (*Pisum sativum*) are legumes containing around 25% crude protein and, similarly to soy protein, are low in methionine and high in arginine. In addition, peas have a higher arginine-to-lysine ratio, which may explain their hypocholesterolemic effect (Sppielmann et al. 2008).

Tong et al. (2021) demonstrated the benefits of pea protein on the lipid profile in a study where hamsters supplemented with pea protein for 30 days showed reduced serum concentrations of total cholesterol and LDL-C when compared to the group receiving pork supplementation.

Few studies have examined the effects of pea protein on human lipid profiles. Sirtori et al. (2012) investigated pea protein and white lupin (*Lupinus albus*) supplementation, as well as combinations with soluble fibers, in adults with hypercholesterolemia. They observed decreased serum cholesterol in patients who consumed pea protein with oat fiber and apple pectin, as well as in those who received casein with apple pectin.

Another legume being studied is the cowpea (*Vigna unguiculata*), a legume native to northeastern Brazil. This legume has a protein content of around 25% and a low lipid content (Onwuliri 2002).

Regarding the effects of cowpeas on hypercholesterolemia, Frota et al. (2008) demonstrated reductions in total cholesterol and non-HDL cholesterol in hypercholesterolemic hamsters fed a diet rich in saturated fats and cholesterol, supplemented with 20% protein from either whole cowpeas or isolated cowpea protein for four weeks. The authors noted that the effects on sterol excretion were better with the use of the whole bean than those found in the group that received the isolated protein. This difference can be explained by the fact that whole grains contain fiber, saponins, and resistant starches, which act synergistically to reduce the intestinal absorption of cholesterol and, consequently, help reduce circulating cholesterol concentrations (Correa and Poltronieri 2016).

A reduction in serum concentrations of total cholesterol, LDL-c, and non-HDL-c cholesterol, accompanied by an increase in HDL-c, was also demonstrated in humans with hypercholesterolemia who consumed 25 g/day of cowpea protein isolate for six weeks, with a four-week washout interval (Frota et al. 2015).

Researchers have also observed that thermal processing can alter the effect of isolated cowpea protein on the intestinal absorption of cholesterol. When comparing the effect of raw and cooked cowpea protein isolate, Marques et al. (2014) found that thermal processing can potentiate the beneficial effect of greater inhibition of the cholesterol micellar solubilization, facilitating peptide release.

It is important to note that plant proteins, especially those from legumes, have beneficial effects on hypercholesterolemia, potentially contributing to reduced cardiometabolic risk. However, the mechanisms involved in these effects have yet to be elucidated.

ANIMAL PROTEIN AND CIRCULATING LIPOPROTEINS

Animal proteins are generally classified as complete and considered standard in terms of their amino acid composition. Important sources of this type of protein are beef, pork, goat, sheep, and poultry, as well as fish, dairy products, and eggs (Cozzolino 2016).

Dairy products include casein and whey proteins, which are highly digestible proteins that not only contain all the essential amino acids but also have important bioactive, nutritional, biological, and functional properties (Auestad and Layman 2021; Rafiq et al. 2016).

Eggs have the highest protein content in the white (50%), with ovalbumin, ovotransferrin, ovomucoid, ovomucin, and lysozyme being the main constituent proteins (Guha et al. 2019). The yolk contains around 40% proteins, especially lipovitellin, which corresponds to a sub-fraction of high-density lipoprotein (HDL), and phosvitin, which acts to protect against the oxidative degradation of lipids by binding to heavy metal ions (Dubin et al. 1994).

Meat proteins provide essential amino acids with the appropriate composition to promote protein biosynthesis, which is necessary for the production and maintenance of structural, enzymatic, contractile, hormonal, transport, and protective functions. In addition, their amino acids and peptides produce metabolites with important bioactive properties (Geiker et al. 2021) (Figure 27.4).

Despite the high biological value of meat protein, it has been suggested that the cardiovascular risk associated with animal protein consumption is higher for red or processed meat than for poultry and fish, which are associated with a lower risk of developing coronary heart disease (Zhubi-Bakija et al. 2021).

In a study incorporating data from the Rotterdam survey and including a meta-analysis, Chen et al. (2020) showed that higher total protein intake was associated with increased all-cause mortality,

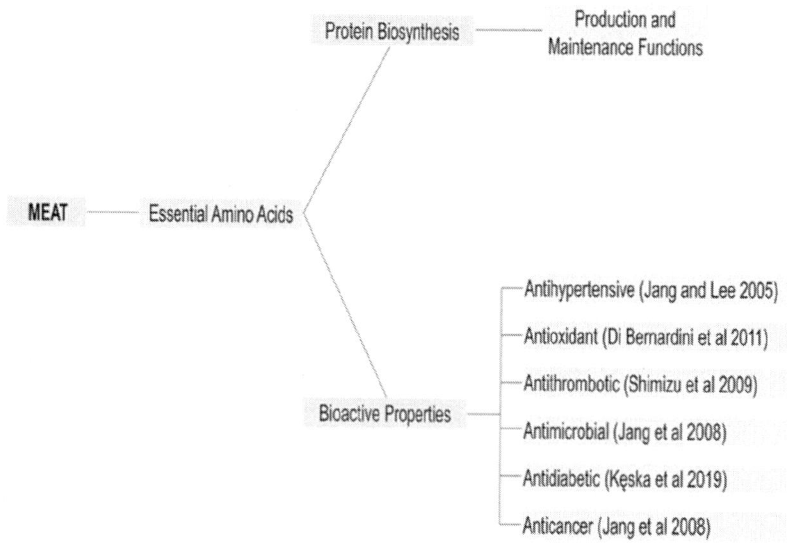

FIGURE 27.4 Properties of essential amino acids present in meat protein (source: compiled by the authors).

with a positive association between animal protein consumption and CVD mortality. The authors suggest that the relationship between animal protein consumption and the risk of CVD may involve the amino acids in the composition, such as branched-chain and aromatic amino acids, as well as other components present in foods of animal origin, such as saturated fatty acids, especially in red and processed meats. As for the types of amino acids, a diet rich in these components is linked to the development of insulin resistance and excess weight, which are considered risk factors for cardiometabolic diseases and, thus, to an increased risk of death from heart disease.

The relationship between branched-chain amino acids and dyslipidemia has been the subject of research, and prospective studies have revealed that increased blood concentrations of branched-chain amino acids are associated with a greater risk of elevated triglyceride concentrations (Mook-Kanamori et al. 2014; Yamakado et al. 2015).

Furthermore, a cross-sectional study investigating the relationship between dietary branched-chain amino acids and dyslipidemia risk found that these amino acids in the diet were positively correlated with total cholesterol and LDL-c. The highest quartiles of intake of this type of amino acid were associated with a greater likelihood of hypercholesterolemia (Yu et al. 2022).

Regarding branched-chain amino acids and HDL-c concentrations, Krishnamurthy et al. (2023) reported a strong negative correlation between blood concentrations of these amino acids and HDL-c, as well as a positive correlation between triglycerides and the amino acids leucine and valine. In addition, a one-way analysis of variance demonstrated the coexistence of increasing concentrations of branched-chain amino acids and HDL-c blood values, with increasing concentrations of leucine, isoleucine, and valine showing a negative trend in relation to blood HDL-c levels.

While the mechanisms explaining the relationship between branched-chain amino acids and lipid metabolism have yet to be fully elucidated, research indicates that adipose tissue and the liver are considered the main sites of interaction between these types of amino acids and lipid metabolism (Yang et al. 2016). Furthermore, 3-hydroxy-3-methylglutaryl-CoA, a by-product of leucine fragmentation, can be used as a substrate for cholesterol biosynthesis (Figure 27.5) (Halama et al. 2016).

The association between aromatic amino acids and lipid profiles has also been explored. Research has shown that the amino acids tryptophan and tyrosine were positively correlated with fasting serum LDL-c concentrations, and that higher concentrations of aromatic amino acids were

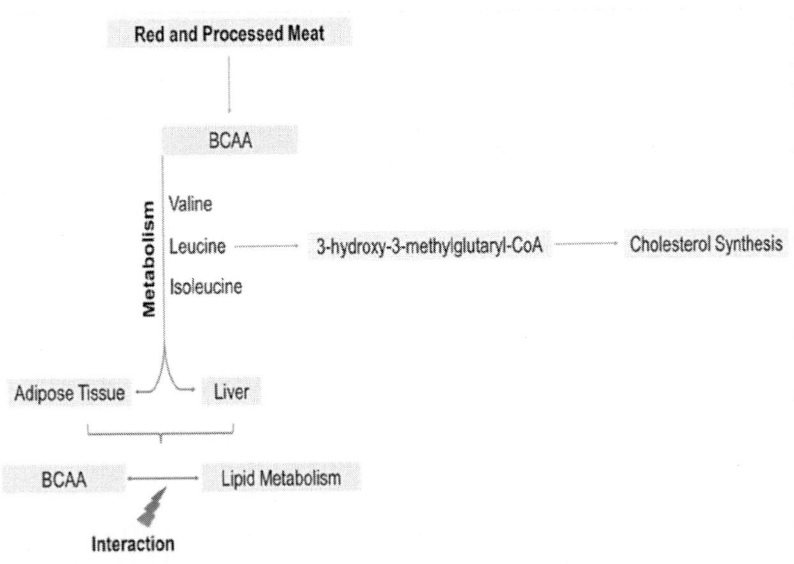

FIGURE 27.5 Relationship between branched-chain amino acids and lipid metabolism. BCAA: branched-chain amino acids (source: compiled by the authors).

present in the serum of individuals with simple steatosis and non-alcoholic steatohepatitis compared to individuals with normal livers (de Mello et al. 2021).

A cross-sectional study of the Chinese population aged 47 to 97 years showed a positive correlation between the concentrations of branched-chain amino acids (valine, isoleucine, leucine) and aromatic amino acids (tyrosine and phenylalanine) with triglycerides and apolipoprotein B/apolipoprotein A-I ratios. Additionally, an inverse relationship was observed with HDL-c and apolipoprotein A-I concentrations. Individuals with increased carotid intima-media thickness had higher concentrations of valine, isoleucine, leucine, total branched-chain amino acids, and phenylalanine (Yang et al. 2014).

Although several studies have investigated the relationship between the consumption of animal proteins and their constituent amino acids and hypercholesterolemia, the results remain inconclusive. Some meta-analyses have shown no effect of animal proteins on the increase in lipid profile parameters (Amirani et al. 2020; Zhang et al. 2016).

CONCLUSION

In summary, the effects of protein-rich diets on cardiovascular health are influenced by the specific protein source. Diets rich in vegetable proteins, especially from legumes such as soy, peas, and beans, have shown benefits in reducing total cholesterol, triglycerides, and LDL-c while increasing HDL-c. On the other hand, excessive consumption of animal proteins, especially from red meat, may be associated with adverse effects on the lipid profile, including increases in the concentrations of total cholesterol, LDL-c, HDL-c, and triglycerides. The complex interactions between amino acids, dietary lipid composition, and hepatic gene expression play a crucial role in these outcomes. A balanced approach that prioritizes healthier protein sources, such as those of plant origin, alongside an overall balanced diet and healthy lifestyle is key to promoting cardiovascular health. However, further research is imperative for a comprehensive understanding of the long-term impacts of high-protein diets. Such studies will provide valuable insights to guide dietary choices and preventive strategies for managing hypercholesterolemia and related diseases.

MINI DICTIONARY OF TERMS

Gastric inhibitory peptides or glucose-dependent insulinotropic polypeptides (GIP): Incretins produced by intestinal endocrine cells, especially in the duodenum and jejunum, in response to food intake. Their primary function is to stimulate pancreatic beta cells to secrete insulin in response to the presence of glucose. GIP also regulates gastrointestinal function by inhibiting stomach acid secretion and optimizing conditions for nutrient digestion and absorption in the small intestine.

Globulins: Proteins present in plasma that are insoluble in water and involved in various processes related to immune function, blood coagulation, and the transport of lipids, vitamins, and hormones.

Peptides: Compounds made up of several amino acids linked by peptide bonds, with an amine and a carboxyl group.

Total cholesterol: It is a measure of the total amount of cholesterol in the blood. It includes both low-density lipoprotein (LDL) cholesterol and high-density lipoprotein (HDL) cholesterol.

SUMMARY POINTS

- Vegetable proteins, such as those from legumes like soybeans, peas, and beans, are associated with lower serum cholesterol concentrations.
- Proteins enriched with isoflavones have a greater effect on reducing LDL-c levels and increasing HDL-c levels.

- Consumption of red or processed meats is linked to a higher cardiovascular risk.
- Branched-chain and aromatic amino acids in animal proteins are associated with an elevated risk of dyslipidemia.
- High-protein diets provide benefits for body weight management and cardiovascular risk factors; however, it is crucial to consider the source of protein. Too much animal protein in these diets can pose challenges for blood lipid levels.
- High-protein diets, particularly when derived from animal sources, may be linked to an increased risk of hypercholesterolemia, possibly due to a concomitant increase in fat intake.

REFERENCES

Amini, P., Maghsoudi, Z., Feizi, A., Ghiasvand, R., Askari, G. 2016. "Effects of high protein and balanced diets on lipid profiles and inflammation biomarkers in obese and overweight women at aerobic clubs: a randomized clinical trial". *Int J Prev Med* 7:110. https://doi.org/10.4103/2008-7802.190608

Amirani, E., Milajerdi, A., Reiner, Ž., Mirzaei, H., Mansournia, M. A., Asemi, Z. 2020. "Effects of whey protein on glycemic control and serum lipoproteins in patients with metabolic syndrome and related conditions: a systematic review and meta-analysis of randomized controlled clinical trials". *Lipids Health Dis* 19, 1:209. https://doi.org/10.1186/s12944-020-01384-7

Auestad, N., Layman, D. K. 2021. "Dairy bioactive proteins and peptides: a narrative review". *Nutr Rev* 79, 2:36–47. https://doi.org/10.1093/nutrit/nuab097

Bhoite, R., Chandrasekaran, A., Pratti, V. L., Satyavrat, V., Aacharya, S., Mane, A., Mehta, S., et al. 2021. "Effect of a high-protein high-fibre nutritional supplement on lipid profile in overweight/obese adults with type 2 diabetes mellitus: A 24-week randomized controlled trial". *J Nutr Metab* 6634225:1–9. https://doi.org/10.1155/2021/6634225

Campbell, T. C. 2017. "A plant-based diet and animal protein: questioning dietary fat and considering animal protein as the main cause of heart disease". *J Geriatr Cardiol* 14, 5:331–337. https://doi.org/10.11909/j.issn.1671-5411.2017.05.011

Caponio, G.R., Wang, D.Q.H., Ciaula, A., De Angelis, M., Portincasa, P. 2021. "Regulation of cholesterol metabolism by bioactive components of soy proteins: novel translational evidence". *Int J Mol Sci* 22, 1:227. https://doi.org/10.3390/ijms22010227.

Chatterjee, C., Gleddie, S., Xiao, C. W. 2018. "Soybean bioactive peptides and their functional properties". *Nutrients* 10, 9:1211. https://doi.org/10.3390/nu10091211.

Chen, Z., Glisic, M., Song, M., Aliahmad, H. A., Zhang, X., Moumdjian, A. C., Gonzalez-Jaramillo, V., et al. 2020. "Dietary protein intake and all-cause and cause-specific mortality: results from the rotterdam study and a meta-analysis of prospective cohort studies". *Eur J Epidemiol* 35, 5:411–429. https://doi.org/10.1007/s10654-020-00607-6

Coelho-Júnior, H. J., Calvani, R., Picca, A., Tosato, M., Savera, G., Landi, F., Marzetti, E. 2023. "Protein Intake Is Associated with Blood Pressure and Cholesterol Levels in Italian Older Adults: A Cross-Sectional Study". *Metabolites* 13, 3:431. https://doi.org/10.3390/metabo13030431

Correa, I. N., Poltonieri, F. 2016. "Ação hipocolesterolêmica das proteínas de leguminosas". RASBRAN 7, 2:79–90.

Cozzolino, S. M. F. 2016. *Biodisponibilidade de nutrientes*. 5ª ed. SP: Manole.

de Mello, V.D., Sehgal, R., Männistö, V., Klåvus, A., Nilsson,E., Perfilyev, A., Kaminska, D., Miao, Z., Pajukanta, P., Ling, C., Hanhineva,K., Pihlajamäki, J. 2021. Serum aromatic and branched-chain amino acids associated with NASH demonstrate divergent associations with serum lipids.Liver Int. Apr; 41(4):754–763. https://onlinelibrary.wiley.com/doi/epdf/10.1111/liv.14743

Diarz, E. J., Leyaro, B. J., Kivuyo, S. L., Ngowi, B. J., Msuya, S. E., Mfinanga, S. G., Bonfoh, B., Mahande, M. J. 2020. "Red meat consumption and its association with hypertension and hyperlipidaemia among adult maasai pastoralists of ngorongoro conservation area, Tanzania". *PLoS One* 15:e0233777. https://doi.org/10.1371/journal.pone.0233777

Díaz-Rúa, R., Keijer, J., Palou, A., Van Schothorst, E. M, Oliver P. 2017. "Long-term intake of a high-protein diet increases liver triacylglycerol deposition pathways and hepatic signs of injury in rats". *J Nutr Biochem* 46:39–48. https://doi.org/10.1016/j.jnutbio.2017.04.008

Dong J.-Y, Zhang, Z.-L., Wang, P.-Y., Qin, L.-Q. 2013. "Effects of high-protein diets on body weight, glycaemic control, blood lipids and blood pressure in type 2 diabetes: meta-analysis of randomised controlled trials". *Br J Nutr* 110:781–789. https://doi.org/10.1017/S0007114513002055

Dubin, S., McKee, K., Battish, S. 1994. "Essential amino acid reference profile affects the evaluation of enteral feeding products". J *Am Diet Assoc* 94:884–887. https://doi.org/10.1016/0002-8223(94)92369-8

Eilander, A., Harika, R. K., Zock, P. L. 2015. " Intake and sources of dietary fatty acids in Europe: are current population intakes of fats aligned with dietary recommendations?". *Eur J Lipid Sci Technol* 117:1370–1377. https://doi.org/10.1002/ejlt.201400513

Eng, J. M., Estall, J. L. 2021. "Diet-induced models of non-alcoholic fatty liver disease: food for thought on sugar, fat, and cholesterol". *Cells* 10, 7:1805. https://doi.org/10.3390/cells1007180

Ference, B. A., Ginsberg, H. N., Graham, I., Ray, K. K., Packard, C. J., Bruckert, E., Hegele, R. A., et al. 2017. "Low density lipoproteins cause atherosclerotic cardiovascular disease. 1. Evidence from genetic, epidemiologic, and clinical studies. A consensus statement from the European Atherosclerosis Society Consensus Panel". *Eur. Heart J* 38:2459–2472. https://doi.org/10.1093/eurheartj/ehx144

Fogelholm, M., Anderssen, S., Gunnarsdottir, I., Lahti-Koski, M. 2012. "Dietary macronutrients and food consumption as determinants of long-term weight change in adult populations: a systematic literature review". *Food Nutr Res* 56:19103. https://doi.org/10.3402/fnr.v56i0.19103

Frota, K. M., dos Santos, F. R. D., Ribeiro, V. Q., Arêas, J. A. 2015. "Cowpea protein reduces LDL-cholesterol and apolipoprotein B concentrations, but does not improve biomarkers of inflammation or endothelial dysfunction in adults with moderate hypercholesterolemia". *Nutr Hosp* 31, 4:1611–1619. https://doi.org/10.3305/nh.2015.31.4.8457

Frota, K. M., Mendonça, S., Saldiva, P. H., Cruz, R. J., Arêas, J. A. 2008. "Cholesterol-lowering properties of whole cowpea seed and its protein isolate in hamsters". *J Food Sci* 73, 9:H235–240. https://doi.org/10.1111/j.1750-3841.2008.00953.x

Geiker, N. R. W., Bertram, H. C., Mejborn, H., Dragsted, L. O., Kristensen, L., Carrascal, J. R., Bügel, S., Astrup, A. 2021. "Meat and human health-current knowledge and research gaps". *Foods* 10, 7:1556. https://doi.org/10.3390/foods10071556

Guasch-Ferré, M., Ambika Satija, A., Blondin, S. A., Janiszewski, M., Emlen, E., O'Connor, L. E., Campbell, W. W., Hu, F. B., Willett, W. C., Stampfer, M. J. 2019. "Meta-analysis of randomized controlled trials of red meat consumption in comparison with various comparison diets on cardiovascular risk factors". *Circulation* 139:1828–1845. https://doi.org/10.1161/CIRCULATIONAHA.118.035225

Guha, S., Majumder, K., Mine, Y. 2019. "Egg Proteins". In Encyclopedia of Food Chemistry 74–84. https://doi.org/10.1016/B978-0-08-100596-5.21603-X

Halama, A., Horsch, M., Kastenmüller, G., Möller, G., Kumar, P., Prehn, C., Laumen, H., et al. 2016. "Metabolic switch during adipogenesis: From branched chain amino acid catabolism to lipid synthesis". *Arch Biochem Biophys* 589:93–107. https://doi.org/10.1016/j.abb.2015.09.013.

Kim, I. S., Kim, C. H., Yang, W. S. 2021. "Physiologically active molecules and functional properties of soybeans in human health—a current perspective". *Int J Mol Sci* 22, 8:4054. https://doi.org/10.3390/ijms22084054.

Krishnamurthy, H. K., Reddy, S., Jayaraman, V., Krishna, K., Song, Q., Wang, T., Bei, K., Rajasekaran, J. J. 2023. "Preliminary study on the association of serum branched-chain amino acids with lipid and hepatic markers". *Cardiol Res* 14, 3:167–175. https://doi.org/10.14740/cr1454

Lamberg-Allardt, C., Bärebring, L., Arnesen, E.K., Nwaru, B. I., Thorisdottir, B., Ramel, A., Söderlund, F., Dierkes, J., Åkesson, A. 2023. "Animal versus plant-based protein and risk of cardiovascular disease and type 2 diabetes: a systematic review of randomized controlled trials and prospective cohort studies". *Food & Nutr Res* 67:9003. https://dx.doi.org/10.29219/fnr.v67.9003

Li, N., Wu, X., `Zhuang, W., Xia, L., Chen, Y., Zhao, R., Yi, M., Wan, Q., Du, L., Zhou, Y. 2020. "Soy and isoflavone consumption and multiple health outcomes: umbrella review of systematic reviews and meta-analyses of observational studies and randomized trials in humans". *Mol Nutr Food Res* 64, 4:e1900751. https://doi.org/10.1002/mnfr.201900751.

Mach, F., Baigent, C., Catapano, A.L., Koskinas, K. C., Casula, M., Badimon, L., Chapman, M. J., et al. 2020. "ESC scientific document group. 2019 ESC/EAS Guidelines for the management of dyslipidaemias: Lipid modification to reduce cardiovascular risk". *Eur Heart J* 41:111–188. https://doi.org/10.1093/eurheartj/ehz455

Marques, M. R., Soares, F. R. A., Corrêa, C. A. C., Siguemoto, É. S., Fontanari, G. G., Arêas, J. A. 2015. "Peptides from cowpea present antioxidant activity, inhibit cholesterol synthesis and its solubilisation into micelles". *Food Chem* 168:288–293. https://doi.org/10.1016/j.foodchem.2014.07.049

Mejia, S. B., Messina, M., Li, S. S., Viguiliouk, E., Chiavaroli, L., Khan, T. A., Srichaikul, K. 2019. "A meta-analysis of 46 studies identified by the FDA demonstrates that soy protein decreases circulating LDL and total cholesterol concentrations in adults". *J Nutr* 149, 6:968–981. https://doi.org/10.1093/jn/nxz020.

Mook-Kanamori, D. O., Römisch-Margl, W., Kastenmüller, G., Prehn, C., Petersen, A. K., Illig, T., Gieger, C., et al. 2014. "Increased amino acids levels and the risk of developing of hypertriglyceridemia in a 7-year follow-up". *J Endocrinol Invest* 37, 4:369–374. https://doi.org/10.1007/s40618-013-0044-7

Mozaffarian, D. 2016. "Dietary and policy priorities for cardiovascular disease, diabetes, and obesity: a comprehensive review". *Circulation* 133:187–225. https://doi.org/10.1161/CIRCULATIONAHA.115.018585

Najjar, R. S., Moore, C. E., Montgomery, B. D. 2018. "A defined, plant-based diet utilized in an outpatient cardiovascular clinic effectively treats hypercholesterolemia and hypertension and reduces medications". *Clin Cardiol* 41:307–313. https://doi.org/10.1002/clc.22863

Onwuliri, A. V., Obu, J. A. 2002. "Lipids and other constituents of Vigna unguiculata and Phaseolus vulgares grown in northern Nigeria". *Food Chem* 78, 1:1–7. https://doi.org/10.1016/S0308-8146(00)00293-4.

Rafiq, S., Huma, N., Pasha, I., Sameen, A., Mukhtar, O., Khan, M. I. 2016. "Chemical Composition, Nitrogen Fractions and Amino Acids Profile of Milk from Different Animal Species". *Asian-Australas J Anim Sci* 29, 7:1022–1028. https://doi.org/10.5713/ajas.15.0452

Sacks, F. M., Lichtenstein, A., Van Horn, L., Harris, W., Kris-Etherton, P., Winston, M. 2006. "American heart association nutrition committee. Soy protein, isoflavones, and cardiovascular health: An American heart association science advisory for professionals from the nutrition committee". *Circulation* 113:1034–1044. https://doi.org/10.1161/CIRCULATIONAHA.106.171052.

Schwarz, J., Tomé, D., Baars, A., Hooiveld, G. J. E. J., Müller, M. 2012. "Dietary protein affects gene expression and prevents lipid accumulation in the liver in mice". *PLoS One* 7, 10:e47303. https://doi.org/10.1371/journal.pone.0047303

Shi, X., Huang, Z., Zhou, G., Li, C. 2021. "Dietary protein from different sources exerted a great impact on lipid metabolism and mitochondrial oxidative phosphorylation in rat liver". *Front Nutr* 26, 8:719144. https://doi.org/10.3389/fnut.2021.719144.

Sirtori, C. R., Triolo, M., Bosisio, R., Bondioli, A., Calabresi, L., De Vergori, V., Gomaraschi, M., et al. 2012. "Hypocholesterolaemic effects of lupin protein and pea protein/fibre combinations in moderately hypercholesterolaemic individuals". *Br J Nutr* 107, 8:1176–1183. https://doi.org/10.1017/S0007114511004120.

Spielmann J, Stangl GI, Eder K. 2008. "Dietary pea protein stimulates> Nutrition bile acid excretion and lowers hepatic cholesterol concentration in rats". *J Anim Physiol Anim (Berl)* 92, 6:683–693. doi: 10.1111/j.1439-0396.2007.00766.x.

Tong, L. T., Xiao, T., Wang, L., Lu, C., Liu, L., Zhou, X., Wang, A., Qin, W., Wang, F. 2021. "Plant protein reduces serum cholesterol levels in hypercholesterolemia hamsters by modulating the compositions of gut microbiota and metabolites". *iScience* 24, 12:103435. https://doi.org/10.1016/j.isci.2021.103435

Trautwein, E. A., McKay, S. 2020. "The role of specific components of a plant-based diet in management of dyslipidemia and the impact on cardiovascular risk". *Nutrients* 12, 9:2671. https://doi.org/10.3390/nu12092671

Velázquez, K. T., Enos, R. T., Bader, J. E., Sougiannis, A. T., Carson, M. S., Chatzistamou, I., Carson, J. A., Nagarkatti, P. S., Nagarkatti, M., Murphy, E. A. 2019. "Prolonged high-fat-diet feeding promotes nonalcoholic fatty liver disease and alters gut microbiota in mice". *World J Hepatol* 11, 8:619–637. https://doi.org/10.4254/wjh.v11.i8.619

Xu, C., Markova, M., Seebeck, N., Loft, A., Hornemann, S., Gantert, T., Kabisch, S., et al. 2020. "High-protein diet more effectively reduces hepatic fat than low-protein diet despite lower autophagy and FGF21 levels". *Liver International* 40:2982–2997. https://doi.org/10.1111/liv.14596

Wang, S., Jung, S., Ko, K. S. 2022. "Effects of amino acids supplementation on lipid and glucose metabolism in HepG2 cells". *Nutrients* 14, 15:3050. https://doi.org/10.3390/nu14153050.

Wang, K. F., Wu, C. H., Chang, C. C., Chen, L. C., Wang, K. L., Lu, T. M., Lin, S. J., Chiang, C. E. 2017. "Determinants of treatment modifcation in hypercholesterolemic patients". *Acta Cardiol Sin* 33:156–164. https://doi.org/10.6515/acs20161215a.

Willett, W., Rockström, J., Loken, B., Springmann, M., Lang, T., Vermeulen, S., Garnett, T., et al. 2019. "Food in the Anthropocene: The EAT-lancet commission on healthy diets from sustainable food systems". *Lancet* 393:447–492. https://doi.org/10.1016/S0140-6736(18)31788-4

Yamakado, M., Nagao, K., Imaizumi, A., Tani, M., Toda, A., Tanaka, T., Jinzu, H., et al. 2015. "Plasma free amino acid profiles predict four-year risk of developing diabetes, metabolic syndrome, dyslipidemia, and hypertension in Japanese population". *Sci Rep* 5:11918. https://doi.org/10.1038/srep11918

Yang, R., Dong, J., Zhao, H., Li, H., Guo, H., Wang, S., Yu, S., Chen, W. 2014. "Association of branched-chain amino acids with carotid intima-media thickness and coronary artery disease risk factors". *PLoS One* 9, 6:e99598. https://doi.org/10.1371/journal.pone.0099598

Yang, P., Hu, W., Fu, Z., Sun, L., Zhou, Y., Gong, Y., Yang, T., Zhou, H. 2016. "The positive association of branched-chain amino acids and metabolic dyslipidemia in Chinese Han population". *Lipids Health Dis* 15:120. https://doi.org/10.1186/s12944-016-0291-7

Yoshikawa, M., Fujita, H., Matoba, N., Takenaka, Y., Yamamoto, T., Yamauchi, R., Tsuruki, H., Takahata, K. 2000. "Bioactive peptides derived from food proteins preventing lifestyle-related diseases". *Biofactors* 12, 1–4:143–146. https://doi.org/10.1002/biof.5520120122.

Yu, L., Zhu, Q., Li, Y., Song, P., Zhang, J. 2022. "Dietary branched-chain amino acids (BCAAs) and risk of dyslipidemia in a chinese population". *Nutrients* 14, 9:1824. https://doi.org/10.3390/nu14091824

Zhang, J. W.,Tong, X., Wan, Z., Wang, Y., Qin, L. Q., Szeto, I. M. 2016. "Effect of whey protein on blood lipid profiles: a meta-analysis of randomized controlled trials". *Eur J Clin Nutr* 70, 8:879–885. https://doi.org/10.1038/ejcn.2016.39

Zhao, H., Song, An., Zheng, C., Wang, M., Song, G. 2020. "Effects of plant protein and animal protein on lipid profile, body weight and body mass index on patients with hypercholesterolemia: a systematic review and meta-analysis". *Acta Diabetol* 57, 10:1169–1180. https://doi.org/10.1007/s00592-020-01534-4.

Zhao, L-G, Zhang, Q-L, Liu, X-L, Wu, H., Zheng, J-L, Xiang, Y-B. 2019. "Dietary protein intake and risk of type 2 diabetes: a dose-response meta-analysis of prospective studies". *Eur J Nutr* 58, 4:1351–1367. https://doi.org/10.1007/s00394-018-1737-7

Zhubi-Bakija, F., Bajraktari, G., Bytyçi, I., Mikhailidis, D. P., Henein, M. Y., Latkovskis, G., Rexhaj, Z., Zhubi, E., Banach, M., International Lipid Expert Panel (ILEP). 2021. "The impact of type of dietary protein, animal versus vegetable, in modifying cardiometabolic risk factors: A position paper from the International Lipid Expert Panel (ILEP)". *Clin Nutr*, 40, 1:255–276. https://doi.org/10.1016/j.clnu.2020.05.017

28 High-Protein Diets in the Prevention and Treatment of Type 2 Diabetes

Faidon Magkos

ABBREVIATIONS

BMI	Body mass index
EPIC	European Prospective Investigation into Cancer and Nutrition
HbA1c	Glycated hemoglobin
HOMA-IR	Homeostasis model assessment of insulin resistance
RCT	Randomized controlled trial
T2D	Type 2 diabetes

INTRODUCTION

Protein is an important component of the human diet, providing both essential and non-essential amino acids that serve as building blocks for the synthesis of cellular organelles, transport systems, signaling molecules, enzymes and hormones, and structural and contractile proteins, among many other biological functions. The average healthy adult body contains approximately 13% to 15% protein by weight, translating to about 8 kg in women and 11 kg in men (Wang et al. 2003). To avoid loss of body protein, international organizations recommend adults consume ~0.8 g of protein per kg of body mass daily, although the acceptable range is 1.0–3.7 g/kg/day, corresponding to 10–35% of all calories consumed (Wolfe et al. 2017).

Higher-than-adequate protein intakes are very popular among the public for various reasons, ranging from weight loss to muscle gain (Magkos 2020b). International expert groups and healthcare professionals often recommend high-protein diets to manage conditions like obesity, age-related muscle loss, and type 2 diabetes (T2D) (Busetto et al. 2011; Mittendorfer et al. 2020). In the context of body weight maintenance (i.e., eucaloric diets), high-protein diets typically provide 18–30% of all calories from protein (120–200 g/day), compared to 10–15% (67–100 g/day) in normal-protein diets (Westerterp-Plantenga et al. 2009). For weight loss (i.e., hypocaloric diets), the protein content may be even higher when expressed as percent of total calories (18–47%), but the absolute amount of protein is not particularly high (52–120 g/day) owing to the reduction in total calorie intake (Westerterp-Plantenga et al. 2009).

Habitual protein intake among adults in developed countries varies widely, ranging from less than 0.8 g/kg/day to more than 1.8 g/kg/day. This corresponds to 10–22% of all calories consumed across both sexes and all body mass index (BMI) categories (Mittendorfer et al. 2020). Accordingly, protein intakes >0.8 g/kg/day are very common in the general population when dietary energy, carbohydrate, and fat intakes are not restricted (Mittendorfer et al. 2020).

DOI: 10.1201/b23328-28

DIETARY PROTEIN INTAKE AND RISK FOR TYPE 2 DIABETES

Numerous large-scale population-based prospective cohort studies have investigated the relationship between dietary protein intake and T2D risk. These studies include the Nurses' Health Study I and II and the Health Professionals Follow-up Study (n ~205,800) (Malik et al. 2016); the European Prospective Investigation into Cancer and Nutrition (EPIC) Netherlands Substudy (n ~38,000) (Sluijs et al. 2010); the EPIC-InterAct Case-Cohort Study (n ~28,500) (van Nielen et al. 2014); the Malmö Diet and Cancer Cohort Study (n ~27,000) (Ericson et al. 2013); the Melbourne Collaborative Cohort Study (n ~21,500) (Shang et al. 2016); and the Rotterdam Study (n ~6,800) (Chen et al. 2020). These studies consistently demonstrate that a habitually high dietary protein intake is associated with an increased risk for T2D (and prediabetes), with an approximately 20–40% increase in risk for every 5% increment in dietary energy from protein, equivalent to an additional 25–30 g of protein per day. The increase in risk persists, albeit attenuated, after adjusting for various confounders including dietary factors (e.g., total energy, fiber, type of fat), demographics (e.g., age, sex, ethnicity), lifestyle habits (e.g., alcohol consumption, smoking, physical activity), and often—but not always—after adjusting for obesity indexes as well (e.g., BMI, waist circumference) (Sluijs et al. 2010; Ericson et al. 2013; van Nielen et al. 2014; Malik et al. 2016; Shang et al. 2016; Chen et al. 2020). This suggests that at least some of the putative adverse effect of high-protein intake is mediated by greater total or regional adiposity.

The association between high-protein intake and T2D risk is driven exclusively by protein intake from animal food sources (e.g., meat, fish, and poultry) because protein intake from plant sources (e.g., legumes, nuts, and grains) is not associated with risk for T2D and may, instead, be protective (Sluijs et al. 2010; Ericson et al. 2013; van Nielen et al. 2014; Malik et al. 2016; Shang et al. 2016; Chen et al. 2020). There is also considerable heterogeneity among different animal foods, with the strongest adverse effects on T2D risk observed for red and processed meat; less pronounced effects for fish, poultry, and eggs; and minimal effects for dairy products, which are similar to plant foods in this regard (Malik et al. 2016). Results from these large-scale population-based longitudinal studies are corroborated by three independent meta-analyses, encompassing 11 to 72 prospective cohort studies with a total of ~450,000 to ~2,600,000 participants followed-up for 2.5 to 24 years (Shang et al. 2016; Tian et al. 2017; Fan et al. 2019). However, the potentially beneficial independent effect of plant protein should be interpreted with caution, as animal food sources in all these studies accounted for more than half and up to three-quarters of total protein intake (Sluijs et al. 2010; van Nielen et al. 2014; Malik et al. 2016; Shang et al. 2016; Chen et al. 2020), resulting in moderate-to-low absolute plant protein intake. In other words, total protein intake itself may be important, independent of the food source (Sluijs et al. 2010). Furthermore, the direction and significance of the relationship between dietary protein intake and T2D risk may depend on both the food source and the actual consumption levels. In another meta-analysis, a high but not a moderate intake of animal protein was positively related to risk, whereas a moderate but not a high intake of plant protein was negatively related to risk (Ye et al. 2019). Despite these nuances, in a typical Western diet, there is little doubt that replacing animal protein with plant protein—for the same total protein intake—decreases the risk for T2D; this holds true also when replacing "unhealthy" with "healthy" plant-based foods (Figure 28.1) (Malik et al. 2016; Virtanen et al. 2017).

HIGH-PROTEIN DIETS AND WEIGHT LOSS IN TYPE 2 DIABETES

Weight loss through lifestyle modification, i.e., diet and exercise, is the cornerstone of T2D management (Magkos et al. 2020). Progressive weight loss reduces the risk of developing T2D among people who do not have T2D at baseline (Hamman et al. 2006), and similarly increases the chances of remission among people who have T2D at baseline (Lean et al. 2019), in a dose-dependent manner. An average weight loss of ~12%—induced by a very low-calorie diet followed by a prudent whole-food diet and increased physical activity—reduces glycated hemoglobin (HbA1c) to

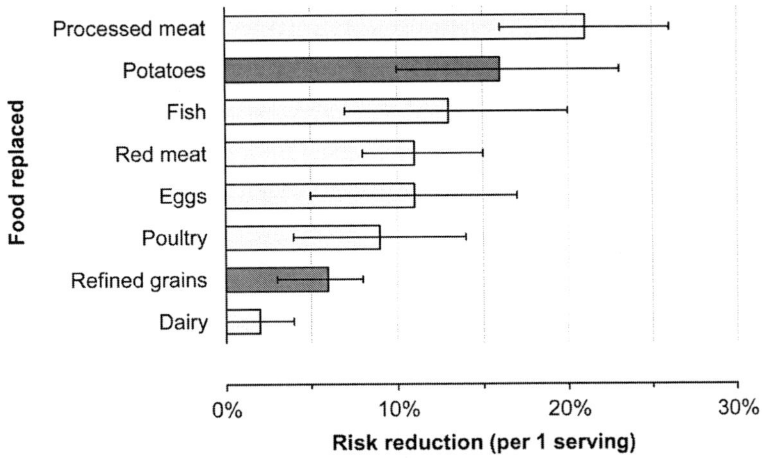

FIGURE 28.1 Reduction in the risk for type 2 diabetes when healthy plant protein foods replace other protein foods in the diet. This figure shows the reduction in type 2 diabetes risk associated with replacement of one serving/day of animal protein foods (dairy, poultry, eggs, red and processed meat; light gray bars) or "unhealthy" plant protein foods (refined grains and potatoes; dark gray bars) with one serving of "healthy" plant protein foods (composite of whole grains, legumes, peanuts, peanut butter, and other nuts). Data are pooled from the Nurses' Health Study I and II in women and the Health Professionals Follow-up Study in men. Results have been adjusted for many demographic, dietary, and lifestyle variables, as well as body mass index, and have been mutually adjusted for each other. Bars indicate mean risk reduction with 95% confidence interval. The figure was created with data provided in Figure 1 in Malik et al. (2016).

non-diabetic levels in two out of three patients with T2D; and of these two patients, one achieves HbA1c within the normoglycemic range (Taheri et al. 2020). Nevertheless, most patients with T2D are unable to maintain large amounts of weight loss over the long term, i.e., after 2–4 years (Franz et al. 2015). However, even without large amounts of weight loss and T2D remission, progressively greater weight loss (from as little as 2–3% of initial body weight) causes dose-dependent reductions in fasting glucose and HbA1c concentrations, the need for antidiabetic medications, fasting triglyceride concentrations, and systolic and diastolic blood pressures, as well as dose-dependent increases in high-density-lipoprotein cholesterol concentration (Wing et al. 2011; Gummesson et al. 2017). Accordingly, the effects of high-protein diets on body weight homeostasis in patients with T2D are critically important for evaluating their utility in T2D management.

Protein-rich diets are promoted for weight loss because of their ability to stimulate diet-induced thermogenesis and suppress hunger (Magkos 2020b). Among people without T2D, high-protein diets—prescribed concurrently with calorie restriction—induce significantly greater weight loss than conventional low-fat or high-carbohydrate diets during the first 3–6 months of dieting, but these differences are not usually apparent after 1–2 years (Magkos 2020b). Likewise, high-protein diets prescribed ad libitum after clinically significant weight loss—to evaluate their utility for weight maintenance—attenuate weight regain by 1–2 kg during the first 3–12 months (Magkos 2020a), but again, this beneficial effect is not apparent after 3 years (Raben et al. 2021). The results from studies among people with T2D are essentially the same. Three of the longest (≥1 year) randomized controlled trials (RCTs) in individuals with overweight or obesity and T2D that compared calorie-restricted high-protein diets (30% of total energy) against isocaloric normal-protein (15–20%), high-carbohydrate diets, matched for fat content, found that weight loss was not significantly different after 1–2 years (Figure 28.2) (Larsen et al. 2011; Krebs et al. 2012; Pedersen et al. 2014). In fact, contrary to observations in non-diabetic individuals (Magkos 2020b), higher dietary protein intakes do not facilitate weight loss in patients with T2D even at earlier time points, i.e., 2–6 months (Luscombe et al. 2002; Larsen et al. 2011; Krebs et al. 2012; Pedersen et al. 2014). One study

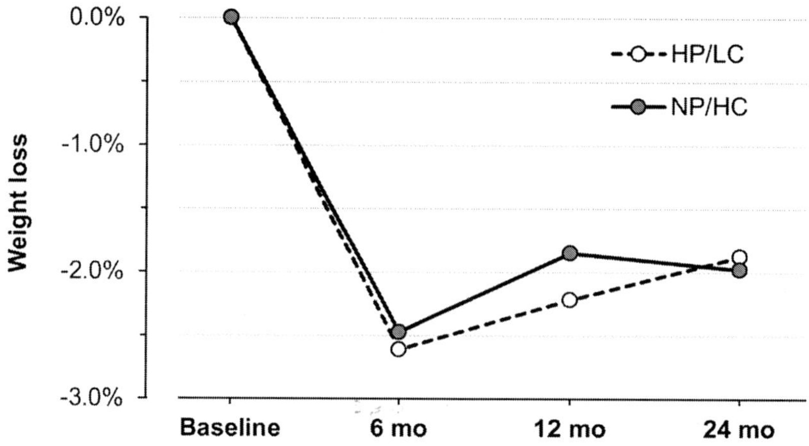

FIGURE 28.2 Effects of a calorie-restricted high-protein diet on body weight in patients with type 2 diabetes. This figure shows the average weight loss (% of initial weight) induced by a high-protein, low-carbohydrate diet (HP/LC: 30% of energy as protein, 40% as carbohydrate, and 30% as fat) or a normal-protein, high-carbohydrate diet (NP/HC: 15%, 55%, and 30%, respectively) in 294 patients with overweight or obesity and type 2 diabetes. Both diets were prescribed in "real-world" settings and aimed at a 500 kcal/day energy deficit. Results have been analyzed with generalized linear mixed models and have been adjusted for baseline. Body weight decreased significantly with time in both diet arms ($P < 0.001$), but there was no time-by-diet interaction ($P = 0.73$). The figure was created with data extracted from Figure 2 in Krebs et al. (2012).

reported greater weight loss 4 months after a protein-rich diet than a carbohydrate-rich diet, but only when the diets were combined with resistance exercise (Wycherley et al. 2010), while another study found no differences after 3 months when similar diets were combined with aerobic exercise (Watson et al. 2016). In a meta-analysis of 18 RCTs in patients with T2D, comparing high-protein against low-protein diets (sometimes differing in both carbohydrate and fat contents) for a duration ranging from 4 weeks to 2 years, no beneficial effects of higher protein intakes on weight loss and body composition were observed (Zhao et al. 2018). This was corroborated in a recent network meta-analysis that included 8 crossover or parallel RCTs reporting on the effects of high-protein diets on body weight (Zeng et al. 2023). Calorie restriction emerged as the best dietary pattern for weight loss in patients with T2D, whereas a high dietary protein content had no independent beneficial effect on body weight homeostasis (Zeng et al. 2023). It is important to note that typical calorie-restricted high-protein diets are well-tolerated and are not accompanied by any significant side effects or adverse changes in kidney function indexes in patients with T2D (Larsen et al. 2011; Krebs et al. 2012), even among those with albuminuria, i.e., an early manifestation of diabetic nephropathy (Pedersen et al. 2014).

EFFECTS OF ACUTE PROTEIN INGESTION ON GLUCOSE HOMEOSTASIS IN TYPE 2 DIABETES

Plasma glucose concentrations reflect the net balance between the rate of glucose entry into the circulation—from both endogenous sources (i.e., hepatic and renal glucose production in the fasted and postprandial states) and exogenous sources (i.e., meal-derived glucose in the postprandial state)—and the rate of glucose removal from the circulation by all tissues of the body (e.g., brain, skeletal muscle, and adipose tissue, but also urinary glucose loss in T2D) (Magkos et al. 2023). After an overnight fast, plasma glucose concentrations are primarily maintained by hepatic glucose production (hepatic glycogenolysis and gluconeogenesis). Ingestion of carbohydrate (but also other nutrients) suppresses endogenous glucose production and stimulates glucose disposal (Magkos et al. 2023). Insulin and glucagon are the primary mediators of these responses: insulin suppresses hepatic gluconeogenesis and glycogenolysis and stimulates glucose uptake in muscle and adipose tissue, whereas glucagon stimulates both gluconeogenesis and glycogenolysis in the liver (Magkos et al. 2023). T2D is characterized by fasting and postprandial hyperglycemia (i.e., higher-than-normal circulating glucose concentrations), owing predominantly to high circulating glucagon, insulin resistance in muscle and liver, and inadequate insulin secretion from the pancreas (Mittendorfer et al. 2020).

Acute ingestion of protein alone, without carbohydrate, increases plasma amino acid concentrations and stimulates both insulin and glucagon secretion (Carr et al. 2008; Claessens et al. 2009; Ang et al. 2019). The net result of this postprandial hormonal milieu is the stimulation of both glucose disposal and glucose production rates to approximately equal extents, so that no changes occur in blood glucose concentration (Carr et al. 2008; Ang et al. 2019). The protein-induced stimulation of endogenous glucose production likely involves both direct actions, mediated by local glucagon signaling, and indirect actions, mediated by the provision of glucogenic amino acids for gluconeogenesis in the liver. The protein-induced stimulation of peripheral glucose uptake is likely the result of impaired insulin signaling in skeletal muscle, as studies using the glucose clamp technique to measure insulin sensitivity (exogenous infusion of insulin at a constant rate to achieve high physiological hyperinsulinemia and infusion of dextrose at a variable rate to maintain euglycemia) have found that hyperaminoacidemia, achieved by infusing amino acid solutions or ingesting protein, impairs insulin-mediated glucose disposal, i.e., causes insulin resistance in muscle (Mittendorfer et al. 2020). Protein, therefore, elicits a coordinated metabolic response that opposes the actions of insulin on the liver—overriding the suppression of glucose production—and on peripheral tissues—blunting the stimulation of glucose uptake—to prevent hypoglycemia in the absence of ingested glucose.

When protein is co-ingested with glucose, a pronounced insulinotropic effect is observed, which is greater than that of glucose alone and depends on the food source of the protein (e.g., animal protein is more potent than plant protein), its rate of absorption (e.g., whey is more potent than casein), and its amino acid composition (e.g., arginine and branched-chain amino acids are among the most potent) (Esteves de Oliveira et al. 2011; Oberoi et al. 2022; Dehnavi et al. 2023). Simultaneously, the typical suppression of glucagon concentrations observed after consuming only glucose is overridden when protein is co-ingested (Oberoi et al. 2022). Various mechanisms may be involved in protein-stimulated insulin secretion from pancreatic beta-cells, including increased gut hormone (incretin) release, changes in intracellular energy production and calcium influx, and activation of the mammalian target of rapamycin (Carr et al. 2008; Yanagisawa 2023). The acute protein-stimulated glucagon secretion from the alpha-cells is thought to be mediated by direct amino acid signaling and an endocrine feedback loop (Dean 2020). Protein-induced insulin secretion is maintained in individuals with T2D who have impaired glucose-induced insulin secretion (Ang et al. 2012). Under these conditions, the balance between protein's effects on pancreatic insulin and glucagon secretion, and its effects on hepatic and peripheral insulin action, will determine protein's net impact on blood glucose homeostasis. Often, but not always, ingestion of protein some time before ("preload") (Figure 28.3) or together with ("co-ingestion") (Figure 28.4) a fixed amount of carbohydrate augments postprandial insulin responses and mitigates postprandial plasma glucose excursions in patients with T2D (Frid et al. 2005; Jakubowicz et al. 2014; Manders et al. 2014; Bae et al. 2018; King et al. 2018). This relative hypoglycemic effect is also evident during isocaloric meals where protein replaces some carbohydrate; however, in these cases, postprandial insulin concentrations are also lower, suggesting that reduced carbohydrate intake is the primary driver behind this response (Samkani et al. 2018).

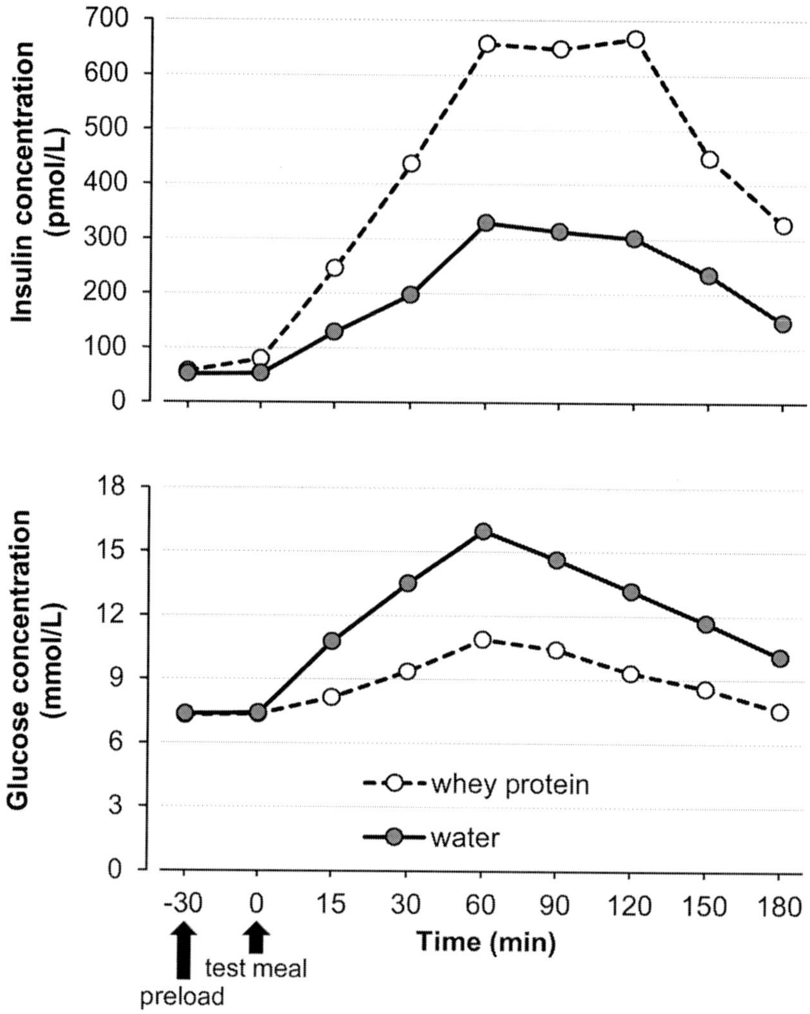

FIGURE 28.3 Effects of a protein "preload" on postprandial insulin and glucose responses in patients with type 2 diabetes. This figure shows the average insulin (top) and glucose (bottom) concentrations before and for 3 hours after ingestion of a high-glycemic-index test meal (~350 kcal; at t = 0 min), preceded by consumption of 50 g of whey protein or plain water (at t = -30 min) in random order, in 15 individuals with type 2 diabetes. The protein "preload" significantly increased postprandial insulin concentrations (total area-under-the-curve, 0–180 min: +105%) and lowered postprandial glucose concentrations (total area-under-the-curve, 0–180 min: -28%). Redrawn with data extracted from Figure 1a and 1b in Jakubowicz et al. (2014).

EFFECTS OF HIGH-PROTEIN DIETS ON GLUCOSE HOMEOSTASIS IN TYPE 2 DIABETES

Contrary to the results from studies examining the responses to one or a few consecutive meals, results from longer-term studies that evaluated the effects of high-protein diets of varying composition and duration on glycemic control, insulin resistance, and beta-cell function in people with (and without) T2D are contradictory and inconclusive. Meta-analyses and systematic and narrative reviews of these studies do not reach definitive conclusions about the effects of high protein intakes on glycemic control and the mechanisms regulating glucose homeostasis, owing mainly to confounding from differences in overall diet composition and the accompanying changes in body weight (Rietman et al. 2014; Mittendorfer et al. 2020; Ancu et al. 2021; Yanagisawa 2023). For

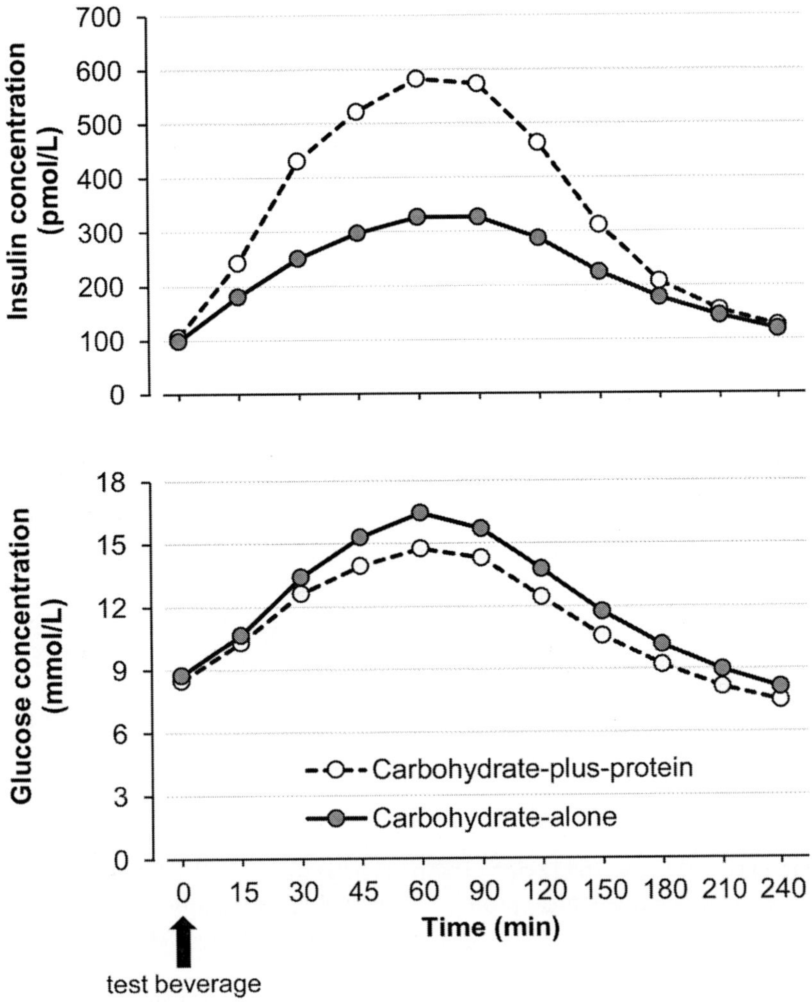

FIGURE 28.4 Effects of protein co-ingestion with carbohydrate on postprandial insulin and glucose responses in patients with type 2 diabetes. This figure shows the average insulin (top) and glucose (bottom) concentrations before and for 4 hours after ingestion of a test beverage containing 0.7 g/kg body weight carbohydrate (50% glucose and 50% maltodextrin), alone or together with 0.3 g/kg body weight intact casein protein, both administered at t = 0 in random order, in 60 individuals with type 2 diabetes. Co-ingestion of protein significantly increased postprandial insulin concentrations (baseline-adjusted area-under-the-curve, 0–240 min: +99%) and lowered postprandial glucose concentrations (baseline-adjusted area-under-the-curve, 0–240 min: -22%). Redrawn with data extracted from Figure 1a and 1b in Manders et al. (2014).

example, in individuals with T2D, increased protein intake coupled with decreased carbohydrate intake for 6 weeks improves glycemia (it lowers fasting and postprandial glucose concentrations and HbA1c) and decreases liver fat compared with a normal-protein, high-carbohydrate diet, during both eucaloric and hypocaloric feeding paradigms (Skytte et al. 2019; Thomsen et al. 2022). However, small differences in body weight between diet arms despite apparent weight stability (Skytte et al. 2019) and differences not only in protein and carbohydrate intakes, but also in dietary fat and possibly other components of the diet (Skytte et al. 2019; Thomsen et al. 2022), make it impossible to dissect the independent contribution of increased dietary protein consumption. Most studies reporting beneficial effects of high-protein diets on glycemic control and insulin resistance in people with T2D do so in the face of calorie restriction and weight loss; therefore, the role of greater protein intake per se remains unclear (Rietman et al. 2014). Furthermore, high-protein meals and diets (20–35% of total energy) combined with carbohydrate restriction produce greater improvements in glycemic control than similarly high-protein meals and diets combined with fat restriction (Huhmann et al. 2018; Hansen et al. 2023). These findings underscore the importance of concomitant changes in body weight and other dietary macronutrients when evaluating high-protein diets in relation to T2D.

In a meta-analysis of 18 RCTs involving 1,099 adults with T2D, who were treated with high-protein or various comparison diets for 1–24 months, no differences were found between diets for the changes in fasting glucose and insulin concentrations and HbA1c (Zhao et al. 2018). Furthermore, greater protein intakes did not affect resting systolic and diastolic blood pressures and cholesterol concentrations, but caused a significant reduction in triglyceride concentration (Zhao et al. 2018). A second meta-analysis of 8 RCTs in patients with T2D did not find any significant beneficial effects of high-protein diets on markers of glycemic control, blood pressure, and the plasma lipid profile, but reported a significant improvement in insulin sensitivity, as evidenced by a reduction in the Homeostasis Model Assessment of insulin resistance (HOMA-IR) index (Zeng et al. 2023). Finally, a third meta-analysis of 13 RCTs lasting 3–52 months, including 1,138 individuals with T2D, found that high-protein diets do not significantly improve glycemic control and blood pressure, but reduce triglyceride and low-density-lipoprotein cholesterol concentrations and the HOMA-IR index (Yu et al. 2020). Despite the lack of consistent beneficial effects of high-protein diets on glucoregulation in T2D, another systematic review and meta-analysis of 13 RCTs found that replacing animal protein with plant protein in the diabetic diet, while maintaining the same total protein intake, modestly but significantly decreases fasting glucose and insulin concentrations and HbA1c (Viguiliouk et al. 2015).

CONCLUSION

Despite the popularity of high-protein diets for managing obesity and T2D, there is very little evidence in favor of higher-than-adequate protein intakes. First, habitually greater protein consumption in the population is strongly associated with an increased risk of T2D. Second, clinical trials targeting weight loss—the cornerstone of T2D management—find no advantage for high-protein diets compared with diets of different macronutrient composition. Third, although protein ingestion acutely decreases postprandial glucose concentrations in response to a carbohydrate challenge in people with T2D, chronically consuming a high-protein diet does not seem to provide any benefits independent of changes in body weight and the concomitant reductions in other dietary macronutrients, particularly carbohydrate. Nevertheless, in the context of most Western dietary patterns, which are more than adequate—if not rich—in protein, replacing animal protein with plant protein will likely be beneficial for the prevention and treatment of diabetes.

MINI DICTIONARY OF TERMS

Diet-induced thermogenesis: The increase in energy expenditure above baseline (resting metabolic rate) that occurs upon ingestion of food and is attributed to the energy required to process and store the ingested nutrients.

Glucose rate of appearance and disappearance: The rate at which glucose enters the circulation from the liver and kidney (and gut, in the postprandial state) is the glucose rate of appearance (Ra), whereas the rate at which glucose exits the circulation via all possible routes (uptake by tissues, loss in urine, etc.) is the glucose rate of disappearance (Ra). Ra and Rd can be measured in vivo in humans by using isotope tracers.

Healthy and unhealthy plant foods: Not all plant foods are created equal. Some plant foods are inversely associated with risk for cardiometabolic disease ("healthy") whereas others are positively associated ("unhealthy").

High-protein diet: The definition varies and can be based on absolute terms (g/day or g/kg/day) or on relative terms (% of energy intake). Operationally defined here as a diet with more protein than the comparison diet.

Hyperinsulinemic-euglycemic clamp: A technique for the direct measurement of insulin sensitivity in human subjects. Insulin is infused intravenously at a constant rate to achieve physiological hyperinsulinemia (which suppresses endogenous glucose production and stimulates glucose uptake), whereas dextrose is infused at a variable rate to maintain plasma glucose concentration steady. When steady-state is reached, the rate of exogenous dextrose infusion (plus any residual endogenous glucose production) is the same as the rate at which glucose is removed from the circulation via all possible routes.

Insulin and glucagon: The two major pancreatic hormones that regulate glucose metabolism. Insulin (secreted from beta-cells) suppresses glucose production (from both gluconeogenesis and glycogenolysis) and stimulates glucose uptake, glycolysis, and glycogenesis. Glucagon (secreted from alpha-cells) has the opposite effects.

Type 2 diabetes: A metabolic disease characterized by higher-than-normal blood glucose concentration in the fasting and postprandial states, owing to a combination of peripheral insulin resistance and pancreatic beta-cell dysfunction.

SUMMARY POINTS

- This chapter examines the role of dietary protein in relation to type 2 diabetes.
- Observational studies indicate that for every 5% increment in dietary energy from protein (25–30 g/day more), the risk of type 2 diabetes increases by 20–40%.
- Replacing animal protein with plant protein typically decreases this risk, although not all animal and plant foods are created equal in this respect.
- Weight loss is the cornerstone of managing type 2 diabetes, but increasing dietary protein does not result in greater weight loss in patients with type 2 diabetes.
- Ingesting protein alone acutely increases the secretion of both insulin and glucagon and does not affect plasma glucose concentrations.
- When protein is ingested just before or together with carbohydrate, the postprandial insulin response is greater and the postprandial glucose response is smaller compared to carbohydrate alone, despite that more calories are consumed.
- The effects of high-protein diets on glucose homeostasis in patients with type 2 diabetes are difficult to interpret because of concomitant changes in body weight and differences in other dietary components.
- Results from meta-analyses indicate no weight-independent effects of high-protein diets on glucoregulation in type 2 diabetes, but replacing animal protein with plant protein is beneficial.
- There is insufficient data to recommend higher-than-adequate protein intakes for the prevention and treatment of type 2 diabetes.

REFERENCES

Ancu, O., M. Mickute, N. D. Guess, N. M. Hurren, N. A. Burd and R. W. Mackenzie. 2021. "Does high dietary protein intake contribute to the increased risk of developing prediabetes and type 2 diabetes?" *Appl Physiol Nutr Metab* 46 (1): 1–9. https://doi.org/10.1139/apnm-2020-0396.

Ang, M., A. S. Muller, F. Wagenlehner, A. Pilatz and T. Linn. 2012. "Combining protein and carbohydrate increases postprandial insulin levels but does not improve glucose response in patients with type 2 diabetes." *Metabolism* 61 (12): 1696–702. https://doi.org/10.1016/j.metabol.2012.05.008.

Ang, T., C. R. Bruce and G. M. Kowalski. 2019. "Postprandial aminogenic insulin and glucagon secretion can stimulate glucose flux in humans." *Diabetes* 68 (5): 939-46. https://doi.org/10.2337/db18-1138.

Bae, J. H., L. K. Kim, S. H. Min, C. H. Ahn and Y. M. Cho. 2018. "Postprandial glucose-lowering effect of premeal consumption of protein-enriched, dietary fiber-fortified bar in individuals with type 2 diabetes mellitus or normal glucose tolerance." *J Diabetes Investig* 9 (5): 1110–8. https://doi.org/10.1111/jdi.12831.

Busetto, L., M. Marangon and F. De Stefano. 2011. "High-protein low-carbohydrate diets: what is the rationale?" *Diabetes Metab Res Rev* 27 (3): 230–2. https://doi.org/10.1002/dmrr.1171.

Carr, R. D., M. O. Larsen, M. S. Winzell, K. Jelic, O. Lindgren, C. F. Deacon and B. Ahren. 2008. "Incretin and islet hormonal responses to fat and protein ingestion in healthy men." *Am J Physiol Endocrinol Metab* 295 (4): E779–84. https://doi.org/10.1152/ajpendo.90233.2008.

Chen, Z., O. H. Franco, S. Lamballais, M. A. Ikram, J. D. Schoufour, T. Muka and T. Voortman. 2020. "Associations of specific dietary protein with longitudinal insulin resistance, prediabetes and type 2 diabetes: The Rotterdam study." *Clin Nutr* 39 (1): 242–9. https://doi.org/10.1016/j.clnu.2019.01.021.

Claessens, M., W. Calame, A. D. Siemensma, M. A. van Baak and W. H. Saris. 2009. "The effect of different protein hydrolysate/carbohydrate mixtures on postprandial glucagon and insulin responses in healthy subjects." *Eur J Clin Nutr* 63 (1): 48–56. https://doi.org/10.1038/sj.ejcn.1602896.

Dean, E. D. 2020. "A primary role for alpha-cells as amino acid sensors." *Diabetes* 69 (4): 542–9. https://doi.org/10.2337/dbi19-0021.

Dehnavi, Z., H. Barghchi, A. J. Esfehani, M. Barati, Z. Khorasanchi, F. Farsi, A. N. Ostad, G. Ranjbar, R. Rezvani, M. R. Gorgani and M. Safarian. 2023. "Animal and plant-based proteins have different postprandial effects on energy expenditure, glycemia, insulinemia, and lipemia: A review of controlled clinical trials." *Food Sci Nutr.* https://doi.org/10.1002/fsn3.3417.

Ericson, U., E. Sonestedt, B. Gullberg, S. Hellstrand, G. Hindy, E. Wirfalt and M. Orho-Melander. 2013. "High intakes of protein and processed meat associate with increased incidence of type 2 diabetes." *Br J Nutr* 109 (6): 1143–53. https://doi.org/10.1017/S0007114512003017.

Esteves de Oliveira, F. C., A. C. Pinheiro Volp and R. C. Alfenas. 2011. "Impact of different protein sources in the glycemic and insulinemic responses." *Nutr Hosp* 26 (4): 669–76. https://doi.org/10.1590/S0212-16112011000400002.

Fan, M., Y. Li, C. Wang, Z. Mao, W. Zhou, L. Zhang, X. Yang, S. Cui and L. Li. 2019. "Dietary protein consumption and the risk of type 2 diabetes: A dose-response meta-analysis of prospective studies." *Nutrients* 11 (11): 2783. https://doi.org/10.3390/nu11112783.

Franz, M. J., J. L. Boucher, S. Rutten-Ramos and J. J. VanWormer. 2015. "Lifestyle weight-loss intervention outcomes in overweight and obese adults with type 2 diabetes: a systematic review and meta-analysis of randomized clinical trials." *J Acad Nutr Diet* 115 (9): 1447–63. https://doi.org/10.1016/j.jand.2015.02.031.

Frid, A. H., M. Nilsson, J. J. Holst and I. M. Bjorck. 2005. "Effect of whey on blood glucose and insulin responses to composite breakfast and lunch meals in type 2 diabetic subjects." *Am J Clin Nutr* 82 (1): 69–75. https://doi.org/10.1093/ajcn.82.1.69.

Gummesson, A., E. Nyman, M. Knutsson and M. Karpefors. 2017. "Effect of weight reduction on glycated haemoglobin in weight loss trials in patients with type 2 diabetes." *Diabetes Obes Metab* 19 (9): 1295–305. https://doi.org/10.1111/dom.12971.

Hamman, R. F., R. R. Wing, S. L. Edelstein, J. M. Lachin, G. A. Bray, L. Delahanty, M. Hoskin, A. M. Kriska, E. J. Mayer-Davis, X. Pi-Sunyer, J. Regensteiner, B. Venditti and J. Wylie-Rosett. 2006. "Effect of weight loss with lifestyle intervention on risk of diabetes." *Diabetes Care* 29 (9): 2102–7. https://doi.org/10.2337/dc06-0560.

Hansen, C. D., E. M. Gram-Kampmann, J. K. Hansen, M. B. Hugger, B. S. Madsen, J. M. Jensen, S. Olesen, N. Torp, D. N. Rasmussen, M. Kjaergaard, S. Johansen, K. P. Lindvig, P. Andersen, K. H. Thorhauge, J. C. Brond, P. Hermann, H. Beck-Nielsen, S. Detlefsen, T. Hansen, K. Hojlund, M. S. Thiele, M. Israelsen and A. Krag. 2023. "Effect of calorie-unrestricted low-carbohydrate, high-fat diet versus high-carbohydrate, low-fat diet on type 2 diabetes and nonalcoholic fatty liver disease : A randomized controlled trial." *Ann Intern Med* 176 (1): 10–21. https://doi.org/10.7326/M22-1787.

Huhmann, M. B., S. Yamamoto, J. M. Neutel, S. S. Cohen and J. B. Ochoa Gautier. 2018. "Very high-protein and low-carbohydrate enteral nutrition formula and plasma glucose control in adults with type 2 diabetes mellitus: a randomized crossover trial." *Nutr Diabetes* 8 (1): 45. https://doi.org/10.1038/s41387-018-0053-x.

Jakubowicz, D., O. Froy, B. Ahren, M. Boaz, Z. Landau, Y. Bar-Dayan, T. Ganz, M. Barnea and J. Wainstein. 2014. "Incretin, insulinotropic and glucose-lowering effects of whey protein pre-load in type 2 diabetes: a randomised clinical trial." *Diabetologia* 57 (9): 1807–11. https://doi.org/10.1007/s00125-014-3305-x.

King, D. G., M. Walker, M. D. Campbell, L. Breen, E. J. Stevenson and D. J. West. 2018. "A small dose of whey protein co-ingested with mixed-macronutrient breakfast and lunch meals improves postprandial glycemia and suppresses appetite in men with type 2 diabetes: a randomized controlled trial." *Am J Clin Nutr* 107 (4): 550–7. https://doi.org/10.1093/ajcn/nqy019.

Krebs, J. D., C. R. Elley, A. Parry-Strong, H. Lunt, P. L. Drury, D. A. Bell, E. Robinson, S. A. Moyes and J. I. Mann. 2012. "The diabetes excess weight loss (DEWL) trial: a randomised controlled trial of high-protein versus high-carbohydrate diets over 2 years in type 2 diabetes." *Diabetologia* 55 (4): 905–14. https://doi.org/10.1007/s00125-012-2461-0.

Larsen, R. N., N. J. Mann, E. Maclean and J. E. Shaw. 2011. "The effect of high-protein, low-carbohydrate diets in the treatment of type 2 diabetes: a 12 month randomised controlled trial." *Diabetologia* 54 (4): 731–40. https://doi.org/10.1007/s00125-010-2027-y.

Lean, M. E. J., W. S. Leslie, A. C. Barnes, N. Brosnahan, G. Thom, L. McCombie, C. Peters, S. Zhyzhneuskaya, A. Al-Mrabeh, K. G. Hollingsworth, A. M. Rodrigues, L. Rehackova, A. J. Adamson, F. F. Sniehotta, J. C. Mathers, H. M. Ross, Y. McIlvenna, P. Welsh, S. Kean, I. Ford, A. McConnachie, C. M. Messow, N. Sattar and R. Taylor. 2019. "Durability of a primary care-led weight-management intervention for remission of type 2 diabetes: 2-year results of the DiRECT open-label, cluster-randomised trial." *Lancet Diabetes Endocrinol* 7 (5): 344–55. https://doi.org/10.1016/S2213-8587(19)30068-3.

Luscombe, N. D., P. M. Clifton, M. Noakes, B. Parker and G. Wittert. 2002. "Effects of energy-restricted diets containing increased protein on weight loss, resting energy expenditure, and the thermic effect of feeding in type 2 diabetes." *Diabetes Care* 25 (4): 652–7. https://doi.org/10.2337/diacare.25.4.652.

Magkos, F. 2020a. "Protein-rich diets for weight loss maintenance." *Curr Obes Rep* 9 (3): 213–8. https://doi.org/10.1007/s13679-020-00391-0.

Magkos, F. 2020b. "The role of dietary protein in obesity." *Rev Endocr Metab Disord* 21 (3): 329–40. https://doi.org/10.1007/s11154-020-09576-3.

Magkos, F., M. F. Hjorth and A. Astrup. 2020. "Diet and exercise in the prevention and treatment of type 2 diabetes mellitus." *Nat Rev Endocrinol* 16 (10): 545–55. https://doi.org/10.1038/s41574-020-0381-5.

Magkos, F., D. N. Reeds and B. Mittendorfer. 2023. "Evolution of the diagnostic value of "the sugar of the blood": hitting the sweet spot to identify alterations in glucose dynamics." *Physiol Rev* 103 (1): 7–30. https://doi.org/10.1152/physrev.00015.2022.

Malik, V. S., Y. Li, D. K. Tobias, A. Pan and F. B. Hu. 2016. "Dietary protein intake and risk of type 2 diabetes in US men and women." *Am J Epidemiol* 183 (8): 715–28. https://doi.org/10.1093/aje/kwv268.

Manders, R. J., D. Hansen, A. H. Zorenc, P. Dendale, J. Kloek, W. H. Saris and L. J. van Loon. 2014. "Protein co-ingestion strongly increases postprandial insulin secretion in type 2 diabetes patients." *J Med Food* 17 (7): 758–63. https://doi.org/10.1089/jmf.2012.0294.

Mittendorfer, B., S. Klein and L. Fontana. 2020. "A word of caution against excessive protein intake." *Nat Rev Endocrinol* 16 (1): 59–66. https://doi.org/10.1038/s41574-019-0274-7.

Oberoi, A., C. Giezenaar, R. S. Rigda, K. Lange, M. Horowitz, K. L. Jones, I. Chapman and S. Soenen. 2022. "Comparative effects of co-ingesting whey protein and glucose alone and combined on blood glucose, plasma insulin and glucagon concentrations in younger and older men." *Nutrients* 14 (15): 3111. https://doi.org/10.3390/nu14153111.

Pedersen, E., D. R. Jesudason and P. M. Clifton. 2014. "High-protein weight loss diets in obese subjects with type 2 diabetes mellitus." *Nutr Metab Cardiovasc Dis* 24 (5): 554–62. https://doi.org/10.1016/j.numecd.2013.11.003.

Raben, A., P. S. Vestentoft, J. Brand-Miller, E. Jalo, M. Drummen, L. Simpson, J. A. Martinez, T. Handjieva-Darlenska, G. Stratton, M. Huttunen-Lenz, T. Lam, J. Sundvall, R. Muirhead, S. Poppitt, C. Ritz, K. H. Pietilainen, M. Westerterp-Plantenga, M. A. Taylor, S. Navas-Carretero, S. Handjiev, M. A. McNarry, S. Hansen, L. Raman, S. Brodie, M. P. Silvestre, T. C. Adam, I. A. Macdonald, R. San-Cristobal, N. Boyadjieva, K. A. Mackintosh, W. Schlicht, A. Liu, T. M. Larsen and M. Fogelholm. 2021. "The PREVIEW intervention study: Results from a 3-year randomized 2 x 2 factorial multinational trial investigating the role of protein, glycaemic index and physical activity for prevention of type 2 diabetes." *Diabetes Obes Metab* 23 (2): 324–37. https://doi.org/10.1111/dom.14219.

Rietman, A., J. Schwarz, D. Tome, F. J. Kok and M. Mensink. 2014. "High dietary protein intake, reducing or eliciting insulin resistance?" *Eur J Clin Nutr* 68 (9): 973–9. https://doi.org/10.1038/ejcn.2014.123.

Samkani, A., M. J. Skytte, D. Kandel, S. Kjaer, A. Astrup, C. F. Deacon, J. J. Holst, S. Madsbad, J. F. Rehfeld, S. B. Haugaard and T. Krarup. 2018. "A carbohydrate-reduced high-protein diet acutely decreases postprandial and diurnal glucose excursions in type 2 diabetes patients." *Br J Nutr* 119 (8): 910–7. https://doi.org/10.1017/S0007114518000521.

Shang, X., D. Scott, A. M. Hodge, D. R. English, G. G. Giles, P. R. Ebeling and K. M. Sanders. 2016. "Dietary protein intake and risk of type 2 diabetes: results from the Melbourne Collaborative Cohort Study and a meta-analysis of prospective studies." *Am J Clin Nutr* 104 (5): 1352–65. https://doi.org/10.3945/ajcn.116.140954.

Skytte, M. J., A. Samkani, A. D. Petersen, M. N. Thomsen, A. Astrup, E. Chabanova, J. Frystyk, J. J. Holst, H. S. Thomsen, S. Madsbad, T. M. Larsen, S. B. Haugaard and T. Krarup. 2019. "A carbohydrate-reduced high-protein diet improves HbA(1c) and liver fat content in weight stable participants with type 2 diabetes: a randomised controlled trial." *Diabetologia* 62 (11): 2066–78. https://doi.org/10.1007/s00125-019-4956-4.

Sluijs, I., J. W. Beulens, A. D. van der, A. M. Spijkerman, D. E. Grobbee and Y. T. van der Schouw. 2010. "Dietary intake of total, animal, and vegetable protein and risk of type 2 diabetes in the European prospective investigation into cancer and nutrition (EPIC)-NL study." *Diabetes Care* 33 (1): 43–8. https://doi.org/10.2337/dc09-1321.

Taheri, S., H. Zaghloul, O. Chagoury, S. Elhadad, S. H. Ahmed, N. El Khatib, R. A. Amona, K. El Nahas, N. Suleiman, A. Alnaama, A. Al-Hamaq, M. Charlson, M. T. Wells, S. Al-Abdulla and A. B. Abou-Samra. 2020. "Effect of intensive lifestyle intervention on bodyweight and glycaemia in early type 2 diabetes (DIADEM-I): an open-label, parallel-group, randomised controlled trial." *Lancet Diabetes Endocrinol* 8 (6): 477–89. https://doi.org/10.1016/S2213-8587(20)30117-0.

Thomsen, M. N., M. J. Skytte, A. Samkani, M. H. Carl, P. Weber, A. Astrup, E. Chabanova, M. Fenger, J. Frystyk, B. Hartmann, J. J. Holst, T. M. Larsen, S. Madsbad, F. Magkos, H. S. Thomsen, S. B. Haugaard and T. Krarup. 2022. "Dietary carbohydrate restriction augments weight loss-induced improvements in glycaemic control and liver fat in individuals with type 2 diabetes: a randomised controlled trial." *Diabetologia* 65 (3): 506–17. https://doi.org/10.1007/s00125-021-05628-8.

Tian, S., Q. Xu, R. Jiang, T. Han, C. Sun and L. Na. 2017. "Dietary protein consumption and the risk of type 2 diabetes: A systematic review and meta-analysis of cohort studies." *Nutrients* 9 (9): 982. https://doi.org/10.3390/nu9090982.

van Nielen, M., E. J. Feskens, M. Mensink, I. Sluijs, E. Molina, P. Amiano, E. Ardanaz, B. Balkau, J. W. Beulens, H. Boeing, F. Clavel-Chapelon, G. Fagherazzi, P. W. Franks, J. Halkjaer, J. M. Huerta, V. Katzke, T. J. Key, K. T. Khaw, V. Krogh, T. Kuhn, V. V. Menendez, P. Nilsson, K. Overvad, D. Palli, S. Panico, O. Rolandsson, I. Romieu, C. Sacerdote, M. J. Sanchez, M. B. Schulze, A. M. Spijkerman, A. Tjonneland, R. Tumino, A. D. van der, A. M. Wurtz, R. Zamora-Ros, C. Langenberg, S. J. Sharp, N. G. Forouhi, E. Riboli, N. J. Wareham and C. InterAct. 2014. "Dietary protein intake and incidence of type 2 diabetes in Europe: the EPIC-interact case-cohort study." *Diabetes Care* 37 (7): 1854–62. https://doi.org/10.2337/dc13-2627.

Viguiliouk, E., S. E. Stewart, V. H. Jayalath, A. P. Ng, A. Mirrahimi, R. J. de Souza, A. J. Hanley, R. P. Bazinet, S. Blanco Mejia, L. A. Leiter, R. G. Josse, C. W. Kendall, D. J. Jenkins and J. L. Sievenpiper. 2015. "Effect of replacing animal protein with plant protein on glycemic control in diabetes: A systematic review and meta-analysis of randomized controlled trials." *Nutrients* 7 (12): 9804–24. https://doi.org/10.3390/nu7125509.

Virtanen, H. E. K., T. T. Koskinen, S. Voutilainen, J. Mursu, T. P. Tuomainen, P. Kokko and J. K. Virtanen. 2017. "Intake of different dietary proteins and risk of type 2 diabetes in men: the Kuopio Ischaemic heart disease risk factor study." *Br J Nutr* 117 (6): 882–93. https://doi.org/10.1017/S0007114517000745.

Wang, Z., W. Shen, D. P. Kotler, S. Heshka, L. Wielopolski, J. F. Aloia, M. E. Nelson, R. N. Pierson, Jr. and S. B. Heymsfield. 2003. "Total body protein: a new cellular level mass and distribution prediction model." *Am J Clin Nutr* 78 (5): 979–84. https://doi.org/10.1093/ajcn/78.5.979.

Watson, N., K. Dyer, J. Buckley, G. Brinkworth, A. Coates, G. Parfitt, P. Howe, M. Noakes and K. Murphy. 2016. "Effects of low-fat diets differing in protein and carbohydrate content on cardiometabolic risk factors during weight loss and weight maintenance in obese adults with type 2 diabetes." *Nutrients* 8 (5): 289. https://doi.org/10.3390/nu8050289.

Westerterp-Plantenga, M. S., A. Nieuwenhuizen, D. Tome, S. Soenen and K. R. Westerterp. 2009. "Dietary protein, weight loss, and weight maintenance." *Annu Rev Nutr* 29: 21–41. https://doi.org/10.1146/annurev-nutr-080508-141056.

Wing, R. R., W. Lang, T. A. Wadden, M. Safford, W. C. Knowler, A. G. Bertoni, J. O. Hill, F. L. Brancati, A. Peters, L. Wagenknecht and A. R. G. Look. 2011. "Benefits of modest weight loss in improving cardiovascular risk factors in overweight and obese individuals with type 2 diabetes." *Diabetes Care* 34 (7): 1481–6. https://doi.org/10.2337/dc10-2415.

Wolfe, R. R., A. M. Cifelli, G. Kostas and I. Y. Kim. 2017. "Optimizing protein intake in adults: Interpretation and application of the recommended dietary allowance compared with the acceptable macronutrient distribution range." *Adv Nutr* 8 (2): 266–75. https://doi.org/10.3945/an.116.013821.

Wycherley, T. P., M. Noakes, P. M. Clifton, X. Cleanthous, J. B. Keogh and G. D. Brinkworth. 2010. "A high-protein diet with resistance exercise training improves weight loss and body composition in overweight and obese patients with type 2 diabetes." *Diabetes Care* 33 (5): 969–76. https://doi.org/10.2337/dc09-1974.

Yanagisawa, Y. 2023. "How dietary amino acids and high-protein diets influence insulin secretion." *Physiol Rep* 11 (2): e15577. https://doi.org/10.14814/phy2.15577.

Ye, J., Q. Yu, W. Mai, P. Liang, X. Liu and Y. Wang. 2019. "Dietary protein intake and subsequent risk of type 2 diabetes: a dose-response meta-analysis of prospective cohort studies." *Acta Diabetol* 56 (8): 851–70. https://doi.org/10.1007/s00592-019-01320-x.

Yu, Z., F. Nan, L. Y. Wang, H. Jiang, W. Chen and Y. Jiang. 2020. "Effects of high-protein diet on glycemic control, insulin resistance and blood pressure in type 2 diabetes: A systematic review and meta-analysis of randomized controlled trials." *Clin Nutr* 39 (6): 1724–34. https://doi.org/10.1016/j.clnu.2019.08.008.

Zeng, B. T., H. Q. Pan, F. D. Li, Z. Y. Ye, Y. Liu and J. W. Du. 2023. "Comparative efficacy of different eating patterns in the management of type 2 diabetes and prediabetes: An arm-based Bayesian network meta-analysis." *J Diabetes Investig* 14 (2): 263–88. https://doi.org/10.1111/jdi.13935.

Zhao, W. T., Y. Luo, Y. Zhang, Y. Zhou and T. T. Zhao. 2018. "High-protein diet is of benefit for patients with type 2 diabetes: An updated meta-analysis." *Medicine (Baltimore)* 97 (46): e13149. https://doi.org/10.1097/MD.0000000000013149.

29 Maternal Low-Protein Diets and Effects on Fetal Programming

Vipin A. Vidyadharan, Chandra Yallampalli, and Chellakkan S. Blesson

ABBREVIATIONS

LP	Low protein
mtDNA	Mitochondrial DNA
T2D	Type 2 diabetes

INTRODUCTION

Nutrition is the sole source of energy and building material for animals from conception to death. Nutritional needs vary at different stages of life, depending on the growth and metabolic status of the organism. Nutrients are classified into two major categories, macronutrients and micronutrients, based on their requirement for growth and metabolism. Proteins occupy a special niche among macronutrients as they function as both a building material and an energy source. Proteins comprise around 20 natural amino acids, of which nine are considered essential because the human body cannot synthesize them. These essential amino acids are obtained from the diet. Proteins are essential for normal growth and development, and their requirements vary during different stages of life. Proteins play a vital role during fetal development and set the metabolic and growth trajectory for the future. Maternal nutrition is the only source of protein for the developing fetus, and it is well known that prenatal protein deficiency leads to low birth weight and in-utero growth restriction. Although maternal and placental factors may contribute to adverse prenatal nutrition, maternal malnutrition is the most important cause of fetal malnutrition and protein deficiency.

A substantial body of research shows an association between a low-protein (LP) diet during fetal development and the occurrence of metabolic diseases in adult life, a phenomenon known as the developmental origin of adult diseases (Vipin et al. 2022; Hales and Barker 2001; Kyle and Pichard 2006; Roseboom et al. 2006). Although this hypothesis has been tested and reproduced over the years, the exact mechanism driving the fetal origin of adult diseases is not clear. Most maternal LP diet studies assume that a mismatch between prenatal and postnatal nutrition and catch-up growth in the early postnatal period drives the metabolic disease phenotype in the offspring. The mismatch concept relies on two major ideas: first, fetal adaptations to adverse in-utero nutrition, and second, that these adaptations are detrimental in nutrient-rich adulthood. Fetal exposure to famine is often reported to have a higher incidence and early onset of metabolic diseases in adulthood than their normal counterparts. The catch-up growth concept is based on the observation that adverse in-utero conditions lead to underweight and smaller babies; however, when exposed to normal dietary conditions after birth, their growth often catches up to match the normal population in early childhood. This dynamic and rapid process of body weight recovery is characterized by hyperinsulinemia and a predisposition towards acquiring metabolic syndrome later in life (Hales and Ozanne 2003).

 DOI: 10.1201/b23328-29

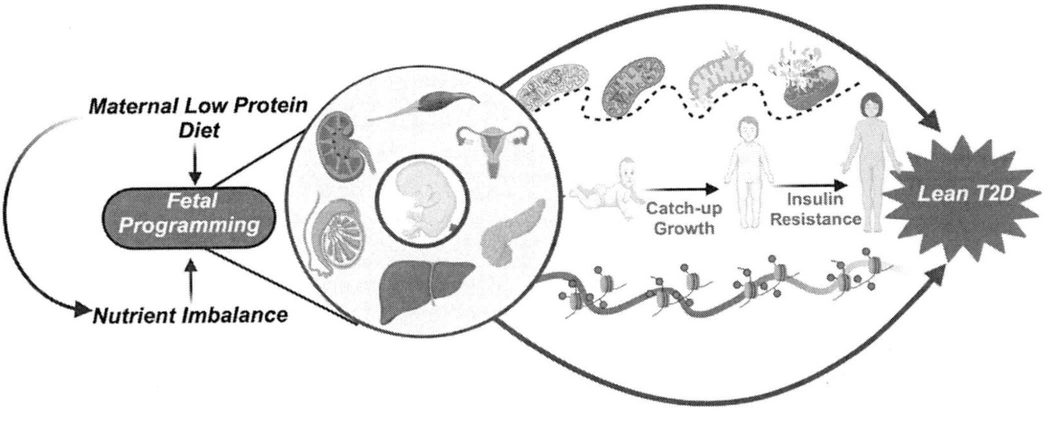

FIGURE 29.1 Schematic representation of maternal low-protein-diet-induced fetal programming of lean T2D.

Considering that cellular nutrient-sensing mechanisms take amino acids as one of the indicators of nutrient availability, the LP diet might be perceived as a cue for starvation by the body, even if the total caloric intake is normal (Shin et al. 2016). Once the body senses a limitation of nutrients, the cells may change the metabolic milieu by activating energy-efficient (higher capacity/efficiency of ATP production) pathways to limit and optimize nutrient consumption (Muller et al. 2015). Often, nervous tissue is least affected (brain sparing) at the expense of other tissues such as skeletal muscle (Sadiq et al. 1999). Glucose is the main energy source for neurons, and in adverse nutritional conditions, the fetus adapts its metabolism in the skeletal muscle and liver to spare glucose to protect the brain, and these fetal metabolic adaptations persist into adulthood. Although these metabolic adaptations are beneficial for fetal survival, they may not be advantageous in the nutrient-rich postnatal period. Thus, in an over-nourished or normally-nourished state, organisms may choose lower bioenergetic efficiency to limit reactive oxygen species (ROS) production and store excess energy-rich nutrients. However, continuous overnutrition and mismatched metabolic adaptations lead to metabolic diseases in adulthood (Cleal et al. 2007) (Figure 29.1).

MATERNAL LOW-PROTEIN DIETS AND METABOLIC DISEASE HYPOTHESIS

The association between prenatal LP diet and the development of metabolic disorders in adulthood is well supported by animal studies (Blesson et al. 2016; Burdge et al. 2007; Barros et al. 2015; Frapin et al. 2020). In humans, the Dutch famine that occurred around the end of World War II provided a unique opportunity to study such an association in humans. Several studies using the Dutch famine data showed a strong association between maternal poor nutrition and adult onset of metabolic diseases (Kyle and Pichard 2006; Roseboom et al. 2006). The first follow-up study showed that early development could be linked to T2D in men from Hertfordshire, UK (Hales 1997). The relationship between poor maternal nutrition and the onset of metabolic diseases in both sexes has been described in several studies (Hales and Barker 2001). The thrifty phenotype hypothesis, also known as Barker's hypothesis, provides a conceptual and mechanistic explanation for these observations and is primarily based on the notion that fetal development is sensitive to the uterine nutritional environment (Hales 1997). Consequently, poor in-utero nutrition prompts the fetus to optimize the growth of certain organs at the expense of others to provide a better chance of fetal survival. However, these fetal adaptations might be detrimental when nutrition is abundant in adult life (Hales and Ozanne 2003). The exact mechanism behind fetal programming or the developmental origins of health and diseases is an active area of research. Decreased fetal growth is generally seen in animal models of severe protein deficiency (6–8% diet) (Blesson et al. 2016). Low birth weight is associated with the occurrence of T2D in several populations in different countries (Hales and Ozanne 2003). The main limitation of this hypothesis is the assumption that severe environmental stress and impaired fetal growth cause fetal programming. The necessity of fetal growth reduction was questioned when it was shown that similar metabolic defects can be seen across normal weight ranges. Further, the model ignored the importance of the pre-conceptional period for fetal programming (Gluckman et al. 2010).

The developmental plasticity hypothesis of fetal programming is the latest model that explains the developmental origin of health and diseases (Bateson et al. 2004). This model argues that organisms are programmed to respond to cues available in early-life environments so that they are well-prepared for the conditions experienced by their parents or grandparents. The developmental plasticity hypothesis separates the necessity of fetal growth restriction from the developmental origin of diseases. Developmental plasticity is the ability of organisms to adapt in response to environmental cues during a sensitive period in development. Often, environmental cues received outside these brief sensitive development periods have little or no effect on the organisms' phenotypic characteristics. These characteristics are induced either by early environments or by the cues to which their parents or grandparents were exposed. It is thought that maintaining developmental plasticity continuously is difficult due to its high energy demand; hence, plasticity is restricted to a specific period of development. The plastic phase in mammals extends from conception through the weaning period. As per the developmental plasticity hypothesis, fetuses exhibit two classes of responses: a phenotypic response for immediate advantages or a response for later fitness benefits. The latter response is called an anticipatory adaptive response. Maternal nutritional or hormonal signals can invoke anticipatory adaptive responses in the fetus and adapt its phenotypic development (Bateson et al. 2004; Gluckman et al. 2010).

ANIMAL MODELS

Due to ethical and practical considerations, laboratory animals are the only system that can be used to prospectively investigate the effects of prenatal nutrition on fetal development. Despite extensive epidemiological studies on the occurrence of metabolic diseases in humans, animal models have been more effective in uncovering the mechanisms of developmental programming. In addition, investigators can easily control the nutritional and genetic influences on in-utero development in these animals. Among different LP animal models, mice and rats are the most preferred, followed by sheep, pigs, and cows. The rodents' short life spans and shorter gestational periods make them the favorite animal models for developmental programming researchers. In recent years, mice have been the preferred choice of animals owing to the ease of genetic manipulation and the availability of different genetic knockouts. However, rats are a better model for cardiovascular and metabolic syndrome studies due to their greater similarity to human physiology compared to mice. The rodent models are not suitable for studying the developmental programming of the central nervous system due to fundamental differences from human brain development. Although primates are suitable models for brain development studies, cost and ethical considerations restrict their use. Rodents and pigs are polytocous (multiple offspring in a single pregnancy), which results in an unequal nutrient supply among fetuses. To overcome this, many researchers have opted for sheep and cows as their preferred animal model.

Although different researchers prefer different animal models depending on the research question and expertise, the main aim is to mimic protein-deficient conditions in the critical periods of gestation and development. Many studies have used an isocaloric diet with a 50–70% reduction in total protein. To maintain the isocaloric nature of the diet, research groups use various ratios of carbohydrates, lipids, and proteins. In addition, the initiation and duration of the LP diet regimens vary among different groups. Most of the studies follow an LP diet during gestation and lactation,

TABLE 29.1
Summary of Rodent Models Used to Investigate Maternal Low-Protein-Associated Insulin Resistance and Glucose Intolerance

	Diet Regimen			
Rodent Species	Protein (%)	Period (Days)	Offspring Age	Mechanisms of T2D Identified
Sprague-Dawley rats	6–10%	12 to 43	Day 0 to 17 weeks	A. Increased Insulin-like Growth (IGF) gene expression B. Sirt3 dysfunction in skeletal muscle C. Dysregulated glucose homeostasis
Wistar rats	6–10%	1 to 43	Fetal Day 20 to 16 weeks	A. Deregulated gluconeogenesis B. Impaired liver structure C. Insulin resistance and glucose Intolerance D. Epigenetic regulation of HNF4A in islets E. Mitochondrial dysfunction in islets F. Changes in IGF axis G. Hepatic proliferative capacity
C57BL/6J mice	8–9	1 to 40	Day 0 to 21 weeks	A. Impaired glucose metabolism B. Up-regulation miR-15b C. Altered PPAR signaling D. Insulin resistance and glucose intolerance E. Altered mitochondrial gene expression in liver and skeletal muscle. F. Increased abdominal adiposity and glucose intolerance

while others manipulate the diets only during the gestation and, in some cases, from pre-conception through gestation periods. Some groups also experiment with a maternal protein-restricted diet only during the early phase or late phase of gestation. Invariably, protein deprivation during these critical periods of development causes intrauterine fetal growth restriction and lower birth weights. The phenotypes of offspring manifesting after different LP diet regimes vary among different groups (Table 29.1).

TISSUE-LEVEL EFFECTS OF LOW-PROTEIN PROGRAMMING

Among the different tissues impacted by nutritional partitioning in adverse maternal nutrition, skeletal muscle is one of the most affected tissues. Skeletal muscle represents 45% of the body mass and participates in 80% of insulin-induced glucose uptake. Therefore, any change in skeletal muscle mass or metabolism may drastically impact whole-body glucose metabolism. Skeletal muscle has a lower priority for nutrient supply during adverse in-utero nutritional conditions while partitioning the limited nutrition compared to other vital tissues and organs. Animals exposed to in-utero LP diets had lower lean mass and insulin resistance in skeletal muscle, and these metabolic alterations led to T2D (Blesson et al. 2016, 2019). LP programming also reduced protein and glycogen synthesis in the skeletal muscle of these animals (de Vasconcelos et al. 2022). Further, impaired insulin signaling and mitochondrial dysfunction are commonly reported in maternal LP diet models.

Altogether, the LP-induced changes in skeletal muscle mass and cellular metabolism play a significant part in the onset of glucose intolerance and insulin resistance in the offspring.

Apart from skeletal muscle, the liver plays a critical role in whole-body metabolic homeostasis. The liver is also central to nutrient sensing and lipid metabolism. A maternal LP diet often perturbs hepatic metabolic functions and dysregulates the hepatic response to changing nutritional states. LP exposure causes hepatic lipid accumulation through autophagic degradation (Ramadan et al. 2013) of APOB proteins and cholesterol metabolism by epigenetic regulation in response to the limited availability of amino acids. Accumulation of lipids in the liver is considered one of the main initial indicators of metabolic diseases in LP-induced T2D. The LP diet also affects genes linked to hepatic glucose production and gluconeogenesis, which are regulated by epigenetic mechanisms (Ramadan et al. 2013).

The gestational LP diet is detrimental to the normal development of the fetal pancreas, reducing the pancreatic beta-cell number and causing insulin insufficiency commonly seen in the lean T2D population (Staimez et al. 2013; Siddiqui et al. 2022; Tan et al. 2015). While skeletal muscle and the liver serve as the primary sites for glucose oxidation and production, insulin produced by pancreatic beta cells acts as the crucial regulator of glucose homeostasis (Staimez et al. 2013; Siddiqui et al. 2022; Tan et al. 2015). Animal studies further demonstrate that the pancreas makes the least recovery when sufficient protein is provided after birth (Serpente et al. 2021). The thrifty phenotype hypothesis emphasizes the impact of the LP diet on pancreatic function and the adult onset of T2D. Protein availability is linked to the reduced beta-cell number and innervation seen in the LP offspring (Petrik et al. 1999; Winkel et al. 2022). In the developing fetus, a lower number of beta cells and insulin insufficiency might aid in glucose transport to the brain by modifying insulin signaling in other organs. However, in adulthood, this adaptation may adversely affect insulin-dependent glucose disposal.

The molecular mechanisms underlying the insulin insufficiency seen in LP offspring are diverse, including lower beta-cell counts, decreased beta-cell proliferation, and increased beta-cell apoptosis. The LP-induced downregulation of Foxo1, Pdx1, or the Reg1 pathway was found to be associated with impaired beta-cell function and number. Similarly, mTOR signaling and epigenetic regulation of HNF4A expression contribute to beta-cell dysfunction (Vipin et al. 2022). Therefore, the lower number of beta cells and insulin insufficiency might aid in preferential glucose transport to the brain during fetal development and result in insufficient glucose for other organs. Thus, dysregulated glucose metabolism in these organs might be the mechanism behind the high prevalence and early onset of T2D in the LP offspring.

Kidney function depends on the synchrony between nephron function, physiological mechanisms, and the renin-angiotensin system. Any asynchrony can lead to renal dysfunction and vascular diseases (Mesquita et al. 2010). The maternal LP diet interrupts adult renal and cardiovascular functions. Several studies have shown that an LP diet reduces nephron number and impairs glomerular morphology in the renal cortex (Langley-Evans et al. 1999). The decreased number of nephrons leads to an increase in functional demand on individual nephrons. These changes are often associated with a dysregulated renin-angiotensin system and increased nitric oxide production (Pedroza et al. 2019; Sahajpal and Ashton 2003). Reduced nephron numbers also cause detrimental gene expression changes aimed at maintaining normal hemodynamic kidney functioning. However, this maladaptation causes the nephron to progressively decline in renal function, leading to increased vascular pressure and cardiovascular dysfunction (Nwagwu et al. 2000; Tappia et al. 2013).

Maternal malnutrition can also impact the gonadal development in offspring. Limited availability of amino acids during embryonic development reduces gonadal size, delays sexual maturation, and impairs reproductive capacity in offspring (Ajuogu et al. 2021; Guzman et al. 2006; Zambrano et al. 2005). Further, male LP offspring exhibit smaller testes with fewer Sertoli cells and lower sperm counts, resulting in reduced fertility (Kotsampasi, Balaskas, et al. 2009). Similarly, female LP offspring show reduced ovarian follicle numbers and decreased expression of genes linked to follicular development (Guzman et al. 2014; Winship et al. 2018). The LP-induced reduction in the microRNAs (miRNAs) such as miR-15b might be the reason for reduced follicular gene expression in these animals (Sui et al. 2014). Additionally, LP-mediated changes in mitochondrial distribution and function can dysregulate the estrous cycle and redox homeostasis (Blesson et al. 2020; Fabozzi et al. 2021). The normal development of gonads and their secretory functions may also depend on the integrity of the hypothalamic-pituitary-gonadal axis (Kotsampasi, Balaskas, et al. 2009; Kotsampasi, Chadio, et al. 2009). This axis is often disrupted in the LP offspring, potentially contributing to impaired gonadal development and function.

MOLECULAR MECHANISMS OF LOW-PROTEIN PROGRAMMING

Epigenetic modifications are considered the main mechanisms of fetal programming. Epigenetic changes that occur during fetal development appear to be the main link between early-life environments and the development of chronic diseases in adulthood. The major epigenetic modifications reported are DNA methylation, histone modifications, and non-coding RNA-mediated changes. The epigenetic mechanism of fetal programming is one potential mechanism behind the development of LP-induced metabolic disease. Changes in DNA methylation are among the most studied epigenetic modifications associated with LP-induced metabolic disease. Transcription factors related to glucose and fatty acid metabolisms have been investigated for DNA methylation changes due to their roles as controlling nodes for different gene networks related to metabolism. For instance, PPARA, PDX1, and HNF4A are differentially regulated via DNA methylation of their promotors (Lillycrop et al. 2010; Abuzgaia et al. 2015; Sandovici et al. 2011; Vaiserman and Lushchak 2019).

Histone modifications associated with programming often appear before any changes in DNA methylation. Changes in the acetylation and methylation profiles of histones H3 and H4 are the common modifications seen in programmed tissues. For example, histone modifications in the regulatory regions of transcription factors such as PDX1 and HNF4A impair pancreatic development and function (Abuzgaia et al. 2015; Sandovici et al. 2011). In recent years, miRNA-mediated epigenetic modification has gained more attention in fetal programming. The miRNAs are small non-coding RNAs that can bind to the 3'-untranslated region of mRNA and regulate post-transcriptional gene expression. Various miRNAs are reported to have a role in the LP-associated onset and development of metabolic diseases (Vaiserman and Lushchak 2019). Among them, the association between increased miR483 levels and decreased lipid storage capacity is well-studied in different tissues (Ferland-McCollough et al. 2012).

Epigenetic programming often depends on the availability of metabolic cofactors. The acetyl group is one of the major cofactors for histone modification, and its levels are controlled by mitochondrial metabolism. Therefore, if the in-utero environment alters normal mitochondrial metabolism, it is likely to change the availability of metabolic cofactors for epigenetic changes during a crucial period of development. It has been reported that mitochondria are the common site of maternal LP diet-induced changes (Vidyadharan et al. 2022, 2023; Fabozzi et al. 2021; Fukunaga 2021). Nutrient availability can influence cellular processes by regulating the availability of the methyl and acetyl groups required for epigenetic modifications (Gut and Verdin 2013). DNA methylation is critical for epigenetic memory (Bird 2002). The methylation process depends on methionine, which is required to produce S-adenosyl-methionine, the universal methyl group donor (Mentch et al. 2015; Schutt et al. 2019). The embryonic demethylation and methylation cycles are both critical in determining the somatic methylation pattern and are often irreversible (Bird 2002).

The human mitochondrial genome (mtDNA), which spans 16.5 kilobases, contains 37 genes essential for the respiratory chain complexes. In humans, mtDNA is exclusively inherited from the mother's oocytes. A single cell may contain 1 to 10,000 mtDNA molecules distributed among a network of mitochondria that continually fuse and divide. The mtDNA is particularly susceptible to mutation due to its exposure to free radicals and the absence of protective histone proteins, resulting in a mixture of both healthy and mutated mtDNA within a cell, a condition known as heteroplasmy (Floros et al. 2018). The significant decrease in mtDNA content within female germ cells, combined with the random distribution of mtDNA during division, results in a distinct separation of mtDNA genetic makeup between the germ cells of the offspring. As a result, only a small portion of the mitochondrial DNA is passed into different tissues (Cao et al. 2007; Shoubridge and Wai 2007). This process creates a bottleneck effect, which becomes more pronounced with advancing maternal age at the time of childbirth and under adverse in-utero conditions. Consequently, the frequencies of mutations can vary significantly between a mother and her offspring. Thus, a child may inherit a disease-causing mutation at a relatively high frequency when subjected to adverse gestational conditions, potentially leading to the development of metabolic diseases in adulthood (Fukunaga 2021; Priliani et al. 2019; Smith et al. 2023; Zaidi et al. 2019). Therefore, mtDNA copy number and mtDNA mutations may independently or synergistically, along with epigenetic changes, preserve temporally distant events, such as maternal LP exposure and metabolic modifications, into adulthood.

Mitochondria can also influence metabolism by making the nucleus respond to mitochondrial signals. This retrograde signaling helps organisms quickly adapt to metabolic demands following environmental changes. Changes in mitochondrial membrane potential, calcium levels, ROS, acetyl coenzyme A, succinate, fumarate, alpha-ketoglutarate, ATP/ADP ratio, and NADH/NAD+ ratio are all triggers of retrograde signaling. Many of these signals have a profound impact on the regulation of transcription factors associated with cellular metabolism and epigenetic modification. Apart from controlling epigenetic changes, mitochondria can influence metabolic health either through changes in mtDNA or mitochondrial retrograde signaling (Matilainen et al. 2017).

PREVENTION AND TREATMENT

Insulin resistance and glucose intolerance are key indicators of T2D, and if left unchecked, they can result in severe complications later in life. The major insulin-dependent tissues responsible for peripheral glucose disposal are skeletal muscles, hepatocytes, and adipocytes. Therefore, the primary focus of T2D treatment strategies should be to enhance muscle sensitivity to insulin, regulate hepatic glucose homeostasis, and manage fat accumulation. In the context of lean T2D management, various drugs have been prescribed, with metformin being one of the most widely used medications, in conjunction with dietary and lifestyle modifications. Notably, over the past two decades, randomized controlled trials conducted across diverse ethnic groups have conclusively demonstrated that prevention is attainable through drug interventions or lifestyle adjustments (Pan et al. 1997; Tuomilehto and Lindstrom 2003; Muilwijk et al. 2018).

Controlling body weight is crucial for managing T2D in overweight or obese patients. Yet, in the context of lean T2D, it appears that individuals with lower body weight often experience more severe beta-cell failure compared to those with normal weight or obesity. Currently, it remains uncertain whether attaining a lower body weight can effectively prevent or reverse specific variants of T2D, such as lean T2D. Numerous studies have investigated the effectiveness of lifestyle modifications and drug therapy. The first trial, initiated in China with a 20-year follow-up, laid the foundation for subsequent research in this area (Li et al. 2008). Other noteworthy studies, including the American Diabetes Prevention Program Outcome Study, the Finnish Diabetes Prevention Study, and the Indian Short Message Service Study, demonstrated the enduring impact of lifestyle modifications even after the active phase of the trial. While lifestyle changes are recognized as an effective and safe strategy for prevention, their relative risk reduction effectiveness varies among different ethnic populations. Notably, in a study involving lean glucose-intolerant individuals in India, lifestyle modifications alone proved effective in preventing diabetes onset, despite the participants' relatively low BMI and high insulin resistance characteristics (Ramachandran et al. 2013). Although the precise pathophysiology of lean T2D remains unclear, research using the maternal LP model has identified potential prevention approaches, particularly related to one-carbon metabolism and associated molecules. Folic acid supplementation has shown promise in countering the adverse effects of fetal programming, and some studies have even suggested its epigenetic reversal potential (Blesson et al. 2018; Clare et al. 2019). Various other amino acid and antioxidant treatments showed mixed outcomes. At present, there are no known cures for LP-induced metabolic abnormalities; mitigation appears to be the best solution.

CONCLUSIONS AND FUTURE DIRECTIONS

A balanced diet with sufficient protein is essential for normal fetal development and the prevention of diseases in adulthood. Proteins act as fundamental building blocks that support fetal growth and development. They play a critical role in energy partitioning among different organs and the formation of vital tissues, including the brain, pancreas, skeletal muscles, and liver. Proteins are essential for the normal functioning of different metabolic processes, including glucose metabolism. Consequently, insufficient protein intake during pregnancy may increase the offspring's risk of developing metabolic diseases such as lean T2D in adulthood. Lean T2D represents a distinct subgroup of the condition, characterized by specific clinical profiles. Understanding the unusual combination of leanness and insulin resistance in lean T2D is crucial for unraveling the disease's etiology and developing effective treatments. The development of T2D in humans is a gradual process, necessitating long-term assessment and prevention strategies.

Understanding the molecular mechanisms underlying the origins of lean T2D is crucial for developing effective prevention strategies. Future research should investigate the roles of epigenetics, mitochondrial function, and changes in mtDNA in the onset of LP-programmed lean T2D. In the global context, obesity-associated T2D is a significant public health issue in both developing and developed nations. Simultaneously, nutritional deficiencies, especially protein deficiency, pose a substantial concern in underdeveloped and developing countries. Studies highlighting the connection between maternal protein intake and T2D risk in offspring underscore the need for further investigation and decisive action to avert a looming global crisis. To tackle these challenges, public health measures must address poverty and improve access to nutritious, protein-rich diets during pregnancy as essential steps to prevent lean T2D. A scientific comprehension of the disease's intricacies, combined with effective health education and well-crafted public policies, can help mitigate the growing global epidemic of T2D, ultimately contributing to healthier populations worldwide.

SUMMARY POINTS

1. An adequate amount of protein in a balanced diet is crucial for normal fetal development and disease prevention in adulthood.
2. Proteins are essential for organ growth, including that of the brain, pancreas, muscles, and liver, as well as for metabolic processes such as glucose regulation.
3. A maternal low-protein diet can result in widespread, irreversible epigenetic and metabolic alterations in fetal energy regulation.
4. Insufficient protein intake during pregnancy increases the risk of metabolic diseases like lean T2D in offspring.
5. A thorough understanding of the molecular mechanisms underlying lean T2D is essential for developing effective prevention and treatment strategies.

REFERENCES

Abuzgaia, A. M., D. B. Hardy, and E. Arany. 2015. "Regulation of postnatal pancreatic Pdx1 and downstream target genes after gestational exposure to protein restriction in rats." *Reproduction* 149 (3): 293–303.

Ajuogu, P. K., M. A. K. Al-Aqbi, R. A. Hart, J. R. McFarlane, and N. A. Smart. 2021. "A low protein maternal diet during gestation has negative effects on male fertility markers in rats - A systematic review and meta-analysis." *J Anim Physiol Anim Nutr (Berl)* 105 (1): 157–66.

Barros, M. A., J. L. De Brito Alves, V. O. Nogueira, A. G. Wanderley, and J. H. Costa-Silva. 2015. "Maternal low-protein diet induces changes in the cardiovascular autonomic modulation in male rat offspring." *Nutr Metab Cardiovasc Dis* 25 (1): 123–30.

Bateson, P., D. Barker, T. Clutton-Brock, D. Deb, B. D'Udine, R. A. Foley, P. Gluckman, K. Godfrey, T. Kirkwood, M. M. Lahr, J. McNamara, N. B. Metcalfe, P. Monaghan, H. G. Spencer, and S. E. Sultan. 2004. "Developmental plasticity and human health." *Nature* 430 (6998): 419–21.

Bird, A. 2002. "DNA methylation patterns and epigenetic memory." *Genes Dev* 16 (1): 6–21.

Blesson, C. S., A. K. Schutt, M. P. Balakrishnan, R. G. Pautler, S. E. Pedersen, P. Sarkar, D. Gonzales, G. Zhu, J. C. Marini, S. K. Chacko, U. Yallampalli, and C. Yallampalli. 2016. "Novel lean type 2 diabetic rat model using gestational low-protein programming." *Am J Obstet Gynecol* 214 (4): 540 e1–540 e7.

Blesson, C. S., A. K. Schutt, S. Chacko, J. C. Marini, P. R. Mathew, D. Tanchico, M. Balakrishnan, and C. Yallampalli. 2019. "Sex dependent dysregulation of hepatic glucose production in lean type 2 diabetic rats." *Front Endocrinol (Lausanne)* 10: 538.

Blesson, C. S., A. K. Schutt, P. R. Mathew, D. Tanchico, M. Balakrishnan, U. Yallampalli, and C. Yallampalli. 2018. "Folate treatment partially reverses gestational low-protein diet-induced glucose intolerance and the magnitude of reversal is age and sex dependent." *Nutrition* 49: 81–9.

Blesson, C. S., A. K. Schutt, V. A. Vipin, D. T. Tanchico, P. R. Mathew, M. Balakrishnan, A. Betancourt, and C. Yallampalli. 2020. "In utero low-protein-diet-programmed type 2 diabetes in adult offspring is mediated by sex hormones in ratsdagger." *Biol Reprod* 103 (5): 1110–20.

Burdge, G. C., J. Slater-Jefferies, C. Torrens, E. S. Phillips, M. A. Hanson, and K. A. Lillycrop. 2007. "Dietary protein restriction of pregnant rats in the F0 generation induces altered methylation of hepatic gene promoters in the adult male offspring in the F1 and F2 generations." *Br J Nutr* 97 (3): 435–9.

Cao, L., H. Shitara, T. Horii, Y. Nagao, H. Imai, K. Abe, T. Hara, J. Hayashi, and H. Yonekawa. 2007. "The mitochondrial bottleneck occurs without reduction of mtDNA content in female mouse germ cells." *Nat Genet* 39 (3): 386–90.

Clare, C. E., A. H. Brassington, W. Y. Kwong, and K. D. Sinclair. 2019. "One-carbon metabolism: linking nutritional biochemistry to epigenetic programming of long-term development." *Annu Rev Anim Biosci* 7: 263–87.

Cleal, J. K., K. R. Poore, J. P. Boullin, O. Khan, R. Chau, O. Hambidge, C. Torrens, J. P. Newman, L. Poston, D. E. Noakes, M. A. Hanson, and L. R. Green. 2007. "Mismatched pre- and postnatal nutrition leads to cardiovascular dysfunction and altered renal function in adulthood." *Proc Natl Acad Sci U S A* 104 (22): 9529–33.

de Vasconcelos, D. A. A., R. T. Nachbar, C. H. Pinheiro, C. L. do Amaral, A. R. Crisma, K. F. Vitzel, P. Abreu, M. I. Alonso-Vale, A. B. Lopes, A. Bento-Santos, F. Falcao-Tebas, D. F. de Santana, E. do Nascimento, R. Curi, T. C. Pithon-Curi, S. M. Hirabara, and C. G. Leandro. 2022. "Maternal low-protein diet reduces skeletal muscle protein synthesis and mass via Akt-mTOR pathway in adult rats." *Front Nutr* 9: 947458.

Fabozzi, G., B. Iussig, D. Cimadomo, A. Vaiarelli, R. Maggiulli, N. Ubaldi, F. M. Ubaldi, and L. Rienzi. 2021. "The impact of unbalanced maternal nutritional intakes on oocyte mitochondrial activity: implications for reproductive function." *Antioxidants (Basel)* 10 (1).

Ferland-McCollough, D., D. S. Fernandez-Twinn, I. G. Cannell, H. David, M. Warner, A. A. Vaag, J. Bork-Jensen, C. Brons, T. W. Gant, A. E. Willis, K. Siddle, M. Bushell, and S. E. Ozanne. 2012. "Programming of adipose tissue miR–483-3p and GDF-3 expression by maternal diet in type 2 diabetes." *Cell Death Differ* 19 (6): 1003–12.

Floros, V. I., A. Pyle, S. Dietmann, W. Wei, W. C. W. Tang, N. Irie, B. Payne, A. Capalbo, L. Noli, J. Coxhead, G. Hudson, M. Crosier, H. Strahl, Y. Khalaf, M. Saitou, D. Ilic, M. A. Surani, and P. F. Chinnery. 2018. "Segregation of mitochondrial DNA heteroplasmy through a developmental genetic bottleneck in human embryos." *Nat Cell Biol* 20 (2): 144–51.

Frapin, M., S. Guignard, D. Meistermann, I. Grit, V. S. Moulle, V. Paille, P. Parnet, and V. Amarger. 2020. "Maternal protein restriction in rats alters the expression of genes involved in mitochondrial metabolism and epitranscriptomics in fetal hypothalamus." *Nutrients* 12 (5).

Fukunaga, H. 2021. "Mitochondrial DNA copy number and developmental origins of health and disease (DOHaD)." *Int J Mol Sci* 22 (12).

Gluckman, P. D., M. A. Hanson, and T. Buklijas. 2010. "A conceptual framework for the developmental origins of health and disease." *J Dev Orig Health Dis* 1 (1): 6–18.

Gut, P., and E. Verdin. 2013. "The nexus of chromatin regulation and intermediary metabolism." *Nature* 502 (7472): 489–98.

Guzman, C., R. Cabrera, M. Cardenas, F. Larrea, P. W. Nathanielsz, and E. Zambrano. 2006. "Protein restriction during fetal and neonatal development in the rat alters reproductive function and accelerates reproductive ageing in female progeny." *J Physiol* 572 (Pt 1): 97–108.

Guzman, C., R. Garcia-Becerra, M. A. Aguilar-Medina, I. Mendez, H. Merchant-Larios, and E. Zambrano. 2014. "Maternal protein restriction during pregnancy and/or lactation negatively affects follicular ovarian development and steroidogenesis in the prepubertal rat offspring." *Arch Med Res* 45 (4): 294–300.

Hales, C. N. 1997. "Fetal and infant growth and impaired glucose tolerance in adulthood: the "thrifty phenotype" hypothesis revisited." *Acta Paediatr Suppl* 422: 73–7.

Hales, C. N., and D. J. Barker. 2001. "The thrifty phenotype hypothesis." *Br Med Bull* 60: 5–20.

Hales, C. N., and S. E. Ozanne. 2003. "The dangerous road of catch-up growth." *J Physiol* 547 (Pt 1): 5–10.

Kotsampasi, B., C. Balaskas, G. Papadomichelakis, and S. E. Chadio. 2009. "Reduced Sertoli cell number and altered pituitary responsiveness in male lambs undernourished in utero." *Anim Reprod Sci* 114 (1–3): 135–47.

Kotsampasi, B., S. Chadio, G. Papadomichelakis, S. Deligeorgis, D. Kalogiannis, I. Menegatos, and G. Zervas. 2009. "Effects of maternal undernutrition on the hypothalamic-pituitary-gonadal axis function in female sheep offspring." *Reprod Domest Anim* 44 (4): 677–84.

Kyle, U. G., and C. Pichard. 2006. "The Dutch Famine of 1944–1945: a pathophysiological model of long-term consequences of wasting disease." *Curr Opin Clin Nutr Metab Care* 9 (4): 388–94.

Langley-Evans, S. C., S. J. Welham, and A. A. Jackson. 1999. "Fetal exposure to a maternal low protein diet impairs nephrogenesis and promotes hypertension in the rat." *Life Sci* 64 (11): 965–74.

Li, G., P. Zhang, J. Wang, E. W. Gregg, W. Yang, Q. Gong, H. Li, H. Li, Y. Jiang, Y. An, Y. Shuai, B. Zhang, J. Zhang, T. J. Thompson, R. B. Gerzoff, G. Roglic, Y. Hu, and P. H. Bennett. 2008. "The long-term effect of lifestyle interventions to prevent diabetes in the China Da Qing Diabetes Prevention Study: a 20-year follow-up study." *Lancet* 371 (9626): 1783–9.

Lillycrop, K. A., J. Rodford, E. S. Garratt, J. L. Slater-Jefferies, K. M. Godfrey, P. D. Gluckman, M. A. Hanson, and G. C. Burdge. 2010. "Maternal protein restriction with or without folic acid supplementation during pregnancy alters the hepatic transcriptome in adult male rats." *Br J Nutr* 103 (12): 1711–9.

Matilainen, O., P. M. Quiros, and J. Auwerx. 2017. "Mitochondria and Epigenetics - Crosstalk in Homeostasis and Stress." *Trends Cell Biol* 27 (6): 453–463.

Mentch, S. J., M. Mehrmohamadi, L. Huang, X. Liu, D. Gupta, D. Mattocks, P. Gomez Padilla, G. Ables, M. M. Bamman, A. E. Thalacker-Mercer, S. N. Nichenametla, and J. W. Locasale. 2015. "Histone Methylation Dynamics and Gene Regulation Occur through the Sensing of One-Carbon Metabolism." *Cell Metab* 22 (5): 861–73.

Mesquita, F. F., J. A. Gontijo, and P. A. Boer. 2010. "Maternal undernutrition and the offspring kidney: from fetal to adult life." *Braz J Med Biol Res* 43 (11): 1010–8.

Muilwijk, M., M. Nicolaou, S. A. Qureshi, C. Celis-Morales, J. M. R. Gill, A. Sheikh, N. Sattar, E. Beune, A. K. Jenum, K. Stronks, and I. G. M. van Valkengoed. 2018. "Dietary and physical activity recommendations to prevent type 2 diabetes in South Asian adults: A systematic review." *PLoS One* 13 (7): e0200681.

Muller, M. J., J. Enderle, M. Pourhassan, W. Braun, B. Eggeling, M. Lagerpusch, C. C. Gluer, J. J. Kehayias, D. Kiosz, and A. Bosy-Westphal. 2015. "Metabolic adaptation to caloric restriction and subsequent refeeding: the Minnesota Starvation Experiment revisited." *Am J Clin Nutr* 102 (4): 807–19.

Nwagwu, M. O., A. Cook, and S. C. Langley-Evans. 2000. "Evidence of progressive deterioration of renal function in rats exposed to a maternal low-protein diet in utero." *Br J Nutr* 83 (1): 79–85.

Pan, X. R., G. W. Li, Y. H. Hu, J. X. Wang, W. Y. Yang, Z. X. An, Z. X. Hu, J. Lin, J. Z. Xiao, H. B. Cao, P. A. Liu, X. G. Jiang, Y. Y. Jiang, J. P. Wang, H. Zheng, H. Zhang, P. H. Bennett, and B. V. Howard. 1997. "Effects of diet and exercise in preventing NIDDM in people with impaired glucose tolerance. The Da Qing IGT and Diabetes Study." *Diabetes Care* 20 (4): 537–44.

Pedroza, A., D. S. Ferreira, D. F. Santana, P. T. da Silva, F. C. A. de Aguiar Junior, D. F. Sellitti, and C. J. Lagranha. 2019. "A maternal low-protein diet and neonatal overnutrition result in similar changes to glomerular morphology and renal cortical oxidative stress measures in male Wistar rats." *Appl Physiol Nutr Metab* 44 (2): 164–171.

Petrik, J., B. Reusens, E. Arany, C. Remacle, C. Coelho, J. J. Hoet, and D. J. Hill. 1999. "A low protein diet alters the balance of islet cell replication and apoptosis in the fetal and neonatal rat and is associated with a reduced pancreatic expression of insulin-like growth factor-II." *Endocrinology* 140 (10): 4861–73.

Priliani, L., E. L. Prado, R. Restuadi, D. E. Waturangi, A. H. Shankar, and S. G. Malik. 2019. "Maternal multiple micronutrient supplementation stabilizes mitochondrial DNA copy number in pregnant women in lombok, Indonesia." *J Nutr* 149 (8): 1309–1316.

Ramachandran, A., C. Snehalatha, J. Ram, S. Selvam, M. Simon, A. Nanditha, A. S. Shetty, I. F. Godsland, N. Chaturvedi, A. Majeed, N. Oliver, C. Toumazou, K. G. Alberti, and D. G. Johnston. 2013. "Effectiveness of mobile phone messaging in prevention of type 2 diabetes by lifestyle modification in men in India: a prospective, parallel-group, randomised controlled trial." *Lancet Diabetes Endocrinol* 1 (3): 191–8.

Ramadan, W. S., I. Alshiraihi, and S. Al-karim. 2013. "Effect of maternal low protein diet during pregnancy on the fetal liver of rats." *Ann Anat* 195 (1): 68–76.

Roseboom, T., S. de Rooij, and R. Painter. 2006. "The Dutch famine and its long-term consequences for adult health." *Early Hum Dev* 82 (8): 485–91.

Sadiq, H. F., U. G. Das, T. F. Tracy, and S. U. Devaskar. 1999. "Intra-uterine growth restriction differentially regulates perinatal brain and skeletal muscle glucose transporters." *Brain Res* 823 (1–2): 96–103.

Sahajpal, V., and N. Ashton. 2003. "Renal function and angiotensin AT1 receptor expression in young rats following intrauterine exposure to a maternal low-protein diet." *Clin Sci (Lond)* 104 (6): 607–14.

Sandovici, I., N. H. Smith, M. D. Nitert, M. Ackers-Johnson, S. Uribe-Lewis, Y. Ito, R. H. Jones, V. E. Marquez, W. Cairns, M. Tadayyon, L. P. O'Neill, A. Murrell, C. Ling, M. Constancia, and S. E. Ozanne. 2011. "Maternal diet and aging alter the epigenetic control of a promoter-enhancer interaction at the Hnf4a gene in rat pancreatic islets." *Proc Natl Acad Sci U S A* 108 (13): 5449–54.

Schutt, A. K., C. S. Blesson, J. W. Hsu, C. T. Valdes, W. E. Gibbons, F. Jahoor, and C. Yallampalli. 2019. "Preovulatory exposure to a protein-restricted diet disrupts amino acid kinetics and alters mitochondrial structure and function in the rat oocyte and is partially rescued by folic acid." *Reprod Biol Endocrinol* 17 (1): 12.

Serpente, P., Y. Zhang, E. Islimye, S. Hart-Johnson, and A. P. Gould. 2021. "Quantification of fetal organ sparing in maternal low-protein dietary models." *Wellcome Open Res* 6: 218.

Shin, H. J., H. Kim, S. Oh, J. G. Lee, M. Kee, H. J. Ko, M. N. Kweon, K. J. Won, and S. H. Baek. 2016. "AMPK-SKP2-CARM1 signalling cascade in transcriptional regulation of autophagy." *Nature* 534 (7608): 553–7.

Shoubridge, E. A., and T. Wai. 2007. "Mitochondrial DNA and the mammalian oocyte." *Curr Top Dev Biol* 77: 87–111.

Siddiqui, M. K., R. M. Anjana, A. Y. Dawed, C. Martoeau, S. Srinivasan, J. Saravanan, S. K. Madanagopal, A. Taylor, S. Bell, A. Veluchamy, R. Pradeepa, N. Sattar, R. Venkatesan, C. N. A. Palmer, E. R. Pearson, and V. Mohan. 2022. "Correction to: Young-onset diabetes in Asian Indians is associated with lower measured and genetically determined beta cell function." *Diabetologia* 65 (7): 1237.

Smith, A. R., A. Hinojosa Briseno, M. Picard, and A. Cardenas. 2023. "The prenatal environment and its influence on maternal and child mitochondrial DNA copy number and methylation: A review of the literature." *Environ Res* 227: 115798.

Staimez, L. R., M. B. Weber, H. Ranjani, M. K. Ali, J. B. Echouffo-Tcheugui, L. S. Phillips, V. Mohan, and K. M. Narayan. 2013. "Evidence of reduced beta-cell function in Asian Indians with mild dysglycemia." *Diabetes Care* 36 (9): 2772–8.

Sui, S., Y. Jia, B. He, R. Li, X. Li, D. Cai, H. Song, R. Zhang, and R. Zhao. 2014. "Maternal low-protein diet alters ovarian expression of folliculogenic and steroidogenic genes and their regulatory MicroRNAs in neonatal piglets." *Asian-Australas J Anim Sci* 27 (12): 1695–704.

Tan, V. M., Y. S. Lee, K. Venkataraman, E. Y. Khoo, E. S. Tai, Y. S. Chong, P. Gluckman, M. K. Leow, and C. M. Khoo. 2015. "Ethnic differences in insulin sensitivity and beta-cell function among Asian men." *Nutr Diabetes* 5 (7): e173.

Tappia, P. S., C. Guzman, L. Dunn, and N. Aroutiounova. 2013. "Adverse cardiac remodeling due to maternal low protein diet is associated with alterations in expression of genes regulating glucose metabolism." *Nutr Metab Cardiovasc Dis* 23 (2): 130–5.

Tuomilehto, J., and J. Lindstrom. 2003. "The major diabetes prevention trials." *Curr Diab Rep* 3 (2): 115–22.

Vaiserman, A., and O. Lushchak. 2019. "Developmental origins of type 2 diabetes: Focus on epigenetics." *Ageing Res Rev* 55: 100957.

Vidyadharan, V. A., A. Betancourt, C. Smith, C. Yallampalli, and C. S. Blesson. 2022. "Prenatal low-protein diet affects mitochondrial structure and function in the skeletal muscle of adult female offspring." *Nutrients* 14 (6).

Vidyadharan, V. A., C. S. Blesson, D. Tanchico, A. Betancourt, C. Smith, and C. Yallampalli. 2023. "Low protein programming causes increased mitochondrial fusion and decreased oxygen consumption in the hepatocytes of female rats." *Nutrients* 15 (7).

Vipin, V. A., C. S. Blesson, and C. Yallampalli. 2022. "Maternal low protein diet and fetal programming of lean type 2 diabetes." *World J Diabetes* 13 (3): 185–202.

Winkel, L., M. Rasmussen, L. Larsen, L. T. Dalgaard, and J. H. Nielsen. 2022. "Fetal programming of the endocrine pancreas: impact of a maternal low-protein diet on gene expression in the perinatal rat pancreas." *Int J Mol Sci* 23 (19).

Winship, A. L., S. E. Gazzard, L. A. Cullen McEwen, J. F. Bertram, and K. J. Hutt. 2018. "Maternal low-protein diet programmes low ovarian reserve in offspring." *Reproduction* 156 (4): 299–311.

Zaidi, A. A., P. R. Wilton, M. S. Su, I. M. Paul, B. Arbeithuber, K. Anthony, A. Nekrutenko, R. Nielsen, and K. D. Makova. 2019. "Bottleneck and selection in the germline and maternal age influence transmission of mitochondrial DNA in human pedigrees." *Proc Natl Acad Sci U S A* 116 (50): 25172–8.

Zambrano, E., G. L. Rodriguez-Gonzalez, C. Guzman, R. Garcia-Becerra, L. Boeck, L. Diaz, M. Menjivar, F. Larrea, and P. W. Nathanielsz. 2005. "A maternal low protein diet during pregnancy and lactation in the rat impairs male reproductive development." *J Physiol* 563 (Pt 1): 275–84.

30 Effects of Maternal Diet and Body Composition on Human Milk Protein

Sandra Abreu, Cecília Morais, and Margarida Liz Martins

ABBREVIATIONS

BMI	Body mass index
HBM	Human breast milk
NW	Normal weight
OW	Overweight

INTRODUCTION

Breastfeeding is recommended exclusively for the first six months of an infant's life, as it provides optimal nutrition that supports infant growth and development (ESPGHAN Committee on Nutrition et al. 2009; World Health Organization 2002). The composition of human breast milk (HBM) is dynamic, changing throughout the day, and varying with the frequency and duration of breastfeeding, as well as across different stages of lactation. Additionally, HBM composition can be influenced by various maternal factors, including diet, nutritional status, genetics, health, and lifestyle, as well as infant factors (Samuel et al. 2020).

Several studies have investigated the impact of maternal dietary intake on HBM composition. Most have focused on the relationship between maternal intake of specific nutrients, particularly fatty acids, and the macronutrient content of HBM. It is well established that the maternal dietary intake of fatty acids, as well as certain fat- and water-soluble vitamins, is reflected in the composition of HBM (Keikha et al. 2017; Bravi et al. 2016; Samuel et al. 2020). However, evidence regarding the impact of maternal diet on other nutrients in HBM, such as protein, is conflicting. Additionally, while evidence suggests that maternal body mass index (BMI) and adiposity are associated with higher levels of fat in HBM (Leghi et al. 2020; Daniel et al. 2021), the impact of maternal body composition on the protein content of HBM is not well understood.

HBM protein comprises a mixture of whey, casein, and mucin proteins, which provide numerous bioactive proteins and peptides with essential functions (Figure 30.1). Proteins make up approximately 1% of HBM composition, which is typically lower than in cow's milk or formula (Lawrence 2015). Despite changes in protein content and composition of HBM after the second week of lactation, it still provides sufficient support for optimal growth, immune function, and overall health in infants (Haschke, Haiden, and Thakkar 2016). Furthermore, considering the role of protein in HBM on metabolism, the immune system, and growth, it can significantly impact long-term health outcomes (Koletzko et al. 2019).

DOI: 10.1201/b23328-30

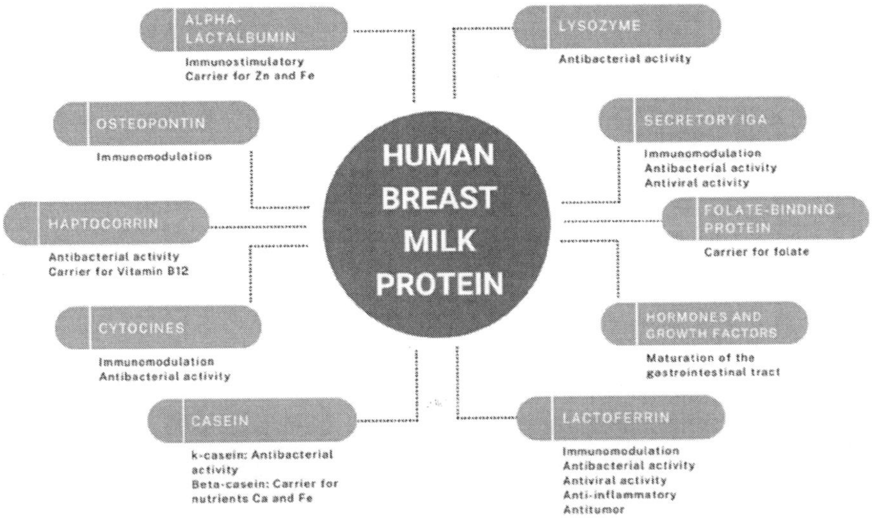

FIGURE 30.1 Overview of biological activity of human breast milk protein. This figure illustrates some of the biological activity of protein found in human breast milk. Ca, calcium; Fe, iron; Zn, zinc. Adapted from Donovan (2019), Lawrence (2015), and Lonnerdal (2016).

MATERNAL NUTRITIONAL INTAKE AND HUMAN MILK PROTEIN

Several studies have investigated the influence of maternal nutritional intake on modulating the quantity and quality of HBM protein. Researchers have observed that maternal nutrient intake can affect the protein content of HBM, along with other factors related to the mother and infant. A systematic review reported that only 35% of the included studies examining the relationship between maternal dietary intake and HBM protein composition found a significant association with at least one macronutrient consumed by lactating women (Adhikari et al. 2022). However, some studies did not find a significant association between dietary intake and the protein content of HBM. Therefore, evidence regarding the impact of maternal diet on the protein content of HBM remains mixed.

Evidence from Observational Studies

A growing body of observational research has investigated the impact of maternal nutritional intake on HBM protein. Table 30.1 lists most of the observational studies that have explored the association between maternal nutritional intake and protein content in HBM.

A study of 77 exclusively breastfeeding mothers during 4–8 weeks of lactation (Bzikowska-Jura et al. 2020) showed that maternal intake of plant-derived protein was positively associated with the true protein content of HBM, but not with the total protein content. No associations were found between maternal macronutrient intake (in grams per day and as a percentage of total energy intake) or intake of animal-derived protein and the true protein or total protein content of HBM. A cross-sectional study (Wati et al. 2023) of 110 exclusively breastfeeding mothers of 1-to-6-month-old infants found that maternal protein intake was positively associated with the total protein content of HBM, whereas no significant associations were observed for fat and carbohydrates. Furthermore, maternal intake of animal-derived and plant-derived protein, as well as amino acids such as isoleucine, leucine, lysine, phenylalanine, methionine, and valine, was found to have a positive correlation with total HBM protein. Another study reported that HBM protein was negatively associated with maternal energy intake from protein and vitamin D among lactating women at 6–8 weeks postpartum (Butts et al. 2018). Similarly, Leelahakul et al. (2009) found a positive correlation between the total protein concentration in HBM and maternal energy intake. Additionally, the concentration of

TABLE 30.1
Characteristics and Main Findings of Observational Studies Examining the Association between Maternal Nutritional Intake and Protein Content in Human Breast Milk

Authors, Year	Study Details	Country	Dietary Assessment	Exposure	Outcome	Adjustment	Main Findings
Boniglia et al. 2003	n = 117 exclusively breastfeeding mothers of term infants (mean age = 31.2 ± 4.5 years) Cross-sectional study Milk collection: ±4 days after the infant's first month	Italy	Two 24-hour recalls	Maternal energy and protein intake Maternal age and BMI	Total nitrogen Protein nitrogen Non-protein nitrogen Free amino acid	None	No correlations were found between maternal energy or protein intake and protein and non-protein nitrogen fractions in HBM.
Leelahakul et al. 2009	n = 39 mothers of term infants (mean age Thai mothers = 26.0 ± 1.4 years; mean age Japanese mothers = 30.5 ± 1.3 years) Cross-sectional study Milk collection: five days postpartum	Japan	24-hour recall	Maternal energy, macronutrient, and micronutrient intake	Total protein Major proteins in the whey	None	There was a positive correlation between the total protein concentration in the HBM and the mothers' daily energy intake ($r = 0.49$, $p < 0.01$). There was a negative correlation between the lactoferrin concentration in the HBM and the mothers' daily fat intake (r = −0.23, p <0.05).
Quinn et al. 2012	n = 102 lactating women (mean age = 23.8 ± 0.3 years) Cross-sectional study Milk collection between 2 weeks and 18 months after birth	Philippines	24-hour recall	Maternal BMI, percent body fat, energy and macronutrient intake, and infant age	Total protein	Breastfeeding duration, infant age, energy	There were no significant associations between any maternal dietary predictors and HBM protein content.

(Continued)

TABLE 30.1 (CONTINUED)
Characteristics and Main Findings of Observational Studies Examining the Association between Maternal Nutritional Intake and Protein Content in Human Breast Milk

Authors, Year	Study Details	Country	Dietary Assessment	Exposure	Outcome	Adjustment	Main Findings
Bzikowska-Jura et al. 2018	n = 40 exclusively breastfeeding mothers of term infants (mean age = 31.1 ± 4.4 years); n = 22 and 15, at third and sixth months of lactation, respectively Longitudinal study Milk collection at first, third, and sixth months of lactation	Poland	Three-day dietary records at first, third, and sixth months of lactation	Maternal energy, macronutrient, and micronutrient intake Maternal body composition	Total protein True protein	None	There is no significant correlation between the protein content of HBM and the nutrient intake of the mother's diet at three different time points.
Butts et al. 2018	n = 78 mothers of term infants (mean age = 31 ± 5 years) Cross-sectional study Milk collection: three breast milk samples over a one-week period 6–8 weeks postpartum	New Zealand	Three-day dietary record	Maternal energy, macronutrient, and micronutrient intake	Total protein	None	There were negative associations between breast milk concentrations of protein and maternal energy intake from protein as well as intake of vitamin D. There were positive associations between breast milk concentrations of protein and maternal starch intake.
Aumeistere et al. 2019	n = 61 lactating women (age range = 23–39 years) Cross-sectional study Milk collection between 1.5 and 21 months postpartum	Latvia	Three-day dietary record	Maternal food groups and micronutrient intake	Total protein	Maternal age, infant age and sex, parity, and feeding pattern	There was no significant association between the protein content of HBM and maternal macronutrient intake.

(Continued)

TABLE 30.1 (CONTINUED)
Characteristics and Main Findings of Observational Studies Examining the Association between Maternal Nutritional Intake and Protein Content in Human Breast Milk

Authors, Year	Study Details	Country	Dietary Assessment	Exposure	Outcome	Adjustment	Main Findings
Bzikowska-Jura et al. 2020	n = 77 exclusively breastfeeding mothers of term infants (mean age = 32.4 ± 4.4 years) Cross-sectional study Milk collection 4–8 weeks postpartum	Poland	Three-day dietary record	Infant sex and anthropometric parameters Maternal macronutrient intake and nutritional status	Total protein True protein	None	The true protein content of HBM was found to be correlated with the amount of plant-derived protein consumed by the mother (Spearman rho = 0.23, p = 0.045). No significant correlations were observed for the total protein content of HBM.
Thakur et al. 2021	n = 60 lactating women of preterm infants (age range = 18–45 years) Longitudinal study Milk collection at 7th, 14th, 21st, and 28th days of lactation	India	Three-day dietary records at 7th, 14th, 21st, and 28th days of lactation	Maternal and infant factors	True protein	None	The content of protein in the HBM was not found to be associated with the mother's dietary components.
Ryoo and Kang 2022	n = 159 lactating women of term infants (mean age = 33.98 ± 4.35 years) Cross-sectional study Milk collection 7–14 days postpartum	Korea	Food intake frequency survey	Maternal macronutrient intake Maternal factors (stress, sleep quality, spousal support, breastfeeding assessment score)	Total Protein	None	There were no significant associations between maternal dietary intake of macronutrients and HBM protein (carbohydrates: r = 0.02; fat: r = −0.04; protein: r = 0.06; p >0.05 for all)

(*Continued*)

TABLE 30.1 (CONTINUED)
Characteristics and Main Findings of Observational Studies Examining the Association between Maternal Nutritional Intake and Protein Content in Human Breast Milk

Authors, Year	Study Details	Country	Dietary Assessment	Exposure	Outcome	Adjustment	Main Findings
Binder et al. 2023	n = 136 lactating women (mean age = 33.98 ± 4.35 years) Cross-sectional study Milk collection: median on the 29th day of lactation	Austria	24-hour recall	Maternal energy and macronutrient intake Maternal BMI	Total protein True protein	None	There was a positive correlation between maternal dietary protein intake (g) and true protein (Spearman's rho = 0.997; p <0.001), and the protein intake (% kcal) and true protein in the HBM (Spearman's rho = 0.995; p <0.001). True and total protein in the HBM were also correlated with fat (g and % kcal), carbohydrate (g and % kcal), and energy.
Komatsu et al. 2023	n = 71 lactating women (mean age = 31.1 ± 4.9 years) Cross-sectional study Milk collection one month postpartum	Japan	Self-administered diet history questionnaires	Maternal macronutrient and micronutrient intake	Total protein	Maternal age, early-pregnancy BMI at the first clinical visit, gestational weeks, parity, mode of delivery, maternal weight gain, infant sex, and infant birth weight	None of the analyzed nutrients were associated with protein in HBM.

(Continued)

TABLE 30.1 (CONTINUED)
Characteristics and Main Findings of Observational Studies Examining the Association between Maternal Nutritional Intake and Protein Content in Human Breast Milk

Authors, Year	Study Details	Country	Dietary Assessment	Exposure	Outcome	Adjustment	Main Findings
Wati et al. 2023	n = 110 exclusively breastfeeding mothers of term infants (age range = 20–35 years) Cross-sectional study Milk collection: 1–6 months postpartum	Indonesia	Three 24-hour recalls	Maternal energy and macronutrient intake Maternal nutritional status	Total protein	Maternal age, education, occupation, parity, and current BMI	Findings from the adjusted regression model showed that maternal protein intake was positively associated with protein in HBM (r = 0.554, p <0.001), while no associations were found for carbohydrates, fat, and energy intake. The intake of plant-based (r = 0.239) and animal protein (r = 0.593) was positively correlated with the protein content of HBM. The amino acids isoleucine, leucine, lysine, phenylalanine, methionine, and valine were positively correlated with protein in HBM.

This table summarizes the evidence from observational studies investigating the relationship between maternal nutritional intake and the protein content of HBM. For the purposes of this chapter, the focus is solely on the identification of the HBM protein.

BMI, body mass index; HBM, human breast milk; r, Pearson correlation.

lactoferrin in HBM was negatively correlated with maternal fat intake. The concentration of lactoferrin in HBM also positively associated with maternal intake of energy, vitamin A, and cholesterol, and negatively associated with protein intake, after adjusting for potential confounders (Liu et al. 2022). Researchers have noted that lactoferrin in HBM plays an important role in iron metabolism. It also inhibits iron-dependent bacteria to prevent gastrointestinal infections and possesses anti-inflammatory, antitumor, and antiviral properties (Donovan 2019; Lawrence 2015; Lonnerdal 2016).

Although the pattern of protein content change during the lactation period is similar between term and preterm infants (Gidrewicz and Fenton 2014), a stronger correlation was observed between maternal protein intake and the protein content of HBM in preterm infants. A study of 136 Austrian lactating women (Binder et al. 2023), 93% of whom had given birth to preterm infants and with lactation days ranging from 14 to 66, found a strong positive association between maternal dietary protein intake and the true protein in HBM. Weaker positive associations were seen with daily dietary carbohydrate, fat, and energy intake. Another noteworthy finding from this study is that mothers' protein intake was positively associated with the adiponectin and leptin content of HBM. There is some evidence suggesting that the presence of these proteins in HBM may influence the growth and development of infants, as well as their risk of being overweight in later life (Kratzsch, Bae, and Kiess 2018).

On the other hand, the protein content of HBM was not found to be associated with maternal nutritional intake in several studies. Bzikowska-Jura et al. (2018) conducted a study to investigate the relationship between maternal diet and the composition of HBM during the first, third, and sixth months of lactation. Their findings revealed no association between maternal energy, macronutrient, or micronutrient intake and the protein content of HBM. Similarly, findings from the SMILE Iwamizawa study (Survey on Mothers, Infants, and Children Lives and Environments in Iwamizawa, Hokkaido, Japan) among 71 Japanese lactating women showed, after adjusting for potential confounders, that protein content in HBM at one month postpartum was not associated with any of the maternal nutrient intakes analyzed (Komatsu et al. 2023). This aligns with the findings of a study of 159 breastfeeding Korean mothers of term infants (Ryoo and Kang 2022), which showed no significant associations between dietary intake of macronutrients and HBM protein at two weeks postpartum. Other observational studies have revealed similar results among both preterm (Thakur et al. 2021) and term infants (Quinn et al. 2012; Aumeistere et al. 2019; Boniglia et al. 2003).

Evidence from Experimental Studies

Early experimental studies have shown that protein intake may impact the protein composition of HBM. These studies found that a short-term protein-restricted diet resulted in a decrease in non-protein nitrogen, specifically free amino acids (Motil et al. 1995; Forsum and Lonnerdal 1980). Additionally, other macronutrients may affect the protein content of HBM. A controlled short-term feeding study conducted over a 12-hour period, with increased sugar or fat intake, found that HBM protein decreased in response to the higher fat diet compared to the control, while no effect was observed for sugar (Ward et al. 2021).

More recently, findings from a two-week dietary intervention trial aimed at reducing fat and sugar intake among women who exclusively breastfed a term infant between 6 and 20 weeks postpartum also found no association between maternal nutritional intake and HBM protein (Leghi et al. 2021). At the end of the intervention, even with energy restriction (at least 500 kcal/day), HBM macronutrient concentrations remained unaffected. This is consistent with another study, which also found no change in the macronutrient composition of HBM after lactating women engaged in a low-fat, high-carbohydrate diet over an eight-day period (Mohammad, Sunehag, and Haymond 2009). Taken together, these results suggest that short-term changes in maternal energy and nutritional intake have no adverse impact on the protein content of HBM.

Further studies have found that the macronutrient content of HBM is not affected, even when lactating women do not meet recommended energy or macronutrient needs (Aumeistere et al. 2019; Leghi et al. 2021) or when they are on short-term or chronic fasting (Allegaert and Smits 2020). Thus, it has been proposed that compensatory physiological mechanisms may buffer variations in human milk composition against short-term alterations in lactating women's nutritional intake, which may explain the weak or nonexistent association between maternal nutrient intake and the protein content of HBM (Keikha et al. 2017; Tigas, Sunehag, and Haymond 2002).

MATERNAL DIETARY PATTERN AND HUMAN MILK PROTEIN

The causal relationship between HBM composition and maternal diet remains controversial (Bravi et al. 2016; Keikha et al. 2017; Xi et al. 2023) due to a lack of comprehensive research on HBM composition under different dietary patterns. Most studies have primarily focused on the association between individual foods or nutrients and the content of HBM. For example, egg consumption during lactation has been positively associated with ovalbumin (Metcalfe et al. 2016) and lactoferrin content in HBM (Liu et al. 2022). However, dietary intake is a combination of multiple foods, nutrients, and their interactions. Thus, analyzing dietary patterns based on factors associated with food consumption, including interactions and synergies between individual nutrients and the entire food matrix, better reflects the complexity of the diet.

Regarding the analysis of different dietary patterns, evidence indicates that the protein content in HBM is higher for mothers who adhere highly to the Mediterranean diet, likely due to their higher intake of plant-derived proteins in the diet. Adhering to a healthy and sustainable dietary pattern such as the Mediterranean diet during breastfeeding can provide better quality HBM for infant nutrition (Karbasi et al. 2023).

Looking at vegetarian dietary patterns, the evidence shows that the composition of HBM is not significantly different between mothers who follow a vegetarian diet and those who do not. It is acknowledged that vegetarian eating patterns can exhibit significant variability, influenced by the degree to which animal products are excluded (Karcz and Krolak-Olejnik 2021), and it is known that continuous dietary restrictions during the lactation period can negatively affect the nutrient content of breast milk (Institute of Medicine 1991). Nevertheless, according to a systematic review that included 13 studies, the protein content of HBM is similar between mothers who follow a meat-free diet and those who have no restrictions. Regarding the amino acid profile, a difference was found in taurine concentrations. The milk of omnivorous mothers was significantly richer in this amino acid than that of vegetarian mothers, which can be attributed to the mother's diet (Karcz and Krolak-Olejnik 2021).

Concerning fasting or short-term dieting in lactating women, research has found no association between fasting and HBM protein composition (Allegaert and Smits 2020; Rakicioglu et al. 2006). Allegaert and Smits (2020) found, using physiology-based pharmacokinetic models, that acute maternal fasting or fluid abstention does not significantly affect the macronutrient composition of HBM, whereas chronic, repetitive fasting during Ramadan has a moderate but not clinically relevant effect.

Different studies have identified various dietary patterns that present inconsistent results regarding the association between maternal dietary patterns and HBM protein composition. The MEDIDIET study, which included Italian breastfeeding mothers, identified five maternal dietary patterns: "Vitamins, Minerals and Fibres", "Proteins and Fatty Acids with Legs", "Fatty Acids with Fins", "Fatty Acids with Leaves", and "Starch and Vegetable Proteins". No association was found between these maternal dietary patterns and the protein content in HBM (Bravi et al. 2021). A meta-analysis conducted by Xi et al. (2023) focused on the relationship between the lactation diet and the composition of mature HBM, considering four dietary patterns, according to the reference proportion of energy from proteins and fats in relation to the total energy value of the diet. The study showed that the protein content of HBM was negatively correlated with fat intake in patterns with

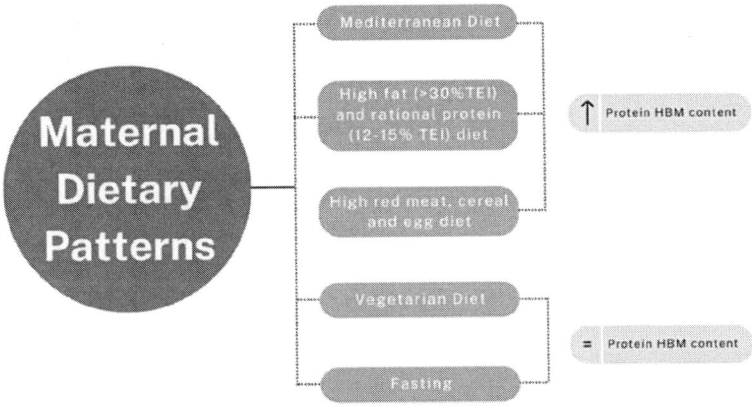

FIGURE 30.2 The influence of maternal dietary patterns on the protein content of human breast milk. This figure summarizes the scientific evidence regarding the influence of different dietary patterns on HBM content. HBM, human breast milk; TEI, total energy intake; ↑, a higher level was found compared to other dietary patterns; =, no significant differences were found compared to other dietary patterns.

high fat and protein content. Additionally, data from a prospective study showed that breast milk from mothers with a maternal dietary pattern defined as "high fat and rational protein" (more than 30% of energy supplied from fat and 12–15% from protein) had the highest protein content, while the lowest protein value was associated with the "rational fat and rational protein" dietary pattern (20–30% of the energy supplied from fat and 12–15% from protein) (Binder et al. 2023). A study conducted among Chinese lactating women classified maternal diets into four dietary patterns: high in animal foods, high in eggs, high in plant foods, and high in fruits. The authors showed that dietary patterns with high proportions of animal (meat/poultry, fish/shrimp, milk/dairy) and plant-based foods (tubers, vegetables, and beans) were associated with lower protein concentration in colostrum (milk collected in the first week postpartum). Conversely, a higher protein concentration in mature milk was found among women consuming high amounts of eggs, fruits, and cereals (Hu et al. 2021). Similarly, another study reported that a dietary pattern characterized by a high consumption of red meat, cereals, and eggs was associated with a higher protein content in breast milk (Huang and Hu 2020). On the other hand, a cross-sectional study involving Iranian breastfeeding women concluded that there were no differences in the protein content of HBM between women with a healthy dietary pattern characterized by high consumption of whole grains, legumes, vegetables, eggs, and red meat and women with an "unhealthy" dietary pattern characterized by a high intake of soft drinks, snacks, honey, seafood, and chicken (Karbasi et al. 2022).

Figure 30.2 summarizes the scientific evidence regarding the impact of different dietary patterns on HBM content.

MATERNAL BODY COMPOSITION AND HUMAN MILK PROTEIN

The influence of maternal body composition on HBM content has been poorly explored by the scientific community, although there is a global understanding that maternal dietary intake and nutritional status may affect some of the HBM components. One of the first difficulties encountered in approaching this topic lies in the parameters evaluated by each research study. Many studies collect information concerning height and weight, thus providing data concerning the postpartum and/or pre-pregnancy BMI, but information on more specific data of the mothers' body composition is scarce. Despite the different methodologies and the diversity of the parameters considered for study, some evidence of the association between maternal body composition and HBM content has been reported in several studies, though it is still controversial.

A systematic review that explored the impact of maternal dietary intake and nutritional status on HBM macronutrient composition found that, out of the 17 studies that investigated the association between maternal nutritional status and HBM protein, more than half did not find a significant association (Adhikari et al. 2022). Former systematic reviews have also indicated that the protein content in HBM is not affected by maternal BMI and adiposity parameters (Daniel et al. 2021; Leghi et al. 2020).

A longitudinal study investigating the association between maternal BMI and the HBM composition, mainly in terms of energy and macronutrients, over a long period (2 weeks to 9 months postpartum) found that the protein content of HBM was higher at some points when comparing overweight (OW) and obese to normal-weight (NW) women who were breastfeeding (Sims et al. 2020). A cohort study involving the collection of HBM at two weeks postpartum did not find significant differences in macronutrients between OW and NW mothers (Lemas et al. 2016). Another research found lower protein content at two weeks but not at four months, concluding that there were no consistent differences between the protein content of HBM of OW and NW mothers at both time points (Young et al. 2017). In a study that divided women into four categories, considering underweight, NW, OW, and obese mothers separately, the protein content was lower in relation to true and crude protein for underweight mothers when compared to NW and OW. The content of true and crude protein was also higher for obese mothers' HBM compared to the underweight mothers' HBM (Binder et al. 2023).

When analyzing BMI as a continuous variable, Bzikowska et al. (2018) found no association between current or pre-pregnancy BMI and the content of protein in HBM, for either total or true protein. Another study also found a significant correlation between total protein and pre-pregnancy BMI, but not for true protein. Additionally, this research evaluated other parameters of body composition and found that the current lean body mass of the mother presented a negative correlation with the total and true protein content of HBM, while the total water content was also inversely correlated with true protein (Bzikowska-Jura et al. 2020). In a prior study, Bzikowska-Jura et al. (2018) observed moderate to strong positive correlations between total protein content and true protein at the third month postpartum, for BMI and for the majority of body composition measures, such as the percentage of fat mass, fat-free mass (in kg), and muscle mass. Furthermore, a negative correlation between total protein and the percentage of total body water was also found.

Despite the conflicting evidence found by different studies (Daniel et al. 2021; Leghi et al. 2020), many of the findings suggest an association between BMI and the protein content of HBM. This relationship is not linear when attempting to evaluate protein as total or true. Moreover, studies that present information on body composition measures are rare, making it difficult to establish more robust associations. Considering different methodologies, it was also evident from reports that the time for HBM collection was considerably variable among all the studies and that evidence for the relation between BMI and body composition parameters was not found consistently and continuously in HBM during the breastfeeding process. Sims et al. (2020) stressed the need for more longitudinal studies aimed at evaluating the impact of BMI on HBM macronutrient composition, pointing out the limitations of studying HBM composition and its relation to maternal factors in cross-sectional studies.

CONCLUSION

There is no doubt that HBM is the optimal nourishment for infants. Several factors may impact the composition of HBM, with maternal dietary intake and body composition holding significant influence. Although the protein content in HBM appears to be more resistant to variations in maternal factors than other nutrients (Keikha et al. 2017), the evidence reviewed suggests that maternal intake of energy and protein, dietary patterns high in protein-rich food sources, and BMI can modulate HBM protein.

However, discrepancies are observed within the existing studies, which may be due mainly to differences in methodological issues, thus collectively challenging the comparability of these studies and the establishment of a causal relationship between HBM protein content and maternal diet or body composition. With regard to maternal dietary assessment, it is important to note that different methodologies have been used, including 24-hour recall, food record, and food frequency questionnaire. The lack of validity of these questionnaires in different populations, recall bias associated with the data collection, overestimation or underestimation of dietary intake, individual variations in population diet, different reference periods of maternal diet, diverse classification of dietary patterns and food sources, and different tools for dietary data analysis may explain some of the less consistent results between studies. Furthermore, other reasons for the disparity of findings arise from methods used to determine the protein content of HBM and the time point of milk collection, since the majority of the studies investigated the composition in the first week postpartum (Adhikari et al. 2022; Keikha et al. 2017). Other limitations that impair comparisons refer to the use of different BMI categories for breastfeeding women and different methodologies to assess body mass composition (Adhikari et al. 2022). The lack of consideration of controlling for relevant confounding factors is another important issue since HBM composition, as well as maternal diet and body composition, can be influenced by several maternal and infant-related factors (Keikha et al. 2017).

While evidence shows that the maternal diet and body composition have little effect on the protein content of HBM, it is generally accepted that an adequate maternal nutritional status and diet are essential to provide newborn infants with their mother's own milk with an optimal content (Binder et al. 2023). Therefore, further research is needed to better understand the impact of maternal diet and body composition on the protein content of HBM over a longer period. Such research could provide valuable insights into the effects of maternal factors on HBM composition and help guide recommendations for breastfeeding mothers.

THE EFFECTS OF OTHER PROTEINS

In this chapter, we reviewed evidence on the association between maternal diet and body composition and HBM protein content. We did not provide an exhaustive review of different proteins, as most studies highlighted the total protein or true content of HBM. Overall, some studies have investigated the influence of maternal dietary intake and adiposity on HBM lactoferrin, leptin, and adiponectin content. Studies have shown that leptin in HBM can be modulated by maternal BMI (Christensen et al. 2022) and diet (Binder et al. 2023). Additionally, maternal BMI has been positively associated with milk adiponectin concentrations (Martin et al. 2006). On the other hand, Christensen et al. (2022) found no association between maternal BMI and adiponectin levels in HBM. Moreover, Binder et al. (2023) reported that increased maternal protein intake correlated positively with adiponectin levels in HBM. Regarding lactoferrin levels in HBM, evidence is conflicting, with studies reporting no to weak associations with maternal nutritional status and diet (Villavicencio et al. 2017; Leelahakul et al. 2009; Liu et al. 2022; Rahfiludin M. Z and Pangestuti D. R 2020).

MINI DICTIONARY OF TERMS

Adiponectin: An adipokine protein with anti-inflammatory and insulin-sensitizing effects.

Lactoferrin: An iron-binding glycoprotein present in breast milk, especially in colostrum. It is the main protein in the whey fraction.

Leptin: An adipokine protein secreted by adipose tissue with orexinergic effects.

Non-protein nitrogen: Components that contain nitrogen. These include milk protein-derived peptides, peptide hormones, amino acids, oligosaccharides, monosaccharides, nucleic acids, free nucleotides, and other small molecules.

Total Protein: Represents the quantity of protein based on the total quantity of nitrogen. Also designated as crude protein.

True protein: Represents the quantity of protein corrected for non-protein nitrogen.

SUMMARY POINTS

- Protein content in human breast milk is around 1%.
- Human breast milk protein is involved in biological activities that promote infant growth and development.
- Maternal dietary intake and body composition may modulate the protein content of human breast milk.
- Maternal energy and protein intake may influence the protein content of human breast milk.
- Dietary patterns high in protein-rich food sources may modulate the protein content of human breast milk.
- The high heterogeneity across studies impairs their comparability.
- Body mass index and adiposity seem to be positively correlated with the content of human breast milk protein.

REFERENCES

Adhikari, S., U. Kudla, J. Nyakayiru, and E. M. Brouwer-Brolsma. 2022. "Maternal dietary intake, nutritional status and macronutrient composition of human breast milk: systematic review." *Br J Nutr* 127 (12):1796–1820. doi: 10.1017/S0007114521002786.

Allegaert, K., and A. Smits. 2020. "Acute maternal fasting or fluid abstention does not significantly affect the macronutrient composition of human milk: clinical and clinical research relevance." *Children (Basel)* 7 (6). doi: 10.3390/children7060060.

Aumeistere, L., I. Ciprovica, D. Zavadska, J. Andersons, V. Volkovs, and K. Celmalniece. 2019. "Impact of maternal diet on human milk composition among lactating women in latvia." *Medicina (Kaunas)* 55 (5). doi: 10.3390/medicina55050173.

Binder, C., S. Baumgartner-Parzer, L. I. Gard, A. Berger, and A. Thajer. 2023. "Maternal diet influences human milk protein concentration and adipose tissue marker." *Nutrients* 15 (2). doi: 10.3390/nu15020433.

Boniglia, C., B. Carratu, F. Chiarotti, S. Giammarioli, and E. Sanzini. 2003. "Influence of maternal protein intake on nitrogen fractions of human milk." *Int J Vitam Nutr Res* 73 (6):447–52. doi: 10.1024/0300-9831.73.6.447.

Bravi, F., M. Di Maso, Srbm Eussen, C. Agostoni, G. Salvatori, C. Profeti, P. Tonetto, P. A. Quitadamo, I. Kazmierska, E. Vacca, A. Decarli, B. Stahl, E. Bertino, G. E. Moro, M. Ferraroni, and Group on Behalf of the Medidiet Working. 2021. "Dietary patterns of breastfeeding mothers and human milk composition: data from the Italian MEDIDIET study." *Nutrients* 13 (5). doi: 10.3390/nu13051722.

Bravi, F., F. Wiens, A. Decarli, A. Dal Pont, C. Agostoni, and M. Ferraroni. 2016. "Impact of maternal nutrition on breast-milk composition: a systematic review." *Am J Clin Nutr* 104 (3):646–62. doi: 10.3945/ajcn.115.120881.

Butts, C. A., D. I. Hedderley, T. D. Herath, G. Paturi, S. Glyn-Jones, F. Wiens, B. Stahl, and P. Gopal. 2018. "Human milk composition and dietary intakes of breastfeeding women of different ethnicity from the manawatu-wanganui region of new zealand." *Nutrients* 10 (9). doi: 10.3390/nu10091231.

Bzikowska, A., A. Czerwonogrodzka-Senczyna, H. Weker, and A. Wesolowska. 2018. "Correlation between human milk composition and maternal nutritional status." *Rocz Panstw Zakl Hig* 69 (4):363–367. doi: 10.32394/rpzh.2018.0041.

Bzikowska-Jura, A., A. Czerwonogrodzka-Senczyna, G. Oledzka, D. Szostak-Wegierek, H. Weker, and A. Wesolowska. 2018. "Maternal nutrition and body composition during breastfeeding: association with human milk composition." *Nutrients* 10 (10). doi: 10.3390/nu10101379.

Bzikowska-Jura, A., P. Sobieraj, D. Szostak-Wegierek, and A. Wesolowska. 2020. "Impact of infant and maternal factors on energy and macronutrient composition of human milk." *Nutrients* 12 (9). doi: 10.3390/nu12092591.

Christensen, S. H., J. I. Lewis, A. Larnkjaer, H. Frokiaer, L. H. Allen, C. Molgaard, and K. F. Michaelsen. 2022. "Associations between maternal adiposity and appetite-regulating hormones in human milk are mediated through maternal circulating concentrations and might affect infant outcomes." *Front Nutr* 9:1025439. doi: 10.3389/fnut.2022.1025439.

Daniel, A. I., S. Shama, S. Ismail, C. Bourdon, A. Kiss, M. Mwangome, R. H. J. Bandsma, and D. L. O'Connor. 2021. "Maternal BMI is positively associated with human milk fat: a systematic review and meta-regression analysis." *Am J Clin Nutr* 113 (4):1009–1022. doi: 10.1093/ajcn/nqaa410.

Donovan, S. M. 2019. "Human milk proteins: composition and physiological significance." *Nestle Nutr Inst Workshop Ser* 90:93–101. doi: 10.1159/000490298.

ESPGHAN Committee on Nutrition, C. Agostoni, C. Braegger, T. Decsi, S. Kolacek, B. Koletzko, K. F. Michaelsen, W. Mihatsch, L. A. Moreno, J. Puntis, R. Shamir, H. Szajewska, D. Turck, and J. van Goudoever. 2009. "Breast-feeding: A commentary by the ESPGHAN committee on nutrition." *J Pediatr Gastroenterol Nutr* 49 (1):112–25. doi: 10.1097/MPG.0b013e31819f1e05.

Forsum, E., and B. Lonnerdal. 1980. "Effect of protein intake on protein and nitrogen composition of breast milk." *Am J Clin Nutr* 33 (8):1809–13. doi: 10.1093/ajcn/33.8.1809.

Gidrewicz, D. A., and T. R. Fenton. 2014. "A systematic review and meta-analysis of the nutrient content of preterm and term breast milk." *BMC Pediatr* 14:216. doi: 10.1186/1471-2431-14-216.

Haschke, F., N. Haiden, and S. K. Thakkar. 2016. "Nutritive and bioactive proteins in breastmilk." *Ann Nutr Metab* 69 Suppl 2:17–26. doi: 10.1159/000452820.

Hu, R., Srbm Eussen, F. P. C. Sijtsma, B. Papi, B. Stahl, Y. Jin, M. Mank, J. Li, and Z. Wang. 2021. "Maternal dietary patterns are associated with human milk composition in Chinese lactating women." *Nutrition* 91–92:111392. doi: 10.1016/j.nut.2021.111392.

Huang, Z., and Y. M. Hu. 2020. "Dietary patterns and their association with breast milk macronutrient composition among lactating women." *Int Breastfeed J* 15 (1):52. doi: 10.1186/s13006-020-00293-w.

Institute of Medicine. 1991. *Nutrition During Lactation*. Washington, DC: The National Academies Press.

Karbasi, S., A. Bahrami, Z. Asadi, F. Shahbeiki, M. Naseri, A. Zarban, and G. A. Ferns. 2022. "The association of maternal dietary quality and the antioxidant-proxidant balance of human milk." *Int Breastfeed J* 17 (1):56. doi: 10.1186/s13006-022-00498-1.

Karbasi, S., M. Mohamadian, M. Naseri, Z. Khorasanchi, A. Zarban, A. Bahrami, and G. A. Ferns. 2023. "A Mediterranean diet is associated with improved total antioxidant content of human breast milk and infant urine." *Nutr J* 22 (1):11. doi: 10.1186/s12937-023-00841-0.

Karcz, K., and B. Krolak-Olejnik. 2021. "Vegan or vegetarian diet and breast milk composition - a systematic review." *Crit Rev Food Sci Nutr* 61 (7):1081–1098. doi: 10.1080/10408398.2020.1753650.

Keikha, M., M. Bahreynian, M. Saleki, and R. Kelishadi. 2017. "Macro- and micronutrients of human milk composition: are they related to maternal diet? A comprehensive systematic review." *Breastfeed Med* 12 (9):517–527. doi: 10.1089/bfm.2017.0048.

Koletzko, Berthold, Hans Demmelmair, Veit Grote, and Martina Totzauer. 2019. "Optimized protein intakes in term infants support physiological growth and promote long-term health." *Seminars in Perinatology* 43 (7):151153. doi: https://doi.org/10.1053/j.semperi.2019.06.001.

Komatsu, Y., Y. Wada, F. Tabata, S. Kawakami, Y. Takeda, K. Nakamura, T. Ayabe, K. Nakamura, T. Kimura, and A. Tamakoshi. 2023. "Associations between maternal diet, human milk macronutrients, and breast-fed infant growth during the first month of life in the SMILE Iwamizawa in Japan." *Nutrients* 15 (3). doi: 10.3390/nu15030654.

Kratzsch, J., Y. J. Bae, and W. Kiess. 2018. "Adipokines in human breast milk." *Best Pract Res Clin Endocrinol Metab* 32 (1):27–38. doi: 10.1016/j.beem.2018.02.001.

Lawrence, Ruth A. 2015. "4 - Biochemistry of human milk." In *Breastfeeding*, edited by Ruth A. Lawrence and Robert M. Lawrence, 8th ed., 93–144. Philadelphia: Elsevier.

Leelahakul, V., F. Tanaka, N. Sinsuksai, K. Vichitsukon, W. Pinyopasakul, N. Kido, and S. Inukai. 2009. "Comparison of the protein composition of breast milk and the nutrient intake between Thai and Japanese mothers." *Nurs Health Sci* 11 (2):180–184. doi: 10.1111/j.1442-2018.2009.00445.x.

Leghi, G. E., M. J. Netting, C. T. Lai, A. Narayanan, M. Dymock, A. Rea, M. E. Wlodek, D. T. Geddes, and B. S. Muhlhausler. 2021. "Reduction in maternal energy intake during lactation decreased maternal body weight and concentrations of leptin, insulin and adiponectin in human milk without affecting milk production, milk macronutrient composition or infant growth." *Nutrients* 13 (6). doi: 10.3390/nu13061892.

Leghi, G. E., M. J. Netting, P. F. Middleton, M. E. Wlodek, D. T. Geddes, and A. B. S. Muhlhausler. 2020. "The impact of maternal obesity on human milk macronutrient composition: A systematic review and meta-analysis." *Nutrients* 12 (4). doi: 10.3390/nu12040934.

Lemas, D. J., B. E. Young, P. R. Baker, A. C. Tomczik, T. K. Soderborg, T. L. Hernandez, B. A. de la Houssaye, C. E. Robertson, M. C. Rudolph, D. Ir, Z. W. Patinkin, N. F. Krebs, S. A. Santorico, T. Weir, L. A. Barbour, D. N. Frank, and J. E. Friedman. 2016. "Alterations in human milk leptin and insulin are associated with early changes in the infant intestinal microbiome." *Am J Clin Nutr* 103 (5):1291–1300. doi: 10.3945/ajcn.115.126375.

Liu, L., Q. Guo, M. Cui, J. Liu, C. Yang, X. Li, P. Liu, and L. Wang. 2022. "Impact of maternal nutrition during early pregnancy and diet during lactation on lactoferrin in mature breast milk." *Nutrition* 93:111500. doi: 10.1016/j.nut.2021.111500.

Lonnerdal, B. 2016. "Bioactive proteins in human milk: health, nutrition, and implications for infant formulas." *J Pediatr* 173 (Suppl):S4–9. doi: 10.1016/j.jpeds.2016.02.070.

Martin, Lisa J., Jessica G. Woo, Sheela R. Geraghty, Mekibib Altaye, Barbara S. Davidson, Walter Banach, Lawrence M. Dolan, Guillermo M. Ruiz-Palacios, and Ardythe L. Morrow. 2006. "Adiponectin is present in human milk and is associated with maternal factors3." *The American Journal of Clinical Nutrition* 83 (5):1106–1111. doi: https://doi.org/10.1093/ajcn/83.5.1106.

Metcalfe, J. R., J. A. Marsh, N. D'Vaz, D. T. Geddes, C. T. Lai, S. L. Prescott, and D. J. Palmer. 2016. "Effects of maternal dietary egg intake during early lactation on human milk ovalbumin concentration: a randomized controlled trial." *Clin Exp Allergy* 46 (12):1605–1613. doi: 10.1111/cea.12806.

Mohammad, M. A., A. L. Sunehag, and M. W. Haymond. 2009. "Effect of dietary macronutrient composition under moderate hypocaloric intake on maternal adaptation during lactation." *Am J Clin Nutr* 89 (6):1821–1827. doi: 10.3945/ajcn.2008.26877.

Motil, K. J., M. Thotathuchery, A. Bahar, and C. M. Montandon. 1995. "Marginal dietary protein restriction reduced nonprotein nitrogen, but not protein nitrogen, components of human milk." *J Am Coll Nutr* 14 (2):184–191. doi: 10.1080/07315724.1995.10718492.

Quinn, E. A., F. Largado, M. Power, and C. W. Kuzawa. 2012. "Predictors of breast milk macronutrient composition in Filipino mothers." *Am J Hum Biol* 24 (4):533–540. doi: 10.1002/ajhb.22266.

Rahfiludin M. Z, and Pangestuti D. R. 2020. "Lactoferrin association with maternal nutritional status and lactation stages." *Curr Res Nutr Food Sci* 8 (1). doi: http://dx.doi.org/10.12944/CRNFSJ.8.1.16.

Rakicioglu, N., G. Samur, A. Topcu, and A. A. Topcu. 2006. "The effect of Ramadan on maternal nutrition and composition of breast milk." *Pediatr Int* 48 (3):278–83. doi: 10.1111/j.1442-200X.2006.02204.x.

Ryoo, C. J., and N. M. Kang. 2022. "Maternal factors affecting the macronutrient composition of transitional human milk." *Int J Environ Res Public Health* 19 (6). doi: 10.3390/ijerph19063308.

Samuel, T. M., Q. Zhou, F. Giuffrida, D. Munblit, V. Verhasselt, and S. K. Thakkar. 2020. "Nutritional and non-nutritional composition of human milk is modulated by maternal, infant, and methodological factors." *Front Nutr* 7:576133. doi: 10.3389/fnut.2020.576133.

Sims, C. R., M. E. Lipsmeyer, D. E. Turner, and A. Andres. 2020. "Human milk composition differs by maternal BMI in the first 9 months postpartum." *Am J Clin Nutr* 112 (3):548–557. doi: 10.1093/ajcn/nqaa098.

Thakur, A., N. Kler, P. Garg, P. Gandhi, and S. Srivastava. 2021. "Macronutrient analysis of human milk and factors associated with its composition in mothers of preterm infants </= 32 weeks." *Eur J Pediatr* 180 (12):3527–3534. doi: 10.1007/s00431-021-04158-0.

Tigas, S., A. Sunehag, and M. W. Haymond. 2002. "Metabolic adaptation to feeding and fasting during lactation in humans." *J Clin Endocrinol Metab* 87 (1):302–7. doi: 10.1210/jcem.87.1.8178.

Villavicencio, Aasith, Maria S. Rueda, Christie G. Turin, and Theresa J. Ochoa. 2017. "Factors affecting lactoferrin concentration in human milk: how much do we know?" 95 (1):12–21. doi: 10.1139/bcb-2016-0060 %M 28075610.

Ward, E., N. Yang, B. S. Muhlhausler, G. E. Leghi, M. J. Netting, M. J. Elmes, and S. C. Langley-Evans. 2021. "Acute changes to breast milk composition following consumption of high-fat and high-sugar meals." *Matern Child Nutr* 17 (3):e13168. doi: 10.1111/mcn.13168.

Wati, L. R., D. Sargowo, T. Nurseta, and L. Zuhriyah. 2023. "The role of protein intake on the total milk protein in lead-exposed lactating mothers." *Nutrients* 15 (11). doi: 10.3390/nu15112584.

World Health Organization. 2002. *The Optimal Duration of Exclusive Breastfeeding: Report of An Expert Consultation*. Geneva: World Health Organization.

Xi, Q., W. Liu, T. Zeng, X. Chen, T. Luo, and Z. Deng. 2023. "Effect of different dietary patterns on macronutrient composition in human breast milk: a systematic review and meta-analysis." *Nutrients* 15 (3). doi: 10.3390/nu15030485.

Young, B. E., Z. W. Patinkin, L. Pyle, B. de la Houssaye, B. S. Davidson, S. Geraghty, A. L. Morrow, and N. Krebs. 2017. "Markers of oxidative stress in human milk do not differ by maternal bmi but are related to infant growth trajectories." *Matern Child Health J* 21 (6):1367–1376. doi: 10.1007/s10995-016-2243-2.

31 The Paradoxical Impact of Dietary Protein on Growth in Early Life

The Role of Different Protein Sources

Kulnipa Kittisakmontri and Mary Fewtrell

ABBREVIATIONS

ABP	Animal-based protein
ASF	Animal-source foods
BCAA	Branched-chain amino acid
BMIZ	Body mass index z-score
CF	Complementary feeding
DBM	Double burden of malnutrition
EAA	Essential amino acid
FFM	Fat-free mass
FM	Fat mass
HAZ	Height-for-age z-score
HICs	High-income countries
HM	Human milk
IF	Infant formula
IGF-1	Insulin-like growth factor 1
IGFBP-3	Insulin-like growth factor binding protein 3
LAZ	Length-for-age z-score
LMICs	Low- and middle-income countries
mTOR	Mammalian target of rapamycin
PBP	Plant-based protein
% PE	Percentage of protein energy to total energy intake
RCT	Randomized controlled trial
SDS	Standard deviation scores
SMD	Standardized mean difference
WAZ	Weight-for-age z-score
WLZ	Weight-for-length z-score

INTRODUCTION

Proteins are essential macromolecules that play a pivotal role in human growth and serve as functional components in many physiological and pathological pathways in the human body. Although

DOI: 10.1201/b23328-31

only twenty amino acids form the basic elements of polymeric structures, these amino acids can form up to 20,000 primary proteins that differ in shape and are involved in a wide range of functions (Millward 2020). Among the numerous biological processes occurring in the human body, growth is among the most important and complex, requiring several proteins to act both as the main building blocks and functional molecules that drive this process. Therefore, dietary protein that provides these amino acids is a key component of the human diet that cannot be replaced by other macronutrients, including carbohydrates and fat (Wu 2016).

The need for dietary protein is especially important during infancy when growth velocity is high, and the period up to age two has been highlighted as the most vulnerable for the onset of faltering growth (Victora et al. 2010). In past centuries, protein deficiency was recognized as one of the main contributors to pediatric undernutrition, and protein consumption was promoted as a major part of public health programs, especially in low- and middle-income countries (LMICs), before losing its prominence to micronutrients in the mid-1970s (Semba 2016). It was not until 1995, when a landmark study from a French research group reported a possible negative impact of high protein intake in early life on increased body fatness in childhood (Rolland-Cachera et al. 1995), that the scientific community started to recognize a potential downside of high protein consumption in this context.

Protein quality, not just quantity, deserves attention. Despite providing the same amount of protein, plant-based proteins (PBP) may not provide sufficient essential amino acids (EAAs) due to their poor digestibility and lower biological value, which reflects the match between absorbed amino acids and cellular demands, compared to animal-based proteins (ABP) (Milward 2020). Therefore, dietary protein cannot be regarded as a single, homogeneous nutrient, and it must be recognized that dietary proteins with different structures and amino acid compositions are not always interconvertible (Parikh et al. 2021). In LMICs, while pediatric populations may consume adequate amounts of protein (Semba 2016), only one in five infants and young children regularly consumes animal-source food (ASF) in their diet (Aguayo 2017). In contrast, infants and young children aged 0–24 months in high-income countries (HICs) receive nearly 2–3 times the daily serving size of ASFs compared to the average intake worldwide, although the gap is narrower in older children (Miller et al. 2023). A study conducted in five European countries reported that the main sources of dietary protein during complementary feeding (CF) were ASFs such as infant/follow-on formula, milk/dairy products, and meats/meat products. The proportions of ABP were typically between 70–80%, of which dairy, including formula and other dairy products, was highest at 6 months old (>80%) and steadily dropped to 40% at 24 months old (Damianidi et al. 2016).

Dietary protein seems to play a paradoxical role in optimizing growth during early life. While a low intake of high-quality protein (e.g., milk and meats) negatively affects growth, particularly linear growth, among children in low-income countries (Millward 2017), lower consumption of ABP may reduce the risk of childhood obesity in HICs (Arnesen et al. 2022) and vice versa. Therefore, recommendations for dietary protein intake in early life cannot be considered from only one perspective without considering protein sources and other specific conditions such as population characteristics, infection rates, and feeding practices (Figure 31.1). Notably, as many countries face nutritional transition and the double burden of malnutrition (DBM), where both undernutrition and overnutrition are prevalent, the situation is even more complicated. Further evidence is needed to better understand how protein quantity and quality in early life influence growth and long-term outcomes, and to investigate underlying mechanisms.

This chapter provides a comprehensive review addressing these issues. It begins with scientific evidence from different contexts (i.e., considering population characteristics and feeding practices) focusing on distinct effects on growth of each protein source, and then proposes possible mechanisms that could explain the link between dietary protein in early life and growth outcomes.

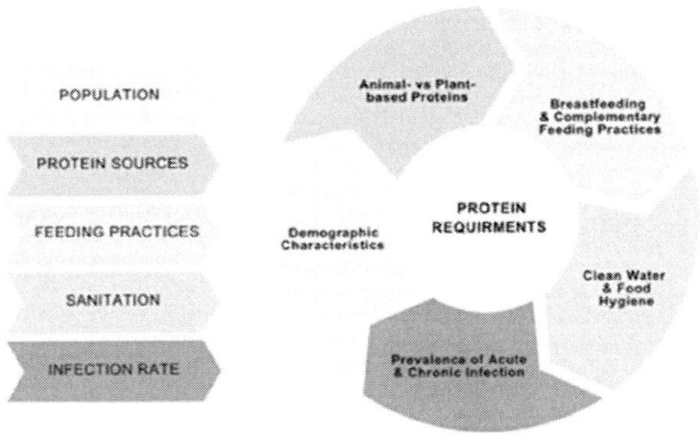

FIGURE 31.1 Factors influencing protein requirement in early life. This figure shows factors that should be considered when recommending protein requirements in early life. Differences in population characteristics, availability and affordability of protein sources, feeding practices, sanitation, and rates of infection/inflammation may result in specific requirements for individual nations.

EFFECTS OF DIETARY PROTEIN ON GROWTH OUTCOMES IN DIFFERENT POPULATIONS

Consistent with current concepts of protein intake and growth outcomes, most clinical studies fall into two categories, focusing either on 1) protein deficiency and undernutrition or 2) high protein consumption in early life and childhood overweight/obesity. While researchers have extensively studied the first in resource-restricted populations, the second is more common in HICs. However, extremely limited evidence incorporates both perspectives, and we still do not know which is the best strategy for countries facing the DBM where all forms of malnutrition co-exist. The following sections provide evidence from different contexts.

Low- and Middle-Income Countries

Evidence from Global Studies

Regarding protein quantity, global evidence suggests that populations living in a majority of the 180 countries studied consumed adequate amounts of total protein without considering its quality. Using this crude protein intake, the estimated risk of protein inadequacy ranged from zero to less than 20%. However, when protein quality was considered, inadequacy could reach more than 30% among populations in Sub-Saharan Africa. More importantly, the risk of stunting in children was associated with protein intake, but only when quality was considered. The authors suggested that protein quality was necessary for assessing protein adequacy and was clinically relevant to the linear growth of these populations (Ghosh et al. 2012).

Similarly, a recent study reported a negative association between ASF consumption and the prevalence of stunting in a pooled analysis of 74,548 infants and young children aged 6–23 months living in LMICs. Those who did not receive any ASF on the previous day had more than 1.4 times higher odds of being stunted compared with their peers who consumed all three types of ASF (i.e., eggs, meats, and dairy) (Krasevec et al. 2017). Another analysis of nearly 200,000 weaned children demonstrated large differences in reported consumption of ABP, including infant formula, yogurt, meats, and other dairy products (61.6% for other dairy products to 311.9% for infant formula) between the lowest and highest household wealth deciles, while only small variations (7.4–23.8%) were found for PBP (i.e., grains, roots, tubers, legumes, and nuts) (Heemann et al. 2022). Although the global evidence shows a positive impact of ABP on linear growth by reducing stunting, a causal relationship cannot be assumed unless these results are confirmed by randomized controlled trials (RCTs).

Evidence from Randomized Controlled Trials

A recent meta-analysis of 15 RCTs mostly conducted in LMICs indicated that food-based interventions in children under 5 years of age can promote linear growth compared to control groups (a mean difference of LAZ: 0.20; 95% CI: 0.04, 0.35), but the effects on WAZ and weight-for-length z-score (WLZ) were not significantly different. The authors suggested that ABP has a greater effect on linear growth than PBP, as evidenced by larger effect sizes in interventions providing at least one type of ABP (e.g., milk, whey, meats, eggs) compared to interventions that included only PBP (Mamun et al. 2023).

Two RCTs were conducted concurrently on Malawian infants and young children to investigate the effects of two PBPs, cowpea and common pea, on growth outcomes compared with a control group receiving an isocaloric corn-soy blend. The study on infants receiving additional protein from either cowpea or common pea from 6–12 months showed a preventive effect on linear growth, demonstrated by the least reduction of LAZ at 9 months, but only in the cowpea group. However, this effect disappeared at 12 months, and none of the intervention foods improved WLZ in these infants (Stephenson et al. 2017). Interestingly, in the RCT conducted on young children, neither cowpea nor common pea supplementation had a significant impact on growth compared with the control diet (Agapova et al. 2018).

A recent meta-analysis of RCTs using ABP in the form of infant/follow-on formula or food-based interventions for infants and young children in LMICs reported a positive effect of the formula trials on both weight and length/height gain, while pooled analyses of the food-based interventions showed a smaller effect on LAZ/height-for-age z-score (HAZ) (standard mean difference (SMD) +0.059) and WLZ (SMD +0.060). Of note, the food-based interventions used several types of ABP, but most primarily contained milk or other dairy products (Pimpin et al. 2019).

Additionally, a study on Bangladeshi infants aged 6 months examined the effect of four food-based interventions, one including milk protein and peanut while the others contained different PBPs (rice-lentil, chickpea, or wheat-soy blend), on growth outcomes and prevalence of undernutrition. The results showed greater effects of the intervention containing milk protein in preventing wasting, underweight, and stunting, as well as reducing the fall in LAZ and WLZ at 18 months. Only the intervention using chickpeas resulted in a non-inferior result compared to the ABP intervention, while the rice-lentil and wheat-soy blend groups were less effective in promoting linear growth and preventing stunting (Christian et al. 2015).

Overall, evidence shows that ABP has more favorable effects than PBP in promoting both linear and ponderal growth of children living in resource-limited countries. Milk protein is the most effective source for improving nutritional status in these populations, while non-dairy ABPs are also effective, especially when consumed regularly, as shown in the RCT using food-based interventions. However, without fortification, milk protein cannot be the sole ABP given to infants and young children during CF due to its low content of iron and zinc. Thus, further research is needed to determine appropriate recommendations for dairy and non-dairy ABP intake to optimize somatic growth and fulfill the requirements for all essential micronutrients.

High-Income Countries

The concept that high protein consumption in early life may contribute to later obesity has been widely documented for at least two decades. In 2005, Koletzko and colleagues proposed the "early protein hypothesis," suggesting that protein intakes exceeding metabolic demands promote weight gain in infancy and increase obesity risk later (Koletzko et al. 2005). Since then, more evidence has accumulated, and a recent meta-analysis suggested that high protein intake in early childhood

(<18 months old), especially ABP, is probably causally linked to higher childhood BMI up to 18 years old. Interestingly, this meta-analysis shows consistent findings across prospective cohorts conducted in several HICs (Arnesen et al. 2022).

Evidence from Observational and Longitudinal Studies

Researchers have observed that formula-fed infants are heavier and longer than breastfed infants starting from 3 months of age. While breastfed infants show a gradual decline in growth rate toward the end of the first year, formula-fed infants progressively gain weight throughout this period. This observation has led to several assumptions about influential factors, but the difference in protein content between breast milk and formula is one of the most convincing hypotheses (Heinig et al. 1993; Agostoni et al. 1999). Unlike infant formula, which has constant and high protein content, protein in breast milk gradually decreases over the lactation period (Kouwenhoven et al. 2022).

The most recent meta-analysis, including more than 300,000 participants, indicates that breastfeeding is associated with a reduced risk of obesity in children aged 2–6 years. More importantly, there is a dose-response relationship between breastfeeding duration and the risk of childhood obesity. In a pooled analysis, the risk of obesity decreased by 3.2% for each additional month of breastfeeding. Although this meta-analysis did not adjust for socioeconomic status, over 80% of the included studies were conducted in HICs (Qiao et al. 2020). Additionally, results from two birth cohorts in France indicated that increased protein intake from infant formula was associated with weight-for-age z-score (WAZ) at 12 months, and infants who consumed formula containing more than 2.8 g/100 kcal of protein had a higher body mass index z-score (BMIZ) than infants receiving lower protein formula (Camier et al. 2021).

During the CF period, infants need more energy and nutrients to fill nutritional gaps and establish their eating skills. At this time, protein intake rapidly changes from roughly 5–15% of total energy consumed, especially in breastfed infants (Michaelsen et al. 2017). The diversity of dietary protein sources during this time greatly increases beyond just breast milk/formula. Several longitudinal cohorts have suggested that ABP consumption in the first few years of life was positively associated with body weight and BMI in later age groups, while none found a significant relationship with PBP (Rolland-Cachera et al. 1995; Thorisdottir et al. 2014; Braun et al. 2016; Voortman et al. 2016). Milk protein is more likely to correlate with rapid weight gain and higher BMI/BMIZ than meat (Hoppe et al. 2004; Hopkins et al. 2015).

Evidence from Randomized Controlled Trials

To test the "early protein hypothesis," a large, multi-center, double-blind RCT including 1,138 formula-fed infants was conducted in five European countries starting in 2002. The study randomly assigned formula-fed infants to two groups receiving either high- or low-protein formula during the first year of life, while breastfed infants were recruited as a reference. This RCT found that body weight and BMI rapidly increased in the high-protein group, while infants fed the low-protein formula were significantly thinner and more closely aligned with the breastfed group at 2 and 6 years of age. Researchers found no significant difference in linear growth between groups (Koletzoko et al. 2009; Weber et al. 2014). At 6 years of age, the group receiving the high-protein formula had 2.43 times higher odds of being obese compared to the low-protein group (Weber et al. 2014). Although most recent meta-analyses including only RCTs have found some supporting evidence for the causal effects of high-protein formulas on greater weight-related parameters compared to low-protein formulas, the pooled analyses unexpectedly showed inconsistent results in growth outcomes (Patro-Golab et al. 2016; Pimpin et al. 2019). Researchers have assumed that variations in the timing of intervention, age of enrollment, protein content, and amino acid profiles of the interventional formulas, as well as the use of different growth parameters and references, possibly lead to inconclusive results (Kouwenhoven et al. 2022).

Regarding RCTs conducted during the CF period, one in a US population showed that breastfed infants who were randomly assigned to receive "pureed meat" as a complementary food at 5 months old showed greater changes in both WAZ and length-for-age z-score (LAZ) at 9 months old (ΔWAZ: +0.24; ΔLAZ: +0.14) compared to those who given isocaloric "micronutrient-fortified cereal" (ΔWAZ: −0.07; ΔLAZ: −0.27). However, infants not only received different types of protein but also consumed different amounts of protein (17% vs. 9% of daily energy intake in the meat and cereal groups, respectively). Although the authors concluded that higher protein intake from meats was associated with higher length and weight gain without excessive gain in adiposity, it is difficult to determine whether these results were attributable to meat or higher protein consumption in the meat group (Tang and Krebs 2014).

Comparing two common ABPs, "meat-based" and "dairy-based" protein, an RCT reported that infants randomly assigned to eat "pureed meat" had higher LAZ compared to those who consumed "dairy-based food" providing similar energy and protein content. The mean changes in LAZ from 5 to 12 months old were +0.33 and −0.30 standard deviation scores (SDS) in the meat and dairy groups, respectively. Conversely, infants in the dairy group showed more rapid weight gain with a substantial increase in their WLZ over time (mean ΔWLZ from 5–12 months: +0.76 vs. +0.30 in the dairy and meat groups, respectively). Of note, all participants in this study were formula-fed infants, and the energy from protein compared to total energy intake (% PE) was 9 ± 1% and 15 ± 4% at baseline and 12 months (Tang, Hendricks, and Krebs 2018). A year after the intervention ended, the average LAZ of participants in the meat group remained significantly higher than in the dairy group, but the mean WLZ of both groups no longer differed either at 24 months or in change over time (Tang et al. 2019).

Recently, researchers conducted a double-blind RCT providing baby foods and nutritional advice to caregivers of infants aged about 4.5 months to follow either a conventional Swedish diet or a protein-reduced Nordic diet enriched with plant foods, fish, and rapeseed oil, but with less added sugar, saturated fat, meat, and meat products. The authors reported similar growth outcomes and iron status at ages 12 and 18 months in both groups, although the Nordic group consumed significantly less protein compared with the control group at both time points. Energy and fat intake did not differ. Interestingly, infants in the Nordic group consumed significantly more fruits and vegetables daily than the control group. However, it should be noted that most infants were breastfed at baseline in both groups, and the study did not report the amount of each protein source (Johansson et al. 2023).

Thus, in HICs, scientific evidence suggests distinctive effects of different protein sources. Regarding weight-related anthropometry, strong evidence indicates a causal relationship between high intake of milk protein, either from formulas or dairy products, and rapid weight gain during infancy and early childhood, potentially increasing the risk of obesity in older children. Non-dairy ABP can promote ponderal growth to some extent, but the impact is less pronounced compared to milk protein. Nevertheless, there is inconclusive evidence for linear growth. Studies report inconsistent effects of ABP, both milk and non-dairy, on linear growth. Notably, PBP seems to play a limited role in growth promotion, which is concordant with studies in LMICs. However, higher PBP consumption in high-income populations seems to result in more favorable growth outcomes and may not lead to undernutrition as reported in resource-limited countries.

Populations Facing the Double Burden of Malnutrition

It has been reported that more than one-third of the population in LMICs is affected by the DBM (Seferidi et al. 2022). Early-life nutrition is widely recognized as one of the main drivers of this public health problem. Nevertheless, when looking for scientific evidence supporting the role of dietary protein in early life and its preventive effect on the DBM, most studies focus on the promotion of ABP, especially milk-based foods, to overcome undernutrition and micronutrient deficiency, while little evidence evaluates high protein intake and childhood obesity in these populations.

The DBM is increasing most rapidly in middle-income countries (Tzioumis et al. 2016). Evidence from a multi-center cohort in Thailand reported that most infants consumed higher dietary protein than the WHO recommendations during the CF period, and nearly 80% of this was ABP. More importantly, milk protein, mostly from formula, showed a greater effect on weight-related parameters than non-dairy ABP, whereas PBP had no impact on any growth parameters (Kittisakmontri et al. 2022). A study in Vietnam, where the DBM is prevalent, also reported an increased risk of being overweight/obese if young children consumed more powdered milk (Nguyen et al. 2021).

Interestingly, a study in Indonesia where more than one-third of households had the DBM, found that a dietary pattern called "nutrition transition," comprising high-fat and animal products/processed foods, was significantly associated with higher protein and energy consumption compared with other dietary patterns, while the "plant-based diet" was associated with stunting. According to the multi-level analysis, the study showed a negative relationship between age and the score of the nutrition transition pattern, implying that younger generations tended to consume a higher amount of high-fat and animal products than older age groups (Lowe et al. 2021).

Unlike countries where only one form of malnutrition is dominant, the DBM causes a more complicated situation and needs to be adequately addressed. To prevent both under- and overnutrition, more research is required in this specific context to ensure proper provision of ABP that optimizes growth and nutritional status while preventing high-risk populations from becoming overweight and obese. When developing recommendations for protein intake, it is important to determine whether all protein sources should be considered equally if the scientific evidence points out their distinctive effects on growth, especially during the infancy and young child period (Figure 31.2). The effect of each protein source on growth in different populations is summarized in Table 31.1.

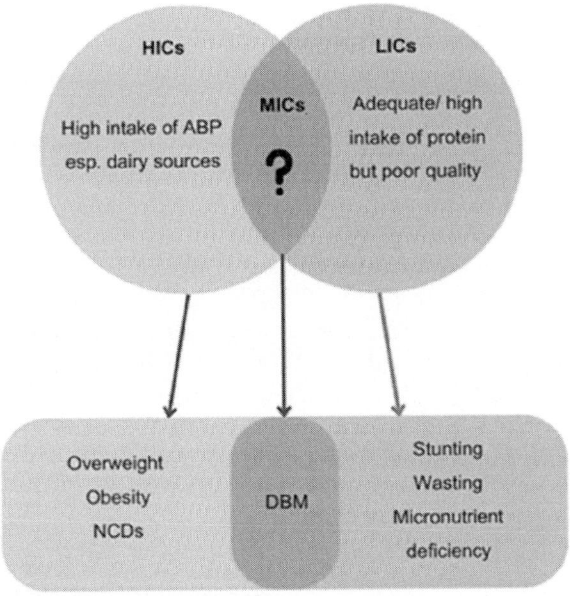

FIGURE 31.2 Paradoxical roles of dietary protein on growth in the early life of populations from different socioeconomic settings. There are two opposite paradigms considering the effect of protein intake on growth outcomes in early life. While evidence from high-income countries (HICs) focuses on concerns that high intakes of animal-based protein (ABP), especially dairy protein, may cause obesity and non-communicable diseases (NCDs) in later life, studies from low-income countries (LICs) indicate that consumption of poor-quality protein, even in adequate amounts, can lead to all forms of malnutrition. However, it is still unknown which paradigm plays a central role in populations facing the double burden of malnutrition (DBM), the majority of which are middle-income countries (MICs).

TABLE 31.1
Summary Effect of Each Protein Source on Growth Outcomes in Various Populations Classified by Socioeconomic Status

Protein Sources	Populations		
	High-Income	**Facing DBM**	**Low-Income**
Dairy protein	High intake increases the risk of childhood obesity	High intake might increase the risk of childhood obesity	Adequate intake may improve growth
Non-dairy ABP	Less effect on growth outcomes compared to dairy protein	Insufficient evidence	Adequate intake may improve growth
PBP	No significant effect on growth outcomes	Insufficient evidence	Inconsistent findings (data favor legumes over cereals)

DBM, double burden of malnutrition; ABP, animal-based protein; PBP, plant-based protein.

EFFECTS OF DIFFERENT PROTEIN SOURCES ON BODY COMPOSITION

Beyond anthropometry and nutritional status, body composition is also an important growth indicator. For infants and young children, body composition is more sensitive than body weight in differentiating whether each protein source promotes the production of fat-free mass (FFM), including structural tissues and bones, or an accumulation of fat mass (FM), which can increase the risk of overweight and obesity in later life. To support an optimal growth trajectory, protein intake should contribute to a balanced body composition.

A recent systematic review concluded that limited but suggestive evidence shows a positive effect of total protein intake in early life on body fatness, while there was a null effect for PBP. Regarding ABP, the study found inconsistent findings and could not support an association between ABP and body composition (Arnesen et al. 2022). However, the large, multi-center RCT in five European countries mentioned previously clearly showed that infants consuming high-protein formula during infancy accumulated more FM at 6 years of age than breastfed infants or those receiving low-protein formula (Totzauer et al. 2018). More significantly, the accumulation of FM was predominantly in the "pre-peritoneal area," rather than subcutaneous, which may result in adverse metabolic or later health outcomes (Gruszfeld et al. 2016). Following infancy, a double-blind RCT providing growing-up milk with high or low protein also found a significant effect of high milk protein on increased percentage of body fatness, although there was no difference in percentage FFM between these two groups (Wall et al. 2019). Although many studies compare infants consuming intervention formulas with breastfed infants, a systematic review reported an inconclusive effect of protein from breast milk on infant body composition (Norrish et al. 2023).

In contrast, a cluster-RCT in rural Bangladesh reported significant increases in both FFM and FM, with FFM increments being twice that of FM in infants randomly assigned to milk- and whey-based food from 6 to 18 months of age. Although the chickpea-based food group also showed higher FFM and FM than the control group, its effect was slightly less than the intervention food containing milk and whey protein. Additionally, post-intervention analyses only supported the effect of milk- and whey-based food on body composition, while the effects of other interventions were absent. Notably, this RCT reported similar effects of the wheat- and soy-blended food on body composition compared to the control group (Shaikh et al. 2020). However, an RCT from Uganda reported that all forms of lipid-based supplements, regardless of dairy protein, increased both FM and FFM compared to unsupplemented young children aged under 59 months (Mbabazi et al. 2023).

Regarding non-dairy ABP, a systematic review focusing on interventions during the CF period found that consuming different amounts of meat/meat products did not contribute to any favorable/

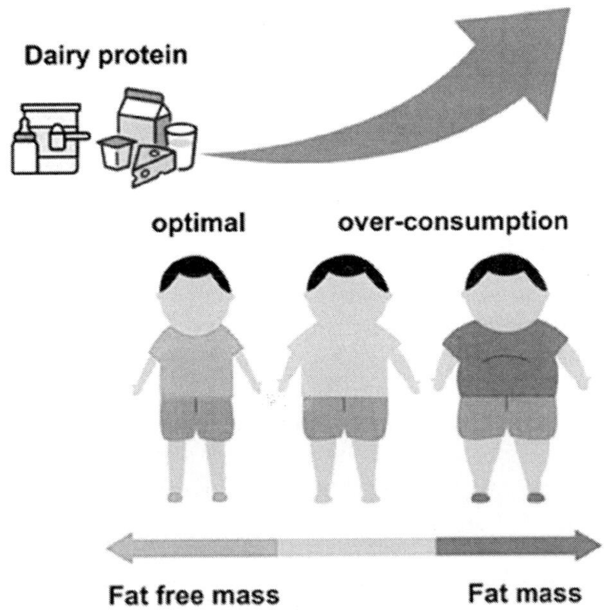

FIGURE 31.3 Effect of dairy protein on body composition. Dairy protein is an essential protein source promoting fat-free mass if consumption meets the physiological needs of infants and young children. However, overconsumption can cause unnecessary accumulation of fat mass, which can result in later obesity.

unfavorable body composition in infants and young children. Notably, none of the included studies used measures of body composition that could differentiate FM/FFM (English et al. 2019). In studies measuring body composition using proper techniques, a birth cohort in Australia found that the intake of non-dairy ABP, including meat, fish, and eggs at 12 months of age, was associated with an increasing FFM index z-score, but not FM index z-score at 2–3 years of age (Smith-Brown et al. 2018). However, an RCT in Swedish infants found no differences in the body composition of infants randomly assigned to the intervention group consuming lower protein from meat while receiving more fruits and vegetables during the CF period compared to the control group (Johansson et al. 2023). Of note, both groups received less than 15% PE, which is considered the upper limit for infants and young children according to the Nordic recommendations (Blomhoff et al. 2023). There is still a lack of evidence to determine the effect of non-dairy ABP on body composition in resource-restricted populations.

In summary, further studies focusing on the impact of each protein source on infant and young child body composition are needed. According to current evidence, dairy protein seems to promote both FM and FFM accumulation. However, if dairy protein consumption exceeds the physiological requirement, it may predominantly enhance body fatness, especially in the pre-peritoneal area, which may consequently result in poorer metabolic health later in life (Figure 31.3). Regarding non-dairy ABP and PBP, available studies do not allow conclusions to be drawn about their effect on body composition.

PROPOSED MECHANISMS UNDERPINNING THE RELATIONSHIP BETWEEN DIETARY PROTEIN AND GROWTH

Determining the underlying mechanisms that explain the link between protein intake and growth outcomes is essential. For decades, researchers have focused on metabolic and hormonal pathways, particularly the growth hormone (GH)–insulin-like growth factor-1 (IGF-1) axis. Recently, more novel mechanisms have been explored as potential factors in this relationship, including the mammalian target of rapamycin (mTOR) signaling pathways and gut microbiota.

Plasma Amino Acids and the GH–IGF-1 Axis

Different protein sources have distinct amino acid patterns. While ABP usually provides all EAAs, most PBP typically lack at least one EAA. Therefore, many researchers hypothesize that the differences in amino acid patterns may stimulate human metabolism and growth-related hormones in distinctive ways.

In LICs, a prospective cohort in rural Malawi reported that stunted children aged 12–59 months had significantly lower serum concentrations of all EAAs, some conditionally EAAs, and nonessential amino acids. The authors suggested that the low levels of serum EAAs could be attributed to the consumption of poor-quality protein, which provides insufficient EAAs to promote linear growth (Semba, Shardell, et al. 2016). Supporting this suggestion, a study in Indonesia showed that stunted young children consumed fewer EAAs from their diets than non-stunted peers, although total dietary protein intakes did not differ between the groups (Maulidiana and Sutjiati 2021). RCTs in Ghana and rural Malawi confirmed this hypothesis, reporting higher plasma EAAs in infants randomly assigned to receive legume-based complementary foods compared to a control group (Furuta et al. 2021; Ordiz et al. 2019).

In contrast, evidence from HICs indicates that high protein intake from infant formula may increase plasma EAA and BCAA levels, which in turn promote weight gain. Data from observational studies and RCTs indicate a positive effect of high-protein formula on plasma BCAAs (Lonnerdal and Chen 1990; Karlsland et al. 1998; Kirchberg et al. 2015). Kirchberg et al. found that BCAAs, including leucine, isoleucine, and valine, were the most discriminant metabolites between infants receiving high- and low-protein formula in metabolomic analyses. However, regarding non-dairy ABP, the RCT in the US did not find any differences in plasma amino acids between formula-fed infants who consumed meat-based complementary foods and those assigned to the dairy-based intervention. However, the provision of protein-rich complementary foods may increase some EAAs over time (Tang et al. 2021). For PBP, the study did not find a significant change in plasma amino acids in formula-fed infants supplemented with cereal-based complementary foods (Lonnerdal and Chen 1990).

In vitro and in vivo studies demonstrate the effect of the GH–IGF-1 axis on growth. Amino acids, especially BCAAs, are among the most potent stimuli promoting this axis (Millward 2017). Evidence from a large RCT suggests that the intake of high-protein formula intake may overstimulate the secretion of plasma IGF-1, which can result in rapid weight gain and accumulation of body fat. The authors also showed a significant increase in urine C-peptides, the main metabolites of insulin, in those infants (Socha et al. 2011). Likewise, a prospective cohort in Thailand found a positive correlation between dairy protein and serum IGF-1, IGFBP-3, and insulin, which could explain a link between high dairy protein and weight-related z-scores. Notably, there was no association between non-dairy ABP or PBP and growth-related hormones in this study (Kittisakmontri et al. 2022).

In summary, dietary protein of better quality seems to promote optimal growth in infants and young children generally by increasing concentrations of EAAs, particularly BCAAs, which stimulate the GH–IGF-1 axis. However, if dietary protein, especially dairy protein, consumption exceeds physiological requirements, these high levels of BCAAs may contribute to overstimulation of the axis as well as induce secretion of insulin that eventually promotes rapid weight gain and accumulation of body fat (Figure 31.4).

Other Novel Mechanisms

Over the past decade, evidence has accumulated on the role of novel mechanisms, including the mTOR signaling pathways and gut microbiome, in early-life growth. It is possible that dietary protein promotes growth through these novel mechanisms.

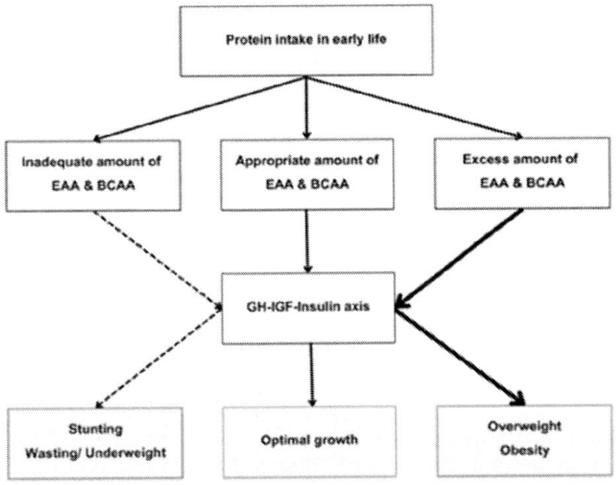

FIGURE 31.4 Impacts of amino acid availability on growth outcomes via the GH–IGF-insulin axis. Dietary protein provides essential amino acids (EAAs) and branched-chain amino acids (BCAAs) that activate the growth hormone (GH)-insulin-like growth factor (IGF) and insulin axis. If EAA and BCAA intake is sufficient, the activation of this axis will promote optimal growth of infants and young children. However, either insufficient or excessive EAAs and BCAAs can cause under- (dashed arrow) or overstimulation (thick arrow) of this axis, respectively. The results of these inappropriate stimulations can contribute to undernutrition or obesity.

mTOR is a conserved serine/threonine protein kinase that plays central roles in regulating metabolism, growth, cellular proliferation, and aging. It is sensitive to growth factors, energy, and nutrients, especially amino acids. In the absence of amino acids, mTOR may not be activated by other signals such as growth factors or energy. Therefore, during periods of rapid growth, researchers assume mTOR is a master regulator as several growth processes, such as skeletal muscle synthesis, bone elongation, myelination of the nervous system, gut homeostasis, and organ size, are stimulated by mTOR (Semba, Trehan, et al. 2016). Although clinical studies demonstrating a relationship between dietary protein consumption and mTOR signaling pathways in infants and young children are lacking, strong evidence from many animal studies indicates that the understimulation of mTOR by amino acid deficiency can impair growth and even cause severe growth restriction in those models (Semba, Trehan, et al. 2016). In humans, supplementation with either whey protein or BCAAs can increase mTOR activity in skeletal muscle in adult women compared to a control group (Smith et al. 2015). Notably, these findings may explain the links between too little/too much protein consumption and DBM.

Emerging evidence links gut microbiome colonization in early life to growth and long-term health outcomes. However, a study in Malawi reported that supplementation of breastfed infants with PBP, cowpea, or common bean did not affect the gut microbiota pattern at age 9 months compared to the control group, even though the cowpea group showed improved length (Ordiz et al. 2020). In contrast, a small RCT with US infants showed that breastfed infants who randomly received high-protein (17% PE), meat-based complementary food had a significantly increased abundance of butyrate-producing bacteria, Clostridia group XIVa, compared to the low-protein (9% PE), iron-fortified cereal group (Krebs et al. 2013). However, this effect may be attributed to other differences, such as high fiber and iron content in the cereal group.

In conclusion, increasing evidence supports novel mechanisms that could explain the link between dietary protein intake and growth in early life (Figure 31.5). However, further research is

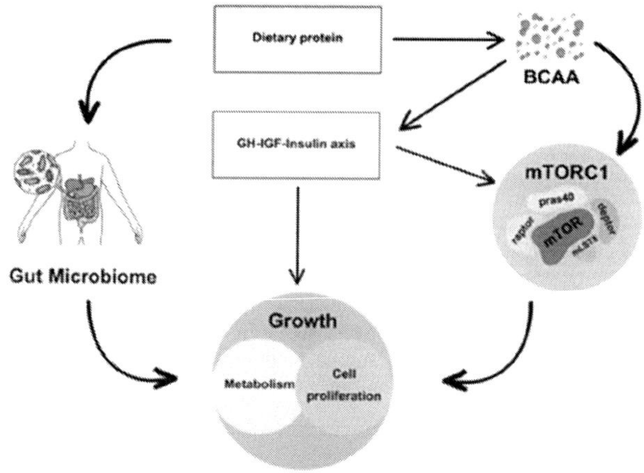

FIGURE 31.5 Novel mechanisms related to the relationship between dietary protein and growth outcomes. There are two mechanisms that could explain a link between dietary protein and growth: 1) The central conductor of growth is the mammalian target of rapamycin complex-1 (mTORC1), which can be activated directly by BCAAs or indirectly through inducing secretion of growth-promoting factors from the GH–IGF-insulin axis; 2) Dietary protein may shape the gut microbiome, which can influence metabolism and growth at an early age.

needed, especially clinical studies. Understanding how dietary protein affects human growth will create more opportunities to prevent both undernutrition and overnutrition, as well as ameliorate many health burdens in later life. To optimize growth, researchers should consider the effects of dietary protein from all perspectives. They should account for baseline characteristics of populations and relevant contexts, such as type of milk fed and food availability, before making recommendations on dietary protein for infants and young children. Moreover, separate recommendations may be necessary for different sources of dietary protein, as they show distinct effects on growth outcomes.

MINI DICTIONARY OF TERMS

Body composition: A term describing the basic components of the human body, typically divided into fat mass and fat-free mass.

Branched-chain amino acids: Three essential amino acids – leucine, isoleucine, and valine – that have branched chains in their structure.

Complementary feeding: The period when infants transition from exclusive milk feeding to consuming other foods in addition to milk.

Essential amino acids: Amino acids required for bodily functions that cannot be produced by the human body. These amino acids are obtained through dietary sources or supplementation.

Fat-free mass: The portion of body composition comprising lean muscle, bones, and other tissues that are free from fat.

SUMMARY POINTS

- Different sources of dietary protein have distinct effects on the growth and body composition of infants and young children.
- Suboptimal intake of animal-based protein (ABP) can lead to all forms of undernutrition, especially stunting.

- Overconsumption of ABP, especially dairy protein, can result in rapid weight gain, accumulation of fat mass, and an increased risk of overweight/obesity in later life.
- Although plant-based protein shows small and inconsistent effects on growth in most clinical studies, some legumes may prevent growth impairment in resource-limited populations.
- Both insufficient and excessive supply of amino acids, especially branched-chain amino acids from different amounts and sources of dietary protein, may result in unfavorable growth through many mechanisms. These include the growth hormone-insulin-like growth factor-1 axis, the mammalian target of rapamycin signaling pathways, and changes in gut microbiota.
- A "one size fits all" recommendation for protein intake may not be appropriate for infants and young children living in different circumstances. Recommendations should take into account the setting, nutritional risks, available and acceptable foods, and different protein sources.

REFERENCES

Agapova, S.E., Stephenson, K.B., Divala, O., et al. 2018. Additional common bean in the diet of Malawian children does not affect linear growth, but reduces intestinal permeability. *J Nutr* 148:267–74.

Agostoni, C., Grandi, F., Gianni, M.L., et al. 1999. Growth patterns of breast fed and formula fed infants in the first 12 months of life: an Italian study. *Arch Dis Child* 81:395–99.

Aguayo, V.M. 2017. Complementary feeding practices for infants and young children in South Asia: a review of evidence for action post–2015. *Matern Child Nutr* 13(suppl2):e12439.

Arnesen, E.K., Thorisdottir, B., Lamberg-Allardt, C., et al. 2022. Protein intake in children and growth and risk of overweight or obesity: a systematic review and meta-analysis. *Food Nutr Res* 21:66.

Blomhoff, R., Andersen, E.K., Arnesen, E.K., et al. 2023. *Nordic nutrition recommendations 2023.* Copenhagen: Nordic Council of Ministers.

Braun, K.V., Erler, N.S., Kiefte-de Jong, J.C., et al. 2016. Dietary intake of protein in early childhood is associated with growth trajectories between 1 and 9 years of age. *J Nutr* 146:2361–7.

Camier, A., Davisse-Paturet, C., Scherdel, P., et al. 2021. Early growth according to protein content of infant formula: results from the EDEN and ELFE birth cohorts. *Pediatr Obes* 16:e12803.

Christian, P., Shaikh, S., Shamim, A.A., et al. 2015. Effect of fortified complementary food supplementation on child growth in rural Bangladesh: a cluster-randomized trial. *Int J Epidemiol* 44:1862–76.

Damianidi, L., Gruszfeld, D., Verduci, F., et al. 2016. Protein intakes and their nutritional sources during the first 2 years of life: secondary data evaluation from the European childhood obesity project. *Eur J Clin Nutr* 70:1291–7.

English, L.K., Obbagy, J.E., Wong, Y.P., et al. 2019. Types and amounts of complementary foods and beverages consumed and growth, size, and body composition: a systematic review. *Am J Clin Nutr* 109: 956S–77S.

Furuta, C., Sato, W., Murakami, H., et al. 2021. Changes of plasma amino acid profiles in infants with a nutrient-fortified complementary food supplement: evidence from a 12-month single-blind cluster-randomized controlled trial. *Front Nutr* 8:606002.

Ghosh, S., Suri, D., Uauy R. 2012. Assessment of protein adequacy in developing countries: quality matters. *Br J Nutr* 108: S77–87.

Gruszfeld, D., Weber, M., Gradowska, K., et al. 2016. Association of early protein intake and pre-peritoneal fat at five years of age: follow-up of a randomized clinical trial. *Nutr Metab Cardiovasc Dis* 26:824–32.

Heemann, M., Kim, R., Sharma, S., et al. 2022. Food groups consumption patterns among children meeting and not meeting WHO's recommended dietary diversity: Evidence from 197,514 children in 59 countries. *Food Policy* 112:102368

Heinig, M.J., Nommsen L.A., Peerson, J.M., et al. 1993. Energy and protein intakes of breast-fed and formula-fed infants during the first year of life and their association with growth velocity: the DARLING study. *Am J Clin Nutr* 58: 152–61.

Hopkins, D., Steer, C.D., Northstone, K., et al. 2015. Effects on childhood body habitus of feeding large volumes of cow or formula milk compared with breastfeeding in the latter part of infancy. *Am J Clin Nutr* 102:1096–103.

Hoppe, C., Udam, T.R., Lauritzen, L., et al. 2004. Animal protein intake, serum insulin0like growth factor I, and growth in healthy 2.5-y-old Danish children. *Am J Clin Nutr* 80: 447–52.

Johansson, U., Ohlund, I., Lindberg, L., et al. 2023. A randomized, controlled trial of a Nordic, protein-reduced complementary diet in infants: effects on body composition, growth, biomarkers, and dietary intake at 12 and 18 months. *Am J Clin Nutr* 117:1219–31.

Karlsland Akeson, P.M., Axelsson, I.E., Raiha, N.C., et al. 1998. Protein and amino acid metabolism in three to twelve-month-old infants fed human milk or formulas with varying protein concentrations. *J Peditr Gastroenterol Nutr* 26:297–304.

Kirchberg, F.F., Harder, U., Weber, M., et al. 2015. Dietary protein intake affects amino acids and acylcarnitine metabolism in infants aged 6 months. *J Clin Endocrinol Metab* 100:149/58.

Kittisakmontri, K., Lanigan, J., Wells, J.C.K., et al. 2022. Quantity and source of protein during complementary feeding and infant growth: evidence from a population facing double burden of malnutrition. *Nutrients* 14:3948.

Koletzko, B., Broekaert, I., Demmelmair, H., et al. 2005. Protein intake in the first year of life: a risk factor for later obesity? The E.U. childhood obesity project. *Adv Exp Med Biol* 569: 69–79.

Koletzko, B., von Kries, R., Closa, R., et al. 2009. Lower protein in infant formula is associated with lower weight up to age 2 y: a randomized clinical trial. *Am J Clin Nutr* 86:1836–45.

Kouwenhoven, S.M.P., Muts, J., Finken, M.J.J., et al. 2022. Low-protein infant formula and obesity risk. *Nutrients* 14:2728.

Krasevec, J., An, X., Kumapley, R., et al. 2017. Diet quality and risk of stunting among infants and young children in low- and middle-income countries. *Matern Child Nutr* 13:e12430.

Krebs, N.F., Sherlock, L.G., Westcott, J., et al. 2013. Effects of complementary feeding regimens on iron status and enteric microbiota in breastfed infants. *J Pediatr* 163:416–23.

Lonnerdal, B. and Chen, C.L. 1990. Effects of formula protein level and ratio on infant growth, plasma amino acids and serum trace elements I. cow's milk formula. *Acta Paediatr Scand* 79:257–65.

Lowe, C., Kelly, M., Sarma, H., et al. 2021. The double burden of malnutrition and dietary patterns in rural central java, Indonesia. *Lancet Reg Health West Pac* 14:100205.

Mamun, A.A., Mahmudiono, T., Yudhastuti, R., et al. 2023. Effectiveness of food-based intervention to improve the linear growth of children underfive: a systematic review and meta-analysis. *Nutrients* 15: 2430.

Mbabazi, J., Pesu, H., Mutumba, R., et al. 2023. Effect of milk protein and whey permeate in large-quantity lipid-based nutrient supplement on linear growth and body composition among stunted children: a randomized 2x2 factorial trial in Uganda. *PLoS Med* 20:e1004227.

Michaelsen, K.F., Grummer-Strawn, L., Begin, F. 2017. Emerging issues in complementary feeding: global aspects. *Matern Child Nutr* 13(suppl 2):e12444

Miller, V., Webb, P., Cudhea, F., et al. 2023. Children's and adolescents' rising animal-source food intakes in 1990-2018 were impacted by age, region, parental education and urbanicity. *Nat Food* 4:305–19.

Millward, D.J. 2017. Nutrition, infection and stunting: the roles of deficiencies of individual nutrients and foods, and of inflammation, as determinants of reduced linear growth of children. *Nutr Res Rev* 30:50–72.

Millward, D.J. 2020. Nutrition and metabolism of proteins and amino acids. In *Introduction to human nutrition*, ed. S.A. Lanham-New, T.R. Hill, A.M. Gallagher and H.H. Vorster, 131–160. West Sussex: John Wiley & Sons Ltd.

Maulidiana, A.R. and Sutjiati, E. 2021. Low intake of essential amino acids and other risk factors of stunting among under-five children in Malang city, East Java, Indonesia. *J Public Health Res* 10:2161.

Nguyen, T., Sokal-Gutierrez, K., Lahiff, M., et al. 2021. Early childhood factors associated with obesity at age 8 in Vietnamese children: the young lives cohort study. *BMC Public Health* 21:301.

Norrish, I., Sindi, A., Sakalidis et al. 2023. Relationships between the Intakes of human milk components and body composition of breastfed infants: A systematic review. *Nutrients* 15(10):2370 doi: 10.3390/nu15102370.

Ordiz, M.I., Janssen, S., Humphrey, G., et al. 2020. The effect of legume supplementation on the gut microbiota in rural Malawian infants aged 6 to 12 months. *Am J Clin Nutr* 111:884–92.

Ordiz, M.I., Semba, R.D., Moaddel, R. et al. 2019. Serum amino acid concentrations in infants from Malawi are associated with linear growth. *Curr Dev Nutr* 3:nzz100.

Parikh, P. Semba, R.D., Manary, M., et al. 2021. Animal source foods, rich in essential amino acids, are important for linear growth and development of young children in low- and middle-income countries. *Matern Child Nutr* 18: e13264.

Patro-Golab B, Zalewski BM, Kouwenhoven SMP, et al. 2016. Protein concentration in milk formula, growth, and later risk of obesity: A systematic review. *J Nutr* 146(3):551–64.doi: 10.3945/jn.115.223651.

Pimpin, L., Kranz, S., Liu, E. et al. 2019. Effects of animal protein supplementation of mothers, preterm infants, and term infants on growth outcomes in childhood: a systematic review and meta-analysis of randomized trials. *Am J Clin Nutr* 110:410–29.

Qiao, J., Dai, L., Zhang, Q., et al. 2020. A meta-analysis of association between breastfeeding and early childhood obesity. *J Pediatr Nurs* 53:57–66.

Rolland-Cachera, M.F., Deheeger, M., Akrout, M., et al. 1995. Influence of macronutrients on adiposity development: a follow up study of nutrition and growth from 10 months to 8 years of age. *Int J Obes Relat Metab Disord* 19:573–8.

Seferidi, P., Hone, T., Duran, A.C., et al. 2022. Global inequalities in the double burden of malnutrition and associations with globalisation: a multilevel analysis of Demographic and Health Surveys from 55 low-income and middle-income countries, 1992–2018. *Lancet Glob Health* 10:e482–90.

Semba, R.D. 2016. The rise and fall of protein malnutrition in global health. *Ann Nutr Metab* 69:79–88.

Semba, R.D., Shardell, M., Sakr Ashour, F.A., et al. 2016. Child stunting is associated with low circulating essential amino acids. *EBiomedicine* 6:246–52.

Semba, R.D., Trehan, I., Gonzalez-Freire, M., et al. 2016. Perspective: the potential role of essential amino acids and the mechanistic target of rapamycin complex 1 (mTORC1) pathway in the pathogenesis of child stunting. *Adv Nutr* 7:853–65.

Shaikh, S., Campbell, R.K., Mehra, S., et al. 2020. Supplementation eith fortified lipid-based and blended complementary foods has variable impact on body composition among rural Bangladeshi children: a cluster-randomized controlled trial. *J Nutr* 150:1924–32.

Smith, G.I., Yoshino, J., Stromsdorfer, K.L., et al. 2015. Protein ingestion induces muscle insulin resistance independent of leucine-mediated mTOR activation. *Diabetes* 64:1555–63.

Smith-Brown, P., Morrison, M., Krause, L., et al. 2018. Growth and protein-rich food intake in infancy is associated with fat-free mass index at 2–3 years of age. *J Paediatr Child Health* 54:770–75.

Socha, P., Grote, V., Gruszfeld, D., et al. 2011. Milk protein intake, the-metabolic-endocrine response, growth in infancy: data from a randomized clinical trial. *Am J Clin Nutr* 94:1776s–84s.

Stephenson, K.B., Agapova, S.E., Divala, O., et al. 2017. Complementary feeding with cowpea reduces growth faltering in rural Malawian infants: a blind, randomized controlled clinical trial. *Am J Clin Nutr* 106:1500–7.

Tang, M., Andersen, V., Hendricks, A.E., et al. 2019. Different growth patterns persist at 24 months in formula-fed infants randomized to consume a meat- or dairy-based complementary diet from 5 to 12 months. *J Pediatr* 206:78–82.

Tang, M., Hendricks, A.E. and Krebs, N.F. 2018. A meat- or dairy-based complementary diet leads to distinct growth patterns in formula-fed infants: a randomized controlled trial. *Am J Clin Nutr* 107:734–42.

Tang, M. and Krebs, N.F. 2014. High protein intake from meat as complementary food increases growth but not adiposity in breastfed infants: a randomized trial. *Am J Clin Nutr* 100:1322–8.

Tang, M., Weaver, N.E., Berman, L.M., et al. 2021. Different blood metabolomics profiles in infants consuming a meat- or dairy-based complementary diet. *Nutrients* 13:388.

Thorisdottir, B., Gunnarsdottir, I., Palsson, G.I., et al. 2014. Animal protein intake at 12 months is associated with growth factors at the age of six. *Acta Pediatr* 103:512–7.

Totzauer, M., Luque, V., Escribano, J., et al. 2018. Effect of lower versus higher protein content in infant formula through the first year on body composition from 1 to 6 tears: follow-up of a randomized clinical trial. *Obesity* (Silver Spring) 26: 1203–10.

Tzioumis, E., Kay, M.C., Bentley, M.E., et al. 2016. Prevalence and trends in the childhood dual burden of malnutrition in low- and middle-income countries, 1990–2012. *Public Health Nutr* 19: 1375–88.

Victora, C.G., de Onis, M., Hallal, P.C. et al. 2010. Worldwide timing of growth faltering: revisiting implications for interventions. *Pediatrics* 125: e473–80.

Voortman, T., Braun, K.V., Kiefte-de Jong, J.C., et al. 2016. Protein intake in early childhood and body composition at the age of 6 years: the generation R study. *Int J Obes* (Lond) 40:1018–25.

Wall, C.R., Hill, R.J., Lovell, A.L., et al. 2019. A multicenter, double-blind, randomized, placebo-controlled trial to evaluate the effect of consuming growing up milk "lite" on body composition in children aged 12–23 mo. *Am J Clin Nutr* 109:576–85.

Weber, M., Grote, V., Closa-Monasterolo, R., et al. 2014. Lower protein content in infant formula reduces BMI and obesity risk at school age: follow-up of a randomized trial. *Am J Clin Nutr* 99:1041–51.

Wu, G. 2016. Dietary protein intake and human health. *Food Funct* 7: 1251–6.

32 Protein Requirements and Quality in Infancy and Early Childhood

Aamer Imdad, Sarah Malik, and Zulfiqar Ahmed Bhutta

ABBREVIATIONS

AAA	Aromatic amino acid
AAP	American Academy of Pediatrics
ASPEN	American Society for Parenteral and Enteral Nutrition
DIAAS	Digestibility indispensable amino acid score
DRI	Dietary reference intake
EAACI	European Academy of Allergy and Clinical Immunology
EAR	Estimated average requirement
EFSA	European Food Safety Authority
FAO	Food and Agriculture Organization
FDA	US Food and Drug Administration
His	Histidine
Ile	Isoleucine
Leu	Leucine
LRSO	Life Science Research Office
Lys	Lysine
NPN	Non-protein nitrogen
PDCAAS	Protein digestibility-corrected amino acid score
PRI	Population reference intake
RDA	Recommended dietary allowance
SAA	Sulfur amino acid
Thr	Threonine
Trp	Tryptophan
UNU	United Nations University
Val	Valine
WHO	World Health Organization

INTRODUCTION

Infancy and early childhood are the most important periods for growth and development. Dietary proteins are an essential component of the diet of infants and children, supporting rapid growth and neurodevelopment. They provide the nitrogen and amino acids needed for the synthesis of the approximately 25,000 proteins encoded in the human genome. In addition to supporting protein synthesis, dietary proteins are used to synthesize other nitrogen-containing biomolecules, such as peptide hormones, neurotransmitters, nucleic acids, glutathione, and creatine (Garlick 2006; WHO/

DOI: 10.1201/b23328-32

FAO 2007; EFSA Panel on Dietetic Products and Allergies 2012). This chapter focuses on protein requirements and protein quality in infants and early childhood.

Definitions

According to the World Health Organization (WHO), the protein requirement for infants and children is defined as the minimum protein intake necessary to achieve nitrogen equilibrium at an appropriate body composition during energy balance with moderate physical activity, plus the tissue growth consistent with good health (WHO/FAO 2007). Two key concepts are associated with the dietary requirement of proteins. The first is the average requirement (or mean requirement), which is a population-level measure representing the mean value needed to satisfy ~50% of the population, assuming a normal distribution. The second concept is the "safe level" (also called recommended dietary allowance or dietary reference intake), defined as the amount of dietary protein meeting the needs of 95% of a population (typically two standard deviations above the mean requirement, assuming a normal distribution). These terms should not be used interchangeably, as they ensure nutritional adequacy for vastly different population segments. Furthermore, in reality, protein intake within a specific population may not follow a normal distribution, meaning an average intake equal to the safe level does not guarantee protein malnutrition will be rare.

Dietary protein requirements account for metabolic demands (maintenance) and the efficiency of protein digestion and bioavailability. Metabolic demands are determined by the nature and extent of metabolic pathways that consume amino acids. These pathways are driven by the need to maintain bodily function and to satisfy special physiological states, such as growth in children, acute weight loss, pregnancy, lactation, and disease (WHO/FAO 2007). Dietary proteins are the body's source of nitrogen and indispensable amino acids. Crude protein content in a dietary source is calculated by multiplying the total nitrogen content by 6.25 (EFSA Panel on Dietetic Products and Allergies 2012). This conversion factor is based on the fact that about 95% of the nitrogen content in the body and diet is from protein, and the rest is from non-protein sources such as free amino acids, urea, or nucleotides (WHO/FAO 2007; EFSA Panel on Dietetic Products and Allergies 2012). Protein quality refers to the effectiveness with which absorbed dietary nitrogen achieves nitrogen balance (Garlick 2006; WHO/FAO 2007; EFSA Panel on Dietetic Products and Allergies 2012).

MEASUREMENT OF THE PROTEIN REQUIREMENT

Protein requirement can be defined as the minimum amount of dietary protein needed to maintain nitrogen balance and sustain age-appropriate tissue growth rates (WHO/FAO 2007). This calculation is based not only on the nitrogen balance but also on the efficiency of digestion of the protein (bioavailability). Human studies demonstrate that in the absence of dietary protein/nitrogen (with adequate non-protein energy and nutrients), the body experiences ongoing losses of nitrogen, identified as obligatory nitrogen losses (WHO/FAO 2007). These obligatory nitrogen losses occur mainly via urine, sweat, feces, skin, and hair (Mojtahedi et al. 2002; WHO/FAO 2007). The minimum protein requirement is the protein intake adequate to enable nitrogen balance. This determines the amount of protein needed for maintenance, a value which, when combined with the protein bioavailability, defines an individual's protein requirement. For children, the calculation also includes protein deposition for growth.

The calculation of protein requirements for children is based on studies of nitrogen balance in infants and children from different parts of the world. The WHO/FAO protein requirement analysis for infants and children drew on data from ten studies encompassing 53 participants aged 9 months to 14 years; these studies investigated protein requirements over a range of dietary protein intakes. Four studies measured nitrogen balance exclusively (Dewey et al. 1996; WHO/FAO 2007). Protein requirements were calculated by plotting a straight line between the nitrogen balance and nitrogen intake and describing it mathematically by the intercept on the intake axis and the gradient of the line. The intercept is the amount of nitrogen intake required to balance the nitrogen losses, and the gradient describes the efficiency of the utilization of dietary protein.

The next step in the calculation of protein requirement for infants is to calculate the rate of growth or protein deposition. Historically, the protein deposition data was calculated from the growth of body mass adjusted for the percentage of deposition as fat. This method, however, was not considered robust, and more direct methods to determine protein deposition were considered, including total body potassium measurement. The WHO/FAO recommendations use data from Butte et al. (2000), which prospectively assessed total body potassium in infants from 2 weeks to 2 years of age. The data for older children comes from a cross-sectional study by Ellis et al., who measured total body potassium for children aged 4–18 years (Ellis et al. 2000). Notably, there was a paucity of data for children between 2 and 4 years of age, and the protein requirement for that age group was extrapolated from the two studies mentioned above and growth data from another study conducted in this age group.

Based on the combined results of the study, the value for maintenance was determined as 110 mg nitrogen per kilogram per day, which is about 0.687 g protein per kilogram per day with a slope of 0.58. However, when the studies using animal protein sources such as milk and egg were considered, the maintenance value was 93 mg nitrogen per kilogram per day (0.58 g protein/kg/day) due to the higher efficiency of animal protein, and the slope was adjusted to 0.66. Consequently, a value of 0.58 g/kg/day was adopted as the maintenance protein requirement for infants under six months of age. Table 32.1 summarizes the values for protein requirement in the first six months of life per WHO/FAO guidelines.

For children aged six months to 18 years, the procedure for calculating the average requirement and safety level followed a similar methodology but used different values for maintenance and utilization efficiency. This adjustment reflects that most children in this age group no longer rely exclusively on breast milk and consume mixed diets. Based on data from pediatric studies, the maintenance requirement is 0.68 g protein/kg/day; however, the data from larger adult studies suggest a value of around 0.66 g/kg/day, which was applied to children (WHO/FAO 2007). Table 32.2 provides the values for protein requirements for children aged 0–10 years.

Protein Quality of Foods in Infancy and Childhood

Protein quality reflects the proportion of amino acids that can be absorbed from the diet and utilized in the body. Assessing protein quality helps predict a food's capacity to satisfy the body's demand for amino acids and nitrogen. Quality varies by food source and can be affected by the mode of processing (Millward 2012).

TABLE 32.1
Protein Requirement in Infants Less than Six Months of Age

Age (months)	Maintenance Requirement (g/kg per day)	Growth Requirement (g/kg per day)	Average Requirement (g/kg per day)	Safe requirement (g/kg per day)
1	0.58	0.83	1.41	1.77
2	0.58	0.65	1.23	1.50
3	0.58	0.55	1.13	1.36
4	0.58	0.49	1.07	1.24
6	0.58	0.40	0.98	1.14

Adapted from WHO/FAO (2007).

TABLE 32.2
Average Protein Requirement and Safe Requirement in Infants and Children

Age (years)	Maintenance Requirement (g/kg per day)	Growth Requirement (g/kg per day)	Average Requirement (g/kg per day)	Safe Requirement (g/kg per day)
0.5	0.66	0.46	1.12	1.31
1	0.66	0.29	0.95	1.14
1.5	0.66	0.19	0.85	1.03
2	0.66	0.13	0.79	0.97
3	0.66	0.07	0.73	0.90
4	0.66	0.03	0.69	0.86
5	0.66	0.06	0.69	0.85
6	0.66	0.04	0.72	0.89
7	0.66	0.08	0.74	0.91
8	0.66	0.09	0.75	0.92
9	0.66	0.09	0.75	0.92
10	0.66	0.09	0.75	0.91

Adapted from WHO/FAO 2007 (WHO/FAO 2007).

Methods to Define Protein Digestibility

Several methods evaluate protein quality, considering protein composition data based on digestibility and amino acid bioavailability. The protein digestibility-corrected amino acid score (PDCAAS), widely used since the 1950s, is not recommended anymore due to three criticisms. It was argued that the PDCAAS score overestimates crude protein digestibility by including the value of fecal protein, even though the proteins in human feces are influenced by the microbiome and may not indicate the true digestibility of the protein. The PDCAAS was also criticized for truncating the estimates and ignoring anti-nutritional factors that overestimate or underestimate individual amino acids.

In 2011, an FAO expert consultation led to a new method for assessing protein quality, the digestible indispensable amino acid score (DIAAS).

The DIAAS is defined as:

DIAAS = 100 × [(mg of digestible dietary indispensable amino acid in 1 g of dietary protein) / (mg of the same dietary indispensable amino acid in 1 g of the reference protein)].

This method is expected to overcome the limitations of the prior methods by providing a better estimate of crude protein digestibility, i.e., decreasing the overestimation relative to PDCAAS. DIAAS should be more accurate because it focuses on the digestibility of individual amino acids as opposed to crude protein and approximates the digestibility at the end of the small intestine, which might be a more exact measure of the number of amino acids absorbed by the body. DIAAS also addresses the issue of truncation, leading to a better scoring system for amino acids as it recognizes the nutritional value of higher-quality proteins. The concern related to bioavailability is addressed by considering the components in protein that can affect the levels of amino acids. For example, immunoassays, such as that for lysine, give a better indication of the individual amino acid components of a protein.

DIAAS is similar to the amino acid scoring pattern, which is also used for protein quality assessments. The amino acid scoring pattern is defined as the dietary indispensable amino acid requirement divided by the safe protein requirement. This scoring pattern can be used to adjust digestible protein intake to assess available protein in specific diets by comparison to reference amino acid scoring patterns. For infants, the reference amino acid scoring pattern is based on the amino acid composition of breast milk. Table 32.3 shows the milligrams of indispensable, dispensable, and conditionally indispensable amino acids per gram of nitrogen and protein, and per 100 kcal in breast milk. For young children (6 months–3 years), the pattern is based on the amino acid profile of a 6-month-old infant. For older children and adolescents, the pattern is based on a 3–10-year-old's profile. Table 32.4 displays amino acid scoring patterns for all three age groups (WHO/FAO 2007).

TABLE 32.3
Indispensable, Dispensable, and Conditionally Indispensable Amino Acids for Humans in Breast Milk

Amino Acid	mg per g Nitrogen	mg per g Protein	mg per 100 kcal
Indispensable			
Histidine	141	23	41
Isoleucine	319	51	92
Leucine	586	94	169
Lysine	395	63	114
Methionine	85	14	24
Phenylalanine	282	45	81
Threonine	268	43	77
Tryptophan	114	18	33
Valine	315	50	90
Dispensable			
Alanine	–	111	–
Aspartic acid	–	110	–
Asparagine	–	–	–
Glutamic acid	–	148	–
Serine	–	74	–
Conditionally Indispensable			
Arginine	–	67	–
Cysteine	131	21	38
Glutamine	–	–	–
Glycine	–	117	–
Proline	–	–	–
Tyrosine	–	36	–

Source: WHO/FAO (2007).

TABLE 32.4
Recommended Amino Acid Scoring Patterns for Infants, Children, and Adolescents

Age Group	His	Ile	Leu	Lys	SAA	AAA	The	Trp	Val
Infant (birth–6 mo)	21	55	96	69	33	94	44	17	55
Child (6 mo–3 yr)	20	32	66	57	27	52	31	8.5	43
Older child and adolescent	16	30	61	48	23	41	25	6.6	40

Adapted from FAO 2013 (WHO 2013).

LIMITATIONS OF METHODS FOR DETERMINATION OF PROTEIN REQUIREMENT

The recommendations for protein requirements in children and adults have remained unchanged over the last four decades (Trumbo et al. 2002; WHO/FAO 2007). The methods used in WHO and Institute of Medicine guidelines to establish nitrogen balance and protein requirements have limitations (Weiler et al. 2023). For example, most of the studies that were aimed at determining nitrogen balance were conducted under conditions where subjects received adequate energy content but no protein. Newer studies, drawing insights from urea and citric acid cycle enzyme kinetics, strongly suggest that changes in amino acid supply and/or concentrations have effects on the kinetics of protein turnover, including amino acid oxidation (Bandegan et al. 2017; Weiler et al. 2023). This means that nitrogen losses noted during zero protein intake may not accurately represent true nitrogen losses. Additionally, amino acids serve roles beyond being used as subunits of protein synthesis, and protein losses may be influenced by amino acid availability, potentially skewing nitrogen balance results (Dalangin et al. 2020; Weiler et al. 2023). Protein requirement studies are mostly based on nitrogen balance data, and functional health outcomes have not been assessed across different levels of protein intake, which is the ultimate goal of establishing the protein requirement for the population (Dewey et al. 1996). Newer studies using advanced techniques may be needed to revise interpretations of nitrogen balance data, accounting for both essential and non-essential amino acid requirements and alternative metrics such as dietary energy percentage or per-meal protein intake (Weiler et al. 2023).

Sources of Protein

Term Infants (Birth to 6 Months)

Human milk is the recommended source of nutrition for the first six months of life; however, only 37% of infants globally are exclusively breastfed during this period (Victora et al. 2016; Meek et al. 2022; WHO 2023). Human milk contains hundreds of proteins with diverse functions, including providing essential amino acids, enhancing immunological defense, promoting intestinal growth and maturation, shaping the gut microbiome, increasing the absorption of micronutrients, and

TABLE 32.5
Levels of Proteins in Breastmilk Based on Stage of Lactation

Protein (g/L ±SD)	Stage of Lactation			
Whey	**Colostrum (0–5 Days)**	**Early (6–10 Days)**	**Transitional (11–30 Days)**	**Mature (31–365 Days)**
Total	20.6	15.7	14.8	11.1
α-Lactalbumin	4.56 ±0.41	4.3 ±0.41	3.52 ±0.27	2.85 ±0.24
Lactoferrin	6.15 ±0.89	3.65 ±1.19	2.46 ±0.27	1.76 ±0.28
Lysozyme	0.32 ±0.01	0.30 ±0.01	0.28 ±0.11	0.38 ±0.15
Secretory IgA	5.45 ±1.7	1.5 ±0.22	1.10 ±0.32	1.14 ±0.21
Casein	**Early (0–10 Days)**		**Transitional (11–30 Days)**	**Mature (31–365 Days)**
Total casein	2.49 ±0.41		2.59 ±0.59	1.92 ±0.72
α-Casein	0.34 ±0.09		0.33 ±0.07	0.33 ±0.18
β-Casein	1.29 ±0.28		1.46 ±0.46	1.03 ±0.53
κ-Casein	0.86 ±0.09		0.80 ±0.10	0.55 ±0.05

Modified from Donovan (2019).

enhancing learning and memory (Donovan 2019). Breast milk proteins are categorized as whey, casein, and mucins. Whey protein makes up 90% of total protein in colostrum, decreasing to around 60% in mature milk. Proteomic analysis identifies 115 unique whey proteins, primarily involved in immune response (35%), cell communication (17%), metabolism/energy production (16%), and transport (12%) (Liao et al. 2011; Donovan 2019). Key whey proteins in human milk include α-lactalbumin, lactoferrin, IgA, osteopontin, and lysozyme (Donovan 2019). Their concentrations decline sharply in the first month of life, except lysozyme, which remains stable (Table 32.5).

Human milk contains α-, β-, and κ-caseins. The concentration of α-casein remains fairly stable (Table 32.5) (Liao et al. 2011; Donovan 2019), while the concentrations of β- and κ-caseins rise slightly from early to mature milk. The human milk fat globule contains about 125 distinct proteins with diverse functions, including metabolism/energy production (21%), cell communication (19%), transport (16%), and immune response (20%) (Lonnerdal 2016; Donovan 2019).

Dietary Guidelines for Americans recommend the use of infant formula when breastfeeding is not possible (U.S. Department of Agriculture and U.S. Department of Health and Human Services 2020). In the US, the FDA regulates infant formula, requiring 1.8–4.5 g protein/100 kcal of infant formula (FDA 2023). The European Food Safety Authority regulates infant formula in Europe and recommends 1.8–2.5 g/100 kcal (EU 2015). Most formulas are cow's milk-based, though goat's milk-, soy-, rice-, and pea-based options are also available (Martinez and Ballew 2011; DiMaggio et al. 2019)

The protein content of cow's milk-based formulas differs from human milk, with a higher casein-to-whey ratio (70:30). Cow's milk contains α-, β-, γ-, and κ-caseins, whereas human milk contains α-, β-, and κ-caseins. β-lactoglobulin, the predominant whey protein in cow's milk, is absent in human milk. The concentrations of α-lactalbumin, lactoferrin, and osteopontin are lower in bovine milk than in human milk (Donovan 2019).

Some formulas have extensively hydrolyzed proteins, where about 90% of the protein is hydrolyzed. Some formulas contain only amino acids and no intact proteins (Martinez and Ballew 2011; Milbrandt 2017). These extensively hydrolyzed and amino acid-based formulas are considered hypoallergenic and are used for treating cow's milk allergy. A recent US report noted that about 5% of infant formulas purchased were extensively hydrolyzed or amino acid-based, a cow's milk allergy prevalence of only 1–3% (Strzalkowski et al. 2022). While hydrolysis of protein mitigates allergies, it increases formula osmolality, raising solute load on kidneys and gastrointestinal losses during diarrheal illness. Thus, its use should be discouraged in infants without cow's milk allergy. Hydrolysis of proteins involves proteolytic enzymes, sometimes derived from porcine sources, which may conflict with halal and kosher principles (Martinez and Ballew 2011; Milbrandt 2017). Formula packaging should clearly disclose this to help families make informed decisions.

Some families prefer to use a plant-based formula, and the available options include soy-, pea-, and rice-based formulas. Soy formulas have comparable protein bioavailability to cow's milk-based formulas but higher total protein, supplemented with amino acids such as L-methionine, taurine, and L-carnitine (Martinez and Ballew 2011). They have been shown to promote similar growth rates to cow's milk-based formulas, although some infants with cow's milk allergies show cross-reactivity to soy protein. Soy formulas are used to treat galactosemia but are not recommended for preterm infants due to their high phytate content, which may interfere with the absorption of phosphorus and calcium (Martinez and Ballew 2011).

Preterm Low-Birth-Weight Infants (0–6 months)

One in ten births in the US is preterm (<37 weeks of gestation) (March of Dimes 2022). Globally, about 13.4 million infants were born preterm in 2020, with 15% very preterm (<32 weeks), requiring neonatal intensive care unit admission (Ohuma et al. 2023). About 20.5 million infants were born with low birth weight (<2500 g) in 2015, mostly in low- and middle-income countries (Blencowe et al. 2019). The mortality rate for low-birth-weight and preterm infants remains high in low-income

TABLE 32.6
Amino Acid Composition of Commonly Used Amino Acid Solutions in the US

Amino Acid	Aminosyn-PF (Abbott) (mg/dL) (10% Solution)	TrophAmine (B. Braun) (mg/dL) (10% Solution)	Primene (Baxter) (mg/dL) (10% Solution)
Histidine	312	480	380
Isoleucine	760	820	670
Leucine	1,200	1,400	1,000
Lysine	677	820	1,100
Methionine	180	340	240
Phenylalanine	427	480	420
Threonine	512	420	370
Tryptophan	180	200	200
Valine	673	780	760
Alanine	698	540	800
Arginine	1,227	1,200	840
Proline	812	680	300
Serine	495	380	400
Taurine	70	25	60
Tyrosine	44	240	45
Glycine	385	360	400
Cysteine	—	<16	189
Glutamic Acid	820	500	1,000
Aspartic Acid	527	320	600

Source: Poindexter and Denne 2003

countries (Ohuma et al. 2023), but survival has improved in high-income countries despite growth challenges (Poindexter and Denne 2003).

Protein requirements for preterm infants account for maintenance (calculation of nitrogen losses), protein deposition (accretion), and digestibility. Preterm infants exhibit higher protein losses and deposition needs than term infants (Hay and Thureen 2010). While there is no consensus on preterm infant protein requirements, the Life Science Research Office (LSRO) recommends 3.0–4.3 g/kg/day (Klein and Heird 2005).

Preterm infants may be fed enterally or parenterally (via a central or peripheral line), or by a combination of both. The feeding route depends on the newborn's gestational age, birth weight, and stability. For example, extremely low-birth-weight babies (<1000 g) often require 3.5–4.0 g/kg/day parenteral nutrition, while heavier infants (>1000 g) need 3.0–3.5 g/kg/day (Poindexter and Denne 2003). Table 32.6 presents the amino acid compositions of commonly used amino acid solutions in the United States. The recipe for these solutions is based on the serum level of breastfed infants and/or cord blood concentration of amino acids. All three solutions contain all the amino acids except cysteine and glutamine. Of the included amino acids, tyrosine has limited bioavailability from these solutions due to low solubility. Cystine and glutamine are not included in the amino acid solutions due to concerns related to the long-term stability of these amino acids, and separate infusion solutions of these amino acids are available. These are important considerations, as tyrosine and cystine are conditionally essential in preterm infants, and glutamine is the most abundant amino acid in plasma and human milk.

Enterally fed preterm infants require 3.4–4.3 g/kg/day protein. Human milk is preferred due to its protective effect against necrotizing enterocolitis, but the protein content of human milk declines over time. Protein content may be as high as 4.1 ±2.1 g/dL in the first three days, decrease to 1.6–3.4

g/dL by days 4–7, and drop to 0.9–2.2 g/dL by day 28. This decrease in protein content might necessitate enrichment with human milk fortifiers to satisfy the increased caloric and protein requirements of growing preterm infants (Gates et al. 2021). Human milk from milk banks typically offers a consistently stable protein content, as the milk is pooled from multiple donors. The composition of preterm formulas differs compared to term formulas, with most preterm formulas available in the US containing 2.2–3.3 g/dL protein (at a ratio of 70:30 whey:casein, which closely resembles human milk). Preterm formulas also typically have a higher cystine content, as it is a conditionally essential amino acid for preterm infants (Ziegler 2016).

INFANTS (6–24 MONTHS)

WHO and AAP recommend continuing breastfeeding after the first six months of life along with appropriate complementary foods (Meek et al. 2022; WHO 2023). For non-breastfed infants between 6 and 11 months of age, WHO recommends either using a follow-up formula or animal milk such as cow's milk (WHO 2023). However, the Dietary Guidelines for Americans recommend against the use of non-formula animal milk during the first year of life. The use of animal milk during infancy has been associated with gastrointestinal blood loss and an increased risk of iron deficiency anemia; however, the growth rates are similar between animal milk-fed infants compared to mixed-fed and formula-fed infants (Ehrlich et al. 2022).

WHO recommends that infants and young children (6–23 months) should consume animal-source foods on a daily basis. Animal-based foods include meat, fish, eggs, and dairy products. Pulses, nuts, and seeds should be consumed frequently, particularly when meat, fish, eggs, or vegetables are limited in the diet. Starchy staple foods should be minimized due to their low protein quality (WHO 2023).

Animal protein sources (meat, poultry, eggs, seafood, dairy) are complete proteins, containing all essential amino acids. Incomplete protein sources lack sufficient indispensable amino acids and include most plant foods (beans, grains, nuts, peas, seeds, vegetables). For example, grains are deficient in lysine and threonine, while legumes lack methionine and tryptophan. Diversifying protein sources through complementary supplementation can address these deficiencies (Garlick 2006; WHO 2023).

Breastfeeding can be continued into the second year of life and beyond, as mutually desired by baby and mother; however, about two-thirds of infants do not receive human milk beyond the first year. WHO and AAP recommend transitioning to cow's milk at one year of age. The use of follow-up formula, toddler milk, or plant-based milk is discouraged (Fuchs et al. 2023; WHO 2023). Plant-based milks generally have lower protein content than cow's milk. Table 32.7 compares their protein content (Merritt et al. 2020).

TABLE 32.7
Protein Comparison of Cow's Milk and Plant-Based "Milk"

Per 1 cup (240 ml)	Cow's Milk	Almond	Cashew	Coconut	Flax Seed	Hemp	Oat	Pea	Rice	Soy
Protein (g)	150	30–100	25–80	45–90	55	70–170	130	115	110	90

Adapted from Merritt et al. (2020).

Children (2 Years and Above)

Protein deposition slows after age two but continues through adolescence. The Dietary Guidelines for Americans recommend diverse protein sources during early childhood, including animal- and plant-based foods. Suboptimal protein intake (low or high) during early childhood is associated with adverse outcomes in later life. A case-control study from Indonesia found stunted children (24–59 months) consumed fewer essential amino acids than healthy peers (Rizky Maulidiana and Sutjiati 2021). Data from high-income countries show that high protein intake during early childhood correlates with later obesity risk, likely via stimulation of insulin-like growth factors, although the exact mechanisms remain unclear (Arnesen et al. 2022).

Protein-Energy Malnutrition

The term protein-energy malnutrition is used synonymously with severe acute malnutrition or wasting (Bhutta et al. 2017). The etiology of protein-energy malnutrition in children is multifactorial, with contributing factors such as maternal malnutrition, low birth weight, lack of breastfeeding, delayed or inappropriate complementary feeding, food insecurity, and lack of clean drinking water, which in turn are associated with social and cultural norms and depend on political stability. In lower-income countries, the "protein gap" – insufficient quantity and quality of protein – is a key factor in protein-energy malnutrition (Semba 2016). Recent data show that protein quantity and quality may be adequate in older children but remain suboptimal in infants aged 6–12 months (Arsenault and Brown 2017). The treatment algorithm for severe acute malnutrition depends on the severity of malnutrition, with different regimens for inpatient and outpatient care. Inpatient care commonly uses F-75 and F-100 liquid formulas administered orally or via a feeding tube. These formulations contain animal-sourced proteins like cow, goat, or camel milk. Outpatient care relies on ready-to-use therapeutic foods with proteins sourced from milk and peanuts. Local sources of proteins, such as soy-maize-sorghum, have also been used (Akomo et al. 2019). The adoption of ready-to-use therapeutic foods has transformed outpatient management of non-complicated severe acute malnutrition, increasing treatment access and reducing mortality (Schoonees et al. 2019).

Protein-Related Allergies during Early Childhood

The prevalence of childhood allergies seems to be increasing in high-income countries (Loh and Tang 2018). The most common allergens in early childhood include cow's milk, eggs, peanuts, tree nuts, and seafood (Halken et al. 2021). The allergies may be IgE-mediated, leading to systemic responses characterized by respiratory, skin, and gastrointestinal reactions that can be life-threatening. Non-IgE-mediated allergies cause mostly gastrointestinal symptoms like diarrhea, mucus, and bloody stools. Studies have investigated whether early exposure to potential allergens in high-risk infants would prevent the incidence and severity of later allergies. A significant breakthrough in this area showed that early introduction of peanuts in high-risk infants could reduce later allergy incidence and severity, prompting the AAP to revise their recommendations from delayed to early introduction of allergenic foods. The Dietary Guidelines for Americans now encourage introducing small amounts of potentially allergenic foods during early childhood (Togias et al. 2017; Abrams et al. 2020). Hydrolyzed cow's milk-based formula has been studied for allergy prevention, but the evidence is not conclusive (Osborn et al. 2018; McGuire and Soll 2019; Halken et al. 2021). Additionally, the European Academy of Allergy and Clinical Immunology (EAACI) recommends introducing well-cooked eggs at 6 months to prevent egg allergy (Halken et al. 2021) and avoiding soy formula for allergy prevention (Halken et al. 2021).

Concluding Remarks

Early infancy and childhood are critical for growth and development. While the protein requirement methodologies have improved, further refinements could enhance precision for protein requirements at the population and individual levels, which could help improve short- and long-term clinical outcomes, particularly for understudied populations like preterm infants.

MINI DICTIONARY OF TERMS

Average protein requirement: The amount of protein that meets the needs of approximately 50% of the population.

Crude protein level: The amount of protein in a dietary source, calculated by multiplying the total nitrogen content by 6.25.

Protein quality: A concept describing the effectiveness with which nitrogen balance can be achieved for a given amount of absorbed dietary nitrogen.

Protein requirement: For infants and children, the minimum protein intake needed to allow nitrogen equilibrium at an appropriate body composition during energy balance at moderate physical activity, plus the growth rate consistent with good health.

Recommended dietary allowance: The amount of dietary protein that meets the requirements of 95% of a population.

SUMMARY POINTS

- This chapter focuses on protein requirements and protein quality during infancy and early childhood.
- Protein requirements per kilogram are higher during childhood than in adulthood.
- The protein requirement during childhood is calculated based on nitrogen balance studies, protein deposition, and the efficiency of protein digestion.
- Protein requirements for preterm infants are higher than those for term infants.
- Animal-based foods are an important source of protein when solid foods are introduced around six months of age.

REFERENCES

Abrams, E. M., E. S. Chan and S. Sicherer (2020). "Peanut allergy: new advances and ongoing controversies." *Pediatrics* **145**(5).

Akomo, P., P. Bahwere, H. Murakami, C. Banda, E. Maganga, S. Kathumba, K. Sadler and S. Collins (2019). "Soya, maize and sorghum ready-to-use therapeutic foods are more effective in correcting anaemia and iron deficiency than the standard ready-to-use therapeutic food: randomized controlled trial." *BMC Public Health* **19**(1): 806.

"Aminosyn-PF Injection Prescribing Information." Available at: https://www.drugs.com/pro/aminosyn-pf-injection.html (Last Accessed January 2, 2024).

Arnesen, E. K., B. Thorisdottir, C. Lamberg-Allardt, L. Barebring, B. Nwaru, J. Dierkes, A. Ramel and A. Akesson (2022). "Protein intake in children and growth and risk of overweight or obesity: a systematic review and meta-analysis." *Food Nutr Res* **66**.

Arsenault, J. E. and K. H. Brown (2017). "Dietary protein intake in young children in selected low-income countries is generally adequate in relation to estimated requirements for healthy children, except when complementary food intake is low." *J Nutr* **147**(5): 932–939.

Bandegan, A., G. Courtney-Martin, M. Rafii, P. B. Pencharz and P. W. Lemon (2017). "Indicator amino acid-derived estimate of dietary protein requirement for male bodybuilders on a nontraining day is several-fold greater than the current recommended dietary allowance. *J Nutr* **147**(5): 850–857.

Bhutta, Z. A., J. A. Berkley, R. H. J. Bandsma, M. Kerac, I. Trehan and A. Briend (2017). "Severe childhood malnutrition." *Nat Rev Dis Primers* **3**: 17067.

Blencowe, H., J. Krasevec, M. de Onis, R. E. Black, X. An, G. A. Stevens, E. Borghi, C. Hayashi, D. Estevez, L. Cegolon, S. Shiekh, V. Ponce Hardy, J. E. Lawn and S. Cousens (2019). "National, regional, and worldwide estimates of low birthweight in 2015, with trends from 2000: a systematic analysis." *Lancet Glob Health* **7**(7): e849–e860.

Butte, N. F., J. M. Hopkinson, W. W. Wong, E. O. Smith and K. J. Ellis (2000). "Body composition during the first 2 years of life: an updated reference." *Pediatr Res* **47**(5): 578–585.

Dalangin, R., A. Kim and R. E. Campbell (2020). "The Role of Amino Acids in Neurotransmission and Fluorescent Tools for Their Detection." *Int J Mol Sci* **21**(17).

Dewey, K. G., G. Beaton, C. Fjeld, B. Lonnerdal and P. Reeds (1996). "Protein requirements of infants and children." *Eur J Clin Nutr* **50**(Suppl 1): S119–147; discussion S147–150.

DiMaggio, D. M., N. Du, C. Scherer, S. Brodlie, V. Shabanova, P. Belamarich and A. F. Porto (2019). "Comparison of imported european and us infant formulas: labeling, nutrient and safety concerns." *J Pediatr Gastroenterol Nutr* **69**(4): 480–486.

Donovan, S. M. (2019). "Human milk proteins: composition and physiological significance." *Nestle Nutr Inst Workshop Ser* **90**: 93–101.

EFSA Panel on Dietetic Products, N. and Allergies (2012). "Scientific opinion on dietary reference values for protein." *EFSA Journal* **10**(2): 2557.

Ehrlich, J. M., J. Catania, M. Zaman, E. T. Smith, A. Smith, O. Tsistinas, Z. A. Bhutta and A. Imdad (2022). "The effect of consumption of animal milk compared to infant formula for non-breastfed/mixed-fed infants 6-11 months of age: a systematic review and meta-analysis." *Nutrients* **14**(3).

Ellis, K. J., R. J. Shypailo, S. A. Abrams and W. W. Wong (2000). "The reference child and adolescent models of body composition: a contemporary comparison." *Ann N Y Acad Sci* **904**: 374–382.

EU (2015). "THE European Commission: Commission Delegated Regulation (EU) 2016/127 of 25 September 2015 supplementing Regulation (EU) No 609/2013 of the European Parliament and of the Council as regards the specific compositional and information requirements for infant formula and follow-on formula and as regards requirements on information relating to infant and young child feeding." http://publications.europa.eu/resource/cellar/95d90c87-c97e-11e5-a4b5-01aa75ed71a1.0006.03/DOC_1 (Last Accessed September 20, 2023).

FDA. (2023). "Infant formula." Available at https://www.fda.gov/food/resources-you-food/infant-formula.

Fuchs, G. J., S. A. Abrams, A. A. Amevor and N. Committee On (2023). "Older infant-young child 'formulas'." *Pediatrics* **152**(5), e2023064050.

Garlick, P. J. (2006). "Protein requirements of infants and children." *Nestle Nutr Workshop Ser Pediatr Program* **58**: 39–47; discussion 47–50.

Gates, A., T. Marin, G. Leo and B. K. Stansfield (2021). "Review of preterm human-milk nutrient composition." *Nutr Clin Pract* **36**(6): 1163–1172.

Halken, S., A. Muraro, D. de Silva, E. Khaleva, E. Angier, S. Arasi, H. Arshad, H. T. Bahnson, K. Beyer, R. Boyle, G. du Toit, M. Ebisawa, P. Eigenmann, K. Grimshaw, A. Hoest, C. Jones, G. Lack, K. Nadeau, L. O'Mahony, H. Szajewska, C. Venter, V. Verhasselt, G. W. K. Wong, G. Roberts, A. European Academy of, A. Clinical Immunology Food and G. Anaphylaxis Guidelines (2021). "EAACI guideline: preventing the development of food allergy in infants and young children (2020 update)." *Pediatr Allergy Immunol* **32**(5): 843–858.

Hay, W. W. and P. Thureen (2010). "Protein for preterm infants: how much is needed? How much is enough? How much is too much?" *Pediatr Neonatol* **51**(4): 198–207.

Klein, J. C. and C. W. Heird (2005). "Life science research office (LRSO): summary and comparison of recommendations for nutrient contents of low-birth-weight infant formulas." Available at: http://www.lsro.org/articles/lowbirthweight_rpt.pdf (Last Accessed January 2, 2024).

Liao, Y., R. Alvarado, B. Phinney and B. Lonnerdal (2011). "Proteomic characterization of human milk whey proteins during a twelve-month lactation period." *J Proteome Res* **10**(4): 1746–1754.

Loh, W. and M. L. K. Tang (2018). "The epidemiology of food allergy in the global context." *Int J Environ Res Public Health* **15**(9).

Lonnerdal, B. (2016). "Bioactive proteins in human milk: health, nutrition, and implications for infant formulas." *J Pediatr* **173**(Suppl): S4–9.

March-of-dimes. (2022). "March of dimes report card: Iowa." Available at file:///C:/Users/aimdad/Downloads/MarchofDimesReportCard-Iowa%20(1).pdf (Last Accessed October 9, 2023).

Martinez, J. A. and M. P. Ballew (2011). "Infant formulas." *Pediatr Rev* **32**(5): 179–189; quiz 189.

McGuire, W. and R. Soll (2019). "Commentary on 'infant formulas containing hydrolysed protein for prevention of allergic disease and food allergy'." *Neonatology* **116**(3): 286–289.

Meek, J. Y., L. Noble and B. Section on (2022). "Policy statement: breastfeeding and the use of human milk." *Pediatrics* **150**(1), e2022057988.

Merritt, R. J., S. E. Fleet, A. Fifi, C. Jump, S. Schwartz, T. Sentongo, D. Duro, J. Rudolph, J. Turner and N. C. O. Nutrition (2020). "North American society for pediatric gastroenterology, hepatology, and nutrition position paper: plant-based milks." *J Pediatr Gastroenterol Nutr* **71**(2): 276–281.

Milbrandt, T. P. (2017). "Specialized infant formulas." *Pediatr Rev* **38**(5): 241–242.

Millward, D. J. (2012). "Amino acid scoring patterns for protein quality assessment." *Br J Nutr* **108**(Suppl 2): S31–43.

Mojtahedi, M., L. C. de Groot, H. A. Boekholt and J. M. van Raaij (2002). "Nitrogen balance of healthy Dutch women before and during pregnancy." *Am J Clin Nutr* **75**(6): 1078–1083.

Ohuma, E. O., A. B. Moller, E. Bradley, S. Chakwera, L. Hussain-Alkhateeb, A. Lewin, Y. B. Okwaraji, W. R. Mahanani, E. W. Johansson, T. Lavin, D. E. Fernandez, G. G. Dominguez, A. de Costa, J. A. Cresswell, J. Krasevec, J. E. Lawn, H. Blencowe, J. Requejo and A. C. Moran (2023). "National, regional, and global estimates of preterm birth in 2020, with trends from 2010: a systematic analysis." *Lancet* **402**(10409): 1261–1271.

Osborn, D. A., J. K. Sinn and L. J. Jones (2018). "Infant formulas containing hydrolysed protein for prevention of allergic disease." *Cochrane Database Syst Rev* **10**(10): CD003664.

Poindexter, B. B. and S. C. Denne (2003). "Protein needs of the preterm infant." *Neoreviews* **4**(2): e52–e59.

"Primene Prescribing Information." Available at https://www.drugs.com/pro/primene.html (Last Accessed January 2, 2024).

Rizky Maulidiana, A. and E. Sutjiati (2021). "Low intake of essential amino acids and other risk factors of stunting among under-five children in Malang City, East Java, Indonesia." *J Public Health Res* **10**(2).

Schoonees, A., M. J. Lombard, A. Musekiwa, E. Nel and J. Volmink (2019). "Ready-to-use therapeutic food (RUTF) for home-based nutritional rehabilitation of severe acute malnutrition in children from six months to five years of age." *Cochrane Database Syst Rev* **5**(5): CD009000.

Semba, R. D. (2016). "The rise and fall of protein malnutrition in global health." *Ann Nutr Metab* **69**(2): 79–88.

Strzalkowski, A. J., K. M. Jarvinen, B. Schmidt and B. E. Young (2022). "Protein and carbohydrate content of infant formula purchased in the United States." *Clin Exp Allergy* **52**(11): 1291–1301.

Togias, A., S. F. Cooper, M. L. Acebal, A. Assa'ad, J. R. Baker, Jr., L. A. Beck, J. Block, C. Byrd-Bredbenner, E. S. Chan, L. F. Eichenfield, D. M. Fleischer, G. J. Fuchs, 3rd, G. T. Furuta, M. J. Greenhawt, R. S. Gupta, M. Habich, S. M. Jones, K. Keaton, A. Muraro, M. Plaut, L. J. Rosenwasser, D. Rotrosen, H. A. Sampson, L. C. Schneider, S. H. Sicherer, R. Sidbury, J. Spergel, D. R. Stukus, C. Venter and J. A. Boyce (2017). "Addendum guidelines for the prevention of peanut allergy in the United States: report of the national institute of allergy and infectious diseases-sponsored expert panel." *J Allergy Clin Immunol* **139**(1): 29–44.

"TrophAmine Prescribing Information." Available at https://www.drugs.com/pro/trophamine.html (Last Accessed January 2, 2024).

Trumbo, P., S. Schlicker, A. A. Yates, M. Poos, Food and T. N. A. Nutrition Board of the Institute of Medicine (2002). "Dietary reference intakes for energy, carbohydrate, fiber, fat, fatty acids, cholesterol, protein and amino acids." *J Am Diet Assoc* **102**(11): 1621–1630.

"U.S. Department of Agriculture and U.S. Department of Health and Human Services." Dietary Guidelines for Americans, 2020–2025. 9th Edition. December 2020. Available at DietaryGuidelines.gov (Last Accessed December 19, 2023). 9th Edition.

Victora, C. G., R. Bahl, A. J. Barros, G. V. Franca, S. Horton, J. Krasevec, S. Murch, M. J. Sankar, N. Walker, N. C. Rollins and G. Lancet Breastfeeding Series (2016). "Breastfeeding in the 21st century: epidemiology, mechanisms, and lifelong effect." *Lancet* **387**(10017): 475–490.

"World Health Organization. Management of severe malnutrition: a manual for physicians and other senior health workers, release 1999." Available at https://apps.who.int/iris/handle/10665/41999 (Last Accessed September 5, 2022). from https://apps.who.int/iris/handle/10665/41999.

Weiler, M., S. R. Hertzler and S. Dvoretskiy (2023). "Is it time to reconsider the U.S. Recommendations for dietary protein and amino acid intake?" *Nutrients* **15**(4).

WHO (2023). *WHO Guideline for Complementary Feeding of Infants and Young Children 6-23 Months of Age*. Geneva: WHO.

WHO/FAO. (2007). "Protein and amino acid requirements in human nutrition: report of a joint FAO/WHO/UNU expert consultation. (WHO technical report series; no. 935)." Available at https://iris.who.int/bitstream/handle/10665/43411/WHO_TRS_935_eng.pdf?ua=1 (Last Accessed December 12, 2023).

Ziegler, E. E. (2016). "Protein needs of preterm infants: why are they so difficult to meet?" *Nestle Nutr Inst Workshop Ser* **86**: 121–128.

33 Protein Quantity and Quality and Links with Mental Health in Adolescents

Bani Tamber Aeri and Harmeet Kaur

ABBREVIATIONS

ADHD	Attention deficit hyperactivity disorder
ATP	Adenosine triphosphate
BBB	Blood-brain barrier
BDNF	Brain-derived neurotrophic factor
CRP	C-reactive protein
EAA	Essential amino acids
HPA	Hypothalamus-pituitary-adrenal axis
ICMR-NIN	Indian Council of Medical Research – National Institute of Nutrition
LNAA	Large neutral amino acid
MDD	Major depressive disorder
NO	Nitric oxide
RCTs	Randomised controlled trials
SAM	S-adenosyl methionine
UPF	Ultra-processed foods
WHO	World Health Organisation

INTRODUCTION

Adolescence is a crucial period of development, marking the transition from childhood to adulthood. During this period (early adolescence to youth, ages 10–24 years), adolescents experience significant physical and cognitive changes, including the onset of puberty, the formation of their identity and the exploration of personal values, beliefs and interests. These changes are associated with unique challenges and opportunities, which profoundly impact all domains of development, including biological, social, cognitive and emotional.

Globally, the number of young people aged 10–19 years is nearly 1.2 billion, accounting for one-sixth of the total population. Notably, 90% of these young people live in developing countries and face unique challenges such as limited access to education, healthcare and employment opportunities (Lee et al. 2016; WHO 2021). The changes adolescents experience, when accompanied by these limitations and exposure to poverty, abuse, violence and neglect, can make them susceptible to physical and mental health issues. Adolescent well-being, which encompasses not just physical health and a body free from disease but also includes emotional and mental health, is thus at risk.

A mental health disorder is characterised by disruption or impairment of an individual's normal personal function. These disorders constitute a significant portion of the global disease burden, impacting millions of people worldwide, including adolescents. Despite this, mental health conditions are often unrecognised and untreated in this age group. If left unaddressed, these disorders can

DOI: 10.1201/b23328-33

cause disability and self-destruction, placing adolescents at a higher risk for social exclusion. They may drop out of school and shy away from seeking help, creating a vicious cycle of stigma, ill health and lack of support (Gaiha et al. 2020).

Therefore, it is essential to protect adolescents from adversity, encourage socio-emotional learning and ensure access to mental health care to promote their overall health and well-being.

COMMON MENTAL HEALTH DISORDERS AMONG ADOLESCENTS: EXTENT OF THE PROBLEM

According to WHO (2021), one in seven adolescents aged 10–19 experiences a mental disorder, accounting for 13% of the global burden of disease in this age group. Despite this, the majority do not seek help or receive care. An estimated 3–5% of 10–19-year-olds experience anxiety and depression disorders, which may cause sudden and unexpected mood changes that can affect school attendance and performance, and social withdrawal that can lead to isolation, loneliness and the risk of suicide. Attention-deficit/hyperactivity disorder (ADHD) affects 2–3% of adolescents, and conduct disorder, characterised by destructive behaviour, affects approximately 3%. These conditions can disrupt education and may lead to criminal behaviour. Eating disorders, such as anorexia nervosa and bulimia nervosa, are also common among adolescents. These involve abnormal eating behaviour and preoccupation with food, often accompanied by concerns about body weight and shape. Anorexia nervosa can lead to premature death if left untreated.

Psychosis, characterised by symptoms such as hallucinations or delusions, and risk-taking behaviours, such as substance abuse, alcohol consumption and sexual risk-taking, can severely impact the mental and physical well-being of adolescents (WHO 2021).

FINDING SOLUTIONS TO THE PROBLEM

Mental well-being is now considered as important as physical well-being in determining the full growth potential of an adolescent. The onset of mental disorders often occurs early in life, and the risk factors for adolescent mental health are multifaceted (Kieling et al. 2011). These range from genetic endowment and prenatal factors to family relationships, social circle, self-perceptions, stressful life events, an adolescent's own thought processes and their health and nutritional status. Addressing each of these factors, whether biological or environmental in origin, is essential to reduce these preventable fatalities. The present discussion, however, focusses on the role of diet in achieving positive mental health among adolescents with an emphasis on protein intake.

DIET AND MENTAL HEALTH OF ADOLESCENTS

The relationship between diet and health seems to be a simple one, as is expressed by the adage "we are what we eat." This relationship extends beyond physical health to mental health as well. Research has established that a better-quality diet is associated with better mental health outcomes among adolescents. Mediterranean diets are linked to a lower risk of depression, while Western-type diets are independently associated with a greater likelihood or risk for poor mental health, including depression. Adolescents who have low consumption of fruits and vegetables, whole grains and low-fat dairy, and increased intake of Western dietary patterns (including sweetened beverages, high-glycaemic-index foods, processed foods and foods rich in saturated fatty acids) have an increased risk of poor mental health (O'Niel et al. 2014; Khanna et al. 2019; Chopra et al. 2021; Zielińska et al. 2022).

A recent analysis found that individuals with major depressive disorder (MDD) had a higher consumption of a pro-inflammatory Western diet, characterised by higher total energy, carbohydrates,

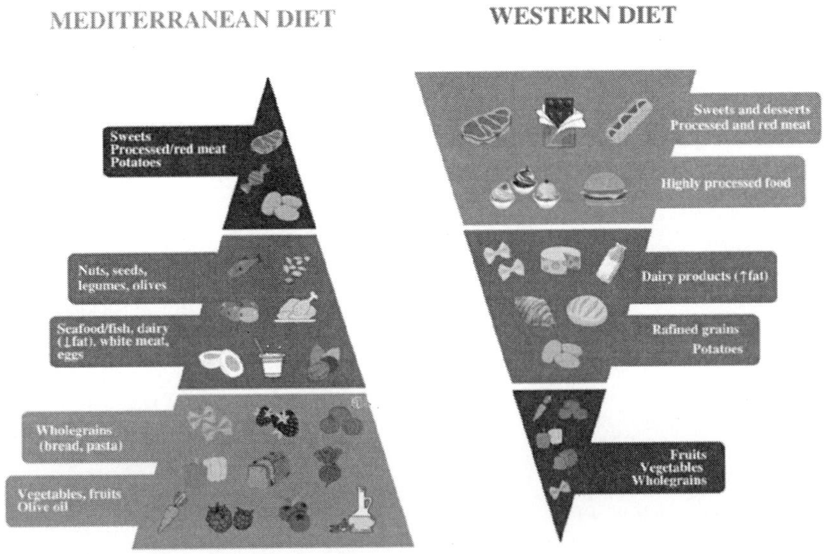

FIGURE 33.1 Characteristics of the Western and Mediterranean diets. The inverse relationship between the Western and Mediterranean diets is illustrated in pyramid form (Zielińska et al. 2022, with permission).

sugar, total fat, and saturated fat intake. These subjects also had a higher C-reactive protein (CRP) level, indicating medium-level inflammation in them (Bernier et al. 2023).

Animal experiments have also shown that poor diets may not just affect the brain but also can result in more systemic problems like increased inflammation, oxidative stress and an up-regulated stress response. This may result in reduced brain-derived neurotrophic factor (BDNF) and impact synaptic plasticity, learning and memory (Jiang et al. 2021).

A Mediterranean-style diet, which emphasises plant-based foods, healthy fats, moderate fish, seafood, dairy and poultry consumption, and low intake of red meat and sweets, is associated with a lower risk of depression (Ventriglio et al. 2020). Thus, promoting the intake of traditional diets among adolescents that are balanced and rich in complex carbohydrates, good quality proteins, a balanced amount of fat (including omega-3 fatty acids) and vitamins and minerals is important for achieving good mental health (Figure 33.1).

THE SPECIFIC ROLE OF PROTEINS IN MENTAL HEALTH

Proteins, the structural and functional components of all living cells, perform several essential functions in the body. These functions, ranging from growth and maintenance to synthesis of enzymes and hormones, as well as functioning as a protective agent, are all vital for human life.

Proteins are present in various foods like eggs, meats, dairy, seafood, legumes, nuts and seeds. While the amount and quality of protein vary considerably among foods, animal sources tend to be superior in both protein quantity and quality compared to plant sources. Meat, eggs and milk are considered excellent sources of high-quality protein. Egg protein is often used as the ideal (complete) protein against which the indispensable/essential amino acid profile of other foodstuffs is compared (Watford and Wu 2018; Puglisi and Fernandez 2022).

Irrespective of the protein source, it is broken down during digestion and absorption processes, and the basic units – amino acids – are utilised for performing several building-up functions in the body. While hundreds of amino acids exist in nature, humans use only 20 of them. One way to classify them is by identifying which amino acids the human body can synthesise on its own and which ones it cannot (Figure 33.2).

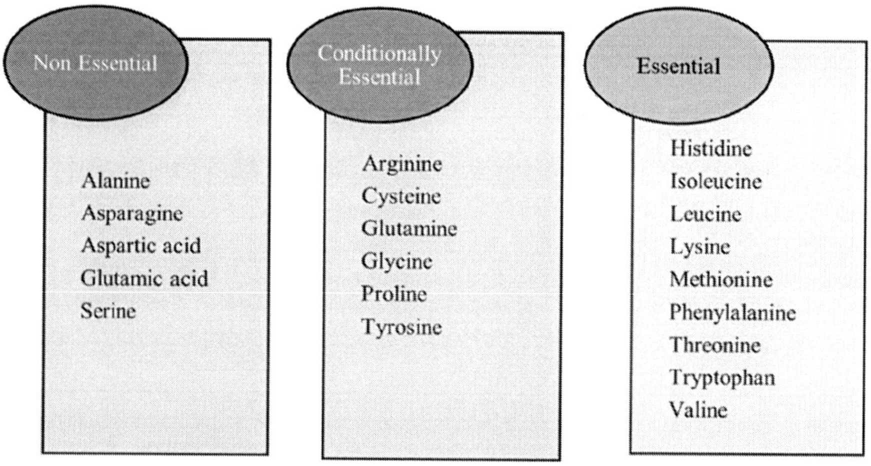

FIGURE 33.2 Classification of amino acids. This figure presents the names of non-essential, conditionally essential and essential amino acids in humans.

The three classes of amino acids are:

Non-essential amino acids: Five amino acids are termed non-essential because they can be obtained from foods and generated within the human body as well. All these amino acids perform important functions in the body; therefore, the term non-essential is a misnomer.

Conditionally essential amino acids: Six amino acids are termed conditionally essential because healthy bodies can generate them under normal physiological conditions. They become essential under certain conditions such as starvation or inborn errors of metabolism.

Essential amino acids (EAA): Nine amino acids are essential because they cannot be generated within the body. Dietary protein, therefore, provides these amino acids, which are needed to make certain hormones and other important molecules in the body.

Consumption of protein-rich diets helps in achieving positive mental health in adolescents in several ways as discussed below.

Proteins and Brain Neurotransmitters

The role of protein in mental health is usually linked to its impact on neurotransmitter synthesis. Research shows that the plasma levels of particular amino acids can regulate central neurotransmission, cognitive performance and mood state in adolescents.

A neurotransmitter is a molecule secreted by neurons that acts as a chemical messenger or signalling molecule between other neurons or target cells such as glands or muscles. The brain utilises amino acids like tryptophan, tyrosine, histidine, arginine and threonine, along with a few others, to synthesise various neurotransmitters and neuromodulators (Figure 33.3). The following discussion focusses on five amino acids and their role in adolescent mental health.

Tryptophan

Tryptophan is the rarest of the EAA found in food and is the precursor of serotonin. Serotonin, a neurotransmitter, is involved in modulating mood, cognition, reward, learning and memory. Evidence suggests that a change in the protein-to-carbohydrate ratio of food consumed can alter brain serotonin levels. Thus, the availability of tryptophan to the brain can alter behaviour owing to its serotonergic effect.

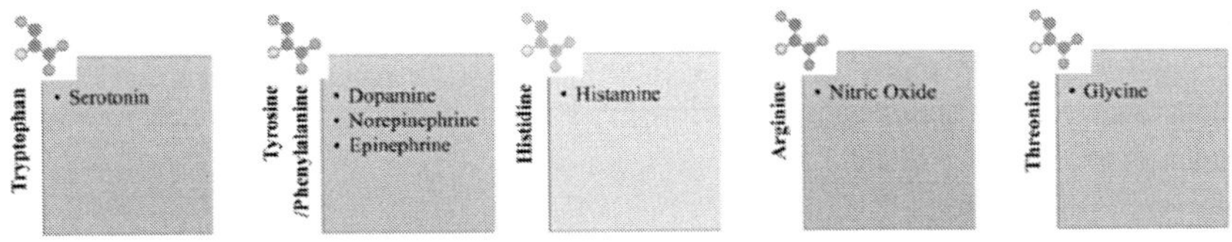

FIGURE 33.3 Specific amino acids and their respective neurotransmitters. This figure details certain amino acids that act as substrates for neurotransmitters.

Tryptophan has been shown to decrease quarrelsome behaviours, increase agreeable behaviours and improve mood (Rot et al. 2006). In never-depressed healthy volunteers with a high risk for depression due to familial risk factors, acute tryptophan depletion produced clear abnormalities in mood control (Feder et al. 2011). In remitted depressed patients, temporarily lowering tryptophan levels resulted in an acute depressive relapse (Booij et al. 2005). Therefore, tryptophan may act as an effective antidepressant in mild-to-moderate depression (Young 2007). This indicates that the administration of tryptophan-rich dietary proteins in adolescents can enhance tryptophan availability to the brain and potentially increase serotonin synthesis. Clinical studies have indicated that tryptophan supplementation improves attention scores and memory and reduces the symptoms of obsessive-compulsive disorders (Jenkins et al. 2016; Lakhan et al. 2008).

Tyrosine

Tyrosine is another amino acid that has been studied for its impact on behaviour. It is a precursor to three neurotransmitters: norepinephrine, epinephrine and dopamine. Research indicates that conditions of acute stress (physiological stress like exposure to extreme weather or psychological stress) increase the release of catecholamines like norepinephrine and dopamine, which are thought to play a critical role in the brain's response to acute stress. Prolonged stress can result in the depletion of catecholamine levels. Tyrosine is a precursor of catecholamines, and any change or reduction in its availability can influence their synthesis. Depletion of norepinephrine has the potential to impair/compromise cognitive and physical performance. Under such stressful situations, tyrosine supplementation has been shown to reduce cognitive decline and some behavioural deficits (Hase et al. 2015).

A review of ten RCTs and four clinical trials also demonstrated the benefit of tyrosine supplementation on cognitive functions, indicating that maintaining an abundant supply of tyrosine may prevent catecholamine depletion in the brain and improve the performance of cognitive tasks in healthy young volunteers (Attipoe et al. 2015).

Histidine

The amino acid histidine is decarboxylated to form histamine, and thus, dietary histidine affects histamine concentrations in immune cells, the stomach and the brain. Histamine is known to regulate the sleep-wake cycle and cognitive functions (an excess of it also causes allergic reactions in the body). Studies suggest that daily ingestion of histidine may improve feelings of fatigue and increase performance during memory tasks due to its effect in enhancing clear thinking and attentiveness (Sasahara et al. 2015).

Increased histidine intake in the form of dried bonito (dried fish, mostly consumed in Japanese cuisine and a good source of histidine) broth improved mood state and mental task performance of subjects. Furthermore, dried bonito broth has been reported to suppress inflammation and inhibit aggressive behaviour in children and adolescents (Nishijo et al. 2021). In animal studies, insufficient histidine intake has been shown to reduce brain histamine content, which has been linked with anxiety-like behaviours (Yoshikawa et al. 2014). Histidine and histidine-containing substances may exert beneficial effects via their antioxidant, anti-inflammatory and chelating properties and may modulate the histamine content in the brain (Holeček 2020).

Arginine

Arginine, a semi-essential amino acid, is a crucial substrate for certain pathways essential for the central nervous system, such as urea and nitric oxide (NO) cycles. A deficiency of arginine or any alteration in the NO pathway may lead to abnormalities and dysregulation of nitric oxide and the urea cycle. This dysregulation has been linked to the pathogenesis of severe depression in both human and animal models. This can be explained by the fact that NO acts as a precursor of signalling molecules and is involved in the vasodilation of blood vessels, learning and memory processing, the plasticity of neurons and modulation of neuronal function during stress and anxiety (Paul and Ekambaram 2011). Depressed patients show higher plasma nitrate concentrations – an index of NO production – but lower circulating L-arginine concentrations (Pinto et al. 2012). In fact, studies have shown that restoring NO production will help an individual feel better, think better, sleep better, have less anxiety and experience fewer symptoms of depression. Therefore, arginine and its metabolites are important in maintaining good mental health (Fan et al. 2021).

Threonine

Threonine is an essential amino acid that gets converted to glycine. Glycine, also an amino acid, has been shown to reduce persistent, unwanted muscle contractions, as seen in certain neurological diseases. Studies also show that changes in serum levels of serine and glycine, both requiring threonine for production, are linked to major depression. Glycine is known to help improve sleep, mental performance, mood and memory. As a precursor to glycine, threonine is often used to alleviate anxiety and depression symptoms. Adequate intake of threonine may support healthy brain function and a positive mental state. Its sleep-enhancing properties may also prove helpful for those who have insomnia and other sleep disorders (Razak et al. 2017).

Disturbances in amino acid metabolism have been implicated in the pathophysiology of schizophrenia, and high doses of glycine have been shown to reduce the more subtle symptoms of schizophrenia, such as social withdrawal, emotional flatness and apathy, which do not respond to medications. Symptoms of schizophrenia are typically present during adolescence (Lakhan et al. 2008).

Along with the five amino acids discussed above, a few other amino acids are also known to impact mental health. Taurine is an amino acid made in the liver from cysteine known to elicit a calming effect in the brain. A deficiency of this amino acid may increase manic episodes in bipolar patients. Methionine, another amino acid, combines with ATP to produce S-adenosylmethionine (SAM), which facilitates the production of neurotransmitters in the brain.

Figure 33.4 provides an overview of the main functions some neurotransmitters perform in the body. While amino acids have a critical role in regulating and maintaining the levels of these neurotransmitters, risk factors for mental health among adolescents are complex. Nonetheless, consuming a protein-rich diet or supplements of specific amino acids can alter brain activity, leading to better mental health among adolescents.

Proteins and Stress

Adolescents are susceptible to stress due to various physiological or psychological factors, which can have deleterious effects on the brain, behaviour and cognitive function. Any form of stress activates the hypothalamic–pituitary–adrenal axis, which leads to the secretion of glucocorticoids from the adrenal glands. Sustained elevation of cortisol due to chronic stress may be detrimental to adolescent health by impairing cognition (due to neuronal loss), suppressing thyroid function, decreasing bone density and even reducing muscle tissue (Lupien et al. 2009). Chronic stress and inflammation (as measured by higher CRP) were associated with enhanced tryptophan breakdown in children and adolescents, which may decrease serotonin production. (Michels et al. 2018).

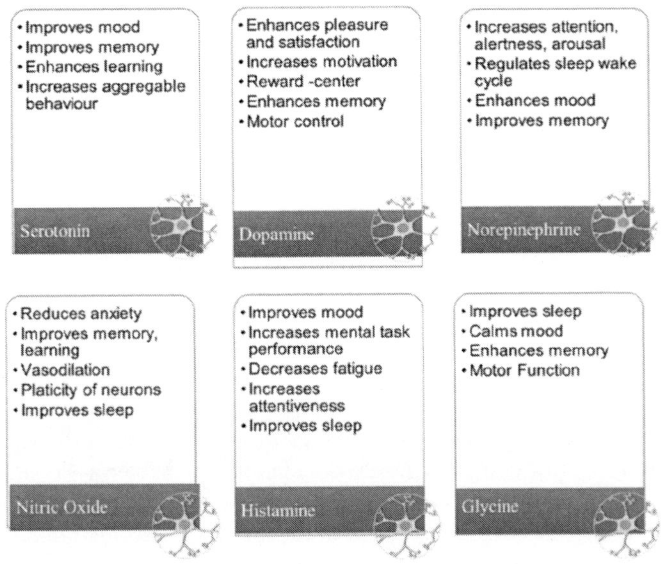

FIGURE 33.4 Neurotransmitters and their role in mental health. This figure describes the main functions regarding mental health of select neurotransmitters in humans.

Prolonged stress induces a catabolic state, which results in increased urinary nitrogen loss and an increased metabolic rate. The main reason for this response is the mobilisation of amino acids for production glucose to provide energy for the cells responding to the stress, which could be an immune response or wound repair. A deficiency of proteins among adolescents in such a situation would lead to the breakdown of stores and further exacerbate losses. To improve coping ability under stress, Dutch university students (aged 17 and above) were administered whey proteins enriched in tryptophan. This enhanced serotonin production, reduced depressive feelings and improved coping ability. Proteins like whey stimulate constructive processes and work antagonistically to cortisol, and are thus needed to manage the catabolic state of metabolic stress (Markus et al. 2000).

Proteins and Reduced Caloric Intake

Moderately high-protein diets promote satiety, reducing the likelihood of adolescents reaching for easily available, often ultra-processed (UPF) snacks high in sugar or fat. UPF consumption is very high among adolescents, and this can result in depression and internalising and externalising behavioural problems. UPF components (e.g., additives, trans fatty acids, sugar, salt, flavour enhancers and preservatives) can impair the hippocampus and cortex, which can have implications for emotional processes (Reales-Moreno et al. 2022).

Data indicates that when the protein fraction of the diet decreases, it leads to a compensatory increase in total energy intake in an attempt to maintain a target amount of absolute protein intake. This increased energy intake might result in weight gain (Hall 2019). Hence, increasing the intake of proteins can help in correcting weight imbalances, which may have implications like distorted body image among adolescents.

Protein-Rich Foods as Sources of Other Important Nutrients

Protein-rich foods are also rich sources of other nutrients, including foods that are typically high in iron. Iron deficiency anaemia, prevalent among adolescents, especially in the developing world, can result in fatigue, irritability, lack of motivation and impaired learning and memory, all of which can contribute to feelings of depression. Anaemic adolescents are counseled to consume iron-rich protein sources in their diet. This would treat anaemia and improve mood, enhance physical work capacity and increase productivity.

Proteins, Gut Microbiota and Mental Health

The gut-brain axis is a bi-directional system between the brain and gastrointestinal tract, linking the emotional and cognitive centres of the brain with the peripheral functioning of the digestive tract. Growing evidence indicates that changes in gut microbiota composition are related to mental health issues, including anxiety and depression. The gut microbiome offers a novel way to alter neurotransmitter modulation in the brain (Aeri 2021).

Research has also established that the manner in which the gut microbiota changes across the lifespan can impact tryptophan metabolism, affecting the serotonergic system. Evidence for direct regulation of tryptophan in the gut comes from studies on germ-free animals that are laboratory-raised and gut microbiota-deficient. These animals show increased levels of circulating tryptophan and decreased serotonin. Introducing tryptophan-metabolising bacteria to their gut reduces circulating levels of tryptophan. Within the brain, an increase in hippocampal serotonin levels was observed, along with a decrease in anxiety-like behaviour. This demonstrated that gut microbiota influence both behaviour and brain neurochemistry (Jenkins et al. 2016).

A review examining the relationship between the microbiome, chronic stress and the Western diet commonly consumed by adolescents implied that changes in the gut microbiome increase the release of microbial lipopolysaccharides, which activate a gut inflammatory response. This stimulates the HPA axis, inducing symptoms associated with depression (Simkin 2019). Individuals consuming higher proportions of protein and fat reported increased well-being and decreased anxiety and depression, mainly because low-carbohydrate diets had a positive influence on the gut microbiome (Martin et al. 2023). Thus, promoting a protein-rich healthy diet and lifestyle among adolescents would be effective in maintaining a healthy microbiome.

PROTEIN AS A COMPONENT OF A WELL-BALANCED DIET FOR ADOLESCENTS: INFLUENCE ON MENTAL HEALTH

Adequate amounts of macro and micronutrients are essential for all functions of the brain. It has been widely documented that adolescents with mental health disorders are more to likely consume nutrient-poor and calorie-dense diets.

Research has shown that consuming high-fat, high-sugar diets can affect the proteins responsible for signalling BDNF. However, when the synthesis of BDNF is increased by consuming a protein-rich diet, depression symptoms have been shown to improve (Bocchio-Chiavetto 2010). This was evident from a study of adolescent elite athletes, which found that higher protein intake was associated with lower depressive symptom severity (Gerber et al. 2023; Figure 33.5).

A study conducted among adolescents in India also affirms that a low intake of proteins (less than EAR) was associated with higher depression scores. Furthermore, a multivariate regression analysis of the data indicated that adolescents not consuming milk for breakfast had higher odds of depression (male odds ratio (OR): 0.28; 95% CI: 0.10–0.77; p = 0.0177 and females OR: NE; 95% CI: NE; p <0.0177; Khanna and Aeri 2020). Similarly, a study on Chinese college students found that dairy consumption was negatively associated with the occurrence of psychological symptoms (Zhao et al. 2023).

Analysis of large nutrition and health survey datasets (N = 60,935) from the United States of America and South Korea examining the association between macronutrients (carbohydrates, proteins and fat) and depression revealed that a 10% increase in the proportion of calorie intake from protein significantly reduced the prevalence of depression in both the United States (OR: 95% CI,

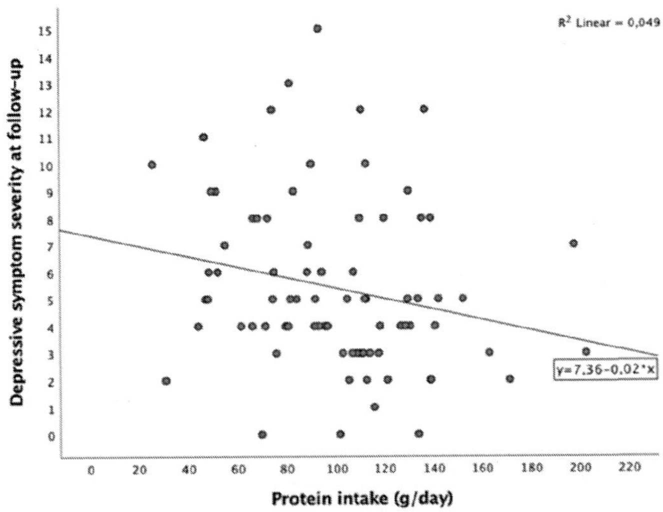

FIGURE 33.5 Association between protein intake and depressive symptoms. This scatterplot and regression line represent the association between protein intake (g/day) and depressive symptoms among adolescent athletes. Protein consumption significantly predicts lower depressive symptoms (Gerber et al. 2023, with permission).

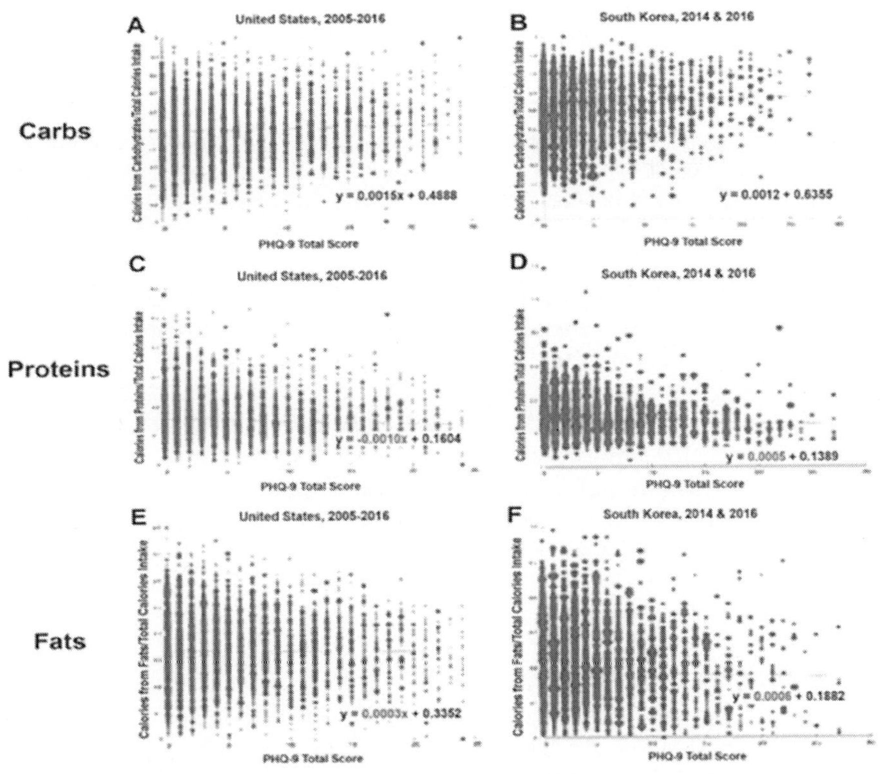

FIGURE 33.6 Scatter plots of depression scores and macronutrient intake. This weighted scatter plot distribution of depression scores (assessed through the Patient Health Questionnaire, PHQ-9) and carbohydrate, protein and fat intake indicates that protein intake is associated with lower depression scores (Oh et al. 2020, under Creative Commons Attribution Licence (CC BY)).

0.621 (0.530–0.728)) and South Korea (0.703 (0.397–0.994)). On the other hand, a 10% increase in the proportion of calories from carbohydrates correlated with increased depression prevalence. Fats showed no association with depression (Oh et al. 2020; Figure 33.6). This again goes on to prove that moving adolescents away from high-carb diets and ensuring adequate protein intake will help reduce the risk of depression.

A study of US adults found that protein intake from milk and milk products might reduce the risk of depressive symptoms (Li et al. 2020). Anxiety symptoms have also been shown to improve after following a high-protein, low-carbohydrate diet (Aucoin 2021).

Considering that mental health disorders also fall under the category of non-communicable diseases, the consumption of balanced diets can be used as an important tool for preventing and delaying (maybe even slowing) the onset of these disorders (Firth et al. 2018; Khanna et al. 2019; Chopra et al. 2021).

SUPPORTING ADOLESCENT MENTAL HEALTH: QUANTITY AND QUALITY OF PROTEIN INTAKE

Balanced, protein-rich diets are key modifiable factors for preventing mental health disorders among adolescents. The recommended protein intake ranges from 35 to 55 g/day among Indian adolescents, providing around 15% of total calorie intake. Appropriate combinations of cereals with pulses (in the ratio of 3:1) plus 250 ml of milk daily can meet EAA requirements. Non-vegetarians consuming animal sources such as meat, poultry, eggs, fish, milk and cheese in appropriate amounts can also achieve EAA intake (ICMR-NIN 2020).

In their enthusiasm for consuming proteins, some adolescents resort to consuming extra in the form of protein supplements. It must be made clear that most recommendations for protein intake across the world are, on average, 0.8 g/kg of body weight/day. This can easily be achieved through a well-balanced, diverse diet. No additional benefits accrue from consuming extra protein, as was shown by Li et al. (2020), who found that while the odds ratio of depressive symptoms decreased when the total protein intake increased, it plateaued at protein intake above 1.0 g/kg/day. Consuming just the right amount of proteins is not difficult to achieve if adolescents consume traditional balanced diets.

The quality of the diet also has a great impact on mental health. As discussed previously, pro-inflammatory eating patterns like the Western diet are associated with an increased risk of mental health problems in adolescents. Adherence to the Mediterranean diet can counteract the effects of stress-induced inflammation (as measured by CRP and IL-6) and reduce the risk of future mental health issues among adolescents.

Animal protein intake has been linked to psychological distress and risk for depression, leading to concerns that it may jeopardise mental health (Zhang et al. 2017). However, this association may stem from the low intake of anti-inflammatory foods in these diets, which is associated with the risk. Animal proteins, in fact, provide the right proportions of essential amino acids (EAA) needed by the body. If animal foods are consumed in appropriate proportions alongside adequate portions of vegetables, fruits and whole foods, they will add high biological value proteins to adolescent diets. However, processed meats may negate these beneficial effects, so their intake should be minimised.

Adolescents who consume vegetarian proteins can be taught to balance the amino acid profile of their diets using cereal-pulse combinations and diversifying their diets. Plant proteins also contain fibre, which may also be good for gut health. Low-fat dairy can serve as a good source of high-quality proteins in vegetarian diets. Many adolescents, especially in developing countries, consume protein-deficient diets due to food scarcity or lack of availability of nutritious foods. On the other hand, the consumption of ultra-processed foods and sugar-sweetened beverages has nudged the nutritive components like proteins out of the diets of these youngsters.

Educating adolescents via various channels, be it schools, social media campaigns, peer leaders, etc., to ensure correct food choices will help in this regard. Reading food labels, choosing fresh foods in place of processed ones, promoting locally available inexpensive foods and diet diversity will help in fighting the onset of mental health disorders. Adolescents with mental health problems should be screened for protein deficiency (and other deficiencies as well), and diet counseling should be provided by a nutritionist to improve protein intake, correct distorted eating patterns and address nutritional deficiencies, if any. Sensitising parents and caregivers to mental health problems so that they seek immediate help if their child shows any signs of changed behaviour is also necessary. Encouraging adolescents to engage in regular physical activity would also instil confidence and boost mental and physical well-being.

CONCLUSION

The risk factors for the onset of mental disorders among adolescents are multifaceted, encompassing psychological, social and genetic influences, as well as inflammation, oxidative stress and gut dysbiosis. Consuming protein-rich balanced diets can influence neurotransmitter synthesis, potentially reduce inflammation and help maintain the microbiota-gut-brain axis, which can contribute to the prevention of the risk of mental health problems among adolescents.

Encouraging adolescents to opt for protein-rich meals and snacks instead of sugary, processed foods can help them steer clear of anxiety and depression triggers. Additionally, a diet rich in protein enhances energy levels, enabling them to feel better and stay active. This positive outlook can improve body image perception and boost confidence.

Adolescence is a crucial stage of human development, both physically and mentally. Adolescents acquire cognitive and social-emotional skills that shape their personalities and prepare them for their future roles in society. Poor mental health and psychosocial development during this period can have long-lasting consequences, limiting opportunities in adulthood. Therefore, adolescent well-being requires a multidimensional and holistic approach, and supporting it is crucial for realising their full potential. Consuming a protein-rich, well-balanced diet can contribute significantly to achieving this well-being.

MINI DICTIONARY OF TERMS

Dopamine: Also called the happy hormone, it provides feelings of reward and pleasure.

Gut-brain axis: The two-way communication between the brain and the gut microbiota, now considered important for understanding mental health.

Neurotransmitter: A chemical messenger that relays signals from one neuron to the next.

Serotonin: A neurotransmitter released in the body that gives a feeling of well-being.

Ultra-processed foods: Highly processed foods typically high in sugar, salt and fat, containing artificial colours and stabilisers.

SUMMARY POINTS

- The rising incidence of mental health disorders among adolescents is a cause for concern.
- Consumption of balanced diets promotes better mental health outcomes in adolescents.
- Protein-rich diets provide the body with amino acids like tryptophan, tyrosine, histidine, arginine, threonine, etc., which act as substrates for the synthesis of neurotransmitters (e.g., tryptophan is a precursor for serotonin).
- These neurotransmitters improve mood, provide pleasure and satisfaction and enhance attention, memory and cognition among adolescents.

- Diets rich in proteins are also anti-inflammatory and are beneficial for maintaining a healthy weight and the microbiota-gut-brain axis.
- Communication between the enteric and central nervous systems is increasingly recognised as significant for maintaining positive mental health.
- Encouraging adolescents to opt for protein-rich meals and snacks instead of sugary, processed foods can help them steer clear of mental health risk factors.
- The move away from Western diets and towards more traditional balanced diets needs to be promoted.
- Addressing adolescent mental health is crucial due to its far-reaching repercussions into adulthood.

REFERENCES

Aeri, Bani Tamber. "Role of Probiotics and Vitamins in Maintaining a Healthy Gut Microbiome: Recent Advances." *Indian Journal of Community Health* 33, no. 3 (September 30, 2021): 543–46. https://doi.org/10.47203/ijch.2021.v33i03.027

Attipoe, Selasi, Stacey A. Zeno, Courtney Lee, Cindy Crawford, Raheleh Khorsan, Avi R. Walter, and Patricia A. Deuster. "Tyrosine for Mitigating Stress and Enhancing Performance in Healthy Adult Humans, a Rapid Evidence Assessment of the Literature." *Military Medicine* 180, no. 7 (July 1, 2015): 754–65. https://doi.org/10.7205/milmed-d-14-00594

Aucoin, Monique, Laura LaChance, Umadevi Naidoo, Daniella Remy, Tanisha Shekdar, Negin Sayar, Valentina Cardozo, Tara Rawana, Irina Chan, and Kieran Cooley. "Diet and Anxiety: A Scoping Review." *Nutrients* 13, no. 12 (December 10, 2021): 4418. https://doi.org/10.3390/nu13124418

Bernier, Veronique, Marie-Hélène Debarge, Matthieu Hein, Sarah Ammendola, Anais Mungo, and Gwenole Loas. "Major Depressive Disorder, Inflammation, and Nutrition: A Tricky Pattern?." *Nutrients* 15, no. 15 (August 3, 2023): 3438. https://doi.org/10.3390/nu15153438

Bocchio-Chiavetto, Luisella, Vincenzo Bagnardi, Roberta Zanardini, Raffaella Molteni, Maria Gabriela Nielsen, Anna Placentino, Caterina Giovannini et al. "Serum and Plasma BDNF Levels in Major Depression: A Replication Study and Meta-analyses." *World Journal of Biological Psychiatry* 11, no. 6 (March 25, 2010): 763–73. https://doi.org/10.3109/15622971003611319.

Booij, Linda, Willem Van Der Does, P. M. J. Haffmans, Wim J. Riedel, Durk Fekkes, and Marc Blom. "The Effects of High-Dose and Low-Dose Tryptophan Depletion on Mood and Cognitive Functions of Remitted Depressed Patients." *Journal of Psychopharmacology* 19, no. 3 (May 1, 2005): 267–75. https://doi.org/10.1177/0269881105051538.

Chopra, Chandni, Subhadra Mandalika, and Narendra Kinger. "Does Diet Play a Role in the Prevention and Management of Depression among Adolescents? A Narrative Review." *Nutrition and Health* 27, no. 2 (January 8, 2021): 243–63. https://doi.org/10.1177/0260106020980532.

Fan, Mingyue, Xiao Gao, Li Li, Zhongyu Ren, Leanna MW Lui, Roger S. McIntyre, Kayla M. Teopiz, Peng Deng, and Bing Cao. "The Association Between Concentrations of Arginine, Ornithine, Citrulline and Major Depressive Disorder: A Meta-Analysis." *Frontiers in Psychiatry* 12 (November 18, 2021): 686973. https://doi.org/10.3389/fpsyt.2021.686973.

Feder, Adriana, Jamie Skipper, James Blair, Katherine R. Buchholz, Sanjay J. Mathew, Markus J. Schwarz, John Doucette, et al. "Tryptophan Depletion and Emotional Processing in Healthy Volunteers at High Risk for Depression." *Biological Psychiatry* 69, no. 8 (April 1, 2011): 804–7. https://doi.org/10.1016/j.biopsych.2010.12.033.

Firth, Joseph, Brendon Stubbs, Scott Teasdale, Philip B. Ward, Nicola Veronese, Nitin Shivappa, James R. Hébert, Michael Berk, Alison R. Yung, and Jerome Sarris. "Diet as a Hot Topic in Psychiatry: A Population-scale Study of Nutritional Intake and Inflammatory Potential in Severe Mental Illness." *World Psychiatry* 17, no. 3 (September 7, 2018): 365–67. https://doi.org/10.1002/wps.20571.

Gaiha, Shivani Mathur, Tatiana Taylor Salisbury, Mirja Koschorke, Usha Raman, and Mark Petticrew. "Stigma Associated with Mental Health Problems among Young People in India: A Systematic Review of Magnitude, Manifestations and Recommendations." *BMC Psychiatry* 20, no. 1 (November 16, 2020): 1–24. https://doi.org/10.1186/s12888-020-02937-x.

Gerber, Markus, Sarah Jakowski, Michael Kellmann, Robyn Cody, Basil Gygax, Sebastian Ludyga, Caspar Müller, Sven Ramseyer, and Johanna Beckmann. "Macronutrient Intake as a Prospective Predictor of Depressive Symptom Severity: An Exploratory Study with Adolescent Elite Athletes." *Psychology of Sport and Exercise* 67 (July 1, 2023): 102442. https://doi.org/10.1016/j.psychsport.2023.102442.

Hall, Kevin D. “The Potential Role of Protein Leverage in the US Obesity Epidemic.” *Obesity* 27, no. 8 (May 16, 2019): 1222–24. https://doi.org/10.1002/oby.22520.

Hase, Adrian, Sophie E. Jung, and Marije Aan Het Rot. “Behavioral and Cognitive Effects of Tyrosine Intake in Healthy Human Adults.” *Pharmacology, Biochemistry and Behavior* 133 (June 1, 2015): 1–6. https://doi.org/10.1016/j.pbb.2015.03.008.

Holeček, Milan. “Histidine in Health and Disease: Metabolism, Physiological Importance, and Use as a Supplement.” *Nutrients* 12, no. 3 (March 22, 2020): 848. https://doi.org/10.3390/nu12030848.

ICMR/NIN. *Nutrient Requirements for Indians*. Hyderabad: ICMR-National Institute of Nutrition, 2020. https://www.nin.res.in/RDA_Full_Report_2020.html

Jenkins, Trisha A., Jason Nguyen, Kate E. Polglaze, and Paul Bertrand. “Influence of Tryptophan and Serotonin on Mood and Cognition with a Possible Role of the Gut-Brain Axis.” *Nutrients* 8, no. 1 (January 20, 2016): 56. https://doi.org/10.3390/nu8010056.

Jiang, Shuai, Hui Liu, and Chunbao Li. “Dietary Regulation of Oxidative Stress in Chronic Metabolic Diseases.” *Foods* 10, no. 8 (August 11, 2021): 1854. https://doi.org/10.3390/foods10081854.

Khanna, Preeti, and Bani Tamber Aeri. “Association of Quantity and Quality of Protein Intake with Depression and Anxiety Symptoms among Adolescent Boys and Girls (13–15 Years) Studying in Public Schools of Delhi.” *Journal of Nutritional Science and Vitaminology* 66, no. Supplement (January 1, 2020): S141–48. https://doi.org/10.3177/jnsv.66.s141.

Khanna, Preeti, Vijay Kumar Chattu, and Bani Tamber Aeri. “Nutritional Aspects of Depression in Adolescents - A Systematic Review.” *International Journal of Preventive Medicine* 10, no. 1 (January 1, 2019): 42. https://doi.org/10.4103/ijpvm.ijpvm_400_18.

Kieling, Christian, Helen Baker-Henningham, Myron L. Belfer, Gabriella Conti, Ilgi Ozturk Ertem, Olayinka Omigbodun, Luis Augusto Rohde, Shoba Srinath, Nurper Ulkuer, and Atif Rahman. “Child and Adolescent Mental Health Worldwide: Evidence for Action.” *The Lancet* 378, no. 9801 (October 1, 2011): 1515–25. https://doi.org/10.1016/s0140-6736(11)60827-1.

Lakhan, Shaheen E., and Karen Vieira. “Nutritional Therapies for Mental Disorders.” *Nutrition Journal* 7, no. 1 (January 21, 2008): 1–8 . https://doi.org/10.1186/1475-2891-7-2.

Lee, Lana, Krishna K. Upadhya, Pamela A. Matson, Hoover Adger, and Maria Trent. “The Status of Adolescent Medicine: Building a Global Adolescent Workforce.” *International Journal of Adolescent Medicine and Health* 28, no. 3 (August 1, 2016): 233–43. https://doi.org/10.1515/ijamh-2016-5003.

Li, Yan, Caixia Zhang, Suyun Li, and Dongfeng Zhang. “Association between Dietary Protein Intake and the Risk of Depressive Symptoms in Adults.” *British Journal of Nutrition* 123, no. 11 (February 20, 2020): 1290–1301. https://doi.org/10.1017/s0007114520000562.

Lupien, Sonia J., Bruce S. McEwen, Megan R. Gunnar, and Christine Heim. “Effects of Stress throughout the Lifespan on the Brain, Behaviour and Cognition.” *Nature Reviews Neuroscience* 10, no. 6 (April 29, 2009): 434–45. https://doi.org/10.1038/nrn2639.

Markus, C. Rob, Berend Olivier, Geert Panhuysen, Jan Van Der Gugten, Alles, Adriaan Tuiten, H. G. M. Westenberg, Durk Fekkes, H. F. Koppeschaar, and E. E. H. F. De Haan. “The Bovine Protein α-Lactalbumin Increases the Plasma Ratio of Tryptophan to the Other Large Neutral Amino Acids, and in Vulnerable Subjects Raises Brain Serotonin Activity, Reduces Cortisol Concentration, and Improves Mood under Stress.” *The American Journal of Clinical Nutrition* 71, no. 6 (June 1, 2000): 1536–44. https://doi.org/10.1093/ajcn/71.6.1536 .

Martin, Sydney E., Colleen Suzanne Kraft, Thomas R. Ziegler, Erin C. Millson, Lavanya Rishishwar, and Greg S. Martin. “The Role of Diet on the Gut Microbiome, Mood and Happiness.” *medRxiv* (2023): 2023. https://doi.org/10.1101/2023.03.18.23287442

Michels, Nathalie, Gerard Clarke, Loreto Olavarría-Ramírez, Sonia Gómez-Martínez, L. E. Díaz, Ascensión Marcos, Kurt Widhalm, and Livia A. Carvalho. “Psychosocial Stress and Inflammation Driving Tryptophan Breakdown in Children and Adolescents: A Cross-Sectional Analysis of Two Cohorts.” *Psychoneuroendocrinology* 94 (August 1, 2018): 104–11. https://doi.org/10.1016/j.psyneuen.2018.05.013.

Nishijo, Muneko, Ngoc Thao Pham, Hai Thai Thu Duong, Ngoc Nghi Tran, Takashi Kondoh, Yoshikazu Nishino, Hiroshi Nishimaru, Quyet Ba, and Hisao Nishijo. “Nutritional Intervention with Dried Bonito Broth for the Amelioration of Aggressive Behaviors in Children with Prenatal Exposure to Dioxins in Vietnam: A Pilot Study.” *Nutrients* 13, no. 5 (April 25, 2021): 1455. https://doi.org/10.3390/nu13051455.

Oh, Jihoon, Kyongsik Yun, Jeong-Ho Chae, and Tae-Suk Kim. “Association between Macronutrients Intake and Depression in the United States and South Korea.” *Frontiers in Psychiatry* 11 (March 17, 2020). https://doi.org/10.3389/fpsyt.2020.00207.

O'Neil, Adrienne, Shae E. Quirk, Siobhan Housden, Sharon L. Brennan, Lana J. Williams, Julie A. Pasco, Michael Berk, and Felice N. Jacka. "Relationship between Diet and Mental Health in Children and Adolescents: A Systematic Review." *American Journal of Public Health* 104, no. 10 (October 1, 2014): e31–42. https://doi.org/10.2105/ajph.2014.302110.

Paul, Vanaja, and Perumal Ekambaram. "Involvement of Nitric Oxide in Learning & Memory Processes." *The Indian Journal of Medical Research* 133, no. 5 (2011): 471. https://pubmed.ncbi.nlm.nih.gov/21623030.

Pinto, Vitor Laerte, Junior, Paula Fontoura Coelho De Souza, Tatiana M.C. Brunini, M. B. L. Oliveira, Monique Bandeira Moss, Mariana Alves De Sá Siqueira, Marcos Rochedo Ferraz, and Antônio Cláudio Mendes-Ribeiro. "Low Plasma Levels of L-Arginine, Impaired Intraplatelet Nitric Oxide and Platelet Hyperaggregability: Implications for Cardiovascular Disease in Depressive Patients." *Journal of Affective Disorders* 140, no. 2 (October 1, 2012): 187–92. https://doi.org/10.1016/j.jad.2012.02.008.

Puglisi, Michael J., and Maria Luz Fernandez. "The Health Benefits of Egg Protein." *Nutrients* 14, no. 14 (July 15, 2022): 2904. https://doi.org/10.3390/nu14142904.

Razak, Meerza Abdul, Pathan Shajahan Begum, Buddolla Viswanath, and Senthilkumar Rajagopal. "Multifarious Beneficial Effect of Nonessential Amino Acid, Glycine: A Review." *Oxidative Medicine and Cellular Longevity* 2017 (January 1, 2017): 1–8. https://doi.org/10.1155/2017/1716701.

Reales-Moreno, Marta, P Tonini, Rosa M. Escorihuela, Montserrat Solanas, Sílvia Fernández-Barrés, Dora Romaguera, and Oren Contreras-Rodríguez. "Ultra-Processed Foods and Drinks Consumption Is Associated with Psychosocial Functioning in Adolescents." *Nutrients* 14, no. 22 (November 15, 2022): 4831. https://doi.org/10.3390/nu14224831.

Rot, Marije Aan Het, D. S. Moskowitz, Gilbert Pinard, and Simon N. Young. "Social Behaviour and Mood in Everyday Life: The Effects of Tryptophan in Quarrelsome Individuals." *Journal of Psychiatry and Neuroscience* 31, no. 4 (July 1, 2006): 253–62. https://pubmed.ncbi.nlm.nih.gov/16862243.

Sasahara, Ikuko, Naoko Fujimura, Yoshinori Nozawa, Yasufumi Furuhata, and Hitoshi Sato. "The Effect of Histidine on Mental Fatigue and Cognitive Performance in Subjects with High Fatigue and Sleep Disruption Scores." *Physiology & Behavior* 147 (August 1, 2015): 238–44. https://doi.org/10.1016/j.physbeh.2015.04.042.

Simkin, Deborah R. "Microbiome and Mental Health, Specifically as It Relates to Adolescents." *Current Psychiatry Reports* 21, no. 9 (September 1, 2019). https://doi.org/10.1007/s11920-019-1075-3.

Ventriglio, Antonio, Federica Sancassiani, Maria Paola Contu, Mariateresa Latorre, Melanie Di Salvatore, Michele Fornaro, and Dinesh Bhugra. "Mediterranean Diet and Its Benefits on Health and Mental Health: A Literature Review." *Clinical Practice & Epidemiology in Mental Health* 16, no. 1 (July 30, 2020): 156–64. https://doi.org/10.2174/1745017902016010156.

Watford, Malcolm, and Guoyao Wu. "Protein." *Advances in Nutrition* 9, no. 5 (September 1, 2018): 651–53. https://doi.org/10.1093/advances/nmy027.

World Health Organization: WHO. "Mental Health of Adolescents." *Www.Who.Int*, November 17, 2021. https://www.who.int/news-room/fact-sheets/detail/adolescent-mental-health.

Yoshikawa, Takeo, Tadaho Nakamura, Tetsuro Shibakusa, Mayu Sugita, Fumito Naganuma, Takahiro Iida, Y. Miura, Attayeb Mohsen, Ryuichi Harada, and Kazuhiko Yanai. "Insufficient Intake of L-Histidine Reduces Brain Histamine and Causes Anxiety-Like Behaviors in Male Mice." *Journal of Nutrition* 144, no. 10 (July 23, 2014): 1637–41. https://doi.org/10.3945/jn.114.196105.

Young, Simon N. "How to Increase Serotonin in the Human Brain without Drugs." *Journal of Psychiatry & Neuroscience: JPN* 32, no. 6 (2007): 394.

Zhang, Yi, Ye Yang, Ming-Sheng Xie, Xiang Ding, Hui Li, Zhi-Chen Lui, and Shi-Fang Peng. "Is Meat Consumption Associated with Depression? A Meta-Analysis of Observational Studies." *BMC Psychiatry* 17, no 1 (December 28, 2017): 409. https://doi:10.1186/s12888-017-1540-7.

Zhao, Zhimin, Ruibao Cai, Yongxing Zhao, Yanyan Hu, Jingzhi Liu, and Minghao Wu. "Association between Dairy Consumption and Psychological Symptoms: Evidence from a Cross-Sectional Study of College Students in the Yangtze River Delta Region of China." *International Journal of Environmental Research and Public Health* 20, no. 4 (February 13, 2023): 3261. https://doi.org/10.3390/ijerph20043261.

Zielińska, Magdalena, Edyta Łuszczki, Izabela Michońska, and Katarzyna Dereń. "The Mediterranean Diet and the Western Diet in Adolescent Depression-Current Reports." *Nutrients* 14, no. 20 (October 19, 2022): 4390. https://doi.org/10.3390/nu14204390.

34 Protein Intake and Weight Loss in Postmenopausal Women

Jamie I. Baum, Yuxiang Sun, and Aubree L. Hawley

ABBREVIATIONS

AA	Amino acid
AMDR	Acceptable macronutrient distribution range
BCAA	Branched-chain amino acid
BMI	Body mass index
BMR	Basal metabolic rate
CVD	Cardiovascular disease
E_2	Estradiol
FFM	Fat-free mass
FSH	Follicle-stimulating hormone
GLP-1	Glucagon-like peptide-1
HDL	High-density lipoprotein
LBM	Lean body mass
LDL	Low-density lipoprotein
LNAA	Large neutral amino acid
PYY	Peptide YY
RCT	Randomized controlled trial
RDA	Recommended dietary allowance
REE	Resting energy expenditure
SWAN	Study of Women's Health Across the Nation
TEF	Thermic effect of food

INTRODUCTION

People worldwide are living longer. The aging population across the world is growing and is expected to be twice that of children by 2050 (Steverson 2022). Women outnumber men at older ages due to increased life expectancy, accounting for nearly 56% of people aged 65 or older (U.N. 2022). Menopause is a natural part of the aging process, with most women experiencing menopause between the ages of 45 and 55 years (WHO 2022). Natural menopause is defined as occurring 12 months after the last menstrual cycle and marks the end of menstrual cycles (WHO 2022). Menopause results in declining endogenous estrogen production caused by the loss of ovarian follicular function. This decrease in estrogen during the menopausal transition is associated with increased central adiposity and loss of fat-free mass (FFM; e.g., skeletal muscle loss), putting postmenopausal women at increased risk for negative health outcomes such as cardiovascular disease and type-2 diabetes mellitus (Donato et al. 2006; Rolland et al. 2007). In addition to cardiometabolic risk, age- and menopause-related reduction in muscle mass and function is associated with decreased well-being such as poor sleep quality and depression (Lee and Lee 2013).

DOI: 10.1201/b23328-34

Over 43% of postmenopausal women have obesity (Knight et al. 2021). Currently, the treatment of obesity in postmenopausal women focuses on lifestyle and dietary changes (Knight et al. 2021). Research suggests that nutritional strategies focused on incorporating high-quality protein are potential methods to mitigate weight gain and improve cardiometabolic health outcomes in postmenopausal women (Silva and Spritzer 2017). This chapter examines body composition and cardiometabolic health changes occurring with menopause and the potential role of dietary protein intake in mitigating these factors.

TRANSITIONING TO MENOPAUSE

The length of menopause typically lasts four to five years; however, the duration of the transition is highly variable, ranging from less than one year to ten years or longer (Harlow 2018). The menopausal transition is generally defined as the time between the onset of irregular menstrual cycles and menopause (Talaulikar 2022). Age at menopause is influenced by multiple factors such as diet, exercise levels, smoking, socioeconomic background, ethnicity, culture, concurrent medical issues, and body mass index (BMI) (Talaulikar 2022). A higher BMI is related to a later onset of the menopausal transition but not its duration (Talaulikar 2022). The transition involves changes in hormones, with women experiencing dramatic decreases in circulating estrogens, particularly estradiol (E_2), and increases in the gonadotropin follicle-stimulating hormone (FSH) (Marlatt et al. 2022). These hormonal changes are linked to many of the metabolic and physiological changes observed with menopause (Figure 34.1) (Motlani et al. 2023).

BODY COMPOSITION AND WEIGHT CHANGES WITH MENOPAUSE

Obesity is a major public health concern that affects women at a disproportionate rate compared to men (Flegal et al. 2016). The postmenopausal state is associated with increases in weight gain, increasing the risk for obesity. Women aged 40–59 years and those 60 and over have an obesity prevalence of 43.3% in (Hales et al. 2020). On average, women gain 5–7 pounds (or 2–3 kg) during the menopause transition, though interindividual variability is substantial (Marlatt et al. 2022). Changes in hormones, along with aging, lifestyle, race, ethnicity, and genetic factors, contribute to the changes in body weight and body composition observed with menopause. These changes in

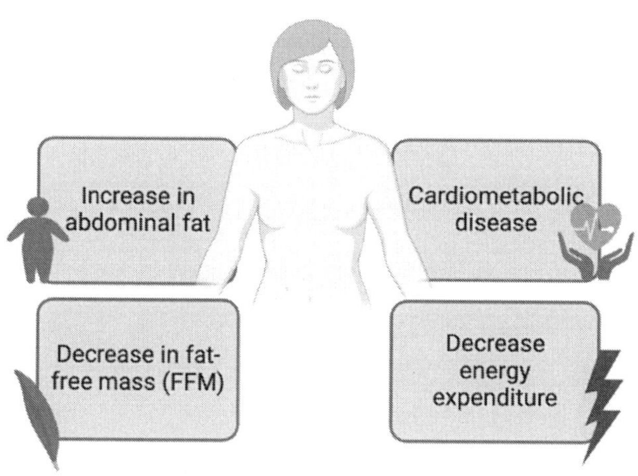

FIGURE 34.1 Physiological changes influencing weight gain in postmenopausal women. Changes in hormones during the menopause transition can lead to shifts in body composition, increasing abdominal fat mass and decreasing muscle mass. These changes may also lead to increased weight gain via a decrease in energy expenditure, changes in sleep, and increased risk for poor cardiometabolic health.

body composition result in decreased energy expenditure and increased energy intake, promoting a positive energy balance that often leads to weight gain and contributes to increased cardiometabolic risk (Marlatt et al. 2022; Greendale et al. 2019). The Study of Women's Health Across the Nation (SWAN) confirms that the accelerated gain in fat mass and loss of FFM were related to the menopause transition rather than aging (Greendale et al. 2019; Pasquali et al. 1994; Marlatt et al. 2022). Additional studies have confirmed the gain in fat mass during menopause, particularly an increase in both subcutaneous and visceral abdominal fat (Lovejoy et al. 2008; Guthrie et al. 2004; Janssen et al. 2010). In addition, as women age, there tends to be an increased amount of fat tissue in the abdomen and a relatively reduced amount of fat in the hip-thigh area (Ko and Jung 2021; Greendale et al. 2019). Increased weight gain, especially in the abdominal area, also influences the severity f vasomotor symptoms (e.g., hot flashes and night sweats) and insomnia, as well as increased fatigue and decreased quality of life (Marlatt et al. 2022).

Sarcopenia refers to the skeletal muscle loss that occurs with age. Skeletal muscle is lost at a rate of 3–8% every 10 years after the age of 30 and accelerates with age, coinciding with the peri- and postmenopausal periods. Sarcopenia is associated with an increased risk of functional disability, falls, fractures, and overall mortality among older adults (Baum and Wolfe 2015; Hawley and Baum 2024). Women develop sarcopenia earlier than men, and the incidence of sarcopenia increases rapidly during the menopausal period (Ko and Kim 2020). Several cross-sectional studies have reported that postmenopausal women exhibit lower FFM in the whole body, trunk, and lower extremity regions than premenopausal women (Ko and Kim 2020; Poehlman, Toth, and Gardner 1995).

FFM accounts for approximately 60–85% of body mass and is a major determinant of basal metabolic rate (BMR) or resting energy expenditure (REE). The loss of FFM with age results in a decrease in REE and may result in changes in energy metabolism (Ko and Kim 2020). The Healthy Transitions longitudinal study supports this, demonstrating changes in body composition that may be attributed to reductions in energy expenditure and fat oxidation in postmenopausal women (Lovejoy et al. 2008). The same study also demonstrated that women who had completed the menopause transition had a greater reduction in 24-hour energy expenditure and sleep energy expenditure compared to premenopausal women (Lovejoy et al. 2008).

CARDIOMETABOLIC HEALTH IN POSTMENOPAUSAL WOMEN

Changes in body composition and reproductive hormones lead to increased cardiometabolic risk in postmenopausal women (Marlatt et al. 2022). It has been suggested that the increase in central adiposity may increase the risk of insulin resistance and of developing metabolic syndrome; however, data supporting this association is mixed (Marlatt et al. 2022). Estrogens, particularly E_2, play a role in maintaining glucose homeostasis through effects on insulin secretion and clearance; however, mechanisms are unclear (Godsland 2005). In addition, the Framingham study was one of the first studies to provide evidence of an increase in cardiovascular events in postmenopausal women compared to premenopausal women of the same age (Kannel et al. 1976; Roa-Diaz et al. 2021).

Cardiovascular disease (CVD) is the leading cause of death in postmenopausal women, and CVD risk factors such as dyslipidemia, hypertension, and cardiac dysfunction increase during menopause (Marlatt et al. 2022). Women are at higher risk of developing CVDs after menopause due to estrogen deficiency and dysregulated lipid metabolism. Estrogens, especially E_2, play a protective role in the cardiovascular system and are produced primarily in the ovaries via a process that uses low-density lipoprotein (LDL) cholesterol as a substrate. However, circulatory LDL cannot be utilized to synthesize estrogen during menopause, thereby resulting in decreased estrogen production (reviewed in Ko and Kim 2020).

The prevalence of sleep disturbances increases from the premenopausal to postmenopausal periods. Sleep quality also decreases in women around the menopause transition due to the decline of E_2 and progesterone. Vasomotor symptoms associated with menopause may further contribute to reduced sleep quality. The SWAN study found sleep disturbances ranged from 16 to 42% in premenopausal women and increased to 39–47% and 35–60% in perimenopausal and postmenopausal women, respectively (Kravitz et al. 2011; Kravitz and Joffe 2011). This is important because poor sleep quality is an additional risk factor for poor cardiometabolic health (Marlatt et al. 2022).

PROTEIN NEEDS OF POSTMENOPAUSAL WOMEN

Nutrition plays an essential role in the health, function, and well-being of postmenopausal women (Hawley and Baum 2024). Among nutrients, dietary protein has been identified as a key supportive nutrient for older adults and is important for maintaining energy balance (Westerterp-Plantenga, Lemmens, and Westerterp 2012), reducing cardiometabolic risk, and promoting overall well-being (Peuhkuri, Sihvola, and Korpela 2012).

Recommendations for protein intake focus on age and/or body weight, but not sex. This means both men and women over the age of 18 years are advised to consume the same amount of protein per kilogram (kg) of body weight to meet basic needs. However, factors like menopause and body composition (e.g., muscle mass) can influence individual protein requirements. The current dietary protein recommendations are based on studies that estimate the minimum protein intake necessary to maintain nitrogen balance (Rand et al. 2003). Currently, the Food and Nutrition Board of the National Academy of Medicine has set the recommended daily allowance (RDA) for protein at 0.8 g/kg/day, covering the minimum requirements of 97–98% of all healthy adults >18 years of age, including older adults (Trumbo et al. 2002). In addition to the RDA, recommendations for protein intake are also provided in the context of a complete diet within the acceptable macronutrient distribution range (AMDR) (Dietary reference intakes for energy, carbohydrate, fiber, fat, fatty acids, cholesterol, protein and amino acids. Institute of Medicine. 2005). The AMDR expresses protein intake recommendations as a percentage of total caloric intake (10–35% of daily energy intake from protein) and is more relevant in the context of a complete diet than the RDA (Wolfe et al. 2017). The AMDR upper value of 35% far exceeds the RDA of 0.8 g/kg/day by approximately four times at ~3.0 g/kg/day. Surprisingly, the current percent daily value (DV%) for protein is based on 50 g of dietary protein in the context of a 2000 kcal/day diet, equating to 10% of daily intake, the minimum amount of daily protein to maintain net protein balance. Contrarily, moderate or high consumption of 1.2–2.0 g/kg/day of dietary protein easily falls within the AMDR and should be considered for older adults, including postmenopausal women (Phillips, Chevalier, and Leidy 2016; Wolfe 2012; Baum and Wolfe 2015; Hawley and Baum 2024).

Recent dietary protein and aging research has focused on the optimal daily and per-meal intake to promote skeletal muscle mass and function in older adults (Paddon-Jones and Leidy 2014; Paddon-Jones and Rasmussen 2009; Volpi et al. 2013). Optimal protein intake, defined in terms of skeletal muscle, is the minimal dose of protein that stimulates a maximal anabolic response and maintains or improves skeletal muscle mass and function over time (Baum, Young, and Wolfe 2016; Symons et al. 2009). According to the PROT-AGE (Dietary Protein Needs with Aging Study Group) recommendations, older adults are recommended to consume 25–30 g of protein (e.g., 0.4 g/kg/meal) (Moore et al. 2014) and 2.5–2.8 g of the branched-chain amino acid (BCAA) leucine per meal, equating to 1.0–1.2 g/kg/day of dietary protein intake (Bauer et al. 2013; Agriculture 2015). In line with these recommendations, a cross-sectional study in healthy older adults found a positive association between daily protein distribution of >25 g per meal and appendicular skeletal muscle (Valenzuela et al. 2013). According to National Health and Nutrition Examination Survey (NHANES) data, older men and women consume 1.01 ± 0.03 g/kg/day and ~0.97 ± 0.04 g/kg/day of dietary protein, respectively (Berryman et al. 2018; Volpi et al. 2013; Bauer et al. 2013). However, 19.21 ± 2.11% and 13.17 ± 1.33% of older men and women, respectively, fell below the RDA for dietary protein (Berryman et al. 2018). In addition, protein consumption in older adults follows a highly skewed distribution pattern with a disproportionate amount of daily protein consumed during the evening meal (>60% of daily protein) and far less at breakfast (Arentson-Lantz et al. 2015).

For example, a cross-sectional analysis found older men and women to consume 11.4 ± 0.4 g and 15.3 ± 0.5 g of protein at breakfast compared to 44.5 ± 1.0 g and 44.8 ± 1.0 g at dinner (Berner et al. 2013).

PROTEIN INTAKE, WEIGHT LOSS, AND BODY COMPOSITION IN POSTMENOPAUSAL WOMEN

While some research suggests higher protein intake benefits weight loss in older adults, there is a gap in comprehensive studies focused solely on postmenopausal women. As mentioned above, most dietary protein guidelines do not differentiate based on sex or menopausal status. This lack of specific research makes it difficult to determine the optimal protein intake for postmenopausal women trying to lose weight and improve body composition (e.g., decrease fat mass and maintain or increase FFM).

In general, diet-induced weight loss results in a decrease in both fat mass and lean body mass (Villareal et al. 2005; Mathus-Vliegen and Obesity Management Task Force of the European Association for the Study of Obesity 2012), which could worsen age-related muscle loss, increase the ratio of fat mass to FFM, and further impair cardiometabolic health (Figure 34.2). Several population-based studies have evaluated the effect of weight loss on mortality in middle-aged to older age groups (Andres, Muller, and Sorkin 1993; Williamson and Pamuk 1993; Lissner et al. 1991; Blair et al. 1993; Villareal et al. 2005). The data from these studies indicate that weight loss in older adults is associated with increased, rather than decreased, mortality. It is important to note that these are purely observational-based studies and not randomized controlled trials (RCTs).

However, several studies, primarily conducted in middle-aged adults, have found that intentional weight loss is associated with reduced mortality rates in people with diabetes, impaired glucose tolerance, and other health conditions (Villareal et al. 2005). Specifically, weight loss diets higher in protein (≥30% of energy intake) and lower in carbohydrates have been shown to promote weight loss, improve body composition (i.e., decrease fat mass and increase FFM), and regulate glycemic response in women (Farnsworth et al. 2003; Gannon et al. 2003; Layman, Shiue, et al. 2003; Layman, Boileau, et al. 2003; Baum, Gray, and Binns 2015).

There are several meta-analyses and systematic reviews that have assessed whether higher protein diets impact weight loss and changes in body composition in adult men and women (Clifton, Condo, and Keogh 2014; Dong et al. 2013a; Wycherley et al. 2013). In studies lasting less than a year, higher protein intake ranging from 1.2–1.6 g/kg/day (i.e., 16–45% of energy intake) demonstrated greater weight loss, fat mass loss, and preservation of lean mass compared with normal protein diets ranging from 5–23% energy from protein. Weight loss effects were also observed with higher protein intake in studies on dietary intervention focused on weight maintenance (Dong et al. 2013a; Phillips, Chevalier, and Leidy 2016). In studies longer than a year, higher protein intake led to greater weight loss and fat loss compared with normal protein diets; however, changes in FFM were not observed (Clifton, Condo, and Keogh 2014; Phillips, Chevalier, and Leidy 2016). These differences could be due to a lack of compliance or difficulties maintaining higher protein intake in the long term.

Another potential mechanism for the impact of higher protein intake on weight loss may be appetite regulation, potentially through the hunger-stimulating hormone ghrelin and the satiety hormones peptide YY (PYY) and glucagon-like peptide-1 (GLP-1). Most of the data related to the effect of higher protein intake on the regulation of appetite stem from single-meal acute studies. The majority of these studies demonstrated improvements in at least one marker of appetite and satiety (e.g., subjective or hormonal measurements) (Phillips, Chevalier, and Leidy 2016; Leidy et al. 2015). Several studies have also shown that consumption of high-protein meals containing at least 30 g of protein led to greater reductions in select cortico-limbic neural responses to food stimuli compared with the consumption of protein meals containing 15–18 g of protein (Leidy et al. 2011).

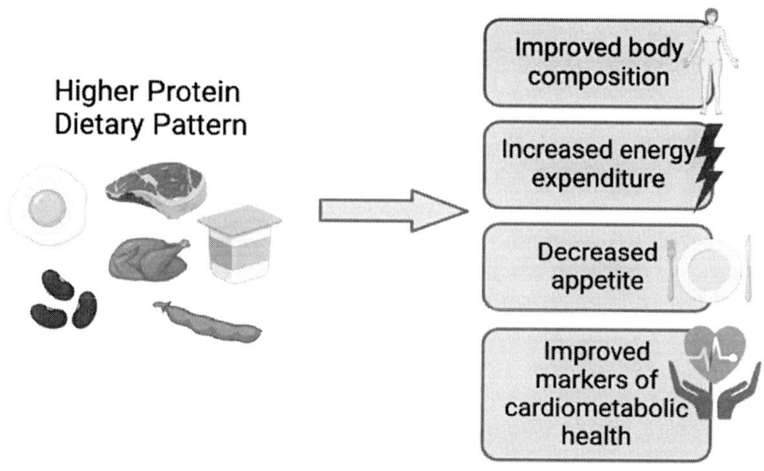

FIGURE 34.2 Benefits of higher protein dietary patterns on weight loss for postmenopausal women. Higher protein dietary patterns (e.g., >1.2 g protein/kg body weight) may improve weight loss in postmenopausal women through preservation of fat-free mass (FFM), increase in energy expenditure, and regulation of appetite. Cardiometabolic health may also be improved through decreases in total cholesterol, increases in HDL cholesterol, and improvements in glycemic regulation.

However, most of this research has been conducted in young adults, and it is unclear if postmenopausal women will elicit the same appetite response in response to protein intake.

Due to changes in body composition, postmenopausal women experience a decrease in energy expenditure, which may be a contributor to weight gain (Lovejoy et al. 2008). Acutely, higher protein diets increase energy expenditure through increased postprandial energy expenditure (e.g., thermic effect of food (TEF)) and REE (Leidy et al. 2015). Dietary protein consistently elicits a greater TEF than either carbohydrates or fat (Halton and Hu 2004) due to the large amount of adenosine triphosphate (ATP) used in the postprandial period following a higher protein (20–30% energy) meal for the process of metabolizing and storing protein. Protein intake is positively associated with TEF (Ravn et al. 2013); however, this effect varies from person to person. TEF has also been suggested as one of the mechanisms that influence appetite sensations, including satiety (Westerterp-Plantenga et al. 1999). Even a small positive association between TEF and satiety could have clinical implications for weight loss and preventing weight gain during the postmenopausal period.

ADDITIONAL HEALTH BENEFITS ASSOCIATED WITH DIETARY PROTEIN INTAKE

Although most older adults fall within the RDA for dietary protein intake (Berryman et al. 2018), there is increasing evidence that diets with greater levels of high-quality protein, especially at the expense of simple carbohydrates, may decrease cardiometabolic risk (Dong et al. 2013b). Proposed therapeutic effects of high-protein diets include lower energy intake associated with increased thermogenesis (Acheson et al. 2011) and preservation of skeletal muscle (Houston et al. 2008) translating to decreased waist circumference, systolic and diastolic blood pressure, triglycerides, and fasting insulin, as well as increased high-density lipoprotein (HDL) (Santesso et al. 2012; Baum and Wolfe 2015; Wolfe 2012, 2015). According to observational data, higher protein diets (1.0–1.5 g/kg/day) are associated with lower BMI and waist circumference and higher HDL cholesterol levels compared to protein intake at the RDA (Pasiakos, Lieberman, and Fulgoni 2015).

Layman et al. (Layman et al. 2009) investigated the effect of following a high-protein diet (high protein: 1.6 g/kg/day, ~30% energy versus RDA: 0.8 g/kg; ~15% energy) after four months of weight loss on long-term weight maintenance. The high-protein diet was more effective for long-term fat loss (38% greater fat loss) and produced sustained reductions in triglycerides and increases in HDL cholesterol (Layman et al. 2009). In line with these findings, a meta-analysis of long-term weight-maintenance diets following higher-protein, low-carbohydrate intake found increases in sustained weight loss and decreases in fasting triglycerides and insulin levels, supporting a cardiometabolic benefit (Clifton, Condo, and Keogh 2014). Contrarily, an RCT examining the effects of a high-protein diet (high protein: 1.4 g/kg/day versus RDA 0.8 g/kg/day) in older adults (70 ± 5 years) observed no differences in outcomes of cardiometabolic health (e.g., HDL, LDL, fasting glucose, blood pressure) nor markers of inflammation including TNF-α, IL-6, and CRP (Wright et al. 2018). Further RCTs are needed to establish the relationship between protein intake and cardiometabolic health in postmenopausal women.

Sleep

Few RCTs have investigated the effect of dietary protein on sleep quality and duration in late-middle-aged and older adults. One RCT conducted by Zhou et al. (Zhou et al. 2016), assessed the effect of a high-protein, energy-restricted diet for 16 weeks on late-middle-aged obese adults (56 ± 3 years) and found high-protein diets to improve sleep quality.

The mechanism of the effect of protein on sleep after acute feeding may be related to tryptophan, tyrosine, and the synthesis of the brain neurotransmitters serotonin, melatonin, and dopamine (Peuhkuri, Sihvola, and Korpela 2012; Espana and Scammell 2011). As BCAAs are transported into the brain across the blood-brain barrier via the same carrier that transports large neutral amino acids (LNAAs), phenylalanine, tyrosine, and tryptophan, the competition between BCAAs and the aromatic amino acids may influence the synthesis of some neurotransmitters, including dopamine, norepinephrine, and serotonin (Holecek 2018). Although higher protein intake results in greater postprandial plasma AAs, high-protein diets do not translate into constant low tryptophan- or tyrosine-to-LNAA ratios. For example, Zhou et al. (2016) found no acute changes in the ratio of tryptophan to LNAAs and tyrosine to LNAAs. The relationship between acclimated protein intake and the body's ability to produce or remove brain tryptophan and serotonin is yet to be elucidated. However, in an intervention conducted in rhesus monkeys, a higher-protein diet increased plasma and cerebrospinal fluid concentrations of tryptophan and serotonin metabolites, indicating a probable beneficial sleep effect (Grimes, Cameron, and Fernstrom 2000). More long-term RCTs are needed to investigate the relationship between tryptophan:LNAA and tyrosine:LNAA ratios in older adults consuming high-protein diets. Therefore, the beneficial effect of dietary protein on sleep has yet to be established.

SUMMARY

Dietary patterns higher in protein have the potential to improve weight loss and cardiometabolic health in postmenopausal women. However, most weight loss interventions focused on increasing dietary protein intake have been conducted in both men and women, and few focus on weight loss in older adults, specifically postmenopausal women. In addition, a majority of the research related to the health benefits of higher protein intake and weight loss used high-quality, animal-based protein sources (Phillips, Chevalier, and Leidy 2016), and limited information is available about the impact of plant-based protein sources on weight loss. Therefore, more research is needed before definitive conclusions can be made about the effect of increased dietary protein intake on weight loss in postmenopausal women.

MINI DICTIONARY OF TERMS

Body composition: The proportion of fat, muscle, bone, and other tissues in the body.
Cardiometabolic health: The relationship between cardiovascular and metabolic health that impacts an individual's risk for developing chronic diseases.

Complete protein: A protein containing all nine essential amino acids in adequate proportions, which the body cannot produce on its own.

Higher-protein dietary pattern: Involves consuming a greater proportion of protein-rich foods relative to other macronutrients, such as carbohydrates and fats.

Postmenopausal: The time in a woman's life when she has not had a menstrual cycle for at least 12 months, marking the end of her reproductive years.

SUMMARY POINTS

- This chapter focuses on the role of protein intake on weight loss in postmenopausal women.
- Over 40% of postmenopausal women have obesity.
- The increase in obesity with menopause increases the risk of chronic disease.
- The decrease in estrogen during the menopausal transition is associated with changes in body composition, including muscle mass loss and increased central adiposity. This results in a decrease in energy expenditure.
- Dietary patterns higher in protein may help mitigate weight gain and changes in body composition that occur with menopause.
- Most dietary protein interventions for weight loss in older adults focus on men and women. There are very few long-term dietary interventions studying the effect of higher protein intake on weight loss in postmenopausal women.

REFERENCES

Acheson, K. J., A. Blondel-Lubrano, S. Oguey-Araymon, M. Beaumont, S. Emady-Azar, C. Ammon-Zufferey, I. Monnard, S. Pinaud, C. Nielsen-Moennoz, and L. Bovetto. 2011. "Protein choices targeting thermogenesis and metabolism." *Am J Clin Nutr* 93 (3): 525–34. https://doi.org/10.3945/ajcn.110.005850. https://www.ncbi.nlm.nih.gov/pubmed/21228266.

Agriculture., U.S. Department of Health and Human Services and U.S. Department of. 2015. Dietary Guidlines for Americans 2015–2020 8th Edition

Andres, R., D. C. Muller, and J. D. Sorkin. 1993. "Long-term effects of change in body weight on all-cause mortality. A review." *Ann Intern Med* 119 (7 Pt 2): 737–43. https://www.ncbi.nlm.nih.gov/pubmed/8363208.

Arentson-Lantz, E., S. Clairmont, D. Paddon-Jones, A. Tremblay, and R. Elango. 2015. "Protein: a nutrient in focus." *Appl Physiol Nutr Metab* 40 (8): 755–61. https://doi.org/10.1139/apnm-2014-0530. https://www.ncbi.nlm.nih.gov/pubmed/26197807.

Bauer, J., G. Biolo, T. Cederholm, M. Cesari, A. J. Cruz-Jentoft, J. E. Morley, S. Phillips, C. Sieber, P. Stehle, D. Teta, R. Visvanathan, E. Volpi, and Y. Boirie. 2013. "Evidence-based recommendations for optimal dietary protein intake in older people: a position paper from the PROT-AGE study group." *J Am Med Dir Assoc* 14 (8): 542–59. https://doi.org/10.1016/j.jamda.2013.05.021. https://www.ncbi.nlm.nih.gov/pubmed/23867520.

Baum, J. I., M. Gray, and A. Binns. 2015. "Breakfasts higher in protein increase postprandial energy expenditure, increase fat oxidation, and reduce hunger in overweight children from 8 to 12 years of age." *J Nutr.* https://doi.org/10.3945/jn.115.214551. http://www.ncbi.nlm.nih.gov/pubmed/26269241.

Baum, J. I., Kim Il-Young, and Robert R. Wolfe. 2016. "Protein consumption and the elderly: what is the optimal level of intake?" *Nutrients* 8 (6): 359. https://doi.org/10.3390/nu8060359. http://0-search.ebscohost.com.library.uark.edu/login.aspx?direct=true&db=ccm&AN=116338130&site=ehost-live&scope=site.

Baum, J. I., and R. R. Wolfe. 2015. "The link between dietary protein intake, skeletal muscle function and health in older adults." *Healthcare (Basel)* 3 (3): 529–43. https://doi.org/10.3390/healthcare3030529. https://www.ncbi.nlm.nih.gov/pubmed/27417778.

Berner, L. A., G. Becker, M. Wise, and J. Doi. 2013. "Characterization of dietary protein among older adults in the United States: amount, animal sources, and meal patterns." *J Acad Nutr Diet* 113 (6): 809–15. https://doi.org/10.1016/j.jand.2013.01.014. https://www.ncbi.nlm.nih.gov/pubmed/23491327.

Berryman, Claire E, Harris R Lieberman, Victor L Fulgoni, III, and Stefan M Pasiakos. 2018. "Protein intake trends and conformity with the dietary reference intakes in the united states: analysis of the national health and nutrition examination survey, 2001–2014." *Am J Clin Nutr* 108 (2): 405–13. https://doi.org/10.1093/ajcn/nqy088. https://doi.org/10.1093/ajcn/nqy088.

Blair, S. N., J. Shaten, K. Brownell, G. Collins, and L. Lissner. 1993. "Body weight change, all-cause mortality, and cause-specific mortality in the multiple risk factor intervention trial." *Ann Intern Med* 119 (7 Pt 2): 749–57. https://www.ncbi.nlm.nih.gov/pubmed/8363210.

Clifton, P. M., D. Condo, and J. B. Keogh. 2014. "Long term weight maintenance after advice to consume low carbohydrate, higher protein diets-a systematic review and meta analysis." *Nutr Metab Cardiovasc Dis* 24 (3): 224–35. https://doi.org/10.1016/j.numecd.2013.11.006. https://www.ncbi.nlm.nih.gov/pubmed/24472635.

Institute of Medicine. 2005. Dietary reference intakes for energy, carbohydrate, fiber, fat, fatty acids, cholesterol, protein and amino acids. Washington, DC: The National Academies Press.

Donato, G. B., S. C. Fuchs, K. Oppermann, C. Bastos, and P. M. Spritzer. 2006. "Association between menopause status and central adiposity measured at different cutoffs of waist circumference and waist-to-hip ratio." *Menopause* 13 (2): 280–5. https://doi.org/10.1097/01.gme.0000177907.32634.ae. https://www.ncbi.nlm.nih.gov/pubmed/16645541.

Dong, J. Y., Z. L. Zhang, P. Y. Wang, and L. Q. Qin. 2013a. "Effects of high-protein diets on body weight, glycaemic control, blood lipids and blood pressure in type 2 diabetes: meta-analysis of randomised controlled trials." *Br J Nutr* 110 (5): 781–9. https://doi.org/10.1017/S0007114513002055. https://www.ncbi.nlm.nih.gov/pubmed/23829939.

Dong, Jia-Yi, Zeng-Li Zhang, Pei-Yu Wang, and Li-Qiang Qin. 2013b. "Effects of high-protein diets on body weight, glycaemic control, blood lipids and blood pressure in type 2 diabetes: meta-analysis of randomised controlled trials." *British Journal of Nutrition* 110 (5): 781–9. https://doi.org/10.1017/S0007114513002055. https://www.cambridge.org/core/article/effects-of-highprotein-diets-on-body-weight-glycaemic-control-blood-lipids-and-blood-pressure-in-type-2-diabetes-metaanalysis-of-randomised-controlled-trials/4D4F7A3943BE752FB061CE60671204B1.

Espana, R. A., and T. E. Scammell. 2011. "Sleep neurobiology from a clinical perspective." *Sleep* 34 (7): 845–58. https://doi.org/10.5665/SLEEP.1112. https://www.ncbi.nlm.nih.gov/pubmed/21731134.

Farnsworth, E., N. D. Luscombe, M. Noakes, G. Wittert, E. Argyiou, and P. M. Clifton. 2003. "Effect of a high-protein, energy-restricted diet on body composition, glycemic control, and lipid concentrations in overweight and obese hyperinsulinemic men and women." *Am J Clin Nutr* 78 (1): 31–9. http://www.ncbi.nlm.nih.gov/pubmed/12816768.

Flegal, K. M., D. Kruszon-Moran, M. D. Carroll, C. D. Fryar, and C. L. Ogden. 2016. "Trends in obesity among adults in the United States, 2005 to 2014." *JAMA* 315 (21): 2284–91. https://doi.org/10.1001/jama.2016.6458. https://www.ncbi.nlm.nih.gov/pubmed/27272580.

Gannon, M. C., F. Q. Nuttall, A. Saeed, K. Jordan, and H. Hoover. 2003. "An increase in dietary protein improves the blood glucose response in persons with type 2 diabetes." *Am J Clin Nutr* 78 (4): 734–41. http://www.ncbi.nlm.nih.gov/pubmed/14522731.

Godsland, I. F. 2005. "Oestrogens and insulin secretion." *Diabetologia* 48 (11): 2213–20. https://doi.org/10.1007/s00125-005-1930-0. https://www.ncbi.nlm.nih.gov/pubmed/16193292.

Greendale, G. A., B. Sternfeld, M. Huang, W. Han, C. Karvonen-Gutierrez, K. Ruppert, J. A. Cauley, J. S. Finkelstein, S. F. Jiang, and A. S. Karlamangla. 2019. "Changes in body composition and weight during the menopause transition." *JCI Insight* 4 (5). https://doi.org/10.1172/jci.insight.124865. https://www.ncbi.nlm.nih.gov/pubmed/30843880.

Grimes, M. A., J. L. Cameron, and J. D. Fernstrom. 2000. "Cerebrospinal fluid concentrations of tryptophan and 5-hydroxyindoleacetic acid in Macaca mulatta: diurnal variations and response to chronic changes in dietary protein intake." *Neurochem Res* 25 (3): 413–22. https://doi.org/10.1023/a:1007557524370. https://www.ncbi.nlm.nih.gov/pubmed/10761988.

Guthrie, J. R., L. Dennerstein, J. R. Taffe, P. Lehert, and H. G. Burger. 2004. "The menopausal transition: a 9-year prospective population-based study. The Melbourne women's midlife health project." *Climacteric* 7 (4): 375–89. https://doi.org/10.1080/13697130400012163. https://www.ncbi.nlm.nih.gov/pubmed/15799609.

Hales, C. M., M. D. Carroll, C. D. Fryar, and C. L. Ogden. 2020. "Prevalence of obesity and severe obesity among adults: United States, 2017–2018." *NCHS Data Brief* (360): 1–8. https://www.ncbi.nlm.nih.gov/pubmed/32487284.

Halton, T. L., and F. B. Hu. 2004. "The effects of high protein diets on thermogenesis, satiety and weight loss: a critical review." *J Am Coll Nutr* 23 (5): 373–85. https://doi.org/10.1080/07315724.2004.10719381. https://www.ncbi.nlm.nih.gov/pubmed/15466943.

Harlow, S. D. 2018. "Menstrual cycle changes as women approach the final menses: what matters?" *Obstet Gynecol Clin North Am* 45 (4): 599–611. https://doi.org/10.1016/j.ogc.2018.07.003. https://www.ncbi.nlm.nih.gov/pubmed/30401545.

Hawley, A. L., and J. I. Baum. 2024. "Nutrition as the foundation for successful aging: a focus on dietary protein and omega-3 polyunsaturated fatty acids." *Nutr Rev* 82 (3): 389–406. https://doi.org/10.1093/nutrit/nuad061. https://www.ncbi.nlm.nih.gov/pubmed/37319363.

Holecek, M. 2018. "Branched-chain amino acids in health and disease: metabolism, alterations in blood plasma, and as supplements." *Nutr Metab (Lond)* 15: 33. https://doi.org/10.1186/s12986-018-0271-1. https://www.ncbi.nlm.nih.gov/pubmed/29755574.

Houston, Denise K., Barbara J. Nicklas, Jingzhong Ding, Tamara B. Harris, Frances A. Tylavsky, Anne B. Newman, Jung Sun Lee, Nadine R. Sahyoun, Marjolein Visser, and Stephen B. Kritchevsky. 2008. "Dietary protein intake is associated with lean mass change in older, community-dwelling adults: the Health, Aging, and Body Composition (Health ABC) Study." *Am J Clin Nutr* 87: 150–5.

Janssen, I., L. H. Powell, R. Kazlauskaite, and S. A. Dugan. 2010. "Testosterone and visceral fat in midlife women: the study of women's health across the nation (SWAN) fat patterning study." *Obesity (Silver Spring)* 18 (3): 604–10. https://doi.org/10.1038/oby.2009.251. https://www.ncbi.nlm.nih.gov/pubmed/19696765.

Kannel, W. B., M. C. Hjortland, P. M. McNamara, and T. Gordon. 1976. "Menopause and risk of cardiovascular disease: the Framingham study." *Ann Intern Med* 85 (4): 447–52. https://doi.org/10.7326/0003-4819-85-4-447. https://www.ncbi.nlm.nih.gov/pubmed/970770.

Knight, M. G., C. Anekwe, K. Washington, E. Y. Akam, E. Wang, and F. C. Stanford. 2021. "Weight regulation in menopause." *Menopause* 28 (8): 960–965. https://doi.org/10.1097/GME.0000000000001792. https://www.ncbi.nlm.nih.gov/pubmed/34033603.

Ko, S. H., and Y. Jung. 2021. "Energy metabolism changes and dysregulated lipid metabolism in postmenopausal women." *Nutrients* 13 (12). https://doi.org/10.3390/nu13124556. https://www.ncbi.nlm.nih.gov/pubmed/34960109.

Ko, S. H., and H. S. Kim. 2020. "Menopause-associated lipid metabolic disorders and foods beneficial for postmenopausal women." *Nutrients* 12 (1). https://doi.org/10.3390/nu12010202. https://www.ncbi.nlm.nih.gov/pubmed/31941004.

Kravitz, H. M., E. Avery, M. Sowers, J. T. Bromberger, J. F. Owens, K. A. Matthews, M. Hall, H. Zheng, E. B. Gold, and D. J. Buysse. 2011. "Relationships between menopausal and mood symptoms and EEG sleep measures in a multi-ethnic sample of middle-aged women: the SWAN sleep study." *Sleep* 34 (9): 1221–32. https://doi.org/10.5665/SLEEP.1244. https://www.ncbi.nlm.nih.gov/pubmed/21886360.

Kravitz, H. M., and H. Joffe. 2011. "Sleep during the perimenopause: a SWAN story." *Obstet Gynecol Clin North Am* 38 (3): 567–86. https://doi.org/10.1016/j.ogc.2011.06.002. https://www.ncbi.nlm.nih.gov/pubmed/21961720.

Layman, D. K., R. A. Boileau, D. J. Erickson, J. E. Painter, H. Shiue, C. Sather, and D. D. Christou. 2003. "A reduced ratio of dietary carbohydrate to protein improves body composition and blood lipid profiles during weight loss in adult women." *J Nutr* 133 (2): 411–7. http://www.ncbi.nlm.nih.gov/pubmed/12566476.

Layman, D. K., E. M. Evans, D. Erickson, J. Seyler, J. Weber, D. Bagshaw, A. Griel, T. Psota, and P. Kris-Etherton. 2009. "A moderate-protein diet produces sustained weight loss and long-term changes in body composition and blood lipids in obese adults." *J Nutr* 139 (3): 514–21. https://doi.org/10.3945/jn.108.099440.

Layman, D. K., H. Shiue, C. Sather, D. J. Erickson, and J. Baum. 2003. "Increased dietary protein modifies glucose and insulin homeostasis in adult women during weight loss." *J Nutr* 133 (2): 405–10. http://www.ncbi.nlm.nih.gov/pubmed/12566475.

Lee, J. Y., and D. C. Lee. 2013. "Muscle strength and quality are associated with severity of menopausal symptoms in peri- and post-menopausal women." *Maturitas* 76 (1): 88–94. https://doi.org/10.1016/j.maturitas.2013.06.007. https://www.ncbi.nlm.nih.gov/pubmed/23838172.

Leidy, H. J., P. M. Clifton, A. Astrup, T. P. Wycherley, M. S. Westerterp-Plantenga, N. D. Luscombe-Marsh, S. C. Woods, and R. D. Mattes. 2015. "The role of protein in weight loss and maintenance." *Am J Clin Nutr* 101 (6): 1320S–29S. https://doi.org/10.3945/ajcn.114.084038. https://www.ncbi.nlm.nih.gov/pubmed/25926512.

Leidy, H. J., R. J. Lepping, C. R. Savage, and C. T. Harris. 2011. "Neural responses to visual food stimuli after a normal vs. higher protein breakfast in breakfast-skipping teens: a pilot fMRI study." *Obesity (Silver Spring)* 19 (10): 2019–25. https://doi.org/10.1038/oby.2011.108. https://www.ncbi.nlm.nih.gov/pubmed/21546927.

Lissner, L., P. M. Odell, R. B. D'Agostino, J. Stokes, 3rd, B. E. Kreger, A. J. Belanger, and K. D. Brownell. 1991. "Variability of body weight and health outcomes in the Framingham population." *N Engl J Med* 324 (26): 1839–44. https://doi.org/10.1056/NEJM199106273242602. https://www.ncbi.nlm.nih.gov/pubmed/2041550.

Lovejoy, J. C., C. M. Champagne, L. de Jonge, H. Xie, and S. R. Smith. 2008. "Increased visceral fat and decreased energy expenditure during the menopausal transition." *Int J Obes (Lond)* 32 (6): 949–58. https://doi.org/10.1038/ijo.2008.25. https://www.ncbi.nlm.nih.gov/pubmed/18332882.

Marlatt, K. L., D. R. Pitynski-Miller, K. M. Gavin, K. L. Moreau, E. L. Melanson, N. Santoro, and W. M. Kohrt. 2022. "Body composition and cardiometabolic health across the menopause transition." *Obesity (Silver Spring)* 30 (1): 14–27. https://doi.org/10.1002/oby.23289. https://www.ncbi.nlm.nih.gov/pubmed/34932890.

Mathus-Vliegen, E. M., and Obesity Obesity Management Task Force of the European Association for the Study of. 2012. "Prevalence, pathophysiology, health consequences and treatment options of obesity in the elderly: a guideline." *Obes Facts* 5 (3): 460–83. https://doi.org/10.1159/000341193. https://www.ncbi.nlm.nih.gov/pubmed/22797374.

Moore, Daniel R., Tyler A. Churchward-Venne, Oliver Witard, Leigh Breen, Nicholas A. Burd, Kevin D. Tipton, and Stuart M. Phillips. 2014. "Protein ingestion to stimulate myofibrillar protein synthesis requires greater relative protein intakes in healthy older versus younger men." *J Gerontol A Biol Sci Med Sci* 70: 57–62.

Motlani, V., G. Motlani, S. Pamnani, A. Sahu, and N. Acharya. 2023. "Endocrine changes in postmenopausal women: a comprehensive view." *Cureus* 15 (12): e51287. https://doi.org/10.7759/cureus.51287. https://www.ncbi.nlm.nih.gov/pubmed/38288203.

Paddon-Jones, D., and H. Leidy. 2014. "Dietary protein and muscle in older persons." *Curr Opin Clin Nutr Metab Care* 17 (1): 5–11. https://doi.org/10.1097/MCO.0000000000000011. https://www.ncbi.nlm.nih.gov/pubmed/24310053.

Paddon-Jones, D., and B. B. Rasmussen. 2009. "Dietary protein recommendations and the prevention of sarcopenia." *Curr Opin Clin Nutr Metab Care* 12 (1): 86–90. https://doi.org/10.1097/MCO.0b013e32831cef8b. https://www.ncbi.nlm.nih.gov/pubmed/19057193.

Pasiakos, S. M., H. R. Lieberman, and V. L. Fulgoni, 3rd. 2015. "Higher-protein diets are associated with higher HDL cholesterol and lower BMI and waist circumference in US adults." *J Nutr* 145 (3): 605–14. https://doi.org/10.3945/jn.114.205203. https://www.ncbi.nlm.nih.gov/pubmed/25733478.

Pasquali, R., F. Casimirri, A. M. Labate, O. Tortelli, G. Pascal, B. Anconetani, M. R. Gatto, R. Flamia, M. Capelli, and L. Barbara. 1994. "Body weight, fat distribution and the menopausal status in women. The VMH collaborative group." *Int J Obes Relat Metab Disord* 18 (9): 614–21. https://www.ncbi.nlm.nih.gov/pubmed/7812415.

Peuhkuri, K., N. Sihvola, and R. Korpela. 2012. "Diet promotes sleep duration and quality." *Nutr Res* 32 (5): 309–19. https://doi.org/10.1016/j.nutres.2012.03.009. https://www.ncbi.nlm.nih.gov/pubmed/22652369.

Phillips, S. M., S. Chevalier, and H. J. Leidy. 2016. "Protein "requirements" beyond the RDA: implications for optimizing health." *Appl Physiol Nutr Metab* 41 (5): 565-72. https://doi.org/10.1139/apnm-2015-0550. https://www.ncbi.nlm.nih.gov/pubmed/26960445.

Poehlman, E. T., M. J. Toth, and A. W. Gardner. 1995. "Changes in energy balance and body composition at menopause: a controlled longitudinal study." *Ann Intern Med* 123 (9): 673–5. https://doi.org/10.7326/0003-4819-123-9-199511010-00005. https://www.ncbi.nlm.nih.gov/pubmed/7574222.

Rand, W. M., Peter L. Pellet, and Vernon R. Young. 2003. "Meta-analysis of nitrogen balance studies for estimating protein requirements in healthy adults." *Am J Clin Nutr.* 77 (1): 109–27. doi: 10.1093/ajcn/77.1.109

Ravn, A. M., N. T. Gregersen, R. Christensen, L. G. Rasmussen, O. Hels, A. Belza, A. Raben, T. M. Larsen, S. Toubro, and A. Astrup. 2013. "Thermic effect of a meal and appetite in adults: an individual participant data meta-analysis of meal-test trials." *Food Nutr Res* 57. https://doi.org/10.3402/fnr.v57i0.19676. https://www.ncbi.nlm.nih.gov/pubmed/24376394.

Roa-Diaz, Z. M., P. F. Raguindin, A. Bano, J. E. Laine, T. Muka, and M. Glisic. 2021. "Menopause and cardiometabolic diseases: what we (don't) know and why it matters." *Maturitas* 152: 48–56. https://doi.org/10.1016/j.maturitas.2021.06.013. https://www.ncbi.nlm.nih.gov/pubmed/34674807.

Rolland, Y. M., H. M. Perry, 3rd, P. Patrick, W. A. Banks, and J. E. Morley. 2007. "Loss of appendicular muscle mass and loss of muscle strength in young postmenopausal women." *J Gerontol A Biol Sci Med Sci* 62 (3): 330–5. https://doi.org/10.1093/gerona/62.3.330. https://www.ncbi.nlm.nih.gov/pubmed/17389732.

Santesso, N., E. A. Akl, M. Bianchi, A. Mente, R. Mustafa, D. Heels-Ansdell, and H. J. Schünemann. 2012. "Effects of higher- versus lower-protein diets on health outcomes: a systematic review and meta-analysis." *Eur J Clin Nutr* 66 (7): 780–88. https://doi.org/10.1038/ejcn.2012.37. https://pubmed.ncbi.nlm.nih.gov/22510792. https://www.ncbi.nlm.nih.gov/pmc/articles/PMC3392894/.

Silva, T. R., and P. M. Spritzer. 2017. "Skeletal muscle mass is associated with higher dietary protein intake and lower body fat in postmenopausal women: a cross-sectional study." *Menopause* 24 (5): 502–9. https://doi.org/10.1097/GME.0000000000000793. https://www.ncbi.nlm.nih.gov/pubmed/27922938.

Steverson, M. 2022. *Ageing and Health*. World Health Organization. Accessed January 27. https://www.who.int/news-room/fact-sheets/detail/ageing-and-health.

Symons, T. B., M. Sheffield-Moore, R. R. Wolfe, and D. Paddon-Jones. 2009. "A moderate serving of high-quality protein maximally stimulates skeletal muscle protein synthesis in young and elderly subjects." *J Am Diet Assoc* 109 (9): 1582–6. https://doi.org/10.1016/j.jada.2009.06.369. https://www.ncbi.nlm.nih.gov/pubmed/19699838.

Talaulikar, V. 2022. "Menopause transition: physiology and symptoms." *Best Pract Res Clin Obstet Gynaecol* 81: 3–7. https://doi.org/10.1016/j.bpobgyn.2022.03.003. https://www.ncbi.nlm.nih.gov/pubmed/35382992.

Trumbo, P., S. Schlicker, A. A. Yates, and M. Poos. 2002. "Dietary reference intakes for energy, carbohydrate, fiber, fat, fatty acids, cholesterol, protein and amino acids." *J Am Diet Assoc* 102 (11): 1621–30. https://doi.org/10.1016/s0002-8223(02)90346-9.

U.N. 2022. *Ageing*. Global Issues. United Nations. Accessed January 26. https://www.un.org/en/global-issues/ageing.

Valenzuela, R. E., J. A. Ponce, G. G. Morales-Figueroa, K. A. Muro, V. R. Carreón, and H. Alemán-Mateo. 2013. "Insufficient amounts and inadequate distribution of dietary protein intake in apparently healthy older adults in a developing country: implications for dietary strategies to prevent sarcopenia." *Clin Interv Aging* 8: 1143–8. https://doi.org/10.2147/cia.S49810.

Villareal, D. T., C. M. Apovian, R. F. Kushner, S. Klein, Nutrition American Society for, and The Obesity Society Naaso. 2005. "Obesity in older adults: technical review and position statement of the American society for nutrition and NAASO, the obesity society." *Obes Res* 13 (11): 1849–63. https://doi.org/10.1038/oby.2005.228. https://www.ncbi.nlm.nih.gov/pubmed/16339115.

Volpi, E., W. W. Campbell, J. T. Dwyer, M. A. Johnson, G. L. Jensen, J. E. Morley, and R. R. Wolfe. 2013. "Is the optimal level of protein intake for older adults greater than the recommended dietary allowance?" *J Gerontol A Biol Sci Med Sci* 68 (6): 677–81. https://doi.org/10.1093/gerona/gls229. https://www.ncbi.nlm.nih.gov/pubmed/23183903.

Westerterp-Plantenga, M. S., S. G. Lemmens, and K. R. Westerterp. 2012. "Dietary protein - its role in satiety, energetics, weight loss and health." *Br J Nutr* 108 Suppl 2: S105–12. https://doi.org/10.1017/S0007114512002589. https://www.ncbi.nlm.nih.gov/pubmed/23107521.

Westerterp-Plantenga, M. S., V. Rolland, S. A. Wilson, and K. R. Westerterp. 1999. "Satiety related to 24 h diet-induced thermogenesis during high protein/carbohydrate vs high fat diets measured in a respiration chamber." *Eur J Clin Nutr* 53 (6): 495–502. https://doi.org/10.1038/sj.ejcn.1600782. https://www.ncbi.nlm.nih.gov/pubmed/10403587.

WHO. 2022. *Menopause*. World Health Organization. Accessed January 24. https://www.who.int/news-room/fact-sheets/detail/menopause.

Williamson, D. F., and E. R. Pamuk. 1993. "The association between weight loss and increased longevity. A review of the evidence." *Ann Intern Med* 119 (7 Pt 2): 731–6. https://www.ncbi.nlm.nih.gov/pubmed/8363207.

Wolfe, R. R. 2012. "The role of dietary protein in optimizing muscle mass, function and health outcomes in older individuals." *Br J Nutr* 108: S88–S93. https://doi.org/10.1017/s0007114512002590. <Go to ISI>://WOS:000310383200010.

Wolfe, R. R. 2015. "Update on protein intake: importance of milk proteins for health status of the elderly." *Nutr Rev* 73: 41–47. https://doi.org/10.1093/nutrit/nuv021. <Go to ISI>://WOS:000359652200008.

Wolfe, R. R., A. M. Cifelli, G. Kostas, and I. Y. Kim. 2017. "Optimizing protein intake in adults: interpretation and application of the recommended dietary allowance compared with the acceptable macronutrient distribution range." *Adv Nutr* 8 (2): 266–75. https://doi.org/10.3945/an.116.013821. https://www.ncbi.nlm.nih.gov/pubmed/28298271.

Wright, C. S., J. Zhou, R. D. Sayer, J. E. Kim, and W. W. Campbell. 2018. "Effects of a high-protein diet including whole eggs on muscle composition and indices of cardiometabolic health and systemic inflammation in older adults with overweight or obesity: a randomized controlled trial." *Nutrients* 10 (7). https://doi.org/10.3390/nu10070946.

Wycherley, T. P., J. D. Buckley, M. Noakes, P. M. Clifton, and G. D. Brinkworth. 2013. "Comparison of the effects of weight loss from a high-protein versus standard-protein energy-restricted diet on strength and aerobic capacity in overweight and obese men." *Eur J Nutr* 52 (1): 317–25. https://doi.org/10.1007/s00394-012-0338-0. https://www.ncbi.nlm.nih.gov/pubmed/22406907.

Zhou, J., J. E. Kim, C. L. Armstrong, N. Chen, and W. W. Campbell. 2016. "Higher-protein diets improve indexes of sleep in energy-restricted overweight and obese adults: results from 2 randomized controlled trials." *Am J Clin Nutr* 103 (3): 766–74. https://doi.org/10.3945/ajcn.115.124669. https://www.ncbi.nlm.nih.gov/pubmed/26864362.

35 The Impact of Plant and Animal Proteins on the Diets of Older Individuals

Eirini Koidou, Christina F. Dolopikou, Fani Thoma, and Sousana Papadopoulou

ABBREVIATIONS

AAs	Amino acids
BC	Body composition
DIAAS	Digestible indispensable amino acid score
EAAs	Essential amino acids
EAR	Estimated average allowance
FFM	Fat-free mass
IAAO	Indicator amino acid oxidation
MPS	Muscle protein synthesis
QoL	Quality of life
RDA	Recommended dietary allowance
WHO	World Health Organization

INTRODUCTION

The gradual increase in average life expectancy and longevity, brought about by advancements in medicine, nutrition, improved living conditions, enhanced hygiene, and public health, has resulted in a rise in the number of individuals aged 60 and over. Aging is an exceptionally complex and multifaceted process that remains not fully understood. However, studies suggest that numerous genetic, epigenetic, stochastic, and environmental factors interact with one another to interpret and decode this dynamic process. Aging is characterized by alterations in the cardiovascular, endocrine, gastrointestinal, respiratory, renal, and musculoskeletal systems, as well as changes in body composition (BC) (Wilmoth et al. 2023). A healthy BC is characterized by a low proportion of body fat and a higher proportion of fat-free mass (FFM). Although body fat plays a fundamental role in maintaining good health, functional ability, and longevity and provides valuable insights into a person's nutritional status, it is well-established that the natural aging process is accompanied by rapid and significant changes in BC, regardless of an individual's sex or race, and these changes are not necessarily linked to shifts in body weight. Changes in BC inevitably affect energy needs, basal metabolic rate, and macronutrient oxidation. Research suggests that these changes in BC translate into an increased risk for various health conditions, including sarcopenia, osteoporosis, sarcopenic obesity, obesity, cachexia, frailty, metabolic syndrome, and coronary diseases. Importantly, these effects are observed irrespective of an individual's sex or race (Pacifico et al. 2020).

 DOI: 10.1201/b23328-35

THE ROLE OF NUTRITION IN OLDER ADULTS

Older adults constitute a diverse population and, regrettably, are among the groups most vulnerable to nutritional deficiencies and a decline in nutritional status. These issues can have significant implications for BC, functionality, clinical outcomes, quality of life (QoL), and oral health. Furthermore, for a population that often lives with chronic diseases, maintaining a balanced diet is vital for effective disease management.

As older individuals often have lower energy needs but similar or even increased nutrient requirements compared to younger adults, dietary recommendations and guidelines for this demographic emphasize the following key principles by Volkert et al. (2022):

- following a regular eating schedule,
- consuming nutrient-dense foods,
- maintaining energy balance,
- hydrating,
- consuming adequate fiber, healthy fats, calcium, vitamins D and B_{12}, and
- reducing salt consumption.

An adequate protein intake seems to be essential for healthy aging. The role of proteins and amino acids in diet and health appears to be more complex, and this complexity can either affect or determine dietary requirements (Table 35.1).

Proteins are the primary constituents of the body's cells and serve multiple essential functions. They act as the main structural components of various tissues, including bones and muscles. They play a crucial role in producing enzymes, hormones, antibodies, and neurotransmitters. In addition, they help maintain fluid and acid-base balance, transport nutrients to cells, and support the regulation and expression of DNA and RNA. In specific situations, such as starvation or malnutrition, a small amount of proteins can be used to meet energy needs. Proteins are composed of long chains of amino acids (AAs). Traditionally, AAs are classified as either essential or non-essential. It's worth noting that there is a third category of AAs known as conditionally essential (Fürst 2009) (Table 35.2).

A recommended method for evaluating protein quality is through amino acid scoring. There are two primary methods for expressing amino acid scores, namely the protein digestibility-corrected amino acid score (PDCAAS) and the digestible indispensable amino acid score (DIAAS). These methods are invaluable for assessing the efficacy of protein sources in terms of supporting muscle mass maintenance, physical function, and healthy aging, as highlighted by Paddon-Jones et al. (2009).

The PDCAAS was previously employed to gauge the quality of a protein source. However, it has since been considered deficient due to its reliance on a single value of fecal crude protein digestibility. This limitation has prompted the adoption of a new method, the DIAAS. Instead of relying on a single fecal crude protein digestibility value, the DIAAS method employs true ileal amino acid digestibility for each individual EAA, providing a more accurate reflection of the bioavailability of EAAs (Figure 35.1), as established by FAO in 2013.

THE ROLE OF PROTEIN IN OLDER INDIVIDUALS

Muscle Protein in the Elderly

Aging is closely linked to an accelerated loss of skeletal muscle mass, strength, and physical function. Notably, the rate of muscle mass loss can increase by up to 15% each decade after the age of 70. Additionally, a decrease in the fractional synthesis rate (FSR) of skeletal muscle proteins has been observed in older individuals (Yarasheski et al. 1993; Wolfe, 2012).

TABLE 35.1
Changes in the Physiology of Various Systems that Affect Nutritional Health and Age-Related Metabolic Consequences, Based on Observational and Rodent Studies

System	"Normal" Aging Changes	Common Problems Relative to Nutrition	Nutritional Conditions
Cardiovascular	Decreased elasticity of blood vessels, blood volume and pulse volume Increased blood pressure, size of heart chambers, and arterial stiffness	Hypertension Atherosclerosis Heat failure Anemia	High saturated fat consumption Low fiber consumption Overweight/obesity Harmful use of alcohol Low consumption of fruit, vegetables, and whole grain High protein consumption High intake of salt
Nervous and cognition	Decreased nerve conduction velocity affecting sensory functions Losing sense of taste, smell, thirst Changes in mastication and swallowing	Malnutrition Weight loss Obesity Constipation Dehydration Wernicke-Korsakoff syndrome Eating disorders Poor appetite/anorexia	Cancer Chronic renal failure Liver disease Deficiencies of niacin, folate, vitamins B12, B6, B1, E, and D, minerals, essential amino acids, and omega-3 fatty acids
Digestive	Decrease in the amount of saliva or a change in the quality of saliva Missing teeth or ill-fitting dentures Dysphagia	Decreased ability to taste, chew, and swallow Lack of interest in food or fluids Denture problems Malnutrition Dehydration	Dehydration, insufficient food intake (vegetables and fruits) Deficiencies of vitamin D, vitamin C, B vitamins and vitamin A, zinc, iron and Protein Energy Malnutrition (PEM) Hyposalivation, Xerostomia and Oral Candidiasis on Oral Dysphagia
Endocrine	Reducing the release of aldosterone Dehydroepiandrosterone levels drop, reducing the release of cortisol Loss of estrogen and progesterone production in women Gradual decline in testosterone in men Reduction in GH and IGF-1 Reduced glucose tolerance DHEA and DHEA-S gradually decline until death in men and women Thyroid hormone axis activity	Dehydration and electrolyte abnormalities Increased body fat mass, waist-to-hip ratio, risk of obesity, type 2 diabetes, metabolic syndrome, visceral adipose mass, and cardiovascular disease Increased risk for reduction of muscle mass, strength, bone density, lean body mass Increased risk of orthostatic hypotension	Gut microbiome Decreased dietary fibers, iodine, iron, zinc, selenium, vitamin D, and probiotic factors Increased medication

(Continued)

TABLE 35.1 (CONTINUED)
Changes in the Physiology of Various Systems that Affect Nutritional Health and Age-Related Metabolic Consequences, Based on Observational and Rodent Studies

Respiratory	Reduced respiratory and work capacity	Decreased muscle strength and metabolic health	Protein-energy malnutrition Dehydration Low consumption of Omega-3 fatty acids, antioxidants, and anti-inflammatory vitamins High consumption of simple and refined carbohydrates, trans and saturated fats, and sodium Osteoporosis Obesity
Renal	Reduced number of nephrons Increased nephrosclerosis and nephron hypertrophy Loss of functional glomeruli Reduced production of klotho protein	Increased risk of type 2 diabetes, high blood pressure, obesity, chronic kidney disease, and cardiovascular diseases	High consumption of fats, sodium, alcohol, and sugar Low consumption of fiber, antioxidants, anti-inflammatory vitamins, and probiotics Dehydration
Musculoskeletal	Muscle anabolic inflexibility Loss of muscle mass, function, and strength Decreased lean body mass, bone density, and basal metabolic rate	Increased risk of sarcopenia, osteoporosis, and osteoarthritis	Malnutrition Protein undernutrition Decreased magnesium and vitamin High consumption of saturated fatty acids

TABLE 35.2
Classification of Amino Acids

Essential Amino Acids	Non-Essential Amino Acids
Arginine	Alanine
Histidine	Arginine
Isoleucine	Asparagine
Leucine	Aspartic acid
Lysine	Cysteine
Methionine	Glutamic acid
Phenylalanine	Glutamine
Threonine	Glycine
Tryptophan	Proline
Valine	Serine
	Tyrosine

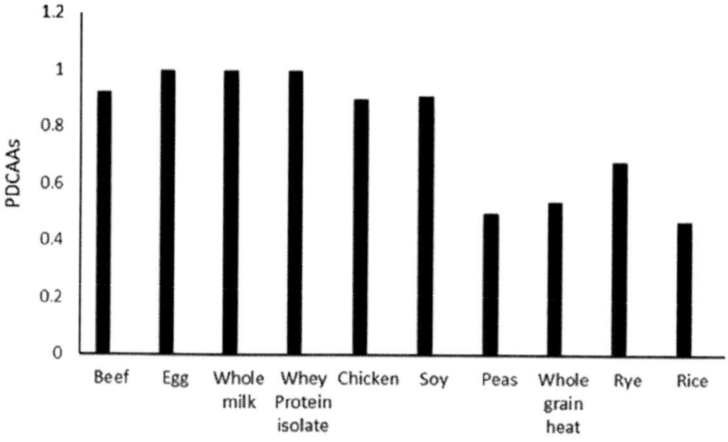

FIGURE 35.1 Protein digestibility-corrected amino acid score (PDCAAS) for different protein sources. From FAO (2013).

Hughes et al. (2001) conducted a longitudinal study that indicated muscle strength declines in older individuals, regardless of whether muscle mass is maintained or even increased through exercise. This finding suggests the need to explore the role of other cellular, neural, and metabolic factors influencing changes in strength in both men and women.

Previous studies, such as the one by Yarasheski et al. (1993), found that the fractional rate of muscle protein synthesis (MPS) was lower in elderly women and men compared to younger individuals. However, the skeletal MPS pathways are activated after resistance exercise in both groups. Welle et al. (1993) noted reduced myofibrillar protein synthesis in healthy older men, even though myofibrillar protein synthesis represented a smaller fraction of whole-body protein synthesis in the older group.

Trappe et al. (2004), in a study involving both young and older men, discovered higher levels of the amino acid 3-methylhistidine in the older group. This suggests increased proteolysis in the rested and fasted states, both before and after 24 hours after following a single bout of resistance exercise, aligning with the decrease in muscle mass associated with aging. Moreover, there is evidence indicating that sarcopenia in aging is partly due to a decrease in the rates of basal post-absorptive myofibrillar MPS and a protein deficit in older individuals.

In a recent study by Stephenson et al. (2023), magnetic resonance spectroscopy was used to examine men with sarcopenia and healthy, age- and sex-matched controls. This study found increased contractile costs during exercise and potential mitochondrial dysfunction in sarcopenic men. The researchers also observed correlations between mitochondrial enzyme activity from muscle biopsies and magnetic resonance spectroscopy bioenergetic markers, with strong associations noted with complex II and complex III activity, as well as citrate synthase and succinate dehydrogenase activity.

These findings highlight the multifaceted nature of age-related muscle loss and the importance of understanding the various factors and processes involved in maintaining muscle mass and strength in older individuals.

Although the precise causes and mechanisms are not yet fully understood, numerous studies suggest that age-related changes in proteolytic activity, reduced protein metabolism, diminished muscle protein synthesis, and alterations in protein digestion and absorption can lead to detrimental health outcomes. These outcomes may include a decreased immune response, autoimmune diseases, susceptibility to bacterial infections, cancer, and viral diseases. Furthermore, inadequate physical activity, a decline in resting metabolic rate, changes in fat distribution, age-related alterations in body composition, hormonal imbalances, malnutrition, insulin resistance, and mild metabolic acidosis may contribute to the development of conditions like sarcopenia and osteoporosis.

Despite these challenges, it's important to recognize that protein intake remains a critical factor, and the age-related changes mentioned contribute to the significance of addressing this issue (Frankowska et al. 2022).

Animal and Plant Protein

Animal protein is generally considered superior to plant protein for building muscle mass due to its higher protein quality, as supported by research studies (Berrazaga et al. 2019; Van Vliet et al. 2015). Animal-based protein is favored for its high leucine content and complete amino acid profile, making it more anabolic than plant-based protein (Figure 35.2).

Conversely, healthcare professionals often encourage the substitution of animal protein with plant-based alternatives to reduce the risk of cardiovascular diseases. Plant-based diets have

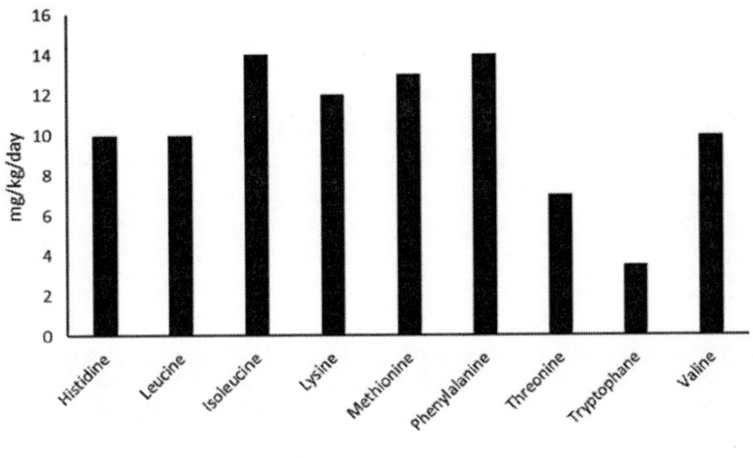

FIGURE 35.2 Amino acid requirements for adults. Data are presented in mg/kg/day. From FAO (2013).

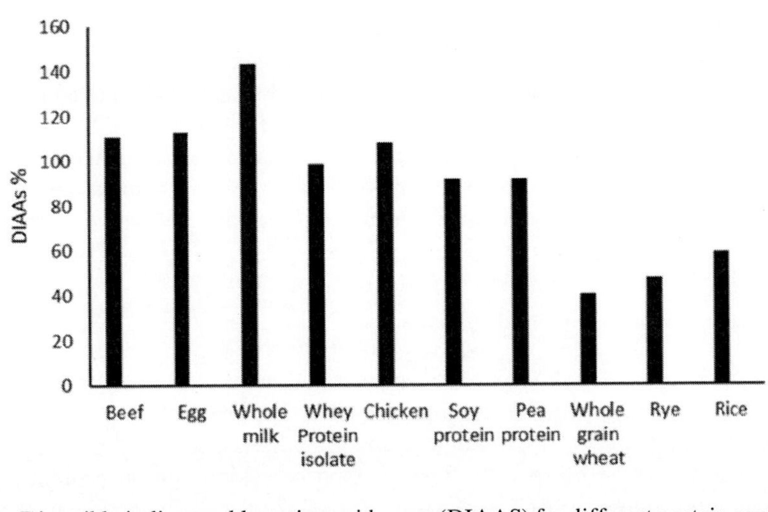

FIGURE 35.3 Digestible indispensable amino acid score (DIAAS) for different protein sources. DIAAS % = 100 x [(mg of digestible dietary indispensable amino acid in 1 g of the dietary protein) / (mg of the same dietary indispensable amino acid in 1 g of the reference protein)]. From FAO (2013).

demonstrated various health benefits, including a lower risk of muscle loss, frailty, type 2 diabetes, and overall mortality (Zhubi-Bakija et al. 2021; Coelho-Júnior et al. 2019; Im et al. 2022).

Plant-based diets promote the preservation of muscle mass and function in older adults due to their alkaline properties and abundance of potassium and magnesium (Ganapathy and Nieves 2020). In contrast, animal-based diets tend to elevate acid production, potentially leading to adverse effects on muscle mass and function through low-grade metabolic acidosis (Carnauba et al. 2017). Maintaining muscle mass with aging is possible when consuming plant-based protein sources alongside a well-balanced diet rich in essential nutrients (Gorissen and Witard 2017).

Given that animal-based proteins are generally richer in essential AAs (EAAs) compared to plant-based sources, larger quantities of plant-based protein are required to achieve a similar impact on muscle size (Van Vliet et al. 2015). Accounting for the energy content of the protein source, meeting EAA requirements from plant sources necessitates a considerably higher caloric intake compared to animal sources. This is a vital consideration, especially in the context of the prevalent public health concern of obesity, particularly among the aging population.

According to Gorissen et al. (2018), consuming an equivalent amount of EAAs would require an adult to ingest approximately 32 g of whey protein, 47 g of casein protein, 55 g of soy protein, or 48 g of pea protein (Figure 35.3)

Finally, animal-based proteins have been found to be more potent stimulators of MPS than plant-based proteins, largely due to their higher digestibility, which directly impacts their anabolic potential (Berrazaga et al. 2019; Van Vliet et al. 2015). Ingesting plant protein has been associated with increased protein oxidation (Yang et al. 2012), indicating that AAs from plant-based proteins are primarily directed toward urea production, leaving fewer AAs available to stimulate MPS (Van Vliet et al. 2015).

Protein Recommendations

In recent years, protein requirements, regardless of life stage and sex, have been widely defined using terms like recommended dietary allowance (RDA) and estimated average requirement (EAR). The RDA is set higher than the EAR (RDA = EAR + 2 SD) to ensure that the nutritional needs of individuals who may have higher requirements are met. In the case of healthy adults, the EAR for protein is set at 0.66 g per kg per day (g/kg/d), and the RDA is set at 0.8 g/kg/d. These values are designed to cover the protein needs of 97.5% of adults (WHO 2007). Notably, these recommendations do not specify the type of proteins to be consumed. It's worth mentioning that these protein intake values for healthy individuals are established based on nitrogen balance studies rather than considerations related to muscle mass maintenance or other specific needs.

A meta-analysis conducted by Rand et al. (2003) provides the most current and widely accepted recommendations for protein intake in healthy adults, and it includes a study that involved older adults. In 2019, the European Society for Clinical Nutrition and Metabolism (ESPEN) guidelines on clinical nutrition and hydration in geriatric populations suggested a protein intake of at least 1 g/kg/d for older adults. It's important to consider factors such as physical activity, health status, and the individual's energy requirements when determining protein intake. As of now, data to establish a tolerable upper intake level for protein remain insufficient (Table 35.3).

The available evidence from various studies, which employ different methodologies and techniques, including advanced "Omic" and systems biology approaches, generally does not reveal significant differences in the rates of MPS between men and women. It's important to note that the majority of research participants in these studies are typically healthy younger men and women, or the comparisons are made between older and younger adults. Thus, based on current knowledge, it appears that throughout the lifespan, both sexes tend to exhibit similar protein turnover rates, considering interactions between sex-associated genetic differences, dietary factors, nutrition, and overall health (Comitato et al. 2015). Markofski and Volpi (2011) have suggested that differences in hormone profiles and changes, which can occur due to variations in BC, might exist and could

TABLE 35.3
Methods and Techniques to Evaluate Protein Metabolism

Methods	Advantages	Disadvantages
Nitrogen balance	Highly accurate method when used for prolonged periods (e.g., 1 week or more)	Errors toward positive nitrogen balance due to overestimation of dietary nitrogen intake and underestimation of nitrogen losses from the body Long periods of adaptation (e.g., 5–7 days) are required to equilibrate the body's nitrogen pool when protein intake is altered Influence of dietary energy/carbohydrate intake can alter utilization of amino acids for energy Difficult and time-consuming method
Urine concentration in urine and sweat	Simple and inexpensive method	Affected by diet (protein intake)
3-methylhistidine in urine	Reliable biomarker for muscle protein breakdown	It does not provide information about actual changes in muscle mass Strict control of meat consumption
Radiolabeled Isotopes	Simple and inexpensive method	Possible health compromise due to the radioactive isotopes
Stable isotopes	No health risk	Expensive method with special equipment requirements
Arterial-venous difference approach	Can reliably predict changes in muscle mass over time	Invasive method The results are highly dependent on the accurate measurement of blood flow rate, which varies in each tissue
Indicator amino acid oxidation (IAAO)	Requires only peripheral tracer infusion and the collection of peripheral blood and breath samples Determination of the response at different levels of protein or EAA intake requires only a few hours	Determination of the rate of total CO_2 production by indirect calorimetry can be quite variable compared to the measurement of CO_2 enrichment
Tag tracer to specific protein	Provides direct information regarding protein synthesis	Expensive method Tissue biopsies Special equipment is required to analyze the results
Fractional composition and degradation rate	Safe method Utilizes stable isotopes	Short-term method Expensive Special equipment is required to analyze the results Does not provide information about the actual changes in muscle mass

potentially result in variations in therapeutic approaches and their outcomes. These distinctions could also influence the development of public health strategies.

Both physical activity and dietary intake are primarily anabolic stimuli that regulate protein synthesis in both men and women. These factors play a crucial role in maintaining and promoting muscle protein synthesis (MPS) and overall health.

PROTEIN REQUIREMENTS IN OLDER INDIVIDUALS

Indeed, there is a growing body of research supporting the notion that older adults may benefit from a higher protein intake than the RDA of 0.8 g/kg/day. A protein intake of 1.2 g/kg/day or more has been suggested to help prevent age-related diseases and promote healthy aging. The ESPEN recommends protein consumption of 1.0–1.2 g/kg/day for healthy elderly individuals and a further increase to 1.2–1.5 g/kg/day for those who are chronically ill or malnourished (Deutz et al. 2014). For elderly individuals with severe illness or malnutrition, protein needs may be as high as 2.0 g/kg/day (Bauer et al. 2013). These higher protein intake recommendations are often made with the aim of preserving muscle mass, supporting overall health, and mitigating the negative effects of aging.

The nutritional recommendations for the treatment and prevention of sarcopenia, as proposed by the Society for Sarcopenia, Cachexia, and Wasting Disease (SCWD), are consistent with the approach to address sarcopenia, which increases the risk of frailty, muscle weakness, difficulties in daily activities, falls, and can impact life expectancy. The SCWD recommends a protein intake of at least 1.0–1.5 g/kg body weight per day in combination with appropriate exercise to prevent the loss of muscle mass and function (Morley et al. 2010).

Increased protein requirements in older adults can be attributed to a phenomenon known as anabolic resistance, where the muscle's response to protein intake and resistance exercise is diminished. As a result, higher protein intake is needed to effectively stimulate and maintain MPS. Supporting this theory, a study found that older adults who consumed 1.5 g/kg/day of protein (nearly twice the RDA) exhibited higher rates of MPS compared to those consuming 0.8 g/kg/day (Kim et al. 2015).

Moreover, protein intake exceeding the RDA may contribute to a reduction in the risk of hip fractures and the loss of bone mineral density, thus improving overall bone health (Wallace et al. 2017). To maximize protein synthesis, it is recommended to consume approximately 25–30 g of protein per meal or 0.35–0.40 g/kg/meal (Nowson and O'Connell 2015). However, if the total protein intake per meal is less than 20 g, the protein synthetic response appears to be impaired in older individuals (Katsanos et al. 2005). Conversely, consuming more than 30 g of protein per meal does not seem to further enhance the stimulation of MPS (Nowson and O'Connell 2015).

Frailty is a significant concern associated with aging, characterized as a state of increased weakness that carries an elevated risk of adverse health outcomes. It is a topic of considerable interest to researchers. In this context, findings claim the relationship between higher protein intake (>1 g/kg/d) and reduced risk of frailty-related health issues such as slow walking speed. These findings underscore the importance of proper protein intake in promoting healthy aging (Rahi et al. 2016; Beasley et al. 2010)

OBESITY AND PROTEIN NEEDS

The prevalence of obesity among the aging population is increasing as the population ages. In light of this, the American Society for Nutrition (ASN) and the North American Association for the Study of Obesity (NAASO) (Villareal et al. 2005) have recommended that the diets of elderly individuals who intend to lose weight should provide 1.0 g/kg/d of high-quality protein. Similarly, ESPEN (Volkert et al. 2022) suggests that moderate caloric restriction, while ensuring a protein intake of at least 1 g/kg/d, may reduce the risk of physical frailty by increasing muscle anabolism.

Studies have explored the protein needs of older adults during weight loss programs, as the risk of losing skeletal muscle mass and bone mass increases. For instance, Weijs et al. (2016) conducted a study with 60 obese older adults who were involved in a 13-week weight loss program, which included a hypocaloric diet and resistance training. They found that protein needs for this population were higher than 1.2 g/kg/d or 1.9 g/kg/FFM, and they recommended an intake of 1.5 g/kg/d based on muscle mass accretion.

Additionally, Coker et al. (2019) conducted a study with eight overweight men and women using stable isotope methodology to compare the anabolic response and protein balance between the consumption of an essential amino acid-enriched meal replacement and a high-protein meal replacement beverage in older individuals aiming to lose weight. Their findings suggested that the essential amino acid-enriched meal replacement had an advantage in promoting anabolic response. Kim et al. (2016) conducted a systematic review and meta-analysis that included two trials involving populations aged over 60 years, but the results did not yield consistent outcomes.

Porter Starr et al. (2016) implemented a 6-month weight loss intervention program in 67 male and female participants, with a meal-based enhancement of protein quantity (1.2 g/kg/d) and quality. They observed similar amounts of weight loss but noted greater functional improvements relative to the control group.

In a 12-week intervention program for men and women, Backx et al. (2016) found similar results in weight loss, with no differences in 400-meter walking speed between the experimental group (protein intake of 1.7 g/kg/d) and the control group (0.9 g/kg/d). They concluded that increasing protein intake in overweight older individuals does not necessarily preserve lean body mass, strength, or physical performance.

It's crucial to identify the need for further investigation into nutrient modification in overweight and obese older adults, recognizing that there is more to learn in this area.

Generally, it's important to note that while anabolic resistance may play a role in protein requirements, it remains unclear whether protein digestion and amino acid absorption kinetics are impaired with aging (Gorissen et al. 2020). Further research is needed to fully understand the mechanisms behind protein metabolism in older adults and to refine dietary recommendations accordingly.

ANIMAL AND PLANT PROTEIN IN OLDER INDIVIDUALS

Observational studies tend to favor animal protein sources over plant-based options concerning parameters related to health. However, it's worth noting that when protein intake is adequate, the specific dietary protein source may become less significant. Moreover, numerous studies indicate that reducing cardiovascular disease (CVD) risk is possible by favoring plant sources alongside unprocessed, low-saturated-fat animal-based protein foods (Campbell et al. 2023).

Older adults who incorporate a variety of high-quality, protein-rich foods into a healthy dietary pattern can lower their risk of physical performance decline and sarcopenia. The proportion of animal and plant proteins in a diverse diet does not significantly impact the chronic response of muscle turnover, provided that the total adequate amounts of protein are consumed. In cases where incomplete plant-based proteins are consumed, they can be combined to create complete EAA profiles resembling those of animal proteins, especially when the relative intake is less than 70% of the total intake (Nishimura et al. 2023).

Collectively, it appears that a balanced proportion of plant and animal protein sources contributes to healthier and more beneficial dietary patterns, as opposed to diets heavily skewed toward high animal or high plant-based protein consumption, as suggested by Fabek et al. in 2021. The specific biological mechanisms explaining the health benefits of vegetable protein over animal protein are currently unknown. Many epidemiological and intervention studies have evaluated their respective health benefits; however, it remains unclear whether these effects are primarily attributed to amino acid composition or other components present in the food sources, making it challenging to attribute the effects solely to protein, as highlighted by Richter et al in 2015.

In conclusion, both animal and plant-based protein sources offer essential nutrients, including high-quality proteins, that can have varying effects on health outcomes depending on factors such as age group or health status. The differences in quality between these two protein sources extend beyond their amino acid profiles, making it necessary to consider the broader context of food matrices and the accompanying nutrients and bioactives present in them. Therefore, making general statements about these protein sources may oversimplify their effects.

STUDIES THAT REPORT NEW APPROACHES TO PROTEIN RECOMMENDATIONS IN OLDER INDIVIDUALS' DIETS

The amount of protein required also depends on factors like the amount of FFM. When protein catabolism is increased, studies have shown high variability in BC, leading to significant variations in what might be considered "optimal and effective" protein intakes. Research by Campbell et al. (2023), Rafii et al. (2015), Geisler et al. (2016), and Carlos et al. (2018) has highlighted this variability in protein requirements.

For instance, Dekker et al. (2022), used three different methods to estimate protein requirements and found large differences in the recommended grams of protein. Notably, protein requirements calculated based on FFM were lower compared to those based on actual or corrected body weight. This suggests that measuring FFM is crucial, particularly in cases where there is substantial variation in BC, as is often the case in elderly individuals and those who are overweight. This topic is currently being investigated by many researchers, and it holds particular importance, especially in older individuals and various clinical populations.

Recently, the indicator amino acid oxidation (IAAO) method has been used to determine protein requirements in older adults under free-living conditions. However, it's important to note that this method has certain limitations. Preliminary results, often derived from studies with a small number of participants and short-term interventions, have shown that current protein recommendations may underestimate the actual protein needs of older individuals.

For example, Tang et al. (2014), using a log-normal assumption model, suggested that the protein requirement was 0.96 g/kg/day, with an adequate protein allowance of 1.39 g/kg/day. Similar results were obtained by Rafii et al. (2015) and by Rafii et al. (2016). The mean protein requirement and RDA for protein were estimated to be higher than the current recommendations. Mao et al. (2020) assessed protein needs in 14 healthy men and women using an isocaloric diet that provided energy at 1.5 times their resting energy expenditure. They determined an EAR of 0.91 g/kg/day and a recommended nutrient intake of 1.17 g/kg/day, again higher than the existing guidelines.

Recent research has also investigated the total sulfur amino acid requirement. The results showed that the requirement in women aligned with current recommendations but was 75% higher in men (Paoletti et al. 2023). The same group also determined the minimum methionine requirement in 15 healthy individuals aged 60 and older. They derived an EAR of 5.4 mg/kg/day for males and 4.6 mg/kg/day for females, with RDAs of 8.8 mg/kg/day and 6.8 mg/kg/day, respectively. These findings support the idea that increasing the consumption of plant proteins may have beneficial effects on BC and metabolic health in older adults.

The study by Nishimura et al. (2023) underscores the importance of a comprehensive and holistic approach to determine the actual and optimal dietary protein requirements and recommendations for healthy older adults. This approach should include the use of various methodologies and concurrent measurements of whole-body protein and skeletal mass. The metabolic and physiological changes that occur with aging significantly influence and may necessitate adjustments in protein requirements for older individuals.

MINI DICTIONARY OF TERMS

Anabolic resistance: The reduced ability of skeletal muscle to respond to anabolic stimuli such as dietary proteins.

Animal protein: A nutrient found in foods such as meat, fish, poultry, eggs, and dairy products.

Deficiency: The state of not having, or not having enough, protein in a diet.

Healthy aging: A process of maintaining functionality, well-being, and independence in the elderly.

Muscle protein breakdown: Loss of muscle mass and, consequently, a reduction in strength, power, and functionality, which is experienced in age-related muscle loss, inactivity, diet, and other catabolic conditions.

Muscle protein synthesis: The metabolic process in which protein is produced to repair muscle damage or facilitate the maintenance of muscle mass due to aging.

Nutritional requirement: The amount of a nutrient that an older person needs to be healthy and functional.

Older individual: A healthy woman and man aged >60 years.

Plant protein: A nutrient found in foods such as beans, vegetables, greens, lentils, nut butter, nuts, seeds, peas, rice, oats, and soy.

Protein intake: The amount of protein that an older person consumes per day.

SUMMARY POINTS

- The world's population is aging.
- A balanced diet is a vital factor in healthy aging.
- Physiological and musculoskeletal changes might increase the protein needs of the elderly without sex differences.
- At least 1.0–1.2 g/kg/d is recommended for adults aged over 65 years to optimize muscle health.
- Based on the available data in the literature, it is recommended that healthy older adults should consume 1.0–1.5 g/kg/day of dietary protein.
- Adequate protein intake has a fundamental role to play in the mental and physical health of older people.
- Low protein intake is correlated with the onset and progression of sarcopenia, frailty, and osteoporosis.
- The quality of consumed protein plays a crucial role in muscle health.
- Plant-based proteins ordinarily contain fewer quantities of EAAs and are less digestible than animal proteins.
- Understanding the protein needs of the elderly should consider not only the quantity but also the quality of protein, its distribution throughout the day, and the role of physical activity in promoting muscle health in aging individuals.

REFERENCES

Backx, Evelien, Michael Tieland, Karin J. Borgonjen-Van Den Berg, P. R. Claessen, Luc J. C. Van Loon, and Lisette C. P. G. M. De Groot. 2016. "Protein Intake and Lean Body Mass Preservation during Energy Intake Restriction in Overweight Older Adults." *International Journal of Obesity* 40 (2): 299–304. https://doi.org/10.1038/ijo.2015.182

Bauer, Jüergen M., Gianni Biolo, Tommy Cederholm, Matteo Cesari, Alfonso J. Cruz-Jentoft, John E. Morley, Stuart M. Phillips, et al. 2013. "Evidence-Based Recommendations for Optimal Dietary Protein Intake in Older People: A Position Paper from the PROT-AGE Study Group." *Journal of the American Medical Directors Association* 14 (8): 542–59. https://doi.org/10.1016/j.jamda.2013.05.021

Beasley, Jeannette M., Andrea Z. LaCroix, Marian L. Neuhouser, Ying Huang, Lesley F. Tinker, Nancy Fúgate Woods, Yvonne Michael, J. David Curb, and Ross L. Prentice. 2010. "Protein Intake and Incident Frailty in the Women's Health Initiative Observational Study." *Journal of the American Geriatrics Society* 58 (6): 1063–71. https://doi.org/10.1111/j.1532-5415.2010.02866.x

Berrazaga, Insaf, Valérie Micard, Marine Gueugneau, and Stéphane Walrand. 2019. "The Role of the Anabolic Properties of Plant- versus Animal-Based Protein Sources in Supporting Muscle Mass Maintenance: A Critical Review." *Nutrients* 11 (8): 1825. https://doi.org/10.3390/nu11081825

Campbell, Wayne W., Nicolaas E. P. Deutz, Elena Volpi, and Caroline M. Apovian. 2023. "Nutritional Interventions: Dietary Protein Needs and Influences on Skeletal Muscle of Older Adults." *The Journals of Gerontology* 78 (Supplement_1): 67–72. https://doi.org/10.1093/gerona/glad038

Carnauba, Renata Alves, Ana Beatriz Baptistella, Valéria Paschoal, and Gilberti Hubscher. 2017. "Diet-Induced Low-Grade Metabolic Acidosis and Clinical Outcomes: A Review." *Nutrients* 9 (6): 538. https://doi.org/10.3390/nu9060538

Celis-Morales, Carlos, Fanny Petermann, Lewis Steell, Jana Anderson, Paul Welsh, Daniel Mackay, Stamatina Iliodromiti, et al. 2018. "Associations of Dietary Protein Intake with Fat-Free Mass and Grip Strength: A Cross-Sectional Study in 146,816 UK Biobank Participants." *American Journal of Epidemiology* 187 (11): 2405–14. https://doi.org/10.1093/aje/kwy134

Coelho-Júnior, Hélio José, Riccardo Calvani, Ivan O. Gonçalves, Bruno Rodrigues, Anna Picca, Francesco Landi, Roberto Bernabei, Masayuki Uchida, and Emanuele Marzetti. 2019. "High Relative Consumption of Vegetable Protein Is Associated with Faster Walking Speed in Well-Functioning Older Adults." *Aging Clinical and Experimental Research* 31 (6): 837–44. https://doi.org/10.1007/s40520-019-01216-4

Coker, Robert H., Kenneth J. Shin, Kristen Scholten, Michelle Johannsen, Jean Tsigonis, Il-Young Kim, Scott Schutzler, and Robert R. Wolfe. 2019. "Essential Amino Acid-Enriched Meal Replacement Promotes Superior Net Protein Balance in Older, Overweight Adults." *Clinical Nutrition* 38 (6): 2821–26. https://doi.org/10.1016/j.clnu.2018.12.013

Comitato, Raffaella, Anna Saba, Aida Turrini, Claudia Arganini, and Fabio Virgili. 2015. "Sex Hormones and Macronutrient Metabolism." *Critical Reviews in Food Science and Nutrition* 55 (2): 227–41. https://doi.org/10.1080/10408398.2011.651177

Dekker, Ingeborg M., Natascha M. Van Rijssen, A. M. Verreijen, P. J. M. Weijs, W. B. De Boer, Dominique Terpstra, and Hinke Kruizenga. 2022. "Calculation of Protein Requirements; a Comparison of Calculations Based on Bodyweight and Fat Free Mass." *Clinical Nutrition ESPEN* 48 (April): 378–85. https://doi.org/10.1016/j.clnesp.2022.01.014

Deutz, N. E. P., Jürgen Bauer, Rocco Barazzoni, Gianni Biolo, Yves Boirie, Anja Bosy-Westphal, Tommy Cederholm, et al. 2014. "Protein Intake and Exercise for Optimal Muscle Function with Aging: Recommendations from the ESPEN Expert Group." *Clinical Nutrition* 33 (6): 929–36. https://doi.org/10.1016/j.clnu.2014.04.007

FAO. 2013. *Dietary Protein Quality Evaluation in Human Nutrition: Report of an FAO Expert Consultation, 31 March-2 April, 2011, Auckland, New Zealand*. FAO.

Frankowska, Natalia, Katarzyna Lisowska, and Jacek M. Witkowski. 2022. "Proteolysis Dysfunction in the Process of Aging and Age-Related Diseases." *Frontiers in Aging* 3 (July). https://doi.org/10.3389/fragi.2022.927630

Fürst, Peter. 2009. "Basics in Clinical Nutrition: Proteins and Amino Acids." *E-SPEN* 4 (2): e62–65. https://doi.org/10.1016/j.eclnm.2008.07.010

Ganapathy, Aravinda, and Jeri W. Nieves. 2020. "Nutrition and Sarcopenia—What Do We Know?" *Nutrients* 12 (6): 1755. https://doi.org/10.3390/nu12061755

Geisler, Corinna, Carla M. Prado, and Manfred J. Müller. 2016. "Inadequacy of Body Weight-Based Recommendations for Individual Protein Intake—Lessons from Body Composition Analysis." *Nutrients* 9 (1): 23. https://doi.org/10.3390/nu9010023

Gorissen, Stefan H. M., Julie J. R. Crombag, Joan M. Senden, W. A. Huub Waterval, Joergen Bierau, Lex B. Verdijk, and Luc J. C. Van Loon. 2018. "Protein Content and Amino Acid Composition of Commercially Available Plant-Based Protein Isolates." *Amino Acids* 50 (12): 1685–95. https://doi.org/10.1007/s00726-018-2640-5

Gorissen, Stefan H. M., and Oliver C. Witard. 2017. "Characterising the Muscle Anabolic Potential of Dairy, Meat and Plant-Based Protein Sources in Older Adults." *Proceedings of the Nutrition Society* 77 (1): 20–31. https://doi.org/10.1017/s002966511700194x

Gorissen, Stefan. H. M., Trommelen J. Kouw, I. W., Pennings, B., Wall, B. T., Churchward-Venne, T. A., ... & van Loon, L. J. 2020. Protein type, protein dose, and age modulate dietary protein digestion and phenylalanine absorption kinetics and plasma phenylalanine availability in humans. *The Journal of nutrition*, *150*(8), 2041-2050. https://doi.org/10.1093/jn/nxaa024

Hughes, Virginia, Walter R. Frontera, Michael Wood, William J. Evans, Gerard E. Dallal, Ronenn Roubenoff, and Maria A. Fiatarone Singh. 2001. "Longitudinal Muscle Strength Changes in Older Adults: Influence of Muscle Mass, Physical Activity, and Health." *The Journals of Gerontology* 56 (5): B209–17. https://doi.org/10.1093/gerona/56.5.b209

Im, Ji-Hyun, Hyoungsu Park, and Kyong Park. 2022. "Dietary Essential Amino Acid Intake Is Associated with High Muscle Strength in Korean Older Adults." *Nutrients* 14 (15): 3104. https://doi.org/10.3390/nu14153104

Katsanos, Christos S., Hisamine Kobayashi, Melinda Sheffield-Moore, Asle Aarsland, and Robert R. Wolfe. 2005. "Aging Is Associated with Diminished Accretion of Muscle Proteins after the Ingestion of a Small Bolus of Essential Amino Acids." *The American Journal of Clinical Nutrition* 82 (5): 1065–73. https://doi.org/10.1093/ajcn/82.5.1065

Kim, Il-Young, Scott Schutzler, Amy Schrader, Horace J. Spencer, Patrick Kortebein, N. E. P. Deutz, Robert R. Wolfe, and Arny A. Ferrando. 2015. "Quantity of Dietary Protein Intake, but Not Pattern of Intake, Affects Net Protein Balance Primarily through Differences in Protein Synthesis in Older Adults." *American Journal of Physiology-Endocrinology and Metabolism* 308 (1): E21–28. https://doi.org/10.1152/ajpendo.00382.2014

Kim, Jung Eun, Lauren O'Connor, Laura P. Sands, Maribeth Slebodnik, and Wayne W. Campbell. 2016. "Effects of Dietary Protein Intake on Body Composition Changes after Weight Loss in Older Adults: A Systematic Review and Meta-Analysis." *Nutrition Reviews* 74 (3): 210–24. https://doi.org/10.1093/nutrit/nuv065

Mao, Deqian, Fengge Chen, Rui Wang, Bai Ping, Yanhong Zhang, Wei Zhao, Jing Chen, Lichen Yang, Xiaoguang Yang, and Min Li. 2020. "Protein Requirements of Elderly Chinese Adults Are Higher than Current Recommendations." *Journal of Nutrition* 150 (5): 1208–13. https://doi.org/10.1093/jn/nxaa031

Markofski, Melissa M., and Elena Volpi. 2011. "Protein Metabolism in Women and Men: Similarities and Disparities." *Current Opinion in Clinical Nutrition and Metabolic Care* 14 (1): 93–97. https://doi.org/10.1097/mco.0b013e3283412343

Morley, John E., Josep M. Argilés, William J. Evans, Shalender Bhasin, David Cella, N. E. P. Deutz, Wolfram Doehner, et al. 2010. "Nutritional Recommendations for the Management of Sarcopenia." *Journal of the American Medical Directors Association* 11 (6): 391–96. https://doi.org/10.1016/j.jamda.2010.04.014

Nishimura, Yusuke, Grith Højfeldt, Leigh Breen, Inge Tetens, and Lars Holm. 2023. "Dietary Protein Requirements and Recommendations for Healthy Older Adults: a Critical Narrative Review of the Scientific Evidence." *Nutrition Research Reviews* 36 (1): 69–85.

Nowson, Caryl, and Stella O'Connell. 2015. "Protein Requirements and Recommendations for Older People: A Review." *Nutrients* 7 (8): 6874–99. https://doi.org/10.3390/nu7085311.

Pacifico, Jacob, Milou A. J. Geerlings, Esmee M. Reijnierse, Christina Phassouliotis, Wen Kwang Lim, and Andrea B. Maier. 2020. "Prevalence of Sarcopenia as a Comorbid Disease: A Systematic Review and Meta-Analysis." *Experimental Gerontology* 131 (March): 110801. https://doi.org/10.1016/j.exger.2019.110801

Paddon-Jones, Douglas, and Blake B. Rasmussen. 2009. "Dietary Protein Recommendations and the Prevention of Sarcopenia." *Current Opinion in Clinical Nutrition and Metabolic Care* 12 (1): 86–90. https://doi.org/10.1097/mco.0b013e32831cef8b

Paoletti, Alyssa, Paul B. Pencharz, Ronald O. Ball, Dehan Kong, Libai Xu, Rajavel Elango, and Glenda Courtney-Martin. 2023. "The Dietary Requirement for Total Sulfur Amino Acids in Adults Aged ≥60 Years Appears to Be Higher in Males than in Females." *The American Journal of Clinical Nutrition* 118 (3): 538–48. https://doi.org/10.1016/j.ajcnut.2023.06.015

Rafii, Mahroukh, Karen Chapman, Rajavel Elango, Wayne W. Campbell, Ronald O. Ball, Paul B. Pencharz, and Glenda Courtney-Martin. 2016. "Dietary Protein Requirement of Men >65 Years Old Determined by the Indicator Amino Acid Oxidation Technique Is Higher than the Current Estimated Average Requirement." *Journal of Nutrition* 146 (4): 681–87. https://doi.org/10.3945/jn.115.225631

Rafii, Mahroukh, Karen Chapman, Jillian Owens, Rajavel Elango, Wayne W. Campbell, Ronald O. Ball, Paul B. Pencharz, and Glenda Courtney-Martin. 2015. "Dietary Protein Requirement of Female Adults >65 Years Determined by the Indicator Amino Acid Oxidation Technique Is Higher Than Current Recommendations." *Journal of Nutrition* 145 (1): 18–24. https://doi.org/10.3945/jn.114.197517

Rahi, Berna, Zoé Colombet, Magali González-Colaço Harmand, Jean-François Dartigues, Yves Boirie, Luc Letenneur, and Catherine Féart. 2016. "Higher Protein but Not Energy Intake Is Associated with a Lower Prevalence of Frailty among Community-Dwelling Older Adults in the French Three-City Cohort." *Journal of the American Medical Directors Association* 17 (7): 672.e7–672.e11. https://doi.org/10.1016/j.jamda.2016.05.005

Rand, William, Peter L. Pellett, and Vernon R. Young. 2003. "Meta-Analysis of Nitrogen Balance Studies for Estimating Protein Requirements in Healthy Adults." *The American Journal of Clinical Nutrition* 77 (1): 109–27. https://doi.org/10.1093/ajcn/77.1.109

Richter, Chesney K., Ann C. Skulas-Ray, Catherine M. Champagne, and Penny M. Kris-Etherton. 2015. "Plant Protein and Animal Proteins: Do They Differentially Affect Cardiovascular Disease Risk?" *Advances in Nutrition* 6 (6): 712–28. https://doi.org/10.3945/an.115.009654

Starr, Kathryn N. Porter, Carl F. Pieper, Melissa C. Orenduff, Shelley McDonald, Luisa B. McClure, Zhou Run, Martha E. Payne, and Connie W. Bales. 2016. "Improved Function with Enhanced Protein Intake per Meal: A Pilot Study of Weight Reduction in Frail, Obese Older Adults." *The Journals of Gerontology* 71 (10): 1369–75. https://doi.org/10.1093/gerona/glv210

Stephenson, Mary C., Jamie Xiu-Mei Ho, Eugenia Migliavacca, Maria Kalimeri, Neerja Karnani, Subhasis Banerji, John J. Totman, Jérôme N. Feige, Reshma Aziz Merchant, and Stacey Kh Tay. 2023. "Evidence for Inefficient Contraction and Abnormal Mitochondrial Activity in Sarcopenia Using Magnetic Resonance Spectroscopy." *Journal of Cachexia, Sarcopenia and Muscle* 14 (3): 1482–94. https://doi.org/10.1002/jcsm.13220

Tang, Minghua, George P. McCabe, Rajavel Elango, Paul B. Pencharz, Ronald O. Ball, and Wayne W. Campbell. 2014. "Assessment of Protein Requirement in Octogenarian Women with Use of the Indicator Amino Acid Oxidation Technique." *The American Journal of Clinical Nutrition* 99 (4): 891–98. https://doi.org/10.3945/ajcn.112.042325

Trappe, Todd A., Rick H. Williams, John A. Carrithers, Ulrika Raue, B. Esmarck, Michael Kjær, and Robert C. Hickner. 2004. "Influence of Age and Resistance Exercise on Human Skeletal Muscle Proteolysis: A Microdialysis Approach." *The Journal of Physiology* 554 (3): 803–13. https://doi.org/10.1113/jphysiol.2003.051755

Van Vliet, Stephan, Nicholas A. Burd, and Luc J. C. Van Loon. 2015. "The Skeletal Muscle Anabolic Response to Plant- versus Animal-Based Protein Consumption." *Journal of Nutrition* 145 (9): 1981–91. https://doi.org/10.3945/jn.114.204305

Villareal, Dennis T., Caroline M. Apovian, Robert F. Kushner, and Samuel Klein. 2005. "Obesity in Older Adults: Technical Review and Position Statement of the American Society for Nutrition and NAASO, The Obesity Society." *The American Journal of Clinical Nutrition* 82 (5): 923–34. https://doi.org/10.1093/ajcn/82.5.923

Volkert, Dorothee, Anne Marie Beck, Tommy Cederholm, A. J. Cruz-Jentoft, Lee Hooper, Eva Kiesswetter, Marcello Maggio, et al. 2022. "ESPEN Practical Guideline: Clinical Nutrition and Hydration in Geriatrics." *Clinical Nutrition* 41 (4): 958–89. https://doi.org/10.1016/j.clnu.2022.01.024

Wallace, Taylor C., and Cara L. Frankenfeld. 2017. "Dietary Protein Intake above the Current RDA and Bone Health: A Systematic Review and Meta-Analysis." *Journal of the American College of Nutrition* 36 (6): 481–96. https://doi.org/10.1080/07315724.2017.1322924

Weijs, P. J. M., and Robert R. Wolfe. 2016. "Exploration of the Protein Requirement during Weight Loss in Obese Older Adults." *Clinical Nutrition* 35 (2): 394–98. https://doi.org/10.1016/j.clnu.2015.02.016

Welle, Stephen, Charles A. Thornton, Ralph F. Józefowicz, and Marcia Statt. 1993. "Myofibrillar Protein Synthesis in Young and Old Men." *American Journal of Physiology-Endocrinology and Metabolism* 264 (5): E693–98. https://doi.org/10.1152/ajpendo.1993.264.5.e693

Wilmoth, John Richard, Daniela Bas, Santanu Mukherjee, and Naveed Hanif. 2023. *World Social Report 2023: Leaving No One Behind in an Ageing World.* United Nations.

Wolfe, Robert R. 2012. "The Role of Dietary Protein in Optimizing Muscle Mass, Function and Health Outcomes in Older Individuals." *British Journal of Nutrition* 108 (S2): S88–93. https://doi.org/10.1017/s0007114512002590

World Health Organization [WHO]. 2007. *Protein and Amino Acid Requirements in Human Nutrition.* PubMed. National Institutes of Health. https://pubmed.ncbi.nlm.nih.gov/18330140

Yang, Yifan, Tyler A. Churchward-Venne, Nicholas A. Burd, Leigh Breen, Mark A. Tarnopolsky, and Stuart M. Phillips. 2012. "Myofibrillar Protein Synthesis Following Ingestion of Soy Protein Isolate at Rest and after Resistance Exercise in Elderly Men." *Nutrition & Metabolism* 9 (1). https://doi.org/10.1186/1743-7075-9-57

Yarasheski, Kevin E., Jeffrey J. Zachwieja, and Dennis M. Bier. 1993. "Acute Effects of Resistance Exercise on Muscle Protein Synthesis Rate in Young and Elderly Men and Women." *American Journal of Physiology-Endocrinology and Metabolism* 265 (2): E210–14. https://doi.org/10.1152/ajpendo.1993.265.2.e210

Zhubi-Bakija, Fjolla, Gani Bajraktari, Ibadete Bytyçi, Dimitri P. Mikhailidis, Michael Y. Henein, Gustavs Latkovskis, Zarife Rexhaj, et al. 2021. "The Impact of Type of Dietary Protein, Animal versus Vegetable, in Modifying Cardiometabolic Risk Factors: A Position Paper from the International Lipid Expert Panel (ILEP)." *Clinical Nutrition* 40 (1): 255–76. https://doi.org/10.1016/j.clnu.2020.05.017

36 Recommended Resources for Protein Intake in Health and Disease

Rajkumar Rajendram, Vinood B. Patel, and Victor R. Preedy

INTRODUCTION

It is well recognized that protein serves as the cornerstone for essentially every biological process within the human body. Thus, adequate protein intake is clearly important to maintain health and promote physical function (Food and Agriculture Organization of the United Nations, 2017; Hruby et al., 2020).

Data from clinical trials and observational studies suggest that higher protein intake is associated with a decreased rate of muscle loss and improved physical performance in older adults (Hruby et al., 2020; Naseeb and Volpe, 2017). Dietary protein also helps maintain physical mobility and cognitive function (Ardisson Korat et al., 2024). However, most of these studies had short follow-up times, and the associations of specific protein sources (animal or plant) were inconclusive for many outcomes (Ardisson Korat et al., 2024).

However, the optimal amount of protein needed can vary depending on a multitude of factors, including age, activity level, and underlying health conditions (Phillips and Wolfe, 2009; Naseeb and Volpe, 2017; World Health Organization, 2007; Food and Agriculture Organization of the United Nations, 2017). Also, there are different ways of measuring amino acid and/or protein requirements, as well as different ways of assessing the amino acid profiles and protein contents of foods.

It is important for those active in the field to keep up to date and abreast of current research into protein intake in health and disease. To address this, we have compiled a list of resources as recommended by researchers, physicians, practitioners, and others who work in this field. Information on over 100 organizations, scientific societies, regulatory bodies, and other web-based resources are included, as well as recommended books.

The list of material in these tables is included to provide general information only. It does not constitute any recommendation or endorsement of the activities of these sites, facilities, or other resources listed in this chapter by the authors or editors of this book.

Tables 36.1–36.5 list the most up-to-date information on the regulatory bodies (Table 36.1), professional societies (Table 36.2), books (Table 36.3), emerging technologies and platforms (Table 36.4), and other resources of interest (Table 36.5) relevant to an evidence-based approach to protein intake in health and disease. Some organizations are listed in more than one table as they occasionally fulfill multiple roles.

DOI: 10.1201/b23328-36

TABLE 36.1
Websites, Regulatory Bodies, and Organizations Dealing with Protein Intake in Health and Disease

Regulatory Body or Organization	Web Address
American Institute for Cancer Research	www.aicr.org/
American Society for Nutrition	https://nutrition.org/
Association for Nutrition	www.associationfornutrition.org/
Belgian Nutrition Society	www.belgiannutritionsociety.be/
Bone Health and Osteoporosis Foundation (BHOF)	www.bonehealthandosteoporosis.org/
Brazilian Society for Food and Nutrition	https://iuns.org/organisations/brazilian-society-for-food-and-nutrition/
British Nutrition Foundation	www.nutrition.org.uk/
Cleveland Clinic	https://my.clevelandclinic.org/
Çocuk Beslenme ve Metabolizma Derneği (CBMD; Children's Nutrition Metabolism Association)	https://cbmd.org.tr/index.php
European Food Safety Authority (EFSA)	www.efsa.europa.eu/en
Federation of African Nutrition Societies	https://fanus.org/
Federation of Asian Nutrition Societies	https://fansnutrition.org/
Federation of European Nutrition Societies	https://fensnutrition.org/
Finnish Society for Nutrition Research	https://suomenravitsemustieteenyhdistys.fi/en/
Food and Agriculture Organization of the United Nations	www.fao.org/home/en
Food and Nutrition Board/National Academies	www.nationalacademies.org/fnb/food-and-nutrition-board
German Nutrition Society	www.dge.de/english/
International Agency for Research on Cancer	www.iarc.who.int/
International Atomic Energy Agency: Information on Nutrition	www.iaea.org/topics/nutrition
International Osteoporosis Foundation	www.osteoporosis.foundation/
Latin American Nutrition Society	www.slaninternacional.org/
National Institute of Health	www.nih.gov/
Nordic Co-operation	https://pub.norden.org/
Nutrition Society	www.nutritionsociety.org/
PKU Family Association	www.pkuaile.com/
Rare Diseases Network	www.nadirhastaliklaragi.org.tr/en/
Spanish Nutrition Society	www.sennutricion.org/es/inicio
Swedish Association for Clinical Nutrition	www.sfkn.se/
The American College of Lifestyle Medicine (ACLM)	https://lifestylemedicine.org/
UNICEF (United Nations International Children's Emergency Fund)	www.unicef.org/
United States Department of Agriculture (USDA)	www.usda.gov/
United States Food and Drug Administration (US FDA)	www.fda.gov/
World Cancer Research Fund International	www.wcrf.org/
World Health Organization (WHO)	www.who.int/

This table lists the regulatory bodies and organizations relevant to protein intake in health and disease. The links were accurate at the time of going to press but may move or alter. In these cases, the use of the "Search" tabs should be explored at the parent address or site. Dual-language sites often have pages for automatic translation into English. In some cases, links direct the reader to pages related to protein intake in health and disease within parent sites. Some societies and organizations prefer shortened terms, such as acronyms and abbreviations. Some organizations provide a dual role of imparting information on specific conditions (from diagnostics to epidemiology) and acting as professional or regulatory bodies. See also Table 36.2.

TABLE 36.2
Professional Societies Dealing with Protein Intake in Health and Disease

Society Name	Web Address
Academy of Nutrition and Dietetics	www.eatright.org/
American Cancer Society	www.cancer.org/
American Society for Nutrition	https://nutrition.org/
American Society for Parenteral and Enteral Nutrition (ASPEN)	www.nutritioncare.org/
British Dietetic Association	www.bda.uk.com/
British Nutrition Foundation	www.nutrition.org.uk/
Clinical Enteral Parenteral Nutrition Association	www.kepan.org.tr/
European Food Safety Authority	www.efsa.europa.eu/en
European Society for Clinical and Economical Aspects of Osteopororosis, Osteoarthritis, and Musculoskeletal Diseases	www.esceo.org/
European Society for Clinical Nutrition and Metabolism (ESPEN)	www.espen.org/
European Society for Paediatric Gastroenterology, Hepatology and Nutrition (ESPGHAN)	www.espghan.org/
Federation of Asian Nutrition Societies (FANS)	https://fansnutrition.org/
Federation of European Nutrition Societies (FENS)	https://fensnutrition.org/
Food Insight	https://foodinsight.org/
Gıda Mühendisleri Derneği (GIDAMUHDER; Turkish Food Engineers Association)	www.gidamuhendisleri.org.tr/
Global Leadership Initiative on Malnutrition (GLIM)	www.espen.org/education/glim
International Society of Sports Nutrition	www.sportsnutritionsociety.org/
International Union of Nutritional Sciences	https://iuns.org/
Latin American Nutrition Society	www.slaninternacional.org/
National Association for Nutrition Professionals	https://nanp.org/
National Institute of Health	www.nih.gov/
Protein Foods and Nutrition Development Association of India	www.pfndai.org/index.html
Royal Osteoporosis Society	https://theros.org.uk/
School Nutrition Association	https://schoolnutrition.org/
Society for Nutrition Education and Behavior (SNEB)	www.sneb.org/
The American Society for Bone and Mineral Research	www.asbmr.org/Default.aspx
The German Nutrition Society	www.dge.de/english/
The Protein Society	www.proteinsociety.org/
Turkish Association of Dietitians (TDD)	www.tdd.org.tr/TDD/tabid/21007/Default.aspx
US Department of Agriculture	www.usda.gov/
World Health Organization	www.who.int/

This table lists the professional societies relevant to protein intake in health and disease. See also Table 36.1.

TABLE 36.3
Books on Protein Intake in Health and Disease

Book Title	Authors or Editors	Publisher	Year of Publication
A Clinician's Guide to Integrative Oncology	O'Brien K, Sali A	Springer	2017
Alternative Protein Sources: Balancing Food Innovation, Sustainability, Nutrition, and Health: Proceedings of a Workshop	National Academies of Sciences, Engineering, and Medicine; Health and Medicine Division; Food and Nutrition Board	National Academy Press	2023
Complete Food and Nutrition Guide. 5th Ed	Duyff RL	Harvest	2017
Dietary Reference Intakes for Energy, Carbohydrate, Fiber, Fat, Fatty Acids, Cholesterol, Protein, and Amino Acids.	National Academy of Sciences, Institute of Medicine, Food and Nutrition Board.	National Academy Press.	2005
Fundamentals of Cancer Prevention	Alberts D, Hess LM	Springer	2013
Future Proteins: Sources, Processing, Applications and the Bioeconomy.	Tiwari BK, Healy LE.	Elsevier Academic Press	2023
Introduction to Human Nutrition, 3rd Ed	Lanham-New SA, Hill TR, Gallagher AM, Vorster HH.	Wiley Blackwell	2019
Modern Nutrition in Health And Disease	Ross AC, Caballero B, Cousins RJ, Turner KL, Ziegler TR	Wolters Kluwer Health/Lippincott Williams and Wilkins	2014
Nestlé Nutrition Institute Workshop Series: Protein In Neonatal and Infant Nutrition: Recent Updates: 86th Nestlé Nutrition Institute Workshop, Beijing, May 2015	Bhatia J, Shamir R, Vanderplas Y	Karger	2016
Nutrigenomics and Proteomics In Health And Disease	Kussmann M, Stover PJ	Wiley	2017
Nutrition Across Life Stages	Bernstein M	Jones and Bartlett Learning	2023
Nutrition and Bone Health	Holick MF, Nieves JW	Humana Press	2015
Nutrition In Pediatrics: Basic Science, Clinical Applications. 5th Ed	Duggan C, Watkins JB, Koletzko B, Walker WA.	PMPH USA Ltd.	2016
Nutritional Health: Strategies for Disease Prevention	Temple NJ, Wilson T, Jacobs DR, Bray GA	Springer	2023
Our Future Proteins: A Diversity of Perspectives.	Pyett S, Jenkins W, van Mierlo B, Trindade L, Welch D, and van Zanten H	VU University Press	2023
Pediatric Bone: Biology and Diseases	Glorieux FH, Pettifor JM	Academic Press	2011
Protein Based Therapeutics	Singh DB, Tripathi T	Springer	2023
Protein Diets: Nutritional Sources, Health Benefits and Intake Recommendations	Hudson M	Nova Science Publishers, Incorporated	2015
Proteins Associated with Neurodevelopmental Disorders	Qoronfleh MW, Essa MM, Babu CS	Springer	2022
The Necessity of Protein Intake For The Body	Stamps R	Self-publisher Open Library	2023

TABLE 36.4
Emerging Techniques, Instruments, and Analytical Platforms, Sources, or Devices for Investigating Protein Intake in Health and Disease

Organization or Company Name	Web Address
Agilent Scientific Instruments	www.agilent.com/
Bio-Rad Laboratories	www.bio-rad.com/
D_3-Creatine Dilution	PMID: 30900400 and PMCID: PMC9745462
Bioelectrical Impedance Analysis Phase Angle	PMID: 37202048
Danaher Corporation	www.danaher.com/
Human Proteome Organization	www.hupo.org/
Mass Spectrometry–Based Protein Identification In Proteomics—A Review	https://academic.oup.com/bib/article/22/2/1620/5733238
Muscle Protein Fractional Synthetic Rate (FSR)*	PMID: 21683485
Olink Proteomics	www.bioxpedia.com/olink-proteomics/
QIAGEN Digital Insights	https://digitalinsights.qiagen.com/
Sage-N	www.sagenresearch.com/
Swiss Bioinformatics Resource Portal	www.expasy.org/

This table lists some emerging technologies relevant to protein intake in health and disease. The links were accurate at the time of going to press but may move or alter. (*) This article offers a comprehensive list of online software resources for the analysis of protein data.

Other Resources

The Wellcome Collection (https://wellcomecollection.org/collections) and the British Library (www.bl.uk) also hold material on topics related to protein in health and disease.

Other chapters on resources relevant to nutrition (recommended by authors and practitioners) may also be relevant to protein in health and disease (Rajendram et al., 2013a, 2013b, 2014, 2015, 2016, 2017, 2019a, 2019b, 2020, 2022a, 2022b; 2023, Alzaid et al., 2015).

TABLE 36.5
Other Resources of Interest or Relevance for Health Care Professionals or Patients Related to Protein Intake in Health and Disease

Name of Resource or Organisation	Web Address
About Herbs, Botanicals and Other Products	www.mskcc.org/cancer-care/diagnosis-treatment/symptom-management/integrative-medicine/herbs
American Society for Nutrition: Information on Nutritional Psychiatry	https://nutrition.org/food-and-mood-what-is-nutritional-psychiatry/
British Nutrition Foundation	www.nutrition.org.uk/
British Nutrition Foundation: Information on Protein	www.nutrition.org.uk/nutritional-information/protein/
Cancer Care Ontario	www.cancercareontario.ca/en
Carer Centres And Services Patients Association	www.nhs.uk/services/service-directory/patients-association/N10498759
Cleveland Clinic: Information on Integrative Medicine	https://my.clevelandclinic.org/health/treatments/21683-integrative-medicine
Deakin University: Information on Food and Mood Centre	https://foodandmoodcentre.com.au/team/felice-jacka/
Food and Nutrition Information Center (FNIC)	www.nal.usda.gov/programs/fnic
Global Communities	https://globalcommunities.org/
Harvard Health Publishing By Harvard Medical School	www.health.harvard.edu/
Harvard Medical School: Information on Nutritional Psychiatry: Your Brain On Food	www.health.harvard.edu/blog/nutritional-psychiatry-your-brain-on-food-201511168626
Harvard T.H. Chan School of Public Health	www.hsph.harvard.edu/
MedlinePlus	https://medlineplus.gov/
MedLinePlus Nutrition	https://medlineplus.gov/nutrition.html
Merck Manual Overview of Nutrition	www.msdmanuals.com/en-gb/professional/nutritional-disorders/nutrition-general-considerations/overview-of-nutrition
National Cancer Institute	www.cancer.gov/
National Health Service (NHS)	https://nhs.uk
Nordic Co-operation: Information on Protein	https://pub.norden.org/nord2023-003/protein-.html

(Continued)

TABLE 36.5 (CONTINUED)
Other Resources of Interest or Relevance for Health Care Professionals or Patients Related to Protein Intake in Health and Disease

Name of Resource or Organisation	Web Address
The European Consumer Organisation	www.beuc.eu/
The European Patients' Forum	www.eu-patient.eu/
United States Department of Agriculture (USDA) Food and Nutrition Service	www.fns.usda.gov/

This table lists some other resources relevant to protein intake in health and disease. The links were accurate at the time of going to press but may move or alter. In these cases, the use of the "Search" tabs should be explored at the parent address or site. In some cases, links direct the reader to pages related to protein intake in health and disease within parent sites.

ACKNOWLEDGMENTS (IN ALPHABETICAL ORDER)

We thank the following authors for their contributions to the development of this resource. We apologize if some of the suggested material was not included in this chapter or has been moved to different tables.

Abreu S
Aeri BT
Jurado-Fasoli L
Kittisakmontri K
Liz Martins M
Morais C
Neyestani T
Ozer I
Pallaro A
Pierzynowska K
Rodriguez-Garcia C
Siraj N
Rosa Bergamo R
Yazar H

REFERENCES

Alzaid, F., Rajendram, R., Patel, V.B., and Preedy, V.R. (2015). Expanding the knowledge base in diet, nutrition and critical care. Electronic and published resources. In Rajendram, R., Preedy, V.R., and Patel, V.B. (eds.) *Diet and Nutrition in Critical Care*. Berlin, Germany: Springer.

Ardisson Korat, A. V., Shea, M. K., Jacques, P. F., Sebastiani, P., Wang, M., Eliassen, A. H., Willett, W. C., and Sun, Q. (2024). Dietary protein intake in midlife in relation to healthy aging - results from the prospective nurses' health study cohort. *The American Journal of Clinical Nutrition*, 119(2), 271–282.

Food and Agriculture Organization of the United Nations. (2017). *Dietary Protein Quality for Human Life*. Rome, Food and Agriculture Organization of the United Nations. FAO Report

Hruby, A., Sahni, S., Bolster, D., and Jacques, P. F. (2020). Protein intake and functional integrity in aging: The framingham heart study offspring. *The Journals of Gerontology. Series A, Biological Sciences and Medical Sciences*, 75(1), 123–130.

Naseeb, M. A., and Volpe, S. L. (2017). Protein and exercise in the prevention of sarcopenia and aging. *Nutrition Research* (New York, N.Y.), 40, 1–20.

Phillips, S. M., and Wolfe, R. R. (2009). Dietary protein requirements and adaptations to dietary restriction. *Applied Physiology, Nutrition, and Metabolism*, 34(3), 899–909. PubMed

Rajendram, R., Gyamfi, D., Patel, V.B., and Preedy, V.R. (2022a). Recommended resources for biomarkers of nutrition. In Preedy, V.R., and Patel, V.B. (eds.) *Biomarkers of Nutrition*. New York: Elsevier.

Rajendram, R., Gyamfi, D., Patel, V.B., and Preedy, V.R. (2022b). Recommended resources for biomarkers in diabetes: Methods, discoveries, and applications. In Patel, V.B., and Preedy, V.R. (eds.) *Biomarkers in Diabetes: Biomarkers in Disease: Methods, Discoveries and Applications*. Berlin, Germany: Springer.

Rajendram, R., Gyamfi, D., Patel, V.B., and Preedy, V.R. (2023). Recommended resources on diabetes in relation to foods, plants, herbs and spices in human health. In Rajendram, R., Patel, V.B., and Preedy, V.R. (eds.) *Ancient and Traditional Foods, Plants, Herbs and Spices Used in Diabetes*. Boca Raton: CRC Press.

Rajendram, R., Patel, V.B., and Preedy, V.R. (2014). Web based resources and suggested reading. In Rajendram, R., Patel, V.B., and Preedy, V.R. (eds.) *Glutamine in Health and Disease* (pp. 527–532). New York: Springer.

Rajendram, R., Patel, V.B., and Preedy, V.R. (2015). Web based resources and suggested reading. In Rajendram, R., Patel, V.B., and Preedy, V.R. (eds.) *Branched Chain Amino Acids in Health and Disease*. New York: Springer.

Rajendram, R., Patel, V.B., and Preedy, V.R. (2016). Recommended resources on biomarkers in cardiovascular disease. In Patel, V., and Preedy, V. (eds.) *Biomarkers in Cardiovascular Disease*. Dordrecht: Springer. https://doi.org/10.1007/978-94-007-7741-5_52-1

Rajendram, R., Patel, V.B., and Preedy, V.R. (2017). Recommended resources on maternal nutrition. In Rajendram, R., Patel, V.B., and Preedy, V.R. (eds.) *Nutrition and Diet in Maternal Diabetes* (pp. 495–500). New York: Springer.

Rajendram, R., Patel, V.B., and Preedy, V.R. (2019a). Resources in famine, starvation, and nutrient deprivation. In Patel, V.B., and Preedy, V.R. (eds.) *Famine, Starvation, and Nutrient Deprivation* (pp. 2399–2406). New York: Springer.

Rajendram, R., Patel, V.B., and Preedy, V.R. (2019b). Resources in diet, nutrition and epigenetics. In Patel, V.B., and Preedy, V.R. (eds.) *Nutrition and Epigenetics* (pp. 2309–2314). New York: Springer.

Rajendram, R., Patel, V.B., and Preedy, V.R. (2020). Recommended resources for nutrition, oxidative stress, and dietary antioxidants. In Martin, C.R., and Preedy, V.R. (eds.) *Nutrition, Oxidative Stress, and Dietary Antioxidants* (pp. 393–397). New York: Elsevier.

Rajendram, R., Rajendram, R., Patel, V.B., and Preedy, V.R. (2013a). Interlinking diet, nutrition, the menopause and recommended resources. In Hollins-Martin, C.J., Watson, R.R., and Preedy, V.R. (eds.) *Nutrition and Diet in Menopause*. Berlin, Germany: Springer.

Rajendram, R., Rajendram, R., Patel, V.B., and Preedy, V.R. (2013b). Diet quality: What more is there to know? In Preedy, V.R., Hunter, L-A., and Patel, V.B. (eds.) *Diet Quality: An Evidence-Based Approach* (pp. 397–401). Berlin, Germany: Springer.

World Health Organization. (2007). Joint FAO/WHO/UNU Expert Consultation on Protein and Amino Acid Requirements in Human Nutrition (2002 : Geneva, Switzerland), Food and Agriculture Organization of the United Nations, World Health Organization and United Nations University. (2007). *Protein and Amino Acid Requirements in Human Nutrition: Report of a Joint FAO/WHO/UNU Expert Consultation*. World Health Organization. Available at https://iris.who.int/handle/10665/43411 (Accessed 14/07/2024)

Index